Next Generation Wireless Systems and Networks

Next Generation Wireless Systems and Networks

Hsiao-Hwa Chen
National Sun Yat-Sen University, Taiwan

Mohsen Guizani
Western Michigan University, USA

John Wiley & Sons, Ltd

Contents

Preface

This book arose from the idea that the next generation wireless communication has a close interplay between the physical layer (system level) and the upper layer (network level) design.

In the last decade, the explosive growth of mobile and wireless communications has brought a fundamental change to the design of wireless systems and networks. The demands on traditional voice-centric services have been quickly overtaken by data-centric applications. The circuit-switched end-to-end connection communication system and network design philosophy has been replaced by all-IP packet-switched connectionless architecture. The traditional layered architecture of wireless communication systems or networks has faced a great challenge from cross-layer optimized design. The previously clearly defined boundaries between the seven Open System Interface (OSI) layers are diminishing. On the other hand, the advancement in microelectronics has made it possible to implement a complex communication end-user terminal in a pocket-sized or a namecard-sized handset, even with sufficiently high intelligence to work adaptively to the changing environment (i.e. cognitive radio). At the same time, the data transmission rate through a wireless air-link has increased tremendously, from 9.6 kbps in 1995 (on GSM) to 2 Mbps in 2005 (on a WCDMA system), increasing by more than 200 times within the last 10 years. The international research community has targeted "Super-3G" or "Beyond-3G" wireless systems and networks with a peak data transmission rate that can reach as high as 500 Mbps, as demonstrated in the very recent field trials made in Japan by NTT DoCoMo. Even more ambitious 4G wireless systems and networks will provide a peak data transmission rate of approximately 1 Gbps. The great demands on the capacity and quality offered over wireless communication links have pushed us hard to innovate new design methodologies and concepts for wireless systems and networks.

This book project was initiated to respond to the evolutional trend in the design of wireless systems and networks. It is written as an attempt to offer a handy reference, which has taken in almost all the essential background of wireless communications on both the system level and the network level, including the fundamental knowledge of wireless communication channels, almost all major 3G mobile cellular standards, wireless local area networks (LANs), wireless personal area networks (PANs), Bluetooth, All-IP wireless networking, B3G wireless, and other emerging technologies, such as ultra-wideband (UWB), orthogonal frequency-division multiplexing (OFDM), multiple-input multiple-output (MIMO), cognitive radio, and evolution UTRAN (E-UTRAN) systems. Inevitably, it was extremely difficult to write this book in the sense that we had to make a great effort to keep a good balance on the completeness of the coverage and limited page budget. We do hope that this project has achieved the goal and will be appreciated by you, the readers.

Altogether, there are 10 chapters discussed in this book. As mentioned earlier, the primary goal is to offer an up-to-date research reference, which provides the readers with almost all the important technological advancements in wireless systems and networks achieved in the last 20 years. The book includes virtually all major third-generation mobile cellular technologies, such as CDMA2000, WCDMA and TD-SCDMA technical standards. The coverage on those 3G mobile cellular technologies has been tuned to a level, at which their working principles, design philosophies, and salient features can be easily understood without the need to refer to other references given at the end of this book. However, the focus of this book has been put on newly emerging technologies, such as UWB, Multi-Carrier Code Division Multiple Access (MC-CDMA), OFDM, MIMO, cognitive radio, and Beyond-3G (B3G) systems.

This book can also serve as supplementary teaching material for the communications-related courses taught for either undergraduate or postgraduate students, whose major is Electrical and Computer Engineering, Computer Science, or Telecommunications Systems. If it is used as teaching material for undergraduate students, the best effects will be achieved if the students have already taken some prerequisites, such as "Signals and Systems" and "Digital Communications," and so on. A good background of engineering mathematics will also be desirable to easily follow the advanced part of the materials presented. In addition, it can also be successfully used as the main teaching material for professional training courses, which may last as long as a full semester/term.

We are all grateful to our families for their consistent support throughout this book project. Hsiao-Hwa Chen would like to thank his wife, Tsuiping, for her patience and compassion during the holidays and weekends spent working on this project. He would also like to thank his daughter, Cindy, and his son, Peter, for their understanding rendered to their father for not being able to play with them on weekends and holidays. Mohsen Guizani would like to thank his wife Saida and his children, Nadra, Fatma, Maher, Zainab, Sara, and Safa for their understanding and patience throughout the duration of this project.

Many people have helped us during the preparation of this book. Hsiao-Hwa Chen, would especially like to thank his students, En-Hung Chou, Ming-Jiun Liu, Yang-Wen Chen, Ho-Tai Lo, Bir-Rong Sue, Kuo-Bin Wang, Hsiang-Yi Shih, Wei-Cheng Huang, Yao-Lin Tsao, Cheng-Lung Wu, Juang-Wei Jang and Yu-Ming Kuo for helping in various ways to collect the data and references, and so on. Some parts of the works given in this book resulted partly from their theses research works. Mohsen Guizani would like to thank many of his students, in particular, Mr. Joe Baird.

Hsiao-Hwa Chen
National Sun Yat-Sen University
Taiwan

Mohsen Guizani
Western Michigan University
USA

About the Authors

Hsiao-Hwa Chen is currently a Professor at the Institute of Communications Engineering, National Sun Yat-Sen University, Taiwan. He received his BSc and MSc degrees from Zhejiang University, China, and a PhD degree from the University of Oulu, Finland, in 1982, 1985 and 1990, respectively, all in Electrical Engineering. He worked with the Academy of Finland as a Research Associate during 1991–1993 and the National University of Singapore as a Lecturer and then a Senior Lecturer from 1992 to 1997. He joined the Department of Electrical Engineering, National Chung Hsing University, Taiwan, as an Associate Professor in 1997 and was promoted to a Full Professor in 2000. In 2001, he moved to National Sun Yat-Sen University, Taiwan, as a founding Director of the Institute of Communications Engineering of the University. Under his leadership, the institute was ranked second in the country in terms of SCI journal publications and National Science Council funding per faculty in 2004. He has been a visiting Professor to the Department of Electrical Engineering, University of Kaiserslautern, Germany, in 1999, the Institute of Applied Physics, Tsukuba University, Japan, in 2000, the Institute of Experimental Mathematics, University of Essen, Germany in 2002, and the Chinese University of Hong Kong in 2004. His current research interests include wireless networking, MIMO systems, next generation CDMA technologies, and B3G wireless. He is a recipient of numerous Research and Teaching Awards from the National Science Council and Ministry of Education, Taiwan. He has authored or co-authored over 140 technical papers in major international journals and conferences, and three books and two book chapters in the areas of communications. He served and is serving as a TPC member and symposium chair of major international conferences, including IEEE VTC 2003 Fall, IEEE ICC 2004, IEEE Globecom 2004, IEEE ICC 2005, IEEE Globecom 2005, IEEE ICC 2006, IEEE Globecom 2006, IEEE VTC 2006 Spring, and IEEE ICC 2007, and so on. He served or is serving as a member of the Editorial Board or Guest Editor of IEEE Communications Magazine, IEEE Wireless Communications Magazine, IEEE JSAC, IEEE Networks Magazine, IEEE Transactions on Wireless Communications, IEEE Vehicular Technology Magazine, Wireless Communications and Mobile Computing (WCMC) Journal and International Journal of Communication Systems, and so on. His original work in CDMA wireless networks, digital communications and radar systems has resulted in five US patents, two Finnish patents, three Taiwanese patents and two Chinese patents, some of which have been licensed to industries for commercial applications. He has been an Honorable Guest Professor of Zhejiang University, China, and Shanghai Jiao Tong University, China, since 2003 and 2005, respectively. For more information about Professor Hsiao Hwa Chen, please visit the web site at http://www.ice.nsysu.edu.tw/hshwchen.html.

Mohsen Guizani is currently a Full Professor and the Chair of the Computer Science Department at Western Michigan University. He served as the Chair of the Computer Science Department at the University of West Florida from 1999 to 2003. He was an Associate Professor of Electrical and Computer Engineering and the Director of Graduate Studies at the University of Missouri-Columbia from 1997 to 1999. Prior to joining the University of Missouri, he was a Research Fellow at the University of Colorado-Boulder. From 1989 to 1996, he held academic positions at the Computer Engineering Department at the University of Petroleum and Minerals, Dhahran, Saudi Arabia. He was also a

Visiting Professor in the Electrical and Computer Engineering Department at Syracuse University, Syracuse, New York during the academic years 1988–1989. He received his B.S. (with distinction) and M.S. degrees in Electrical Engineering; M.S. and Ph.D. degrees in Computer Engineering in 1984, 1986, 1987, and 1990, respectively, all from Syracuse University, Syracuse, New York. His research interests include Computer Networks, Design and Analysis of Computer Systems, Wireless Communications and Computing, and Optical Networking. He currently serves on the editorial boards of many national and international journals, such as the IEEE Transaction on Wireless Communications, IEEE Transaction on Vehicular Technology, IEEE Communications Magazine, the Journal of Parallel and Distributed Systems and Networks, and the International Journal of Computer Research to name a few. He served as a Guest Editor in the IEEE Communication Magazine, IEEE Journal on Selected Areas in Communications, Journal of Communications and Networks, The Simulation Transaction, International Journal of Computer Systems and Networks, International Journal of Communication Systems, International Journal of Computing Research, and Journal of Cluster Computing. Dr. Guizani is the founder and Editor-In-Chief of "Wireless Communications and Mobile Computing," journal published by John Wiley (http://www.interscience.wiley.com/jpages/1530–8669/). He is the author of four books: Designing ATM Switching Networks, by McGraw-Hill 1999 (http://www.pbg.mcgrawhill. com/computing/authors/guizani.html), Wireless Systems and Mobile Computing, by Nova Science Publishers 2001, Optical Networking and Computing for Multimedia Systems, by Marcel Dekker, June 2002, and Wireless Communications Systems and Networks, by Kluwer, June 2004. He served as a Keynote Speaker for many international conferences and has also presented a number of Tutorials and Workshops. He served as the General Chair for the Parallel and Distributed Computer Systems (PDCS 2002), IEEE Vehicular Technology Conference 2003 (VTC2003), PDCS 2003 and IEEE WirelessCom 2005. He also served as the program chair for many conferences, such as Parallel and Distributed Computer Systems, Wireless Networking Symposium (VTC2000), Annual Computer Simulation Systems Conference, Optical Networking Symposium (Globecom 2002), Collaborative Technologies Symposium 2002 (in conjunction with Western Multi-conference on Simulation and Modeling), and the General Conference of IEEE Globecom 2003. He has more than 140 publications in refereed journals and conferences in the areas of High-Speed Networking, Optical Networking, and Wireless Networking and Communications. Dr. Guizani is the Co-Chair of the IEEE Communications Society Technical Committee on Transmissions, Access, and Optical Systems (IEEE TAOS), Conference Coordinator of the IEEE Communications Society Technical Committee on Computer Communications (IEEE TCCC), a member of the IEEE Communications Society of Optical Networking (IEEE ONTC), the Secretary for the IEEE Communications Society of Personal Communications (IEEE TCPC), and a member of the Computer Network Security Sub-Committee. He is designated by the IEEE Computer Society as a Distinguished National Speaker until December 2005. He is also ABET Accreditation Evaluator for Computer Science and Information Technology Programs. He received both the Best Teaching Award and the Excellence in Research Award from the University of Missouri-Columbia in 1999 (a college wide competition). He won the best Research Award from KFUPM in 1995 (a university wide competition). He was selected as the Best Teaching Assistant for two consecutive years at Syracuse University, 1988 and 1989. He is a senior member of IEEE, a member of IEEE Communication Society, IEEE Computer Society, ASEE, ACM, OSA, SCS, and Tau Beta Pi. For more details, please visit: http://www.cs.wmich.edu/mguizani/.

1

Introduction

As everybody is undoubtedly aware, we are living in the midst of rapid information renovation and innovation. The changing world of information technology (IT) can be challenging or even frightening to all of us. For any IT engineer who stays out of the technological advancement for even a very short period of time, he/she will quickly find himself/herself an outsider to the technological transitions, with difficulty in understanding hundreds and thousands of new terminologies that are invented every year for newly emerging technologies.

In the last 30 years, the IT industries have witnessed two big waves of revolution, one being the invention of the Internet, and the other the wide applications of wireless technologies. The Internet technologies have for the first time in human history provided us with a high-speed information-dissemination infrastructure via its global optical fiber webs that cover virtually every corner of the world. If the time could be flashed back to 30 years earlier, people would have hardly believed that all the information contained in an enormous number of books in libraries can be accessed without going there personally. In addition, the readiness of two way data transactions on the Internet have triggered fundamental changes in many sectors of our life. For instance, intercontinental telephone calls will no longer be considered as a symbol of a lifestyle of luxury. Very soon everybody will be given the privilege that all voice telephone calls (either demotic or international) will be free of charge, thanks to the wide accessibility of the Internet throughout the world. The Internet operates on an all-IP based networking architecture, and thus the network level design and performance-ensuring mechanism play a critical role in all Internet-related applications. Table 1.1 shows the top 20 countries with most Internet subscribers in the world as recorded in 2005.[1]

On the other hand, the revolution of wireless technologies fuels the advancements in modern telecommunication systems through its cordless and mobile extension of wired networks, such as the Internet. Mobility is one of the most important characteristics of modern society. Everything and everyone are in motion. Therefore, the information-dispatching facilities should also be made available while people are on the move. The explosive increase in mobile cellular telephone services around the world has reflected the great demand for mobile communications. The availability of mobile cellular communications has exerted a strong influence on the lifestyle, the business models, as well as on the sense of value, distance, and time. The wireless technologies work on radio frequency (RF) to establish the data connection paths (or radio links) via electromagnetic radiation waves. The invisible RF air links connect users' end-terminals through *base stations* or *access points* with fixed or wired network infrastructures, such as the Internet. Therefore, the wireless technologies have a lot to do with the physical layer design and architecture.

[1]It is amazing to note that China has contributed around 11% of the world's total Internet subscribers in 2005, and has become second only to the United States in terms of the percentage of the world's total Internet subscribers.

Next Generation Wireless Systems and Networks   Hsiao-Hwa Chen and Mohsen Guizani
© 2006 John Wiley & Sons, Ltd

Table 1.1   Top 20 countries with most Internet subscribers in the world as recorded in 2005.

Country or region	Number of subscribers	Population in 2005	Penetration rate (%)	World percentage (%)
United States	202,888,307	296,208,476	68.5	21.6
China	103,000,000	1,282,198,289	7.9	11.0
Japan	78,050,000	128,137,485	60.9	8.3
Germany	47,127,725	82,726,188	57.0	5.0
India	39,200,000	1,094,870,677	3.6	4.2
United Kingdom	35,807,929	59,889,407	59.8	3.8
South Korea	31,600,000	49,929,293	63.3	3.4
Italy	28,610,000	58,608,565	48.8	3.0
France	25,614,899	60,619,718	42.3	2.7
Brazil	22,320,000	181,823,645	12.3	2.4
Russia	22,300,000	144,003,901	15.5	2.4
Canada	20,450,000	32,050,369	63.8	2.2
Spain	15,565,138	43,435,136	35.8	1.7
Indonesia	15,300,000	219,307,147	7.0	1.6
Mexico	14,901,687	103,872,328	14.3	1.6
Taiwan	13,800,000	22,794,795	60.5	1.5
Australia	13,784,966	20,507,264	67.2	1.5
Netherlands	10,806,328	16,316,019	66.2	1.2
Poland	10,600,000	38,133,891	27.8	1.1
Malaysia	9,513,100	26,500,699	37.9	1.1
Rest of the World	176,943,950	2,444,250,712	7.2	18.8
World total	938,710,929	6,420,102,722	14.6	100.0

The Internet in the absence of wireless technologies' support cannot offer the end users such convenience and readiness; while the wireless systems without the backup of the Internet infrastructure will limit its diversity in services and content. The combination of the Internet and wireless technologies will provide us access to information services at any time, in any place, and to any one. The combination of the Internet and wireless technologies has also created many challenging issues, such as the joint optimization of software and hardware implementations, cross-layer design for network solutions, all-IP wireless platforms, intelligent radios, and so on. Therefore, a modern wireless communication architecture can always be viewed from two different aspects: that is, the system level view, which is based mainly on the hardware and physical layer implementations on a local scale, and the network level, which is observed from the topological configuration and upper layer design on a global scale. We have observed a trend where wireless designs on the system and network levels come together. Investigations on both the system level (in a local scale) and the network level (in a global scale) helps to better understand any wireless communication entity of today.

This book was written in an effort to give the readers an up-to-date research reference containing almost all major technological advancements on both the system and the network levels that have happened in the last 20 years. The contents of this book can be divided into four major parts. We give a brief introduction for each part as follows.

1.1   Part I: Background Knowledge

The first part of this book was written by Professor Hsiao-Hwa Chen and it deals with the fundamentals and background knowledge of wireless communications. This part consists of only one chapter, that is, Chapter 2, titled "Fundamentals of Wireless Communications," in which there are six sections altogether.

The first section introduces the theory of radio communication channels, which is part of the most important background knowledge needed to understand why and how a wireless communication system or network suffers various problems and bottlenecks when it works in a particular application scenario. The importance of the knowledge of the wireless channels lies in the fact that, no matter how advanced a future wireless communication system or network might be, we have to deal with the same set of problems associated with the radio propagation channels, such as delay spread, Doppler spread, coherent time, coherent bandwidth, and so on, which will always be there.

With the necessary information about radio channels, we can proceed to introduce various spread spectrum (SS) techniques in the second section. It has to be noted that the SS techniques are the basis of code division multiple access (CDMA) technology, which was first applied to the IS-95 standard [317–326] and has become the primary multiple access scheme chosen by almost all third generation (3G) mobile cellular systems, including CDMA2000 [345–359], WCDMA [425–431], TD-SCDMA [432–439], and so on. In this section, three major SS techniques are discussed, including direct sequence spread spectrum (DSSS), frequency hopping spread spectrum (FHSS), and time hopping spread spectrum (THSS) techniques. It is to be noted that the basic concepts of ultra-wideband (UWB) technologies are introduced while discussing the THSS, which has been applied to many emerging UWB systems, in particular, for those based on pulse position modulation (PPM). Nevertheless, it must be admitted that the most commonly used SS techniques are the DS and FH, rather than the TH technique. The wireless communication systems based on the SS techniques can often be called *wideband* wireless applications, in contrast to those that do not use the SS techniques.

It is amazing to note that the technological evolution happens so rapidly that used-to-be advanced technologies quickly become common background knowledge for all. In the late 1990s one of the coauthors of this book, Professor Hsiao-Hwa Chen, worked in the Telecommunications Laboratory, University of Oulu, Finland, which used to be the largest research group focusing on SS techniques in the whole of Europe. Then, most of the SS techniques had not been unclassified and there were very few publications on the SS techniques applied to civilian applications. The SS techniques were considered to be one of the most advanced know-hows at that time. Therefore, we had to resort to many technical reports and patents for reference information. Nowadays, knowledge of the SS techniques has become a must for all electrical engineers working in telecommunication areas.

Chapter 2 continues with the fundamentals of wireless communications to discuss the issues on multiple access technologies. Three major commonly used multiple access technologies are included in the discussions given in this section, namely, *frequency division multiple access* (FDMA), *time division multiple access* (TDMA), and CDMA technologies. Being different from CDMA, both FDMA and TDMA are always considered to be the traditional *narrow band* multiple access technologies. On the other hand, the CDMA technology was developed from the SS techniques and is often referred to as a *wideband* multiple access technology. As mentioned earlier, CDMA technology has become the main multiple access technology used in all major 3G mobile cellular standards owing to its relatively high bandwidth efficiency and robustness against time-dispersive frequency-selective fading and other external interferences. The first successful application of CDMA technology in commercial communication systems is IS-95A [317–326], which was developed by Qualcomm Inc., USA, and works based on DSSS techniques. The IS-95A system has been deployed in many countries in the world today and its reliability and stability have been confirmed from its long time operations in many countries. Now we have entered the era of 3G mobile communications. All major 3G systems operate on CDMA technologies with no exception, indicating the great popularity of these technologies. An interesting question arises: Can CDMA be still used as the primary multiple access technology in B3G wireless communications? Some people have suggested that the CDMA was developed in the later 1980s and is suitable only for slow-speed, continuous-time traffic, and voice-centric applications. It has also been suggested that the current CDMA technologies may not be well suited for those applications where dominant traffic will carry high-speed, bursty, and data-centric services. To answer this challenging question, we have more discussions on this issue in Chapter 7 of this book.

The fourth section of Chapter 2 consists of three subsections, which are "Multiuser joint detection against MAI," "Pilot-aided CDMA signal detection," and "beam-forming techniques against cochannel interference." Obviously, all these three subsections deal with issues on how to suppress or eliminate the interferences. The *multiuser detection* (MUD) techniques used to be a very popular research topic a couple of years ago, due to the widespread application of CDMA technologies. As indicated by the title of the subsection, the MUD techniques are used only for overcoming the *multiple access interference* (MAI) problem. The MUD concept is smart in terms of the fact that it treats all MAI as a whole and proceeds with the detection through de-correlating the MAI components from the useful signals. Thus, it treats all signals, whether useful or not, as indispensable parts in the entire detection process. This entails a very high detection efficiency when compared to many other traditional interference suppression techniques, which treat unwanted signals individually as a component that should be suppressed as much as possible.

Pilot-added signal detection is another important issue covered in this subsection, where we introduce many useful conclusions from others' and our own research results. Pilot-added detection is used for overcoming the interferences introduced mainly by dispersive channels, where time dispersion (caused by multipath effect) and frequency dispersion (caused by Doppler effect) may exist individually or jointly. Unlike the MUD schemes, the pilot-added signal detection cannot solve the problems associated with MAI. Therefore, the pilot-added detection should always work along with the MUD to enable a wireless receiver with the capability to work successfully under various channel conditions. In this subsection, we summarize the experience gained in designing pilot-added detection schemes by the "three-same condition," which states that the pilot signal should be constructed at the same time, with the same frequency, and the same code as those used for data-carrying signals, to ensure an accurate estimation of the channel condition.

Section 2.4 ends with the subsection titled "Beam-forming techniques against cochannel interference." Actually, the beam-forming technique is a type of *aperture-synthesizing* technique, which uses multiple antennas to transmit or receive the same signal to achieve a certain array gain through narrowed beams. By using the beam-forming technique in transmitter (Tx) antennas we can pinpoint the transmitting signal to a particular directional angle to facilitate the signal reception at the receiver of interest.[2] On the other hand, we can also use receiver antenna beam-forming to reduce the cochannel interference coming from the angle-of-arrival outside the beam.[3] It has to be pointed out that the major difference between the traditional beam-forming techniques and the emerging multiple-in-multiple-out (MIMO) systems lies in the fact that the signals transmitted or received in multiple antennas in beam-forming techniques are exactly the same replicas (except for different delays), while those in a MIMO system are subject to different coding processes in different antennas to achieve spatial diversity gain or multiplexing capability.

The last three sections of Chapter 2 discuss the issues on "OSI reference model," "switching techniques," and "IP-based networking." Section 2.5 discusses an important concept on the Open System Interconnection (OSI) layered networking model. Section 2.6 concentrates on the discussions on two major network switching architectures, that is, circuit switching and packet switching, both of which play very important roles in wired and wireless networks. It is to be noted that circuit switching as a traditional switching technique has been revitalized recently because of its emerging applications in high-capacity fiber-optical trunk networks. Section 2.7 gives a brief introduction to IP-based

---

[2]This is also called the *beam-steering* algorithm.

[3]This sometimes is also called the *null-steering* algorithm.

networking and its development, which has gained great attention due to its wide applications in the Internet-related applications and systems.

1.2 Part II: 3G Mobile Cellular Standards

The second part of this book covers the major 3G mobile cellular standards, including CDMA2000, WCDMA, and TD-SCDMA, which are discussed in Sections 3.1, 3.2, and 3.3, respectively, to form Chapter 3, which was written by Professor Hsiao-Hwa Chen.

As mentioned earlier, it is a great challenge to cover all these major 3G mobile cellular standards within the limited space available in this book, which also discusses many other up-to-date wireless technologies. Obviously, it is not desirable to give only a very brief introduction to each of them, as they also provide important information about the current state-of-the-art wireless technologies, which is the foundation for further discussions on more advanced beyond 3G (B3G) wireless technologies in this book. A relatively informative discussion on these important 3G systems will also be a useful reference to the readers. On the other hand, we are not allowed to spend too much of the space in this book to address the issues on 3G mobile cellular systems only. Therefore, we have to keep a very careful balance on the contents covered here. For this purpose, the discussions given in Chapter 3 have been made as informative as possible, while focusing mainly on their key technical features. Some detail specifications given in the long standard documentations have been omitted for conciseness of the discussions. Therefore, the discussions about the major 3G mobile cellular standards should not be considered complete as it is utterly impossible to condense a standard written in several thousands of pages into a section with only a few tens of pages.

The inclusion of the three major 3G mobile cellular standards, such as CDMA2000, WCDMA, and TD-SCDMA, takes into account the fact that they have been deployed in many countries in the world. The CDMA2000 is the standard that originated from the United States; whereas the WCDMA was the one proposed by Europe.[4] It is to be noted that the Japanese 3G system, ARIB WCDMA, bears a great similarity to the European 3G system, ETSI UMTS-UTRA. So, we do not discuss them individually in two different sections due to the limitation of space in this book. Instead, we put the two into the same section (i.e., Section 3.2), and their differences are discussed and explained in Section 3.2.2. In fact, ARIB has committed to make the Japanese 3G standard fully compatible with ETSI UMTS-UTRA system eventually, although there still are some differences in their specifications at the moment. At the time when this book was written, the roaming between Japanese 3G networks and European 3G networks has not been widely implemented.

Chapter 3 begins with the North American standard, or CDMA2000 standard, a 3G mobile cellular standard proposed by the TIA/EIA of the United States.[5] The discussion of CDMA2000 allows us to understand better how the evolutional change from 2G to 3G mobile cellular systems happens. The CDMA2000 technology is always referred to as the successor of its 2G solution, IS-95. CDMA2000 is one of the IMT-2000 candidate submissions to the International Telecommunication Union (ITU).

As early as 1985, ITU regulators had a vision that the future of mobile cellular systems would be multimedia – involving voice, video, and data services. Thus, in 1985 the ITU, the world's governing telecommunication body, began planning for the next generation digital cellular – "Future Public Land Mobile Telecommunications Systems" (FPLMTS) – later known as *IMT-2000*.[6] The goal of FPLMTS was to provide broadband multimedia wireless services via a single global frequency band

---

[4]It is to be noted that the WCDMA standard was initially proposed by ETSI, Europe, and ARIB, Japan, jointly. Although there still are some minor differences in their network operations, both committed to make the standards fully compatible under the framework of 3GPP.

[5]The abbreviations of TIA and EIA stand for "Telecommunications Industry Association" and "Electronics Industry Association," respectively. Both the organizations are based in the United States.

[6]The abbreviation "IMT-2000" stands for "International Mobile Telecommunications in the year 2000," which would have been standardized in the year 2000 according to the ITU's vision of the 1990s. Unfortunately, the

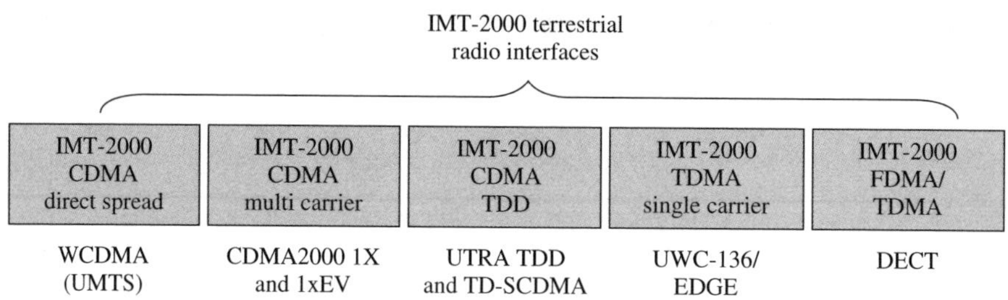

Figure 1.1 Five radio interfaces for IMT-2000 standards as a part of the ITU-R M.1457 Recommendation.

allocation and standardized, interoperable technologies. The frequency range allocated would be around 2000 MHz.

The ITU requires that IMT-2000 (or 3G) networks, among other capabilities, deliver improved system capacity and spectrum utilization efficiency over the 2G systems and support data services at minimum transmission rates of 144 kbps in mobiles (outdoor) and 2 Mbps in fixed (indoor) environments. On the basis of these requirements, in 1999 ITU approved five radio interfaces for IMT-2000 standards as a part of the ITU-R M.1457 Recommendation. CDMA2000 is one of the five standards. It is also known by its ITU name *IMT-CDMA Multi-Carrier*. Figure 1.1 shows five radio interfaces for IMT-2000 standards as a part of the ITU-R M.1457 Recommendation, where WCDMA (Universal Mobile Telephone System (UMTS)) was submitted jointly by Europe and Japan, CDMA2000 1x and 1xEV was proposed by the United States, UTRA time division duplex (TDD), and TD-SCDMA were proposed by Europe and China, UWC-136 and EDGE were proposed by the United States, and DECT was submitted by Europe.

The development of the CDMA2000 standard was driven mainly by North American technology developers with an invested interest in the progression of IS-95, or its later version cdmaOne, as the global standard for next generation mobile cellular systems. The share for CDMA remains to be roughly same, if not reduced. Obviously, CDMA2000 is not a single standard in itself. From IS-95 (the 2G equivalent of Global System for Mobile Communication (GSM)) through CDMA2000 1xRTT, which increases the voice capacity of the former by approximately 40% and allows data transfer speeds up to a peak of 144 kbps, to CDMA2000 1xEV-DO, which has a theoretical bit rate of 2 Mbps, CDMA2000 1xEV-DO should be considered a full-fledged 3G standard.

CDMA2000 represents a family of technologies that includes CDMA2000 1x and CDMA2000 1xEV. CDMA2000 1x can double the voice capacity of cdmaOne (formerly known as *IS-95*) networks and can deliver peak packet data speeds up to 307 kbps in mobile environments. CDMA2000 1xEV includes the following:

- CDMA2000 1xEV-DO, which delivers peak data speeds of 2.4 Mbps and supports applications such as MP3 transfers and video conferences.

- CDMA2000 1xEV-DV, which provides integrated voice and simultaneous high-speed packet data multimedia services at speeds of up to 3.09 Mbps, which has already exceeded the peak data rate specified by IMT-2000 (or 3G) specifications.

1xEV-DO and 1xEV-DV are both backward compatible with CDMA2000 1x and cdmaOne. It is noted that the world's first CDMA2000 1x commercial system was launched by SK Telecom (Korea)

ITU's effort to reach a consensus on the IMT-2000 standard in the year 2000 was not successful because of obvious reasons.

in October 2000. Since then, CDMA2000 1x has been deployed in Asia, North and South America, and Europe, and it was estimated that the subscriber base is growing at 700,000 subscribers per day. CDMA2000 1xEV-DO was launched in 2002 by SK Telecom (Korea) and KT Freetel (Korea).

Section 3.1 discusses the CDMA2000 standard in a progressive way so that it is easy to understand for a person with a minimum background knowledge of wireless communications. Section 3.1 is made up of altogether 13 subsections, in which we give an introduction to the various technical aspects of the CDMA2000 standard. In order to make discussions up-to-date and consistent throughout Section 3.1, we concentrate on the CDMA2000 1xEV (or IS-856) standard based mainly on the specifications given in the following standard documentations [360–361]:

- CDMA2000 High Rate Packet Data Air Interface Specification, 3GPP2 C.S20024 Version 2.0, Date: October 2000.

- CDMA2000 High Rate Packet Data Air Interface Specification, 3GPP2 C.S20024-A, Version 1.0, Date: March 2004

the former of which is known as *Release 0* and the later is Revision A of CDMA2000 1xEV or IS-856 standard. There are also numerous references to CDMA2000 1xEV and their evolutional versions CDMA2000 1xEV-DO and CDMA2000 1xEV-DV, and readers may refer [360–367] for more information about them. In particular, the April 2005 issue of IEEE Communications Magazine has published a Special Issue on CDMA2000 1xEV-DV and seven papers appeared in this issue [368–374], which indicated that the CDMA2000 1xEV-DV will gain greater popularity around the world.

While focusing on the discussion on CDMA2000 1xEV, we also refer to the CDMA2000 1x standard from time to time in Section 3.1. Release 0 of the CDMA2000 1x standard [348–353] consists of the following 3GPP2 documents:

- C.S0001-0 Introduction to CDMA2000 Standards for Spread Spectrum Systems

- C.S0002-0 Physical Layer Standard for CDMA2000 Spread Spectrum Systems

- C.S0003-0 Medium Access Control (MAC) Standard for CDMA2000 Spread Spectrum Systems

- C.S0004-0 Signaling Link Access Control (LAC) Standard for CDMA2000 Spread Spectrum Systems

- C.S0005-0 Upper Layer (Layer 3) Signaling Standard for CDMA2000 Spread Spectrum Systems

- C.S0006-0 Analog Signaling Standard for CDMA2000 Spread Spectrum Systems

All final revisions (or Revision D) of CDMA2000 1x standard can be found in the reference list [359], given at the end of this book.

After having discussed CDMA2000 standards proposed by the United States, we will move on to Section 3.2, which covers the European 3G standard, the WCDMA system. The WCDMA is also one of the IMT-2000 candidate proposals, as shown in Figure 1.1, proposed by ETSI, Europe, and ARIB, Japan, jointly. The discussions given in Section 3.2 focus on the ETSI UMTS system; whereas the difference between European UMTS-FDD and Japanese ARIB WCDMA systems are explained, in particular, in Section 3.2.2 titled "ETSI UMTS versus ARIB WCDMA," to save space in this book. The conclusion will be drawn as a result from the comparison between ETSI UMTS-FDD and ARIB WCDMA that the two should be made fully compatible in mid-2003 under the time frame specified in UMTS Realease'99. [7] under the time frame specified in UMTS Release'99.

---

[7]It has to be noted that some delay occurred in the compatibility time frame between the two, and up to the time when this book was written the roaming between Japanese and European 3G networks has not been widely implemented.

Japanese mobile services operator NTTDoCoMo launched the world's first commercial WCDMA network in 2001, although its operation was limited to only the great Tokyo area initially. When compared to the world's first CDMA2000 1x commercial system launched by SK Telecom (Korea) in October 2000, the NTTDoCoMo WCDMA system had suffered many technical problems during the initial phase of its services in Japan, including both its FOMA terminals[8] and networking. Even today, both Korea and Japan claim that they had the first 3G network in the world, and it is always a very tough issue as to who is number one if the two progress neck and neck in the development of new technologies. Nevertheless, the competition between Japan and Korea in the 3G mobile cellular communication field has been very serious for a long time. That is why many people really doubt the claim that Europe or the United States is the serious pusher for 3G development and standardization. Japan has been very worried for a long time about the fact that Korea has obtained the core CDMA intellectual property rights (IPRs) transfer from Qualcomm Inc., and consequently has grasped the know-hows in many key CDMA technologies. By following the United States's suit in developing its TDMA-based second-generation (2G) cellular technology, Japanese Digital Cellular (JDC) (which was not compatible with any of the 2G systems operating in the world), Japan virtually had no access to the lucrative world of the mobile cellar market. This sad experience made Japan determined to be a must-winner in the race for 3G technology, and motivated the country to work closely with Europe for developing the WCDMA technology. It seems that Japan has got the right bid, as clearly seen from its big share in terms of total WCDMA subscribers in the world. The number of 3G subscribers has so far topped more than 100 million, including four-million WCDMA users (mainly from Japan). With the growing maturity of WCDMA-related products and technology, its commercial-user networks are undergoing a dramatic development. Today, more than 70 3G/UMTS networks using WCDMA technology are operating commercially in 25 countries, supported by a choice of over 100 terminal designs from Asian, European, and US manufacturers.

The last section, Section 3.3, in Chapter 3 talks about the TD-SCDMA standard, which was proposed by China as one of the five IMT-2000 candidate proposals, as shown in Figure 1.1. Currently, very few books have covered the TD-SCDMA standard in their chapters related to 3G technologies. The importance of this Chinese-owned 3G standard will become clearer with the increase in the leverage weight of China's role in the world telecommunication market. China undoubtedly is the largest single market for mobile communications. The number of its mobile service subscribers surpassed the United States a few years ago, making it the most influential mobile cellular market in the world. It will be considered very silly if a mobile communication vendor/manufacturer has no presence in China today. Everybody wants to catch a bite of big China's mobile communication market. On the other hand, China clearly knows that it is stupid for its service providers to buy hundreds of thousands cellular equipments from outsiders with their hard-earned money. Even for those made-in-China equipments, they have to pay a large amount of loyalty fees to the foreign players due to the use of their IPRs. This harsh reality has motivated China to develop its own mobile communication standard in an effort to reduce or even eliminate the heavy reliance on these imported mobile cellular technologies.

However, the path toward the development of its own 3G system is not smooth either. The TD-SCDMA standard was proposed originally by CATT, which is one of the largest institutes[9] that specializes in telecommunication research in China. However, many people criticized the TD-SCDMA as standard lacks novelty technically and thus may still face a heavy licence fee payable to foreign companies even after its commercialization. It was also suggested that the TD-SCDMA used too many core technologies, such as power control (IS-95), RAKE receiver (IS-95), orthogonal variable spreading factor (OVSF) code for channelization (WCDMA), time division duplex (TDD) (UTRA-TDD), and so on, which are borrowed from Qualcomm as well as other companies.

---

[8]The name "FOMA" is the abbreviation for "Freedom Of Mobile multimedia Access."

[9]CATT used to be a very large research institute that belonged to the former Ministry of Post and Telecommunications, China, and is now a privatized company specialized for mobile communications. For more information, please refer to the web site at http://www.catt.ac.cn.

Also, the use of TDD techniques has made it very hard to operate the TD-SCMA in a large cell, thus making it necessary for a service operator to have to offer a dual-system, one for small cells in cities and the other for large cells in the countryside or in suburb areas. This will not make sense economically for any operator. China, also being the largest free-market country, no longer suits the governing mode that worked before when centralized economy played a key role. Now, nobody even from the central government can order a mobile service provider to buy the equipments made only in China. The question then becomes who will care about TD-SCDMA if it does not work well technically. Maybe time will play its role in China again: TD-SCDMA needs some more time to get ready for its practical applications in China, and therefore the time when its 3G licences will be issued has become the focus of the world mobile cellular market now.

Observers widely expect that the Chinese government will make decisions on 3G licensing when it is certain that the homegrown TD-SCDMA is a viable option. However, analysts are urging China to roll out its 3G licensing soon. China cannot afford more delays in the licensing of 3G wireless communications telephony, they say. Delays could harm the development of the country's telecommunications industry, as they may be against the interests of the whole value chain. A clear timetable will help all players, both foreign and domestic, prepare resources planning, manufacturing, and research and development (R&D), they say.

According to current market expectations, China Mobile, the world's biggest mobile service carrier for a number of subscribers, would build a 3G system on the WCDMA standard, which is based on the GSM technology popular in Europe. China Unicom would build a system based on the CDMA2000 standard developed by Qualcomm Inc of the United States. For the homegrown TD-SCDMA system, all six domestic telecom operators are doing network trial tests based on the system. Analysts also expect that the major fixed-line operators China Telecom and China Netcom will receive licences, as will the two existing mobile operators, China Mobile and China Unicom. Some experts on TD-SCDMA predicted that if the 3G licences can be released in 2006, China's 3G mobile telecommunications revenue would likely reach 300 billion yuan (US$36 billion) in 2010. Revenue from 3G between 2005 to 2010 will accumulate to one trillion yuan (US$120 billion).

The discussions on TD-SCDMA in Section 3.3 span 12 subsections, dealing with various aspects of its technical features, including an overview of TD-SCDMA, frame structure, smart antenna applications, adaptive beam patterns, uplink synchronization control, intercell synchronization, baton handover, intercell dynamic channel allocation, flexibility in network deployment, and so on. In particular, we spend quite a bit of space to address the salient features of the TD-SCDMA technology, such as uplink synchronization control, baton handover, and so on. Before the end of Section 3.3, we also touch on the issue of the technical limitations of the TD-SCDMA, where we point out the possible technical problems that can arise when a TD-SCDMA system works on practical applications, as well as the global impact of TD-SCDMA technology.

We hope that the coverage of the TD-SCDMA standard in this book will be useful to those who are particularly interested in the TD-SCDMA technology as well as the Chinese 3G market.

1.3   Part III: Wireless Networking

The third part of this book is about wireless networks, which will be dealt with in two chapters, that is, Chapter 4, titled "Wireless Data Networks", and Chapter 5, titled "All-IP Wireless Networking". Both these chapters were written by Professor Mohsen Guizani.

In Chapter 4, we discuss wireless data networks. We cover all the different standards (including IEEE 802.11b, IEEE 802.11a, IEEE 802.11g, IEEE 802.15, IEEE 802.16, ETSI HIPERLAN and HIPERLAN/2, Japan's MMAC, etc.) and explain their purpose. Then, we discuss the architecture and functionality of the *MAC* sublayer. Next, we summarize the functionalities of *FHSS* and *DSSS*.

Then, we discuss security, authentication and Wired Equivalent Privacy (WEP). We conclude this chapter with a discussion on Bluetooth technologies and related security issues.

In Chapter 5, we discuss All-IP Wireless Networking, including mobile IP, IPv6 versus IPv4, Mobile IPv6, *wireless application protocol* (WAP), and few important routing protocols.

It is to be noted that the research on all-IP wireless is now very active around the world. Many new research topics are emerging everyday, such as voice over wireless local area networks (VoWLANs)[10], mobility support in heterogenous networking environment, security issues on all-IP wireless, and so on. Therefore, the discussions given in Chapter 5 only cover very fundamental knowledge for further research in this area.

Similar to discussions related to networking issues, Chapters 4 and 5 have a lot to do with the upper layer architecture/design of wireless networks. Obviously, to understand the discussions given in these chapters, it is desirable to gain enough knowledge on the *OSI* seven-layer model[11], which has been discussed in Section 2.5.

It should also be noted that the layered network architecture is a traditional network design methodology, which has been used extensively in many different wired and wireless communication systems/networks including all 1G to 3G mobile cellular standards and other networks, such as IEEE 802.11x WLANs. Because of the regularity in its modular structure, the OSI seven-layer model can help network designers to concentrate on the functionalities provided by each layer; while the interface between layers will be taken care of by the standard message exchange formality specified in the OSI-model. However, the reliance on extensive message exchanges happening in the boundaries of different layers results in long processing delays. This problem becomes even more acute where a wireless system is working in a high-speed burst-type traffic mode.

Motivated to solve this problem, a new concept called *cross-layer network design* approach has been proposed recently in an effort to greatly reduce the processing delay in the OSI layered model. The basic idea of cross-layer design is to achieve a global optimization design across different layers by tearing down the boundaries between different layers. In most cases, it might happen that the cross-layer design method is only applied to a few but not all layers. Because of the limited space, we will not elaborate more on the cross-layer design approach in this book.

1.4   Part IV: B3G and Emerging Wireless Technologies

The fourth (also the last) part of this book addresses the issues on B3G wireless communications and other emerging technologies, including Chapter 6, Chapter 7, Chapter 8, Chapter 9, and Chapter 10, which are titled "Architecture of B3G Wireless Systems", "Multiple Access Technologies for B3G Wireless", "MIMO Systems", "Cognitive Radio Technology", and "E-UTRAN: 3GPP's Evolutional Path to 4G", respectively. Chapter 6 was written by Professor Mohsen Guizani, and the other four chapters, that is, Chapters 7 to 10, were written by Professor Hsiao-Hwa Chen.

In Chapter 6, we concentrate on the architecture of B3G Wireless Systems. The discussion in this chapter includes spectrum allocation issues, high-speed data, multimode and reconfigurable platforms, ad hoc mobile networking, and satellite systems in B3G wireless. Since many issues are still in research labs, we conclude this chapter listing some challenging research topics of interest to our readers.

---

[10]Voice over IP has been a research topic for some time; while *voice over wireless local area networks* (VoWLANs) is still a relatively new topic that emerged very recently. The main objective for VoWLANs research is to provide voice communications via any WLAN-compatible terminals, such as PDAs, notebook PCs, and so on.

[11]The OSI seven-layer architecture has become a standard layered architecture for any (either wired or wireless) network system. The seven layers in a downward order include "Application layer," "Presentation layer," "Session layer," "Transport layer," "Network layer," "Data link layer," and "Physical layer."

Chapter 7 titled "Multiple Access Technologies for B3G Wireless" consists of six sections, covering issues on various aspects of the multiple access technologies, which might be applied to futuristic B3G wireless communication systems. Section 7.1 gives a brief introduction on the required characteristics that a multiple access technology should have to suit the applications in B3G wireless. It is pointed out in the section that two major challenging issues have to be taken into account to architect future B3G wireless systems. One is the extremely high data rate that a B3G wireless system or network should provide to the subscribers when compared to that of the current 2G to 3G mobile cellular systems. The peak data rate in a future B3G system is suggested to be at least 500 Mbps. A field trial of a prototype 4G wireless system has been conducted very recently in the suburb area of Tokyo by NTTDoCoMo with a peak data transmission rate of about 1 Gbps being reached in its downlink transmission channels. This data transmission rate was achieved with a transceiver mounted in a slow moving vehicle that communicated with a base station without a constant line-of-sight path. The area where the field trial was conducted had a lot of high-rise buildings, and therefore it has to be considered as a very urbanlike environment with very rich multipath propagation components. This pioneer field test for a 4G system carried out by NTTDoCoMo has set up an example for the target data transmission rate for all futuristic B3G wireless systems. At such a high transmission rate, many new problems that are not visible in a slow-speed wireless communication system will surface, and even technically become a serious bottleneck to the whole wireless system design. Because of the relatively poor bandwidth efficiency in the currently available multiple access technologies (either FDMA, TDMA, or CDMA), it is simply impossible to provide all wireless subscribers with such a high data rate (i.e., about 1 Gbps). Some viable new technological solutions have to be found and tested thoroughly before they can be applied to future B3G wireless systems. It is indeed a tough but an interesting research topic.

Another critical issue that should be taken into account in designing or searching for B3G multiple access technologies is that the dominant traffic in future wireless channels will be virtually all burst-type owing to the wide use of all-IP wireless architecture in all future B3G systems. There will be no continuous transmissions originating from normal mobile subscribers, as it will be extremely wasteful in terms of bandwidth utilization. Continuous transmissions may happen only in some wireless trunk loops. Therefore, all subscribers will use high-speed packet-switched data transmission with a lot of on-and-off pauses in their transactions. The bursty nature of the B3G wireless communications will completely change the design philosophy of a wireless transceiver, which has to work at a very high speed to conduct channel estimation and signal detection on a packet basis, instead of on a frame basis as is the case with the current 2-3G wireless systems.[12] The high-speed bursty transmission in the B3G wireless systems has also made many currently available technologies obsolete and some new ones should be used to replace them. For instance, CDMA technology has been used in all current 3G mobile cellular standards, becoming a defacto standard multiple access technology for them. All 3G mobile standards were designed based basically on the circuit-switched traffic load with a relatively low peak transmission rate, typically below 2 Mbps. However, if the same CDMA technology was used in a 4G wireless system with a peak transmission rate being as high as 1 Gbps, big problems will definitely emerge. An obvious fact is that the conventional CDMA codes used in the 3G wireless systems were designed without considering at all their partial cross-correlation and partial auto-correlation functions, implying that an unprecedented high MAI and multipath interference (MI) will appear if a system needs to capture a lot of short bursts, each of which may contain only a small number of bits. Therefore, in this sense, the current CDMA technologies have to be greatly improved before they can be applied to any of the B3G wireless applications.

In Section 7.2 we introduce, in particular, a special issue titled "Multiple Access Technologies for B3G Wireless Communications," published in the 2005 February issue of the IEEE Communications Magazine [562], which was edited by both the authors of this book, Professor Hsiao-Hwa Chen and Professor Mohsen Guizani. This IEEE Communications Magazine special issue contains altogether

---

[12]Here, we assume that each frame is always much longer than a packet.

eight papers contributed by very well known experts working in these areas worldwide, some of which have been included in the top 10 most popularly cited references in this particular research area in 2005. The significance of this special issue in the IEEE Communications Magazine [562] is also reflected in the timing of its publication, when the research community is working hard to start 4G wireless system design, and when multiple access technologies are always the core of its fundamental architecture. Also, the performance of a B3G wireless system will have very much to do with the multiple access technology it uses.

Section 7.3 is titled "Next Generation CDMA Technologies," whose focus has been indicated clearly in the title itself. As mentioned earlier, the current CDMA technologies are not suitable for B3G wireless applications, where high-speed bursty traffic will be dominant in the channels. This section tries to give possible answers to the questions, such as what the next generation CDMA technologies will look like and how to implement them, and so on. Several important aspects in innovating the current CDMA technologies have been addressed in Section 7.3. They include the design of better CDMA codes, the consideration of spreading and carrier modulation schemes[13], and so on. In particular, we introduce a new CDMA code design approach, namely, the Real Environment Adaptation Linearization (REAL) approach, as an effort to find some optimized CDMA codes with unique MAI-free and MI-free properties. In the REAL approach, a CDMA code set has been generalized as a generic complementary code set with its element code length, the flock size and set size being N, M, and K. If $M = 1$, the code becomes a normal unitary code, such as Gold codes, Walsh-Hadamard codes, OVSF codes, and so on. If $M > 1$, a complementary code is considered.

Two very important conclusions have been drawn from the discussions given in Subsection 7.3.5. The first conclusion is that it is impossible to implement an MAI-free and MI-free CDMA system if conventional unitary CDMA codes are used.[14] The second conclusion is that, theoretically speaking, such an MAI-free and MI-free CDMA system can only support maximum M users, where M is the flock size of the complementary codes used in this CDMA system. The two conclusions made in this subsection may guide us to search for more suitable CDMA codes for application in future CDMA-based B3G wireless communication systems. The significance of the REAL approach is that it is the first time in the history of the literature that the theory of the validity of the two aforementioned conclusions, telling us that we have to use complementary codes as the CDMA codes for the next generation CDMA technologies, is proven. Therefore, the complementary codes will become an essential part of the next generation CDMA technologies. Do not waste time in designing a CDMA-based B3G wireless system using any of the unitary codes.

Section 7.4 discusses multicarrier (MC) CDMA technology, which is another very popular air-interface architecture proposed recently, and has been found to have many applications in wireless systems. Intuitively, the MC-CDMA technologies were developed from single-carrier CDMA technologies. The major difference between a traditional single-carrier CDMA and MC-CDMA systems lies in the fact that the latter can take the advantage of transmitting multiple data streams in parallel via different carriers to make each subcarrier channel a flat fading channel. The errors caused in these subcarrier channels that coincide, deep fades in a frequency-selective fading channel and can be corrected via interleaving plus error-correction coding techniques, which have been widely used in many other wireless subsystems.

We introduce four major MC architectures in Section 7.4, including duplicated time-spreading MC-CDMA, duplicated frequency-spreading MC-CDMA, multiplexed time-spreading MC-CDMA, and multiplexed frequency-spreading MC-CDMA scheme. Their performances are compared in the section. It has to be noted that the importance of the MC-CDMA architecture has also been

---

[13]In Subsection 7.3.3, we proposed a novel spreading modulation scheme, that is, offset stacking (OS) spreading modulation, for its possible application in next generation CDMA systems.

[14]Here, we define a unitary CDMA code as the one that works on a one-code-per-user basis. On the other hand, a complementary code works the other way: each user is assigned a flock (which is normally an even number) of element codes for CDMA purpose.

reflected in the fact that it is undoubtedly the foundation for another emerging technology, namely, the *orthogonal frequency division multiplex (OFDM) technology*, which is addressed in detail in Section 7.5.

Section 7.5 is dedicated to the discussions on OFDM technology, which has been given a lot of attention very recently because of its successful application in several commercial wireless systems, such as IEEE802.11a, IEEE802.11g, Digital Audio Broadcasting (DAB), Digital Video Broadcasting (DVB), and so on. As mentioned earlier, an OFDM scheme can be considered as an economical implementation of a MC system. One of the salient features of OFDM is its simplified base band implementation of parallel carrier modulation (via IFFT algorithm) and demodulation (via FFT algorithm) modules, which otherwise would have to be implemented by analogue components in a MC system. As both IFFT and FFT can be effectively realized using software solutions, it will make an OFDM module extremely flexible to fit into different applications with a different number of subcarrier channels by simply adjusting the number of FFT points.

It is noted that OFDM can use overlapped subcarriers to greatly improve its bandwidth efficiency because of the fact that it sends information by the combinations of tones (instead of a signal always occupying a certain bandwidth). Thus, the output from the IFFT unit will be a linear combination of multiple tones, depending on the input data patterns at the IFFT unit. Obviously, if there is only one single nonzero input with the rest being zero, the output signal from the IFFT unit will be a single sinusoidal waveform (or tone) at a certain frequency. Otherwise, a mixture of several sinusoidal waveforms (or tones) with different frequencies will be the result. Section 7.5 also addresses some implementation problems particularly associated with an OFDM system, such as cyclic prefix (CP) in Subsection 7.5.2, peak-to-average-power (PAPR) issues in Subsection 7.5.3, and Orthogonal Frequency Division Multiple Access (OFDMA) issues in Subsection 7.5.4, and so on. It has to be noted that OFDMA[15] is a rising star as a novel multiple access technology with a great potential for its possible application in future wireless communication systems.

Section 7.6 deals with the issues on UWB technology. In this section, in addition to the introduction of UWB technologies, we will also show how to use an analytical method to study a DS-CDMA UWB system, where both flat fading and frequency-selective fading channels will be taken into account.[16]

Chapter 8 titled "MIMO Systems" was written by Professor Hsiao-Hwa Chen. The MIMO system is a very important emerging technology, which was proposed initially in the later 1990s. The extreme importance of the MIMO technology has been reflected in the fact that it creates the "third" dimension of the diversity mechanism on top of the existing two, that is, frequency diversity (e.g., via MC transmission) and time diversity (e.g., via RAKE reception) techniques. It should be noted that only the MIMO system can provide such a spatial diversity gain to a wireless communication system without consuming other precious radio resources (such as the frequency and the time), which are nonreplaceable, of course with some price to pay in terms of complexity. This is partly the reason that some people call the MIMO technology as one of the most important technological breakthroughs in the wireless communications arena in the last 20 years.

There are two advantages in using MIMO technologies in a wireless communication system. One is it provides a scalable spatial diversity gain depending on the number of antennas used in the Tx and the receiver (Rx). This spatial diversity gain is obtained from the statistical independence existing in the channels over different Tx–Rx Rx antenna pairs, each of which can be viewed as an independent signal replica of others. Thus, combining these independent replicas can achieve substantially high

---

[15]OFDMA treats frequency (or tone) and time as two signal spaces, different combinations of which can be assigned to different users in the same network for multiple access purpose. On the other hand, please note that OFDM is not a multiple access technology. Instead, similar to a MC scheme, OFDM can only provide multiplexing capability to a particular user.

[16]In the analysis that considers frequency-selective fading channels, a modified Saleh-Valenzuela (S-V) channel model will be used to fit a typical operational environment, such as an indoor channel.

diversity gain at a Rx without consuming any of the previous bandwidth resources, such as time and frequency. Another advantage that a MIMO system can exploit is its multiplexing capability provided through multiple transmission paths formed on different Tx–Rx antenna pairs, each of which can be considered as an independent data pipe. Thus, the use of more Tx–Rx antenna pairs will create more transmission paths in parallel, contributing to a great increase of the total data transmission rate in a particular point-to-point wireless air link. In fact, the advantage in the diversity gain and the multiplexing capability can be traded off to fit a particular application scenario. This is a very powerful leverage for a MIMO technology to offer, thus making it an indispensable part of future wireless communication systems. In Chapter 8, the theories on both spatial diversity and multiplexing schemes for a MIMO system are covered.

The MIMO technology has been treated as another focal point in this book, as clearly seen from the number of pages and the breadth of the topics it covers. In addition to the introduction to varied background knowledge of SISO, SIMO, MISO, and MIMO systems and the diversity versus the multiplexing capability in Sections 8.1, 8.2, and 8.3, we also provide an analytical example to evaluate the performance of a space–time block coded (STBC) CDMA system based on complementary codes, as explained in detail from Sections 8.4 to 8.8. The proposed complementary-coded STBC-CDMA scheme is the result of our very recent research activities. The purpose of this new scheme is to try to combine the desirable features of a MIMO system with the complementary-coded CDMA system with unique MAI-free and MI-free properties. This is a part of our effort to search for a suitable system architecture for futuristic B3G wireless communications. The results shown in Sections 8.4 to 8.8 have demonstrated that the STBC-CDMA system based on complementary codes is very promising in terms of its bit error rate performance, which can also be translated into capacity and bandwidth efficiency gains.

Chapter 9 covers the topic on "Cognitive Radio Technology," which is a new wireless technology proposed very recently as a solution to reuse the fallow spectrum that has not been fully used. The importance of the cognitive radio technology has been reflected in the remarks made by Ed Thomas, former Chief Engineer of the Federal Communication Commission (FCC): "If you look at the entire RF frequency up to 100 GHz, and take a snapshot at any given time, you'll see that only five to ten percent of it is being used. So there's 90 GHz of available bandwidth." This tells us that the utilization of the current radio spectrum is severely inefficient, and thus the cognitive radio can find its way to exploit the unused spectrum from time to time, as long as the vacancy appears in the spectrum.

In terms of its operation mode, cognitive radios are intelligent cell phones or smart radios that determine the best way to operate in any given channel situation. Instead of following a set of predefined protocols, as regular radios do, cognitive radios configure to their environment and their user's needs. Cognitive radios are similar to living creatures in that they are aware of their surroundings and understand their own and their user's capabilities and the governing social constraints. A radio's actions arise from a rational process that predicts probable consequences and remembers all of its failures and successes. The radios are treated like animals that learn to evolve over time with their changing environment. Basically, the cognitive engine is a brain that reads the radio's meters and turns the radio's knobs in order to get the desired outcome.

Cognitive radios can be applied to mainly two wireless application environments, one being the situation where cognitive radios should work on licensed bands, in which incumbent users are working; and the other being the environment where cognitive radios work on unlicensed bands. In the first situation, one or more cognitive radios should work with licensed primary users, which are never cooperative. In the first case, avoiding interference with these primary users while keeping reliable communications for cognitive radios, is the issue of concern. In the second scenario, it is possible to implement cognitive radios in all wireless terminals, which work in the same unlicensed bands. In this case a close cooperation among all cognitive radio terminals will be the key for successful operation of the cognitive radio network.

In this chapter, there are 11 sections, starting from the very basic concept of cognitive radio technology to various possible applications of cognitive radios in WPANs, WLANs, WMANs, WWANs, and WRANs. Several new IEEE 802 standards, such as IEEE 802.11h, IEEE 802.22, IEEE 802.16h, and so on, which are related to the applications of cognitive radio technologies, are introduced. We also introduce a few available cognitive radio products as examples in this chapter, to allow the readers to have a feeling about the real world of the cognitive radio.

Chapter 10 is titled "E-UTRAN: 3GPP's Evolutional Path to 4G"; "E-UTRAN" stands for Evolved UTRAN, which is a new technology owing to UTRA's long term evolution. E-UTRAN is also called *Super-3G* or simply 4G technology, and is still a technical standard under discussion in many 3GPP TSG RAN and SA Working Groups meetings. The primary targets for the development of 3GPP E-UTRAN are explained as follows.

To ensure competitiveness of the 3GPP systems in a time frame of the next 10 years and beyond, a long-term evolution of the 3GPP access technology needs to be considered. In particular, to enhance the capability of the 3GPP system to cope with the rapid growth in IP data traffic, the packet-switched technology utilized within 3G mobile networks requires further enhancement. A continued evolution and optimization of the system concept is also necessary in order to maintain a competitive edge in terms of both performance and cost.

Important parts of such a long-term evolution include reduced latency, higher user data rates, improved system capacity and coverage, and reduced overall cost for the operator. Additionally, it is expected that IP-based 3GPP services will be provided through various access technologies. A mechanism to support seamless mobility between heterogenous access networks, for example WLANs and 3GPP access systems, is a useful feature for future network evolution. In order to achieve this, an evolution or migration of the network architecture, as well as an evolution of the radio interface, should be considered. Architectural considerations will include end-to-end systems aspects, including core network aspects and the study of a variety of IP connectivity access networks (e.g., fixed broadband access).

In Chapter 10, there are in total eight sections, which cover the issues ranging from the organization of 3GPP TSG WGs for E-UTRAN, the origin of E-UTRAN, general technical features of E-UTRAN, introduction of E-UTRAN radio interface protocols, E-UTRAN physical layer aspects, and a summary remark. In this chapter, we include, in particular, a section (Section 10.7) to discuss a specific physical layer architecture design based on downlink and uplink FDD-OFDMA technology, which consists of many cutting-edge techniques.

It is noted that the coverage on the 3GPP E-UTRAN technology can be rarely seen in currently available books. Therefore, the information contained in this chapter is very timely and will be a useful reference to the people working in next generation wireless systems and networks.

1.5 Suggestions for Using This Book

As mentioned earlier, this book can be treated primarily as an up-to-date research reference, including many cutting-edge technologies for the people working in wireless communication systems and networks. In addition, this book can also be used as a teaching material to serve different teaching purposes, such as ordinary undergraduate/postgraduate courses in universities/colleges or short courses in professional training or continued education classes, and so on. In order to exploit maximum benefit from this book as either a supplementary teaching material (for undergraduate or postgraduate classes) or as the main text for a professional training course, we to provide some guidance in this section to the instructors who will conduct the lectures in the aforementioned courses.

There are a few paradigms for the course teaching to proceed, depending on the titles and the duration of the courses, as well as the background knowledge level of the attendees of the courses.

If the course is focused more on "Wireless Networks," we suggest that the instructors use the following chapters, that is, Chapters 2, 4 and 5, from this book to form the major part of the lecturing material, as shown in Figure 1.2(a).

If the course is concentrated mainly on "Wireless Communication Systems," we suggest that the flow diagram shown in Figure 1.2(b) should be used for the course teaching. On the other hand, if a course about "3G Mobile Cellular Communications" is given, the teaching flow diagram

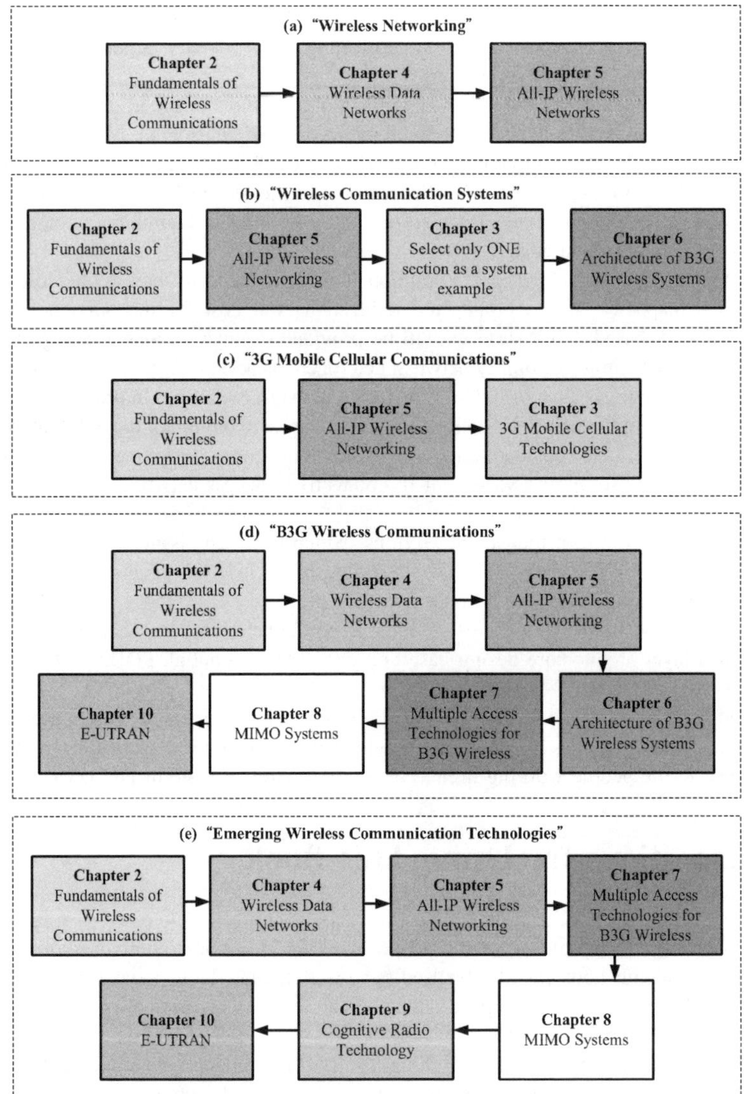

Figure 1.2 The teaching flow diagrams for different courses concentrated on (a) wireless networking, (b) wireless communication systems, (c) 3G mobile cellular communications, (d) B3G wireless communications, and (e) emerging wireless communication technologies.

shown in Figure 1.2(c) could be used. The book can also be used to conduct a teaching course on a more advanced level for senior postgraduate students. For instance, a course titled "B3G Wireless Communications" can be taught on the basis of the materials given in this book. An instructor can use the paradigm shown in Figure 1.2(d) to conduct the course in a time frame of one term. Another possible course using this book is "Emerging Wireless Communication Technologies," which can be taught within one semester by using the paradigm illustrated in Figure 1.2(e).

Therefore, it can be seen that this book can be used for teaching different courses for either undergraduate or postgraduate students in universities. It is also suitable for conducting many other different training classes dedicated to telecommunication engineers for continuing education purposes. Because of the time constraint, it is unfortunate that we could not provide exercise problems at the end of each chapter in this edition. This will be an important part when the next edition is brought out. We will, in particular, welcome any comments on how to use this book. The comments can be sent via email to either of the authors, Professor Hsiao-Hwa Chen (hshwchen@ieee.org), or Professor Mohsen Guizani (mguizani@cs.wmich.edu).

2

Fundamentals of Wireless Communications

In this chapter some important prerequisite concepts necessary to understand the other parts of the book are discussed, such as theory of mobile and wireless channels, SS communication, multiple access technologies, switching techniques and so on. This chapter aims to provide the essential knowledge for readers with limited background of wireless networks and communications to follow the major contents covered in this book easily. This is done in a fairly introductory way, and only limited mathematics is used. More in-depth information discussed in this chapter can also be found in open literature, such as [1–40].

Wireless networks and communications are not new technologies, it being about 30 years ago when mobile cellular telephony was first introduced into our daily lives. Some pioneer work in this area should be acknowledged. Here, we need not trace them back to as early as the time when Popov of Russia [42] and Marconi of Italy [4, 41] did their important experiments on long-distance radio telegram transmissions in 1895 and 1902, respectively. However, we would like to look at what happened in the development of wireless networks and communications in the last 50 years.

In the mid-1960s, primitive data packet technology was developed for its application in the Advanced Research Projects Agency Network (ARPANET) [43], which was established in 1969 by the US Department of Defense. Initiated in 1970, the ALOHANET, based at the University of Hawaii, was the first large-scale packet radio project, an effort to link its different campuses scattered over several isles using a so called *packet radio* system to solve the problems of data link connections between the campuses. This interesting experiment not only solved its own data connection problems, but also triggered a new wave of research on telecommunications with connectionless transmissions, which was probably the first revolutionary trial on modern packet-switched wireless networks.

The experiments made in the 1960s and 1970s in both ARPANET and ALOHANET were of great significance, as they incubated two most important technological ingredients of modern all-IP–based wireless communications, the Internet, and packet switching wireless networks. *Packet radio in motion* has provided a generic model for today's cutting-edge research, high-speed burst traffic wireless networks and communications, a topic that has to be dealt with in all wireless systems beyond 3G (B3G).[1] Therefore, we can see the synergy between the effort made in all pioneer experiments

[1] Mobile communication has experienced three generations up to now: The first generation was based on analog voice-centric technologies; the second generation was developed by digital voice-centric transmission techniques; the third generation is focused on multimedia applications.

Next Generation Wireless Systems and Networks Hsiao-Hwa Chen and Mohsen Guizani
© 2006 John Wiley & Sons, Ltd

and modern wireless networks and communications research. The understanding of its history will
definitely help us to gain great interest in our current studies as well as to envisage its future trends.

2.1   Theory of Radio Communication Channels

The success in the development of the modern wireless networks and communications technologies
is attributed to research breakthrough in wireless and mobile communication channels. Before the
1960s the main research effort in telecommunications was made for the development of various
techniques to overcome noise and external interferences in radio channels. The introduction of mobile
cellular systems, exemplified by the first field trial in Chicago City in 1978, followed by the world-
first commercial cellular services also in the Chicago region in the 1980s, made people realize the
importance of investigating the negative impact of mobile terminals in motion with multipath effect
on the overall performance of a mobile cellular communication system as a whole. Now, we have
entered the era of B3G wireless communications, which should deliver a data transmission rate of
the order of one gigabit per second, but we still face challenges from the unpredictability of channel
conditions due to the high transmission rate compounded by mobility. The existence of multipath
propagation has further complicated the issues that we have to deal with in the design of B3G wireless
systems.

   Therefore, the theory of wireless communication channels is probably one of the most important
and difficult parts in modern wireless networks and communications. Many books in this area are
already available on the shelves. Unfortunately, most of them tend to use too many equations and
acronyms with too few illustrations related to the real world. In this section, we will try to introduce
some basic concepts about wireless communication channels, followed by discussions on the various
major phenomena encountered in wireless and mobile cellular environments.

2.1.1   Radio Signal Propagation

Before going to some complex details of the theory on wireless or mobile communication channels,
we would like to introduce some fundamental concepts about radio propagation channels.

Free-space propagation model

Radio signals will attenuate with the increase of their propagation distances. Assume that an omnidi-
rectional transmitter antenna is concerned here. With a fixed transmitting power from the transmitter,
the signal strength picked up at a receiver antenna will reduce as the distance between the transmitter
and receiver increases due to the conservation law of the power dissipated from the source, or the
transmitter antenna. This gives us a *free-space propagation model*, which can be used to predict the
received signal strength if a transmitter and a receiver have an unobstructed line-of-sight (LOS) path
between them. There are many communication scenarios that support such a free-space propagation
model, including satellite communication systems, microwave relay radio links, deep space communication systems, and so on. The free-space propagation model tells us that received power decays as
a function of the transmitter–receiver separation distance raised to some power (i.e., as a power law
function). The power received in a free-space model by a receiver antenna that is separated from a
radiating transmitter antenna by a distance d can be calculated by the *Friis free-space equation*, or

$$P_r(d) = \frac{P_t G_t G_r \lambda^2}{16\pi^2 d^2 L} \tag{2.1}$$

where $P_r(d)$ is the received power in Watt and is a function of the transmitter-receiver separation
distance in meters, P_t is the transmitting power, G_t is the transmitting antenna gain, G_r is the receiver

antenna gain, d is the distance in meters between the transmitter and the receiver, L is the system loss factor that is not associated with propagation ($L \geq 1$), and λ is the wavelength of the radio signal in meters.

The gain of an antenna is associated with its *effective aperture*, A_e, or

$$G = \frac{4\pi A_e}{\lambda^2} \tag{2.2}$$

On the other hand, the effective aperture A_e is related to the physical size of an antenna and λ is derived from the carrier frequency as

$$\lambda = \frac{c}{f} = \frac{2\pi c}{\omega_c} \tag{2.3}$$

where f is the carrier frequency in Hertz, c is the speed of light in meters per second, and ω_c is the carrier frequency in radians per second.

It should be noted from Equation (2.1) that the values for P_r and P_t must be expressed in the same unit and G_r and G_t are dimensionless variables. The miscellaneous loss L (≥ 1) is usually because of transmission line losses, filter attenuation and antenna losses in the whole communication system. Setting $L = 1$, we imply that there is no loss in the entire transmission system.

It is to be noted that, in a non–free-space operational environment, the order of the power of d^2 in Equation (2.1) depends on the environment that the radio link is operating in and can change from two to five. The higher the power, the faster the received power will decay.

Reflection, diffraction, and scattering

We will discuss three major radio signal propagation mechanisms, that is, *reflection*, *diffraction* and *scattering*.

In a real communication environment, free-space propagation seldom happens. Instead, a great amount of dissipated power from a transmitter antenna will be absorbed by the above three basic propagation mechanisms, especially when the radio signals are sent into an area where a lot of human architecture exists, such as cities, suburban areas, villages, and so on. Even if LOS transmission does occur, the combination of signal components from reflection, diffraction, and scattering mechanisms under the LOS transmission will make it impossible for the received signal to obey the *free-space propagation law*.

Reflection occurs when a propagation radio wave hits an object that is very large in size when compared to the wavelength of the transmitted signal. Reflection occurs from the surfaces of rocks, buildings, big glass windows, cliffs, and walls, and so on. Reflection is the major cause of *multipath effect* in wireless communication channels.

Diffraction will take place when the electromagnetic wave propagation path is obstructed by a surface that consists of a lot of sharp irregularities (sharp edges or bumps). The waves diffracted from the obstructing surfaces are present throughout the space and sometimes behind the obstacle, resulting in a bending of waves around the obstacle, even when an LOS path does not exist between a transmitter and a receiver. At a high frequency range, diffraction depends on the geometry of the obstacles, as well as the amplitude, phase, and polarization of the incident wave at the point of diffraction. Usually, diffraction takes place less often when carrier frequency is higher, as the electromagnetic waves behave more like particles rather than waves. Diffraction is the main cause of the *shadowing effect* in wireless communication channels.

Scattering or *Diffusion* occurs when the radio medium, through which the wave passes, consists of objects with dimensions that are small when compared to the wavelength of the radio signal, and

---

[2]d^2 is assumed in Equation (2.1) due to the free-space propagation; otherwise the power can be larger than 2, depending on the propagation environments of concern.

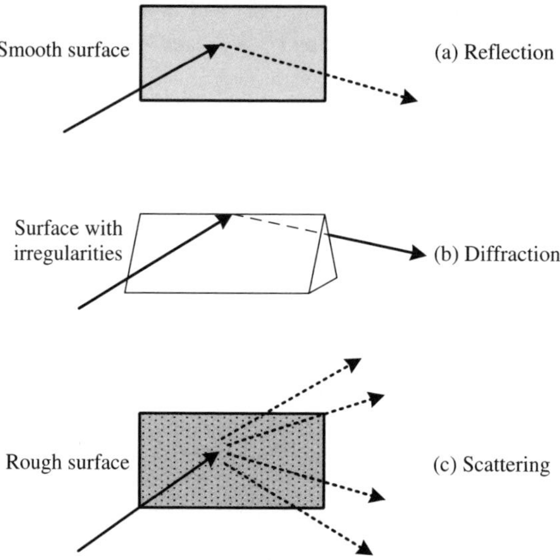

Figure 2.1 The illustration of the three major radio propagation mechanisms.

where the number of obstacles per unit volume is relatively large. Scattered or diffused radio waves are usually generated by rough surfaces, small objects, or other irregularities in the radio channel. Normally, foliage, street sign poles and lamp posts induce scattering in a radio propagation channel.

The three main radio propagation mechanisms are illustrated in Figure 2.1. However, it should be pointed out that the aforementioned three major propagation mechanisms always come together instead of individually. The combined effect of the three radio wave propagation mechanisms will make the signal received at a real receiver behave like a complex random process, discussed in the following text.

2.1.2 Fading Channel Models

Fading is a serious impairing effect introduced by the radio wave's propagation through the channel and causes a big problem to the signal detection process at the receiver.

When the signal experiences fading in the channel, both its envelope and phase will fluctuate over time. Where a coherent modulation scheme is concerned, the fading effects on the signal phase can seriously impair performance, unless some necessary measures are taken to compensate for them at the receiving end even at the cost of complexity of the receiver. In many cases, analysis of systems using such coherent modulation schemes assumes that the phase effects due to fading are perfectly corrected by the receiver. This assumption is also referred to as *ideal coherent demodulation scheme*. On the other hand, for noncoherent modulation schemes, phase information is not needed at the receiver and thus the fading in phase will not affect the demodulation process at a noncoherent receiver. Therefore, the performance analysis for both the ideal coherent and noncoherent modulation schemes over a fading channel requires only information about the fading envelope statistics. Also, in the case of slow fading[3] (to be discussed later in this chapter), where the fading is considered

---

[3]Usually we assume that a fading is slow if its changing cycle is longer than the smallest time interval in which a receiver should process; otherwise a fast fading should be assumed. For instance, the smallest time interval in a code division multiple access (CDMA) system is often the chip interval, if each chip is sampled only once.

to be constant during the duration of a symbol time, the fading envelope random process can be represented by a random variable over the symbol duration to simplify the analysis. Therefore, in most cases the fading effects on the envelope plays a more important role than the effects on phase in determining the performance of a wireless communication system.

As mentioned in the earlier subsection, there are mainly three different radio wave propagation mechanisms in a wireless or mobile communication channel, that is, reflection, diffraction, and scattering, which always act simultaneously on a signal traveling through the channel until it reaches a receiver. Therefore, from the receiver's point of view, the received signal is always the result of a combination of reflected, diffracted, and scattered signals from different obstacles along the propagation path. In addition, a wireless or mobile communication channel is usually shared by many users, and thus the received signals will be a mixture of signals from different sources through different propagation paths. This introduces great unpredictability or randomness to the received signal.

In order to describe this randomness in a received signal, let us look at the issue from two different angles, one being the *local point of view* and the other being the *global point of view* of the radio wave propagation mechanisms. For the local point of view, let us assume that the transmitted signal impinges onto a large plate or a wall-like obstacle with a rough surface, such as big rocks and concrete walls. One single impinged signal will yield a bundle of outgoing rays with their amplitudes and phases being slightly different from one another. Assume that the transmitted signal takes a complex expression, as it always has amplitude and phase components. Also assume that the LOS propagation path does not exist in the received signals. When those large number of outgoing complex signals are combined together at a receiver, we can use the *large number law* or *central limit theory* from *probability theory* to assert that the received signal must be a random variable obeying the complex Gaussian distribution.

On the other hand, the situation can be examined from a global point of view. Assume that there are many transmitters sending their signals into the channel, each of which will propagate through different paths and will experience reflections, diffractions, and scatterings in the channel before arriving at a receiver. Also assume that the signal sent by each transmitter is a complex signal with its independent amplitude and phase components. Therefore, observed from the receiver, the received signal consists of the components from different signals sent by distinct transmitters. Thus, again it is heuristically right for us to assert that the combined signal viewed at the receiver is a complex Gaussian random variable [49, 50].

Rayleigh fading

Therefore, no matter how the problem is viewed, the received composite signal from a channel without the LOS path, but with reflection, diffraction and scattering, will be a complex Gaussian random variable. The amplitude of this complex Gaussian random variable obeys a Rayleigh distribution with its variance being the same as that of the complex Gaussian random variable and its probability density function (pdf) being

$$f_{Rayleigh}(r) = \frac{r}{\sigma^2}\, e^{\left(-\frac{r^2}{2\sigma^2}\right)} \qquad (0 \leq r < \infty) \qquad (2.4)$$

where σ is the root mean squared value of the received signal before detection, and σ^2 is the average power of the received fading signal.

The phase of this complex Gaussian random signal obeys a uniform distribution defined over $[0, \pi]$ with its probability density function being

$$f_{Uniform}(\theta) = \frac{1}{2\pi} \qquad (0 \leq \theta \leq 2\pi) \qquad (2.5)$$

Rician fading

Assume again that the signal sent from a transmitter is a complex signal with its real and imaginary parts, which is the case with most quadrature modulated carrier signals, such as QPSK, MSK, and QAM. If there exists a strong LOS ray along the propagation path, which is a relatively stationary or unfaded signal component, the envelope of the signal received from a mobile channel at a receiver will be a Rician distributed random variable and the phase will be again a uniformly distributed random variable. In such a situation, the reflected, diffracted, and scattered signals arriving at different angles will be superimposed on the strong stationary component to form the Rician random envelope and a uniformly distributed phase as a whole. As the dominant LOS component becomes weaker, the envelope of the composite signal looks more like a Rayleigh distributed random variable. Therefore, the Rician distribution will decay to a Rayleigh distribution when the strong LOS component decreases to a certain level.

The Rician random variable, like the Rayleigh distributed random variable, is also continuous and its probability density function is defined as

$$f_{Rician}(r) = \frac{r}{\sigma^2} \, e^{-\frac{r^2+A^2}{2\sigma^2}} \, I_0\left(\frac{Ar}{\sigma^2}\right) \qquad (0 \le r < \infty, \ 0 \le A < \infty) \qquad (2.6)$$

where σ^2 is the variance of the received fading signal, A is the peak envelope level of the dominant LOS signal component and $I_0(x)$ denotes the modified Bessel function of the first kind of zero order with x being the variable. The phase of the received signal in a Rician channel still obeys uniform distribution as shown in Equation (2.5).

Nakagami-*m* fading

Another important fading signal distribution is the Nakagami-m distribution, which can be used to characterize the signal received in a land mobile and many other indoor-mobile environments. Also, the Nakagami-m model can be employed to describe the statistics of signals received in scintillating ionospheric radio links.[4] However, the most important application of the Nakagami-m channel model is its versatility to express other random fading distributions, such as one-sided Gaussian distribution $\left(m = \frac{1}{2}\right)$ and Rayleigh distribution ($m = 1$). When $m \to \infty$, the Nakagami-m distributed fading channel will converge to a nonfading *additive white Gaussian noisy* (AWGN) channel. In addition, the Nakagami-m channel model can also be used to approximate Rician and several other random distributions with the help of some appropriate one-to-one parameter mapping algorithms.

The importance of the Nakagami-m fading channel model is also reflected in that fact that it gives the widest span of *amount of fading* (AF), which is also sometimes called *fading figure*, among all commonly referred fading channel models. The AF of Nakagami-m fading channel model spans from zero to two via its parameter m, where the AF is defined by

$$AF = \frac{var(r^2)}{\{E[r^2]\}^2} = \frac{E[(r^2 - 2\sigma^2)^2]}{4\sigma^2} \qquad (2.7)$$

where r denotes the received fading signal level, σ^2 stands for the variance of the fading signal, $E[\cdot]$ is the statistical average and $var[\cdot]$ represents the variance. The use of Equation (2.7) was first proposed by [59]. It is to be noted that the AF is typically independent of the average fading power $2\sigma^2$ and thus can be used perfectly as a unified measure of the severity of the channel fading. With the widest span of the AF value, the Nakagami-m fading channel model is indeed a general fading model to predict fading severity in different types of wireless communication channels.

---

[4]Many researches have suggested that the Nakagami-m distribution often gives the best fit to land mobile and indoor wireless multipath propagation, as well as scintillating ionospheric radio links.

Assume that r is a Nakagami-m random variable, then its corresponding pdf is described by

$$f_{Nakagami\text{-}m}(r) = \frac{2r^{2m-1}}{\Gamma(m)\Omega^m}\, e^{-\frac{r^2}{\Omega}} \qquad (0 \le r < \infty) \tag{2.8}$$

where $\Gamma(\cdot)$ is the Gamma function, $\Omega = \frac{\overline{r^2}}{m}$ with $\overline{r^2}$ being the average received signal power and m representing the inverse normalized variance r^2, which has to satisfy the condition of $m \ge \frac{1}{2}$, describing the fading severity. In addition, it has been well known that r can be considered as the square root of the sum of squares of m independent Rayleigh or $2m$ independent Gaussian random variables [60–62].

Figure 2.2 plots the probability density functions of Nakagami-m random variable with different parameters m. It can be seen that the Nakagami-m distribution will become Rician, Rayleigh, half-Gaussian when the parameter m takes $\frac{3}{2} \sim 3$, 1 and $\frac{1}{2}$, respectively.

Log-normal fading

In terrestrial communication and satellite land mobile communication systems, the link quality can also be affected by some slow variation in its mean signal level because of the *shadowing effect* caused by terrain, buildings, trees, and other large-sized obstacles right in the radio signal propagation paths. Based on many previous studies, which provided a lot of valuable experimental results, there is a general consensus that shadowing effect can be modeled effectively by a log-normal distribution for various outdoor and indoor environments [63–68]. The log-normal fading model can also be widely

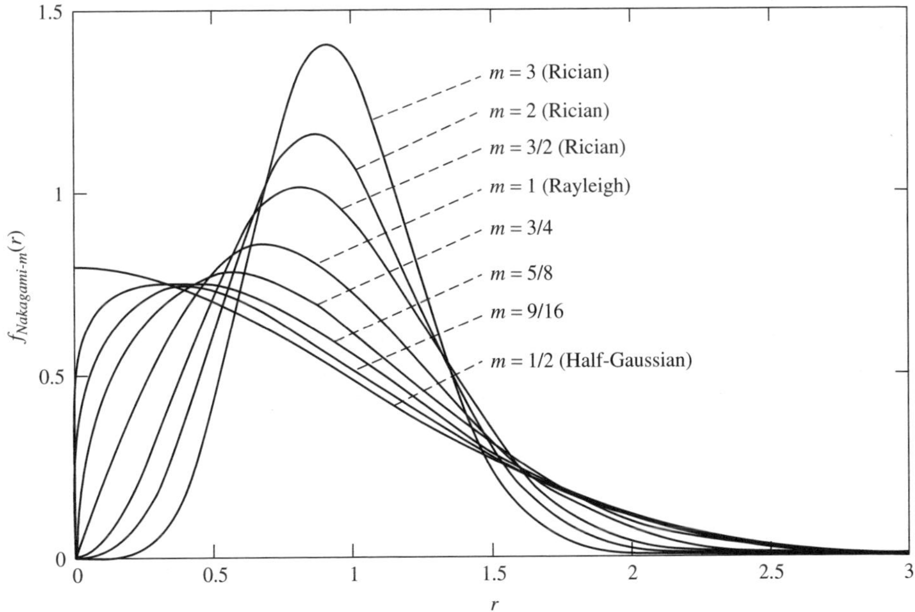

Figure 2.2 The probability density function of Nakagami-m random variable with m being its parameter. It can be seen that the Nakagami-m distribution will become Rician, Rayleigh, half-Gaussian when the parameter m takes $\frac{3}{2} \sim 3$, 1 and $\frac{1}{2}$, respectively.

used to characterize indoor non-LOS propagation path loss, especially when the signal will penetrate through walls, windows (especially with steel frames), partition boards and other large obstacles.

The log-normal distributed random variable in fact obeys a normal or Gaussian distribution when the values are measured in decibels. If x is a Gaussian random variable with its mean μ and variance σ^2, and if r is related to x by the relation $r = \ln x$, then r is a log-normally distributed random variable with its probability density function as

$$f_{log\text{-}normal}(r) = \frac{1}{r\sqrt{2\pi\sigma^2}}\, e^{-\frac{(\ln r - \mu)^2}{2\sigma^2}} \qquad (0 \le r < \infty) \tag{2.9}$$

Provided that the fading signal level is measured in decibels (as a tradition in all propagation measurements), the pdf function usually dealt with is in fact a Gaussian probability density function of x, or

$$f_{log\text{-}normal}^{(dB)}(x) = \frac{1}{\sqrt{2\pi\sigma^2}}\, e^{-\frac{(x - \mu)^2}{2\sigma^2}} \qquad (0 \le x < \infty) \tag{2.10}$$

where μ is the mean path loss measured in decibels (either measured directly or predicted with the help of other ways) and σ is the standard deviation of the path loss. It is noted that the variable σ is usually a function of terrain and antenna heights, and so on.

2.1.3   Narrowband and Frequency-Domain Characteristics

In this subsection and the one that follows, we will use an experimental approach to discuss the characteristics of wireless or mobile channels. Our experimental approach is based on a generic channel measurement system setup and will proceed in terms of two different perspectives, that is,

- narrowband/frequency-domain channel characteristics;

- wideband/time-domain channel characteristics.

Our approach will start with some very simple and purified measurement system setups for both narrowband characteristics and wideband characteristics of the channels. Through those channel measurement examples, we will explain step by step how the signals will penetrate the channels to reach a particular receiver. After having studied these different facets of the wireless or mobile channels, we will gain a vivid multidimensional image on how a channel will interact with different signals generated from a transmitter. Another benefit of using the studying approach is the conveying of a great number of abstract concepts used to describe complex wireless communication channels, and thus allowing a reader with some knowledge of communication systems to understand easily how a wireless channel will react to different types of transmitting signals. The most important parameters of a wireless channel, such as Doppler spread, delay spread, coherent bandwidth and coherent time, and so on, is explained under the framework of this approach. Therefore, the readers will be able to anticipate correctly the consequences of changing channel conditions to the signals received at a terminal.

A wireless channel can be examined with the help of some narrowband signals, a typical example of which is sinusoidal tones at different frequencies. A channel sounding experiment setup can be used here to help us understand how the approach works. For description simplicity, let us first confine our experimental environment to an indoor space only, such as a classroom, although our discussion could also be easily extended to any outdoor channels in general. It is to be noted that the narrowband channel measurement technique is attractive due to its relatively low cost in acquiring all needed equipments and instruments.

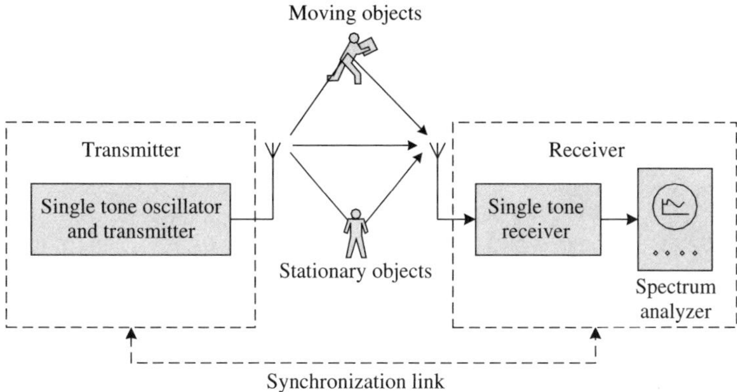

Figure 2.3 A generic system setup of a narrowband channel sounding experiment system, which is made up of a single tone generator cum transmitter and a tone receiver followed by a spectrum analyzer.

This simple channel sounding experiment setup consists of a single tone generator cum transmitter and a tone receiver followed by a spectrum analyzer. The tone generator can transmit a series of single tone continuous sinusoidal waves with different carrier frequencies. The receiver can capture the single tone signals sent from the transmitter with perfect synchronization, such that the receiver will know the exact time the transmitter will take to shift from one frequency to another. The frequencies of the received signals will be displayed in the spectrum analyzer. Figure 2.3 shows the system setup of the narrowband channel sounding experiment set.

With this simple measurement system, we can proceed to study the narrowband characteristics of an indoor channel as follows. First, let us place the transmitter at one corner of a room and the receiver at another corner. The transmitter will send a series of tones with different carrier frequencies. The receiver will record all those tones and display them in a spectrum analyzer. For each transmitted tone, the spectrum analyzer will show it as a spectral line at a particular frequency equal to the carrier frequency of the tone sent. In this way, if we can record all spectral lines in the spectrum analyzer corresponding to all the different tones sent from the transmitter, the *frequency response* or *channel transfer function* of the wireless channel is obtained. Therefore, the narrowband characteristics of a wireless channel and also its frequency-domain response can be studied at the same time with the help of the narrowband channel sounding system setup as shown in Figure 2.3.

Assume that there is absolutely no moving object in the room and that both the transmitter and receiver are fixed in their locations. We will see a very *clean* vertical spectral line in the spectrum analyzer for each tone sent from the transmitter. This result is true even under the multipath effect due to the reflections from walls, ceiling, and floor, and so on. To illustrate it more clearly, let us think about a scenario where the transmitter is sending a tone as $r_1(t) = \sin \omega_1 t$. If only two multipath returns exist, the receiver will get $r_2(t) = \sin \omega_1 t + \sin \omega_1(t + \tau)$, where the propagation delay of the LOS ray is omitted and τ is the relative propagation delay between the two rays. Obviously, we cannot distinguish the spectrum of $r_1(t)$ and $r_2(t)$ in terms of their amplitudes shown in the spectrum analyzer due to the fact that the difference in the spectra of $r_2(t)r_1(t)$ lies only on their phases, not on their amplitudes. Therefore, the multipath effect will not alter the appearance of the spectral lines shown in the spectrum analyzer in a narrowband channel sounding system setup.

On the other hand, now let us introduce some mobility in the indoor channel by either making people move around (as shown in Figure 2.3) or moving or rotating the antennas (of either the transmitter or receiver) directly. Assume that the transmitter is sending a tone with a fixed frequency.

We will observe that the spectral line corresponding to the carrier frequency shown in the spectrum analyzer now becomes spread out or blurred, instead of a clean line as displayed earlier without mobility in the indoor channel. If we examine carefully the width of the spread-out shown in each spectral line, we can see that the width is proportional to the highest speed of the mobility present in the channel. The faster the velocity of the moving objects, the wider the spectral line will be spread out. We call the width of the spectral line spread-out as the *Doppler spread* of the channel. The Doppler spread is a very important characteristic parameter of a wireless or mobile channel and will determine the rapidity, with which the wireless channel varies with time.

The cause of the Doppler spread can also be explained perfectly by the following analysis. Assume the transmitted signal from the transmitter is again a single tone of $r_1(t) = \sin \omega_c t$. When this signal hits a moving object (such as the running person shown in Figure 2.3), the signal reflected from the moving object will carry an extra frequency component called *Doppler frequency*, which is calculated by

$$f_d = \frac{v}{\lambda} \cos \theta = \frac{v f_c}{c} \cos \theta \tag{2.11}$$

where v stands for the moving speed in kilometers per hour, λ is the wavelength of the carrier frequency in meters, f_c is the carrier frequency in Hertz, c is the speed of light in kilometers per hour, and θ is the angle in radians between the moving direction and the LOS from the transmitter to the receiver. f_d defined in Equation (2.11) is also called *Doppler shift* or *Doppler frequency*. The Doppler spread can be expressed by $\Delta f_d = 2 f_d$. In most cases we can use the following approximation to estimate the maximal Doppler shift caused by a moving object as $f_d \cong 1.852v$, where the unit of v is kilometers per hour and the carrier frequency is assumed to be 2 GHz.

Therefore, the received signal at the receiver becomes

$$r_3(t) = \sin \omega_c t + sin \omega_c(t + \tau) + \sin(\omega_c + 2\pi f_d)(t + \tau) \tag{2.12}$$

where it is assumed that there are three reflection paths in a narrowband channel as shown in Figure 2.3. Obviously, only the term $\sin(\omega_c + 2\pi f_d)$ in 2.12 will contribute to the spread-out of the spectral line in the spectrum analyzer, and the other two terms ($\sin \omega_c t$ and $\sin \omega_c(t + \tau)$) will appear as the same spectral line. If there are many reflection rays from many different moving objects with different moving velocities, we will see a series of continuously spread-out spectral lines in the spectrum analyzer, causing the Doppler spread.

Figure 2.4 shows an example of the Doppler spread observed in a spectrum analyzer. It is to be noted that the widest range of the Doppler spread is equal to $\Delta f_d = 2 f_d$, which is associated with the highest speed of the moving objects in the channel and is in fact very small if compared to the carrier frequency f_c.

To have a numerical impression on how large the Doppler spread will be in different application scenarios, we give Table 2.1, in which the Doppler spreads caused by different moving objects in daily life are given. It can be seen that a bullet train moving at a speed of 300 km/h will cause a maximal Doppler spread of about $\Delta f_d = 2 \times 555$ Hz if a $f_c = 2$ GHz carrier frequency is considered here.

The reciprocal of the Doppler spread is defined as *coherent time* of the channel, or

$$T_{co} \cong \frac{1}{\Delta f_d} \tag{2.13}$$

which gives information about how fast the channel will change with time. The coherent time of a channel in fact tells us how much time is needed for the signal to finish changing a complete cycle, that is, from high to low and back to high.

The relation given in Equation (2.13) is significant and it gives us a way to measure the rapidity of change or fade in channel characteristics. With the coherent time, we can differentiate *slow fading*

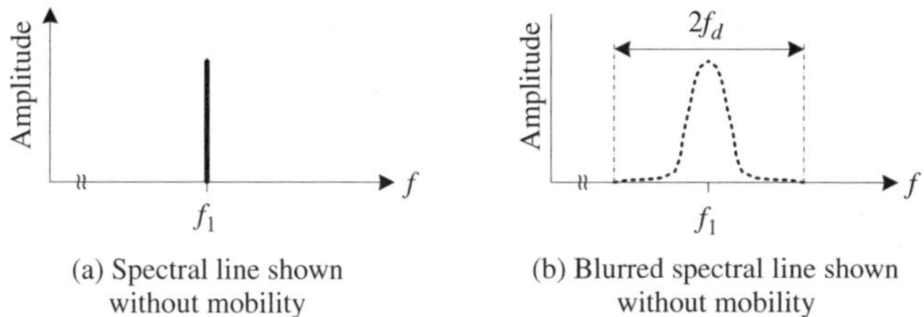

Figure 2.4 The Doppler spread Δf_d caused by mobility in a wireless channel. (a) The spectral line shown in the spectrum analyzer when no mobility exists. (b) The blurred spectral line caused by the mobility in the channel.

Table 2.1 Maximal Doppler spread caused by different moving objects in daily life. Assume that the carrier frequency is 2 GHz here.

Velocity of moving objects (km/h)	Maximal Doppler spread Δf_d (Hz)
Aeroplane (800)	2963
Bullet train (300)	1111
Normal train (120)	444.5
Car (100)	370
Bus (50)	185
Bicycle (10)	37
Walking person (5)	18.5

channel and *fast fading channel*. The fading is said to be slow if the signal symbol time duration T_s is smaller than the coherent time of the channel T_{co}; otherwise it is considered to be a fast fading channel. In a slow fading situation, a particular deep fade will affect many consecutive symbols, which leads to so called *burst errors*. In many communication applications, the fading speed is of great importance to determine the right detection strategy. For instance, when receiver decisions are made based on the observation of received signals over two or more consecutive symbols (such as differentially coded communication schemes), it is always necessary to assume that variation of the fading channel is slow from one symbol to the next; otherwise the differentially coded schemes will not work properly without some extra measures. In the other words, the differentially coded schemes are suitable only for the applications in slow fading channels, rather than the fast fading channels. It is to be noted that the study on the impact of fast fading channels on the performance of a communication system is much more complex and thus more difficult than that of slow fading channels.

Using the narrowband techniques, we can analyze the frequency-domain characteristics of a wireless or mobile communication channel. In particular, it helps us to understand the time-variant properties of a channel in terms of the Doppler spread Δf_d and channel coherent time T_{co}, which forms a reciprocal dual. If a symbol duration of a transmitted signal is longer than the coherent time of the channel, it is said that the signal will suffer a *time selective fading*, which is in contrast with the *frequency selective fading* explained in the following subsection. The time selective fading is the direct consequence of the fast time variation of a channel due to the high mobility of either terminals or moving objects around the receiver. The channel with the Doppler spread caused by mobility of the channel is also referred to as *frequency dispersive channel*.

2.1.4   Wideband and Time-Domain Characteristics

After the discussion on narrowband characteristics of a wireless channel, as given in the previous subsection, we now proceed to study wideband characteristics of the channel again using an experimental approach. Let us think about a channel measurement system which is made up of the following instruments- an impulse generator, a wideband radio transmitter with a wideband antenna, and a wideband receiver followed by a wideband oscilloscope. Assume that a perfect time synchronization is achieved with the help of the synchronization link between the transmitter and the receiver. The wideband measurement system setup is illustrated in Figure 2.5, where it is assumed that the measurement will take place in an indoor environment for simplicity of illustration. Understandably, the wideband channel measurement described in this subsection requires some equipments and instruments, much more expensive than what was needed for the narrowband channel measurements, as discussed in the earlier subsection.

Assume that the transmitter is placed at one corner of the room and the receiver is stationed away from the transmitter, and that there is a LOS path between the transmitter and the receiver. The impulse generator will produce a series of wideband impulses, whose periods should be made longer than the time interval between the first and the last multipath returns, which is also called *delay spread* of the channel, denoted by τ_d. Now, let us make the impulse generator generate the first impulse that is sent out into the channel by the wideband transmitter. When the first impulse arrives at the receiver along the LOS path, we will see the first impulse appearing on the time axis of the oscilloscope, followed by a train of decaying impulses, which travel along the reflection, diffraction, and scattering paths that are always longer than the LOS path. In this way, we have collected one frame of the *delay profile* of the multipath channel. By repeating the same procedure (sending a series of impulses and collecting corresponding delay profiles) over and over again, we can draw a three-dimensional plot with its x-y horizontal plan representing signal propagation delay τ and observation time t, respectively. We should be very careful here to distinguish the propagation time τ from the observation time t, as the former corresponds to multipath return index and the latter stands for the delay profile frame index (for each sent impulse from the transmitter). Figure 2.6 illustrates the appearance of the five frames of delay profiles captured by sending five consecutive impulses into the channel. The delay spread of the channel is measured by the widest delay span across the τ axis, beyond which the channel impulse response becomes lower than a threshold. Therefore, the delay spread of the channel is larger than $\tau_5 - \tau_0$ shown in Figure 2.6. A channel with a nonzero delay spread is also called a *time dispersive channel*, which is in contrast to the frequency dispersive channel (due to the Doppler spread caused by the mobility in the channel) discussed in the previous subsection.

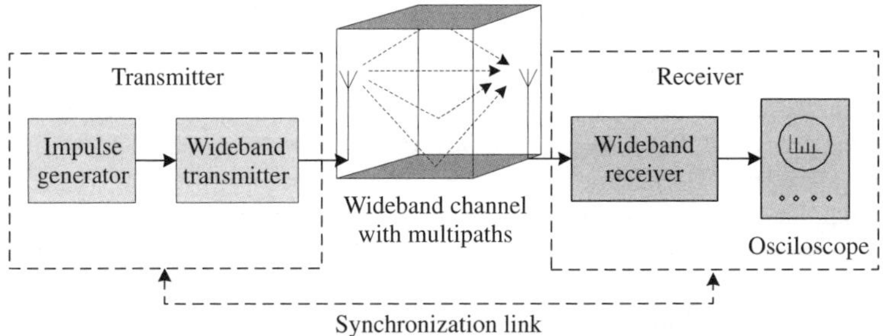

Figure 2.5   Wideband channel measurement system setup, which consists of an impulse generator with a wideband transmitter and a wideband receiver connected by a wideband sampling oscilloscope.

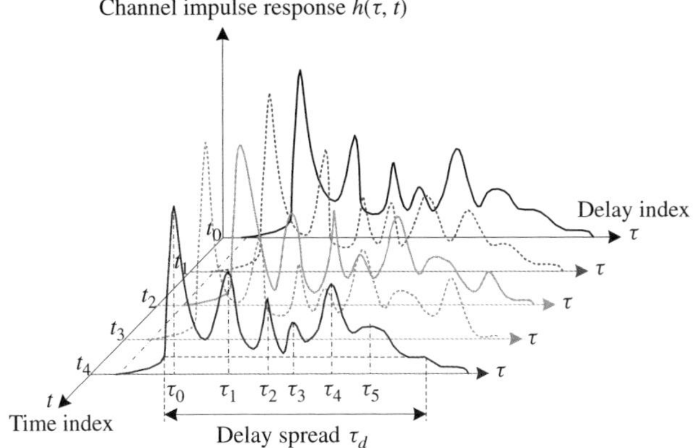

Figure 2.6 Illustration of the three-dimensional impulse response of a multipath channel, where five delay profile frames are captured to show the changes in different measurement times.

With the help of the wideband impulse channel measurement technique, we can obtain another important channel information, that is, channel impulse response, with which we can further collect almost all the essential characteristic parameters of a wireless or mobile communication channel.

On the basis of the information given in the channel impulse response, as shown in Figure 2.6, we can proceed with our study by marking two very interesting cross sections of the channel impulse response, one along the delay index variable τ with a fixed time, for instance, $t = t_2$, and the other along the time index variable t with a particular delay index variable, for example, $\tau = \tau_0$, respectively. The marking of the two cross sections is illustrated by Figure 2.7(a), which is derived directly from Figure 2.6. For illustration clarity, we put these two cross sections in Figure 2.7(b) and Figure 2.7(c), which are named as *delay profile* $h(\tau, t_2)$ and *time profile* $h(\tau_0, t)$, respectively. The notation of $h(\tau, t)$ stands for the two dimensional impulse response function of the channel.

With the delay profile $h(\tau, t_2)$ and the time profile $h(\tau_0, t)$ in our hand, we can easily calculate the delay spread τ_d and the coherent time T_{co} of the channel, as indicated in Figure 2.7(b) and Figure 2.7(c).

From the theory of signals and systems, we know that the Fourier transform of the channel impulse response will yield the frequency-domain transfer function of the channel, if the channel itself is viewed as a particular system with its input being transmitter antenna and the output being the receiver antenna. Having understood this, we readily obtain the frequency response of the channel, $H(f, t_2)$, which is obtained from the Fourier transform of the channel impulse response $h(\tau, t_2)$ with respect to τ. The channel frequency response is shown in Figure 2.7(d), from which we can define another important channel parameter, that is, the *coherent bandwidth* B_{co}, which is usually approximated by the width between the zero frequency and the first null in the channel frequency response. Later we will show using a very simple two-path channel model that the reciprocal of the coherent bandwidth is just equal to the delay spread of the channel, or $\tau_d = \frac{1}{B_{co}}$.

From another cross section of the channel impulse response, or the time profile function $h(\tau_0, t)$, we can study the time-variant properties of a wireless or mobile communication channel. It is clear that τ_0 is used just for explanation and simplicity without losing generality. Any other delay index variable can be used instead, such as $\tau_1, \ldots, \tau_4$, and so on. Again, we can use the Fourier transform to convert the time profile onto the Doppler frequency domain with respect to time index t, resulting in the *Doppler transfer function* of the channel, or $H(\tau_0, f_d)$, which forms a

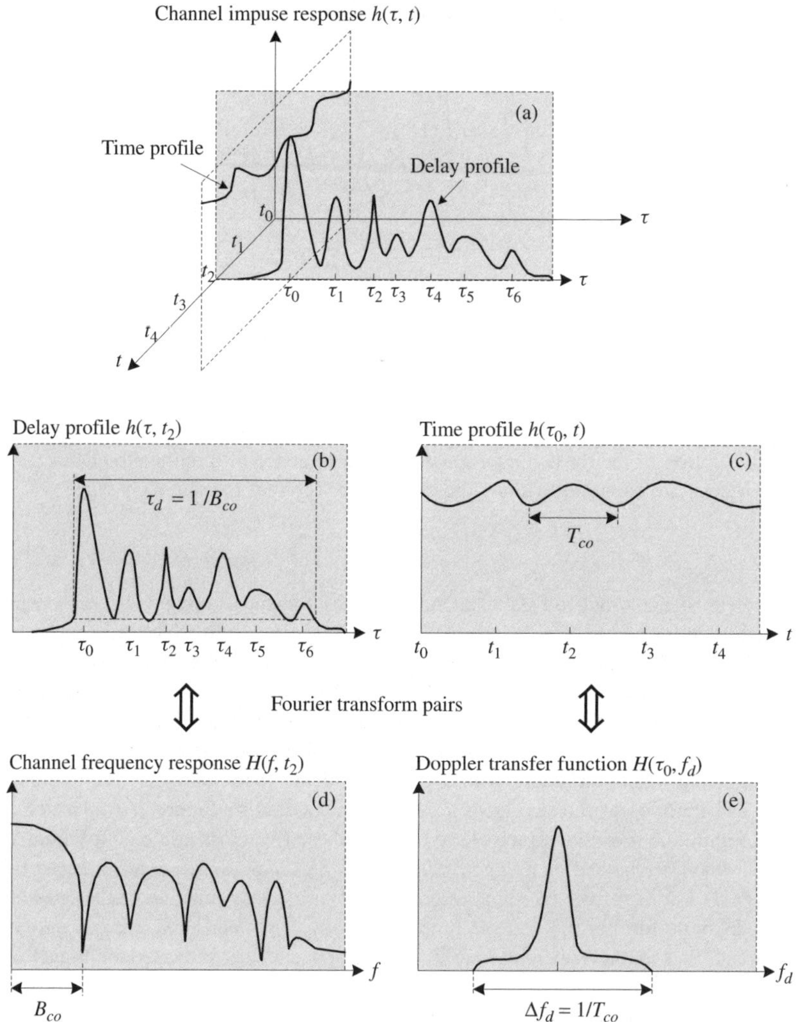

Figure 2.7 Relations among four most important characteristic parameters of a wireless or mobile channel, that is, delay spread τ_d, coherent bandwidth B_{co}, Doppler spread Δf_d and coherent time T_{co}, all of which can be obtained from a series of snapshots at different times from the wideband impulse response of the channel.

Fourier transform pair with the time profile function $h(\tau_0, t)$. As $h(\tau_0, t)$ reflects the characteristics of the channel changing with time, it basically does not change drastically if compared with the delay profile function $h(\tau, t_2)$. Therefore, its Fourier transform (as shown in Figure 2.7(e)) will give a relatively small Doppler frequency span compared to what we can see from the channel frequency response $H(f, t_2)$, shown in Figure 2.7(d). The width of the Doppler spectrum in the Doppler transfer function $H(\tau_0, f_d)$ is defined as the *Doppler spread* of the channel, which is also approximately equal to the reciprocal of the coherent time of the channel shown in Figure 2.7(c), or $T_{co} = \frac{1}{\Delta f_d}$.

The four parameters, that is, delay spread τ_d, coherence bandwidth B_{co}, Doppler spread Δf_d and coherent time T_{co}, are extremely important for us to understand wireless and mobile communication channels. Therefore, their definitions and physical meanings can be summarized as follows;

- The delay spread τ_d is defined as the widest delay span, over which all multipath returns are higher than a certain threshold. The delay spread is approximately equal to the reciprocal of the coherent bandwidth B_{co}.

- The coherence bandwidth B_{co} is defined as the smallest frequency range, within which all signals can pass without suffering serious frequency-selective fading. It is also equal to the reciprocal of the delay spread τ_d.

- The Doppler spread Δf_d is defined as the width of Doppler spectrum caused by mobility in the channel. The Doppler spread is equal to the reciprocal of the coherent time T_{co}.

- The coherent time T_{co} is defined as the time duration, beyond which two signal samples separated longer than T_{co} can usually be considered independent of each other. The coherent time can also be obtained by measuring the average cycle of the signal change in the time profile function $h(\tau_0, t)$, and is equal to the reciprocal of the Doppler spread Δf_d.

Before ending this subsection, we would like to give a real example of channel impulse response obtained from field measurement. The environment considered here is an urban area with high-rise buildings. The carrier frequency is 915 MHz, PN code bit rate is 10 Mbps with a total bandwidth of 20 MHz null-to-null bandwidth. The maximum delay is 51.1 µs. The measurement results are shown in Table 2.2.

For further reading on channel measurement techniques, the readers may refer to [51–58].

Flat fading

After having gone through the discussions in the previous two subsections on narrowband and wideband channel measurements, especially with all the important channel characteristic parameters already in our mind, we are ready to go ahead to characterize different types of fading effects that often occur in wireless or mobile cellular communication channels.

A channel is said to be a *flat fading* channel, or sometimes also called *frequency nonselective fading* channel, if the transmitted signal occupies a bandwidth smaller than the *coherent bandwidth* of the channel or B_{co}, which is approximately equal to the bandwidth spanned by the first null beside the main lobe of the frequency response function of the channel, as illustrated in Figure 2.7(d). It

Table 2.2 Maximal delay spreads for different wireless/mobile cellular environments around the world.

Environment	Carrier frequency (MHz)	Bandwidth (MHz)	Maximal delay spread: τ_d (µs)
Boulder, CO, suburban	1920	20	51.1
Phoenix, AZ, urban	915	20	51.1
Los Angeles, CA, urban	915	10	102.2
Hong Kong, urban	830	2.488	411
Washington State, urban	905–925	20/2.488	51.1/411
Colorado, flat rural	1850–1990	20	51.1
Philadelphia, PA, semirural	902–928, 1850–1990	20	51.1
Indoor Wireless LAN	5800	500	508 ns

should be noted that the multipath effect can also exist in a flat fading channel. However, in a flat fading channel, the spectral characteristics of the sent signal are preserved at a receiver due to the fact that the sent signal has a bandwidth narrower than the coherent bandwidth. The strength of the received signal will change with time owing to fluctuation of the channel gain caused by multipath effect. The statistics of the received signal strength can obey different distributions, such as Rayleigh, Rician, Nakagami-m, log-normal and so on, which have been discussed in the previous subsections.

We can also take a look at the flat fading effect in the time domain. When a flat fading occurs, the symbol duration of transmitted signals is much longer than the delay spread of the flat fading channel,[5] such that the inter-symbol interference will not happen because delayed replicas of the current symbol will not overlap with the next symbol.

The flat fading channels are also known as *amplitude varying channels*, and are sometimes referred to as *narrowband channels* as the bandwidth of the transmitted signals is relatively narrow with respect to the coherent bandwidth of the channels. It should be noted that a flat fading channel can also cause deep fades from time to time, and thus it may require 20 to 30 dB more transmitting power to compensate the losses due to the deep fades. Therefore, even for a channel suffering only flat fading, we should be prepared to employ necessary techniques to mitigate its negative effects on the overall performance of a radio communication system.

Frequency-selective fading

After having discussed flat fading in radio channels, we can define *frequency-selective fading* in a multipath channel. Obviously, different from flat fading, the frequency-selective fading will cause different attenuations in the received signal at different frequencies. Thus, it not only distorts the original signal appearance in both time and frequency domains, but also introduces serious inter-symbol interference (ISI), which is also called *multipath interference* (MI).

Usually, the multipath returns will not come to a receiver as isolated rays. Instead, they arrive at a receiver as a cluster or on a bundle basis, and each cluster of the multipath returns may consist of a great number of delayed replicas, whose phases and amplitudes can change randomly. This forms the *multipath fading*. Therefore, multipath fading is due to the constructive and destructive combination of randomly delayed, reflected, diffracted, and scattered signal components. This type of *fading* changes relatively fast and is therefore responsible for the *short-term signal variation*.

The multipath effect is caused mainly by the reflections of the radio waves by large dimensional and smooth surfaced obstacles in the radio signal propagation path. Thus, the *multipath effect* will become phenomenal when the signal carrier frequency goes up to at least a couple of hundreds MHz, in which the radio wave propagation will experience more reflections and scattering but less diffractions.

If the channel has a constant gain and a linear phase response over a bandwidth that is smaller than the bandwidth of the transmitted signal, the channel will suffer frequency-selective fading on the received signal. Under such a circumstance, the channel impulse response has a multipath delay spread τ_d, which is greater than the symbol duration of the transmitted signal. If this occurs, the received signal includes multiple versions of the sent symbol waveform that are attenuated (or faded) and delayed in time, and then the original transmitted signal is corrupted by those faded and delayed replicas to induce the ISI.

The multipath effect will cause serious interference problems to the signal detection process at a receiver, if no other effective means is used to mitigate the MI. The MI occurs when several delayed replicas arrive at the same receiver and add together to form ISI, which could be either destructive or constructive with equal probability, depending on the signs of the consecutive symbols. However, the constructive effect of the ISI caused by MI is meaningless here due to the impact of its destructive

---

[5]In this case, multipath propagation can still exist, but its coherent bandwidth is wider than the signal bandwidth.

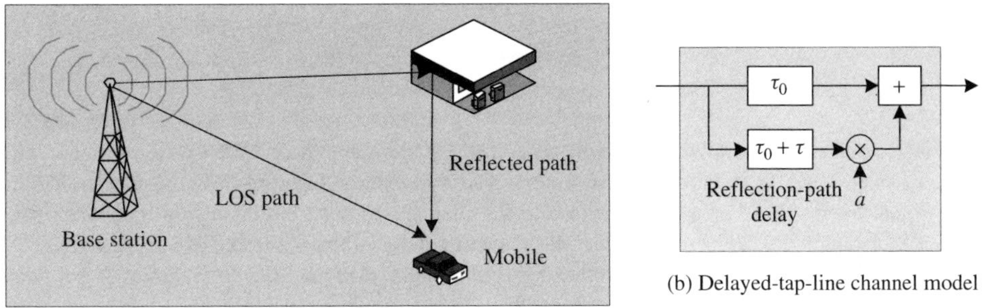

(a) A generic model for two-path mobile channel

(b) Delayed-tap-line channel model

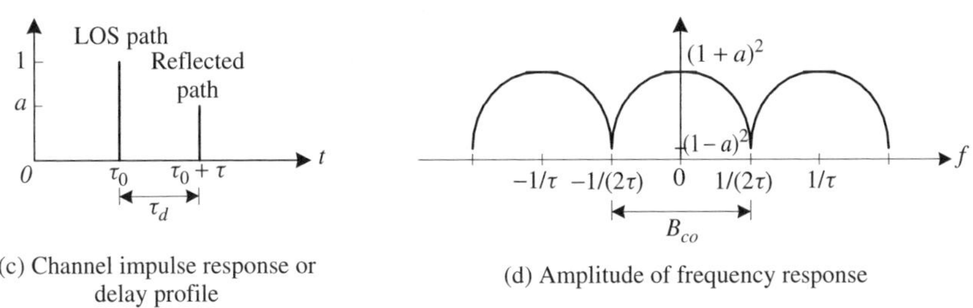

(c) Channel impulse response or
delay profile

(d) Amplitude of frequency response

Figure 2.8 A generic two-path multipath channel model (a). Delayed-tap-line channel model (b). The *delay spread* of the channel is shown in (c). The frequency-selectivity and thus the *coherent bandwidth* of the channel can be seen from the amplitude of frequency response shown in (d).

effect on the signal detection process. Therefore, the consequence of MI poses a great challenge to successful signal detection based on any algorithm at a receiver.

As an effort to illustrate the MI in both the time domain and frequency domain, let us consider a generic two-path channel model, as shown in Figure 2.8. First, let us examine the two-path channel in the time domain, as shown in Figure 2.8(c), which gives the *channel impulse response* or *channel delay profile*. The *delay spread* of this simple channel model is just equal to the interpath delay τ between the line-of-sight and reflected paths.

The MI can also be demonstrated in the frequency domain. It is relatively easy to understand the frequency selected fading in the frequency domain as its name suggests: certain frequency components in the received signal spectrum have greater (or smaller) gains than others. Without losing generality a multipath channel can be well modeled by a delayed tap-line filter with its coefficient in each tap representing the path gain of a particular multipath return and its delay element being the interpath delay. For simplicity, let us look at a two-path channel model shown in Figure 2.8. The amplitude of the transfer function of the impulse response for this two-path channel model is shown in Figure 2.8(d), where a comb-like frequency-domain transfer function can be observed, showing the frequency-selectivity of this particular two-path multipath channel model. Similar observations can also be made for a multipath channel with more than one delayed returns. Therefore, it has been demonstrated once again that an MI can also be referred to as a *frequency-selective fading*.

It is very interesting and important to note that the main lobe of the frequency response amplitude in Figure 2.8(d) is equal to $\frac{1}{\tau}$, which is just equal to the reciprocal of the delay spread of the channel, or τ. Therefore, we obtain

$$\tau_d \cong \frac{1}{B_{co}} \tag{2.14}$$

which states that a multipath channel always specifies a particular coherent bandwidth, beyond which the signal may suffer frequency-selective fading. The *frequency-selectivity* of the channel is relative to the signal bandwidth. For a particular multipath channel, its coherent bandwidth is always fixed. If the signal bandwidth B_s is lager than the coherent bandwidth of the multipath channel B_{co}, a frequency-selective fading takes place; otherwise only a non-frequency-selective fading or *flat fading* will occur.

The study of the techniques to mitigate MI has become a very popular research topic in the last 20 years, driven by a great demand for mobile cellular and wireless communications. Many papers [44–48] have been published as a result of great effort made by both academia and industry. Due to limited space, we will not dwell more on the theory of radio communication channels in this book. For more reading on this subject, the readers may refer to the following publications in the open literature [69–118].

2.2   Spread Spectrum Techniques

Spread spectrum (SS) techniques originated from the development of modern *radar systems* [119–131] at the end of the *second world war* in the 1950s. The earlier radar systems employed *continuous waves* (CW) that were sent as a series of short bursts into the air. The delayed and attenuated echoes received from those short CW bursts were used to measure the distances and directions of the objects in the air as well as in the sea.

Constrained by the maximal available peak power in the CW radio transmission, the earlier radar systems could not detect the incoming objects further than 100 kilometers, depending on the conditions in their operational environments. In order to improve the maximal detection range, *pulse-compression* techniques were introduced to the radar systems so that the detection range could be greatly extended beyond that range without increasing the peak transmitting power. The pulse-compression technique works on the principle that, instead of sending the CW radio signal directly, the carrier signal is first modulated by a coded waveform at the transmitter. When returned to radar receiver, this coded carrier waveform will then be *matched filtered* with the local coded waveform matched to what is used by the transmitter, yielding an autocorrelation peak that is very narrow in time and high in amplitude for easy detection. The commonly used pulse-compression waveforms in pulse radar systems include m-sequences [139–152], Barker codes [153–184], Kronecker sequences [185–189], GMW sequences [191], and so on.

The operation of a communication system is different from that of a radar system in a way that the former works with the transmitter and receiver placed in different locations, whereas the latter works with them in the same place. Another distinction is that a communication system or network always works in a multiple-user environment, in which many users share a common air-link to communicate with one another without introducing excessive interference to them, while a radar system does not have such a problem as it usually works alone. Bearing these differences in mind, we can readily understand that the design concept for the coded pulse-compression radar systems can be borrowed to a communication system to improve signal detection efficiency by the use of *pulse-compression* and *matched-filtering* processes. Therefore, the application of pulse-compression techniques in a communication system yielded *spread spectrum* techniques.

The introduction of the SS techniques in the later 1950s was due to the necessity to overcome some problems in communication systems, which appeared to be very hard to deal with when using conventional noise and interference suppression approaches. Although some communication channels can be accurately modeled by AWGN channels, there are many other channels that do not fit this model. A typical example can be a battle field communication link that might be jammed by a continuous wave tone close to the signal's center frequency or by a distorted retransmission of the enemy's own transmitted signal. In this case, the interference cannot be modeled by a stationary AWGN process. Another possible jamming scenario is that the jammer just transmits wideband pulsed AWGN, which may not necessarily be stationary.

In the later 1950s and early 1960s, many studies showed that there were other types of interferences, which were not caused by enemy's jamming signals or third party transmissions, but by its own transmitted signals, called *self-interference*. This type of self-interference is induced by multipath propagation in the channels, and does not fit the stationary AWGN model either. The receiver can be interfered by the sent signal itself via delayed reception of its own transmitted signal. This phenomenon is called *multipath effect* or *Multipath Interference* and was a problem first found in the LOS microwave digital radio relay transmission systems, such as those used for earlier long-haul telephone trunk transmission as well as in urban mobile radio system in the later 1960s.

In extensive research on mitigating those interference problems which could not be reduced by the typical AWGN channel model, it was found out that SS techniques were extremely effective in dealing with the non-AWGN channel interferences in communication systems. Therefore, the invention and further development of the SS techniques were driven mainly by the applications of the then emerging communication systems and services, such as long-haul microwave relay systems, satellite, and terrestrial land mobile communications, and so on.

Before defining a spread spectrum system, we would like to make sure that we understand what the spectrum of a signal is. Any modulation scheme in a communication system carries two most important characteristic parameters, one being the center frequency at which the signal is modulated; the other the bandwidth of the signal modulated by the carrier waveform. A spectrum, as we are discussing here, is the frequency-domain representation of the signal and especially the modulated signal. We most often see signals presented in the time domain (that is, as the functions of time). Any signal, however, can also be presented in the frequency domain, and different transforms (mathematical operators) are available for converting frequency-domain or time-domain functions from one domain to the other and vise versa. The most frequently used transform operation is the *Fourier transform*, for which the relationship between the time and frequency domains is defined by the pair of Fourier integrals defined as

$$F(\omega) = \int_{-\infty}^{\infty} f(t) e^{-i\omega t} dt \tag{2.15}$$

where $f(t)$ is a time-domain representation of the signal, $F(\omega)$ is defined as the spectrum of the signal $f(t)$, and ω is the radian frequency of the spectral index variable. The Equation (2.15) is always called the *Fourier transform* of the signal $f(t)$. The *inverse Fourier transform* also exists, which can be used to convert the frequency-domain spectral expression $F(\omega)$ back to the time-domain signal representation $f(t)$, or

$$f(t) = \frac{1}{2\pi i} \int_{-\infty}^{\infty} F(\omega) e^{-i\omega t} d\omega \tag{2.16}$$

Therefore, the spectrum of a time-domain signal $f(t)$ is defined as the width and shape of its spectral occupancy in the frequency domain, defined by Equation (2.15).

Tables of Fourier and Laplace transform pairs for different types of time-domain functions can be found in many references (e.g., see [1–3]). Here, we just mention some of the Fourier transforms that are most commonly employed in our analysis of SS systems. For instance, we will be, in particular, interested in the spectra of carriers modulated by pseudorandom binary data streams. Also of interest will be the spectra of frequency hopped carriers, especially where those carriers are to be used in multiple access applications, and it is necessary to restrict any interference between multiple users working in the same band of frequencies. Note that the frequency spectrum produced by modulation with a time-domain square pulse waveform is a $\frac{\sin x}{x}$ function, while modulation with a $\frac{\sin x}{x}$ envelope produces a square-shaped spectrum. The Fourier transform pair for square time waveform and $\frac{\sin x}{x}$ function can be written as

$$f(t) = \frac{\sin\left(\frac{\pi t}{\tau}\right)}{\frac{\pi t}{\tau}} \Leftrightarrow F(\omega) = \begin{cases} \tau, & |\omega| \leq \dfrac{\pi}{\tau} \\ 0, & |\omega| > \dfrac{\pi}{\tau} \end{cases} \tag{2.17}$$

where the notation $\Leftrightarrow$ indicates the Fourier-transform-pair relation between $f(t)$ and $F(\omega)$, and τ is the first zero points beside the main lobe of the function $\frac{\sin\left(\frac{\pi t}{\tau}\right)}{\frac{\pi t}{\tau}}$.

Other spectra that will be of special interest to us are those that are produced by square, triangular, and Gaussian-shaped waveforms, whose Fourier transforms are given as

$$f(t) = \begin{cases} 1, & |t| \leq \dfrac{\tau}{2} \\ 0, & |t| > \dfrac{\tau}{2} \end{cases} \Leftrightarrow F(\omega) = \tau \frac{\sin\left(\frac{\omega\tau}{2}\right)}{\frac{\omega\tau}{2}} \tag{2.18}$$

where τ is the width of the square waveform, and

$$f(t) = \begin{cases} 1 - \dfrac{|t|}{\tau}, & |t| \leq \tau \\ 0, & |t| > \tau \end{cases} \Leftrightarrow F(\omega) = \tau \frac{\sin^2\left(\frac{\omega\tau}{2}\right)}{\left(\frac{\omega\tau}{2}\right)^2} \tag{2.19}$$

where the base of the triangular pulse is 2τ, and

$$f(t) = e^{-\frac{\left(\frac{t}{\tau}\right)^2}{2}} \Leftrightarrow F(\omega) = \tau\sqrt{2\pi}\, e^{\frac{(\tau\omega)^2}{2}} \tag{2.20}$$

respectively. More signal-spectrum Fourier transform pairs can be found in the literature [1–3]. To give some visual examples of real Fourier transforms, we have shown three most commonly referred time waveforms and their Fourier transforms in Figure 2.9, where the square function, the triangular function and the Gaussian function and their frequency-domain Fourier transforms are illustrated. As they were generated from real spectrum plots from MATLAB, they are shown exactly as they should appear in real applications.

Just as an oscilloscope is like a window in the time domain for observing signal waveforms, a spectrum analyzer is a window in the frequency domain, generated by sweeping a filter across the band of interest and detecting the power falling within the filter as it is swept. This power level can then be plotted on the display of an oscilloscope. Usually, all spectra referred to in this book are *power spectra density* (PSD) *functions* of the signals concerned. The relation between the PSD of a signal and its Fourier transform can be written as

$$P(\omega) = |F(\omega)|^2 \tag{2.21}$$

It is to be noted that the power of a signal can be calculated from either the time domain or the frequency domain, and the results should be exactly the same as a consequence of *power conservation law*. That is

$$\frac{1}{2\pi} \int_{-\infty}^{\infty} P(\omega)\,d\omega = \int_{-\infty}^{\infty} |f(t)|^2\,dt \tag{2.22}$$

It is to be noted that the change of physical appearance in the time domain goes in the opposite direction to that in the frequency domain. For instance, if we extend the duration of a signal waveform

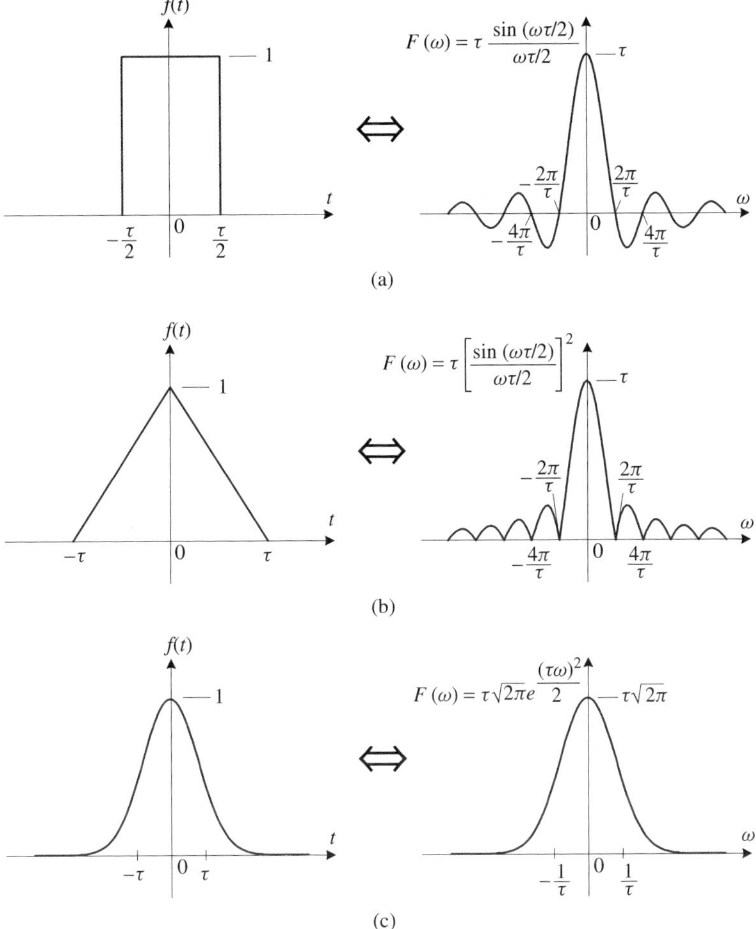

Figure 2.9 Fourier transform pairs for three commonly referred time waveforms. (a) Square waveform. (b) triangular waveform. (c) Gaussian waveform.

in the time domain, its Fourier transform will be compressed in the frequency domain, as stated exactly in the *scaling property* of Fourier transform as follows:

$$f(at) \Leftrightarrow \frac{1}{2\pi |a|} F\left(\frac{\omega}{2\pi a}\right) \tag{2.23}$$

where a is a scaling factor of time index variable t. The scaling factor a can be either more or less than one, resulting in either compressed or extended original signal waveform of $f(t)$ in the time domain.

The spectral bandwidth of a time-domain signal $f(t)$ can be perfectly defined by the width in frequency, at which its power is distributed. Therefore, the signal bandwidth is very much related to the shape or appearance of its power spectra density function. Based on how much power is included in its bandwidth, we have several different definitions of the signal bandwidth. The most commonly used signal bandwidth is 3 dB signal bandwidth, which is defined as the width, over which the power spectral density function falls from its peak value to a level 3 dB lower than the peak. This bandwidth is also called *3-dB bandwidth*. The signal bandwidth can also be defined as the spectral width, over which the included signal power becomes a fixed percentage of the total signal power. This can be easily shown using the following expression as

$$99\% \times \frac{1}{2\pi} \int_{-\infty}^{\infty} P(\omega)\mathrm{d}\omega = \frac{1}{2\pi} \int_{-B_{99\%}}^{B_{99\%}} P(\omega)\mathrm{d}\omega \tag{2.24}$$

where $B_{99\%}$ is the *99 percentage power bandwidth*. Similarly, we can also define other *percentage power bandwidths*, such as *90 percentage power bandwidth, 50 percentage power bandwidth*, and so on.

After having defined the signal bandwidth, we are ready to describe what an SS communication system is in the sequel. Literally, an SS technique can be defined as any method for a transmitter to spread the signal spectrum to a much wider extent than necessary to send the baseband signal itself in a channel. At the receiver side, an SS receiver will be able to effectively collect most, if not all, of the signal energy in a bandwidth spanned by the sent SS signal for effective detection. For instance, a voice signal can be sent with amplitude modulation in a bandwidth roughly twice that of the voice information itself. Other forms of modulation, such as low deviation FM or single sideband AM, also permit information to be transmitted in a bandwidth comparable to the bandwidth of the sent information itself. An SS system, however, often takes a baseband signal (e.g., a voice channel) with a bandwidth of only a few kilohertz, and distributes it over a band that may span many megahertz width in frequency. This is accomplished by modulating the information to be sent together with a wideband encoding waveform, also called *spread modulating signal*. The most familiar example of spectrum spreading is observed in conventional frequency modulation (FM), in which *deviation ratios* greater than one are used. As a result, the bandwidth occupied by an FM modulated signal is dependent on not only the information bandwidth but also the amount of modulation. As in all other spectrum spreading systems, a signal-to-noise[6] advantage is gained by the modulation and demodulation process. To measure the magnitude of this gained advantage, the terminology of *process gain* is always used in an SS system.

Wideband FM could be considered as an SS technique from the standpoint that the carrier spectrum produced in the frequency modulation process is much wider than the transmitted information. However, in the context of this section only those techniques are of interest in which some signal or operation, other than the information being sent, is used for spreading the transmitted signal.

Many different spread spectrum techniques exist, in which the spreading codes or spreading sequences will be used to control the frequency or time of transmission of the data-modulated carrier, thus *indirectly* modulating the data-modulated carrier by the spreading codes or spreading sequences. Several basic spread spectrum techniques available to the communications system designer will be described and discussed in a general way in this part of Chapter 2. This section gives some detailed descriptions of the various techniques and the signals generated. In addition to the most important (or at least most prevalent) forms of SS modulation schemes (i.e., direct sequence (DS) spreading schemes), other useful techniques such as frequency hopping (FH), time hopping, chirping and various hybrid combinations of modulation forms will be described. Each is important in the sense that each has useful applications. The historical tendency has been to confine each form to a particular application scenario. Direct-sequence spreading, for instance, has been found most commonly used in civilian applications. FH is more widely employed in military communication systems. Chirp modulation has been used almost exclusively in radar. These systems will be discussed in the later subsections. The digital codes or sequences used for the spreading signal will also be discussed in detail in Section 2.3 in this chapter.

There are four major techniques that will be accepted here as examples of SS signaling methods:

- Modulation of a carrier by a digital code sequence whose chip rate is much higher than the information signal bandwidth. Such systems are called *direct-sequence* modulated systems.

---

[6]Here, what we mean in "signal-to-noise" ratio is in fact "signal-to-interference" ratio, as no SS technique will help to suppress noise, it will help suppress only interferences.

- Carrier frequency shifting in discrete increments in a pattern determined by a code sequence. This technique is called *frequency hopping* spread spectrum. The transmitter jumps from one frequency to another in some predetermined sequence; the order of appearance of the frequencies is determined by a controlling code sequence.

- The transmitted signal appears in different time slots within a fixed time frame, resulting in the so called *time hopping* spread spectrum technique.[7]

- Pulsed-FM or *chirp modulation* technique, in which a carrier is swept over a wide band during a given pulse interval.

On the basis of the above four different SS techniques, many hybrid versions can be derived, such as *time-frequency hopping* system, where the code sequence determines both the transmitted frequency and the time of transmission, instead of only one as in the case of either FH or time hopping. Also, it is to be noted that the pulse-FM or chirp modulation scheme was a direct derivation from the earlier radar applications and not many applications have been found in modern communication networks and systems due to its relatively low processing gain (PG) achievable and hard to use digital technique for its signal processing.

Recently emerging *UWB* technologies have a lot in common with a time hopping SS system. The UWB techniques will also use *PPM* to modulate digital signal (usually binary) with very narrow pulses. Therefore, the UWB technology is a further development of traditional SS systems. More detail discussions on UWB technologies can be found in Section 7.6. Obviously, spread spectrum techniques form a foundation for modern *CDMA* technologies, which have been playing an extremely important role in current 3G (and maybe beyond 3G as well) wireless networks and communications.

In the following subsections, we will discuss the three major spread spectrum techniques, namely, DS, FH, and time hopping techniques.

2.2.1 Direct-Sequence Spread Spectrum Techniques

The simplest method to spread the spectrum of a data-modulated signal is to modulate the signal a second time using a wideband spreading signal, which always takes some forms of sequences, that is, a *pseudorandom sequence* or PN sequence for short. This second modulation usually takes some form of digital phase modulation, although analog amplitude or phase modulation is conceptually possible. This spread spectrum (SS) scheme is called the *direct sequence* spread spectrum (DSSS) system, (or, more exactly, directly carrier-modulated, code sequence modulation system) which is the best known and most widely used spread spectrum system. This is because of their relative simplicity from the point of view that they do not require a high-speed, fast-settling frequency synthesizer. Nowadays, DS modulation has been used for commercial communication systems and measurement instruments, and even laboratory test equipments that are capable of producing a choice of a number of code sequences or operating modes. It is reasonable to expect that DS modulation will become a familiar form of the spreading modulation scheme in many areas in the years to come due to its unique and desirable features. Even now, commercial applications of DSSS systems are being explored. Characteristics of DS spreading modulation is exactly the modulation of a carrier by a code sequence. In the general case, the format may be AM, FM, or any other amplitude- or angle-modulation form. Very often, however, the *binary phase-shift keying* (BPSK) is used, because it can be implemented at a very low cost: only two balanced multiplication units are required, plus a low-pass filter followed by a decision

---

[7]It has to be noted that one type of emerging ultra-wideband (also called *UWB*) technology works in a very similar way as a time hopping SS. It is also called *TH-UWB* technology. Most commonly used modulation scheme in the TH-UWB is pulse position modulation (PPM).

device. The basic form of a DS signal is that produced by a simple and biphase-modulated (BPSK) carrier. The details about the BPSK DSSS system will be introduced later.

The selection of spreading signals is of great importance in a DSSS system as it should have certain properties that facilitate demodulation of the transmitted data signal by the intended receiver, and make demodulation by an unintended receiver as hard as possible. These same properties will also make it possible for the intended receiver to discriminate between the intended signal and jamming, which usually appears quite differently from what is used for spreading the signal at the transmitter. If the bandwidth of the spreading signal is much larger than the original data signal bandwidth, the SS transmitting signal bandwidth will be dominated by the spreading signal and is nearly independent of the original data signal. Each element of the spreading sequences or codes is usually called a *chip*; its width will determine the bandwidth of the signal after spreading modulation.

Before discussing any DSSS communication systems, we have to introduce the most important characteristic parameter, namely, *PG*, which is defined as a function of the RF bandwidth of the DS signal transmitted, compared with the bandwidth of its data information before carrier modulation. The PG is exhibited as a signal-to-interference improvement resulting from the RF-to-information bandwidth trade-off. It will also govern its capability to mitigate many other undesirable factors appearing in the communication medium and signal detection processes, such as antijamming property, and so on. The usual assumption is that the RF bandwidth is assumed to be equal to the main lobe of the DS spectrum, which is always a $\frac{\sin x}{x}$ function. In many practical applications, the ratio between the chip rate and original data information rate can also be used as the PG. Therefore, for a DSSS system having a 10 Mcps chip rate and a 1 kbps information rate the PG will be $(10^7)/(10^3) = 10^4$ or about 40 dB. A more strict definition of the PG is given as

$$PG_{DS} = \frac{\text{RF bandwidth of DS/SS signal}}{\text{Baseband bandwidth of user data signal}} \tag{2.25}$$

$$\cong \frac{\text{Chip rate of DS/SS signal}}{\text{User data rate}}$$

The question arises then, whether the PG can be raised to a very high level to improve the performance of a DSSS system. This question can be answered best by addressing the limitations that exist with respect to expanding the bandwidth ratio to a arbitrarily large value so that the PG may be increased indefinitely. Obviously, two parameters are available to adjust PG. The first is the RF bandwidth, which depends on the chip rate used. For instance, if we have an RF (null-to-null) bandwidth 100 MHz wide, the chip rate should be at least 50 Mcps. On this basis, how wide should we make the system RF bandwidth and how much benefit can we obtain from the increase of the chip rate? To double the RF bandwidth defined by the chip rate, we can only increase 3 dB PG. However, the price is in its system complexity. With double the chip rate, the sampling rate at a digital receiver has to be at least doubled. This will substantially increase the signal processing load at a DSP chip or CPU. It is to be noted that the increase in the computation load is not linear with the increase in the sampling rate. In other words, the doubling chip rate will probably result in trebling, quadrupling or yielding an even higher computation load in a DSP chip. This imposes a great challenge to implement real-time based communication applications, such as multimedia services. With the decrease in chip duration (or increase in the chip rate) the smallest interval to make a decision at a receiver is also reduced, leaving a result that the hardware and software have to catch up with the data rate to make a sensible decision for each received bit on the basis of the chips. We should remember that the channel characteristics never change with the increase of chip rate, as discussed in Section 2.1. With each chip received at a receiver, all necessary algorithms, such as channel estimation, decision feedback, equalization, and so on, have to be carried out and finished in time before the end of the chip in question. It is still a great challenge to implement a full digital receiver at a chip rate 10 Gcps using the state-of-the-art microelectronics technology. Thus, it is not a wise approach to increase the PG by using a higher chip rate. On the other hand, we can easily understand that it is not sensible to increase the PG by reducing the user data rate either.

The most commonly used techniques for DS spreading are discussed below.

BPSK direct-sequence spread spectrum

The simplest form of DSSS employs BPSK as the spreading modulation. It has to be noted that here we are talking about two modulations, that is, the *spreading modulation* and *carrier modulation*. The former denotes the modulation of data information with a predetermined spreading code or sequence to result in a bandwidth spreading, and the latter stands for modulating the baseband signal with a high frequency radio carrier, only shifting the spectrum of the original baseband signal to a certain RF frequency without yielding any bandwidth spreading. Therefore, for a BPSK DSSS system we imply that the spreading modulation must be done using a BPSK modem. However, it is not certain whether the carrier modulation in a BPSK DSSS system also employs the BPSK modem. As a matter of fact, a BPSK DSSS system can also use any modem, such as BPSK, QPSK, MSK, and so on, for its carrier modulation purpose.

Yet another important point we have to mention here is that the order of the spreading modulation and carrier modulation is irreversible in most cases, and usually the spreading modulation happens before the carrier modulation. In other words, the data signal should first be modulated by a spreading signal, and then the spread signal will be further modulated by a radio frequency (RF) carrier before being fed into the antenna for transmission. However, if both spreading modulation and carrier modulation use BPSK modems, the order of the two become interchangeable.

Ideal BPSK modulation yields instantaneous phase shifts of the carrier by zero or 180 degrees according to the signs of the binary data signal as a modulating signal. It can be mathematically expressed by a multiplication of the carrier by a function $c(i)$ that takes on the values ± 1. Let us consider a constant-envelope data-modulated carrier with power P, carrier radian frequency ω_c, defined by

$$f_d(t) = \sqrt{2P} \cos\left[\omega_c t + \phi_d(t)\right] \tag{2.26}$$

where $\phi_d(t)$ stands for the data-modulated phase, which should take two different values, either zero or 180 degrees depending on the signs (either $+1$ or -1) of binary data information, and the term $\sqrt{2P}$ is to give an average power P.

This signal occupies a bandwidth typically between one-half and twice the data rate prior to DS spreading modulation, depending on the details of the data modulation and the pulse shapes used in shaping the original data pulses. The BPSK spreading is accomplished by simply multiplying $f_d(t)$ by a time-domain signal $c(i)$ that is also called the *spreading signal* or spreading sequence, as illustrated in Figure 2.10.

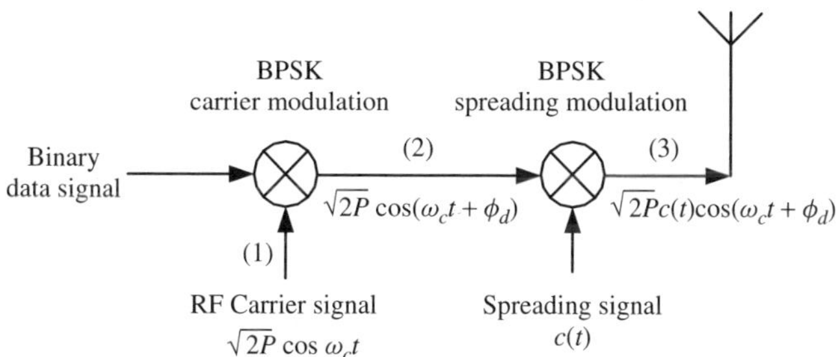

Figure 2.10 Illustration of a BPSK DSSS transmitter.

The transmitted signal after spreading modulation becomes

$$f_s(t) = \sqrt{2P}c(t)\cos\left[\omega_c t + \phi_d(t)\right] \tag{2.27}$$

whose bandwidth is basically determined by the spectral span of the spreading signal $c(t)$, which usually is a wideband spreading sequence. It is to be noted that the process of multiplication of $c(t)$ with $f_d(t)$ will not alter the power of the $f_d(t)$, but only extend the bandwidth of $f_d(t)$. This is what an SS signal means. Then, we look back at the scaling property of the Fourier transform, which tells us that the extension of the spectral span of a signal will equivalently make its time-domain waveform shrink, just as expressed in Equation (2.23). From the power conservation law (Equation (2.22)), the expansion in the bandwidth span of a signal in the frequency domain will reduce its peak amplitude if the total power remains the same. This effect makes an SS signal appear like a wideband noise-like interference to an unintended receiver. It is obvious that a conventional (non-spread-spectrum) receiver would not be useful for detecting the wideband noise-like signal here because it is well below the level of the real noise observed at the receiver.

The signal given in Equation (2.27) is transmitted into an AWGN channel with a transmission delay τ_d. The signal is received and contaminated by interference and channel AWGN noise. Demodulation is accomplished in part by demodulating or remodulating with the spreading code locally generated and appropriately delayed, $c(t - \tilde{\tau}_d)$, as shown in Figure 2.11. This demodulation or correlation of the received signal with the delayed spreading waveform is called the *despreading process* and is an important function in any SS system. The signal after despreading the module in Figure 2.11 will become

$$r_1(t) = \sqrt{2P}c(t - \tau_d)c(t - \tilde{\tau}_d)\cos\left[\omega_c t + \phi_d(t - \tau_d) + \theta\right] \tag{2.28}$$

where $\tilde{\tau}_d$ is the estimated delay at the receiver, τ_d is the propagation delay that the transmitted signal experienced, and θ is the phase delay caused by the propagation delay.

If the estimated delay at the receiver is exactly the same as the real delay, or $\tilde{\tau}_d = \tau_d$, Equation (2.28) will yield

$$\sqrt{2P}\cos\left[\omega_c t + \phi_d(t - \tau_d) + \theta\right] \tag{2.29}$$

as $c(t - \tau_d)c(t - \tilde{\tau}_d) = 1$ if $\tilde{\tau}_d = \tau_d$. This despread signal has been restored into a narrowband signal, which is very similar to the original transmitted phase modulated data signal with only some difference in the delay τ_d and an extra phase θ caused by the propagation delay from the transmitter to the receiver. This despreading process plays a crucial role here to transform the received wideband signal into its original narrowband data signal.

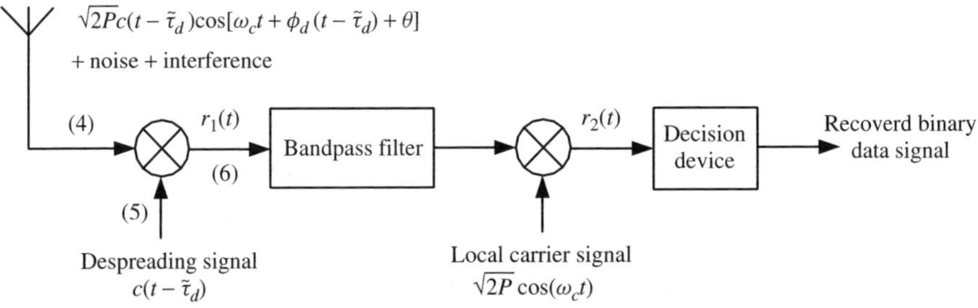

Figure 2.11   Illustration of a BPSK DSSS receiver.

On the other hand, if the receiver uses a wrong spreading signal or spreading sequence, say $c'(t - \tilde{\tau}_d)$, to despread the received wideband signal $\sqrt{2P}c(t - \tau_d)\cos[\omega_c t + \phi_d(t - \tau_d) + \theta]$, it will never accomplish the despreading process to restore the narrowband signal correctly, because $c(t - \tau_d)c'(t - \tilde{\tau}_d)$ will be another wideband sequence no matter whether $\tilde{\tau}_d = \tau_d$ or not and thus the signal

$$\sqrt{2P}c(t - \tau_d)c'(t - \tilde{\tau}_d)\cos[\omega_c t + \phi_d(t - \tau_d) + \theta] \qquad (2.30)$$

will remain a wideband modulated signal. Therefore, the spreading signal $c(t)$ is usually also called the *signature sequence* or *signature code* as it behaves like a *key* to decode or despread the received signal for recovering the original sent narrowband data signal.

There are six different time-domain waveforms observed at the transmitter and the receiver, as shown in (1) to (6) in Figure 2.12. We can also allocate the corresponding observation points from Figure 2.10 and Figure 2.11 accordingly, assuming that the binary data information in this case (shown in Figure 2.12) is a constant value of $+1$ for illustration simplicity. We can then see how a BPSK DSSS communication transceiver works step by step from the time-domain perspective.

The block diagrams shown in Figure 2.10 and Figure 2.11 illustrate a typical DSSS communications transceiver structure. It shows that a DSSS system can be viewed as a conventional AM or FM communications link with only an extra part added to implement spreading modulation and demodulation functionalities. In real applications the carrier modulation usually does not happen before spreading modulation. The baseband information is digitized and added to the spreading sequence first. For the discussion given in this section, however, we assume that the RF carrier has already been data modulated before spreading modulation, because this can simplify the discussion of the modulation-demodulation process in a BPSK DSSS system. After having been amplified, a received signal is multiplied by a reference sequence generated at the receiver locally and, given that the transmitter's sequence and receiver's sequence are synchronous and the same, the carrier inversion phases (as shown in (3) and (4) in Figure 2.12) will be removed successfully and the original carrier waveform will be restored. This narrowband restored carrier can then pass through a bandpass filter designed to pass only the original data-modulated carrier.

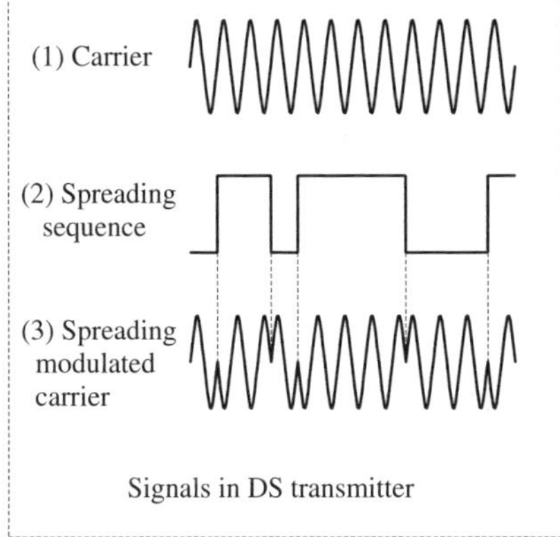

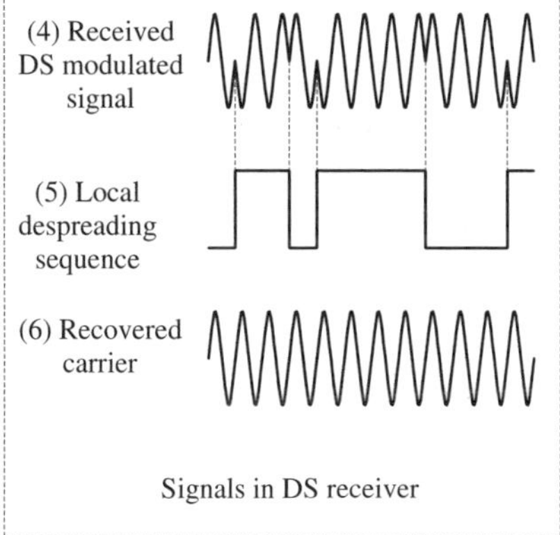

Figure 2.12 Conceptual illustration of time-domain signal waveforms for a BPSK DSSS transceiver. The waveforms shown in this graph correspond to the observation points (1) to (3) in Figure 2.10 and the points (4) to (6) in Figure 2.11, respectively.

All unwanted received signals are also treated by the same process at the receiver as the desired signal, multiplying the received DS signal with a locally generated reference sequence. Any incoming signal not synchronous with the receiver's local reference sequence (a wideband signal) is spread to a bandwidth still equal to the bandwidth of the received signal, because an unsynchronized input signal is mapped into a bandwidth at least as wide as the receiver's reference, such that the band-pass filter can reject almost all the power of these undesired signals. This is the mechanism, by which *process gain* is realized in a DSSS system; that is, the receiver transforms synchronous input signals from the sequence-modulated bandwidth (wideband) to the data-modulated bandwidth (narrowband). At the same time nonsynchronous input signals are spread at least over the spreading sequence-modulated bandwidth. The data-modulated bandwidth specifies the bandwidth of a bandpass filter followed by the decision device, and this bandpass filter in turn effectively controls the amount of power from an unsynchronized or unwanted signal, which reaches the data demodulator. We can see from the discussion here that the multiplication-and-filtering process before data detection at the receiver provides the desired signal with an advantage or *process gain*. In fact, the RF bandwidth in a DSSS system, as discussed earlier, directly affects many capabilities of the system, such as how effectively it can reject external jamming. For instance, if a maximal 10-MHz bandwidth is available, the PG possible is also limited by that 10 MHz. Several practical approaches are available in choosing the proper bandwidth in an anti-interception application; the main interest is to minimize the power transmitted in terms of watts per Hertz. When a maximum PG for interference rejection is needed, the bandwidth again should be made large enough. If either frequency allocation or the propagation medium does not permit the use of a wide RF bandwidth, some restraint must be applied. A prime consideration in SS systems (and in particular, DS systems) is the bandwidth of the system with respect to the interference generated by other systems (that may not necessarily be SS systems) operating in the same or adjacent channels.

A conceptual spectral diagram of this type of DSSS signal format is shown in Figure 2.13, where we only show the envelope of the PSD function of BPSK-modulated DSSS signal for illustration clarity. The main lobe bandwidth (null-to-null) of the signal shown is usually equal to twice the clock rate of the code sequence used as a spreading modulation signal. Each of the sidelobes has a null-to-null bandwidth that is equal to the clock rate; that is, if the code sequence being used as a modulating waveform has a 5 Mcps operating rate,[8] the main lobe of the null-to-null bandwidth will be 10 MHz and each sidelobe will be 5 MHz wide. This is exactly the case in Figure 2.13. On the other hand, in the time domain the BPSK-modulated DSSS carrier looks like the signal shown in Figure 2.12, where the carrier is sent with zero phase shift when the code sequence is a $+1$, and a 180 degree phase shift when the code sequence is a -1.

To illustrate how a DSSS system works in the frequency domain, we would also like to look at the issue from the perspective of power spectral density function as follows. Assume that the input data information stream, as shown in Figure 2.10, is a random sequence with a transmission rate of

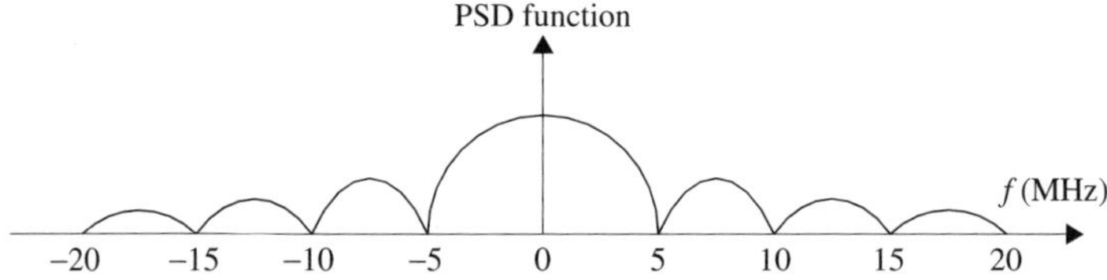

Figure 2.13 Conceptual illustration of power spectral density function for BPSK DSSS signal. The chip rate for this system is 5 Mcps and the null-to-null bandwidth of this DSSS system us 10 MHz.

---

[8]Here, "Mcps" stands for *mega chips per second*.

$\frac{1}{T}$ bit per second (bps), or

$$d(t) = \sum_{k=-\infty}^{\infty} d_k \, p_T(t - kT) \tag{2.31}$$

where $d_k = \pm 1$, the bit duration is T and $p_T(t)$ is the bit pulse waveform function. Thus, its power spectral density (PSD) function $\varphi_d(f)$ can be written into

$$\varphi_d(f) = T \left(\frac{\sin fT}{fT} \right)^2 \tag{2.32}$$

whose shape is illustrated in Figure 2.14(a). Thus, it is seen from the figure that its bandwidth is just equal to $\frac{1}{T}$ Hz. Assume that the spreading sequence is also a random sequence and its chip rate is $\frac{1}{T_c}$. Therefore, its PSD function $\varphi_c(f)$ can be expressed by

$$\varphi_c(f) = T_c \left(\frac{\sin fT_c}{fT_c} \right)^2 \tag{2.33}$$

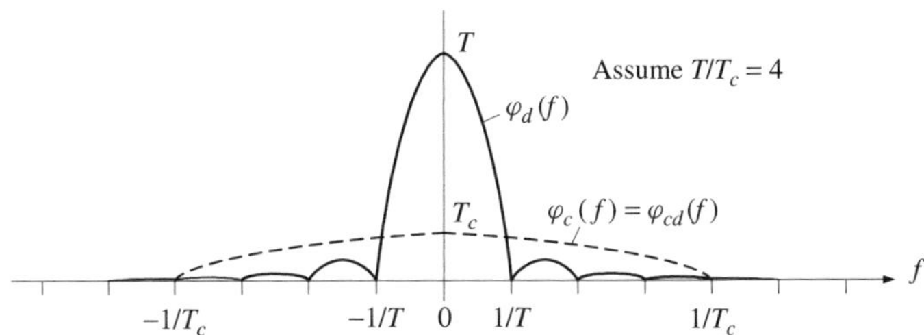

(a) PSD functions for spreading sequence and data signal

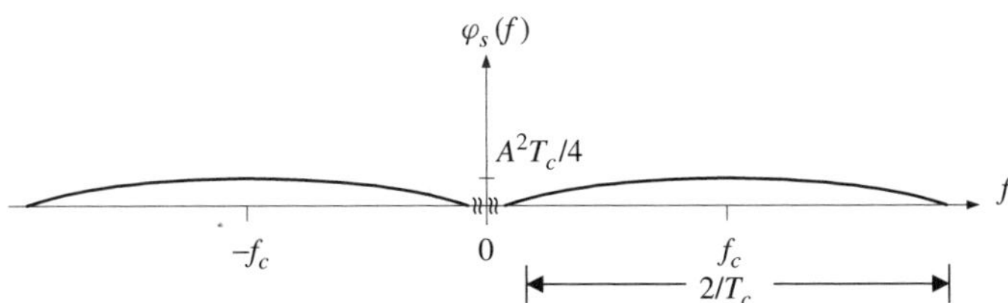

(b) PSD function for BPSK spreading modulated signal

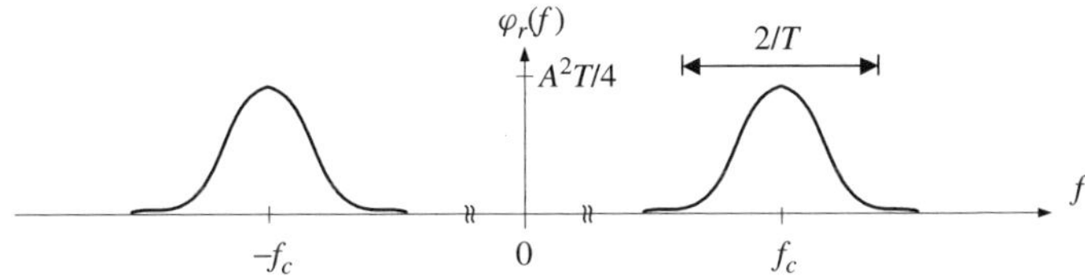

(c) The PSD for the carrier signal after despreading ($r_1(t)$ as shown in DS SS receiver)

Figure 2.14 The PSD functions for (a) original data and spreading sequence, (b) BPSK spreading modulated signal, and (c) the carrier signal after despreading, where it is assumed that $T = 4T_c$ for illustration clarity.

which forms exactly the same expression as Equation (2.32) except for the interchange of bit duration T and chip width T_c. The PSD function for the spreading sequence $\varphi_c(f)$ has also been drawn together with the PSD function of data sequence $\varphi_d(f)$ for easy comparison in Figure 2.14(a), where it is assumed that $T = 4T_c$ for illustration clarity. Obviously, the bandwidth of spreading sequence is equal to $\frac{1}{T_c}$ Hz.

Now, let us consider the spreading modulation process as a simple multiplication between the data signal and spreading sequence, resulting in a PSD function expressed by

$$\varphi_{cd}(f) = T_c \left(\frac{\sin f T_c}{f T_c} \right)^2 \tag{2.34}$$

which takes exactly the same expression as Equation (2.33) and occupies the same bandwidth as that of $\varphi_c(f)$. In this way, the spreading modulation has extended the signal bandwidth to $\frac{T}{T_c} = N$ (N is assumed to be 4 in Figure 2.14) times. N is just equal to the PG of this DSSS system and is usually a fairly large number. The carrier modulation after the spreading modulation will only shift the spectrum $\varphi_{cd}(f)$ to the center frequency f_c, but never changes the physical appearance of $\varphi_{cd}(f)$, as shown in Figure 2.14(b).

The PSD function of transmitted signals from the antenna of the transmitter becomes

$$\varphi_s(f) = \frac{PT_c}{2} \left\{ \left[\frac{\sin(f - f_c)T_c}{(f - f_c)T_c} \right]^2 + \left[\frac{\sin(f + f_c)T_c}{(f + f_c)T_c} \right]^2 \right\} \tag{2.35}$$

which is a bandpass signal and its bandwidth is $\frac{2}{T_c}$ Hz, as shown in Figure 2.14(b). It is observed from the figure that the amplitude of $\varphi_s(f)$ is reduced by $\frac{2T}{PT_c}$ if compared with $\varphi_d(f)$; whereas the width of $\varphi_s(f)$ increases $N = \frac{T}{T_c}$ (which is just the PG value) times if compared with $\varphi_d(f)$.

At the DSSS receiver, the PSD function of the received signal has the same PSD function of the transmitted signal, with only a delay and some extra phase also caused by propagation delay, as shown in Figure 2.11. The delay will never change the shape of the PSD function. It is easy to show that the PSD function of the signal after the despreading process, or signal $r_1(t)$ as indicated in Figure 2.11, can be written as

$$\varphi_r(f) = \frac{PT}{2} \left\{ \left[\frac{\sin(f - f_c)T}{(f - f_c)T} \right]^2 + \left[\frac{\sin(f + f_c)T}{(f + f_c)T} \right]^2 \right\} \tag{2.36}$$

which has been plotted in Figure 2.14(c). It is to be noted that the Equation (2.36) has exactly the same expression as $\varphi_s(f)$, as written in Equation (2.35), except for the interchange of T_c and T. It is not surprising to us as the despreading process at the receiver will restore the original data signal bandwidth, such that most of its power can pass easily through the bandpass filter, as shown in Figure 2.11. It is seen from the figure that, similar to the $\varphi_d(f)$, the PSD function $\varphi_r(f)$ spans also a narrowband spectrum with its bandwidth being $\frac{2}{T}$, which is just the double of that for signal $d(t)$. The spectrum $\varphi_r(f)$ will be restored into a narrowband baseband PSD function after the carrier demodulation, which just shifts its center frequency from f_c back to zero.

QPSK direct-sequence spread spectrum

It is a well-known fact that the use of quadrature modulation scheme can effectively improve the bandwidth efficiency of a digital modem without sacrificing the power efficiency. The two quadrature carriers, that is, $\sin(\omega_c t)$ and $\cos(\omega_c t)$, are perfectly orthogonal due to the simple fact that

$$\int_{-\infty}^{\infty} \sin(\omega_c t) \cos(\omega_c t)\mathrm{d}t = \int_{0}^{2\pi} \sin(\omega_c t) \cos(\omega_c t)\mathrm{d}t = 0 \tag{2.37}$$

It is interesting that we cannot find more carriers than these two, that is, $\sin(\omega_c t)$ and $\cos(\omega_c t)$, which possess such an ideal orthogonality. For instance, in a orthogonal frequency division multiplexing (OFDM) system we can use many subcarriers to send data information in parallel. However, those subcarriers are not orthogonal in a strict sense as each subcarrier is always overlapped by half with its two neighboring subcarriers and this half-overlapping in the same signal space will introduce serious interferences in many circumstances, such as in the case of being under the influence of the multipath effect and Doppler effect. However, the two quadrature carriers, $\sin(\omega_c t)$ and $\cos(\omega_c t)$, can work in a much more robust way against many channel impairments without affecting their perfect orthogonality due to the property that their orthogonality is not established in the same signal space. Instead, their orthogonality is based in a two-dimensional space, that is, in-phase and quadrature spaces, which are vertical with each other,[9] as shown in Figure 2.15. Therefore, it is seen that the carriers $\sin(\omega_c t)$ and $\cos(\omega_c t)$ can always keep their orthogonality even under many undesirable operational conditions because they move in different signal spaces, which are already perfectly orthogonal to each other. The use of QPSK modulation in a digital modem can double the bandwidth efficiency with its power efficiency kept unchanged.

The same idea can be applied to a DSSS system to improve its bandwidth efficiency when compared with a BPSK DSSS system. However, it has to be noted that the use of in-phase and quadrature channels in a QPSK DSSS system should consider the issues on spreading codes or sequences assignment problem, that is, should we assign two different codes to In-phase and quadrature channels, or use the same code for the two channels? Therefore, a QPSK DSSS system should not be considered equivalently as a normal QPSK digital modulation system.

To illustrate the issue clearly, let us consider a generic QPSK DSSS transmitter and a receiver, as shown in Figure 2.16 and Figure 2.17, respectively. $d(t)$ is the input information data stream defined in Equation (2.31) with its duration being T, $c_1(t)$ and $c_2(t)$ are two spreading sequences generated in the transmitter for I and Q channel spreading modulations, $A\sin(2\pi f_c t + \theta)$ and $A\cos(2\pi f_c t + \theta)$ are in-phase and quadrature carriers for QPSK modulation, where the average power of the carrier is $P = \frac{A^2}{2}$ and θ is the initial phase of the carriers.

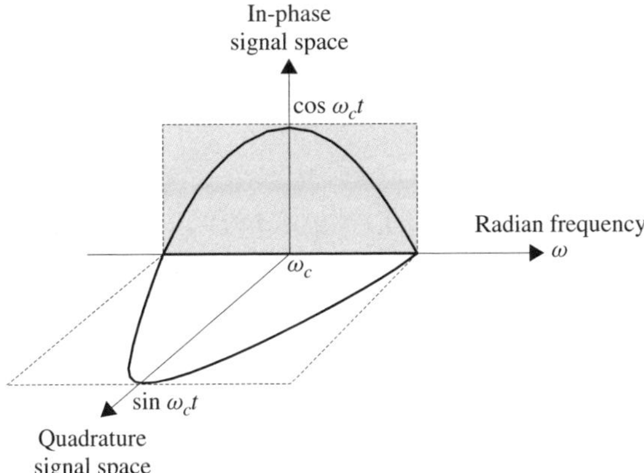

Figure 2.15 Orthogonality of $\sin(\omega_c t)$ and $\cos(\omega_c t)$ carriers in the in-phase and quadrature signal spaces in QPSK digital modulation.

---

[9]We say the two signal spaces are vertical to each other to imply that they have $\pi/2$ phase difference.

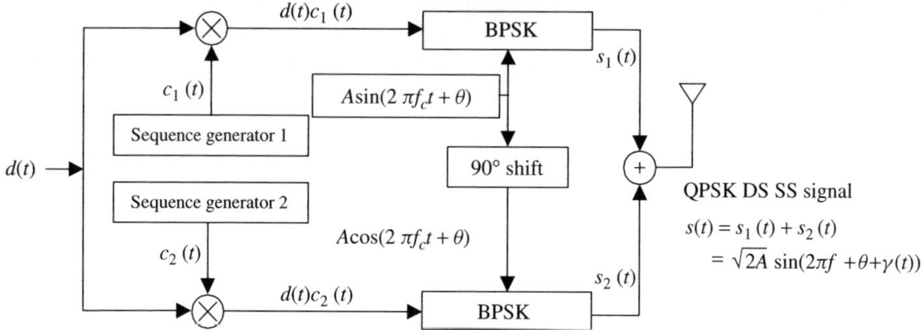

Figure 2.16 A generic QPSK DSSS transmitter.

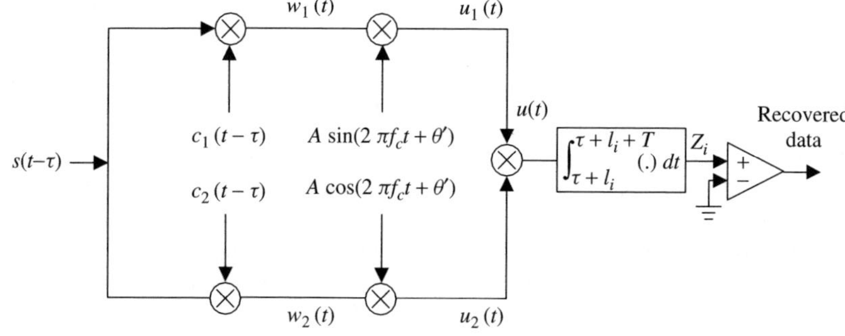

Figure 2.17 A generic QPSK DSSS receiver.

From Figure 2.16, the QPSK DSSS signal can be expressed as

$$s(t) = s_1(t) + s_2(t)$$

$$= Ad(t)c_1(t)\sin(2\pi f_c t + \theta) + Ad(t)c_2(t)\cos(2\pi f_c t + \theta)$$

$$= \sqrt{2}A\sin(2\pi f_c t + \theta + \gamma(t)) \tag{2.38}$$

where the phase modulated component can be written into

$$\gamma(t) = \arctan\frac{c_2(t)d(t)}{c_1(t)d(t)}$$

$$= \begin{cases} \dfrac{\pi}{4}, & \text{if } c_1(t)d(t) = +1 \text{ and } c_2(t)d(t) = +1 \\[2mm] \dfrac{3\pi}{4}, & \text{if } c_1(t)d(t) = -1 \text{ and } c_2(t)d(t) = +1 \\[2mm] \dfrac{5\pi}{4}, & \text{if } c_1(t)d(t) = -1 \text{ and } c_2(t)d(t) = -1 \\[2mm] \dfrac{7\pi}{4}, & \text{if } c_1(t)d(t) = +1 \text{ and } c_2(t)d(t) = -1 \end{cases} \tag{2.39}$$

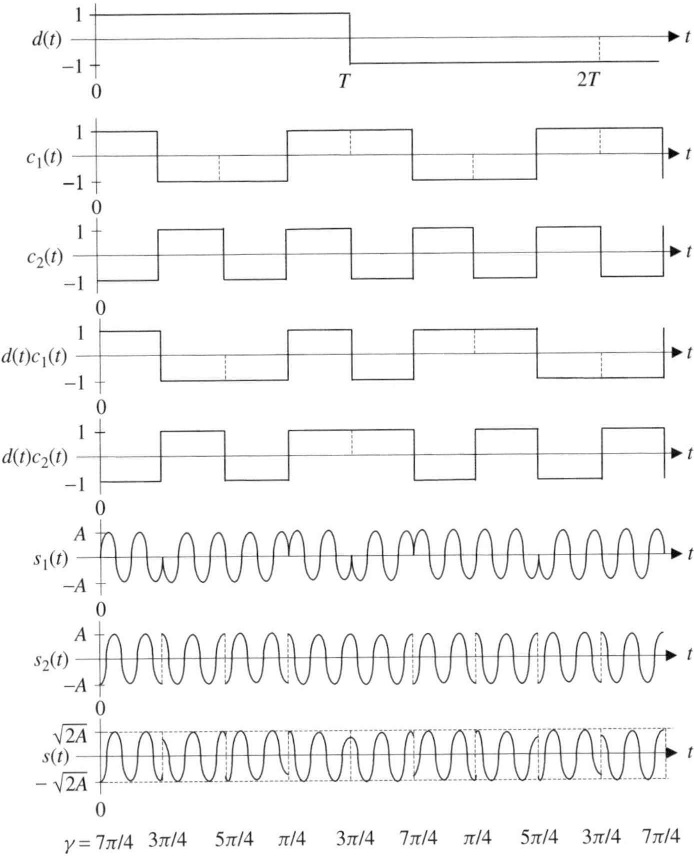

Figure 2.18 Signal waveforms in a generic QPSK DSSS transceiver.

It is seen from Equation (2.39) that $s(t)$ will yield four different phases: $\theta + \frac{\pi}{4}$, $\theta + \frac{3\pi}{4}$, $\theta + \frac{5\pi}{4}$ and $\theta + \frac{7\pi}{4}$, according to different combinations of $d(t)c_1(t)$ and $d(t)c_2(t)$. Figure 2.18 illustrates the signal waveforms in different points of a QPSK DSSS transceiver. It is noted that two different spreading sequences $c_1(t)$ and $c_2(t)$ have been used here to plot Figure 2.18. Of course, there are other alternatives for the assignments of the two spreading sequences in both I and Q channels, resulting in a very different overall performance, implementation complexity, and other characteristic features of the QPSK DSSS system in question. For instance, we can also choose to use the same spreading sequence for spreading modulations in both I and Q channels. In doing so, we will have the advantage that less sequences will be needed for each user in order to support more users in the same *spread spectrum multiple access* (SSMA) network.[10] The use of the same spreading sequence in both I and Q channel spreading modulations is also allowed since the in-phase and quadrature channels employ two orthogonal carriers, $\sin(2\pi f_c t + \theta)$ and $\cos(2\pi f_c t + \theta)$, which have already ensured a good isolation between the two channels. However, extra protection will be given if in-phase and quadrature channels use two different spreading sequences, in case some cross-talk between the I

---

[10]It is to be noted that the two acronyms, SSMA and CDMA, are interchangeable in some cases. The former emphasizes the wideband nature of SS techniques; whereas the later the user division mechanism by codes.

and Q channels exists due to nonideal operational effects, such as the frequency or phase estimation inaccuracy or jitter in the local oscillator of a QPSK DSSS receiver. The price paid to have this extra protection is that the number of spreading sequences needed for the whole SSMA network will be doubled, and in many cases the family size of an appropriate spreading sequences suitable for such a multiple access application is always limited. This issue will be addressed in detail in Section 2.3.3.

The receiver for this generic QPSK DSSS system is shown in Figure 2.17, where it is assumed that the receiver knows the exact propagation delay from the transmitter and receiver or τ and will generate two different spreading sequences $c_1(t - \tau)$ and $c_2(t - \tau)$ accordingly. In this case, the receiver should carry out despreading before carrier demodulation, corresponding to the order of spreading modulation and carrier modulation carried out in the transmitter. As a QPSK modem can be viewed as two BPSK modems working in parallel, we can understand that the order of spreading modulation and carrier modulation can be interchanged in both transmitter and receiver at the same time. It means in this case that we can also first perform carrier modulation before spreading modulation at the transmitter, and thus first have carrier demodulation before the despreading operation at the receiver.

We also assume that the receiver knows the initial phases θ' of the in-phase and quadrature carriers in the received signal $s(t - \tau)$ such that it can regenerate the local carrier references $A \sin(2\pi f_c t + \theta')$ and $A \cos(2\pi f_c t + \theta')$ that are in-phase with the received signal, resulting in a coherent QPSK DSSS signal reception. It is to be noted that the difference between the initial phases θ of the in-phase and quadrature carriers at the transmitter and the initial phases θ' at the receiver is due to the signal propagation delay through the channel.

After the despreading and carrier demodulation process, the signals from the I and Q channels will be combined and undergo integration in a unit, which functions like a low-pass filter to remove higher frequency harmonics generated in the carrier demodulation process. The integration will take place within the duration of the ith data bit of interest from $\tau + t_i$ to $\tau + t_i + T$, where t_i is the starting time of the ith bit and τ is the propagation delay. The output from the integrator will form a decision variable Z_i.

In the following illustration of basic operation of a DSSS receiver we will only concern ourselves with a simple LOS propagation path and will not take into account other channel impairing factors, such as multipath effect, Doppler effect, and so on, for illustration simplicity. Therefore, the received signal can be written as

$$s(t - \tau) = Ad(t - \tau)c_1(t - \tau) \sin\left(2\pi f_c t + \theta'\right)$$

$$+ Ad(t - \tau)c_2(t - \tau) \cos\left(2\pi f_c t + \theta'\right) \tag{2.40}$$

where the initial phase can be also expressed by $\theta' = \theta - 2\pi f_c \tau$ and τ is the propagation delay in the transmission path from the transmitter to the receiver. The signals in the I and Q channels after despreading and carrier demodulation become

$$u_1(t) = Ad(t - \tau) \sin^2\left(2\pi f_c t + \theta'\right)$$

$$+ Ad(t - \tau)c_1(t - \tau)c_2(t - \tau) \sin\left(2\pi f_c t + \theta'\right) \cos\left(2\pi f_c t + \theta'\right)$$

$$= \frac{A}{2}d(t - \tau)\left[1 - \cos(4\pi f_c t + 2\theta')\right]$$

$$+ \frac{A}{2}d(t - \tau)c_1(t - \tau)c_2(t - \tau) \sin(4\pi f_c t + 2\theta') \tag{2.41}$$

and

$$u_2(t) = Ad(t - \tau) \cos^2\left(2\pi f_c t + \theta'\right)$$

$$+ Ad(t - \tau)c_1(t - \tau)c_2(t - \tau) \sin\left(2\pi f_c t + \theta'\right) \cos\left(2\pi f_c t + \theta'\right)$$

$$= \frac{A}{2}d(t - \tau)\left[1 + \cos(4\pi f_c t + 2\theta')\right]$$

$$+ \frac{A}{2}d(t - \tau)c_1(t - \tau)c_2(t - \tau)\sin(4\pi f_c t + 2\theta') \tag{2.42}$$

respectively. The summation of the signals from the I and Q channels will become

$$u(t) = u_1(t) + u_2(t)$$

$$= Ad(t - \tau) + Ad(t - \tau)c_1(t - \tau)c_2(t - \tau)\sin(4\pi f_c t + 2\theta') \tag{2.43}$$

Obviously, after the low-pass filtering, the second term in Equation (2.43) will vanish and only the term $Ad(t - \tau)$ reflecting the data information remains, yielding the decision variable $Z_i = AT$ if $+1$ is sent or $Z_i = -AT$ if -1 is sent. Therefore, the strength of the decision variable generated from a QPSK DSSS receiver is just equal to the twice that generated from a single BPSK DSSS receiver if they work under the same condition (i.e., most importantly, their data transmission rates should be kept the same), implying a 3 dB increase in the *signal-to-noise-ratio* (SNR). It is to be noted that this gain in the SNR does not pay any price in bandwidth efficiency as both the I and Q channels occupy the same bandwidth, which is exactly the same as the bandwidth for a BPSK DSSS system. This is really wonderful and can happen only using the two unique orthogonal carriers $\sin(2\pi f_c t)$ and $\cos(2\pi f_c t)$. It is a pity that we cannot find any more such ideal orthogonal carriers.

The two spreading sequences $c_1(t)$ and $c_2(t)$ applied to the I and Q channels can be two different ones or split up from one same sequence $c(t)$, as shown in Figure 2.19, where the chip duration of $c(t)$ is half of that for either $c_1(t)$ or $c_2(t)$, and thus the length of either $c_1(t)$ or $c_2(t)$ is only the half of that of $c(t)$.

Basically, the *bit error rate* (BER) performance of a QPSK DSSS system is the same as that of a BPSK DSSS system. In fact, either the I or the Q channel can be viewed effectively as a single BPSK DSSS system and each of them possesses the same BER as a normal BPSK DSSS system. Therefore, two BPSK systems (in the QPSK DSSS system in question) working together will yield the same BER as a single BPSK system.

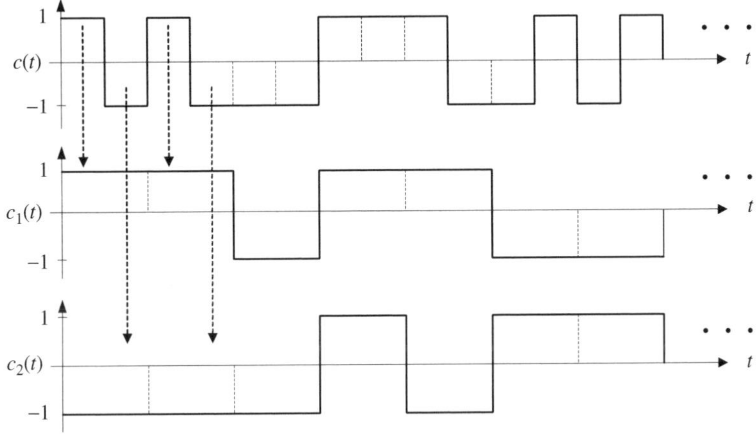

Figure 2.19 Split up of one sequence into two spreading sequences for their use in a QPSK DSSS system.

However, the bandwidth efficiency of a QPSK DSSS system is double that of a single BPSK DSSS system, and it can be explained as follows. Assume that T_c is the chip width for both $c_1(t)$ and $c_2(t)$. Thus, $s_1(t)$ and $s_2(t)$ will have the same bandwidth equal to $\frac{2}{T_c}$. This QPSK DSSS system has its data transmission rate of $\frac{1}{T}$ and the PG of $PG = \frac{T}{T_c}$. The bandwidth of this QPSK DSSS system is determined by the chip width of $c_1(t)$ and $c_2(t)$.

It is to be noted that the I and Q channels in the transmitter shown in Figure 2.16 send the same information bit stream with its data rate of $\frac{1}{T}$. However, we can also use the I and Q channels to deliver different data information to increase the transmission rate to $\frac{2}{T}$, if the bit duration in the I and Q channels is kept unchanged. In this case, the receiver structure should be modified to detect the data information in the I and Q channels separately. The modified block diagrams for a QPSK DSSS system with double data rate are shown in Figure 2.20 and Figure 2.21, respectively.

There are the following factors that will affect the performance of a QPSK DSSS system: transmission rate or bandwidth, PG and SNR (or transmission power). In order to compare the performance of two DSSS systems, such as a BPSK system and a QPSK DSSS system, we have to concentrate on *one* particular parameter, with the other two fixed, to make the comparison easily and objectively. For instance, if we want to compare BPSK and QPSK DSSS systems shown in Figures 2.10 and 2.20, we should fix the data rate and PG first (thus the bandwidth), allowing us to make a fair comparison on SNR values for the two schemes. Since the same data rate is concerned, the bit duration in either the I or Q channel in Figure 2.20 will be twice as wide as that in Figure 2.10. Also due to

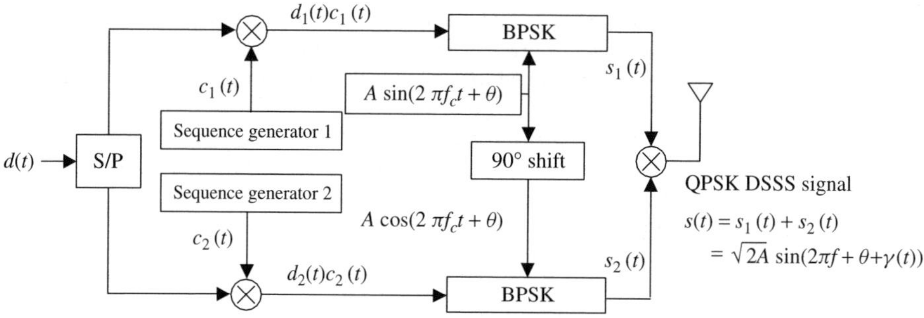

Figure 2.20 An alternative structure of QPSK DSSS transmitter with double transmission rate.

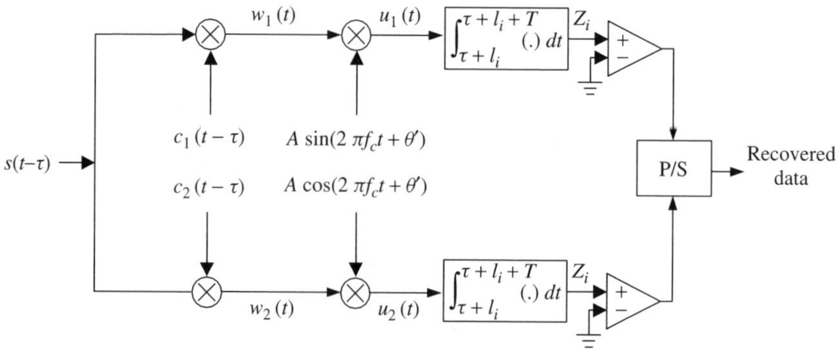

Figure 2.21 An alternative structure of QPSK DSSS receiver with double transmission rate.

the fact that the same PG value is assumed for the two schemes, we can readily conclude that the bandwidth efficiency (defined by bit/s/Hz) for the two schemes should be the same. However, the power efficiency of the QPSK DSSS system shown in Figure 2.20 is double that of a BPSK DSSS system shown in Figure 2.10, due to its wider bit duration and high signal power available for the detection of each bit in the QPSK scheme, implying a higher SNR value.

The DSSS systems using other modulation schemes, such as MSK, QAM and so on, can also be studied using a similar methodology as illustrated in this subsection. Those who are interested in them can refer to these references [255–268].

As a final remark before the end of this subsection, it is to be noted that one of the most successful applications for the QPSK DSSS techniques is the GPS system [244–254], which was launched initially in the United States for positioning applications in military operations. Nowadays, the GPS has found a worldwide applications in various practical systems, most of which are civilian applications and services.

2.2.2 Frequency Hopping Spread Spectrum Techniques

Another method to spread the spectrum of a data-modulated carrier is to switch the carrier frequency from one to another periodically. Usually, each carrier frequency is selected from a set of frequencies, which are spaced approximately as the same width of the data modulation bandwidth. The spreading code in this case does not directly modulate the data-modulated carrier but is instead used to control the appearance sequence of carrier frequencies. Because the transmitted signal appears as a data-modulated carrier which is hopping from one frequency to another, this type of spread spectrum is called *frequency hopping spread spectrum* (FHSS). In the receiver, the FH is removed by mixing (down-converting) with a local oscillator signal that is hopping synchronously with the received signal.

Based on its function and behavior, the *FHSS* technique is more accurately termed as, *code-controlled multifrequency*-FSK modulation. It works very similar to a conventional *frequency shift keying* (FSK) modulation scheme, except that the set of frequencies is very large. On the other hand, a normal FSK modem often uses only two frequencies. For instance, f_1 is sent to denote a *mark*, f_2 is to signify a *space*. In the FH schemes, there will be thousands of frequencies available. The number at a few hundreds to thousands is normal in a real system, which makes discrete frequency selections randomly on the basis of a predetermined sequence in combination with the data information conveyed. The number of frequencies and the rate of hopping from frequency to frequency in an FHSS system is determined by operational requirements for a particular communication application.

The basic structure of an FHSS system can be described as follows. Usually, a FH system must have a sequence generator and a frequency synthesizer, which is capable of generating the corresponding frequencies according to the sequence generator. It is difficult to develop an FHSS system to design a fast-settling frequency synthesizer with a sufficient large number of carrier frequencies. Theoretically speaking, the instantaneous frequency output, the frequency synthesizers generate, must be a single frequency.[11] However, a practical system may produce an output spectrum, which can be a composite of the desired frequency, sidebands generated by hopping, as well as some other spurious frequencies generated as by-products.

Figure 2.22 shows a conceptual block diagram of a FHSS transmitter. The receiver of an FHSS system is given in Figure 2.23. The waveforms generated by this simple FHSS system (in both transmitter and receiver) are shown in Figure 2.24, where it is assumed that the data information is kept at the same level (here all bits are $+1$ constantly) for simplicity of illustration.

---

[11]This is one of the reasons that make a FH system costly and very difficult to implement. In particular, the frequency synthesizer in a fast hopping FH system has to work to switch from one frequency to another in a very fast and stable way, especially when the data rate is very high.

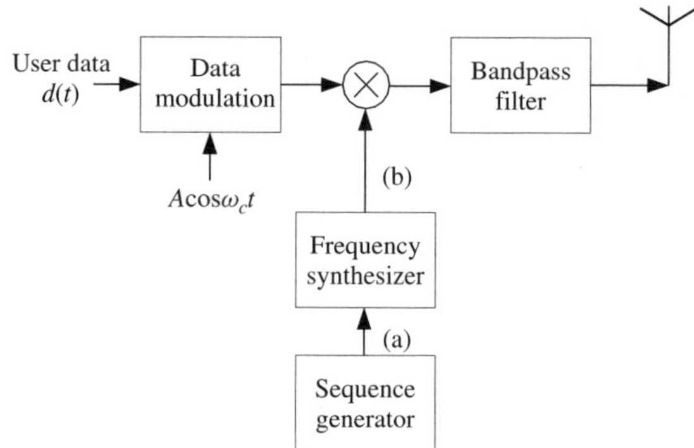

Figure 2.22 Conceptual block diagram of a FHSS transmitter.

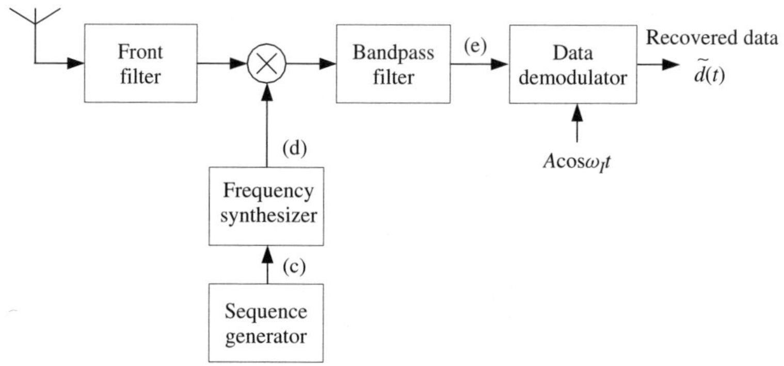

Figure 2.23 Conceptual block diagram of a FHSS receiver.

The FHSS transmitter shown in Figure 2.22 consists of the following basic blocks: a data modulator, a mixer (denoted simply by a multiplier in the figure), a FH patter sequence generator, a frequency synthesizer, a bandpass filter and an antenna. The data modulator will perform the digital modulation between the user data $d(t)$ and a carrier $A\cos\omega_c t$, where A is its amplitude. The frequency synthesizer will work according to the hopping sequences generated by the sequence generator. Usually, the sequence generator can produce a great number of different patterns, each of which will be used by the frequency synthesizer to generate a particular carrier, which will be multiplied with the data-modulated signal in the mixer to produce an up-converted transmitting signal from the antenna. Therefore, the carrier frequency of the transmission signal is under the control of the sequence generator, which can also control the FH rate from one frequency to another. The hopping rate is a very important parameter in a FHSS system, which will determine if it is a *fast hopping* or *slow hopping* FH system.

At the FHSS receiver, as shown in Figure 2.23, the received signal should first go through a *front-end filter*, which will be used to reject the image of the carrier frequency produced in the mixer. For the same purpose, the sequence generator will produce a replica of the sequence used by the transmitter and will yield a FH pattern, which should be exactly the same as that used in the transmitter, in the output of the frequency synthesizer. The locally generated FH pattern will be

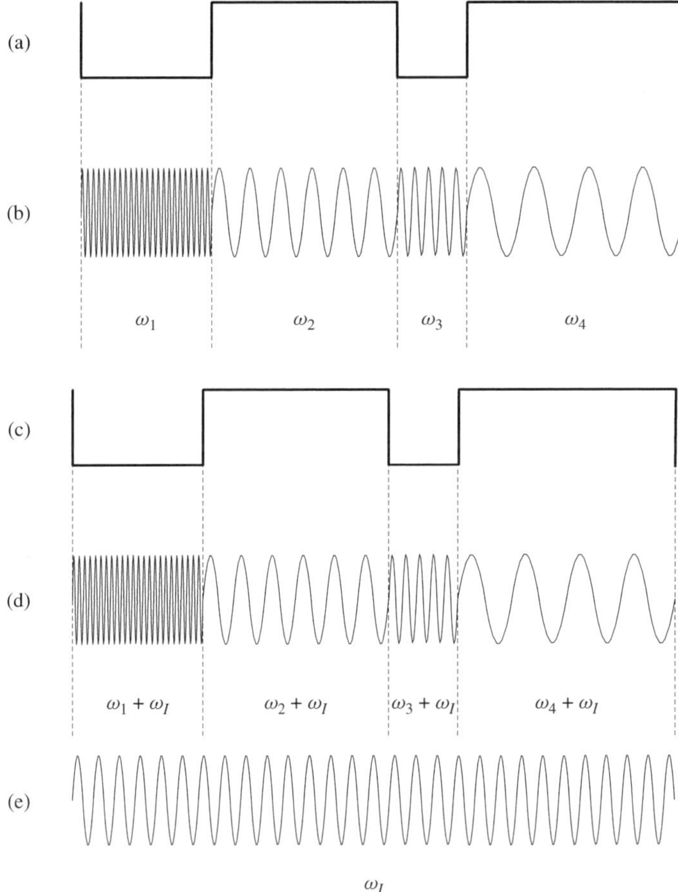

Figure 2.24   Waveforms generated in different points of the FHSS transmitter (as shown in Figure 2.22) and the receiver (as shown in Figure 2.23). (a) The output sequence generated in hopping pattern sequence generator; (b) the output signal from the frequency synthesizer; (c) sequence generated in the local sequence generator at the receiver; (d) the local carrier waveforms generated by the local frequency synthesizer; (e) The carrier output from the mixer of the receiver.

mixed with the received signal to produce a narrowband data-modulated signal with a fixed carrier frequency, which should be equal to the *intermediate frequency* (IF) ω_I. The output IF signal will be demodulated by a data demodulator to recover the transmitted data information or $\tilde{d}(t)$.

Ideally, the spectrum generated from a FH system should be perfectly rectangular, with spectral lines distributed evenly in every predetermined frequency channel. The transmitter should also be designed to send the same amount of power in each frequency. Otherwise, the detection efficiency on different frequencies will be uneven, causing decision errors at the receiver.

As shown in Figure 2.23, the received frequency hopping signal is mixed with a locally generated replica, which is offset by a fixed amount (which is equal to a carrier frequency) suitable for reception process at the receiver, ω_I) such that the output from the mixer in the receiver will produce a constant difference frequency ω_I if the transmitter and receiver code sequences are synchronous.

As in the case of the DSSS system discussed in Section 2.2.1, any signal that is not a replica of the local reference is spread by multiplication with the local reference, not being restored into its original narrowband waveform. Bandwidth of an undesired signal after multiplication with the local reference is approximately equal to the bandwidth before despreading. For instance, an external sinusoidal signal received at a FH receiver will be converted into a signal that will change in the same way as the local reference (a FH carrier), and thus it will never pass the bandpass filter, which is tuned to a fixed carrier frequency or IF, say ω_I. On the other hand, if a desirable signal appears at the input side of the receiver, the output signal from the FH despreading unit will be a narrowband signal modulated by a fixed carrier ω_I, which will undergo a data demodulation process to recover the original data information sent or $\tilde{d}(t)$, as shown in Figure 2.23.

Processing gain of a FH system

The IF mixer and the bandpass filter in the transmitter are effective to reject undesired signal power that lies outside its bandwidth defined by the useful data signal bandwidth. Because this IF bandwidth is only a fraction of the bandwidth of the local FH carrier reference, it can be seen that almost all the undesired signal's power is rejected, whereas a desirable signal is enhanced by being correlated with the local FH carrier reference. In Section 2.2.1, it was illustrated that a DSSS system operates identically from the viewpoint of undesired signal rejection and restoration of the desired signal. From this general point of view, DS and FH systems are similar. However, they are different in the details of their operation. Like the PG defined in the DSSS system, we should also define the PG for the FHSS system, and this value will also play an extremely important role in determining the overall performance of a FH system.

The PG of the FH systems should be defined for two different cases. One case is that all generated carrier frequencies are contiguous, and the other is noncontiguous. The noncontiguous carrier frequencies are common for applications when it is hard to find enough spectrum allocation for the FHSS communications. On the other hand, the contiguous carrier frequencies are the ideal situations, which can simplify the calculation of the PG. By the contiguous FH spectrum, we mean that all the carrier frequencies generated by the synthesizer are evenly spaced in the frequency domain.

The PG of a FH system with a contiguous spectrum can be calculated in the same way as a DSSS system. That is,

$$PG_{FH} = \frac{BW_{RF}}{BW_{information}} \tag{2.44}$$

where BW_{RF} is the bandwidth spanned by all the carrier frequencies generated by the synthesizer collectively, and $BW_{nformation}$ is the bandwidth given by the original baseband information signal, which is determined by the data signal $d(t)$.

On the other hand, if the carrier frequencies generated by the synthesizer are not contiguous, an objective measure of PG can be

$$PG_{FH} = \text{the total number of available frequencies} \tag{2.45}$$

which can also be used to calculate the PG value for a FH system with contiguous carrier channels. For example, a FH system containing 1000 frequencies can have 30 dB available PG. The approximation has been used in the simple calculation for PG, given in Equation (2.44), because all *guard bands* in between two carrier channels are neglected in the formula. If the guard bands in the IF bandwidth cannot be omitted, Equation (2.45) should be used instead.

Hopping rate of a FH system

The FH rate and number of carrier frequencies are determined by the sequence generator. The minimum frequency switching rate of a FHSS system is determined by the following system parameters:

- The bandwidth of information to be sent and its importance.

- The amount of redundancy needed.

- The environment where the FH system will work in terms of severity of the interferences.

Information in a FH system can be sent in any way available to the other systems. Usually, however, some form of digital signal is preferred, whether the information is a digitized analog signal or data. Assume for the present that some digital rate is prescribed and that FH has been chosen as the SS technique. Then, the question is how to determine the FH rate or the *chip* rate. It is to be noted that, in contrast to a DS system, there is no *chip* in a FH system. The term *chip* is used here just for conceptual analogy.

A FH system should possess a sufficiently large number of frequencies on demand. The number required depends on the operational environment of the system. For example, two thousand frequencies will provide satisfactory operation when interference and noise are evenly distributed at every available frequency channel. For equal distribution of interference or jammers in every channel, the interference power required to block communications has to approach two thousand times that of the desired signal power. In other words, the achievable jamming margin is about 33 dB in this case. Unless some sort of redundancy that allows for bit decisions based on more than one frequency is needed, a single narrowband interferer will cause an error rate of 0.5×10^{-3}, which is generally satisfactory for normal digital data transmission. For a simple FH system without any form of transmitted data redundancy (one hop exists in each bit duration), the expected error probability can be approximated by $\frac{J}{N}$, where J is equal to the number of CW interferers whose power is greater than or equal to signal power, and N is the number of frequencies available to the FH system. Error rate probability for a FH system, in which we assume that binary FSK modulation is used (here two frequencies denote binary symbols or $f_1 = +1$ and $f_0 = -1$), can be approximated by the following expansion

$$P_e = \sum_{n=r}^{N_c} \binom{N_c}{r} p^n q^{N_c - n} \tag{2.46}$$

where p is the error probability of a single jamming trial, which is J/N, J denotes the number of jamming carriers, N is the number of channels available to the FH system, q is the probability of no error for a single jamming trial (we always have $q = 1 - p$), N_c is the number of hops in each bit,[12] r is the number of wrong chip decisions necessary to cause a bit error. A chip decision error is defined as the situation where interference power in a "+1" channel exceeds the power in an intended "−1" channel (or vice versa) by some amount ε that is sufficient to cause a decision error.

A FH system is called a *slow frequency hopping* spread spectrum system if only one or less than one hop happens in each bit/symbol duration. Otherwise, if more than one hopping exists, a *fast frequency hopping* system results.

A fast hopping FH system can offer a much better performance than that of a slow hopping FH system at the price of system implementation complexity. For instance, the implementation cost for a fast hopping FH system with three hops per bit will be at least three times more complex than that of a slow hopping FH system with one hop per bit. The synthesizer should work at least three times faster than that of a one-hop-per-bit system. The amount of data a transceiver should process should also be at least three times more than that of an one-hop-per-bit system.

The performance of a three-hop-per-bit FH system can be evaluated by the following method. Assume that the decision rule for this system is on the basis of the two out of three decision rule.

---

[12]The value of N_c will determine fast FH or slow FH. Often, N_c can be a noninteger. For instance, $N_c = 1/2$ means that one hopping happens in every two bits, implying a slow FH system. On the other hand, if $N_c > 1$, a fast FH system is the resultant.

That is if at least two frequencies are correct, the decision will be made in favor of the symbol or bit representing the FH patterns containing the two correct frequencies. Thus, a single jamming trial will cause not more than

$$\binom{3}{2} p^2 q^{3-2} = 3p^2 q \tag{2.47}$$

where $\binom{n}{m}$ denotes the number of combinations of taking m from n items, p and q represent the probability of error caused by a single jamming trial and the probability of no error caused by a single jamming trial, respectively, and $p = 1 - q$. Consider a FH system with totally two thousands carrier frequencies, p will be $\frac{1}{2000}$ and $q = 1 - p = 0.9995$. Thus, the error probability will become $3\left(\frac{1}{3000}\right)^2\left(1 - \frac{1}{3000}\right) \cong 7.5 \times 10^{-7}$, which is much better than 0.5×10^{-3} given by the previous one hop -per bit FH system.

Due to the limited space, we will not discuss the FHSS Systems further. For more information on FHSS, please refer to the references [269–307].

2.2.3   Time Hopping Spread Spectrum and Ultra-Wideband Techniques

After having discussed the two popular SS techniques, DS and FH, we would also like to give a brief introduction of the third SS technique, *time hopping* (TH) technique, in this subsection.

The TH technique, in fact, works in a very similar way as a digital modulation scheme called *pulse position modulation* (PPM). In other words, time hopping is nothing but a type of pulse position modulation in a sense that a code sequence is used to key the transmitter on and off, as shown in Figure 2.25, where the times for the transmitter to switch on and off follow a specific pseudorandom code sequence. The average duty cycle of "on" and "off" in the transmitter can become as large as 0.5. The major difference between the PPM and the TH lies in the fact that the former uses pulse position patterns to represent the data information symbols, whereas the latter denotes a particular code sequence, which acts as a secret key to further decode the data information hidden therein. A conceptual block diagram of a TH system is shown in Figure 2.25, where the *on–off switch logic* unit in the transmitter is used to control the positions that the sent pulse will hop from one to the other. The receiver in Figure 2.25(b) should also follow exactly the same hopping pattern to capture all transmitted power from the transmitter. Obviously, an unintended receiver will not be able to receive all power from a

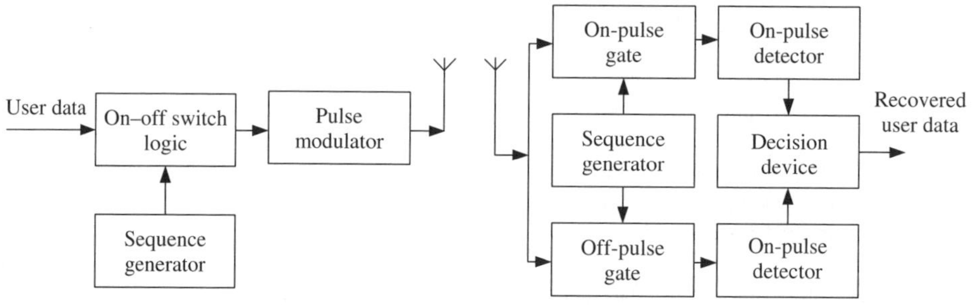

Figure 2.25   Block diagram of a THSS transceiver, where the *sequence generator* is used to control the TH patterns that should change synchronously in both the transmitter and the receiver.

particular TH transmitter due to its unknown pulse hopping pattern. On the other hand, if a hostile jammer wants to block the communications established based on a THSS system, it will be impossible to block all transmitted pulses without the knowledge of transmitter's hopping patterns unless it keeps transmitting all the time, requiring a much higher transmission power than that of a TH transmitter.

The THSS technologies are not as popular as other two SS techniques, that is, DSSS and FHSS techniques. The main reason is its implementation difficulty, especially for the pulse generator, which is the core of a THSS system and should be able to produce a train of very narrow impulses of the order of a nanosecond.[13] The geraniol of impulses train should also provide very good timing accuracy, such that the PPM can be effectively applied to different SS code sequences for multiple access. It remains a challenging task to make such a pulse generator, even today.

The THSS technique seldom works independently in an SS system (except in the case of a UWB system, a technology developed on the basis of the TH technique). Instead, it usually works with some other SS modulation schemes, in particular, the FH technique, discussed in the previous subsection, to result in the time-frequency hopping SS scheme.

The difference separating time-frequency hopping and pure FH lies in the fact that in FH systems the transmitted frequency is changed at each code chip time, whereas a time-frequency hopping system may change frequency and/or amplitude only at one or zero transition instant in the code sequence. The simplicity of the modulator is obvious as is seen from Figure 2.25. Any pulse-modulating signal source that is capable of following code sequences is eligible as a time hopping modulator. TH may be used to aid in reducing interference between systems in *time division multiplexing* (TDM). However, stringent timing requirements must be placed on the overall system to ensure minimum overlap between transmitters. This is one of the reasons that makes a TH system much harder to implement than other SS systems. Also, as in any other SS system, the spreading sequences must be designed or selected carefully from the viewpoint of their cross-correlation properties. As mentioned earlier, a simple THSS system can be blocked by a jammer that uses a continuous carrier at the signal center frequency. The primary advantage offered is in the reduced duty cycle. In other words, to be an effective jammer an interfering transmitter has to be forced to transmit continuously (assuming the TH sequence used by the time hopper is unknown to the interferer). The power required by a legitimate time hopper will be less than that of an interfering transmitter by a factor that should be equal to the PG of this TH system. Because of this relative vulnerability to interferences, a simple TH transmission should not be used for antijamming unless combined with FH to prevent single frequency interferers from causing significant losses.

However, the THSS techniques have been found useful in ranging, multiple access, or other special applications introduced later. A typical example for such applications is the UWB technology, which has had tremendous attention recently, due to its many attractive properties, such as its unique capability to mitigate multipath propagation problems based on its very high time resolution. The issues on the UWB technology will be addressed in more detail in Section 7.6 and will not be discussed further here.

Before ending this subsection, we would like to give a brief summary on the SS techniques we have discussed.

- DSSS technique by far is the most popular technique in all SS communication applications. There are several reasons for its popularity. First, it is the simplest form of the SS techniques and can be implemented at a relatively low cost when compared with other SS techniques, such as FH and TH. All currently available second generation (2G) and third generation (3G) mobile communication standards are based on DSSS technique with almost no exception. Second, a DSSS system can also be designed to operate compatibly with many existing communication networks that operate based on other multiple access technologies, such as *time division multiple access* (TDMA), to achieve so called smooth upgrading. The proposal of a 3G system or

---

[13]Nanosecond is equal to 10 to the power of minus nine, or 10^{-9}, whose bandwidth spans at least 1 GHz or more.

WCDMA [425, 431] was based mainly on this clause as an effort to achieve smooth upgrading from its 2G system named *Global System for Mobile Communications* (GSM) [375–399] working on TDMA technology.

- The FHSS technique is less widely used in civilian communication systems when compared with the DSSS technique. The FH technique has been applied to GSM networks [375–399] as an option to mitigate the *frequency-selective fading* that may happen in some places, especially in urban and downtown areas, where MI may be serious. The implementation of an FH system relies on an accurate and fast-settling frequency synthesizer, which can be costly if using current state-of-the-art RF and microelectronics technology. However, the FHSS techniques have been widely used in military communication systems, such as the battle field commanding system.

- The THSS technique is much less widely used than either DSSS or FHSS technique. The reason is partly because of the fact that it may suffer from serious interference problems if there exists a continuous transmission in the coverage area, as the TH system only works in an on-and-off fashion in a frame. For this reason, the TH technique usually works with other SS techniques, in particular the FH technique, forming a hybrid TH-FH system.

The underlying importance of the SS techniques is that it forms a technical foundation for *CDMA* systems, which have been widely used in 3G mobile cellular communications and may continue to be an important multiple access technology in wireless communications beyond 3G.

Before ending this section, we would also like to throw several questions to the readers as an effort to stimulate some more research interest. We have been discussing, in particular, "spread spectrum" techniques. Then, what can we get from "spread time"? As time and frequency are "in a duel" with each other, can a "spread time" system provide similar PG to mitigate various impairing factors in the channels, just as SS does? If yes, then how can such a "spread time" system be implemented? There has been some research work reported in this particular area, which unfortunately remains very preliminary on "spread time" techniques.

2.3   Multiple Access Technologies

Most communication applications involve more than one communication party, to form a communication system or network, where many users should share a common communication medium to communicate with one another simultaneously. In such a multiuser communication environment, *multiple access technology* will play a pivotal role to determine basically how efficiently the whole communication network can work collectively. A generic multiuser communication model is shown in Figure 2.26, where K users share a common communication medium or channel that may experience different types of impairment in the transmission paths, such as noise, interferences and so on. The receiver in the figure is intended for the kth transmission.

There are three major multiple access technologies, which have found many applications in real communication systems and networks. They are *frequency division multiple access* (FDMA), *TDMA*, and *CDMA*, which will be discussed individually in the following subsections.

2.3.1   Frequency Division Multiple Access

FDMA scheme is the first multiple access technology ever introduced into multiuser communication applications. It can be traced back to the early 1960s when several radio transmissions coexisted in time and space without mutual interference by using different carrier frequencies.

In a FDMA system or network, as its name suggests, all users operating in this system or network will be assigned different carrier frequencies, f_k, where $1 \leq k \leq K$, with a certain bandwidth for each

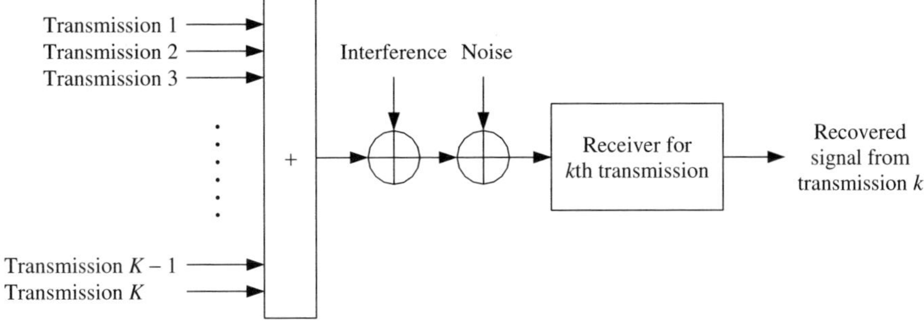

Figure 2.26 A generic multiple access communication system with K users being present. The receiver is for reception of the kth transmission.

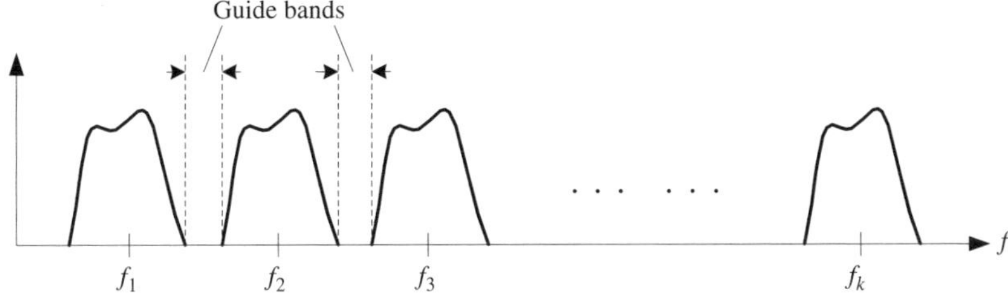

Figure 2.27 Signal spectral channel allocation in a FDMA system.

carrier assuming that there are in total K different users in the network. Therefore, all users will be divided by their distinct carriers when their signals are transmitted in the network. Each carrier together with its associated bandwidth is assigned to a user, also called a *channel*, and thus all user channels should never be overlapped in frequency domain. In most cases, there should be a *guard band* between two consecutive frequency channels to avoid possible interference between the two channels in question, as shown in Figure 2.27.

It is seen from Figure 2.27 that each user has to be allocated a dedicated channel with a certain bandwidth (depending on the modulation scheme used). Basically, the more users in a system, the more frequency channels that are needed. The channels assigned to a specific group of users cannot be easily reallocated to others. Therefore, the channel allocation scheme in a FDMA system is rather rigid in the sense that they cannot be effectively used even if some channels are idle at some instant.

As a precious resource, the frequency bands or channels should be made use of as wisely as possible to support as many users as possible simultaneously. In fact, it has been found that the radio spectrum usable for wireless communication applications is very limited, from a few hundreds Mega Hertz (MHz) to a couple of tens of Giga Hertz (GHz). In addition, most of the available radio spectra have been allocated for certain radio communication applications or services, and not many of them have been left unused. Recent research has clearly indicated that dividing users' transmissions directly among the frequency channels will hardly make effective use of the frequency spectrum. This is why people are working hard now to find more efficient ways to use those vacant spectral bandwidth for new wireless communications.

The FDMA technology has been used in earlier multiuser communication systems, such as Advanced Mobile Phone System (AMPS), which is analogous to the mobile cellular telephony system

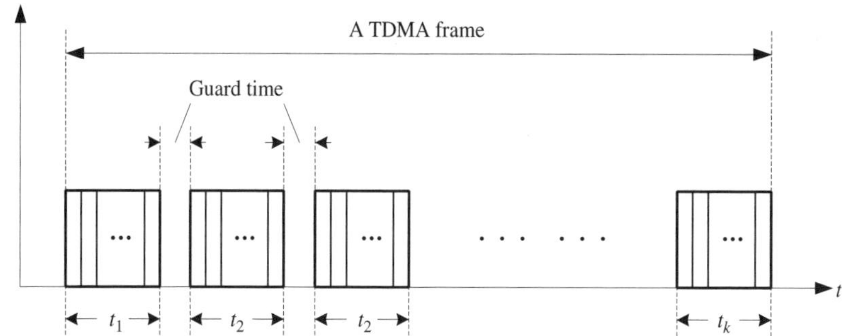

Figure 2.28 Illustration of a time frame structure in a TDMA scheme, where K users share a TDMA frame for their transmissions in their individual time slots, and a guide time (GT) is inserted in between every two consecutive time slots to prevent possible interference caused by transceiver time jitter or synchronization errors, and so on.

developed in North America in the 1980s by AT&T's Bell Laboratories. It operates in 800–900 MHz frequency band and is one of numerous first generation (1G) mobile cellular standards. Even now, some parts of the United States and Canada are still under the coverage of the services of AMPS.

It is necessary for us to distinguish two terminologies, that is, FDMA and *frequency division duplex* (FDD). The FDMA technology is used to separate different users using different frequencies in a communication network; whereas the FDD technique is used to divide *uplink* and *downlink* channels by different carrier frequencies in a point-to-point communication link, usually in mobile cellular or satellite communication networks. Under this context, what we mean by *uplink* is the communication direction from a mobile or ground terminal to the base station (BS) or the satellite in a mobile cellular system or satellite communication system. The downlink will serve transmission in the opposite direction in the systems.

2.3.2 Time Division Multiple Access

Another approach to distinguish different users in a multiuser communication system is to separate different transmissions by different *time slots* in a frame, resulting in a *TDMA* scheme. Each frame is defined as a fixed time interval, in which all users' transmissions should take turns to appear. Therefore, each transmission can only send a small portion of its data in a time slot of the frame, and it has to wait until its next turn (or time slot) comes in the following frame. Thus, the TDMA scheme works like transmissions by taking turns in time, and in each frame all transmissions from different users will proceed in a predetermined sequential order, strictly one by one. Figure 2.28 shows a typical TDMA frame shared by K users' transmissions.

As in the FDMA scheme, a GT has to be present between two consecutive time slots to avoid possible interference caused by any inaccuracy in timing or synchronization control, which play a crucial role in a TDMA system.

Unlike the FDMA scheme, the TDMA scheme is suitable only for transmission of digital signals. The TDMA scheme is hardly suited for sending analog signals, whereas the FDMA scheme can support both analog and digital signal transmissions. For this reason, the kth time slot shown in Figure 2.28 contains a train of bits sent from the kth user.

The TDMA scheme has been successfully applied to many commercial communication systems. A typical example is the *GSM Communications* standard [375–399], which was proposed by ETSI, Europe, and has been deployed in many countries around the world. In fact, the GSM system is one of the most popular mobile cellular systems in the world in terms of the number of subscribers.

It is understandable that a TDMA system needs a precision clock reference in both transmitters and receivers throughout the whole communication network. This will inevitably increase the implementation complexity of a TDMA system when compared to a FDMA system. The recent study shows that sharing channel resource in time is, in particular, well suited for those communication applications where all transmissions happen in a burst or transmission proceeds via a packet of time, and then the next one, and so forth. This is due to the fact that each time slot can naturally fit one or several periods in a TDMA system. This is the reason why many people show great interest in revitalizing the TDMA system for futuristic high-speed all-IP–based wireless communications.

There are some technical limitations for the TDMA scheme as a multiple access solution for future wireless communication systems. Because a TDMA system requires that each user sends data by taking turns in time slots, a transmitter can only send a very small amount of data within a particular time slot allocated to it. When the next slot comes, it has to make a fresh channel estimation and other operations, which are related to the signal detection algorithms, thus consuming a lot of time and hardware/software load, due to the change of channel conditions from slot to slot (spanning a complete frame length). This situation will become worse off if more users should be accommodated at the same time in a TDMA system, as the time slot size will shrink if the frame length is kept unchanged. The GT has been used to avoid possible interference caused by the transceiver time jitter or other synchronization problems in a TDMA system, as shown in Figure 2.28. Obviously, the GT as an overhead will also consume a lot of valuable transmission time resource as all GTs will never yield any data throughput. However, the GT is always indispensable for a TDMA system as it is impossible to achieve a perfect synchronization without it in either transmitters or receivers. To improve the accuracy in synchronization systems of a TDMA network, some sophisticated technology, such as a GPS or other expensive clock references, should be used. This will again increase the overall implementation complexity and cost as a whole.

As in the case of the FDMA versus FDD schemes, a TDMA system can inherently work very well with the *time division duplex* (TDD) scheme, which is another way (other than FDD) to separate uplink and downlink transmissions in a communication system, such as mobile, cellular and satellite communication systems. In the TDD scheme, a frame is also divided into different time slots, some of which can be allocated for uplink transmissions and the others for the downlink transmissions. The obvious advantage for the TDD scheme is its great flexibility to assign a different number of time slots for uplink and downlink transmissions, achieving asymmetric transmissions in uplink and downlink channels, which are required in many real applications, such as Internet applications (where downlink traffic volume is always much higher than that in uplink). Another advantage of the TDD scheme is that it makes use of a relatively narrow spectrum to implement duplexing in uplink and downlink transmissions, without a need to involve the pair-wise spectrum allocation problem as in any FDD scheme.

TDMA can work with FDMA to form a hybrid *TFDMA* schemas, which can create a large number of partitions in time-frequency two-dimensional space, as shown in Figure 2.29, where there are M different frequency bands and N different time slots, supporting all together $M \times N$ different users in the system. Of course, both GT and *guard bands* are needed to protect time slots and frequency channels from possible overlapping due to imperfect operation of transceivers.

Similar to the FDMA scheme, the capacity in a TDMA system is strictly limited by the total number of time slots available in a frame. Therefore, the capacity in a TDMA system is *hard* or very rigid in the sense that it will not be possible to accommodate even one more user when all slots have been used up. The same situation happens to a FDMA system, where its capacity is also strictly constrained by the total number of different frequency bands or channels. The hard capacity problem in both FDMA and TDMA systems makes it very difficult to optimize the utilization of radio resources in a whole network globally. Think about the fact that the communication quality for a user in a TDMA or FDMA system is basically nonadjustable even if only ONE single user is active, because the slots or channels assigned to other users can hardly be reallocated to the single active

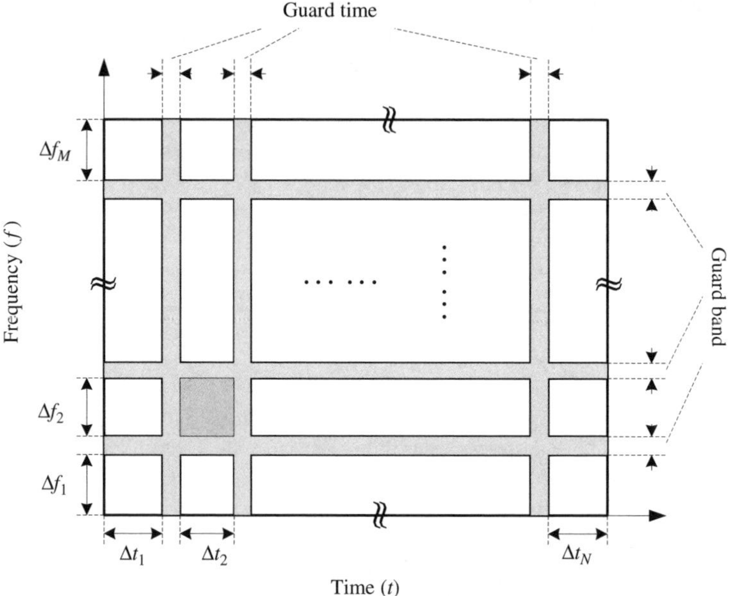

Figure 2.29 Illustration of a time-frequency division multiple access (TFDMA) system, where there are M different frequency bands and N different time slots, supporting all together $M \times N$ different users in the system.

user because of the fixed resource allocation scheme. To overcome this hard capacity problem,[14] a new multiple access technique, called *code division multiple access* technique, can be used and will be discussed in the next subsection.

2.3.3 Code Division Multiple Access

Based on the discussions on SS techniques given in Section 2.2, we will address the issues on CDMA, which has become the most important multiple access technology in contemporary mobile cellular communication systems, as reflected by the fact that almost all 3G standards have been engineered based on CDMA.

It is to be noted that the SS techniques, discussed in Section 2.2, provide only a means to extend the bandwidth of transmitting signals to obtain some other extra operational benefits, which may not be possible if using only a bandwidth comparable to that spanned by the original information signals. However, by spreading the spectrum of the original data signals, we can also obtain other operational advantages, such as allowing more users to share the same communication medium simultaneously in time and frequency without introducing considerable interference.

CDMA has emerged as the most important multiple access technology for the second and third generations (2–3G) wireless communication systems, exemplified by its applications in many important standards, such as IS-95 [317–344], CDMA2000 [345], UMTS-UTRA [425], WCDMA [431] and TD-SCDMA [432], which were proposed by TIA/EIA of the United States, ETSI of Europe, ARIB of Japan and CATT of China, respectively. It is possible that the CDMA will continue to be a primary air-link architecture for the future or *beyond 3G* (B3G) wireless communications, although some

---

[14]Of course, there are many other advantages to use CDMA instead of FDMA or TDMA, such as high bandwidth efficiency, superior multipath diversity due to the use of RAKE receiver, etc.

other multiple access technologies have also gained attention in the community, such as *orthogonal frequency division multiple access* (OFDMA)[15] and even renovated versions of TDMA.

CDMA is a multiple access technology that divides users based on orthogonality or quasi-orthogonality of their signature codes or simply CDMA codes. There are three primarily different types of CDMA technologies that have been extensively investigated in the last two decades, DS-CDMA, FH-CDMA and time hopping (TH) CDMA. Each user in a DS-CDMA system should use a code to spread its information bit stream directly by multiplication or modular-two addition operation, which is also the simplest and most popular CDMA scheme among the three. The FH-CDMA uses a multitone oscillator to generate multiple discrete carrier frequencies and each user in the system will choose a particular frequency hopping pattern among those carriers that are governed by a specific sequence, which should be orthogonal or quasi-orthogonal to the others. Depending on the hopping rate relative to the data rate, FH-CDMA can also be classified into two subcategories: slow hopping and fast hopping techniques. The majority of the currently available FH-CDMA systems are using slow hopping scheme. The third type of CDMA or TH-CDMA is found to be much less widely used than the previous two mainly because of difficulty in implementation and hardware cost associated with a transmitter that should provide an extremely high dynamic range and very high switching speed. As mentioned earlier, the UWB technique can, in one way or another, also be viewed as a type of TH-CDMA systems [594–673].

In addition, there are also many different types of hybrid CDMA schemes, which can be formed by various combinations of DS, FH, and TH, together with multicarrier (MC) and multitone (MT) techniques, as shown in Figure 2.30, where the family tree of various forms of CDMA technologies is depicted, where CC-CDMA and DS/CC-CDMA are two new CDMA schemes to be introduced and then discussed in the latter part of this chapter and *offset stacking* is a new spreading modulation technique used in the CC-CDMA architecture. It is stressed that this chapter will only deal with the DS-based CDMA systems and its evolution issues. However, the conclusions drawn here may also be found equally relevant to the other CDMA schemes.

One of the most important characteristics of a CDMA system is that it allows all users to send their information at the same frequency band and the same time duration simultaneously but at different codes. Therefore, it is obvious that the orthogonality or quasi-orthogonality among the codes or sequences plays an extremely important role. In fact, we should define the two important roles of the

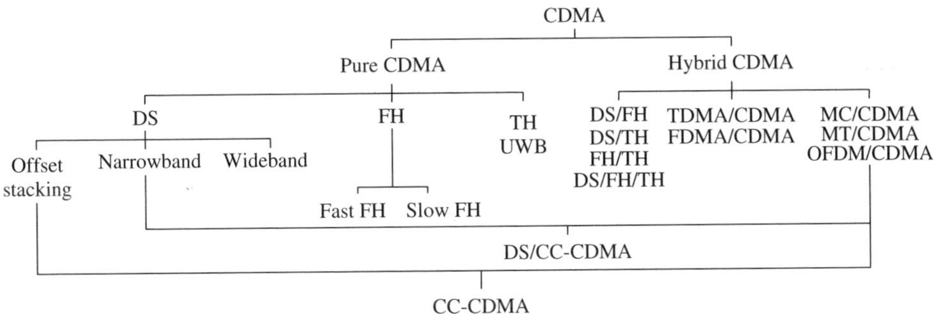

Figure 2.30 Family tree of various CDMA technologies. Offset stacking is a bandwidth-efficient spreading technique used in an orthogonal complementary (OC) coded CDMA.

---

[15]It has to be noted that OFDM and OFDMA are different. The former is only a multiplexing technique that works on a single user basis; whereas the latter is a multiple access scheme that can be applied to a wireless network. In OFDM scheme, different symbols are represented by different combinations of tones. In OFDMA scheme, both time slots and tones are assigned to different users for multiple access.

codes or sequences involved in a CDMA system: one is to act as the signature codes (to accomplish CDMA) and the other is to spread the data bits (to spread signal bandwidth to achieve a certain PG). It should be emphasized that the roles of the former and the latter are not necessarily given to the same code in a particular CDMA air-link architecture. For instance in the uplink channels of IS-95A/B [317–344], the signature codes are long M-sequences and the spreading codes are 64-ary Walsh-Hadamard functions. On the other hand, the downlink of the IS-95A/B standards uses 64-ary Walsh-Hadamard sequences as both the spreading codes and the signature codes.

The question that arises is why has CDMA become the most popular air-interface technology for the current 2G and 3G, possibly also for B3G wireless communications. The main reasons can be summarized as follows. First, so far CDMA is the only technology that can mitigate MI in a very cost effective way, which otherwise would have to be be tackled by using other much more complicated subsystems in FDMA and TDMA systems. Second, the current CDMA technology can offer on the average a far better capacity than its counterparts, such as FDMA and TDMA systems, to meet the increasing demand for mobile cellular applications in the world. Third, the overall bandwidth efficiency of a CDMA system is much higher than that with conventional multiple access technologies, thus giving an operator much bigger incentive to adopt it due to the extremely high price of spectra. Finally, relatively low peak emission power level of a CDMA transmitter offers a unique capability for CDMA-based systems to overlay the existing radio services currently in operation without introducing noticeable interference with each other.

However, we have to admit that current CDMA systems are still far from perfect. It is a well-known fact that a CDMA system is always considered to be interference-limited because of the the existence of multiple access interference (MAI) and MI, which are the two major contributors to the limitation of capacity and performance in any CDMA-based system currently in operation, including all mature 2G and 3G architectures [317–344, 448]. The following questions may come to the mind of anyone who has learned the basics of CDMA:

- Will CDMA systems always have only interference-limited performance?

- Why does a CDMA system have to work with so many complicated auxiliary subsystems, such as close-loop and open-loop power control, RAKE receiver, rate-matching algorithms, uplink synchronization control, multiuser detection, to name just a few examples?

- Can we get rid of all of those complicated subsystems to make a simple and yet well-performing CDMA?

Many people may think it is only a dream to make an ideal CDMA[16] but we would like to offer our different views in this chapter by addressing the issues related to the evolution of CDMA technologies from currently available 2G and 3G systems to the new generation CDMA technology for the future. Here we will also present some of our thoughts to engineer a new CDMA architecture with an enhanced capability to mitigate MAI and MI, a critical issue associated with an interference-free CDMA.

Several assumptions should be made to facilitate the discussions in this chapter. First, we should limit our discussions to DS-CDMA systems only and we will not address the issues related to other CDMA schemes, such as FH-CDMA or TH-CDMA. Second, in such a DS-CDMA system of interest to us, data signal spreading will be performed using short codes (with the chip width being T_c), whose length is exactly the same as the duration of one entire data bit (T_b), or the PG of such a DS-CDMA system is equal to $N = T_b/T_c$. In other words, we will not deal with the situation where long spreading codes, whose length is longer than the width of one data bit, may be used to spread data bit stream. Third, we will concern ourselves with a wireless system with full-duplex operation,

---

[16]An ideal CDMA system can be defined as the one whose performance should not be strictly limited by interferences, such as MAI and MI.

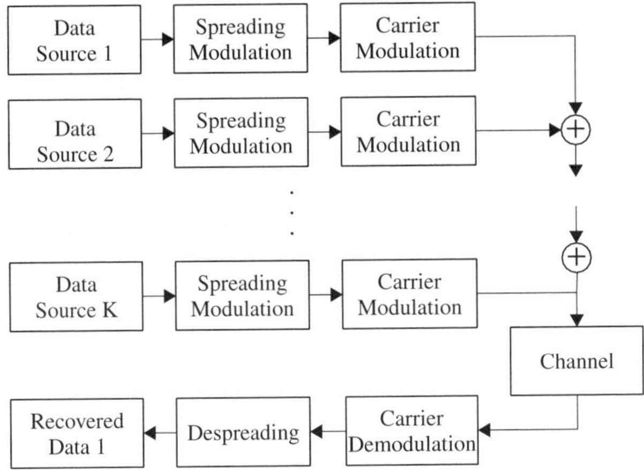

Figure 2.31 A generic K-user DS-CDMA system model, where only uplink channels of the system are concerned.

consisting of mobile terminals and a BS. The transmission link from mobiles to a BS is referred to as *uplink*, and the transmission link in the reverse direction is called *downlink*. The block diagram of a generic DS-CDMA system concerned in this chapter is shown in Figure 2.31, where we are interested only in a DS-CDMA system with K users, each of which is assigned *one* unique code[17] for CDMA purpose, and the signal in question is data source 1. Thus, any form of M-ary CDMA schemes [704–706] is not of interest here.

About the CDMA codes

Clearly, the CDMA codes, whose characteristics will govern the advantages and limitations of a CDMA system, play an essential role in CDMA system architecture. For instance, the use of orthogonal variable spreading factor (OVSF) codes in UMTS-UTRA [425] and WCDMA [431] standards requires that a dedicated rate-matching algorithm has to be carried out in the transceivers involved, whenever user data transmission rate changes to match a specific spreading factor or the system wants to admit as many users as possible in a cell. In addition, the rate change in UMTS-UTRA and WCDMA can be made only at multiples of two, meaning that continuous rate change is impossible. This requirement is a direct consequence from the OVSF code generation tree structure, where the codes in the upper layers bear a lower spreading factor; whereas those in the lower layers offer a higher spreading factor. Therefore, occupancy of a node in the upper layers effectively blocks all nodes in the lower layers, meaning that fewer users can be accommodated in a cell. The rate-matching algorithms [425–432] indeed consume a great amount of hardware and software resource and affect the overall performance, such as increased computation load and processing latency, and so on. Therefore, the choice of CDMA codes is extremely important and should be exercised very carefully at a very early stage of a CDMA system design; otherwise the short-comings of the system architecture due to the use of unsuitable CDMA codes will carry on for ever as the standard.

There are many different ways to characterize the CDMA codes, but nothing can be more effective and intuitive than the autocorrelation function (ACF) and cross-correlation function (CCF), which are discussed in detail in the following text.

---

[17]It has to be noted that here "one" code means either one single code per user or one flock of element codes per user. The former case stands for the traditional unitary codes; whereas the latter represents a CDMA system based on complementary codes.

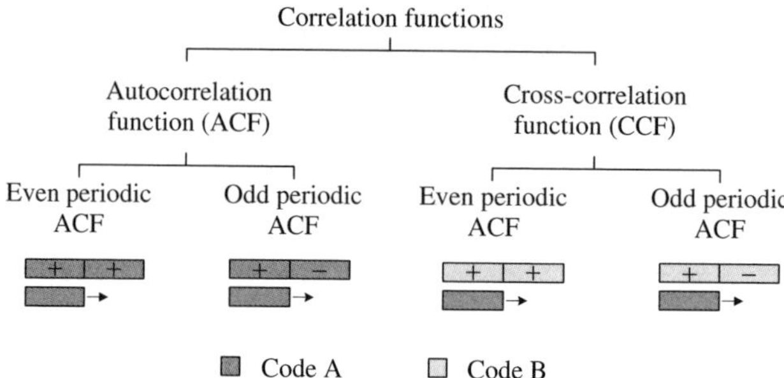

Figure 2.32 Classification of correlation functions of CDMA codes.

Autocorrelation function

The ACF is defined as the result of chip-wise convolution or simply correlation operation between two time-shifted versions of the same code, which can be further classified into two subcategories: periodic ACF and aperiodic ACF, depending on the same and different signs of two consecutive bits, respectively, as shown in the left-sided two branches in Figure 2.32. In a practical CDMA system, usually the periodic ACFs and aperiodic ACFs appear equal likely due to the fact that "+1" and "−1" appear equally probable in the received binary bit stream.

The in-phase ACF or ACF peak, which is often equal to its length or PG value (N), affects detection efficiency of the desirable signal in a CDMA receiver, where a correlator or matched filter is used. On the other hand, the out-of-phase ACFs of a CDMA code will be harmless if no multipath effect is present. However, they will contribute to MAI and will seriously affect system performance under the influence of multiple effect.

Cross-correlation function

The CCF is defined as the result of chip-wise convolution operation between two different spreading codes in a family of the codes. Because of a similar reason as mentioned earlier for the ACF, there are also two different types of CCF, that is, periodic CCF and aperiodic CCF. The former is mainly found in synchronous transmission channels, such as downlink channels in a wireless system, and the latter can appear in either synchronous (if MI is present) or asynchronous channels. In contrast to out-of-phase ACF that will contribute to MAI only under multipath channels, the CCF always contributes to MAI, no matter whether the multipath effect is present or not. On the other hand, the out-of-phase ACF will become harmful if, and only if, a multipath channel is concerned; otherwise it will never yield MAI at a correlating receiver. Obviously, the MAI is one of the most serious threats to jeopardize detection efficiency of a CDMA receiver using either correlator or RAKE and thus has to be kept under a sufficiently low level to ensure a satisfactory performance. Table 2.3 and Table 2.4 list all correlation functions of a CDMA code and their merit behavior in a CDMA system.

Traditional CDMA codes

The search for promising CDMA codes or sequences used to be a very active research topic, which can be traced back to as early as the 1960s. Numerous CDMA codes have been proposed [153–210] and their performance and possible applications in a CDMA system were investigated in the literature [221–228]. In the following text, we briefly discuss some of the frequently referred CDMA codes, including both quasi-orthogonal codes (such as M-sequences [139–152], Gold codes [195], Kasami

Table 2.3   Autocorrelation functions (ACFs) of CDMA codes and their merit behavior in a CDMA system

ACF	IPEP[a]-ACF	OPEP[b]-ACF	IPOP[c]-ACF	OPOP[d]-ACF
Cause	Correlator	MPC[e]	Correlator	MPC
Frequency	Once a bit	High	Once a bit	High
Behavior	Wanted signal	MI	Wanted signal	MI
ITD[f]	Enhance	Impair	Enhance	Impair

[a]IPEP: In-phase even periodic.
[b]OPEP: Out-of-phase even periodic.
[c]IPOP: In-phase odd periodic.
[d]OPOP: Out-of-phase odd periodic.
[e]MPC: Multipath channel.
[f]ITD: Impact to detection.

Table 2.4   Cross-correlation functions (CCFs) of CDMA codes and their merit behavior in a CDMA system

CCF	IPEP[a]-CCF	OPEP[b]-CCF	IPOP[c]-CCF	OPOP[d]-CCF
Cause	Syn. channel	Asyn. channel	Syn. channel	Asyn. channel
Frequency	Once a bit	High	Once a bit	High
Behavior	MAI	MAI	MAI	MAI
ITD[e]	Impair	Impair	Impair	Impair

[a]IPEP: In-phase even periodic.
[b]OPEP: Out-of-phase even periodic.
[c]IPOP: In-phase odd aperiodic.
[d]OPOP: Out-of-phase odd aperiodic.
[e]ITD: Impact to detection.

codes [196], etc.) and orthogonal codes[18] (such as Walsh-Hadamard sequences [200], OVSF codes, complementary codes [206–210], etc.), with their fundamental characteristics. In addition to these commonly referred CDMA codes listed in the preceding text, there are many other less widely quoted ones, such as GMW codes [191], No codes [194], Bent codes [204], and so on, which will not be discussed further in this chapter. For a more comprehensive treatment, the readers may refer to the literature given in [153–210].

M-sequences

M-sequences [132–152] are also called *maximum-length sequences*, whose name reflects the fact that they can be generated by using a primitive polynomial in GF(2)[19] with a specific degree n and have the longest possible length using any polynomial in GF(2) with the same degree. The simplicity of the M-sequences is also reflected in their sequence generator structure, where only a single shift-register is required along with a feedback logic that depends on the primitive polynomial of concern. Basically,

---

[18]It is to be noted that Walsh-Hadamard and OVSF codes should not be considered as "orthogonal codes" in a strict sense, as they are not orthogonal at all with each other if used in asynchronous transmissions, such as the uplink channels in a mobile cellular system. Their orthogonality appears only in synchronous channels.

[19]The acronym GF stands for "Galois Filed," which is a research area in number theory and modern algebra. E. Galois (1811–1832) was a great French Mathematician and one of the inventors of the "Group Theory." He died from a duel in 1832.

M-sequences are not suitable for CDMA applications owing to their uncontrollable CCFs. However, their out-of-phase ACFs are always "-1," making them suitable for some special applications such as radar and synchronization-control systems, where a high time resolution is of ultimate importance. Nevertheless, it should also be pointed out that it is still possible for us to find a *relatively small* group of M-sequences that do maintain reasonably low CCFs among them, such that they can be used as CDMA codes.

Gold codes

One of the most popular quasi-orthogonal CDMA codes is the Gold code [195], which was first studied by R. Gold in 1968, and has been extensively used in many commercial CDMA systems, including IS-93 and WCDMA standards. The popularity of the Gold codes stems from the two main reasons explained below. First, any pair of Gold code families offer a uniquely controllable three-leveled CCF, and so do their out-of-phase ACFs, the maximal value of which is equal to $2^{(n+1)/2} + 1$, where n is the degree of the polynomial generating the Gold codes. This maximal CCF value of the Gold codes is relatively low when compared to those of the other CDMA codes. This characteristic feature of the Gold codes make them, in particular, suitable for CDMA applications with a predictable performance in terms of MAI. Second, a family of Gold codes can be generated by a very simple logic with a pair of shift registers, each of which bears the same structure as that used to generate an M-sequence. In other words, a pair of primitive polynomials are required to generate a complete family of Gold codes, whose size can be as large as $2n + 1$, where n stands for the degree of the primitive polynomials used for generation of the Gold codes. Therefore, the relatively large family size makes Gold codes a popular choice as CDMA codes, being able to support many users in a cell. It should be pointed out that the family size of CDMA codes should also be considered as an important resource, which could be utilized to increase effective transmission rate in a CDMA channel. For instance, if the family size is large enough, each user can be assigned multiple codes, instead of only one as is the case in a conventional CDMA system, such that the use of distinct codes from the same user can effectively deliver more information than that in the case with single code assignment, given the same total bandwidth. This is just what has been done in M-ary CDMA systems [704–706]. In this sense, a large family size is definitely a plus to any CDMA code.

Kasami codes

There are two different types of Kasami codes [196–199], small-set Kasami codes and large-set Kasami codes. The major difference between the two lies in their CCF, the former of which is lower than the latter. The maximal value of the CCF for the small-set Kasami codes is equal to $2^{n/2} + 1$ (where n is the degree of the polynomials generating the small-set Kasami codes), which is already very close to the Welch bound [221]. In fact, a family of the large-set Kasami codes can be divided into two subsets, one of which is the family of small-set Kasami codes and the other is in fact a family of Gold codes with the same degree. Thus, in this sense, the Kasami codes have many features similar to the Gold codes discussed earlier. It should be pointed out that, although the small-set Kasami codes have a rather low peak CCF that is important for MAI reduction, they form a relatively small family size, which limits its wide application as signature codes in a CDMA system.

Walsh-Hadamard sequences

Walsh-Hadamard sequences [200–203] can be obtained from a Walsh-Hadamard matrix, either rows or columns of which can be taken as spreading sequences that are perfectly orthogonal with one another if, and only if, they are used in a synchronous transmission mode. Walsh-Hadamard sequences have been applied to IS-95A/B [317–344] as well as CDMA2000 [345] standards as either channelization codes for downlinks channels or spreading sequences for uplink applications. It is necessary to emphasize that, although Walsh-Hadamard sequences are referred to as *orthogonal codes*, their out-of-phase ACFs are very high, which will seriously affect the asynchronous uplink signal reception in a CDMA system under the multipath effect, where the received signal from a particular mobile consists of multiple replicas with different delays, and thus the out-of-phase ACFs will contribute substantially as a part of the MAI at the receiver. In addition, the out-of-phase CCFs of Walsh-Hadamard sequences are also rather high, giving rise to a substantial increase of MAI. Therefore in this sense, we have a strong reason to doubt the suitability of using Walsh-Hadamard sequences as signature or spreading codes in any CDMA system, which inevitably have to work in a multipath environment.

OVSF codes

Other important orthogonal codes extensively reported are OVSF codes, which have been made famous due to their application in three major 3G standards: UMTS-UTRA [425], WCDMA [431] and TD-SCDMA [432], proposed by ETSI (Europe), ARIB (Japan) and CATT (China), respectively, as their IMT-2000 candidate proposals to International Telecommunication Union (ITU) roughly at about the same time in 1998. As its name suggests, the OVSF codes are generated to fit variable spreading factors (SFs) or the code lengths under a special code generation tree, on which the codes with larger SFs form the lower layers and those with smaller SFs form the upper layers. The SFs can be made variable in multiples of two from 4 to 256 on the basis of a so called *rate-matching* algorithm, as specified in the standards. Thus, possible data rates can also be made variable only in multiples of two. For instance, if a user requires a data rate of 5 units, the system has to assign it a bandwidth associated with a data rate of 8 units, wasting about 37.5% bandwidth resource. The characteristics of ACFs and CCFs of OVSF codes are almost exactly the same as those of Walsh-Hadamard sequences with identical length. Again similar to the Walsh-Hadamard codes, the OVSF codes perform badly under asynchronous uplink channels, where the orthogonality among the codes virtually does not exist. In fact, if all OVSF codes with a fixed SF are arranged into a matrix, we will readily find that it gives a Walsh-Hadamard matrix. In this sense, the OVSF codes should not be considered as new CDMA codes at all.

Because of the fact that both Walsh-Hadamard codes and OVSF codes possess very high out-of-phase ACFs and CCFs, they should not be considered as orthogonal codes, which are supposed to offer ideal ACFs and CCFs with all possible time-shifts. However, there exist genuine orthogonal codes called *orthogonal complementary codes* (OCC), which are of great interest to us in this chapter and are discussed in detail in the following text.

Orthogonal complementary codes

The study of OCC can be traced back to the 1960s, when Golay [206] and Turyn [207] first studied pairs of binary complementary codes whose ACF was zero for all even shifts except the zero shift. However, the main interest on complementary codes at that time was not for their possible applications in CDMA systems but in radar systems. Later, Suehiro [208, 209] extended the concept to the generation of the OCC families whose ACF is zero for all even and odd shifts except the zero shift and whose CCF for any pair is zero for all possible shifts. The work carried out in [209] had paved the way for practical applications of the OCC in modern CDMA systems, one possible architecture of which has been proposed and studied in [210].

There exist several fundamental distinctions between the traditional CDMA codes (such as Gold codes, M-sequences, Walsh-Hadamard codes, etc.) and the OCC. First, the orthogonality of the OCC is based on a *flock* of element codes jointly, instead of a single code as the traditional CDMA code. In other words, every user in a OCC-based CDMA (or simply OC-CDMA) system will be assigned a flock of element codes as its signature code, which ought to be transmitted via different channels (either in FDM or TDM mode) to arrive at a correlator receiver at the same time to produce an autocorrelation peak. Thus, all conventional spreading codes, either quasi-orthogonal or orthogonal codes, are also called *unitary* codes,[20] because they work simply on a one-code-per-user basis. Second, the PG of the OCC is equal to the *congregated length* of a *flock* of element codes. For the OCC of lengths $L = 4$ and $L = 16$ as shown in Table (2.5), the PGs are equal to $4 \times 2 = 8$ and $16 \times 4 = 64$, respectively. Third, zero CCFs and zero out-of-phase ACFs are ensured for any relative shifts between two OCC, which has made the OCC different from the conventional orthogonal codes, such as Walsh-Hadamard sequences and OVSF codes, whose out-of-phase ACFs can never be zero. Also because of this fact, the OCC could be referred to as *truly perfect orthogonal codes*. This property is extremely important as it gives us great hope to further enhance the performance of a CDMA system.

Table 2.5 gives two examples of the OCC in question in the proposed OC-CDMA system model to be introduced in the following text. More examples of OCC (whose PG values are from 8 to 512) can be found in Appendix A given at the end of this book. Table 2.6 shows the *flock* and family sizes for various OCC with different element code lengths (L). For more detailed

Table 2.5   Two examples of orthogonal complementary codes with element code lengths $L = 4$ and $L = 16$

Element code: $L = 4$		Element code: $L = 16$
		A_0: $+ + + + + - + - + + - - + - - +$
		A_1: $+ - + - + + + + + - - + + + - -$
A_0: $+ + + -$	Flock 1	A_2: $+ + - - + - - + + + + + + - + -$
		A_3: $+ - - + + + - - + - + - + + + +$
Flock 1		B_0: $+ + + + - + - + + + - - - + + -$
		B_1: $+ - + - - - - - + - - + - - + +$
A_1: $+ - + +$	Flock 2	B_2: $+ + - - - + + - + + + + - + - +$
		B_3: $+ - - + - - + + + - + - - - - -$
		C_0: $+ + + + + - + - - - + + - + + -$
		C_1: $+ - + - + + + + - + + - - - + +$
B_0: $+ + - +$	Flock 3	C_2: $+ + - - + - - + - - - - - + - +$
		C_3: $+ - - + + + - - - + - + - - - -$
Flock 2		D_0: $+ + + + - + - + - - + + + - - +$
		D_1: $+ - + - - - - - - + + - + + - -$
B_1: $+ - - -$	Flock 4	D_2: $+ + - - - + + - - - - - + - + -$
		D_3: $+ - - + - - + + - + - + + + + +$

Table 2.6   Family sizes and flock sizes for orthogonal complementary codes with various element code lengths L

Element code length ($L = 4^n$)[a]	4	16	64	256	1024	4096
PG ($L\sqrt{L}$)	8	64	512	4096	32768	262144
Family size ($\sqrt{L}$)	2	4	8	16	32	64
Flock size ($\sqrt{L}$)	2	4	8	16	32	64

[a]n can be any integer.

---

[20]The name of "unitary codes" was introduced by us to differentiate them from complementary codes, which work on a flock-basis, where each flock consists of several (usually an even number) element codes.

information on the code generation procedure and other properties of the OCC, the readers may refer to [206–210].

The reason we have paid so much attention to the OCC lies in the fact that the correlation properties of the OCC are based on several (always an even number) element codes, instead of only one single code as is the case in conventional spreading codes, such as Gold codes and OVSF codes, and so on. This observation is significant to make the OCC different from the conventional codes. While it is very difficult to ensure ideal ACFs and CCFs of conventional or *unitary* spreading codes whose correlation function is based on a single code, it may be possible for us to formulate those that are based on the sum of the individual ACFs and CCFs of a flock (always an even number) of element codes, as long as the nonzero values of out-of-phase ACFs and CCFs for different element codes within the same flock can be canceled in the sum operation. Therefore, this gives us a much greater degree of freedom in the code design process and it is no longer necessary for us to strictly require zero out-of-phase ACFs and CCFs for each individual element code.

The examples of both ACF and CCF of a particular OCC of PG equal to $16 \times 4 = 64$ are shown in Figure 2.33 and Figure 2.34, respectively, from which it can be clearly seen that, although the autocorrelation functions of individual element codes are not ideal (there are many nonzero side lobes, as shown in Figure 2.33), the sum of them yields, $R_A(\tau)$, an ideal ACF, which is just what we want. The same observation can also be made with regard to the CCF of some particular OCC or $R_{A,B}(\tau)$, as shown in Figure 2.34. This characteristic feature can also be found in any other OCC, as given in Appendix A. It has to be noted that this desirable feature of the OCC has not been found in any other conventional or unitary CDMA code available so far, including all so-called orthogonal codes, such as Walsh-Hadamard sequences and OVSF codes, and so on.

One example for the application of the OCC can already be found in TD-LAS system [441–447], which has been approved by 3GPP2 as an enhanced standard [440], in which pair-wise OCC, called *LS codes* in the TD-LAS specification,[21] have been used as spreading codes of the users. In other words, each user in the TD-LAS system is assigned two element codes (namely, *S* section and *C* section), which are sent in the same time slot in a time-division-multiplex mode or TDM mode. Combined with LA codes, the use of LS codes in TD-LAS creates a unique interference-free window (IFW), which spans about a few to a few tens of chips beside the ACF main lobe, making it possible for the system to employ some multilevel digital modems, such as 16-QAM, to further improve the bandwidth efficiency in its air-link sector. In April 2002, the TD-LAS system carried out a successful field trial in Shanghai, demonstrating its great potential for the OCC in future wireless communications. Unfortunately, the TD-LAS system can only offer an IFW, which is still much smaller than the code length. In Section 7.3 of this book, we will discuss more on the generation of the OCC, which can offer an MAI-free window as large as the code length. It will be further illustrated that the application of the OCC can offer a new dimension to implement the B3G wireless.

Multiple access interference (MAI)

There are two major sources of interference in a CDMA system; one is MAI and the other is the MI. The former is because of imperfect CCFs of CDMA codes and the latter is caused by the combined effect of out-of-phase ACFs and CCFs of spreading codes used by the system in a multipath channel. The different transmission modes in synchronous downlink and asynchronous uplink channels can further complicate their impact on the performance of a CDMA system.

The downlink channels in a wireless system are referred to as the *transmission direction* from a BS or access point (AP) to mobile terminals within the cell, and are usually synchronous in the way

---

[21]The acronym "LAS" in the TD-LAS system stands for "Large-Area Synchronous" due to the claim that the TD-LAS system can cover a relatively large area and it works in a synchronous way.

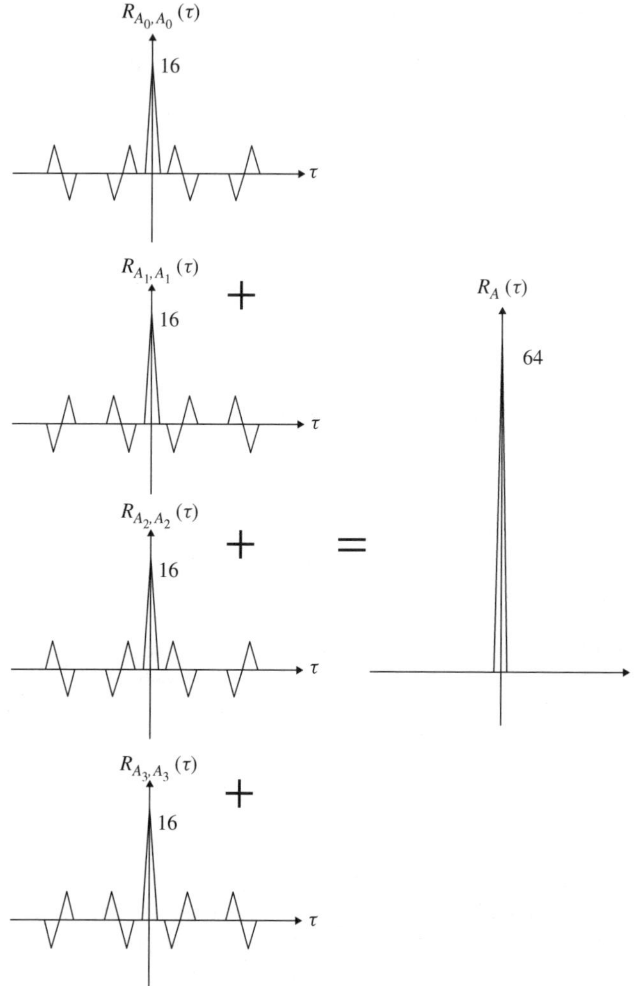

Figure 2.33 Autocorrelation function of an orthogonal complementary code with its element code length being 16 and PG $16 \times 4 = 64$. Detailed information of this particular CC code is given in Table 2.5, where $R_{A_0,A_0}(\tau)$ is the ACF of element code A_0, $R_{A_1,A_1}(\tau)$ is the ACF of element code A_1, $R_{A_2,A_2}(\tau)$ is the ACF of element code A_2, $R_{A_3,A_3}(\tau)$ is the ACF of element code A_3, and $R_A(\tau)$ is the sum of $R_{A_0,A_0}(\tau)$, $R_{A_1,A_1}(\tau)$, $R_{A_2,A_2}(\tau)$, and $R_{A_3,A_3}(\tau)$.

that the bit-streams from the same BS arriving at a particular mobile are aligned bit-by-bit in time. To reflect this characteristic feature of the downlink channels, the delays for the transmissions from a BS to a particular mobile should be considered constant. Without multipath effect, the MAI in the downlink channels will be caused by the periodic CCFs of all CDMA codes active in the system. However, because of the multipath propagation, the MAI will consist of two parts, one part being the CCFs between the code of interest and all other active codes as well as their multipath returns, and the other part being the out-of-phase ACFs of the code concerned at a receiver because of the fact that the receiver will receive several replicas of the code with different delays. Because of the possibility

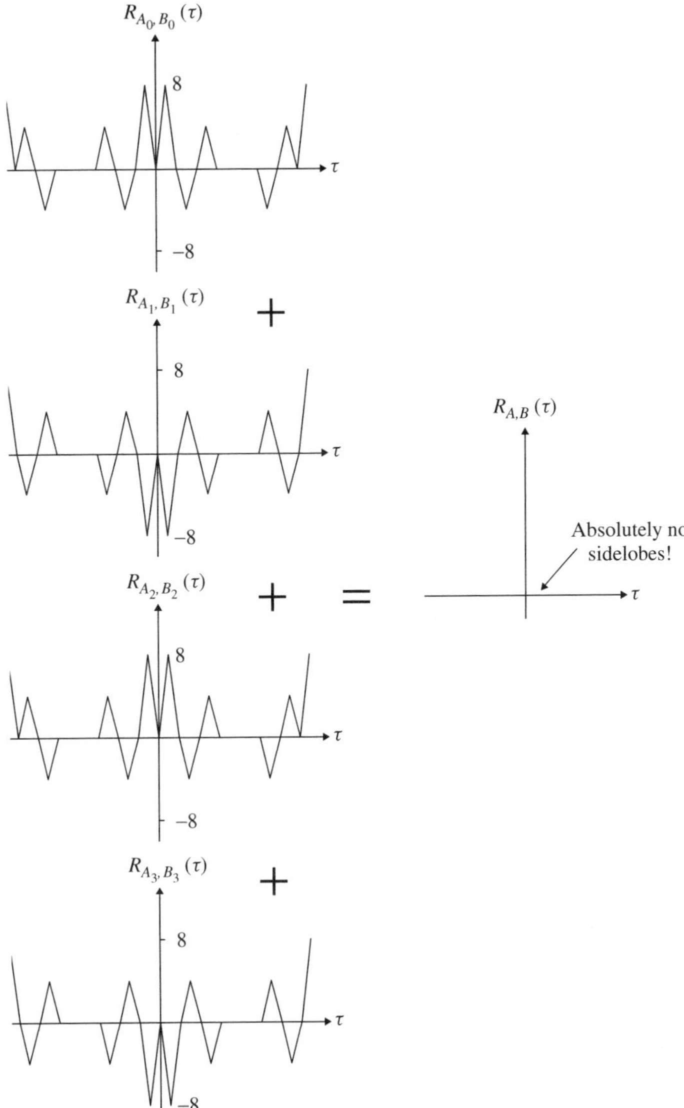

Figure 2.34 Cross-correlation function of an orthogonal complementary code with its element code length being 16 and PG $16 \times 4 = 64$. A detailed information of this particular CC code is given in Table 2.5 where $R_{A_0,B_0}(\tau)$ is the CCF of element codes A_0 and B_0, $R_{A_1,B_1}(\tau)$ is the CCF of element codes A_1 and B_1, $R_{A_2,B_2}(\tau)$ is the CCF of element codes A_2 and B_2, $R_{A_3,B_3}(\tau)$ is the CCF of element codes A_3 and B_3, and $R_{A,B}(\tau)$ is the sum of $R_{A_0,B_0}(\tau)$, $R_{A_1,B_1}(\tau)$, $R_{A_2,B_2}(\tau)$, and $R_{A_3,B_3}(\tau)$.

that two consecutive bits have either the same or different signs, both aperiodic out-of-phase ACFs and periodic out-of-phase ACFs will contribute to MAI with equal probability.

MAI in uplink channels will also consist of two different parts: the CCFs among different user codes and the out-of-phase ACFs due to the multipath effect. In either an AWGN uplink channel or a multipath uplink channel, both even periodic and odd periodic CCFs will be involved in MAI.

Multipath interference (MI)

MI is another major impairing factor in a CDMA system, which usually causes serious ISI at a conventional CDMA receiver based on matched filter or RAKE. In a multipath (either synchronous or asynchronous) channel, transmitted signals will undergo different propagation paths to reach a specific receiver, in which all those multipath returns will be involved either destructively or constructively in the signal detection process.

The multipath effect can be best illustrated in both the time domain and the frequency domain. To do so, let us first define the relative delay between two consecutive paths as *inter-path delay*. If the interpath delay is smaller than one chip width, a traditional matched filter (or correlator) or even a RAKE (with only one sample per chip) will not be able to distinguish individual multipath returns. In this case, if the number of multipath returns is very small, such as only two or three rays, they usually pose little threat to the signal detection process at a receiver and they can be treated as a single multipath return if each chip will be sampled only once. However, if the number of multipath returns (or delay spread) is relatively large, all those closely located multipath returns will span a time duration longer than a chip or even a symbol, causing a serious ISI. On the other hand, if the interpath delay of a multipath channel is larger than the chip width, all multipath returns will be considered as resolvable by using a correlator receiver, and care should also be taken not to let them interfere with one another, especially under the near-far effect. The capability for a correlator to capture individual multipath returns is one of the most important reasons why CDMA technology has become a popular choice in 2–3G wireless communication systems.

As stated in Section 2.1, the multipath effect can also be clearly illustrated in the frequency domain. A multipath channel can be well modeled by a delayed-tap-line filter with the coefficient in each tap representing the path gain of a particular multipath return and its delay element being the interpath delay, as defined earlier, without losing generality. For simplicity, let us look at the two-path channel model shown in Figure 2.8a, and thus only one delay element and one coefficient are involved in such a simple delayed-tap-line channel model, as shown in Figure 2.8b. The transfer function of the impulse response for this two-path channel model is shown in Figure 2.8c, where a comb-like frequency-domain transfer function can be observed, clearly showing the frequency-selectivity of the multipath channel. A similar observation can be made for a multipath channel with more than one delayed return. Therefore, a multipath channel can also be called a *frequency-selective channel*.

Obviously, the cause of MI at a receiver is different from that of MAI from the viewpoint of CDMA codes. As mentioned earlier, MAI is mainly caused by nonideal CCFs of the CDMA codes. However, the MI is mainly due to imperfect or nonzero out-of-phase ACFs of the CDMA codes. Assume that the interpath delay is larger than the chip width. Then, the multipath returns will not be harmful to the signal detection if every CDMA code possesses an ideal ACF, meaning that all (both periodic and aperiodic) out-of-phase ACFs are zero. The output from a correlator will consist of clearly separated autocorrelation peaks, each of which represents an individual multipath return, and can be detected one by one if symbol synchronization is achieved. However, if the out-of-phase ACFs of the CDMA codes concerned are not zero, the ACF side lobes following the earlier autocorrelation peak will seriously interfere with the detection of the later autocorrelation peaks, especially if the earlier peak is much stronger than the later ones owing to the near-far effect in a CDMA system. Both periodic and aperiodic out-of-phase ACFs will be involved in MI owing to the fact that data symbols could equally likely appear either as $+1$ or -1.

Techniques to combat MAI and MI

Combined effect of MAI and MI poses a great threat to successful signal detection at a traditional CDMA receiver. Therefore, many feasible techniques have been developed and studied in open literature to combat them [222–243], some of which will be discussed in the following text and which have been found to have many applications in various commercial CDMA systems. The following techniques are often used to combat MAI and MI in a CDMA system:

- Power control vs. near-far effect;

- RAKE receiver to overcome MI;

- Multiuser joint detection against MAI;

- Pilot-aided CDMA signal detection;

- Beam-forming techniques against co-channel interference;

- Isotropic CDMA air-link technologies;

- Uplink synchronization control.

However, here we will only discuss the first two most widely used schemes for MAI and MI mitigation, that is, *power control* schemes and *RAKE receiver*. The other techniques used for MAI and MI mitigation are discussed in the other chapters in this book to avoid redundant coverage of the same content. For instance, the multiuser joint detection, pilot-aided CDMA signal detection, and beam-forming techniques against co-channel interference are discussed in Section 2.4; the issues on isotropic CDMA air-link technologies and uplink synchronization control are covered in Section 3.3.6.

Power control versus near-far effect

In the development of early CDMA systems, such as Qualcomm's effort to develop the IS-95 standard, a lot of attention has been given to solve the near-far effect in a CDMA system. The near-far effect stems from the fact that some near-by unwanted transmissions (appearing as CCFs between the desirable and unwanted codes) overwhelm a far-away desirable transmission (appearing as an ACF peak of desirable code), jeopardizing the user separation process at a CDMA receiver. Therefore, the near-far effect is because of nonideal CCFs of the CDMA code family in question and can be considered as one of the direct consequences of MAI. In particular, it should be noted that the near-far effect usually has a stronger influence on uplink channels than downlink channels owing to asynchronous transmissions in the uplink channels, where mobile users are distributed in different places in a cell and their received power levels at a BS may vary greatly from one another. On the other hand, the synchronous transmission from the same source (or BS) in the downlink makes it possible for a mobile to receive all signals from the BS at almost equal power. Therefore, there is little near-far effect in the downlink channels when only intracell transmissions are concerned. In other words, the near-far effect in the downlink channels is mainly caused by the transmissions from different cells or BSs, if they are not separated by different frequency bands.

The common practice to combat the near-far effect is to use precision power control techniques, which usually consist of two sectors: one is open-loop power control and the other is closed-loop power control. The former always proceeds before the latter and they should work jointly to ensure that all signals from different mobiles in a cell will be received by a BS with almost equal power. With the help of power control techniques, the transmission power level of every mobile in a cell is constantly and effectively controlled by the BS. Another advantage of using the power control is to limit unnecessary power emission in the whole cellular system to reduce the average intercell and intracell interference, contributing to the overall capacity improvement of the CDMA system.

RAKE receiver to mitigate MI

A salient feature of the CDMA system is its capability to offer a high time resolution to distinguish different multipath waves, thus making it possible to recombine them in a constructive way after separating them, which yields an attractive multipath-diversity gain with the help of a RAKE receiver. While it is still possible for a TDMA or FDMA system to combat MI using one way or another, the great complexity involved in implementing similar capability in a TDMA or FDMA system is a major concern.

A RAKE receiver works with several parallel correlators or *fingers*, each of which is to capture different multipath returns that might undergo different propagation delays to make a constructive recombination, yielding an enhanced decision variable at the output. There are two major recombining schemes in a RAKE; one is called *equal gain combining* (EGC) and the other is maximal ratio combining (MRC), though there is the third option called *selective combining* (SC), which works simply to choose the best signal as the output and is obviously not as popular as the EGC and MRC. In an EGC-RAKE, all captured multipath components are equally weighted before the delay and summation operations, and thus it does not require the information about the path gains, reducing the complexity of the receiver. On the other hand, an MRC-RAKE offers an optimum performance in terms of the maximal SNR achievable at the output by giving different weights to different captured multipath returns according to the known delay profile of the channel. Therefore, the successful operation of an MRC-RAKE virtually requires full information of the multipath channel of concern, which can be obtained only by resorting to other complicated channel estimation techniques, such as a dedicated pilot signaling, and so on. In this sense, it is very difficult to make an MRC-RAKE work adaptively, especially in a fast fading multipath channel with a high Doppler spread due to great mobility of terminals, without the assistance of an independent channel sounding system. Figure 2.35 shows a generic MRC-RAKE receiver model, where only three fingers are presented for illustration simplicity.

In addition to the complexity of a RAKE receiver, its performance can also be sensitively affected by the MAI characteristics of a CDMA system. As mentioned earlier, each finger in a RAKE receiver

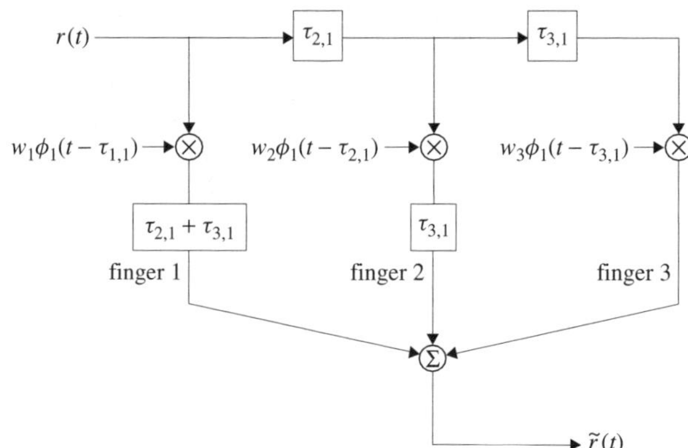

Figure 2.35 A generic model for a three-finger RAKE receiver, where $\tau_{i,j}$ stands for the interpath delay between the i-th and j-th paths, w_i is the path weight for the i-th multipath return, and $\phi_1(t)$ is the spreading code of interest.

should capture a particular multipath signal encoded by a specific code of interest, in addition to which all other unwanted transmissions together with their multipath components (contributing to MAI due to nonideal CCFs) as well as the other multipath returns of the wanted signal (contributing to MI because of nonzero out-of-phase ACFs) will also come into the finger to act jointly as the interference to the wanted particular multipath return signal. Obviously, if the characteristics of CCFs and ACFs of the CDMA codes concerned are not good enough, the effectiveness of a RAKE using either EGC or MRC combining will be severely affected, and sometimes it performs very badly especially if there are many strong later-coming multipath returns followed by the first path. For a more analytical treatment of this subject, the readers may refer to reference [737].

2.3.4 Random Multiple Access Technologies

Having discussed three major multiple access technologies, that is FDMA, TDMA, and CDMA, we would like to take a look at random multiple access techniques, which have been used in wireless communication systems and networks. Different from all three aforementioned multiple access schemes, random multiple access techniques work on an opportunity multiple access basis. The channel resource allocation is not fixed and active users should contend for channel access whenever a packet arrives at the transmitter.

The random multiple access schemes have been proposed, in particular, taking into consideration their suitability for applications in packet-switched networks, and will not function well in a circuit-switched network when compared to any of the three fixed channel access schemes, that is, FDMA, TDMA, and CDMA.

Pure ALOHA

Random multiple access techniques have been widely used in various communication systems and networks where the traffic will take the form of bursts or *packets*. The initial research on random multiple access techniques can be traced to two pioneer research projects, one of which was carried out in ARPANET in the Department of Defense of the United States in the late 1960s [43], and the other was called ALOHANET and was carried out in the University of Hawaii in the early 1970s. The results obtained from these two independent research projects laid an important foundation for the development of many later random multiple access protocols. The ARPANET was a successful experimental model for the modern Internet, which spans a wired packet-switched *World Wide Web* (WWW), and the ALOHANET worked like a test bed for a wireless *packet radio* network, which formed the basis for today's all-IP wireless networks. Obviously, without these two pioneer research projects, we can hardly imagine what today's world would look like.

In principle, random multiple access techniques work like the CDMA scheme in terms of its capacity to allow all users to share the communication resources in both time and frequency domains. The difference between the two lies in the fact that the former usually will not guarantee successful simultaneous transmissions[22] (due to the existence of *collision problems*), even if the number of active users is less than the system capacity; whereas in the latter all simultaneous transmissions will be successful as long as the number of active users is limited within the system capacity. The most important characteristic feature of random multiple access technique is its suitability for supporting burst-type traffic, that is, all traffic will be delivered in short packets, whose transmissions do not need an end-to-end circuit connection, or so-called *connectionless information delivery*. This, in particular, is well suited for *Internet protocol* (IP)–based wired and wireless networking applications.

The simplest random multiple access scheme is *pure ALOHA*, which originated from the ALOHANET developed by the University of Hawaii in the 1970s, as mentioned earlier. It works under

---

[22]Therefore, all random multiple access based networks offer services to users with so-called *best effort*, in which not all traffic will result in successful results.

a very simple rule that can be described in the sequel. When a packet comes to a transmitter in the ALOHA system, it will just send it into the channel. The fate of the sent packet will depend on the status of the receiver that should be located within the effective radio coverage of the transmitter. If the receiver is busy (either busy transmitting or busy receiving from the other transmitter), the sent packet will be ignored and the transmitter will realize the failure of the transmission after a fixed delay, which should be made longer than any possible turn-around time of the channel (which should be longer than two times of the one-way propagation delay between the transmitter and the receiver). In this case, the transmitter will resend the same packet after a random delay called *back-off time*, which is used to avoid a series of collisions due to the possible transmissions from two or more than two transmitters that are active, and thus retransmit also at the same time. The transmitter will continue the same process until it succeeds in sending this packet. The packets transmission in a pure ALOHA system is illustrated in Figure 2.36. On the other hand, if the intended receiver does receive the packet from the transmitter successfully, the receiver should send out an acknowledgment, which will be used as a *token of successful transmission* by the transmitter.

In order to identify the receiver that each packet will address, each packet will start with a *header* containing the address or identification information of the receiver. Of course, the identification of the transmitter can also be given in the header of a packet. The simplest structure of a packet can consist of only two portions, the header and the data portion, if without any protection of *error-correction coding*. To offer some more protection, a packet can contain a tail, which gives the receiver some useful information used for error-correction, such as *parity check codes, cyclic redundancy check* (CRC) codes, and so on.

Therefore, the ALOHA random access protocol can be implemented in a wireless or wired network easily, at an extremely low cost. However, its network performance in terms of *throughput* and *delay* is relatively poor when compared with other refined protocols to be discussed later, such as *slotted ALOHA, carrier sense multiple access* (CSMA), and *carrier sense multiple access with collision detection* CSMA/CD, and so on.[23]

The performance of a pure ALOHA system can be analyzed as follows. In particular, we want to examine the throughput characteristics of a *packet radio network* using pure ALOHA. Some assumptions should be made for analysis simplicity. However, it is to be noted that the designers of a wireless network must usually first model the systems and do extensive simulations to determine actual performance. The simplified model highlights the major issues related to packet network performance and allows us to find some effective methods for improving performance. Many works have been published [738–748] extensively on the performance modeling of packet networks.

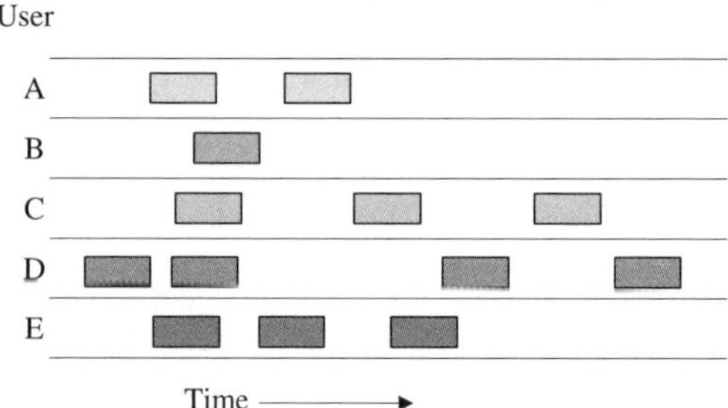

Figure 2.36   Illustration of random or uncoordinated transmissions of packets in a pure ALOHA system. Any packets overlapped in time are collided.

---

[23]The CSMA/CD is a standard random access protocol for Ethernet LAN and many IEEE 802.11 WLAN standards.

Let us consider several packet radio stations where the packets take fixed length, that is τ seconds, and packets enter the system at the rate of G packets per second. If two or more packets in the system collide, then we assume that all packets interfere with each other destructively and none can survive. Whenever a packet collision occurs, the transmitting station does not receive an acknowledgment and should retransmit the same packet. If it retransmits immediately (as very likely other transmitters will also do the same), a collision would occur again and no data would ever get through the packet radio link. Therefore, whenever a collision occurs, the station will delay its retransmission for a random time, X, with average value X_{avg}, which is also called a *random back-off* time. In this case, no packets will ever leave the system and they only get delayed. For analysis simplicity, we assume that X_{avg} is much greater than τ. We can then say that G is the combined arrival rate for new packets and retransmitted packets as a whole from all attempting transmitters. The collision problem in a pure ALOHA system is illustrated in Figure 2.37.

The capacity of the channel, S, is defined as the number of packets per second that are successfully transmitted and must be lower than G, the composite arrival rate, for obvious reasons. If we could stack the packets perfectly without overlap in time, then collisions would not occur and the channel would reach its highest capacity, or

$$C_{max} = \frac{1}{\tau} \quad (packets/second) \tag{2.48}$$

since the channel uses all available transmission time in this case. In a real packet radio network, the packets arrive at transmitters randomly and collisions happen from time to time depending on the traffic load. Thus the capacity C_{max} should always be less than the arrival rate G. To simplify our analysis we would like to assume that all packets arrive independently and follow a Poisson arrival process. It is a well-known fact that the Poisson arrival process will give an exponential distributed interarrival time, that has a probability density function (pdf) as

$$p\{t_{n+1} - t_n < t\} = Ge^{-Gt} \tag{2.49}$$

where t_n and t_{n+1} are two time instances that two packets are transmitted.

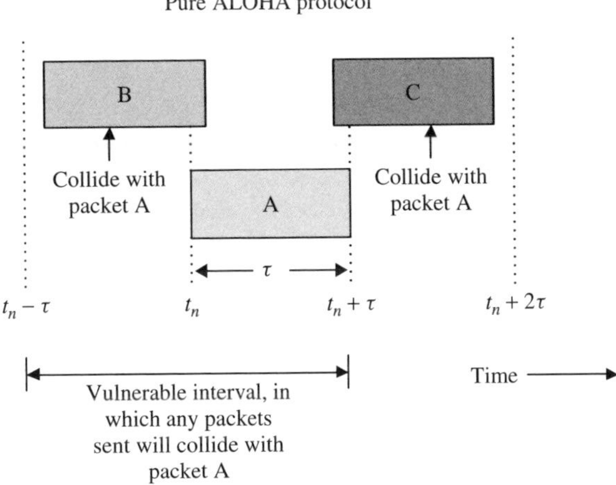

Figure 2.37 Illustration of packet collisions in a pure ALOHA system, where packet A is the packet of interest, and t_n is the transmission time of packet A. There are two other packets transmitted in the vulnerable interval such that they collide with packet A.

Let us consider a packet that starts transmission at time t_n. Its transmission will be successful if, and only if, no other packet transmissions happen during the intervals from t_n to $t_n + \tau$ and from $t_n - \tau$ to t_n. Therefore, the probability of success of a packet transmission will become the product of these two probabilities, or

$$P_s = p\{t_{n+1} - t_n > \tau\}\, p\{t_n - t_{n-1} > \tau\}$$

$$= \left\{\int_\tau^\infty Ge^{-Gt}\,dt\right\}\left\{\int_\tau^\infty Ge^{-Gt}\,dt\right\}$$

$$= e^{-2G\tau} \tag{2.50}$$

The number of successful packets, S, is the packet rate times the probability of success for each packet, that is

$$S = Ge^{-2G\tau}$$

$$= C_{max}\tau Ge^{-2G\tau} \tag{2.51}$$

where we have used the relation $C_{max} = \frac{1}{\tau}$ derived earlier. If we normalize with respect to G by replacing $G\tau$ with G (thus letting $\tau = 1$, and $C_{max} = 1$), we get the standard normalized capacity formula for a packet radio network or an ALOHA system as

$$S = Ge^{-2G} \tag{2.52}$$

from which we can easily see that the maximum capacity of the channel is equal to $\frac{1}{2e}$ when $G = 0.5$. If the offered channel load is high, the capacity of the channel is close to zero. On the other hand, if a very low traffic load occurs, most packets will be successful and the channel capacity is only slightly lower than the offered load. As traffic increases, collisions will take away some of the capacity of the channel. On the other hand, at very high offered loads, almost every packet transmission will experience a collision and very few packets can get through. Figure 2.38 shows the throughput performance of a pure ALOHA system. It can be seen from the figure that the peak throughput is about $\frac{1}{2e}$.

As the traffic load increases, packet delay will also increase since packets may need to be retransmitted several times before a success. We can proceed to calculate the channel delay performance for a pure ALOHA system as follows. Assume that the one-way propagation delay of the channel

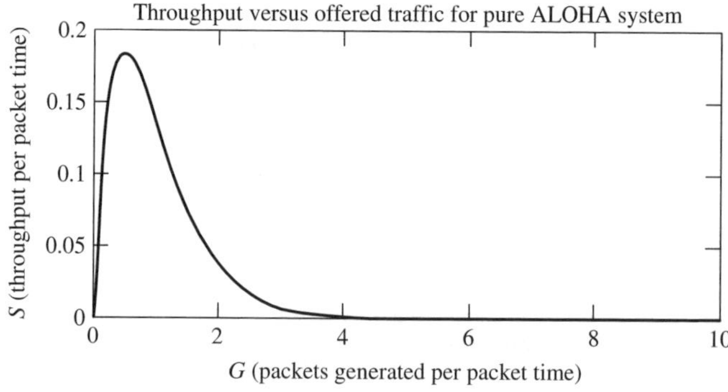

Figure 2.38 Throughput performance of a pure ALOHA system, where the peak throughput will be $\frac{1}{2e}$.

is $\alpha\tau$, the acknowledgment packet length is $a\tau$, and the average back-off duration or the average retransmission delay is $X_{avg} = \delta\tau$. Let the average time between retransmitted packets be $R\tau$, where R is the average delay measured in packet lengths. Then, we obtain

$$R\tau = \tau + \alpha\tau + a\tau + \delta\tau \tag{2.53}$$

or

$$R = 1 + \alpha + a + \delta \tag{2.54}$$

For an offered load of G and a channel capacity of S, the probability of a successful transmission of a packet becomes

$$P_s = \frac{S}{G} \tag{2.55}$$

A packet will be successful on the nth attempt if it failed on all previous $n - 1$ transmission attempts and succeeded on the nth trial.

$$P_s^{(n)} = \left(1 - \frac{S}{G}\right)^{n-1} \left(\frac{S}{G}\right) \tag{2.56}$$

The average number of transmission is

$$N_{avg} = \sum_{n=1}^{\infty} \left(1 - \frac{S}{G}\right)^{n-1} \left(\frac{S}{G}\right) \tag{2.57}$$

Therefore, the average number of retransmission can be calculated as $N_{avg} - 1 = \frac{G}{S} - 1$. The average delay, $D\tau$, is determined by summing over the following intervals:

- the time to transmit one packet (which is τ);

- the average time between retransmitted packets multiplied by the average number of retransmission;

- the propagation delay.

Thus, the average delay can be calculated as

$$D\tau = \tau + \left(\frac{G}{S} - 1\right) R\tau + \alpha\tau \tag{2.58}$$

or

$$D = 1 + \left(\frac{G}{S} - 1\right) R + \alpha \tag{2.59}$$

The average delay performance of a pure ALOHA system is shown in Figure 2.39.

With the throughput and delay performance of a pure ALOHA system, we can easily obtain its backlog performance in units of delayed packet transmissions.

$$Backlog = G - S (packets/terminal) \tag{2.60}$$

which has been plotted in Figure 2.40.

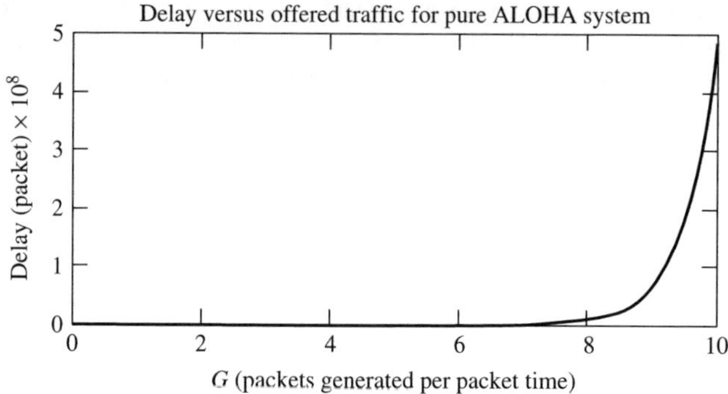

Figure 2.39 Average delay performance of a pure ALOHA system.

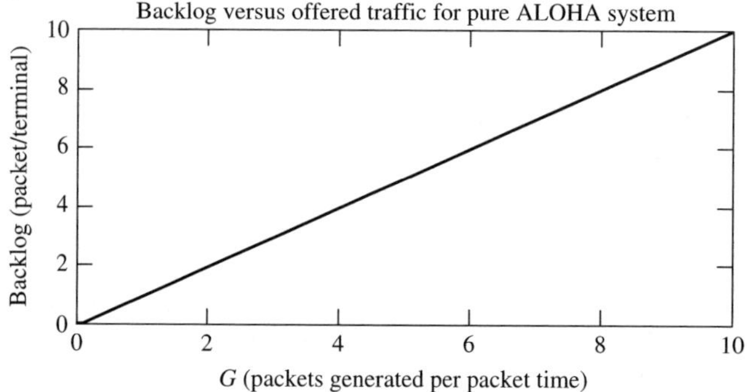

Figure 2.40 Average backlog performance of a pure ALOHA system.

Slotted ALOHA and CSMA

It is seen from the analysis given in the previous text that the pure ALOHA system can only provide a peak *throughput* also called *channel capacity* equal to $\frac{1}{2e}$, which is merely about 0.1839, which is far from satisfactory. There are many approaches to improve the capacity of a packet radio network. Here we will discuss a scheme, that is, slotted ALOHA, to improve the channel capacity by segmenting the transmission times into slots of length τ (as shown in Figure 2.41). It is also required that the transmitter can send a packet only at the beginning of a slot. Thus, any transmitters that are ready to transmit during a slot must wait for the next slot to transmit. Kleinrock and Tobagi [743] showed that the normalized channel capacity or throughput for a slotted ALOHA system is

$$S = Ge^{-G} \tag{2.61}$$

where again G is the offered traffic in the channel. Figure 2.42 shows the capacity of a slotted ALOHA channel as a function of the offered load. Notice that the capacity peak appears at $G = 1.0$.

In both the ALOHA and the slotted ALOHA channels, the transmitter sends a packet without checking channel status (busy or idle). In many ALOHA systems, the transmitter cannot determine if the channel is being used. To improve on this, CSMA systems was proposed, in which the transmitter senses the channel state before transmitting. If the channel is busy, then the transmitter waits until

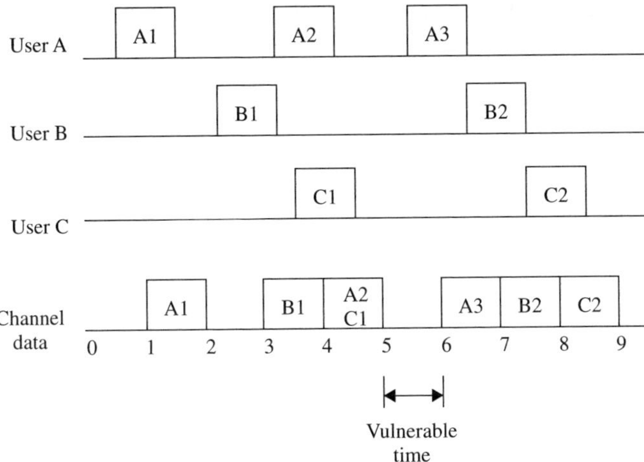

Figure 2.41   Illustration of a slotted ALOHA system, where in total three users (A, B, and C) are contending for transmissions. In the slotted ALOHA the packet that arrived in the previous slot will be sent out at the beginning of the next slot. Users A and C collide in the fifth slot due to the arrivals of their packets (A2 and C1) in the same slot.

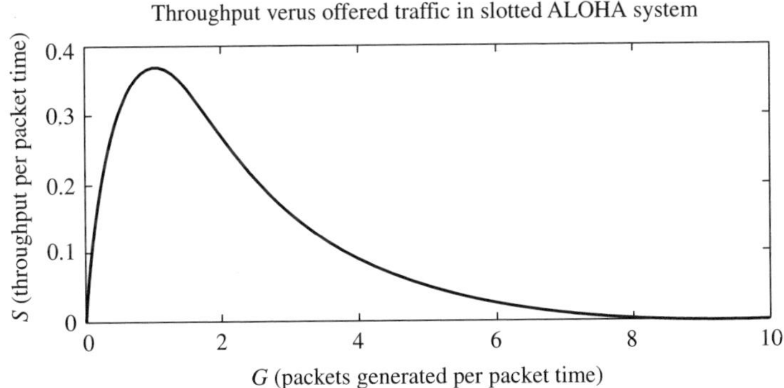

Figure 2.42   Throughput of a slotted ALOHA system, where the capacity peak $\left(S = \frac{1}{e} \simeq 0.3678\right)$ happens when the offered traffic is $G = 1$.

the next slot. Thus, collisions during a transmission are avoided but not at the start of a slot, and the capacity of the channel improves noticeably. When a CSMA system does not work because some of the transmitters are hidden (so-called *hidden terminal problem*), then some of the capacity of the channel must be used to send the status of the reverse channel. Tobagi and Kleinrock [747] called this a *busy tone* solution. This has been used in cellular and PCS systems that send busy-idle bits on the forward control channel to indicate reverse channel status. It is to be noted that the hidden terminal problem exists in many wireless networking scenarios, such as wireless local area networks (WLAN), packet radio networks, and so on. The hidden terminal problem occurs when a transmitter, say A, wants to send packets to a receiver, say B, which is busy receiving packets sent from another transmitter, say C, located at the other end of the network. As the transmitter C is outside A's detection range, A will not know the existence of C, as well as the busy status of the receiver B. In this case, terminal C is called the *hidden terminal* for A. Obviously, the communication between A and B fails because B is already in a busy receiving state. The busy tone can be used in terminal B to overcome the problem.

If all transmitters delay by a random delay before transmitting, the traffic spreads out and the capacity of the channel improves. Kleinrock and Tobagi call this channel a *nonslotted, nonpersistent channel* and calculate the capacity of the channel as

$$S = \frac{G\tau e^{-\alpha G\tau}}{G\tau(1 + 2\alpha) + e^{-\alpha G\tau}} \tag{2.62}$$

where $\alpha\tau$ is again the one-way propagation delay of the channel. For the *slotted, nonpersistent channel*, they assert that the capacity can be calculated as

$$S = \frac{\alpha G\tau e^{-\alpha G\tau}}{\alpha + 1 - e^{-\alpha G\tau}} \tag{2.63}$$

For both channels when the propagation delay is zero, that is, limit $\alpha \to 0$, then the capacity of the channel is

$$S = \frac{G\tau}{1 + G\tau} \tag{2.64}$$

The nonpersistent channel can therefore approach a capacity of one as the offered load increases. This is the ideal approach. The optimum values of the initial delay and the retransmission delay are functions of the offered load. Therefore, at high offered load, the central control of the system must send information to all transmitters to notify the channel status. We have already seen this control capability on the control channels in cellular and PCS systems.

Spreading code protocols

The random multiple access techniques can also work jointly with conventional FDMA, TDMA, and CDMA to form different hybrid versions of multiple access techniques. A popular combination is the joint application of pure ALOHA or slotted ALOHA with CDMA, in which every user will be assigned one or two signature codes for sending their packets [749]. With the joint application of ALOHA and CDMA, a packet radio network can support much more users simultaneously and the *collision* and *hidden terminal* problems can be improved to a large extent.

One of the major design issues in a CDMA-based packet radio network is the architecture of *spreading code protocols*, which specify the way in which spreading codes to different terminals (acting as either a transmitter or receiver) are assigned. Depending on the schemes on the spreading code assignments, basically there are five different spreading code protocols [749]:

- *Common spreading code protocol*: All users use the same spreading code to spread its outgoing packets.

- *Receiver-based spreading code protocol* (R code protocol): Each terminal is assigned a unique spreading code, which will be used only by others to address packets to it.

- *Transmitter-based spreading code protocol* (T code protocol): Each terminal is assigned a unique spreading code, which will be used only to address its own outgoing packets to other terminals.

- *Receiver–Transmitter based spreading code protocol* (R-T code protocol): Each terminal in the network is assigned two codes, one is the receiver-based (R) code and the other the transmitter-based (T) code, respectively. A transmitter should first use the R code to send a request packet to the target and should wait for the confirmation packet (encoded by T code) from the receiver before initiating data packets encoded by the T code.

The common spreading code protocol works in a way very similar to a pure ALOHA system. All users in a packet radio network under the common-code protocol will be using the same spreading code to spread their outgoing packets. Any intended receiver should always check the channel for the packets encoded by the common code. Therefore, the same collision mechanism as existed in a pure ALOHA system is present. It is noted that the use of the spreading code in outgoing packets will bring some operational advantages pertaining to any SS system, such as antijamming, interference-mitigating, and so on, which a pure ALOHA system does not offer. The proposal of the R code, T code, and R-T code protocols is aimed to further improve the performance of a common spreading coded packet radio network.

It is to be noted that all aforementioned spreading code protocols do not provide any *busy-code sensing* capability. Incorporated with code sensing for the target code before transmission, the robustness of the R code protocol can be noticeably improved [772–775]. However, the most vulnerable part of the R code protocol even with the code-sensing is in the initial phase of the pairing-up stage when two or more transmitters may sense the target code free in the channel and thus send packets to the same target simultaneously, resulting in a destructive collision. The receiver-transmitter (R-T) code protocol was also proposed by Sousa and Silvestre [749] to reduce the possible collisions that exist in the R code protocol by giving two codes to each user, in which a transmitter should first use the R code to send a request packet to the target and should wait for the confirmation packet (encoded by T code) from the receiver before initiating data packets encoded by the T code. As the T code will be used only by the transmitter itself, the presence of the T codes in the channel will never bother the activities of any other node, even if the data packet is very long. However, excessive use of spreading codes increases MAI pollution. To address the problem, Chen and Lim [772] proposed the triple-R protocol, in which pairing-up of any two nodes should go through three hand-shaking phases, all using receiver-based code protocol. The study given in [774] tried to solve the *blind-transmission problem* existing in the triple-R protocol by introducing *busy-code broadcasting* to make other transmitters attempting to send packets aware of the active users's busy status to avoid addressing packets to them.

Basically, all the above-mentioned protocols operate in a distributed fashion, and their advantages include the low cost of implementation and flexibility in the network deployment. The major problem with these distributive protocols is the high collision probability, which attributes to long access delay, low average throughput, and network instability especially in a highly loaded scenario, owing to the lack of an effective node-coordination mechanism. In general, the performance of all afore-mentioned protocols [749, 772, 774] is still far from being satisfactory, as illustrated in Figure 2.46 and Figure 2.47 for their performance comparison.

Hierarchy schedule sensing (HSS) protocol

The HSS protocol [750–763, 767, 768] adopts the request scheduling technique incorporated with a slotted permission frame (PF), which is broadcasted by a central scheduler (CS) in a common code C known to all users in the local network. The PF is slotted to differentiate the time slots for different nodes to initiate their request packets. Nodes are assigned different numerical terminal identification (TID) numbers, which appear as a cyclic sequence in the PF. Each node wishing to start a request packet has the obligation to first look up the PF for the right slot under its own TID and may transmit a request packet only at the beginning of the slot. As an attempt to further reduce the waiting time on the PF, a cell may be split up into groups to lessen the number of unique TIDs and thus the length of the PF. Therefore, each node in a group bears the same TID as one other node in each of the other groups. Possible collisions due to the same TIDs in different groups are avoided to some extent by using group IDs. In fact, the TID slots in the PF beacon can also carry some other useful information about the nodes (such as node status, node signature code, node logical names, and so on), which is accessible to all others in the cell due to the use of the common code to encode the PF. Figure 2.43, Figure 2.44, and Figure 2.45 show the pilot frame structure, the pairing-up process, and the hierarchical grouping for the HSS protocol.

The throughput and delay performance of the HSS protocol when compared to other proposed spreading code protocols are given in Figure 2.46 and Figure 2.47, respectively.

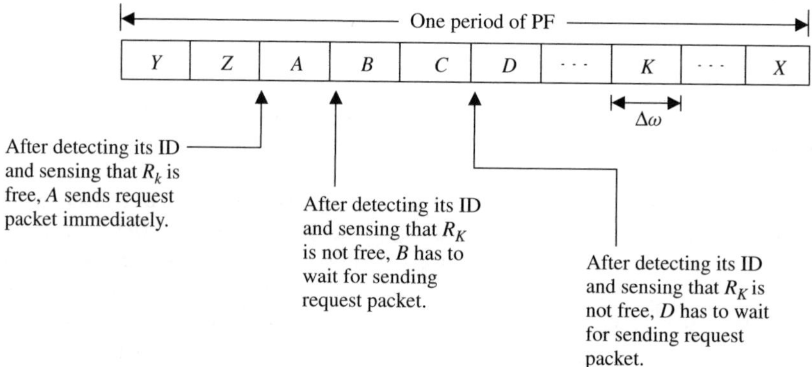

Figure 2.43 Illustration of a period of PF beacon and the ID slots used in the HSS protocol (A, B, and C are contending for sending a request packet to K, and A, B, ..., Z all are numerical numbers).

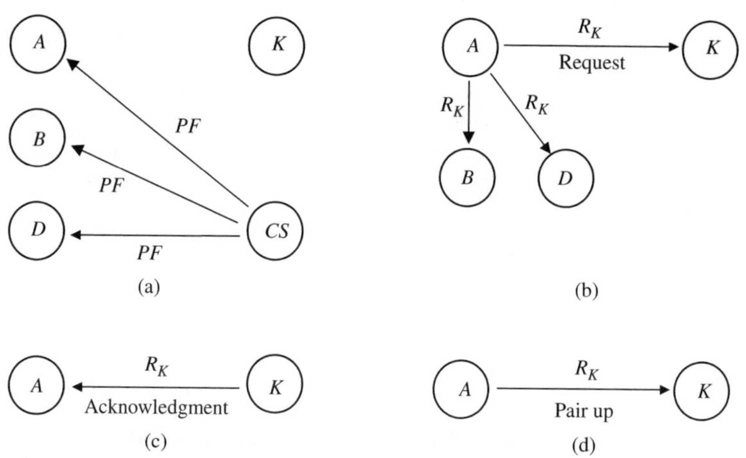

Figure 2.44 Pairing-up procedure between A (a transmitter) and K (a receiver) in the HSS protocol with B and D being contenders. (a) CS broadcasts PF; (b) A initiates request to K; (c) K acknowledges to A; and (d) A pairs up with K.

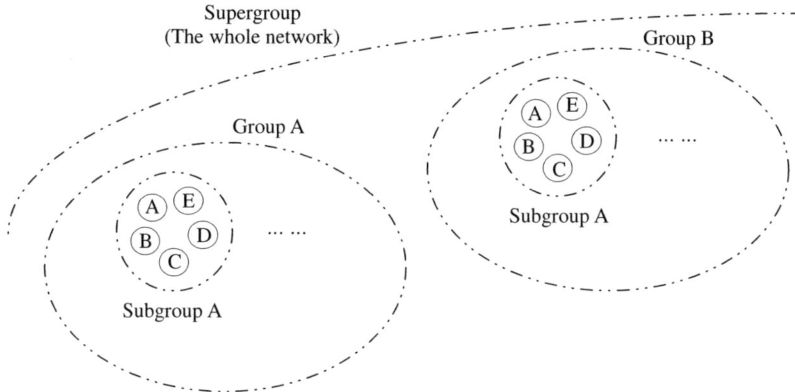

Figure 2.45 Hierarchical grouping in the HSS protocol for a cell with a large number of nodes.

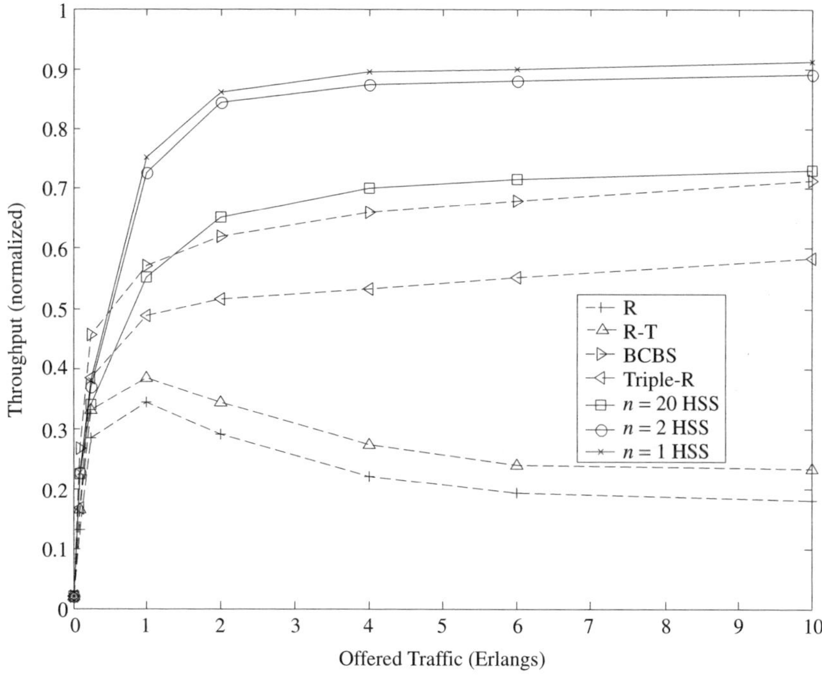

Figure 2.46 Comparison of throughput versus offered traffic of a data network using the HSS protocol with different protocols, where the cell size for R [749], $R - T$ [749], Triple-R [772] and BCBS [774] protocols is $N = 20$.

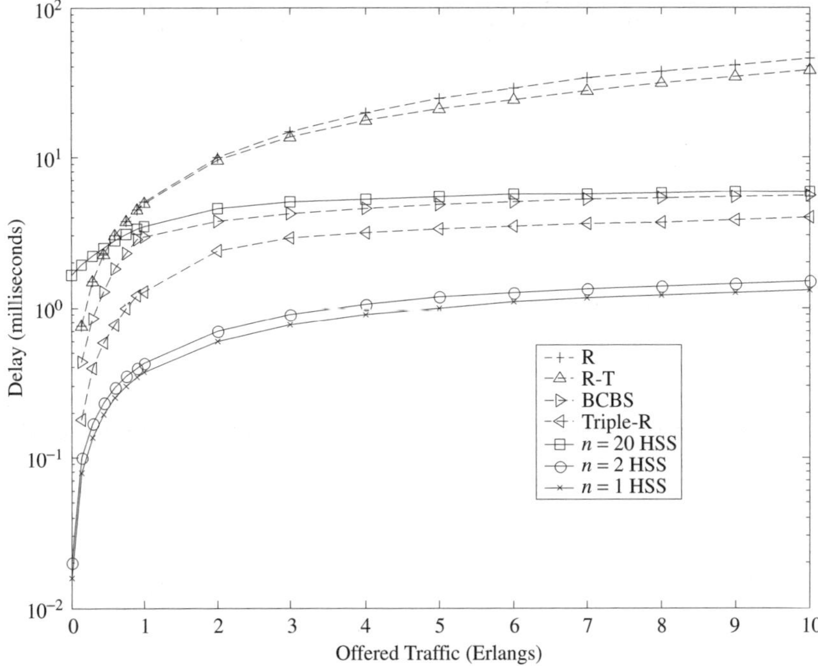

Figure 2.47 Comparison of delay versus offered traffic of a data network using the HSS protocol with different protocols, where the cell size for R [749], $R - T$ [749], Triple-R [772] and BCBS [774] protocols is $N = 20$.

Many more publications can be found for the research work done on random multiple access techniques [738, 776].

2.4   Multiple User Signal Processing

In this section, we will discuss issues on multiple user signal processing in a wireless communication system. In particular, we will concentrate on the following three topics, that is, CDMA multiuser joint detection, pilot-aided CDMA signal reception, and beam-forming techniques for co-channel interference suppression. It is to be noted that another important subject on multiple user signal processing is *multiple-in-multiple-out* (MIMO) system, which is discussed in detail in Chapter 8 of this book.

The multiple user signal processing techniques can be found extremely important in all communication systems based on any form of multiple access techniques. However, because of the great popularity of CDMA techniques, which have been widely used in 2G and 3G wireless communication systems [345−440], we will focus the discussions in this section mainly on multiple user signal processing for a CDMA-based system, although the ideas and principles of analysis can also be applied to any other system based on either FDMA or TDMA [15, 20].

2.4.1 Multiuser Joint Detection against MAI

It is well known that an effective way to combat the MAI in a CDMA system is the use of multiuser detection (MUD), which has become an extremely active research topic in the last 10 to 15 years [708–736]. The basic idea of the MUD was motivated by the fact that a single user–based receiver, such as a matched filter correlator or a RAKE, always treats other transmissions as unwanted interference in the form of MAI that should be suppressed as much as possible in the detection process therein. Such detection methodologies simply ignore the correlation characteristics given by the information coded by different CDMA codes (or MAI) appearing as a whole and all of that correlation among the users have not been utilized as useful information to assist the detection of different signals jointly. On the other hand, the MUD algorithms take the correlation among the users (or MAI) into account in a positive manner and user signal detection proceeds one by one in a certain order as an effort to maximize the detection efficiency as a whole. Some MUD schemes (not all of them), such as the decorrelating detector (DD) [712, 713], have an ideal near-far resistance property in a nonmultipath channel, and thus they can be also used as a countermeasure against the near-far problem in a CDMA system to replace or save the complex power control system that has to be used otherwise. However, it has to be pointed out that in the presence of the multipath effect almost none of the MUD schemes, including the DD, can offer perfect near-far resistance.

There are two major categories of MUDs: linear schemes and nonlinear schemes. It has been widely acknowledged in the literature [708–736] that the linear MUD schemes have a relatively simple structure than the nonlinear schemes and thus they have been given much more attention for their potential application in a practical CDMA system for the simplicity of implementation. In most current 3G standards, such as CDMA2000 [345], UMTS-UTRA [425, 448], WCDMA [431] and TD-SCDMA [432, 433], the MUD has been specified as an important option. However, because of the issue of complexity, this option will remain an option in real systems as most mobile network operators are still reluctant to activate it at this moment.

Two important linear MUD schemes have to be addressed briefly in this subsection; one is the DD [712] and the other is the MMSE detector [713]. DD, as its name suggests, performs MUD via correlation, decorrelating among user signals by using a simple correlation matrix inversion operation. Some of the important properties of the scheme can be summarized as follows. First, it can eliminate MAI completely and thus offer a perfect near-far resistance in the AWGN channel, which is important for its applications, particularly, in uplink channels. Second, it needs correlation matrix inversion operation, which may produce some undesirable side-effects, one of which is the noise-enhancement problem due partly to the ill-conditioned correlation matrix and partly to the fact that it never takes the noise term into account in its decorrelating process. On the other hand, a MMSE detector takes both MAI and noise into account in its objective function to minimize the mean square detection error and thus it offers a better performance than DD especially when signal-to-noise ratio is relatively low in the channel. It should be pointed out that a MUD in the multipath channel behaves very differently when compared with that in the AWGN channel. Usually a successful operation of a MUD in a multipath channel requires full information of the channel, such as the impulse response of the channel in the time domain, and so on. Therefore, a MUD working in multipath channels can be very complex. To overcome this problem, many adaptive MUD schemes [714, 715] have been proposed such that they can perform joint signal detection with only very little or even no channel state information (CSI).

The analysis of a MUD scheme in a downlink channel is much simpler than that in an uplink channel, where all user transmissions are asynchronous. However, with the help of an extended correlation matrix, an asynchronous system can be treated as an enlarged equivalent synchronous system only adding more *virtual user signals* in its dimension-extended correlation matrix. Thus, theoretically speaking, any asynchronous MUD problem can always be solved by this method without losing generality.

Quasi-decorrelating detector (QDD)

Being an important topic of research, many papers on the CDMA MUD have been published and many different forms of MUD schemes have been proposed in the literature [708–736]. *Quasi-Decorrelating Detector* (QDD) [718, 719] is one of the proposed schemes.

The QDD is a nonmatrix inversion–based algorithm for implementing DD. The QDD uses a truncated matrix power expansion instead of the inverted correlation matrix to overcome the problems associated with the inversion transformation in DD, such as noise enhancement, computational complexity, matrix singularity, and so on. Two alternative QDD implementation schemes were presented in [718]; one is to use multistage feed-forward filters and the other is to use an nth order single matrix filter (neither involves matrix inversion). In addition to significantly reduced computational complexity when compared with DD, the QDD algorithm offers a unique flexibility to trade among MAI suppression, near-far resistance, and noise enhancement according to varying system setups. The obtained results show that the QDD outperforms the DD in either AWGN or the multipath channel if the number of feed-forward stages is chosen properly. In the paper [718] the impact of correlation statistics of spreading codes on the QDDs performance was also studied with the help of a performance-determining factor derived explicitly therein, which offers a code-selection guideline for the optimal performance of the QDD algorithm.

It is to be noted that the QDD is also a linear detector but its decorrelating algorithm can be performed without matrix inversion transformation, as an effort to overcome the problems associated with the DD. Similar to the DD, the operation of the QDD does not need the explicit knowledge of the users' signal power, and it can achieve desirable near-far resistance. While retaining many preferable properties of the DD, the QDD also adds several of its own attractive features. The QDD can be implemented by a multistage feed-forward filter, the number of which can be made adjustable to trade MAI suppression for noise enhancement according to varying channel conditions. On the contrary, the DD has a relatively rigid structure and is unable to adapt to a changing operational environment. It can be shown that under varying conditions a fine-tuned QDD (with a carefully chosen number of feed-forward stages) can always outperform the DD in terms of bit error probability (BEP).

Because of the fact that the QDDs performance is closely related to the cross-correlation level (CCLs) statistics of spreading codes, the impact of the CCLs on its performance was also studied in [718] to search for the spreading codes most suitable for the QDD algorithm. The work of Chen [718] deals with the QDD for a synchronous CDMA system in either an AWGN or a multipath channel. In fact, an asynchronous system can be viewed as an equivalent enlarged synchronous one (with more effective users) and thus can be treated in a similar way.

In [718, 719], the study was concentrated on two salient issues: one being the code-dependent analysis of a QDD with the help of performance-determining factors based on the statistical features of the signature codes; and the other being the performance analysis of such a multiuser detector under frequency-selective fading channels, which has been a most serious concern in a wireless or mobile communication system.

Figure 2.48 and Figure 2.49 show the two different implementation schemes for a QDD MUD respectively, one being implemented by multistage feed-forward matrix filters and the other being implemented by an l-order single stage matrix transformation.

Figure 2.50 illustrates the *BEP* of the QDD and the DD in a 3-ray multipath channel with normalized delay profile [0.9275,0.3710,0.0464] and the interpath delay being four chips using EGC and MRC-RAKE receivers. The Gold code length is $N = 31$ and the generation polynomials are [0,0,1,0,1] and [1,0,1,1,1] with their initial state [0,0,0,0,1]. The number of users is $K = 13$ and the detection block size is $M = 5$. Figure 2.51 compares the near-far resistance for both the QDD and the DD in 3-ray multipath channels with different delay profile patterns with interpath delay being four

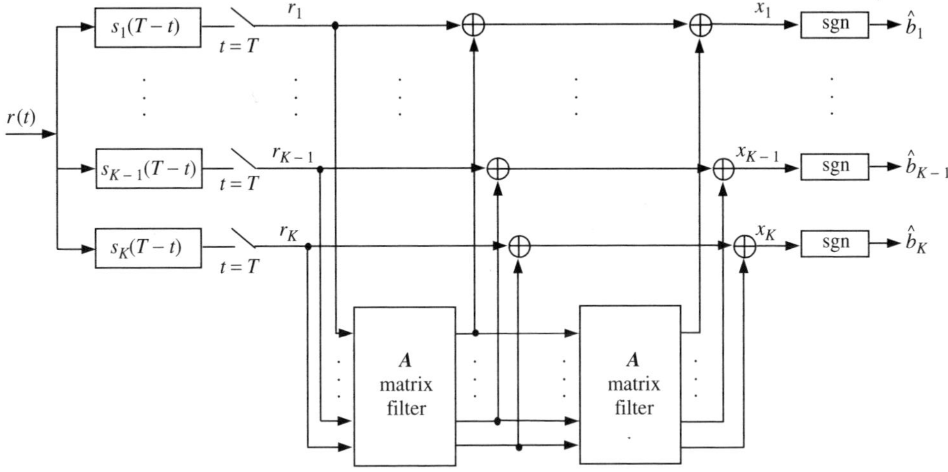

Figure 2.48 QDD scheme implemented by multistage feed-forward matrix filters in the AWGN channel with the front end being a matched filter bank.

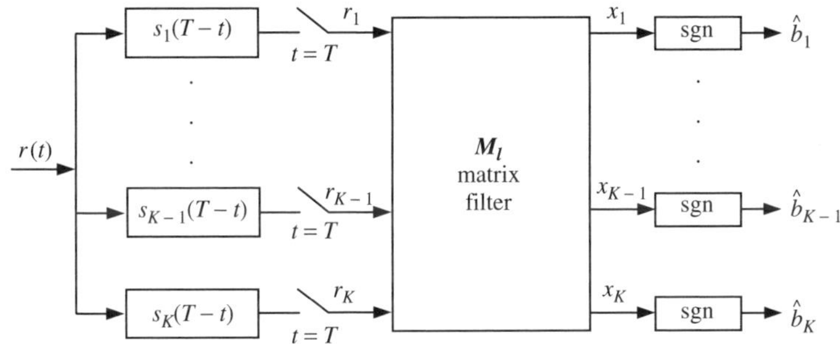

Figure 2.49 QDD implemented by an l-order single stage matrix transformation in AWGN channel with the front end being a matched filter bank.

chips using matched filter, EGC, and MRC-RAKE receivers; Detection block size $M = 5$; number of users $K = 7$; Gold code length is $N = 31$ and generation polynomials are [0,0,1,0,1] and [1,0,1,1,1] for initial state [0,0,0,0,1].

It is seen from Figures 2.50 and 2.51 that the QDD offers a better performance in the multipath channel in terms of its bit error probability and near-far resistance to make it a suitable candidate for its applications in various CDMA wireless systems.

Orthogonal decision-feedback detector (ODFD)

The *orthogonal decision-feedback detector* (ODFD) [720, 721] was proposed to overcome some problems that exist in the *decorrelating decision-feedback detector* (DDFD) [728–730].

Chen and Sim [720] introduced an asynchronous orthogonal decision-feedback detector (AODFD) for asynchronous CDMA multiuser detection. The AODFD based on entire message-length detection

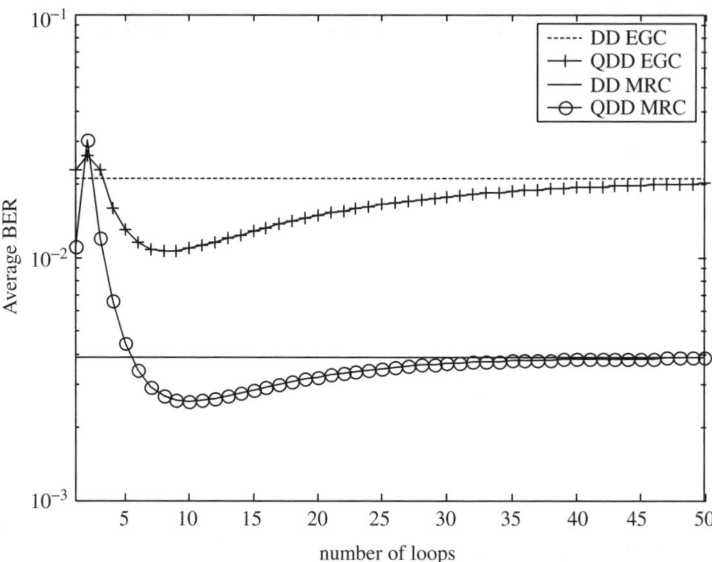

Figure 2.50 BEP of QDD in a 3-ray multipath channel with normalized delay profile being [0.9275,0.3710,0.0464] and interpath delay being four chips using EGC and MRC-RAKE receivers. Gold code length is $N = 31$ and generation polynomials are [0,0,1,0,1] and [1,0,1,1,1] for initial state [0,0,0,0,1]. Number of users is $K = 13$. Detection block size is $M = 5$.

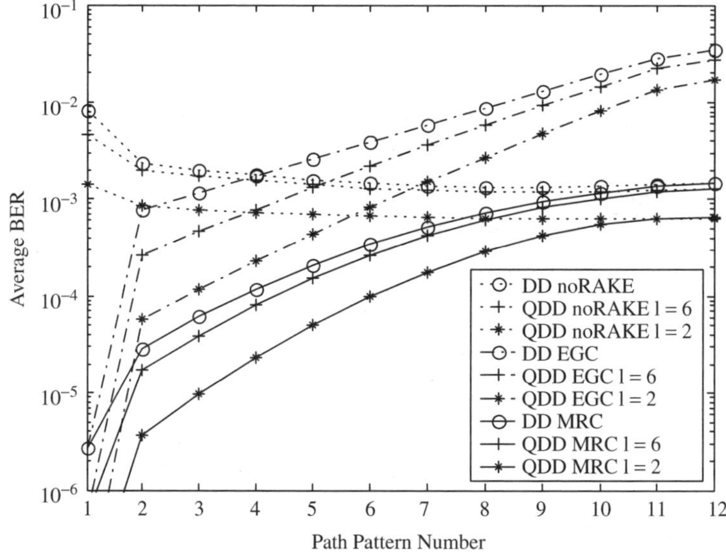

Figure 2.51 BEP of QDD in 3-ray multipath channels with different delay profile patterns and with interpath delay being four chips using matched filter, EGC, and MRC-RAKE receivers; Detection block size $M = 5$; number of users $K = 7$; Gold code length is $N = 31$ and generation polynomials are [0,0,1,0,1] and [1,0,1,1,1] for initial state [0,0,0,0,1].

was studied first. A realizable scheme, sliding-window AODFD, was then proposed and its performance was analyzed. In spite of its simple structure, the sliding-window AODFD performs as good as the asynchronous decorrelating decision-feedback detector (ADDFD) [728–730], which has a much higher complexity. The reduced complexity of the sliding-window AODFD is due to the use of orthogonal matched-filtering and a short window size. Unlike the ADDFD that requires computational intensive z-transformed matrix inversion and spectral factorization, the AODFD uses the agile Gram-Schmidt procedure. It is possible for the AODFD to adopt a simple updating algorithm and parameter updating is no longer always necessary when users leave the system. The comparisons were also made with other orthogonal-based detectors and the BEP results showed that the AODFD is an attractive multiuser detector.

It is well known that a DDFD [728–730] consists of a decorrelating first stage followed by a decision-feedback stage. Decisions are usually made in the order of decreasing power. The complexity of the DDFD grows linearly with the number of users, but the complexity of its algorithm in calculating the linear transformation matrix is of the order of $O(K^3)$, where K is the number of users. When the system setup or received signal power changes (thus, reordering of the users according to their power levels is necessary), the matrix has to be recalculated. In addition, the hardware implementation of the inverse matrix filter is also complicated. Chen and Sim [720] proposed the *orthogonal decision-feedback detector* (ODFD), which is able to overcome most problems associated with the DDFD. The ODFD combines matched filters and the decorrelating matrix filter into a single orthogonal matched filter. Instead of performing match filtering to the users' spreading codes, the ODFDs orthogonal matched filters match to a set of ortho-normal sequences, which span the signal space of all spreading codes. The ODFD can also use soft-decision to further improve its performance (just like improved DDFD (IDDFD) [729]). In fact, implementation complexity is a serious concern with ADDFD, which relies on a noncausal doubly infinite feed-forward filter and has to be truncated for hardware realization. The sliding-window method is one of the most cost-effective ways to make the feed-forward filter realizable. In the paper [720] a sliding-window method was applied to the AODFD to reduce its complexity. To calculate the decorrelating matrix, the ADDFD should perform multidimensional spectral factorization and spectrum matrix inversion, which is a very computationally intensive operation [731]. The AODFD only requires the Gram-Schmidt orthogonalizing procedure to derive the orthogonal matched filter, which plays a pivotal role in simplifying the updating of parameters.

We would also like to discuss some related works done previously by others. Forney has pointed out in his paper [732] that the whitening matched filter can be an orthogonal filter although he did not specifically address the issues related to CDMA multiuser detection. Wei and Rasmussen [733] applied a sliding-window method to a near ideal noise-whitening filter. In their proposed scheme, a matched filter bank is cascaded with the whitening filters followed by an M-algorithm detector. Schlegel *et al.* [734] introduced a multiuser projection receiver to achieve interference cancellation through projecting unwanted user signals onto a space spanned by the desired users' signal vectors, followed by a RLS detector. In this scheme, an independent chip-matched filter bank is still required before the projection filter. In K. B. Lee's paper [735], an orthogonal transformation preprocessing unit, which generates a partially decorrelated output, was used before the LMS or the RLS algorithm for estimating the desired signal. The method does not need a priori knowledge of interfering signal parameters, but the LMS algorithm requires training sequence. Thus, the adaptive algorithm stability will be a concern. Unfortunately, the paper did not provide the analysis on neither BER nor *near-far resistance* performance.

The concept of the ODFD can be easily interpreted using signal vector representation. Consider a two-user system with spreading codes $S_1(t)$ and $S_2(t)$ (as shown in Figure 2.52). As the spreading codes are linearly independent, they form a two-dimensional signal space. There are many pairs of orthogonal functions that can span this signal space, but if the set of orthogonal functions, $\phi_1(t)$ and $\phi_2(t)$ (with normalized energy) as shown in Figure 2.52, are selected, successive decoding can proceed immediately. Suppose that the received signal is matched to $\phi_2(t)$. Then, the output, $S_{2,2}$, is independent of $S_1(t)$ denoting user 1. Therefore, $S_{2,2}$ can be decoded immediately to yield the bit

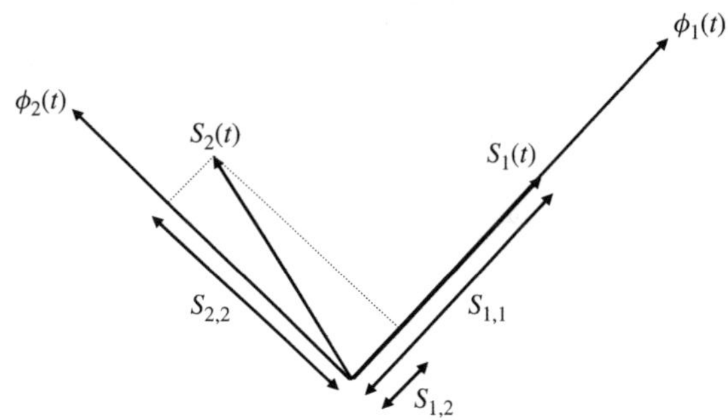

Figure 2.52 Signal space vector representation of two spreading codes and their orthogonal functions in ODFD MUD scheme.

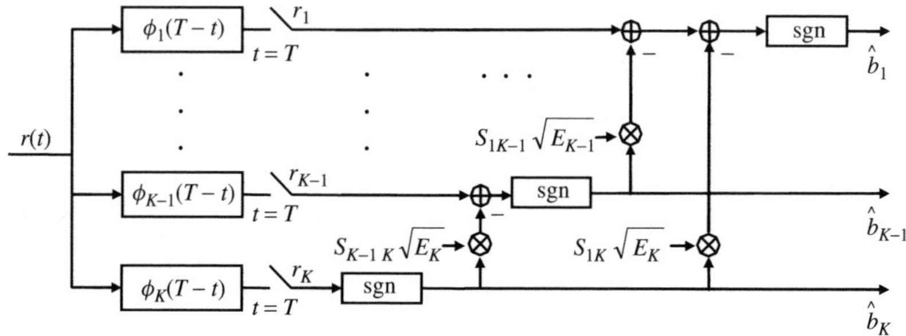

Figure 2.53 Block diagram of a synchronous ODFD MUD scheme.

information for user 2. The other matched filter is matched to $\phi_1(t)$ and the output is $S_{1,1} + S_{1,2}$, in which the bit information for user 1 is corrupted by $S_{1,2}$ but it can be canceled through regeneration since user 2 has already been decoded. It is noted that matched filtering with orthogonal functions instead of the spreading codes may result in some loss in signal-to-noise ratio (SNR). However, the output $S_{2,2}$ is free of interference and its detection can be made more accurate than that of the output from a pure matched filter.

The Gram-Schmidt procedure can generate the set of orthogonal functions and their corresponding coefficients, $S_{1,1}$, $S_{1,2}$, and $S_{2,2}$, which are used in the decision-feedback stage. The block diagram of the ODFD is shown in Figure 2.53, where $r(t)$ is the received signal, $\phi_1, \phi_2, \ldots, \phi_K$ are the orthogonal functions, $s_{m,k}$ $(k, m = 1, 2, \ldots, K)$ are the coefficients generated from the orthogonalization procedure and E_k is the energy of the kth user signal. Figure 2.54 shows the block diagram of an asynchronous ODFD MUD scheme.

In the paper [720], synchronous ODFD, asynchronous ODFD and sliding-window AODFD were studied. The explicit analysis for all three ODFD MUD schemes should not be discussed here because of limited space. Figure 2.55 illustrates the bit error probability of the AODFD MUD scheme, where the detection proceeds at a decreasing delay ordering. It is seen from the figure that the performance of the AODFD scheme is comparable to that of the ADDFD scheme, but with a greatly reduced implementation complexity. More detailed information about the AODFD MUD scheme can be found in [720].

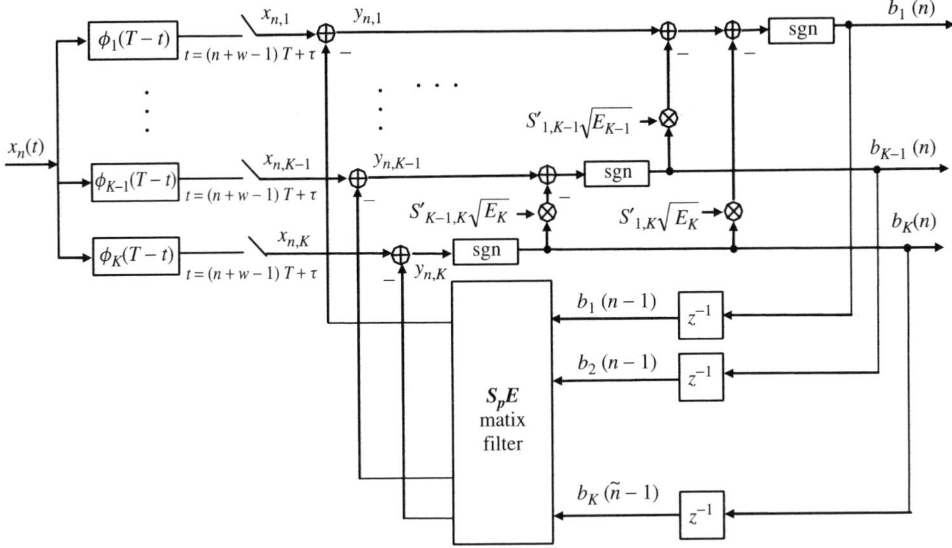

Figure 2.54 Block diagram of the asynchronous ODFD MUD scheme.

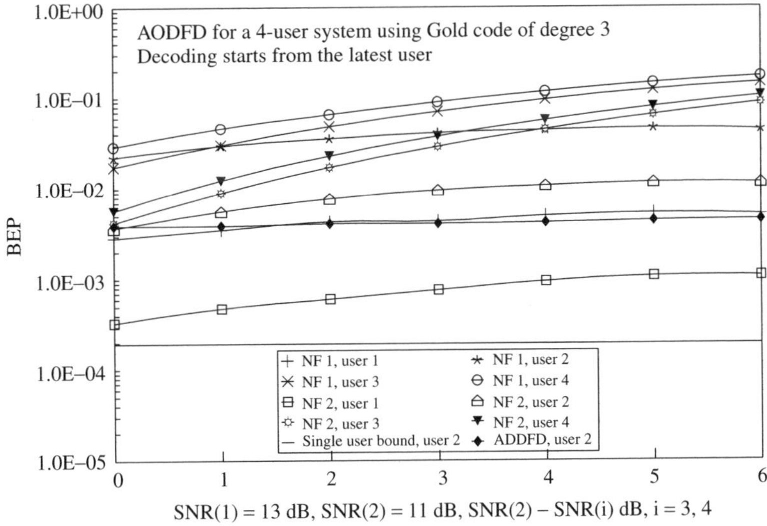

Figure 2.55 Bit error probability of AODFD, where the detection proceeds at a decreasing delay ordering.

2.4.2 Pilot-Aided CDMA Signal Detection

As mentioned earlier, sometimes CDMA signal detection may need the knowledge of the multipath channel, such as the MRC-RAKE and MUD algorithms, and so on. Therefore, the channel estimation becomes a necessity in these CDMA applications. Either a dedicated pilot signal or interleaving a pilot signal with a data signal can be used for channel estimation. Usually, a dedicated pilot channel is feasible only for downlink channel signal detection due to the need to simplify signal detection process at mobiles and the relatively easy allocation of the resources (such as available power and synchronous transmissions, and so on) in a BS.

In order to improve the accuracy of channel estimation, an ideal pilot-signaling design should preferably bear the following three important characteristic features (or called *three-same conditions* [715]):

- The pilot signal should be sent at the *same frequency* as that of the data signal to ensure reliable channel estimation in frequency-selective channels.

- The channel estimation should be carried out at the *same time*[24] as (or as close as possible to) that of data detection to combat fast fading of the channel due to the mobility of the terminals. For a similar reason, time duration of the pilot signal should be made as short as possible to facilitate the real-time channel estimation due to the concern on latency in processing the pilot signal.

- The pilot channel in a CDMA system should share the *same code* as that for data channels to ensure the availability of identical MAI statistics.

In the aforementioned *three-same conditions*, it is noted that the first two conditions are more important than the third one, as the use of different codes is still feasible in many cases to estimate the channel information. However, using the third condition will help to extract the right MAI statistics information, which in some cases is also useful in facilitating the multiuser detection, as discussed in the previous text. It should also be stressed that the need for a pilot signal is because of the requirement for channel estimation, which originally pertains to the specific CDMA signal detection schemes concerned, such as MRC-RAKE and MUD, to mitigate MAI and MI. Obviously, if there is neither MAI nor MI, the complicated pilot signaling is not necessary for a CDMA system.

A typical pilot signal design for an orthogonal complementary code (OCC)–based CDMA system was proposed in [210], as shown in Figure 2.56, where the downlink channel uses a dedicated pilot code with its pilot burst duration and repeating period being T_{d1} and T_{d2}, respectively; and the uplink channel adopts interleaving data signal with the pilot bursts, whose duration and repeating period are T_{u1} and T_{u2}, respectively. It is to be noted that the choice of pilot burst durations and repeating periods for both downlink and uplink channels is of ultimate importance to ensure that the pilot signal works effectively. Usually, we have to select the pilot burst repeating period, that is, T_{d2} and T_{u2}, such that they should be shorter than the channel coherent time T_{co} (which was defined in Equation 2.13 in Section 2.1.3), which is determined by the reciprocal of the Doppler spread (Δf_d) of the mobile channel. On the other hand, the choice of pilot burst duration for both downlink and uplink channels, that is, T_{d1} and T_{u1}, can be made according to the synchronization capture and the tracking properties of a receiver. It should not be made too long so as not to add too much overhead to the data transmission channel and extra interference to other data channels. It should not be too short either to allow a receiver to capture and process the pilot bursts easily.

A specific algorithm for pilot-aided signal detection in an OCC-based CDMA system was proposed in [777]. The motivation of this pilot-aided algorithm was due to the fact that an OCC-CDMA system [210] cannot use the RAKE receiver for signal detection in multipath channels, mainly because

---

[24]Usually, it is required that the time difference between CSI estimation and signal detection must be made shorter than the channel coherent time.

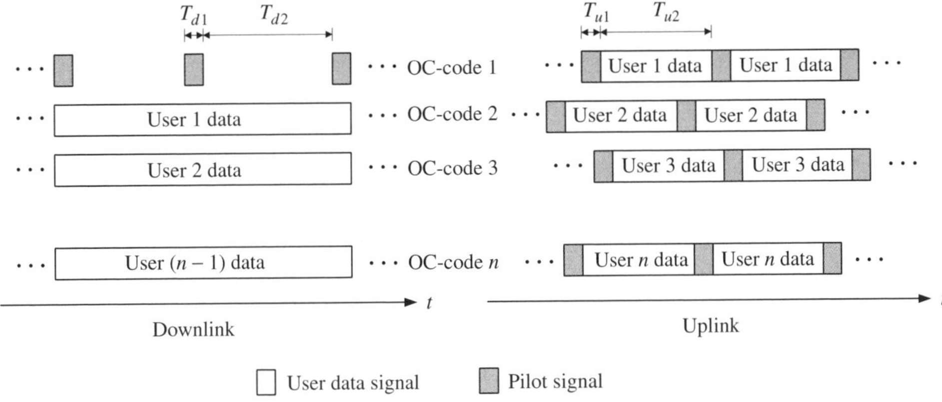

Figure 2.56 Illustration of uplink and downlink pilot frames for an OCC-CDMA system, where T_{d1} and T_{d2} denote downlink pilot burst duration and repeating period, respectively; and T_{u1} and T_{u2} stand for the uplink pilot burst duration and repeating period, respectively.

of its offset stacking spreading modulation [210], which introduces overlapping among different data bits that interfere with one another in the time domain. Thus, other alternative receiver structures have to be found to replace RAKE for signal reception in multipath channels. In [777], we proposed a pilot-added detection scheme for its application in an OCC-CDMA system. The scheme works on a channel matrix estimation algorithm based on the pilot, followed by a matrix inversion operation to obtain the estimates of transmitted data information. It is simple and straightforward to offer satisfactory detection efficiency for the multipath signal reception. In addition, it suits the applications well not only for an OCC-CDMA system but also for other CDMA systems. Figure 2.57 shows the BER versus SNR for an OCC-CDMA system with the pilot-aided detection algorithm. In this figure, the third parameter, that is, the pilot-to-noise ratio (PNR), has been used to plot four different curves. It is seen from the figure that an OCC-CDMA system can work fairly well as long as the PNR value can be made above 18 dB.

Another pilot signal-aided scheme for a CDMA multiuser signal detector was presented in [778], in which a new pilot-aided MUD scheme, single-code cyclic shift (SCCS) detector, was proposed for synchronous CDMA multiuser signal reception. The unique feature of the proposed SCCS detector is that a receiver can decode multiuser signals even without the explicit knowledge of all the signature codes active in the system. The transmitting signal from a BS to a mobile contains two separated channels: the pilot and data channels; the former consists of periodically repeated pilot symbols encoded by the same signature codes as the one spreading the latter. Both pilot and data signals for a specific mobile are sent by a base station using quadrature and in-phase carriers at the same frequency with the QPSK modulation. A matched filter bank, consisting of M correlators that match to distinct cyclic-shifted versions of a "single" signature code, is employed for "channel cyclic shift correlation function" estimation, followed by the MUD algorithm based on the channel information obtained earlier. The performance of the proposed SCCS detector was evaluated and compared to the DD by computer simulations considering various multipath channels with different profiles. The results demonstrated that a synchronous CDMA joint detection can be implemented successfully without necessarily knowing all signature codes of the system.

Figure 2.58 shows a conceptual block diagram of the transceiver using the pilot-aided SCCS MUD detector, consisting of both the transmitter and the receiver. Figure 2.59 illustrates baseband data and the pilot-signaling frame for different users in the system. It is seen from the figures that a BS transmitter will send its data via I channel and its pilot via Q channel in a quadrature modulator,

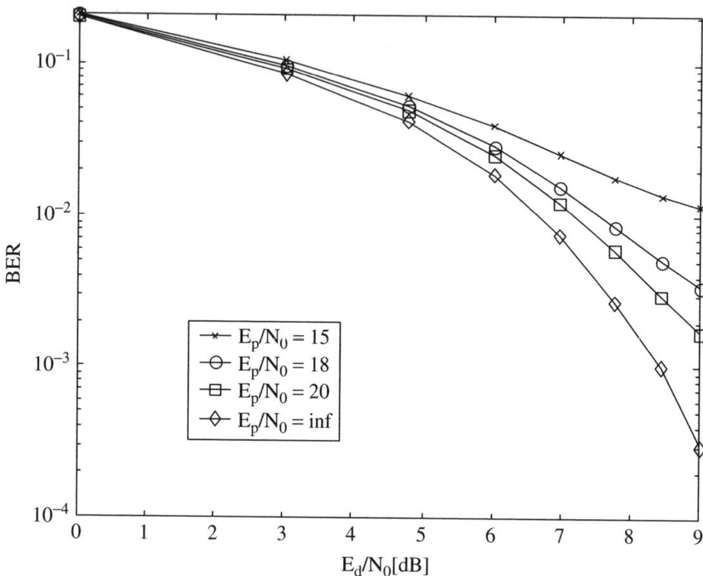

Figure 2.57 BER for OCC-CDMA in uplink multipath channel using pilot-aided detection algorithm. Delay profile = [0.5774, 0.5774, 0.5774], interpath delay = one chip, interuser delay = three chip, PG of CC code = 4×16, and user number = 4.

such as QPSK and and so on. Because of the synchronous transmission in the downlink channel, the pilot bursts for different users will not overlap with one another, allowing a mobile receiver to extract the individual pilot signal easily.

The mean BER averaged for all users is plotted in Figure 2.60, where the pilot-aided SCCS detector is compared with the conventional DD with the PNR as a parameter.

2.4.3   Beam-Forming against Co-Channel Interference

The Beam-forming technique is another effective way to combat MAI or commonly called *co-channel interference* under the context of antenna-array techniques. What we refer here with respect to the antenna-array techniques is either a smart antenna or switched beam system, both of which have been used in some CDMA-based systems to improve the system performance under co-channel interference. The principle of antenna-array techniques is based on various beam-forming algorithms, whose conceptual block diagram is shown in Figure 2.61. It should be clarified that we are concerned here only with the way in which some suitable beam-pattern at either a transmitter or receiver is achieved, and thus is different from what is called *space–time coded MIMO systems*, which is treated in Chapter 8 explicitly.

Generally speaking, beam-forming techniques also belong to the category of multiple-user signal processing as it will take all received signals into account when implementing various beam-forming patterns. An antenna-array system consists of several antenna elements, whose space should be made large enough (usually at least 0.5 to 1 wavelength is required), in order to obtain a statistical independence among the signals received at different elements. An antenna-array can be used by either a transmitter or a receiver. By using various beam-forming algorithms (such as MVDR, MUSIC, LMS, RLS algorithms, etc.) [18, 20], an antenna-array system will generate a directional beam pattern, whose

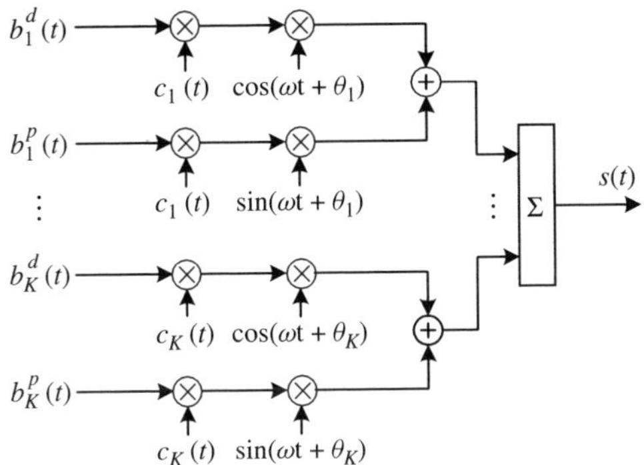

(a) Base-station transmitter

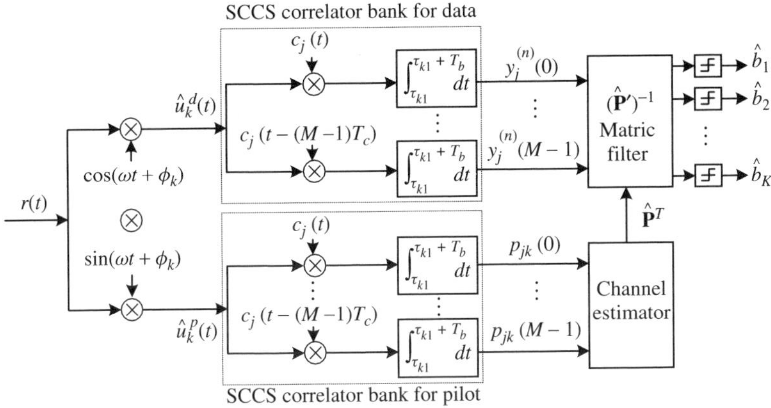

(b) The k-th mobile receiver

Figure 2.58 The conceptual models for transmitter (in BS) and receiver (in mobile) using the SCCS multiuser detector for synchronous CDMA signal reception.

width is dependent on the number of elements used and the beam-forming algorithms in question. In general, the use of more antenna elements will yield a narrower main lobe of the beam-patterns. With such a very narrow directional beam-pattern, a transceiver using an antenna-array system can effectively reduce the co-channel interference generated outside the beam width. If an antenna-array is used as a transmitter antenna in a BS, it can help to project or direct BS signals to some specific mobiles to reduce interference to other mobiles. If an antenna-array is used as a receiver antenna at a BS, it can assist the BS to focus on to the mobiles whose signals it wants to receive to suppress other possible interference outside the beam width. Similar to the aforementioned MUD algorithms, a beam-forming algorithm can also be made adaptive [20] to follow the change of the direction of arrival (DOA) of the target signal with the help of a specific training signal sent by the target, which should be repeated within a time duration shorter than that of a substantial time-related change in DOA of the target signal.

However, the usefulness of antenna-array systems is because of the existence of MAI due to imperfect CCFs among the spreading codes in a CDMA system. Otherwise, the co-channel interference will no longer be a threat to the detection of useful signals and then the antenna-array may not be needed.

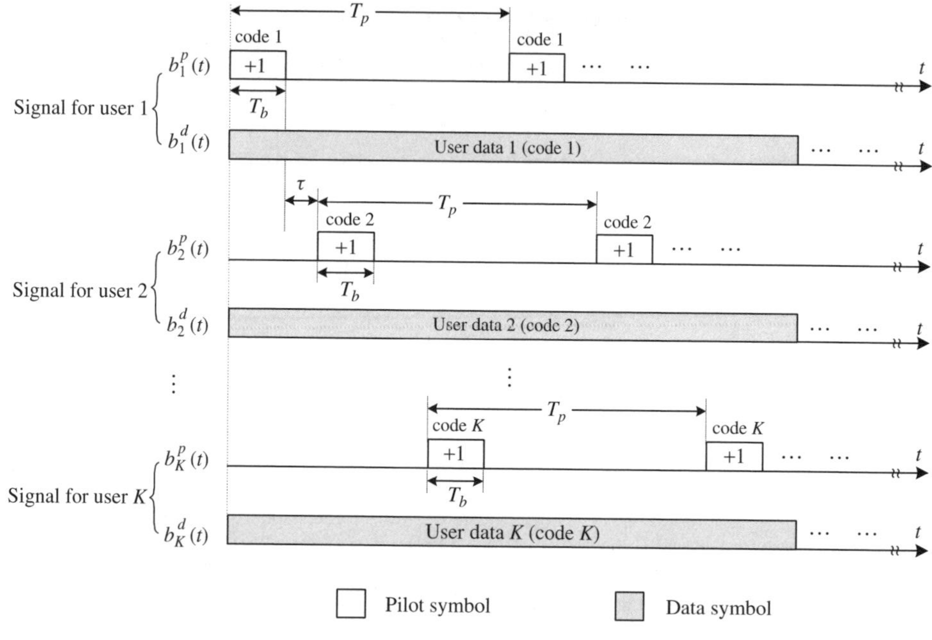

τ : Guard time > Maximum channel delay spread

T_p : Pilot signal renew duration < Channel coherent time

Figure 2.59 Baseband data and pilot signaling structure of a synchronous CDMA system using the SCCS multiuser detector. The composite transmitted signal for each mobile consists of quadrature and in-phase channels to convey pilot and data signals respectively.

In some applications, a beam-forming algorithm can be used jointly with some multiuser detection schemes to form a more effective joint detection solution. An adaptive joint beam-forming and multiuser detection scheme called *B-MMSE* was proposed in [779, 780]. In fact, the combination of antenna-array beam-forming with MUD can effectively improve the detection efficiency of wireless communications under MI, especially for the applications in fast fading channels. Chen and Jen-Siu Lee [779] studied the performance of an adaptive beam-former incorporated with a B-MMSE detector, which works on a unique signal frame characterized by training sequence preamble and data blocks segmented by zero-bits. Both beam-former weights updating and B-MMSE detection are carried out by either the LMS or the RLS algorithm. The comparison of the two adaptive algorithms applied to both the beam-former and the B-MMSE detector was made in terms of the convergence behavior and the estimation mean square error. The final performance in error probability has been given. Various multipath patterns were considered to test the receiver's responding rapidity to changing MI. The performance of the adaptive B-MMSE detector was also compared with that of the nonadaptive version (i.e., through the matrix inversion). The obtained results suggested that the adaptive beam-former should use the RLS algorithm for its fast and robust convergence property; while the B-MMSE filter can choose either the LMS or the RLS algorithm depending on the antenna-array size, multipath severity and complexity.

For more treatments about the beam-forming algorithms and their applications in various wireless systems, the readers may refer to [20].

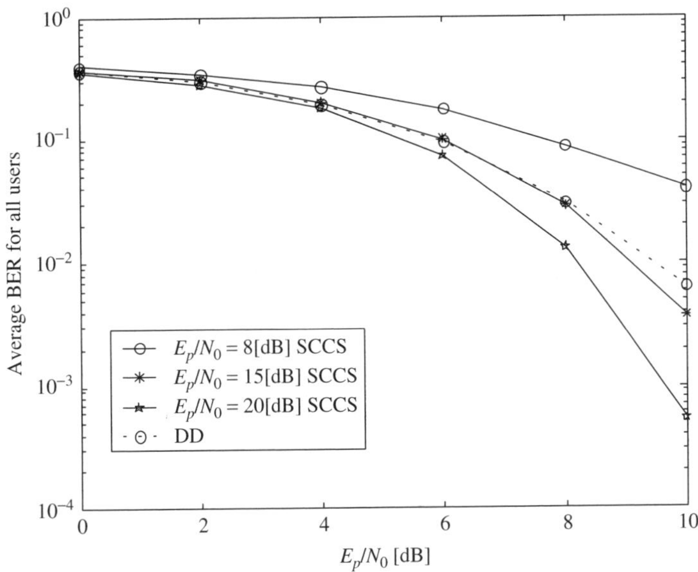

Figure 2.60 The average BER over all users versus pilot channel SNR E_p/N_0 for DD and SCCS detectors in a synchronous CDMA system. Multipath number is $L = 3$, multipath pattern is [1, 0.6, 0.6/8], interpath delay is five chips, the local code is $c_1(t)$, the length of Gold codes is $N = 31$ with the generating polynomials and initial loadings being $p_1 = [1, 1, 0, 1, 1, 1]$; $p_2 = [1, 0, 0, 1, 0, 1]$, $v_1 = [0, 0, 0, 0, 0, 1]$; $v_2 = [0, 0, 0, 0, 0, 1]$.

2.5 OSI Reference Model

The open System Interconnection (OSI) network reference model was proposed by the International Standardization Organization (ISO) as an effort to achieve international standardization of various network protocols. The standardization of different network protocols will also help facilitate the design process of all network architecture based on an open system model. The model is called the *ISO OSI* reference model as it deals with the issue of how to interconnect different network components in an open way, that is, the systems that are open for communication with other systems. We usually just call it the OSI model.

As per its initial definition, the OSI model consists of seven different layers. The principle ideas that were applied to propose the OSI model are as follows:

- A layer should be made where distinct level of abstraction is necessary.

- Each layer should perform a well-defined function.

- The functions of each layer should be defined with enough emphasis on defining internationally standardized network protocols.

- The boundaries between different layers should be chosen to minimize the information exchange across the boundaries.

- The number of layers should be large enough to fit different network protocols, and small enough that the architecture does not become unwieldy.

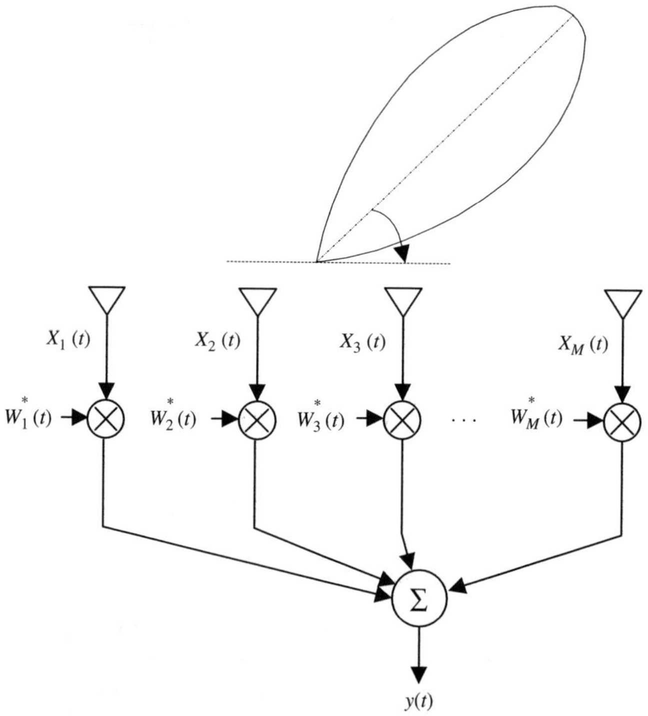

Figure 2.61   A conceptual block diagram of an antenna-array system with M weighted antenna elements.

It is to be noted that the OSI model itself is not a network architecture because it does not specify the exact services and protocols to be used in different layers. The OSI model just gives what each layer should do.

Physical layer

The lowest layer in the OSI model is the physical layer, which is concerned with the transmission of raw bit streams over a physical communication channel regardless of the types of data. The fundamental concern in the physical layer is that when a bit is sent from one side, it is received by the other side as a 1 bit, not a 0 bit. This means that no error should occur. Many specific questions should be answered when designing a physical layer architecture, such as how many volts should be used to represent a 1 and how many volts for a 0; how many microseconds a bit should last; whether transmission may proceed simultaneously in both directions; how the initial connection can be established; and how it will be torn down when communication ends. These design problems in the physical layer have a lot to do with the mechanical, electrical, and procedural interfaces, and the physical transmission medium.

Data link layer

The major function of the data link layer, which is located right above the physical layer, is to take a raw transmission facility and transform it into a stream that appears free of transmission errors to the network layer, which is located right above the data link layer. The data link layer accomplishes this task by breaking the initial long data stream into data frames (typically a few hundred bytes), transmitting the frames sequentially and processing the acknowledgement frames sent back by the receiver. As the physical layer merely sends a stream of bits without any regard to its meaning or structure, it is up to the data link layer to establish and recognize the boundaries of data frames. This is done by attaching a special header and trailer to the beginning and the end of each data frame.

The bits sent through the physical channels can be corrupted by interference and noise, and thus errors are inevitable. It is up to the data link layer to solve the problems caused by damaged, lost, and duplicate frames.

Another function of the data link layer is to keep a fast transmitter from drowning a slow receiver. It can be done by using some traffic regulation mechanism to let the transmitter know the buffer status of the receiver.

Network layer

The network layer is responsible for controlling the operation of the subnet. One of the most important issues is to determine how packets are routed from the source to the destination. Routes can be based on static routing tables or adjusted adaptively according to the network situation.

The network layer should also take care of congestion problems, which will be created if too many packets are present in the subnet at the same time, and they will get in each other's way, forming bottlenecks. The network layer will also count how many packets or bits are sent by each customer to produce billing information.

If a packet has to travel from one network to another to reach its destination, many problems may arise. The addressing used by the other networks may be different from the initiating one. The other networks may not accept the packet simply because it is too large, and so on. It is up to the network layer to solve all those problems.

Transport layer

The transport layer is the fourth layer in the OSI model. The basic function of the transport layer is to accept data from the session layer (the layer right above the transport layer), and to split it up into smaller units if necessary, to pass them to the network layer, and to ensure that all the pieces arrive correctly at the other end of network.

Normally, the transport layer creates a different network connection for each transport connection required by the session layer. If the transport connection requires a high throughput, the transport layer should create multiple network connections, dividing the data among different network connections to improve throughput. On the other hand, if necessary, the transport layer might multiplex several transport connections onto the same network connection to reduce the cost.

The transport layer also decides what type of service is to be provided to the session layer, and ultimately, the users of the network. The most commonly used type of transport connection is an error-free point-to-point channel that delivers messages in the order in which they were sent. The reordering is necessary when the message is broadcasted to multiple destinations.

The transport layer is also called *source-to-destination* or *end-to-end layer*. In other words, a program on the source machine carries on a conversation with a similar program on the destination machine, using the message headers and control messages.

Session layer

The session layer is located right above the transport layer and below the presentation layer. The session layer allows users on different machines to establish sessions between them. One of the services of the session layer is to manage dialogue control. Sessions can allow traffic to go in both directions at the same time, or in only one direction at a time.

A related session service is token management. In some protocol, it is not allowed that both sides do the same operation at the same time. To manage these activities, the session layer provides tokens that can be exchanged. Only the side holding the token may perform the operation.

Another function of the session layer is synchronization. This becomes very important especially when a long data transfer happens. The session layer will provide a method for inserting checkpoints into the long data stream, so that after a crash only the data after the last checkpoint have to be repeated.

Presentation layer

The presentation layer performs certain functions that are requested very often to make it necessary to find a general solution for them, rather than letting every user solve the problems. Different from the other layers in the OSI model, the presentation layer deals with the syntax and semantics of the information transferred through the networks, such as the format of character strings, including ASCII[25] and EBCDIC[26] formats.

The presentation layer is also responsible for other aspects of information representation. For instance, data compression can be used to reduce the number of bits that have to be sent and cryptography is frequently required for privacy and authentication.

Application layer

The upmost layer in the OSI model is the application layer, which contains a variety of protocols that are commonly needed. For instance, there are many different types of incompatible terminals that are in use in the world, each with different screen layouts, escape sequences for inserting and deleting text, moving the cursor, and so on.

One way to solve the problem is to define an abstract network virtual terminal for which other programs can be written by editors. To handle each terminal type, a particular software should be written to map the functions of the network virtual terminal onto the real terminal.

Another application layer function is file transfer. Different networks have distinct file naming conventions, different ways to represent text lines, and so forth. The application layer is responsible for handling the conversion between different filing systems.

Figure 2.62 shows how the data transmission happens in the OSI model. The headers are used in the figure to illustrate how a data message goes through different layers subject to different header addition and deletion processes at the sender and the receiver sides.

2.6   Switching Techniques

In order to establish a communication link between a *source* and a *destination* that may span not necessarily only one single hop, there is a need to route the traffic load over the communication link, which can be built on the basis of either wired or wireless medium.

Traffic routing or switching in wireless networks can be a very complex process. In making a telephone call we never realize how complex the switching process is from the time we dial a number to the instant that we hear the voice from the other end. This end-to-end connection should

---

[25]ASCII stands for American Standard Code for Information Interchange. Computers can only understand numbers, so an ASCII code is the numerical representation of a character such as 'a' or '@' or an action of some sort. ASCII was developed a long time ago and now the nonprinting characters are rarely used for their original purpose.

[26]IBM adopted EBCDIC (Extended Binary Coded Decimal Interchange Code) developed for punched cards in the early 1960s and still uses it on mainframes today. It is probably the next most well-known character set due to the proliferation of IBM mainframes.

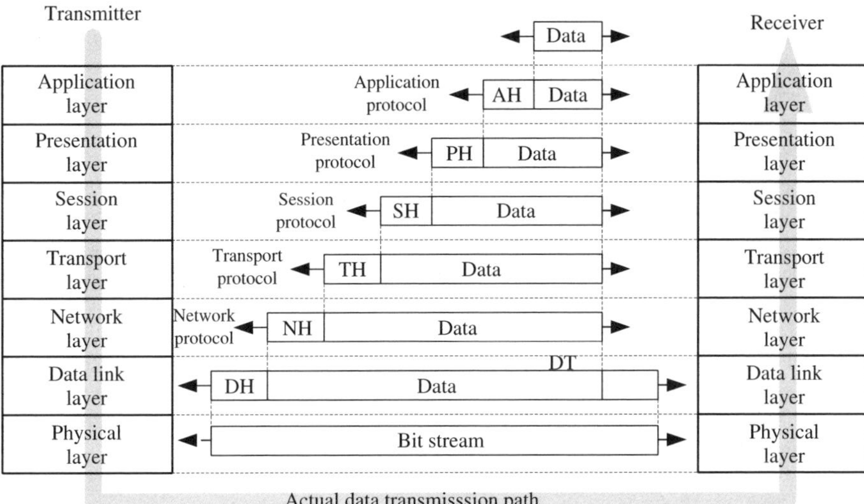

Figure 2.62 Data transmission in the OSI model. Some of the headers may be null.

be done through many intermediate switching offices, each of which will carry out the switching functions automatically with little manual intervention today. The other extreme for traffic routing or switching is sending an e-mail through the Internet. In this case, an entire e-mail message should be first encoded into separate short data groups called *packets*, each of which will contain both the source and destination addresses in its *header* or *trailer*. All of these packets will be sent out in a certain sequential order, one by one. The Internet is just like a huge wired web in the world and has a huge number of nodes in it. Once all those encoded packets belonging to the same e-mail message are sent into the Internet, they will be subject to relays many times by intermediate nodes according to the address information given in the packets until they reach the final destination, but possibly with a wrong sequence due to unexpected random delays for different packets.

As far as a wireless network is concerned, the amount of traffic capacity required in a wireless network is highly dependent upon the type of traffic carried. For instance, a subscriber's telephone call (i.e., voice traffic) requires dedicated network access to provide end-to-end real-time communications, whereas control and signaling traffic may be bursty in nature and may be able to share network resources with other bursty users. Alternatively, some traffic may have an urgent delivery schedule while some may have no need to be sent in real time. The type of traffic carried by a network determines the routing services, protocols, and call-handling techniques that must be employed. Two general routing services are provided by networks. These are connection-oriented services (i.e., virtual circuit routing), and connectionless services (i.e., datagram services). In connection-oriented routing (also called *circuit switching*), the communication path between the message source and destination has to be established for the entire duration of the message, and a call setup procedure is required to dedicate network resources to both the called and calling parties. This is the case with normal telephone calls, as mentioned earlier. Since the path through the network is fixed, the traffic in connection-oriented routing arrives at the receiver in exactly the same order as it was transmitted. A connection-oriented service relies heavily on error control coding to provide data protection in case the network connection becomes noisy. If coding is not sufficient to protect the traffic, the communication is broken, and the entire message must be retransmitted from the beginning. Connectionless routing (also called *packet switching*), on the other hand, does not establish a firm connection for the traffic before the transmission starts, and instead relies on packet-based transmissions. Usually a large number of packets form a message and each individual packet in a connectionless service should be routed separately. Successive packets within the same message might travel completely different routes and encounter widely varying delays throughout the network. Packets sent using connectionless routing do not necessarily arrive at the destination in the order of transmission and must be reordered at the receiver. Because packets take different routes in a connectionless service, some packets may be lost because of network or link failure. However, others may get through with sufficient redundancy to enable the entire message to be recreated at the receiver. Thus, connectionless routing often avoids having to retransmit an entire message, but requires more overhead information for each packet due to some embedded extra information (such as addresses of destination and source, etc.) in addition to the data itself. Typical packet overhead information includes the packet source address, the destination address, the routing information, and information needed to properly order packets at the receiver. In a connectionless service, a call setup procedure is not required at the beginning of a service request, and each message is treated independently by the network.

A detailed description on both circuit switching and packet switching networks will be given in the following subsections.

2.6.1 Circuit Switching Networks

A simple application example of the circuit switching technique is the first generation cellular systems (such as AMPS [316], European Total Access Communication System (ETACS), Nordic Mobile Phone (NMP), etc.), which provide connection-oriented or circuit switching services for each voice subscriber. Voice channels are dedicated for users at a serving base station, and a certain amount of network resource is dedicated to the voice traffic upon initiation of a call. That is, the *mobile switching center* (MSC) dedicates a voice channel connection between the BS and the *public switched telephone network* (PSTN) for the duration of a cellular telephone call. Furthermore, a call initiation sequence is required to connect the called and calling parties on a cellular system. When used in conjunction with radio channels, connection-oriented services are provided by a technique called *circuit switching*, since a physical radio channel is dedicated (or switched in to use) for a particular two-way traffic between the mobile user and the MSC, and the PSTN dedicates a voice circuit between the MSC and the end-user. As calls are initiated and completed, different radio circuits and dedicated PSTN voice circuits are switched in and out to handle the traffic. Circuit switching establishes a dedicated connection (a radio channel between the BS and a mobile, and a dedicated phone line between the MSC and the PSTN) for the entire duration of a call. Despite the fact that a mobile user may be handed over to different BSs, there is always a dedicated radio channel to provide service to the user, and the MSC dedicates a fixed, full-duplex phone connection to the PSTN.

The applications of the circuit switching techniques in old fixed or mobile voice services were based on the then available technological advancement. It worked in a simple way but was far less efficient in terms of the channel resource utilization. The arrival of data communication services motivated the development of the connectionless or packet switching operation mode. As far as data communication is concerned, clearly, circuit switching operation mode is only well suited for the continuous transmission of very long sessions of data transmission. However, if data communication happens in a bursty fashion, circuit switching is never a good solution. Modern wireless data networks, such as WLANs and so on, are not well supported by circuit switching either, due to their short, bursty transmissions, which are often followed by quiet periods. Very often, the time required to establish an end-to-end circuit connection will exceed the duration of the data transmission, resulting in a very inefficient use of precious wireless channel resource. Circuit switching is best suited for dedicated voice-only traffic, or for instances where data is continuously sent over long periods of time.

Table 2.7 shows all popular circuit switching methods in the course of evolution of the circuit switching techniques since 1878.

Table 2.7   The evolutional history of circuit switching techniques.

	Operation	Switching method	Control type	Network type
1878	Manual	Space/analog	Human	Plug/cord/jack
1892	Electromechanical	Space/analog	Distributed stage-by-stage	Stepping switch train
1918	Cross-bar electromechanical	Space/analog	Common control	X-bar switch
1960	ESS-1st generation semielectronic	Space/analog	Common control	Reed switch
1972	ESS-2nd generation semielectronic	Space/analog	Stored program control	Reed switch
1976-	ESS-3rd generation electronic	Time/digital	Stored program common control	Pulse code modulation

2.6.2   Packet Switching Networks

Packet switching refers to the protocols in which messages are broken up into small bursts or packets before they are sent. Each packet is transmitted individually across the networks, and they may even follow different routes to the final destination. Thus, each packet has a header information about the source, destination, packet numbering, and so on. At the destination the packets are reassembled into the original message. Most modern Wide Area Network (WAN) protocols,[27] such as TCP/IP, X.25, and Frame Relay, are based on packet switching technologies.

The main difference between *Packet switching* and *Circuit Switching* lies in the fact that the communication lines are dedicated to passing messages from the source to the destination. In Packet Switching, different messages (and even different packets) can pass through different routes, and when there is a "quiet time" in the communication between the source and the destination, the lines can be used by other routers.

Circuit Switching is ideal when data must be transmitted quickly, must arrive in sequencing order and at a constant arrival rate. Thus, when transmitting real-time data, such as audio and video, Circuit Switching networks will be used. Packet Switching is more efficient and robust for data that is bursty in nature, and can withstand delays in transmission, such as email messages, and Web pages.

Two basic approaches are common to Packet Switching, that is, *Virtual Circuit Packet Switching* and *Datagram Switching*.

In *Virtual Circuit Packet Switching Networks*, an initial setup phase is used to establish a route between the intermediate nodes for all the packets passed during the session between the two end nodes. In each intermediate node, an entry is registered in a table to indicate the route for the connection that has been set up. Thus, packets passed through this route, can have short headers, containing only a *virtual circuit identifier* (VCI), and not their destination. Each intermediate node passes the packets according to the information that was stored in it in the setup phase. In this way, packets arrive at the destination in the correct sequence, and it is guaranteed that essentially there will not be errors. This approach is slower than Circuit Switching, since different virtual circuits may compete over the same resources, and an initial setup phase is needed to initiate the circuit. As in Circuit Switching, if an intermediate node fails, all virtual circuits that pass through it are lost. The most common forms of *Virtual Circuit* networks are X.25 and Frame Relay, which are commonly used for *public data networks* (PDNs).

---

[27]WANs usually cover much large areas, such as the whole region or country, and so on, when compared with wireless metropolitan area networks (WMANs), WLANs, wireless personal area networks (WPANs), and so on.

Datagram Packet Switching Networks, on the other hand, is an approach that uses a different, more dynamic scheme, to determine the route through the network links. Each packet is treated as an independent entity, and its header contains full information about the destination of the packet. The intermediate nodes examine the header of the packet, and decide to which node the packet should be sent so that it will reach its destination. To make a good routing decision, two factors should be taken into account: (1) The shortest way to pass the packet to its destination. The protocols such as RIP/OSPF are used to determine the shortest path to the destination. (2) Finding a free node to pass the packet to. In this way, bottlenecks are eliminated, since packets can reach the destination in alternate routes. Thus, in this scheme, the packets do not follow a preestablished route, and the intermediate nodes (the routers) do not have predefined knowledge of the routes that the packets should be passed through. Packets can follow different routes to the destination, and delivery is not guaranteed (although packets usually do follow the same route, and are reliably sent). Because of the nature of this method, the packets can reach the destination in a different order than they were sent, thus they must be sorted at the destination to form the original message. This approach is time consuming since every router has to decide where to send each packet. The main implementation of the Datagram Switching network is the Internet, which uses the IP network protocol.

As mentioned earlier, packet switching is for providing connectionless services exploiting the fact that dedicated resources are not required for message transmission. Packet switching (also called *virtual switching*) is the most common technique used to implement connectionless services and allows a large number of data users to remain virtually connected to the same physical channel in the network. Since all users may access the network randomly and at will (as discussed in Section 2.3.4), call setup procedures are not needed to dedicate specific circuits when a particular user needs to send data. Packet switching breaks each message into smaller data units for transmission and recovery. When a message is broken into packets or bursts, a certain amount of control information is added to each packet or burst to provide source and destination identification, as well as error recovery provisions.

Figure 2.63 illustrates the sequential format of a packet transmission. The packet consists of header information, user data, and a trailer. The header specifies the beginning of a new packet and contains the source address, destination address, packet sequence number, and other routing and billing information. The user data contains information that is generally protected with error control coding. The trailer contains a CRC that is used for error detection at the receiver.

Figure 2.64 shows the field structure of a transmitted packet, which typically consists of five fields: *flag bits, address field, control field, information field*, and *frame check sequence field*. The flag bits are specific (or reserved) bits that indicate the beginning and end of each packet. The address field contains the source and the destination addresses for transmitting messages and for receiving acknowledgments. The control field defines functions such as transfer of acknowledgments, *automatic repeat requests* (ARQ), and packet sequencing order. The information field contains the user data and may have variable length. The final field is the frame check sequence field or the CRC that is used for error detection of the packet.

In contrast to circuit switching, packet switching (also called *virtual switching* or connectionless switching) provides excellent channel efficiency for bursty data transmissions of short packets. An advantage of packet-switched data is that the channel is utilized only when sending or receiving bursts of information. This benefit is valuable in the case of mobile services where the available bandwidth is limited. The packet radio approach (as discussed in Section 2.3.4) supports intelligent protocols

Header	Data	Trailer

Figure 2.63 A generic data packet format, in which three portions are included, that is, packet header, user data, and packet trailer.

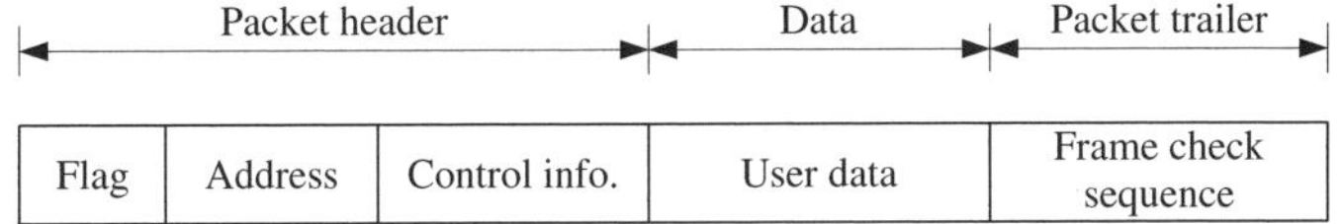

Figure 2.64 Illustration of different fields in a generic data packet format.

for data flow control and retransmission, which can provide highly reliable data transfer in degraded channel conditions.

2.7 IP-Based Networking

It can never be an exaggeration to say that the Internet is one of the most important inventions in human history. The penetration of the Internet in our daily life today can be seen anywhere and anytime, from schools to shops, from developed countries to many developing countries, from commercial service sectors to military operations. The Internet is just like a huge information super-highway, spanning a global *World Wide Web* (WWW) to deliver various *information-on-demand* services to every home around the world. It has exerted a great impact on the human-being's lifestyle today. Without the Internet, the world will be totally different from what it is today.

The explosive demand on the Internet service in the world is also fueled by another important technological advancement in the last twenty years, wireless technologies, which help people to remove the wired barriers to get connected with the outside world. With the help of wireless technologies, a true communication revolution is on its way: the information is at your finger-tips and is there on demand no matter where you stay and how you move. The marriage of the Internet and the wireless technologies yields the *IP-based wireless networking*.

As a more in-depth technical discussion on IP-based wireless networking will be given in Chapter 5, we would like to discuss only the past and present of the Internet technology for both wired and wireless applications, in general. It is very interesting for us to first take a look at the major milestones in the development of the Internet in the history, as illustrated in Table 2.8, from which it is of particular significance for us to note that it took only about 16 years for the Internet node number to increase from 1000 to 35 million. Here, an Internet node means either a FTP site or web-based server connected to the Internet to provide data or information retrieval services. The most recent survey indicated that the e-mail services provided from the Internet has become one of the two most important communication means in our daily life. The other one is still ordinary PSTN telephone services. The survey has also shown that the PSTN telephone services will soon be replaced by *voice over IP* (VoIP) telephone services, which are another important technology developed on the basis of the Internet applications.

Contrary to public perception, the Internet is not one huge cable that links all the major cities of different countries in the world. Instead, it is a mesh-like network, like a web. As mentioned earlier about the packet switching network in Subsection 2.6.2, the Internet links millions of computers via an array of network equipments called *routing devices* and *communications protocols*, in particular, *Transmission Control Protocol/Internet Protocol* (TCP/IP). It provides information delivery and retrieval services to governments, universities, companies, and individuals. In fact, the estimates on the exact number of computers connected to the Internet vary widely. A recent survey of unique information on the World Wide Web (WWW) estimates over one billion pages. Moreover, once connected to the Internet, most of the information can be accessed free of charge. For less than about US$20 per month (as a typical rate available to the US residents), access to this huge information network is readily available. As a matter of fact, several *Internet service providers* (ISPs) have been offering

Table 2.8   The milestones for the development of Internet.

Time	Milestone
1957	ARPA[a] created
1960	Research activities on packet switching
1968	BBN[b] wins ARPANET[c] contract
1968–1969	BBN develops and deploys ARPANET
1970	5 ARPANET sites active
1971	12 ARPANET, sites active
1972	E-mail created and becomes most used application on ARPANET
1974	TCP/IP[d] developed
1975	ARPANET transferred to Defense Comm. Agency
1982	TCP/IP adopted as official protocols on APARNET
1983	MILNET[e] created, Internet created
1984	Over 1000 nodes on Internet
1984	NSFNET[f] created
1985–1990	Commercial networks proliferate
1986	Over 5000 nodes on Internet
1988	Over 60,000 nodes on Internet
1990	ANSNET[g] created upgraded from NSFNET
1990	ARPANET officially retired
1991	Over 600,000 nodes on Internet
1992	Over 1 M nodes on Internet
1993	Over 2 M nodes on Internet
1995	NAPS[h] initiated
1995	ANSNET was sold to America Online
1996	Over 9 M nodes on Internet
1997	Over 16 M nodes on Internet
2000	Over 35 M nodes on Internet

[a] ARPA: Advanced Research Projects Agency, USA
[b] BBN: Bolt, Baranak, and Newman, a firm based in Massachusetts
[c] ARPANET: Advanced Research Projects Agency Network
[d] TCP/IP: Transmission Control Protocol/Internet Protocol
[e] MILNET: Military Network, USA
[f] NSFNET: National Science Foundation Network, USA
[g] ANSNET: Advanced Networks and Services (ANS) Network upgraded from NSFNET
[h] NAPS: Network access points

free Internet access. Only about one million of these people represent paid subscribers. At some moment in the past two years (from 2000 to 2001), the United States passed an important threshold: A majority of households now have access to e-mail services. Free or inexpensive Internet service, however, is not limited to the United States. It is common in the United Kingdom and Brazil as well. Internet access in areas of high competitiveness is becoming less expensive, but this may be changing. There is a retrenching of free ISPs because advertising dollars, intended to subsidize the free ISP service to subscribers, is not proving to be the financial panacea originally promised. It remains to be seen what will transpire in this arena. In some rural areas even in the United States, where access is limited to one ISP, the cost may exceed US$25 per month. In other areas, residents must also pay long-distance charges in addition to the ISP fees because they are too far from a local AP (the local number established by the ISP for a user to dial into in order to use their service) associated with any ISP. The growth of the Internet has been phenomenal as shown in Table 2.8. In Brazil, Internet use is increasing at over 80% per year. Specifically, the Internet is a network of over 30,000 networks. It had 1.7 million computers interconnected in the summer of 1993, with an estimated 30 million users worldwide. The rate at which new computers are being added to the Internet now exceeds one per second. Some interesting data exist on this phenomenon. As of 1995, Norway, for example, had almost five Internet hosts per every 1000 persons. The United States was second with 3.8 per 1000. China only trails the United States in the actual number of people with Internet access at home, according to a recent study (China is in the second position). According to this study, the number of Chinese with Internet access at home is up to more than 56 million. A higher percentage of people in many developed nations can access the Internet at home, but China's huge population placed it above Japan, Germany, and Great Britain in terms of overall at-home Internet access. The United States leads the world with more than 166 million Americans able to access the Internet from home. Extrapolating from previous studies would mean that there will be more than two billion people accessing the Internet by 2006. Many areas of the world depend on access to information via the Internet. In Japan, for example, more than two million people use their wireless handset to send e-mails and to access the Internet. In Finland, where 90% of the teenagers have wireless phones, a wireless user can purchase everything from soft drinks to car washes to bubble gum. This market penetration is much more extensive than in the United States, where wireless access to the Internet is still a big industry. Moreover, this phenomenon is not unique to Finland. There are over 300 million wireless handsets in use throughout the world, and this number is expected to grow to one billion by 2005.

All of the above data have clearly shown that information delivery and retrieval services provided by the Internet or IP-based networking has become an indispensable part of people's life today. Besides, it seems that this trend will not change in a foreseeable time ahead.

3

3G Mobile Cellular Technologies

The great success of mobile cellular communication systems is probably one of the most celebrated events in the history of the telecommunication industry over the past 100 years. The convenience of mobile cellular telephony has finally made a dream come true: people can get in touch with anyone else on earth at any time and at any place. The modern microelectronics industry has made it possible to produce a cell phone is small enough to be carried in a pocket, to put on a necklace, or even to be worn as a watch. In fact, it is hard to imagine the outcome if all mobile cellular services around us disappeared.

Mobile cellular systems have been developed over three key generations. The services of the first generation (1G) systems began when analogue technology-based mobile telephony was first introduced in Chicago City in the 1980s. This 1G analogue mobile cellular system in the United States is also called *Advanced Mobile Phone System* (AMPS) system [316], which operated in 800 MHz and used 30 kHz bandwidth under the *frequency division multiple access* (FDMA) scheme. There are several other 1G systems which were in service in other countries/regions, such as *Total Access Communications System* (TACS) in the United Kingdom, the Nordic Mobile Telephone (NMP) in the Scandinavian countries, and so on. The common characteristic features for all 1G systems around the world can be summarized as follows. First, they operated in the FDMA scheme, in which all voice channels are separated by different frequency carriers with a relatively narrow bandwidth (usually about 30–50 kHz). Second, they were based on analogue transmission and processing technologies. Third, each system only covered a country or a relatively small region. Finally, the capacity of all those 1G mobile cellular systems was small due to their low bandwidth efficiency. Different 1G mobile cellular systems or standards are listed in Table 3.1, and brief descriptions of all 1G mobile cellular systems or standards are given in Table 3.2.

The characteristic features of the 1G mobile cellular systems have motivated the research and development of the second generation (2G) mobile cellular standards, mainly initiated by two different groups, one in the United States and the other in Europe.

The 2G systems proposed by the United States took two different approaches, one leading to a new Time Division Multiple Access (TDMA) based technology, digital AMPS (D-AMPS) standard (the IS-54B and its enhanced version, IS-136 standard); and the other using the Code Division Multiple Access (CDMA) technology, the IS-95 standard [317–344]. The D-AMPS was designed to be compatible with the earlier analog AMPS technology, which was widely deployed in the United States. TDMA is used as an enhancement to the AMPS network by the use of dual-band AMPS/TDMA (or D-AMPS) phones. Use of these phones gives the widespread coverage of the AMPS networks along with some advantages of digital systems in the areas where TDMA networks are available. On

Next Generation Wireless Systems and Networks   Hsiao-Hwa Chen and Mohsen Guizani
© 2006 John Wiley & Sons, Ltd

Table 3.1   The 1G mobile cellular systems or standards worldwide

System or standard	Service start date	Country of origin or region it operated
AMPS	1979 trial, 1983 commercial	United States, then worldwide
AURORA-400	1983	Alberta, Canada
C-Netz and C-Netz C-450	Begins in 81, upgraded in 1988	Germany, Austria, Portugal, South Africa
Comvik	August, 1981	Sweden
ETACS	1987	United Kingdom, now worldwide
JTACS	June, 1991	Japan
N-AMPS (Narrowband Advanced Mobile Phone Service)	1993	United States, Israel
NMT 450	1981	Sweden, Norway, Denmark
NMT 900	1986	Finland, Oman; NMT exists in 30 countries
NTACS/JTACS	June, 1991	Japan
NTT	December, 1979	Japan
NTT Hi Cap	December, 1988	Japan
RadioCom (RadioCom2000)	November, 1985	France
RTMS (Radio Telephone Mobile System)	September, 1985	Italy
TACS (Total Access Communications System)	1985	United Kingdom, Italy, Spain, Austria, Ireland

the other hand, the IS-95 standard proposed by Qualcomm Inc. is based on CDMA technology and offers numerous new capabilities that the former multiple access technologies, such as TDMA and FDMA, can never possess.

At almost the same time, Europe proposed a pan-European 2G mobile phone standard, *Group Special Mobile* which was later called *Global System for Mobile Communications* (GSM), in order to bring the same standard to all major European countries under the strong leadership of *European Telecommunications Standards Institute* (ETSI), an agency based in Sophia Antipolis, in the south of France. This was first undertaken by the European Commission (EC) and later by the European Union (EU). The GSM system uses TDMA technology and can operate in three different bands (800 MHz, 900 MHz and 1.8 GHz) around the world. The very different marketing approaches used by Europe has made the GSM system the single most popular digital mobile cellular system in the world today, and its worldwide subscribers have reached nearly 0.3 billion, according to the latest statistics. Without a doubt, the GSM is the most popular 2G mobile cellular system if measured by its market share in the world.

In addition to the IS-54B, IS-136 and GSM standards, there are many other 2G wireless and cordless digital telephone systems that have been proposed and adopted by different countries and regions in the world. Their major characteristics and brief descriptions have been given in Table 3.3.

The evolution of mobile cellular telephone systems from 1- to 2G has clearly indicated that the Americans were very successful in the development and application of the 1G analog system, AMPS system, not only in North America but also in many other regions in the world. Although the United States was still technologically dominant in 2G mobile cellular systems, Europe was the final winner to grasp the largest mobile cellular market share in the world, that is manifested in the number

Table 3.2   Main Features of the 1G mobile cellular systems or standards worldwide

AMPS	Advanced Mobile Phone System. Developed by Bell Labs in the 1970s and first used commercially in the United States in 1983. It operates in the 800- and 1900 MHz band in the United States and is the most widely distributed analog cellular standard. Close to being defunct.
C-450	Installed in South Africa during the 1980s. Almost like C-Netz. Now known as *Motorphone System 512* and run by Vodacom SA.
C-Netz	Older cellular technology found mainly in Germany and Austria. Operates at 450 MHz. May no longer be working.
Comvik	Launched in Sweden in August 1981 by the Comvik network, lasted until March 31, 1996.
N-AMPS	Narrowband Advanced Mobile Phone System. Developed by Motorola as an interim technology between Analog and digital. It has some 3 times greater capacity than AMPS and operates in the 800 MHz range. Now defunct.
NMT450	Nordic Mobile Telephones/450. Developed specially by Ericsson and Nokia to service the rugged terrain that characterises the Nordic countries. The first multinational celllullar network. Operates at 450 MHz.
NMT900	Nordic Mobile Telephones/900. The 900 MHz upgrade to NMT 450 developed by the Nordic countries to accommodate higher capacities and handheld portables.
NMT-F	French version of NMT900.
NTT	Nippon Telegraph and Telephone. The old Japanese Analog standard. A high-capacity version is called *HICAP*.
RC2000	Radiocom 2000. French system launched November 1985.
TACS	Total Access Communications System. Developed by Motorola. and is similar to AMPS. It was first used in the United Kingdom in 1985, although in Japan it is called *JTAC*. It operates in the 900 MHz frequency range.

Table 3.3   Brief description of the 2G mobile cellular systems or standards worldwide

A1-Net	Austrian Name for GSM 900 networks.
CDMA	Code Division Multiple Access. IS-95. Developed by Qualcomm, characterized by high capacity and small cell radius. It uses the same frequency bands as AMPS and supports AMPS operation, employing spread-spectrum technology and a special coding scheme. It was adopted by the Telecommunications Industry Association (TIA) in 1993. The first CDMA-based networks are now operational.
cdmaOne	Wide ranging wireless specification involving IS-95, IS-96, IS-98, IS-99, IS-634 and IS-41. AT&T, Motorola, Lucent, ALPS, GSIC, Prime Co, Qualcomm, Samsung, Sony, United States West, Sprint, Bell Atlantic, and Time Warner are sponsors.
CDPD	Cellular digital packet data. Overlays existing cellular networks to provide faster data transfer. Bell Atlantic Mobile offers it in the New York metropolitan area, New Jersey, Connecticut, Massachusetts, Pittsburgh, the greater Philadelphia area, the Washington and Baltimore metropolitan areas, and North and South Carolina.
CT-2	A second generation digital cordless telephone standard. CT2 has 40 carriers × 1 duplex bearer per carrier = 40 voice channels. Supposedly withdrawn in Canada.
CT-3	A third generation digital cordless telephone, which is very similar and a precursor to DECT.

(*continued overleaf*)

Table 3.3   (*continued*)

D-AMPS	Digital AMPS. Designed to use existing channels more efficiently, D-AMPS (IS-136) employs the same 30 kHz channel spacing and frequency bands (824–849 and 869–894 MHz) as AMPS. By using TDMA instead of FDMA, IS-136 increases the number of users from 1 to 3 per channel. An AMPS/D-AMPS infrastructure can support either an Analog AMPS phone or digital AMPS phones. (The Federal Communications Commission mandated that digital cellular in the United States must act in a dual-mode capacity with analog). Operates in the 800- and 1900 MHz bands.
DCS	DCS also stands for Digital Communications Systems, another word for American GSM.
DECT	Digital European Cordless Telephony. This started off as Ericsson's CT-3, but developed into the ETSI Digital European Cordless Standard. It is intended to be a far more flexible standard than the CT2 standard, in that, it has more RF channels (10 RF carriers × 12 duplex bearers per carrier = 120 duplex voice channels). It also has better multimedia performance since 32 kbit/s bearers can be concatenated. Ericsson is developing a dual GSM/DECT handset that will be piloted by Deutsche Telekom.
E-Netz	The German name for GSM 1800 networks.
GSM	Global System for Mobile Communications. The first European digital standard, developed to establish cellular compatibility throughout Europe. Its success has spread to all parts of the world and over 80 GSM networks are now operational. It operates at 900- and 1800 MHz in many parts of Europe and in England. Works at 1900 MHz in some parts of the United States. TDMA-based. See below.
PCS	Personal Communications Service. The PCS frequency band in the United States is 1850- to 1990 MHz, encompassing a wide range of new digital cellular standards like N-CDMA and GSM 1900. Singleband GSM 900 phones cannot be used on PCS networks. PCS networks operate throughout the United States.
IS-54	TDMA-based technology used by the D-AMPS system at 800 MHz.
IS-95	CDMA-based technology used at 800 MHz.
IS-136	TDMA-based technology offered at both 800- and 1800 MHz. Should be referred to as cellular. AT&T's choice to offer PCS like services.
JS-008	CDMA-based standard for 1900 MHz.
Nextel	Direct connect service offers point to point communication as well as a TDMA-based cellular telephone in a single handset.
PDC	Personal Digital Cellular is a TDMA-based Japanese standard operating in the 800- and 1500 MHz bands.
PHS	Personal Handy System. A Japanese-centric system that offers high-speed data services and good voice clarity.
TDMA	Time Division Multiple Access. The first U.S. digital standard to be developed. It was adopted by the TIA in 1992. The first TDMA commercial system began in 1993. Called *IS-54* at first and now known as *IS-136*.
TETRA	Trans European Trunked Radio Systems, designed to support both voice and data. Very new. Mostly used in trucks. Allows roaming. Not yet fully implemented.

of worldwide GSM system subscribers. Under these circumstances, Europe, which was enjoying a great amount of benefit from technological transfer as well as 2G-related equipments sales (even today), was not in a hurry to push for the third generation (3G) mobile communication systems. In the meantime, Japan captured little market share in the worldwide 1- to 2G mobile cellular business, partly due to its long time conservative telecommunication policies and regulations and partly to its unique 2G mobile cellular systems, which were incompatible to the others. Japan's urgent desire to join Europe in developing 3G mobile communication systems was also motivated by its technical competition with Korea, which acquired key CDMA technologies from Qualcomm in the mid 1990s and was able to develop and manufacture its own CDMA-based mobile telephone systems in the late 1990s. Seen from Japan's point of view it seemed to be completely unacceptable if it were to lag behind Korea in CDMA-based mobile communication technologies. Therefore, it has to be acknowledged that Japan played an important role in the development of the current 3G mobile communication standards, especially the WCDMA standard proposed by the Association of Radio Industries and Business (ARIB) of Japan and the Universal Mobile Telephone System (UMTS) standard proposed by ETSI.

On the other hand, the initiation of the 3G mobile communication systems should also be attributed to the active involvement of the International Telecommunications Union (ITU), a special technical agency under the United Nations (UN) in the year 2000. The initiative program was

Table 3.4   The comparison of major physical layer parameters of TD-SCDMA, WCDMA, and CDMA2000 standards

Parameters	CDMA2000	WCDMA	TD-SCDMA
Multiple access	DS-CDMA/MC-CDMA	DS-CDMA	TDMA/DS-CDMA
CLPCF	800 Hz	1600 Hz	200 Hz
PCSS	0.25, 1.5 dB	0.25, 0.5, 1.0 dB	1, 2, 3 dB
Channel coding	Convolutional or Turbo	Convolutional, RS or Turbo	Convolutional or Turbo
Spreading code	DL: Walsh, UL: M-ary Walsh	OVSF	OVSF
VSF	4···256	4···256	1···16
Carrier	2 GHz	2 GHz	2 GHz
Modulation	DL: QPSK, UL: BPSK	DL: QPSK, UL: BPSK	QPSK, 8PSK (2 Mbps)
Bandwidth	1.25*2/3.75*2 MHz	5*2 MHz	1.6 MHz
UL-DL spectrum	paired	paired	unpaired
Chip rate	1.2288/3.6864 Mcps	3.84 Mcps	1.28 Mcps
Frame length	20 ms, 5 ms	10 ms	10 ms
Interleaving periods	5/20/40/80 ms	10/20/40/80 ms	10/20/40/80 ms
Maximum data rate	2.4 Mbps	2 Mbps (low mobility)	2 Mbps
Pilot structure	DL: CCMP, UL: DTMP	DL: DTMP, UL: DTMP	CCMP
Detection	PSBC	PCBC	PSBC
Inter-BS timing	Synchronous	Asynchronous/ Synchronous	Synchronous

CCMP: common channel multiplexing pilot
VSF: Variable spreading factor
PCSS: Power control step size
PSBC: Pilot symbol-based coherent
UL: upper-link

DTMP: dedicated time multiplexing pilot
CLPCF: Close loop power control frequency
DL: downlink
PCBC: Pilot channel-based coherent

called *International Mobile Telephony* (IMT-2000), which aimed to put forward a unified worldwide standard that allowed people the use of only one handphone for any place. Although this original objective of the ITU was not fulfilled, the numerous 3G mobile communication proposals submitted by different countries and regions laid the solid foundation for the development of all current major 3G standards, such as WCDMA [431], UMTS UTRA [425], CDMA2000 [345], TD-SCDMA [432, 433], and so on. The basic technical parameters for three major 3G standards, that is, WCDMA, CDMA2000 and TD-SCDMA, are compared in Table 3.4.

In this chapter we are limited to discussing several major 3G standards, such as CDMA2000, WCDMA, UTRA-FDD, UTRA-TDD, and TD-SCDMA systems. However, it is to be noted that the discussions on 3G wireless communication technologies/standards given in this chapter should not be considered as a complete collection of all those standards. We can only offer a general description for each of them by focusing on the most important aspects of the standards concerned. For more comprehensive coverage of the technical details of those standards, readers may check http://www.3gpp.org/ and http://www.3gpp2.org/, which provide the most authoritative and up-to-date information of all those major standards. Finally, we are particularly thankful to the generosity of 3GPP and 3GPP2 to allow us free access to their important standards documentations.

3.1 CDMA2000

Before we start to talk about the CDMA2000 standard, it will be beneficial to take a brief look at the historical background of worldwide 3G mobile cellular systems. The *third generation wireless* is a term used to describe next generation mobile services, which provide better quality voice and high-speed Internet and multimedia services. In contrast, the 2G systems (such as IS-95, GSM, etc.) were basically oriented toward voice-centric applications. While there are many interpretations of what 3G represents, the only universally accepted definition is the one published by the ITU [308], which defines and approves technical requirements and standards, as well as the use of spectra for 3G systems under the IMT-2000 (International Telecommunication Union-2000) program [308]. The ITU requires that IMT-2000 (3G) networks, among other capabilities, deliver improved system capacity and spectrum efficiency over the 2G systems and support data services at the minimum transmission rates of 144 kbps in mobile (outdoor) and 2 Mbps in fixed (indoor) environments. Based on these requirements, the ITU approved five candidate radio interfaces for IMT-2000 standards in 1999 as a part of the ITU-R M.1457 Recommendation. CDMA2000 is one of the five standards. It is also known by its ITU name, or *IMT-CDMA Multi-carrier*. The five radio interfaces for IMT-2000 standards approved by ITU in 1999 as a part of the ITU-R M.1457 Recommendation include:

- IMT-2000 CDMA Direct Spread (also called *WCDMA-UMTS*)

- IMT-2000 CDMA Multi-carrier (also called *CDMA2000 1x and 1xEV*)

- IMT-2000 CDMA TDD (also called *UTRA-TDD and TD-SCDMA*)

- IMT-2000 TDMA Single Carrier (also called *UWC-136/EDGE*)

- IMT-2000 FDMA/TDMA (also called *DECT*)

All of the five approved 3G standards have become important technical standards developed by different countries or regions. The WCDMA-UMTS was jointly proposed by ARIB, Japan, and ETSI, Europe; the CDMA2000 was proposed by Telecommunications Industry Association (TIA)/EIA of the United States; the UTRA-TDD was proposed by ETSI of Europe and TD-SCDMA was proposed by CATT of China, respectively. UWC-136/EDGE was also proposed by TIA/EIA of the United States, and finally, DECT was proposed by ETSI of Europe.

With this background knowledge about the development of worldwide 3G standards, we would like to come back to the topic of interest in this section, or CDMA2000 [345], which is a direct evolution from cdmaOne technology and provides a set of specifications that offer enhanced voice and data capacity. The CDMA2000 family includes: CDMA2000 1x, CDMA2000 1xEV-DO and CDMA2000 1xEV-DV standards. CDMA2000 is also known as *IS-2000*.

CDMA2000 1x (which once carried many other names, such as CDMA2000 phase 1, 1xRTT, 3G CDMA 1x, etc.) is commonly referred to as *1x* or sometimes as *1x RTT*. CDMA2000 1x is a 3G technology that is commercially available today and is 21 times more efficient than analog cellular networks and 4 times more efficient than TDMA networks. Typical CDMA2000 1x networks provide peak rates of 144 kbps for packet data and an average throughput range of 60–90 kbps on a loaded network. It doubles the voice capacity of cdmaOne networks and delivers peak packet data speeds of 307 kbps in mobile environments.

CDMA2000 1xEV[1] includes the two directive technologies, CDMA2000 1xEV-DO [346][2] and CDMA2000 1xEV-DV [347][3]. In fact, the 1xEV name, which stands for "Evolution" was coined in the standards process. The standard was balloted and adopted by 3GPP2, as C.S0024 [361–367], and by TIA/EIA, as IS-856, in October 2000.

1xEV-DO [346] is the short form of "First Evolution," "Data Optimized." CDMA2000 1xEV-DO provides peak data rates of up to 2.4 Mbps in a standard 1.25 MHz channel used exclusively for data. 1xEV-DO provides an average throughput of over 700 kbps, equivalent to cable modem speeds, and fast enough to support applications such as streaming video and large file downloads, and so on. Future releases will increase to 3.08 Mbps for the forward link.

1xEV-DV [347] is the short form of "1x Evolution, Data and Voice." The CDMA2000 1xEV-DV standard is still under development and is expected to be deployed commercially in late 2005 or early 2006. CDMA2000 1xEV-DV can support voice as well as data. Release C of the CDMA2000 1xEV-DV standard supports a forward link of 3.08 Mbps and a reverse link of 153 kbps. Release D supports a forward link of 3.08 Mbps and a reverse link of approximately 1.0 Mbps.

Both CDMA2000 1xEV-DO and CDMA2000 1xEV-DV are backward compatible with CDMA2000 1x and cdmaOne. The CDMA2000 1xEV (also known as *High Data Rate* (HDR) technology or IS-856) working group was established in 3GPP2, TSG-C in March 2000.

The world's first CDMA2000 1x commercial system was launched by SK Telecom (Korea) in October 2000. Since then, CDMA2000 1x has been deployed in Asia, North and South America, and Europe, and the subscriber base is growing at 700,000 subscribers per day. CDMA2000 1xEV-DO was launched in 2002 by SK Telecom and KT Freetel. There are already more than 12 million 1xEV-DO users today, with some more networks expected to be launched this year (2005) and the next. The commercial success of CDMA2000 has made the IMT-2000 vision a reality. The milestones of development and deployment of CDMA2000 systems in the world is tabulated in Table 3.5. The CDMA2000 subscriber growth history from March 2001 through December 2004 is shown in Figure 3.1.

3.1.1 Operational Advantages

CDMA2000 benefited from the extensive experience acquired through several years of operation of IS-95 or cdmaOne systems. As a result, CDMA2000 is more efficient and robust than its predecessor, IS-95 systems. Supporting both voice and data, the standard was devised and tested in various spectrum bands, including the new IMT-2000 allocations. There is tremendous demand for new services and operators are to provide these to many more subscribers at reasonable prices. The unique features, benefits, and performance of CDMA2000 make it a mature technology for high-voice capacity and high-speed packet data. The fact that CDMA2000 1x has the ability to support both voice and data services

---

[1]"EV" in CDMA2000 1xEV stands for "evolution."
[2]"DO" in CDMA2000 1xEV-DO stands for "data optimized."
[3]"DV" in CDMA2000 1xEV-DV stands for "data and voice."

Table 3.5   The Milestones for CDMA2000 system development and deployment worldwide

Time	Events
Aug 2004	Eurotel Praha (Czech Republic) launches the world's first CDMA2000 1xEV-DO network at 450 MHz (CDMA450)
June 2004	100 million commercial CDMA2000 subscribers
April 2004	CDMA2000 1xEV-DO Revision A approved by Third Generation Partnership Project 2 (3GPP2)
March 2004	5 million commercial CDMA2000 1xEV-DO subscribers
March 2004	CDMA2000 1xEV-DV Revision D approved by Third Generation Partnership Project 2 (3GPP2)
Oct 2003	Verizon Wireless (United States) launches the first CDMA2000 1xEV-DO service in North America
May 2003	1 million commercial CDMA2000 1xEV-DO subscribers
May 2003	50 million commercial CDMA2000 subscribers
April 2003	Vesper (Brazil) launches the first CDMA2000 1xEV-DO network in Latin America
Sept 2002	Pelephone Communications Ltd. (Israel) launches the first CDMA2000 1x network in Africa–Middle East region
June 2002	Third Generation Partnership Project 2 (3GPP2) and Telecommunications Industry Association (TIA) approve CDMA2000 1xEV-DV (Data Voice) for publication
May 2002	10 million commercial CDMA2000 subscribers
Jan 2002	SK Telecom (Korea) launches the world's first CDMA2000 1xEV-DO network
Dec 2001	VIVO (Brazil) launches the first CDMA2000 1x network in Latin America
Dec 2001	Telemobil (Romania) launches the world's first CDMA2000 network at 450 MHz (CDMA450)
Aug 2001	1 million commercial CDMA2000 subscribers
July 2001	Western Wireless (United States) deploys the first CDMA2000 1x network in North America
June 2001	ITU recognizes CDMA2000 1xEV-DO as part of the 3G IMT-2000 standard
Oct 2000	SK Telecom and LG Telecom (South Korea) launch the world's first 3G commercial CDMA2000 networks
Nov 1999	ITU-R Task Group 8/1 endorses CDMA2000 standards (three modes) for IMT-2000
July 1999	Phase 1 CDMA2000 standard complete and approved for publication

on the same carrier makes it cost-effective for wireless operators. Because of its optimized radio technology, CDMA2000 enables operators to invest in fewer cell sites and deploy them faster, allowing the service providers to increase their revenues with faster Return On Investment (ROI). Increased revenues, along with a wider array of services, make CDMA2000 the technology of choice for service providers. The major operational advantages of CDMA2000 are summarized in the subsequent text.

Increased voice capacity

Voice is the major source of traffic and revenue for wireless operators, but packet data will emerge as an important source of incremental revenue in the years to come. CDMA2000 delivers high-voice capacity and packet data throughput using a relatively small amount of spectrum for a lower cost.

CDMA2000 1x supports 35 traffic channels per sector per RF (26 Erlangs/sector/RF) using the EVRC vocoder,[4] which became commercial in 1999.

---

[4]Acronym "EVRC" stands for Enhanced Variable Rate Speech Coder, which performs better than the QCELP-13 vocoder. The EVRC vocoder has been specified in 3GPP2 CS0014.

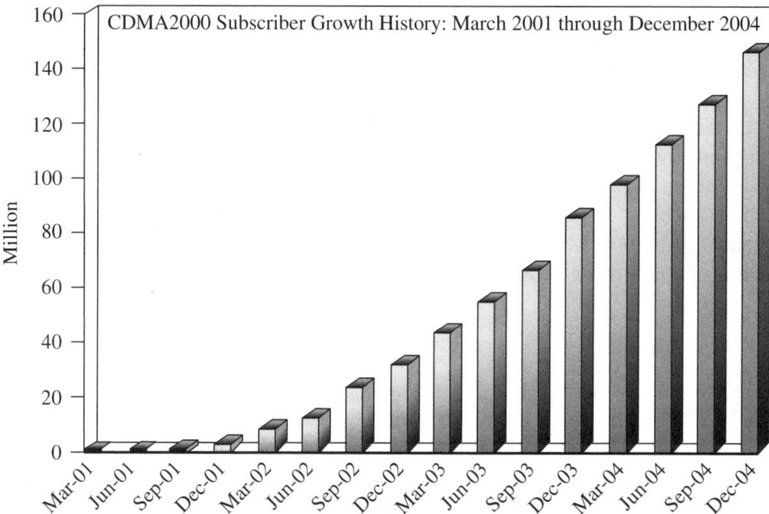

Figure 3.1 The CDMA2000 subscriber growth history from March 2001 through December 2004.

Voice capacity improvement in the forward link is attributed to faster power control, lower code rates (1/4 rate), and transmit diversity (for single path Rayleigh fading). In the reverse link, capacity improvement is primarily due to coherent reverse link.

Higher data throughput

Today's commercial CDMA2000 1x networks (phase 1) support a peak data rate of 153.6 kbps. CDMA2000 1xEV-DO, which has been commercialized in Korea, enables peak rates of up to 2.4 Mbps and CDMA2000 1xEV-DV will be capable of delivering data of 3.09 Mbps.

Frequency band flexibility

CDMA2000 can be deployed in all cellular and PCS spectrums, as shown in Figure 3.15. CDMA2000 networks have already been deployed in the 450 MHz, 800 MHz, 1700 MHz, and 1900 MHz bands; deployments in 2100 MHz and other bands are expected in 2004–2005. CDMA2000 can also be implemented in other frequencies such as 900 MHz, 1800 MHz and 2100 MHz. The high spectral efficiency of CDMA2000 permits high traffic deployments in any 1.25 MHz channel of spectrum.

Increased battery life

CDMA2000 enhances battery performance. Benefits include: (1) Quick paging channel operation with improved reverse link performance; (2) New common channel structure and operation; (3) Reverse link gated transmission; and (4) New Medium Access Control (MAC) states for efficient and ubiquitous idle time operation.

Synchronization

CDMA2000 is synchronized with the Universal Coordinated Time (UCT). The forward link transmission timing of all CDMA2000 base stations worldwide is synchronized within a few microseconds. Base station synchronization can be achieved through different techniques including self-synchronization, radio beep, or through satellite-based systems such as GPS, Galileo, or GLObal NAvigation Satellite System (GLONASS).[5] Reverse link timing is based on the received timing derived from the first multipath component used by the terminal.

There are several benefits to having all the base stations in a network synchronized. First, the common time reference improves the acquisition of channels and handoff procedures since there is no time ambiguity when looking for and adding a new cell in the active set. Second, it also enables the system to operate some of the common channels in soft handoff, which improves the efficiency of the common channel operation. Third, the common network time reference allows the implementation of very efficient *position location* techniques.

Power control

The CDMA2000 basic frame length is 20 ms that is divided into 16 equal power control groups. In addition, CDMA2000 defines a 5 ms frame structure, essential for the support of signaling bursts, as well as 40 and 80 ms frames, which offer additional interleaving depth and diversity gains for data services. Unlike IS-95 where *Fast Closed-Loop Power Control* was applied only to the reverse link, CDMA2000 channels can be power controlled at up to 800 Hz in both the reverse and forward links. The reverse link power control command bits are punctured into the F-FCH or the F-DCCH (explained in later sections) depending on the service configuration. The forward link power control command bits are punctured in the last quarter of the R-PICH power control slot.

In the reverse link, during gated transmission, the power control rate is reduced to 400 or 200 Hz on both links. The reverse link power control subchannel may also be divided into two independent power control streams, either both at 400 bps, or one at 200 bps and the other at 600 bps, allowing for the independent power control of forward link channels.

In addition to the closed-loop power control, the power on the reverse link of CDMA2000 is also controlled through an *Open Loop Power Control* mechanism. This mechanism inverses the slow fading effect due to path loss and shadowing. It also acts as a safety fuse when the fast power control fails. When the forward link is lost, the closed-loop reverse link power control is "freewheeling" and the terminal disruptively interferes with neighboring. In such a case, the open loop reduces the terminal output power and limits the impact to the system. Finally, the Outer Loop Power drives the closed-loop power control to the desired set point based on error statistics that it collects from the forward link or reverse link. Because of the expanded data rate range and various QoS requirements, different users will have different outer loop thresholds; thus, different users will receive different power levels at the base station. In the reverse link, CDMA2000 defines some nominal gain offsets based on various channel frame format and coding schemes. The remaining differences will be corrected by the outer loop itself.

Soft handoff

Even with dedicated channel operation, the terminal keeps searching for new cells as it moves across the network. In addition to the active set, neighbor set, and remaining set, the terminal also maintains a candidate set.

When a terminal is traveling in a network, the pilot from a new BTS (P2) strength exceeds the minimum threshold TADD for its addition in the active set. However, initially, its relative contribution to the total received signal strength is not sufficient and the terminal moves P2 to the candidate set. The decision threshold for adding a new pilot to the active set is defined by a linear function of signal

---

[5]"Galileo" is the global satellite navigation system launched by the European Commission. "GLONASS" is the abbreviation of "GLObal NAvigation Satellite System," which was developed by the former Soviet Union to function in the same way as GPS, developed by the United States.

strength of the total active set. The network defines the slope and cross point of the function. When the strength of P2 is detected to be above the dynamic threshold, the terminal signals this event to the network. The terminal then receives a handoff direction message from the network, requesting the addition of P2 in the active set. The terminal now operates in soft handoff.

The strength of serving BTS (P1) drops below the active set threshold, meaning P1 contribution to the total received signal strength does not justify the cost of transmitting P1. The terminal starts a handoff drop timer. The timer expires and the terminal notifies the network that P1 dropped below the threshold. The terminal receives a handoff message from the network, moving P1 from the active set to the candidate set. P1 strength then drops below TDROP and the terminal starts a handoff drop timer, which expires after a set time. P1 is then moved from the candidate set to the neighbor set. This step-by-step procedure with multiple thresholds and timers ensures that the resource is only used when it is beneficial to the link, and pilots are not constantly added and removed from the various lists, therefore limiting the associated signaling.

In addition to intrasystem and intrafrequency monitoring, the network may direct the terminal to look for base stations on a different frequency or a different system. CDMA2000 provides a framework to the terminal in support of the interfrequency handover measurements consisting of identity and system parameters to be measured. The terminal performs required measurements as allowed by its hardware capability.

In the event of a terminal having dual receiver structure, the measurement can be done in parallel. When a terminal has a single receiver, the channel reception will be interrupted when performing the measurement. In this instance, a certain portion of a frame will be lost during the measurement. To improve the chance of successful decoding, the terminal is allowed to bias the FL power control loop and boost the RL transmit power before performing the measurement. This method increases the energy per information bit and reduces the risk of losing the link in the interval. Based on measurement reports provided by the terminal, the network then decides whether or not to handoff a given terminal to a different frequency system. It does not release the resource until it receives confirmation that the handoff was successful or the timer expires. This enables the terminal to come back in case it could not acquire the new frequency or the new system.

Transmit diversity

Transmit diversity consists of de-multiplexing and modulating data into two orthogonal signals, each of them transmitted from a different antenna at the same frequency. The two orthogonal signals are generated using either Orthogonal Transmit Diversity (OTD) or Space-Time Spreading (STS). The receiver reconstructs the original signal using the diversity signals, thus taking advantage of the additional space and/or frequency diversity.

Another transmission option is directive transmission. The base station directs a beam toward a single user or a group of users in a specific location, thus providing space separation in addition to code separation. Depending on the radio environment, transmit diversity techniques may improve the link performance by up to 5 dB.

Voice and data channels

The CDMA2000 forward traffic channel structure (FTC) may include several physical channels:

- The Fundamental Channel (F-FCH) is equivalent to functionality Traffic Channel (TCH) for IS-95. It can support data, voice, or signaling multiplexed with one another at any rate from 750 bps to 14.4 kbps.

- The Supplemental Channel (F-SCH) supports high-rate data services. The network may schedule transmission on the F-SCH on a frame-by-frame basis, if desired.

- The Dedicated Control Channel (F-DCCH) is used for signaling or bursty data sessions. This channel allows the sending of signaling information without any impact on the parallel data stream.

The reverse TCH structure is similar to the forward TCH. It may include a Paging Indicator Channel (R-PICH), a Fundamental Channel (R-FCH), and/or a Dedicated Control Channel (R-DCCH), and one or several Supplemental Channels (R-SCH). Their functionality and encoding structure is the same as that of the forward link with data rates ranging from 1 kbps to 1 Mbps.[6]

Traffic channel

The TCH structure and frame format is very flexible in CDMA2000. In order to limit the signaling load that would be associated with a full frame format parameter negotiation, CDMA2000 specifics a set of channel configurations. It defines a spreading rate and an associated set of frames for each configuration.

The FTC always includes either a fundamental channel or a dedicated control channel. The main benefit of this multichannel forward traffic structure is the flexibility to independently set up and tear down new services without any complicated multiplexing reconfiguration or code channel juggling. The structure also allows different handoff configurations for different channels. For example, the F-DCCH, which carries critical signaling information, may be in soft handoff, while the associated F-SCH operation could be based on a best cell strategy.

Supplemental channels

One of the key CDMA2000 1x features is its ability to support both voice and data services on the same carrier. CDMA2000 operates at up to 16 or 32 times the FCH rate (also referred to as 16x or 32x in Release 0 and A, respectively). In contrast to voice calls, the traffic generated by packet data calls is bursty, with small durations of high traffic separated by larger durations of no traffic. It is very inefficient to dedicate a permanent traffic channel to a packet data call. This burstiness impacts the amount of available power to the voice calls, possibly degrading their quality if the system is not engineered correctly. Hence, a key CDMA2000 design issue is to ensure that a CDMA channel carrying voice and data calls do so simultaneously with negligible impact to the QoS of both.

Supplemental Channels (SCHs) can be assigned and de-assigned at any time by the base station. The SCH has the additional benefit of improved modulation, coding, and power control schemes. This allows a single SCH to provide a data rate of up to 16 FCH in CDMA2000 Release 0 (or 153.6 kbps for Rate Set 1 rates), and up to 32 FCH in CDMA2000 Release A (or 307.2 kbps for Rate Set 1 rates). Note that each sector of a base station may transmit multiple SCHs simultaneously if it has sufficient transmit power and Walsh codes. The CDMA2000 standard limits the number of SCHs a mobile station can simultaneously support to two. This is in addition to the FCH or DCCH, which are set up for the entire duration of the call since they are used to carrying signaling and control frames as well as data. Two approaches are possible: (1) individually assigned SCHs, with either finite or infinite assignments, or (2) shared SCHs with infinite assignments.

For bursty and delay-tolerant traffic, assigning a few scheduled big pipes is preferable to dedicating many thin or slow pipes. The big-pipe approach exploits variations in the channel conditions of different users to maximize sector throughput. The more sensitive the traffic becomes to delay, such as voice, the more appropriate the dedicated traffic channel (DTCH) approach becomes.

---

[6]It is important to note that while the CDMA2000 standard supports a maximum data rate of 1 Mbps, existing products are supporting a peak data rate of 307 kbps.

Turbo coding

CDMA2000 provides the option of using either turbo or convolutional coding on the forward and reverse SCHs. Both coding schemes are optional for the base station and the mobile station, and the capability of each is communicated through signaling messages prior to the set up of the call. In addition to peak rate increase and improved rate granularity, the major improvement to the traffic channel coding in CDMA2000 is the support of turbo coding at rate 1/2, 1/3, or 1/4. The turbo code is based on 1/8 state parallel structure and can only be used for supplemental channels and frames with more than 360 bits. Turbo coding provides a very efficient scheme for data transmission and leads to better link performance and system capacity improvements. In general, turbo coding provides a performance gain in terms of power savings over convolutional coding. This gain is a function of the data rate, with higher data rates generally providing more turbo coding gain.

In the following subsections, we present a brief introduction to the technical aspects of the CDMA2000 standard. In order to make up-to-date and consistent discussions throughout this section, we concentrate on CDMA2000 1xEV (or IS-856) standard based mainly on the specifications given in the following standard documentations [360, 361]:

- CDMA2000 High Rate Packet Data Air Interface Specification, 3GPP2 C.S20024 v2.0, October 2000.

- CDMA2000 High Rate Packet Data Air Interface Specification, 3GPP2 C.S20024-A, v1.0, March 2004.

The former is known as *Release 0* and the latter Revision A of CDMA2000 1xEV or IS-856 standard. There are also numerous references on CDMA2000 1xEV and their evolutional versions CDMA2000 1xEV-DO and CDMA2000 1xEV-DV, and readers may refer to [360–367] for more information about them. In particular, the recent issue (April 2005) of IEEE Communications Magazine published a Special Issue on CDMA2000 1xEV-DV and seven papers appear in this issue [368–374], which indicate that the CDMA2000 1xEV-DV will gain greater popularity around the world.

However, we also refer to CDMA2000 1x standard from time to time in the discussions on CDMA2000 1xEV technology. Therefore, to understand the CDMA2000 1xEV, it is strongly recommended that readers gain sufficient background knowledge about CDMA2000 1x standard. Release 0 of CDMA2000 1x standard [348–353] consists of the following 3GPP2 documents:

- C.S0001-0 Introduction to CDMA2000 Standards for Spread Spectrum Systems

- C.S0002-0 Physical Layer Standard for CDMA2000 Spread Spectrum Systems

- C.S0003-0 Medium Access Control (MAC) Standard for CDMA2000 Spread Spectrum Systems

- C.S0004-0 Signaling Link Access Control (LAC) Standard for CDMA2000 Spread Spectrum Systems

- C.S0005-0 Upper Layer (Layer 3) Signaling Standard for CDMA2000 Spread Spectrum Systems

- C.S0006-0 Analog Signaling Standard for CDMA2000 Spread Spectrum Systems

Readers can find all the final revisions (or Revision D) of the CDMA2000 1x standard from the reference list [359] provided at the end of this book.

3.1.2   General Architecture

The technical requirements contained in CDMA2000 standard [348–353] form a compatibility standard for the 800 MHz cdmaOne (or IS-95A/B) system and the 1.8 to 2.0 GHz CDMA Personal Communications Services (PCS) systems, also called *JSTD-008* systems. It is required that a mobile station can obtain service in a cellular or PCS system manufactured in accordance with the CDMA2000 standards. The requirements do not address the quality or reliability of that service, nor do they cover equipment performance or measurement procedures. Compatibility, as far as CDMA2000 systems are concerned, is to imply that any mobile station is able to place and receive calls in any 800 MHz IS-95 cellular mobile telecommunication system or in any 1.8 to 2.0 GHz CDMA JSTD-008 system. Under such compatibility, all CDMA systems (CDMA2000, IS-95 and JSTD-008 systems) are able to place and receive calls for any CDMA mobile station. To ensure compatibility, both radio system parameters and call processing procedures should be carefully specified. The sequence of call processing steps that the MSs and BSs execute to establish connections is specified, along with digital control messages and, for dual-mode systems, the analog signals that are exchanged between the two stations. The BS is subject to different compatibility requirements from that of the MS. Radiated power levels, both desired and undesired, are fully specified for MSs, in order to control the radio interference that one MS can cause to another.

BSs are fixed in location and their interference is controlled by proper layout and operation of the system where the station operates. Detailed call processing procedures are specified for MSs to ensure a uniform response to all BSs. The BS procedures that do not affect the MSs' operation are left to the designers of the overall network. This approach to writing the compatibility specification provides the network designer with sufficient flexibility to respond to local service needs and to account for particular topography and propagation conditions. CDMA2000 includes provisions for future service additions and expansion of system capabilities.

Channel naming conventions

To facilitate the understanding of the abbreviations in the illustrations and text given in the discussions followed, we would like to explain the channel naming convention here. The following naming conventions apply to all CDMA2000 standards.

A logical channel name consists of three lower case letters followed by "ch" (channel). A hyphen is used after the first letter. Table 3.6 shows the naming conventions for the logical channels that are used in this family of standards. For example, the logical channel name for the Forward Dedicated Traffic Channel is "f-dtch."

On the other hand, physical channels are named by upper case abbreviations. As in the case of logical channels, the first letters in the names of the channels indicate the direction of the channel (i.e., forward or reverse). Table 3.7 shows the names and meanings of all the physical channels designated in CDMA2000. For example, the physical channel name for the Forward Fundamental Channel is "F-FCH."

Table 3.6   Naming conventions for logical channels in CDMA2000 standard [354]

First letter	Second letter	Third letter
f = Forward	**d** = Dedicated	**t** = Traffic
r = Reverse	**c** = Common	**s** = Signaling

Table 3.7 Naming conventions for physical channels in CDMA2000 standard [354]

Channel name	Physical channel
F/R-FCH	Forward/Reverse Fundamental Channel
F/R-DCCH	Forward/Reverse Dedicated Control Channel
F/R-SCCH	Forward/Reverse Supplemental Code Channel
F/R-SCH	Forward/Reverse Supplemental Channel
F-PCH	Paging Channel
F-QPCH	Quick Paging Channel
RACH	Access Channel
F/R-CCCH	Forward/Reverse Common Control Channel
F/R-PICH	Forward/Reverse Pilot Channel
F-APICH	Dedicated Auxiliary Pilot Channel
F-TDPICH	Transmit Diversity Pilot Channel
F-ATDPICH	Auxiliary Transmit Diversity Pilot Channel
F-SYNCH	Sync Channel
F-CPCCH	Common Power Control Channel
F-CACH	Common Assignment Channel
R-EACH	Enhanced Access Channel
F-BCCH	Broadcast Control Channel
F-PDCH	Forward Packet Date Channel
F-PDCCH	Forward Packet Data Control Channel
R-ACKCH	Reverse Acknowledgment Channel
R-CQICH	Reverse Channel Quality Indicator Channel
F-ACKCH	Forward Acknowledgment Channel
F-GCH	Forward Grant Channel
F-RCCH	Forward Rate Control Channel
R-PDCH	Reverse Packet Data Channel
R-PDCCH	Reverse Packet Data Control Channel
R-REQCH	Reverse Request Channel

Layered architecture

The release of the CDMA2000 family of standards that support Spreading Rate 1 operation has been given in [349] and [355] in the reference list. Figure 3.2 depicts the general system architecture of CDMA2000, and Figure 3.3 shows the layered architecture in mobile stations.

The development of the CDMA2000 family of standards has, to the greatest extent possible, adhered to the architecture by specifying different layers in different standards. The physical layer is specified in [349] and [355], the MAC in [350] and [356], the link access control (LAC) in [351] and [357], and upper layer signaling architecture in [352] and [358].

CDMA2000 provides full backward compatibility with the cdmaOne system. Backward compatibility permits the support of cdmaOne mobile stations by the CDMA2000 infrastructure and also permits the operation of CDMA2000 mobile stations in cdmaOne networks. The CDMA2000 family also supports the reuse of existing cdmaOne service standards, such as those that define speech services, data services, Short Message Services (SMSs), and Over-the-Air Provisioning and Activation services, with the CDMA2000 physical layer.

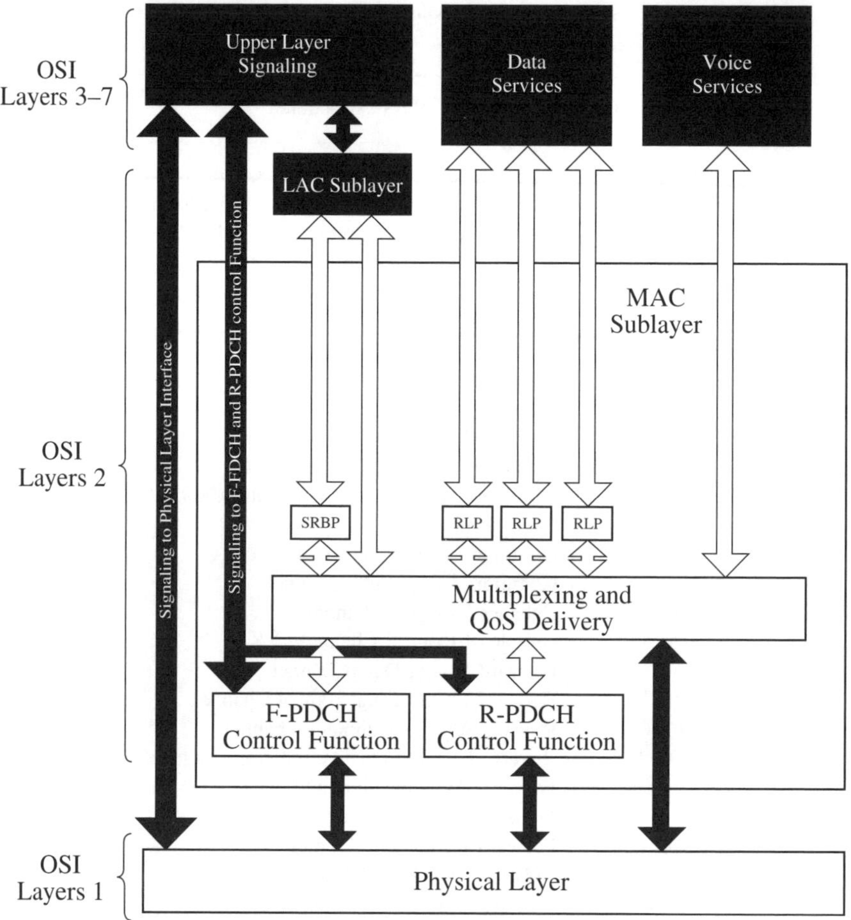

Figure 3.2 General CDMA2000 architecture.

3.1.3 Airlink Design

The 1xEV airlink is designed for packet data optimization. It is spectrally efficient and provides 7.4 Mbps/cell (3 sectors) aggregate forward peak throughput with a single CDMA frequency carrier (1.25 MHz). One of the key premises of 1xEV is that voice and data have very different requirements and there will be inefficiencies any time the two services are combined. With that in mind, the 1xEV design requires a separate CDMA carrier. It is however important to note that the 1xEV waveform retains 100% compatibility with IS-95/1x from the RF viewpoint. The 1xEV waveform uses the same 1.228 Mcps chip rate, link budgets, network plans, and RF designs on both Access Terminals (ATs) and infrastructure. Furthermore, optimizing voice and data on different carriers is advantageous for both services: it simplifies system software development and avoids difficult load-balancing tasks. 1xEV over-the-air system, available since September 1998, demonstrated features and data rates that are proven from real field data experience. 1xEV's forward link uses power efficiently. The access terminal continually updates the Access Network with the data rate it can receive. With this

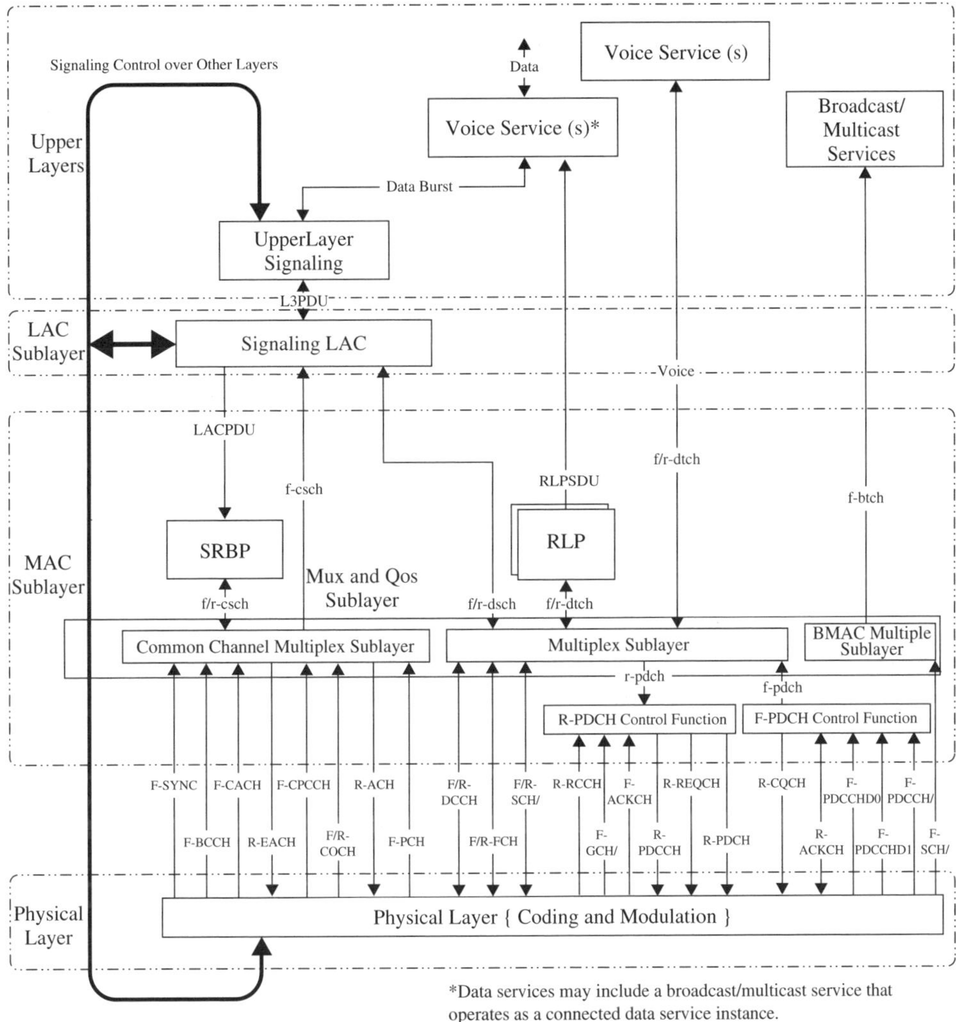

Figure 3.3 CDMA2000 layered architecture in mobile stations.

information the system can service a single user at any instant. This rate control feature enables the 1xEV Access Point (AP) to always transmit at full power, thereby achieving very high peak rates for users that are in good coverage area. The Access Terminal and the AP jointly determine each user's forward link data. The AT measures the pilot strength, and continuously requests an appropriate data rate based on the channel conditions. The Access Point encodes the forward link at precisely the highest rate that the subscriber's wireless channel can support at any instant.

3.1.4 Data Throughput

The majority of data applications receive a larger amount of data from the wireless network's infrastructure through the forward link (also known as *downlink*), than they transmit in the reverse direction

Table 3.8   CDMA2000 1xEV's peak data transmission rates in forward and reverse links

Air-link	Peak data rate
Forward link	2.457 Mbps/sector
Reverse link	153.6 kbps/sector

Table 3.9   CDMA2000 1xEV's average forward and reverse links throughput in a 3-sector cell

Air-link		1-Rx-Antenna	2-Rx-Antenna
	Pedestrian	3.1 Mbps/cell	4.0 Mbps/cell
Forward link	Low mobility	2.0 Mbps/cell	3.1 Mbps/cell
	High-mobility	1.3 Mbps/cell	2.5 Mbps/cell
Reverse link		600 kbps/cell	

(in reverse link or uplink). Therefore, 1xEV provides asymmetric data rates on the forward and reverse links. The forward and reverse links peak data rate in CDMA2000 1xEV system is shown in Table 3.8.

CDMA2000 1xEV airlink makes use of its network resources very efficiently by providing high performance average data throughput with only 1.25 MHz of spectrum, which is only about one quarter of that needed for WCDMA. Given a loaded 1xEV sector, with a number of users distributed uniformly across the coverage area, the average forward and reverse links throughput in a 3-sector cell is shown in Table 3.9.

The above throughput figures do not consider the transmit diversity that may be applied to the cell site. ATs with dual receiver antennas and receiver diversity reception have the benefit of more accurately decoding the received information. Overall, this decreases retransmissions and improves overall system efficiency.

3.1.5   Turbo Coding

CDMA2000 1xEV takes advantage of parallel codes and turbo decoding techniques. Hence, frame sizes are longer than cdmaOne frames.

Forward link code rate

The code rates (R = 1/5 and 1/3) are used on the forward channels. In fact, the code and modulation rates are chosen to be consistent with the required spectral efficiency. Table 3.10 shows the spectral efficiency of each complete packet type distributed over its code rate, and repetition factor.

1xEV limits the lowest code rate to 1/5 because very low code rates increase decoding complexity, while providing only a small increase in the coding gain. A rate 1/5 code with QPSK modulation achieves a spectral efficiency of 2/5 bits/chip. For packet types that require a spectral efficiency lower than 2/5 bits/chip, the rate 1/5 code with QPSK modulation is used. Complete or partial sequence repetition of the modulation symbols is employed to achieve the desired spectral efficiency. For packet types that require spectral efficiency higher than 2/5 bits/chip, higher rate codes and possible higher order modulations are used. Higher rate codes are obtained by indirectly puncturing the basic rate 1/5 turbo code. Accordingly, the output of the rate 1/5 turbo code is first interleaved, followed by a simple truncation of the interleaver output. This design facilitates hybrid-automatic repeat request (ARQ) and provides near-optimal performance for the truncated packet types, because the early termination of

Table 3.10 CDMA2000 1xEV's Turbo coding rate, spectral efficiency, and repetition rate in the forward link

Complete packet type	Spectral efficiency (bits/chip)	Coding rate	Repetition rate
38K	1/24	1/5	9.6
76K	1/12	1/5	4.8
153K	1/6	1/5	2.4
307K-2S	1/3	1/5	1.2
307K-4S	16/49	16/49	2
614K-1S	2/3	1/3	1
614K-2S	32/49	16/49	1
912K	48/49	16/49	1
1.2M-1S	4/3	2/3	1
1.2M-2S	64/49	16/49	1
1.8M	2	2/3	1
2.4M	8/3	2/3	1

Table 3.11 CDMA2000 1xEV's turbo coding rate in the reverse link

Data Rate (kbps)	9.6	19.2	38.4	76.8	153.6
Encoder input block length	250	506	1024	2048	4096
Turbo encoder code rate	1/4	1/4	1/4	1/4	1/2
Encoder output block length	1024	2048	4096	8192	8192

a packet achieves the same effect as puncturing the underlying code. Such a design provides high performance adaptive rate coding for both complete and truncated packet types while maintaining a simple encoder structure.

Reverse link code rate

Code rates ($R = 1/4$ and $1/2$) are used on the reverse channels. The Data Channel supports data rates from 9.6 to 153.6 kbps with 16-slot packets (26.66 ms). The packet is encoded using either rate 1/2 or rate 1/4 Parallel Turbo Code as specified in [360, 361]. The parameters of the reverse link encoder for different data rates are described in Table 3.11. The code symbols are bit-reversal interleaved and block repeated to achieve the 307.2 ksps modulation symbol rate.

3.1.6 Forward Link

The CDMA2000 1xEV channel structure is shown in Figure 3.4, where both forward link and reverse link channels are given. In this subsection we only focus our discussions on forward link. The CDMA2000 1xEV forward link consists of the following time-division multiplexed channels: the Pilot Channel, the MAC Channel, the Forward Traffic Channel and Control Channel. The MAC Channel consists of three subchannels: the Reverse Activity (RA) Channel, the Data Rate Control (DRC) Lock Channel and the Reverse Power Control (RPC) Channel. The Forward Channel hierarchies are shown in Figure 3.4.

When compared to the forward link channels in IS-95A that consists of pilot, sync, paging, and TCHs, the forward link channels in CDMA2000 1xEV has a relatively similar configuration and consists of pilot, medium access control, traffic, and control channels, as shown in Figure 3.4.

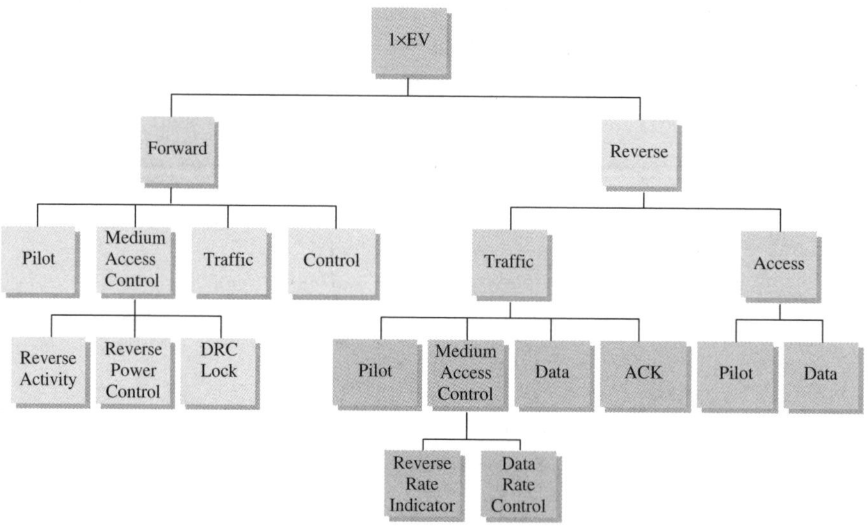

Figure 3.4 CDMA2000 1xEV channels structure for both forward link and reverse link channels.

Channel configuration

A CDMA2000 1xEV forward link carrier is allocated 1.25 MHz of bandwidth and is direct-sequence modulated at the rate of 1.2288 Mcps. The forward link transmission consists of time slots of length 2048 chips (1.66 ms). Groups of 16 slots, referred to as frames, are aligned to the CDMA system time. Within each slot, the Pilot, MAC and Traffic or Control Channels are time-division multiplexed (as shown in Figure 3.5) and are transmitted at the same power level. A slot during which no traffic or control data is transmitted is referred to as an idle slot. During an idle slot, the sector transmits the Pilot and MAC Channels only, thus reducing interference to other sectors. The overall channel structure of the forward link physical layer is shown in Figure 3.6.

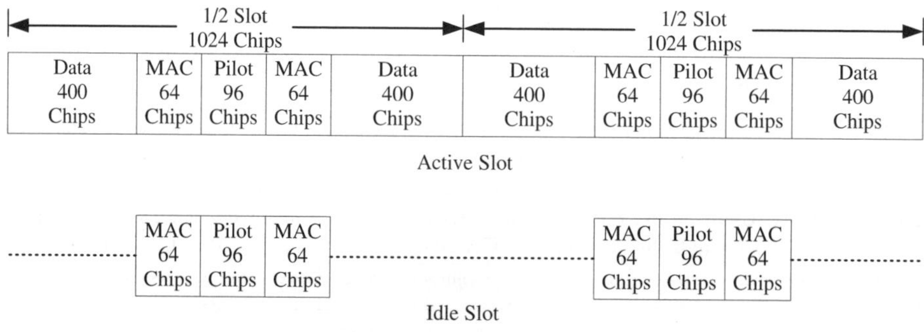

Figure 3.5 CDMA2000 1xEV forward link slot structure.

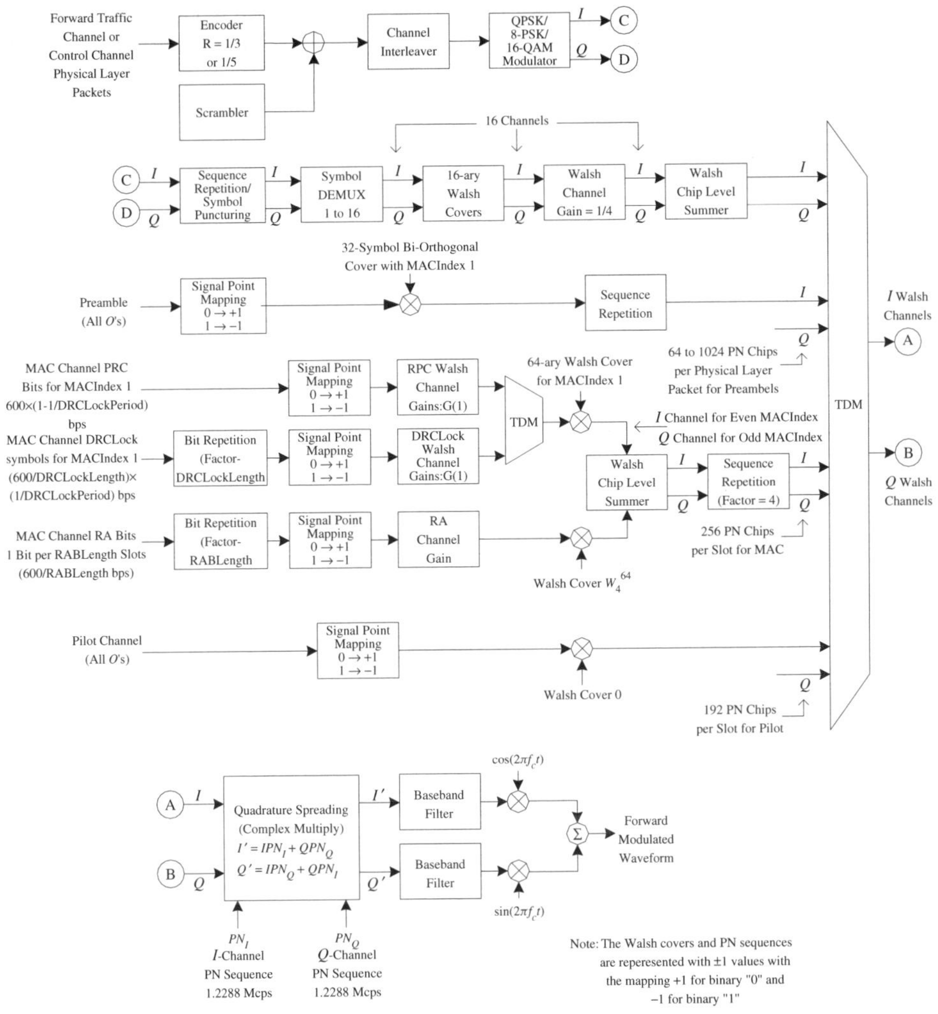

Figure 3.6 CDMA2000 1xEV overall channel structure for the forward link physical layer.

The CDMA2000 1xEV uses a shared forward link and can serve a user at any instant. When the service is performed, an AT receives the full power of the cell transmitter. The AT calculates its C/I ratio and tells the AP the highest data rate at which it can receive information. This allows the AP transmitter to operate at full power and transmit data at the highest data rate that each AT requests. There are additional benefits of such a shared forward link. The scheduling algorithm takes advantage of multiple users and optimizes the data transmission on the forward link. As more subscribers access the 1xEV system, the scheduler assists in improving the traffic flow by fairly scheduling data to each subscriber. The improved efficiency actually increases the subscriber's average throughput.

In the CDMA2000 1xEV system a maximum of 60 simultaneous users in the "connected" state, that is actively requesting and receiving packets, can be served in a sector at any given time. Note that this number represents active connected users (who are not in a dormant state). For example, if the users in a sector are using applications with 10% activity factor, then 600 users per sector can be served during the busy hour, in actual fact. CDMA2000 1xEV also supports both high-speed and high-capacity applications. In the case of high-speed applications with tens of active users per sector at any given time, it ensures the users achieve high throughput. In the case of low data rate applications, where the users are receiving small amounts of information (e.g. stock quotes, telemetry, etc.), the users will not use the channel for a long period of time. The users will receive their short bursts of information and will then release the Forward Link channel. In this case, a much larger number of users (as in hundreds of users) can take turns being one of the 60 active users at a given instant. The 60 active users in a sector are power controlled by the RPC bits. Each power controlled access terminal is assigned a Walsh cover that is used to cover each user's forward link packets. The Walsh code W64 is being used and hence 64 possible Walsh covers are available to be assigned to users. Four out of these 64 are used for other purposes, which leaves 60 RPC bits that can actually be assigned to users. A large number of users are served by the system at any given time, leveraging the bursty nature of the traffic. When a user is not actively using the link, his/her airlink goes dormant. When one starts using it again, the airlink will automatically come up without any special intervention on the part of the user.

Figure 3.7 shows the CDMA2000 1xEV forward channel structure.

As shown in Figure 3.7, the 1xEV forward link is structured to maximize the overall data throughput of a given sector. Therefore, the APs are always transmitting at full power, serving one user at a time. There are no predetermined time slots; the time the user is on the FTC depends on the channel condition or the C/I ratio in that particular time.

As shown in Figure 3.4, the 1xEV Forward Channel consists of the following time multiplexed channels: (1) the Pilot Channel; (2) the Forward MAC Channel; (3) the FTC; and (4) the Control Channel.

The TCH carries user data packets. The Control Channel carries control messages, and it may also carry user traffic if necessary.

The forward link is defined in terms of frames of length 26.67 ms, aligned to the PN rolls of the zero offset PN sequences. Within a frame, there are 16 slots, each with a length of 2048 chips or 1.67 ms duration, as shown in Figure 3.7. Each frame is composed of two half-frame units of eight slots. Each slot is further divided into two half-slots, each of which contains a pilot burst. Each pilot

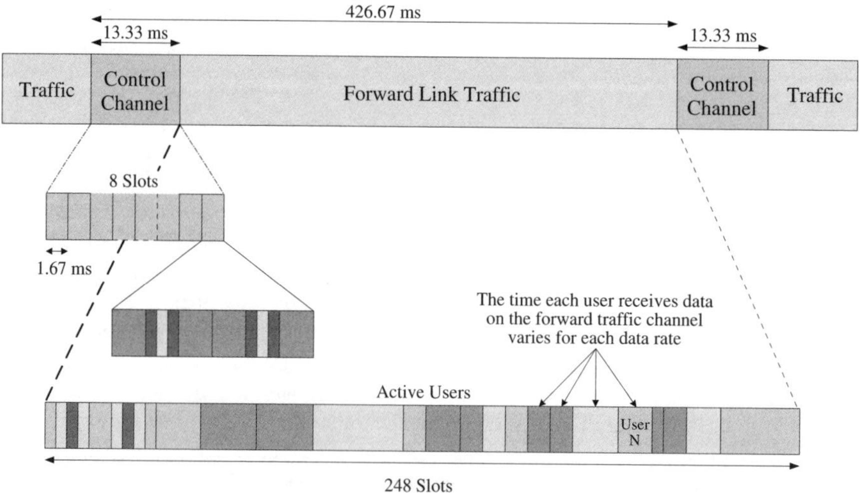

Figure 3.7 CDMA2000 1xEV forward channel architecture.

burst has a duration of 96 chips, and is centered at the midpoint of the half-slot. Within each slot, the Pilot, MAC, and Traffic or Control Channels are time multiplexed. All time-division multiplexed channels are transmitted at the maximum power of the sector. Note that no predetermined time slots are allocated to users.

The MAC Channel consists of two subchannels: the *RPC* Channel and the *RA* Channel. The RA Channel transmits a *reverse-link activity bit* (RAB) stream.

The Control Channel is transmitted at a data rate of 38.4 kbps or 76.8 kbps. The modulation characteristics for the Control Channel transmitted at 38.4 kbps are the same as those of the FTC transmitted at 38.4 kbps. The modulation characteristics for the Control Channel transmitted at 76.8 kbps are the same as those of the Control Channel transmitted at 38.4 kbps except that the packet is transmitted in eight slots.

Burst pilot signal

Figure 3.8 illustrates the structure of CDMA2000 1xEV forward link burst pilot.

The CDMA2000 1xEV system uses a burst pilot, which is optimal for bursty packet data services. The burst pilot is not transmitted on a separate code channel as it is in a cdmaOne system, but is punctured into the forward link waveform at predetermined intervals. The CDMA2000 1xEV burst pilot is transmitted at the maximum power that the cell is enabled to transmit. Using the full power of the cell for the pilot provides the highest possible pilot signal-to-noise ratio so that an accurate channel estimate can be obtained quickly, even during dynamic channel conditions.

The burst pilot is received only in the presence of pilots from other APs and is not affected by other transmissions of data. Since the pilot is transmitted at the full power of the AP, the AT recognizes the burst strictly as a pilot signal. The AT does not need to subtract the effect of data transmissions that may be occurring at the same time as the pilot. This results in a high signal-to-noise ratio for the pilot signal, which aids in rapid C/I estimation.

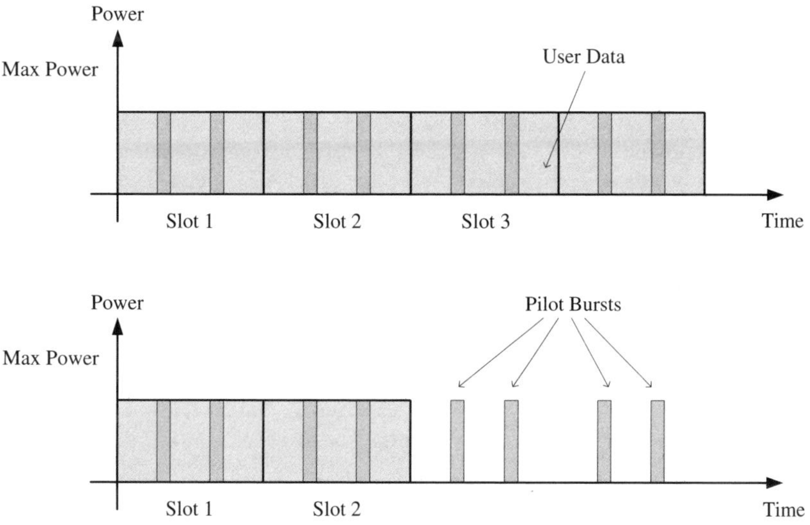

Figure 3.8 Structure of CDMA2000 1xEV forward link burst pilot.

Link adaptive modulation

The link adaptive modulation scheme is another important feature introduced in the CDMA2000 1xEV system. In fact, the 1xEV forward link offers a range of different data rates. The data rates match the range of channel conditions experienced in a typical cellular or PCS networks. QPSK modulation is used to achieve 38.4 kbps through 1228.8 kbps data rates (with the exception of 921.6 kbps), 8PSK for 921.6 kbps and 1843.2 kbps and 16QAM for 1228.8 kbps and 2457.6 kbps.

Table 3.12 shows the correspondence of adaptive modulation schemes, code rate, transmission rate, number of slots, and so on, in the CDMA2000 1xEV Forward Link.

The CDMA2000 1xEV forward link supports dynamic data transmission rates. The AT constantly measures the channel carrier to the interference (C/I) ratio, and then requests the appropriate data rate for the channel conditions every 1.67 ms. The AP receives the AT's request for a particular data rate, and encodes the forward link data at exactly the highest rate that the wireless channel can support at the requested instant. Just enough margin is included to allow the AT to decode the data with a low erasure rate. In this way, as the subscriber's application needs and channel conditions change, the optimum data rate is determined and served dynamically to the user. In summary, the following steps are performed:

- Accurate and rapid measurement of the received C/I ratio from the set of best serving sectors

- Selection of the best serving sector

- Request of transmission at the highest possible data rate that can be received with high reliability given the measured C/I

- Transmission from the selected sector, and only from the selected sector, at the requested data rate.

The AT continuously updates the AP on the DRC channel, indicating a specified data rate to be used on the forward link. The DRC is sent with a Walsh Cover, which indicates which sector should transmit. CDMA2000 1xEV combines the functions of the cdmaOne Sync and Paging overhead channels into a single Control Channel, which is transmitted once every 413.17 ms for a duration of 13.33 ms. This forward link control channel creates notable efficiencies. FTC and Control Channel can be transmitted in a span of 1 to 16 slots. When more than one slot is used, the transmit slots use a 4-slot interlacing technique to further enhance forward link efficiency, as shown in Figure 3.7. For example, data sent at 153.6 kbps is sent in four slots and each slot of data is sent twice to increase the probability of receiving the data. By interlacing the data with every fourth slot, the AT can notify the AP of each slot of data it receives. If the AT is able to decode the data on the first attempt, then it transmits an ACK to the AP. The AP cancels the second slot if the ACK is received prior to its

Table 3.12   CDMA2000 1xEV adaptive modulation schemes, code rate, transmission rate, number of slots in forward link

Data rate (kb/s)	38.4	76.8	153.6	307.2	307.2	614.4	614.4	921.6	1228.8	1228.8	1843.2	2457.6
Modem	QPSK	QPSK	QPSK	QPSK	QPSK	QPSK	QPSK	8PSK	QPSK	16QAM	8PSK	16QAM
Encoded packet length (bits/ms)	1024/ 26.67	1024/ 13.33	1024/ 6.67	1024/ 3.33	2048/ 6.67	1024/ 1.67	2048/ 3.33	3072/ 3.33	2048/ 1.67	4096/ 3.33	3072/ 1.67	4096/ 1.67
Code rate	1/5	1/5	1/5	1/5	1/3	1/3	1/3	1/3	1/3	1/3	1/3	1/3
No. of slots	16	8	4	2	4	1	2	2	1	2	1	1

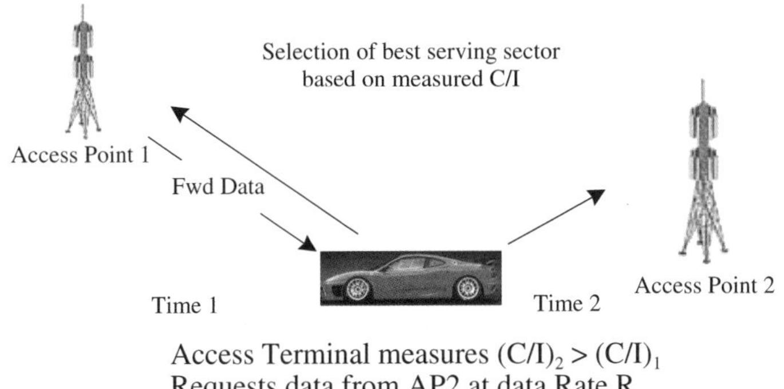

Figure 3.9 Dynamic data rate served based on real-time C/I measurement achieved in CDMA2000 1xEV air-link.

transmission. The system has now increased the throughput to the user and may use the additional slots to serve other users. The combination of these features and the ability to transmit two bits per Hertz in a 1.25 MHz band increases bandwidth efficiency and overall system capacity.

Figure 3.9 shows a conceptual diagram for CDMA2000 1xEV to achieve dynamic data rate served based on real-time C/I measurement.

3.1.7 Scheduling

It is to be noted that CDMA2000 1xEV is optimized for packet data services, in which all terminals do not necessarily demand equal service. Some applications require higher data rates, while others do not. The user's channel condition (i.e., the carrier to interference ratio) is also an important factor in determining the data rate that a given user can attain. The 1xEV system takes advantage of the wireless channel variability, which results in variations of the requested rate over a period of time. The scheduler resides at the BS and takes the data rates requested by different MSs into account. The scheduling algorithm decides which MS is served with the requested data rate at any given instant. The scheduler is weighted to serve users that are improving their signal quality and weighted against users that are experiencing signal degradation. Occasionally, the users may not be served for periods of milliseconds when their requested rates are lower. By the scheduler selecting the optimal time to transmit data to a user, the user's overall moving average throughput is higher than if they were served on a first-in-first-out basis. Please note that the priority in the scheduler is based on a combination of the following: the C/I as well as the duration since the last time a user has been served. Disadvantaged users with low C/I accumulate credits with the scheduler, thereby increasing their priority in the system and improving their throughput.

CDMA2000 1xEV uses *Proportional Fairness Scheduling* for packet scheduling. This algorithm uses a different notion of fairness known as *proportional fairness*. The Proportional Fairness Scheduler maximizes the user's moving average throughput, which improves their experience. The algorithm used by the Proportional Fairness Scheduler takes advantage of variable bit rate that 1xEV uses to deliver data. The algorithm maintains a running average of each user's RF conditions and attempts to deliver data at the requested peak rates, avoiding delivering data when the requested rates are at their lowest points. For example, a particular user has RF conditions that support an average of 614.4 kbps. The changing RF environment surrounding the user causes the RF conditions to oscillate between low and HDRs, with the average being 614.4 kbps. The scheduler's histogram of each user calculates the moving average and serves data when the DRC is equal to or greater than 614.4 kbps, and not at the lower short term rates. The result is that the user's actual data throughput is higher than the running average of the requested data rates. In summary, the Proportional Fairness Scheduler takes advantage of the channel variation over a short period to increase throughput and maintain the grade of service fairness over longer periods of time.

3.1.8   Reverse Link

The 1xEV reverse link structure consists of fixed size physical layer packets (16 slots, 26.67 ms duration). Each slot is just a unit of time. The Reverse Link is different from the forward link physical layer, which has variable modulation schemes in 1.67 ms units of time.

1xEV uses a pilot-aided, coherently demodulated reverse link. Traditional cdmaOne power control mechanisms and soft handoffs (SHOs) are supported on the reverse link. A 1xEV AT may transmit at rates from 9.6 kbps to 153.6 kbps on the reverse link.

The 1xEV Reverse Channel consists of the Access Channel and the TCH. The Access channel consists of a Pilot Channel and a Data Channel. The TCH consists of a Pilot Channel, a MAC Channel, an Acknowledgment (ACK) Channel, and a Data Channel. The Traffic MAC Channel contains a Reverse Rate Indicator (RRI) Channel and a DRC Channel.

The Access Channel is used by the AT to initiate communication with the Access Network or to respond to an AT directed message. The Access Channel consists of a Pilot Channel and a Data Channel. An access probe consists of a preamble followed by an Access Channel data packet. During the preamble transmission, only the Pilot Channel is transmitted. During the Access Channel data packet transmission, both the Pilot Channel and the Data Channel are transmitted.

The reverse link TCH is used by the AT to transmit user specific traffic or signaling information to the Access Network. The reverse link TCH consists of a Pilot Channel, a MAC Channel, an ACK Channel, and a Data Channel. The MAC Channel contains a DRC Channel and an RRI Channel. The ACK Channel is used by the AT to inform the Access Network whether the data packet transmitted on the FTC has been successfully received or not.

The total reverse link capacity is 200 kbps/sector (2.2 times that of IS-95A). This increased capacity is achieved by taking advantage of turbo coding, gaining diversity from the longer packet size (26.67 ms), and the pilot channel.

Reverse link channel structure

Figure 3.10 shows the Reverse Traffic Channel structure of the 1xEV standard. There are four orthogonal code-division multiplexed channels. As shown in Figure 3.10, the Pilot/RRI Channel is time multiplexed so that the RRI channel is transmitted during 256 chips at the beginning of every slot (1.66 ms). The 3-bit RRI symbol transmitted every frame (16 slots), is encoded using a 7-bit simplex codeword. Each codeword is repeated 37 times over the duration of the frame, while the last three code symbols are not transmitted. The DRC symbols (four bits indicating the desired rate) are encoded using 16-ary biorthogonal code. Each code symbol is further spread by one of the 8-ary Walsh functions in order to indicate the desired transmitting sector on the forward link. The DRC message is transmitted with half-slot offset relative to the slot boundary. The reason is to minimize prediction delay while providing enough time for processing at the desired sector before transmission on the forward link starts on the next slot. A DRC message indicating the desired forward link data rate and transmitting sector may be repeated over DRC Length slots, a user-specific parameter set by the access network.

The ACK Channel is BPSK modulated in the first half-slot (1024 chips) of an active slot. A "0" bit is transmitted on the ACK Channel if a data packet has been successfully received on the FTC; otherwise a "1" bit is transmitted. Transmissions on the ACK Channel only occur if the AT detects a data packet directed to it on the FTC. For a Forward Traffic Channel data packet transmitted in slot

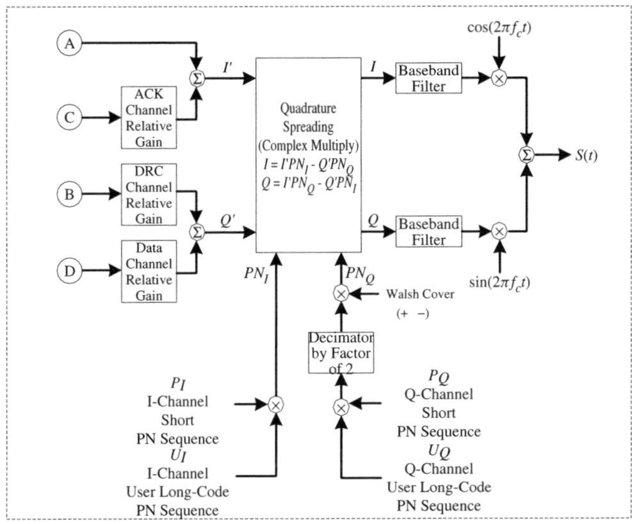

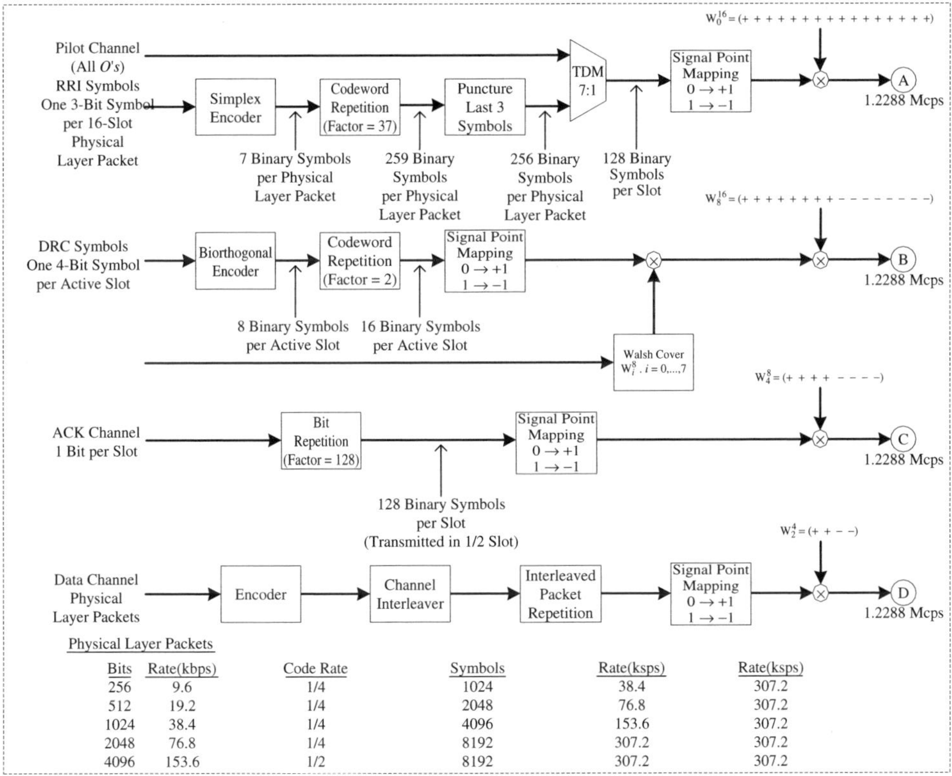

Figure 3.10 Reverse traffic channel structure in CDMA2000 1xEV air-link.

n, the corresponding ACK Channel bit is transmitted in slot $n + 3$ on the Reverse Traffic Channel. The three slots of delay allow the terminal to demodulate and decode the received packet before transmitting on the ACK Channel.

The Data Channel supports data rates from 9.6 to 153.6 kbps with 16-slot packets (26.66 ms). The packet is encoded using either rate 1/2 or rate 1/4 Parallel Turbo Code as specified in 1xEV. The code symbols are bit-reversal interleaved and block repeated to achieve the 307.2 ksps modulation symbol rate.

The Pilot/RRI, DRC, ACK, and Data Channel modulation symbols are each spread by an appropriate orthogonal Walsh function as shown in Figure 3.10. Before quadrature spreading (see Figure 3.10), the Pilot/RRI and ACK Channels are scaled and combined to form the in-phase component. Similarly, the Data and DRC Channels are scaled and combined to form the quadrature component of the baseband signal.

Reverse link power control (both open and closed loops) is applied to the Pilot/RRI Channel only. The powers allocated to the DRC, ACK and Data channels are adjusted by a fixed gain relative to the Pilot/RRI Channel in order to guarantee the desired performance of these channels. For example, the relative gain of the Data Channel increases with the data rate so that the received Eb/Nt is adjusted to achieve the required packet error rate (PER).

The reverse link provides an *RRI*, which aids the AP in determining the rate at which the reverse link is sending data. The RRI is included as the preamble for reverse link frames, indicating the rate at which the data was sent. Figure 3.11 shows the 1xEV reverse channel structure. The data rates supported in CDMA2000 1xEV reverse link are listed in Table 3.13, which actually shows the physical layer parameters of the reverse link channels.

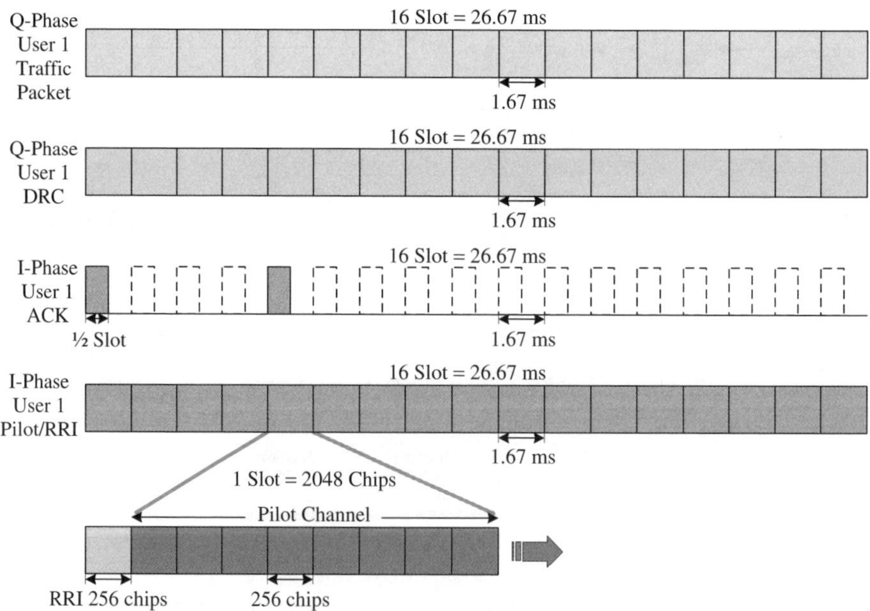

Figure 3.11 Dynamic data rate served based on real-time C/I measurement achieved in CDMA2000 1xEV air-link.

Table 3.13 CDMA2000 1xEV reverse link modulation schemes, code rate, encoded packet length, number of slots

Data Rates (kbps)	9.6	19.2	38.4	76.8	153.6
Modulation	BPSK	BPSK	BPSK	BPSK	BPSK
Encoded packet length (bits)/(ms)	256/26.67	512/26.67	1024/26.67	2048/26.67	4096/26.67
Code rate	1/4	1/4	1/4	1/4	1/2
No. of slots	16	16	16	16	16

3.1.9 CDMA2000 1xEV Signaling

The CDMA2000 1xEV layered architecture enables a modular design that allows partial updates to protocols, software, and independent protocol negotiation. The following are the CDMA2000 1xEV protocol stack layers:

- Physical Layer: The Physical Layer provides the channel structure, frequency, power output, modulation, and encoding specifications for the Forward and Reverse link channels.

- MAC Layer: The Medium Access Control layer defines the procedures used to receive and transmit over the Physical Layer.

- Security Layer: The Security Layer provides authentication and encryption services.

- Connection Layer: The Connection Layer provides air-link connection establishment and maintenance services.

- Session Layer: The Session Layer provides protocol negotiation, protocol configuration, and session state maintenance services.

- Stream Layer: The Stream Layer provides multiplexing of distinct application streams.

- Application Layer: The Application Layer provides the Default Signaling Application for transporting 1xEV protocol messages and the Default Packet Application for transporting user data.

The detail configuration of all different layers in CDMA2000 1xEV standard is shown in Figure 3.12. It is to be noted from the figure that the overall structure of the CDMA2000 1xEV layered architecture was designed according to the general OSI reference model of seven-layer architecture.[7] However, we can see some differences between the standard OSI reference model and CDMA2000 1xEv layered architecture. First of all, the MAC layer has been extracted from the data link layer in the OSI model to become a stand-alone layer. The security layer in the CDMA2000 1xEV is a newly added layer, which does not enjoy the similar emphasis in the OSI reference model. Similarly, both the Connection layer and Stream layer in CDMA2000 1xEV layered architecture do not appear in the OSI reference model as independent layers, although part of their functionalities has been included in either the Transport layer or the Presentation layer.

Next, we will explain the major functions of different layers in CDMA2000 1xEV standard.

---

[7]More detailed discussions on the OSI reference model of seven-layer architecture is given in Section 2.5.

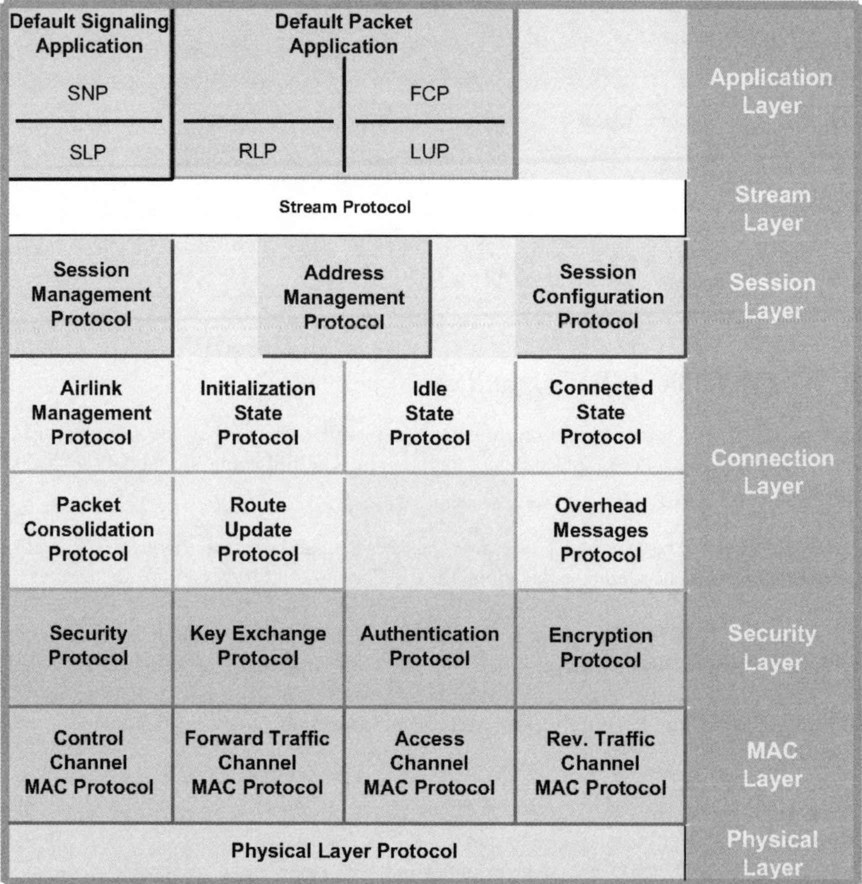

Figure 3.12 Layered network architecture in CDMA2000 1xEV standard.

Physical layer

The functionalities of the physical layer are obvious, delivering physical signaling through the air-link channels without refering to the detailed interpretation of the digital signals. For those descriptions, readers may go back to the previous Subsection 3.1.8 and Subsection 3.1.6.

MAC layer

The MAC Layer is a key component to optimizing the efficiency of the airlink and allowing multiple access to the network in a most cost-effective way. It is comprised of four component protocols, each of which play a part in the transmission of data and system information over the air-link channels, as explained below:

- Control Channel MAC Protocol: It governs the transmission by the Access Network and the subsequent reception by the AT of information on the Control Channel. The Control Channel packets are constructed from the Security Layer packets, and contain information controlling the Access Network transmission and packet scheduling, the AT acquisition, and AT packet reception on the Control Channel. This protocol also adds the AT address to transmitted packets. The rules for Control Channel supervision are part of this protocol as well.

- Access Channel MAC Protocol: It specifies the rules for sending messages on the Access Channel by the AT. This includes the timing as well as power requirements for the transmission. The AT communicates with the Access Network via the Access Channel prior to setting up a traffic connection.

- FTC MAC Protocol: It enables the system to send a user's data packets at optimal efficiency, by utilizing variable and fixed transmission rates and ARQ interlacing. The ARQ interlacing coupled with the DRC and ACK Channel provides the handshake to increase the AT's data throughput performance, resulting in increased capacity of the system. The FTC MAC Protocol also provides the rules that the Access Network uses to interpret the DRC Channel and the rules the AT uses for DRC supervision.

- Reverse Traffic Channel MAC Protocol: It is very similar to the traditional CDMA 1x MAC layer. The protocol transports the information sent by the AT to enable the Access Network in acquiring the Reverse Traffic Channel; and the Reverse Traffic Channel data rate selection.

Security layer

The Security Layer ensures the security of the connection between the AT and the Access Network. It utilizes the Diffie–Hellman key exchange[8] to ensure the intended device is authenticated on the Access Network, and that the connection is not hijacked. It is not intended to encrypt the user's data. For complete security of the user's data it is best to use an end-to-end method, that is, IP Security (IPSEC). IPSEC is a set of protocols developed by the IETF to support the secure exchange of packets at the IP layer. IPSEC has been widely deployed in order to implement Virtual Private Networks (VPNs). IPSEC supports two encryption modes: Transport and Tunnel. The Transport mode only encrypts the data portion (payload) of each packet, but leaves the header untouched. The more secure Tunnel mode encrypts both the header and the payload. On the receiving side, an IPSEC-compliant device decrypts each packet. The majority of today's VPN services utilize IPSEC to encrypt and protect information end-to-end.

The Security Layer provides the following functions:

- Key Exchange: It provides the procedures followed by the Access Network and the AT to exchange security keys for authentication and encryption. The system uses the Diffie–Hellman Key Exchange method.

- Authentication: It provides the procedures followed by the Access Network and the AT for authenticating traffic.

- Encryption: It provides the procedures followed by the Access Network and the AT for encrypting traffic.

Connection layer

The Connection Layer consists of several protocols that are optimized for packet data processing. When they are combined they efficiently manage the 1xEV airlink, reserve resources, and prioritize each user's traffic. They are designed to enhance the user's experience while at the same time bringing

---

[8]Diffie–Hellman key exchange is a cryptographic protocol which allows two parties that have no prior knowledge of each other to jointly establish a shared secret key over an insecure communications channel.

efficiency to the carrier network. Each protocol in the Connection Layer is introduced individually as follows:

- AirLink Management Protocol activates one of the below mentioned three State Protocols based on the AT state.

- Initialization State Protocol (AT has not yet acquired the network) performs the actions associated with acquiring the 1xEV network. This includes network determination, pilot acquisition and system synchronization.

- Idle State Protocol (AT has acquired the network, however it is not sending or receiving any data) monitors the location of the AT via the Route Update Protocol, provides procedures for the opening of a connection, and supports AT power conservation.

- "Suspend Mode" is a new addition to the Idle State Protocol. Suspend Mode expedites the connection setup process. In the suspend mode period, the AT advertises to the network that it will be monitoring the Control Channel before going into slotted mode for a certain period of time; so that the Access Network can quickly assign a TCH to the AT, if needed, rather than going through the usual paging and assignment procedure.

- Connected State Protocol (AT has an open connection with the network) performs the actions of managing the radio link between the AT and the Access Network (handoffs controlled by the Route Update Protocol), and the procedures leading to the close of the connection.

- Route Update Protocol plays a key part in enabling soft and softer handoffs. The AT's Route Update Protocol constantly reports to the Access Network, which AP and sector it is using, as well as potential neighboring sectors. This information is used by the Access Network in maintaining a stable and good quality radio link as the AT moves throughout the network.

- Overhead Messages Protocol is unique owing to the fact that it is used by multiple protocols. It broadcasts essential parameters pertaining to the operation of other protocols over the Control Channel. It also specifies rules for supervision of these messages over the Control Channel.

- Packet Consolidation Protocol is a key element to providing effective QoS to the user. It is responsible for consolidating packets and properly prioritizing them, according to their assigned QoS, for the forward link, and de-multiplexing them on the reverse link. The priority tagging is done at the Stream Layer. It is capable of prioritizing for multiple streams to a single user and multiple streams to many users.

Session layer

The Session Layer protocols provide a support system for the lower layers in the protocol stack. It enables the assignment of the UATI to the AT and configuration information that supports the lower layers. The negotiation of a set of protocols and their configurations for communication between the AT and the Access Network are controlled by this protocol. The Session Layer contains the following protocols:

- Session Management Protocol provides the means to control the ordered activation of the other Session Layer protocols. In addition, this protocol ensures the session is still valid and manages closing the session, resulting in the efficient use of spectrum.

- Address Management Protocol specifies procedures for the initial UATI assignment and maintains the AT addresses.

- Session Configuration Protocol provides the means to negotiate and provision the protocols used during the session, and negotiates the configuration parameters for these protocols.

Stream layer

The Stream Layer tags all the information that is transmitted over the airlink. This includes user traffic as well as signaling traffic. Lower in the stack, these values are read by the Connection Layer's Packet Consolidation Protocol. The two protocols jointly provide effective prioritization of signaling and user traffic. The Stream layer maps the various applications to the appropriate stream and multiplexes the streams for one AT. Stream 0 is always assigned to the Signaling Application. The other streams can be assigned to applications with different QoS requirements or other applications.

Application layer

The Application Layer is the top layer and is a suite of protocols that ensure reliability and low erasure rate over the airlink. The underlining principle of this layer is to increase the robustness of the 1xEV protocol stack. The Application layer has two sublayers, which are the Default Signaling Application that provides best effort and reliable transmission of signaling messages, and the Default Packet Application that provides reliable and efficient transmission of the user's data. The *Default Signaling Application Protocol* has two sublayers:

- Signaling Network Protocol (SNP) provides a message transmission service for signaling messages. These messages are initiated by other protocols, which indicate the appropriate message to be transmitted for a specific function.

- Signaling Link Protocol (SLP) is the transport for the SNP messages. SLP provides a fragmentation mechanism for signaling messages, along with reliable and best-effort delivery services. The fragmentation mechanism increases the efficiency of sending signaling messages that may be larger than a single frame.

Default Packet Application Protocol provides reliable and efficient delivery of the user's data at a low PER, suitable for higher layers (e.g., TCP, UDP), along with mobility management that allows the Access Network to know the location of a mobile at any instance.

Default Packet Application Protocol is comprised of two protocols:

- Radio Link Protocol (RLP): Data applications are not as delay sensitive as voice applications; therefore wireless Internet systems provide various mechanisms for error detection and data retransmission. The RLP layer delivers a frame error rate in the order of 10^{-4}. The combination of RLP and TCP layers deliver an extremely low frame error rate, which is comparable with most land-line data systems today. The RLP protocol uses a NAK-based scheme, thereby reducing the amount of signaling. In addition, the 1xEV enhanced RLP provides a more efficient retransmission mechanism due to the sequencing of octets, rather than the sequencing of frames. This approach eliminates complex segmentation and reassembly issues, in the case that a retransmitted frame cannot fit into the payload available at the time of retransmission.

- Location Update Protocol: This protocol is used to provide mobility management, which enables the Access Network to know the location of an AT at any instance. This service is critical in providing seamless packet transport service to the user through PDSN selection and handover.

- Point-to-Point Protocol (PPP): This protocol is not part of the 1xEV specification, however, it is a key protocol that 3G technologies leverage to provide end-to-end connectivity between the PDSN and each AT. Therefore, it is worth mentioning its role in the 1xEV system. The PPP is a robust tunneling protocol, which sets up a tunnel between the PDSN and AT. The PDSN will maintain each AT's PPP tunnel and forward the user's traffic through its assigned tunnel. The mobile terminal may move in and out of coverage and the PDSN will maintain the PPP state, thus providing a reliable tunnel and an "always on" experience.

3.1.10   Handoffs

The CDMA2000 1xEV AT receives data from not more than one AP at any given time. Instead of combining transmit energy from multiple APs, the AT is able to rapidly switch from communicating with one AP to the other. The AT measures the channel C/I from all the measurable Pilot channels and requests service from the AP with the strongest Pilot signal. This follows the best server rule, where the AT communicates with the requested AP at any given time. The forward link pilot allows the AT to obtain a rapid and accurate C/I estimate. The 1xEV reverse link makes use of soft handoff mechanisms. The AT's transmissions may be received by more than one AP, and frame selection is hence made. The Location Update Message enables the Access Network to connect to the PDSN maintaining the PPP state to the AT; therefore it can reroute traffic to the AT immediately upon receiving the AT's Location Update Message. This method allows the AT to maintain its same IP address and same PPP connection, therefore allowing a seamless handoff.

Handoffs from CDMA2000 1x to cdmaOne systems

CDMA2000 supports the handoff of voice and data calls and other services from a cdmaOne system to a CDMA2000 system, such that the handoffs could happen in the following different scenarios: (1) At a handoff boundary and within a single frequency band; (2) At a handoff boundary and between frequency bands (assuming the mobile station has multiband capability); (3) Within the same cell footprint and within a single frequency band; and (4) Within the same cell footprint and between frequency bands (assuming the mobile station has multiband capability).

CDMA2000 supports the handoff of voice and data calls and other services from a CDMA2000 system to a cdmaOne system in the following situations: (1) At a handoff boundary and within a single frequency band; (2) At a handoff boundary and between frequency bands (assuming the mobile station has multiband capability); (3) Within the same cell footprint and within a single frequency band; and (4) Within the same cell footprint and between frequency bands (assuming the mobile station has multiband capability).

Handoffs from CDMA2000 1x-EV to cdmaOne/CDMA2000 1x systems

The interoperability between 1x and 1xEV Networks are covered in the TIA Standard, IS-878. The following are examples of the handoff scenarios that are possible between 1xEV and 1x systems:

- AT establishes a data session in 1xEV Radio Access Network (RAN). While the AT is dormant, it performs idle handoff from a 1xEV RAN to another 1xEV RAN.

- While the AT is exchanging data in a 1xEV system, it receives a page for an incoming voice service instance from the 1x system. Since the AT is monitoring 1x Forward Common Channel periodically, it is able to receive the page for the voice service instance. In this scenario, the AT can be configured to; (1) continue the data call on the 1x system, (2) to abandon the 1xEV data service instance handoff to the 1x system, and continue with voice only.

- AT is able to receive an SMS while it is exchanging data in the 1xEV system: SMS is received during the AT's assigned paging slot or during a broadcast slot.

- While the AT is exchanging data in a 1xEV system, it decides to initiate a voice call in the 1x system. In this scenario, the AT can be configured to; (1) continue the data call on the 1x system, (2) to abandon the 1xEV data service instance handoff to the 1x system, and continue with voice only.

- AT moves away from the coverage area of the 1xEV system into the coverage area of a 1x system. AT performs an Access Network Change from a 1xEV system to a 1x system.

3.1.11 Summary of CDMA2000 1x-EV

The major characteristic features obtainable from CDMA2000 1x-EV can be summarized as follows:

- CDMA2000 1xEV technology provides cost-effectiveness to the wireless operators. The technology enables operators to offer advanced data services, make very economical use of their spectrum and other network resources, and offer packet data services somewhat earlier than alternative technologies, such as WCDMA (UMTS), and so on. The experience gained from worldwide operators has shown its operational benefits.

- 1xEV leverages from existing hardware and software design, thus providing significant benefits to the equipment manufacturers. The technology offers short development cycles by supporting a quick production turn-around. 1xEV enables the subscriber manufacturers a strategic differentiation by being the first to offer cutting-edge user devices.

- 1xEV unleashes the Internet for the end users by simplifying the use and implementation of mobile wireless devices, and enabling a variety of mainstream ATs for mobile, portable, and fixed applications. Wireless Web lifestyles, the next Internet revolution, will have a lasting positive effect on 1xEV users by increasing their productivity. 1xEV provides an evolution with industry support by using a standardization path under a CDMA2000 umbrella.

3.1.12 CDMA2000 1xEV-DO

CDMA2000 1xEV-DO [346] is short for First Evolution, Data Optimized. CDMA2000 1xEV-DO technology offers near-broadband packet data speeds for wireless access to the Internet. A well-engineered 1xEV-DO network delivers average download data rates between 600 kbps and 1.2 Mbps during off-peak hours, and between 150 kbps and 300 kbps during peak hours. Instantaneous data rates are as high as 2.4 Mbps. These data rates are achieved using only 1.25 MHz of spectrum, one quarter of what is required for WCDMA. 1xEV-DO provides average throughput speeds of over 700 kbps, equivalent to cable modem speeds, and fast enough to support applications such as streaming video and large file downloads. Future releases will increase to 3.08 Mbps for the forward link. A conceptual diagram of a CDMA2000 1x-EV-DO network is shown in Figure 3.13.

1xEV-DO takes advantage of the recent advancement in mobile wireless communications, such as adaptive modulation system, which lets radio nodes optimize their transmission rates based on instantaneous channel feedback received from terminals. This, coupled with advanced turbo coding, multilevel modulation, and macrodiversity via sector selection, lets 1xEV-DO achieve download speeds that are near the theoretical limits of the mobile wireless channel.

1xEV-DO also uses a new concept called *multiuser diversity*. This allows more efficient sharing of available resources among multiple, simultaneously active data users. Multiuser diversity combines packet scheduling with adaptive channel feedback to optimize total user throughput.

CDMA2000 1xEV-D0

CDMA2000 1xEV-D0 technology provides high-speed wireless access
to the Internet over all-IP network.

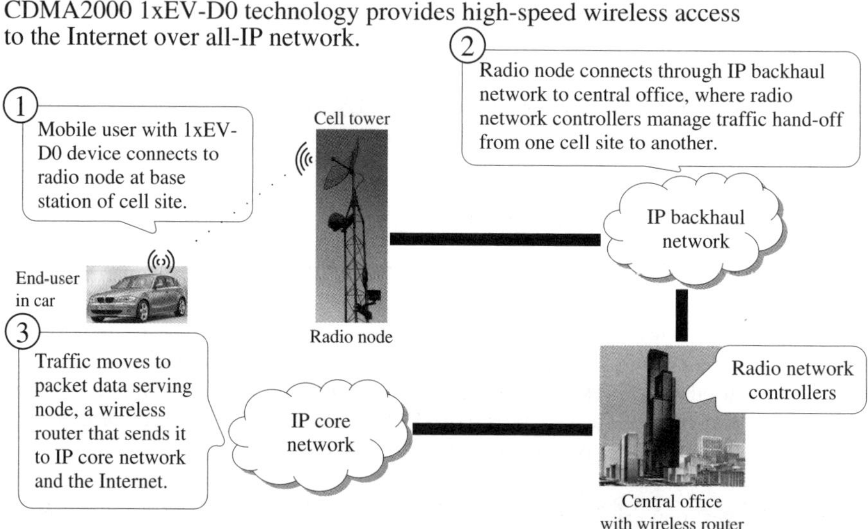

Figure 3.13 Configuration of CDMA2000 1x-EV-DO network.

A 1xEV-DO network is distinguishable from other 3G networks in that it is completely decoupled
from the legacy circuit-switched wireless voice network.[9] This has let some vendors build their 1xEV-
DO networks based entirely on IP technologies. Using IP transport between radio nodes and Radio
Network Controllers (RNCs) lowers backhaul costs by giving operators a choice of backhaul services,
including frame relay, router networks, metropolitan Ethernet and wireless backhaul. IP-based 1xEV-
DO networks take advantage of off-the-shelf IP equipment such as routers and servers, and use open
standards for network management.

1xEV-DO networks have the flexibility to support both user- and application-level QoS. User-
level QoS lets providers offer premium services. Application-level QoS lets operators allocate precious
network resources in accordance with applications' needs. Combined with Differentiated Services-
based QoS mechanisms, flexible 1xEV-DO packet schedulers can enable QoS within an entire wireless
network.

Multimode 1xEV-DO terminals that support CDMA2000 1x voice will let subscribers receive
incoming voice calls even while actively downloading data using 1xEV-DO. While 1xEV-DO is
capable of supporting high-speed Internet access at pedestrian or vehicle speeds, it can also be used
at homes, hotels, and airports.

3.1.13   CDMA2000 1xEV-DV

As CDMA2000 1x networks are being deployed in various countries/regions around the world to a
greater extent, the evolution of 1x is actively being developed within the industry. After the introduction of CDMA2000 1xEV-DO [346], CDMA2000 1xEV-DV is a natural evolution of CDMA2000 1x
family enabling operators to smoothly evolve their networks and provide continuity for their existing

---

[9]The difference between "circuit-switched" and "packet-switched" networks has been discussed in Section 2.6.

services. Key services such as support for voice and data on the same CDMA carrier will continue to be supported while allowing operators to leverage their investments in CDMA2000 1x.

As mentioned earlier, CDMA2000 1xEV-DV [347] is short for "1x Evolution, Data, and Voice." The CDMA2000 1xEV-DV standard is still under development and is expected to be commercially available in 2005. CDMA2000 1xEV-DV can support voice as well as data. Release C of the CDMA2000 1xEV-DV standard supports a forward link of 3.08 Mbps and a reverse link of 153 kbps. Release D supports a forward link of 3.08 Mbps and a reverse link of approximately 1.0 Mbps.

In 2002, 3GPP2 TSG-C has approved CDMA2000 Release C (commonly referred to as 1xEV-DV) for TIA publication. In addition, the ITU has approved 1xEV-DV as a world recognized 3G standard also in 2002. With the completion of 1xEV-DV specifications – both in the CDMA2000 air interface standards and the IOS standards in the end of 2002, we have already seen that initial 1xEV-DV commercial products have begun to be rolled out across various markets of late. Figure 3.14 shows a diagram that summarizes the evolution of CDMA technology. With regards to the evolution of CDMA2000 1x, 1xEV-DV is backward compatible to cdmaOne and CDMA2000 1x; it will enable a smooth migration to 1xEV-DV from 1x networks while preserving the existing services offered by operators, including voice and data services on the same carrier, and simultaneous voice and data.

CDMA2000 1xEV-DV focuses on the enhancement of CDMA2000's data carrying capability to provide higher data transmission rates on the forward link, pertaining to Internet applications such as web browsers, e-mail applications, and so on. The 1xEV-DV system was designed to maintain backward compatibility to all the previous versions of cdmaOne and CDMA2000 families, including the existing channels and signaling structures. An equally important feature of 1xEV-DV is that it does not require new base stations, that is, the same coverage footprint is retained and consequently it will save operators a huge sum of infrastructure costs for upgrading, which might be necessary otherwise. The enhancements occur at the physical layer of the system specifications and are controlled by the upper layers. For the limited space in this subsection, only those physical layer enhancements, that is, forward link enhancements, reverse link enhancements, and so on, different from those described in the previous subsections on CDMA2000 1xEV system will be summarized.

Forward link enhancement

CDMA2000 1xEV-DV incorporates several new features built on its time division multiplexing (TDM) and code-division multiplexing (CDM) capabilities [564]. 1xEV-DV incorporates a number of features that combine to provide an increase in forward link data rates up to 3.1 Mbps and average sector throughput of 1 Mbps.

The data-bearing TCH is referred to as the *Forward Packet Data Channel (F-PDCH or PDCH channel)*. The PDCH is shared by the packet data users and cannot undergo SHO. Depending on system loading, the PDCH consists of 1 to 28 CDM quadrature Walsh subchannels, each spread by 32-ary Walsh function. It can transmit any of a set of fixed packet sizes of 408, 792, 1560, 2328, 3096,

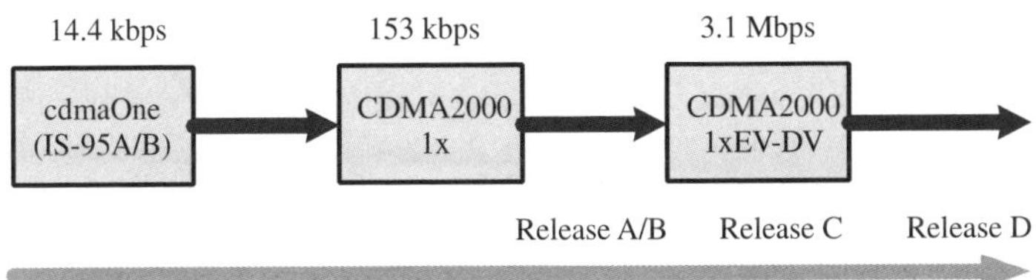

Figure 3.14 A diagram to show the evolution from cdmaOne, CDMA2000 1x to CDMA2000 1xEV-DV.

and 3864 bits. The system has variable packet durations of 1.25, 2.5 and 5 ms. The system also uses channel-sensitive scheduling via *adaptive modulation and coding* (AMC) schemes with higher order modulation of QPSK, 8PSK and 16QAM. The system makes use of a concatenation of *Forward Error Correction* (FEC) coding scheme and an *ARQ* protocol known as *Hybrid-ARQ (HARQ)*. The HARQ operating at the physical layer facilitates shorter round trip delays as compared to those associated with higher-layer retransmission schemes employed in the *RLP*. This important attribute of 1xEV-DV reduces the probability of a data session timeout (e.g. TCP/IP) as compared to RLP retransmission delays. The system has variable CDM common control channels of 1.25, 2.5 and 5 ms with a basic user packet scheduling granularity of 1.25 ms. The control channel, which carries the user's MAC ID, Encoder packet size, HARQ control information, and broadcast of available Walsh codes, is referred to as the *Forward Packet Data Control Channel* (F-PDCCH or PDCCH). The system may use up to two PDCCHs to enable data-bearing services to two different users simultaneously.

The addition of these features provides both the operator and the subscriber with the benefit of higher rate data services. With the addition of 1xEV-DV, subscribers now have access to services that are not available in earlier CDMA technologies, such as cdmaOne and CDMA2000 1x systems.

Reverse link enhancement

As many of the mobile data applications in the near future are expected to be forward link intensive, the majority of the effort in designing 1xEV-DV focused on enhancing the forward link. Although a subsequent release will enhance the data-bearing capability on the reverse link, only minor additions were made to the reverse link in order to be able to support the enhanced forward link.

To support HARQ functionality, the *Reverse Acknowledgment Channel* (R-ACKCH) is added to provide synchronous acknowledgements to the received forward link data packet transmissions. The *Reverse Channel Quality Indicator Channel* (R-CQICH) is used by the MS to indicate to the BS the channel quality measurements of the best serving sector. The MS selects the best serving sector by applying a Walsh cover corresponding to the selected serving sector. In determining the 1xEV-DV design, a significant effort has been undertaken to evaluate the system performance with mixed data and voice services.

Concurrent voice/data support

The CDMA2000 1xEV-DV air interface supports both voice and data services in both forward and reverse links. This provides the operator with a very flexible means of using spectrum. With this feature, the operators can share spectrum between voice and data services, providing concurrent voice and data services. This capability provides the operator with a very flexible method of controlling how spectrum is allocated. By taking advantage of the different usage patterns of voice and data, an operator that shares voice and data on a single carrier can optimize spectrum utilization.

Multiple concurrent traffic types

The 1xEV-DV specifications support both the multiplexing of signaling and user data over the F-PDCH and multiple concurrent data sessions. This provides a benefit to both the operator and the subscriber since this capability supports Personal Computer (PC)–based applications. The subscriber can now operate multiple PC applications simultaneously. The operator can gain revenue from these multiple applications without allocating a fundamental channel to each application.

Backward compatibility

One of the goals of the CDMA2000 1xEV-DV specifications is to offer smooth support for voice and legacy services. This is accomplished by reusing existing CDMA2000 standards wherever possible.

Examples of this reuse include the recycling of the 1x reverse link channels, IS-2000 MAC and signaling layer procedures, support for handoffs between 1xEV-DV radio channels, and other CDMA2000 radio channels and interoperability based on IOS. This benefits the operator by providing a smooth migration path from their deployed CDMA2000 1x infrastructure. This feature also minimizes impacts to the existing infrastructure as the operator upgrades their network to 1xEV-DV. Finally, the subscriber has a surefire guarantee of owning a mobile device that can support both 1x and 1xEV-DV air interfaces, providing a single terminal that can operate over the operator's entire network. An operator has the option of overlaying 1xEV-DV on the same carrier which supports cdmaOne or CDMA2000 1x. This allows the operator to control the migration and customize spectrum usage.

Support of all data services

CDMA2000 1xEV-DV allows the flexibility of both TDM and CDM scheduling, favoring TDM where TDM works best (e.g. services which are akin to the infinite queue best-effort data model, such as FTP, etc.), and allowing CDM to efficiently serve data for other services (e.g. WAP, VoIP, streaming video, etc.). TDM/CDM multiplexing is a powerful as well as a unique feature in 1xEV-DV. It maximizes system throughput by providing optimal modulation and coding rate assignments on a nondiscriminatory basis to all services, thereby providing the operator with the flexibility necessary in a dynamic market environment.

When the authors had almost completed this chapter, a new Feature Topic on CDMA2000 Evolution: 1xEV-DV, was published in IEEE Communications Magazine in the April 2005 issue. There were seven papers published in the Feature Topic in total [368–374], which give the most up-to-date information on the CDMA2000 1xEV-DV.

3.2 WCDMA

WCDMA system, also called *UMTS* [425], is a 3G mobile cellular standard proposed by the ETSI. As discussed in the previous section, the UMTS is one of the Third Generation (3G) mobile systems being developed within the ITU's IMT-2000 framework. It is a realization of a new generation of wideband multimedia mobile telecommunications technology. The coverage area of service provision is to be worldwide in the form of *Future Land Mobile Telecommunications Services* (FLMTS) and now called *IMT-2000*. The coverage will be provided by a combination of cell sizes ranging from in-building pico-cells to global cells covered by satellites, giving services to the remote regions of the world. It is expected that the UMTS is not a replacement of 2G technologies (e.g. GSM, DCS1800, CDMA, DECT etc.), which will continue to evolve to their full potential.

UMTS was mainly developed for countries where 2G GSM networks have been deployed, because these countries have agreed to free new frequency ranges for UMTS networks. As a matter of fact, UMTS is a new technology and operates in a new frequency band, and thus whole new RAN had to be built. This is obviously a disadvantage if compared to the relatively smooth upgrading path from IS-95 to CDMA2000 1x, as discussed in Section 3.1. The advantage of the UMTS system is that the new frequency range gives plenty of new capacity for operators. 3GPP is overseeing the standard development and has wisely kept the Core Network (CN) as close to GSM CN as possible. It is noted that UMTS phones are not meant to be backward compatible with GSM systems, but subscriptions (or SIM cards) can be. It is hoped that dual-mode phones will solve the compatibility problems. UMTS also has two flavors, or FDD (which is also named as *UMTS-FDD*) and TDD (which is also named as *UMTS TDD*). It is quite sure that the former has gained much attention and will be implemented first. Some harmonization has been done between systems, such as chip rate and pilot issues, and so on.

The CDMA technology used by the UMTS system is commonly called *wideband CDMA* or simply WCDMA. 3G WCDMA systems have 5 MHz bandwidth (in either uplink or downlink channels).

In fact, a 5 MHz bandwidth is neither wide nor narrow; it is just a bandwidth. Nevertheless, the new 3G WCDMA systems indeed have a wider bandwidth than the existing 2G CDMA systems (i.e. 1.25 MHz bandwidth in IS-95), which is why it is called *wideband*. It should be noted that the name of WCDMA is true in a relative sense, as there are commercially available CDMA systems operating over a 20 MHz bandwidth.

At this moment, it is significant for us to take a brief look at the different 3G standards in the world. There are FIVE major 3G air interface technologies specified by the ITU Recommendation ITU-R M.1457:

- IMT-2000 CDMA Direct Spread is also known as *UTRA-FDD*, and called WCDMA in Japan; recommended by ARIB/DoCoMo. UMTS is developed by 3GPP.

- IMT-2000 CDMA Multi-carrier, is also known as *CDMA2000* and developed by 3GPP2. IMT-2000 CDMA2000 includes 1x components, like CDMA2000 1x EV-DO.

- IMT-2000 CDMA TDD, is also known as *UTRA-TDD* and TD-SCDMA. TD-SCDMA is developed by China and supported by TD-SCDMA Forum.

- IMT-2000 TDMA Single Carrier, is also known as *UWC-136 (EDGE)* and is supported by UWCC.

- IMT-2000 DECT is supported by the DECT Forum.

3G is a generic name for a set of mobile technologies, which are designed for multimedia communication. Defined by ITU, 3G systems must provide: (1) Backward compatibility with 2G systems; (2) Multimedia support; (3) Improved system capacity compared to 2G and 2.5G cellular systems; and (4) High-speed packet data services ranging from 144 kbps in wide-area mobile environments to 2 Mbps in fixed or in-building environments. The standardization of 3G systems was conducted in several regions through their respective standard organizations:

- ETSI: European Telecommunications Standards Institute

- T1: Standardization Committee-Telecommunications (United States)

- TIA: Telecommunications Industry Association (North America)

- ARIB: Association of Radio Industries and Business (Japan)

- TTC: Telecommunications Technology Committee (Japan)

- TTA: Telecommunications Technology Association (Korea)

- CWTS: China Wireless Telecommunications Standard group

International Mobile Telecommunications-2000 (IMT-2000), initiated by ITU, is the global standard for 3G wireless communications, defined by a set of interdependent ITU Recommendations. IMT-2000 provides a framework for worldwide wireless access. Out of the ITU's IMT-2000 initiative, the Third Generation Partnership Project (3GPP) and the 3GPP2 were born.

3GPP is a collaboration agreement that was established in December 1998. The collaboration agreement brings together a number of telecommunications standards bodies, which are known as *Organizational Partners*. The current organizational partners are ARIB (Japan) and TTC (Japan), CCSA (China), ETSI (Europe), T1 (United States of America) and TTA (Korea). 3GPP is focused on WCDMA-based technology and its derivative and upgraded versions. Refer to the web site at http://www.3gpp.org/ for more information. On the other hand, 3GPP2 is another collaborative effort between five officially recognized standards bodies (ARIB, CCSA, TIA, TTA, and TTC) (as shown in http://www.3gpp2.org/), whose activities are focused on CDMA2000-based technologies.

The proposal of ETSI submitted to 3GPP is called *UMTS*. The terrestrial version of UMTS is called *UMTS Terrestrial Radio Access* (UTRA). The proposal of 3GPP is also called *UTRA*, which stands for *Universal Terrestrial Radio Access*, which has two modes: (1) *Frequency Division Duplex* (FDD) (2) *Time Division Duplex* (TDD). There are salient features for FDD and TDD operation modes, which is summarized below.

FDD operation mode provides simultaneous radio transmission channels for mobiles and base stations. Separate transmit and receive antennas are used at the base station in order to accommodate separate uplink and downlink channels. At the mobile unit, a single antenna is used for both the transmission to and the reception from the base station, and a duplexer is used to enable the use of the same antenna for simultaneous transmission and reception. It is necessary to separate the transmit and receive frequencies so that the duplexer can be given sufficient isolation while being inexpensively manufactured. It is noted that FDD has been exclusively used in earlier analog mobile radio systems. On the other hand, TDD mode shares a single radio channel in time so that a portion of the time is used to transmit from the base station to the mobile, and the rest time is used to transmit from the mobile to the base station. If the data transmission rate is much greater than the end-user's data rate, it is possible to store information bursts and provide the appearance of full-duplex operation to a user, even though two simultaneous radio transmissions exist at any instance of time. TDD is only feasible with digital transmission formats and digital modulation, and is very sensitive to timing.

Table 3.14 compares the difference in major air interface parameters for UMTS UTRA-FDD, UMTS UTRA-TDD and TD-SCDMA systems. Table 3.15 gives major system parameters for UMTS WCDMA and CDMA2000 systems. Table 3.16 makes a comparison among different 2.5–3G technologies in terms of their capabilities.

A UMTS network consists of three interacting domains; CN, UMTS Terrestrial Radio Access Network (UTRAN) and UE. The main function of the CN is to provide switching for user traffic. CN also contains the databases and network management functions. The basic CN architecture for

Table 3.14   Comparison of major system parameters for UMTS UTRA-FDD, UMTS UTRA-TDD and TD-SCDMA systems

	FDD scheme	TDD schemes	
Multiplex technology	WCDMA	TD-CDMA	TD-SCDMA
Bandwidth	2 × 5 MHz paired	1 × 5 MHz unpaired	1 × 1,6 MHz unpaired
Frequency Reuse	1	1	1 (or 3)
Handover	Soft, softer (Interfreq.: hard)	Hard	Hard
Modulation	QPSK	QPSK	QPSK and 8-PSK
Receiver	Rake	Joint Detection Rake (Mobile Station)	Joint Detection Rake (Mobile Station)
Chip Rate	3.84 Mcps	3.84 Mcps	1.28 Mcps
Spreading Factor	4–256	1, 2, 4, 8, 16	1, 2, 4, 8, 16
Power Control*	Fast: every 667 μs**	Slow: 100 cycles/s***	Slow: 200 cycles/s***
Frame organization	0.667/10 ms	0.667/10 ms	0.675/5 ms
Timeslots/Frame	N.A.	15	7

*Range: 80 dB (UL), 30 dB (DL) in step of **0.25 to 1.5 dB; ***1, 2 or 3 dB.

Table 3.15   Comparison of major system parameters for UMTS WCDMA and CDMA2000 systems

Parameter	WCDMA	CDMA2000
Carrier spacing	5 MHz	3.75 MHz
Chip rate	4.096 MHz	3.6864 MHz
Data modulation	BPSK	FW-QPSK; RV-BPSK
Spreading	Complex (OQPSK)	Complex (OQPSK)
Power control frequency	1500 Hz	800 Hz
Variable data rate implement.	Variable SF; multicode	Repeat., puncturing, multicode
Frame duration	10 ms	20 ms (also 5, 30, 40)
Coding	Turbo and convolutional	Turbo and convolutional
Base stations synchronized?	Asynchronous	Synchronous
Base station acquisition/detect	3 step; slot, frame, code	Time shifted PN correlation
Forward link pilot	TDM dedicated pilot	CDM common pilot
Antenna beam-forming	TDM dedicated pilot	Auxiliary pilot

UMTS is based on GSM network with GPRS.[10] All the equipment has to be modified for the UMTS operation and services. The UTRAN provides the air interface access method for UE. The Base Station is referred to as *Node B* and the control equipment for Node Bs is called *Radio Network Controller* (RNC).

The spectrum allocation in Europe, Japan, and Korea for the FDD mode is 1920–1980 MHz for the uplink and 2110–2170 MHz for the downlink, with the bands 1980–2010 MHz and 2170–2200 MHz intended for the satellite part of the future systems. The UTRA-TDD mode utilizes two frequency bands in Europe, the 1900–1920 MHz and the 2010–2025 MHz band. In both modes each carrier has a bandwidth of approximately 5 MHz. In the FDD mode, separate 5 MHz carrier frequencies are used for the uplink and downlink respectively. On the other hand, only one 5 MHz is shared between the uplink and the downlink in TDD. Each operator, subject to its offered licence, can deploy multiple 5 MHz carriers in order to increase capacity. Figure 3.15 shows the UMTS frequency spectrum allocation after the World Radio Conference (WRC) in 2002.

Figure 3.16 compares the voice capacity per 5 MHz spectrum for different 2–3G systems. It is seen from the figure that WCDMA offers a performance that still lags behind CDMA 2000 1x and TD-SCDMA systems. Figure 3.17 shows the handset sale comparison for different 2–3G systems from 2001 to 2007. Table 3.17 lists the 3G networks, the number of licences and the deployment requirements in different countries.

3.2.1   History of UMTS WCDMA

The inception of UMTS standard can be traced back to the early 1990s when ETSI initiated one UMTS research project in RACE1, seven projects in RACE2 and 14 projects in the ACTS Program. RACE projects were funded by *Commission of European Communities* (CEC). ETSI also organized *Future Advanced MObile Universal Telecommunications Systems* (FAMOUS) meetings 3 times a year between Europe, the United States and Japan.

From 1991 to 1995, two CEC funded research projects called *Code Division Testbed* (CODIT) and *Advanced Time Division Multiple Access* (ATDMA) were carried out by the major European telecom manufacturers and network operators. The CODIT and ATDMA projects investigated the

---

[10]"GPRS" stands for General Packet Radio Service, which is a nonvoice value-added service that allows information to be sent and received across a mobile telephone network. It supplements today's circuit-switched data and Short Message Service (SMS).

Table 3.16   Comparison of capabilities for different 2.5–3G technologies

	Peak network downlink speed	Average user through-put for file downloads	Capacity	Other features
GPRS	115 kbps	30–40 kbps		
EDGE	473 kbps	100–130 kbps	Double of that for GPRS	GPRS backward compatible
UMTS WCDMA	2 Mbps	220–320 kbps	Increased over EDGE for high-bandwidth applications	Simultaneous voice and data operation, enhanced security, QoS, multimedia support and reduced delay
UMTS-HSDPA*	10 Mbps	550–1100 kbps	Two and a half to three and a half times that of WCDMA	Backward compatible with WCDMA
CDMA2000 1xRTT	153 kbps	50–70 kbps		
CDMA2000 1xEV-DO	2.4 Mbps	300–500 kbps		Optimized for data, VoIP in development

High speed Downlink Packet Access (HSDPA) is in actual fact an extension of UMTS and can offer a data rate of up to 10 Mbps on downlink channel. HSDPA is a new 3GPP standard to increase the downlink throughput by replacing QPSK in UMTS by 16QAM in HSDPA. It works to offer a combination of channel bundling (TDMA), code multiplex (CDM) and improved coding (adaptive modulation and coding). It also introduces a separate control channel in order to facilitate the data transmission speed. Similar techniques will be available later for uplink with HSUPA. The major reference source for 3GPP HSDPA can be found from: (1) 3GPP TS 25.855 HSDPA; Overall UTRAN description; (2) 3GPP TS 25.856 HSDPA; Layer 2 and 3 aspects; (3) 3GPP TS 25.876 Multiple-Input Multiple-Output Antenna Processing for HSDPA; (4) 3GPP TS 25.877 HSDPA – Iub/Iur Protocol Aspects; (5) 3GPP TS 25.890 HSDPA; User Equipment (UE) radio transmission and reception (FDD).

suitability of wideband CDMA and TDMA-based radio access technologies for 3G systems. This work was later continued in the *Future Radio Wideband Multiple Access System* (FRAMES) project and became the basis of the further ETSI UMTS work until decisions were taken in 1998.

In February 1992 the *WRC* in Malaga, Spain, allocated frequencies for future UMTS use. Frequencies 1885–2025 MHz and 2110–2200 MHz were identified for IMT-2000. The UMTS Task Force was established in February 1995, issuing "The Road to UMTS" report.

The UMTS Forum was established at the inaugural meeting, held in Zurich, Switzerland, in December 1996. Since then, the planned "European" WCDMA standard has been known as the *UMTS*. In June 1997 the UMTS Forum produced its first report entitled *A regulatory Framework for UMTS*. The UMTS core band was decided in October 1997.

In January 1998 ETSI SMG meeting in Paris, both WCDMA and TD-CDMA proposals were combined to UMTS air interface specification. In June 1998, Terrestrial air interface proposals (UTRAN, WCDMA, CDMA2000, EDGE, EP-DECT, TD-SCDMA) were handed into the ITU-R as possible

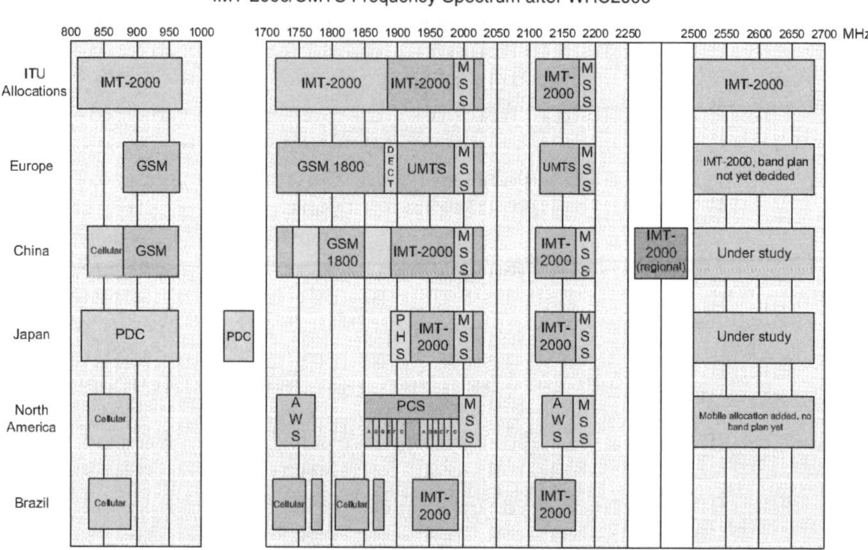

Figure 3.15   IMT-2000/UMTS spectrum allocation for different regions in the world, which was decided in the World Radio Conference (WRC) in 2002.

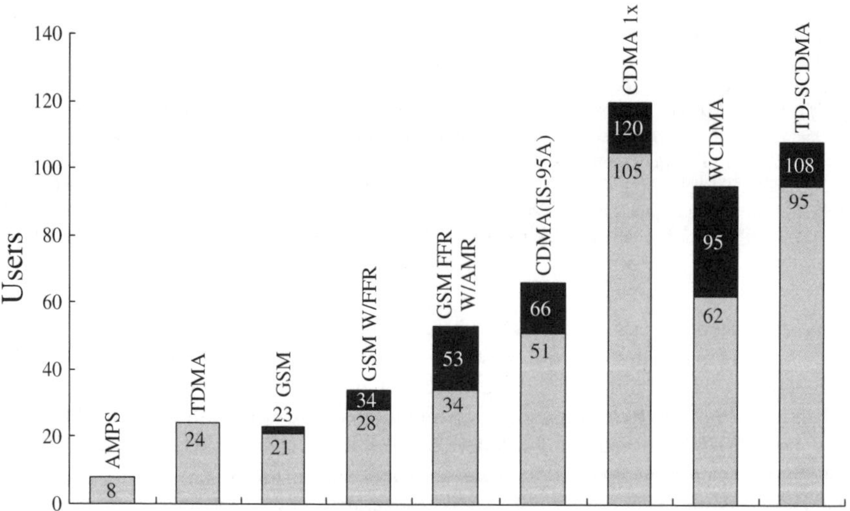

Figure 3.16 Voice capacity comparison in terms of per 5 MHz spectrum for different 2–3G systems, where the dark regions on the bars show the capacity variation among applications with variable link set-ups.

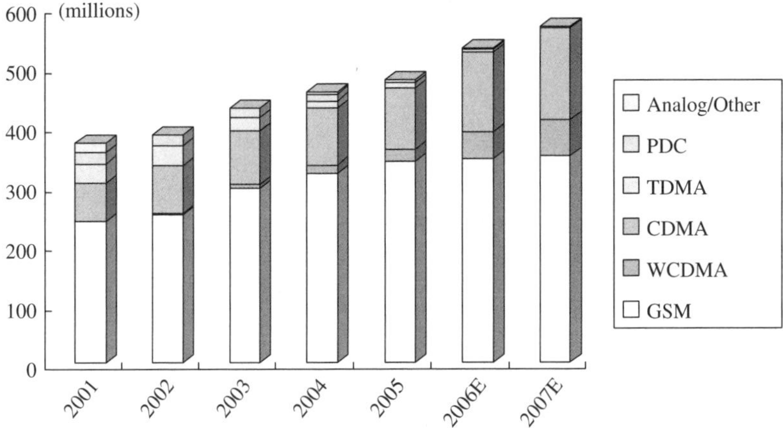

Figure 3.17 Handset sale comparison for different 2–3G systems.

IMT-2000 candidate proposals. The first call using a Nokia WCDMA terminal in DoCoMo's trial network was completed in September 1998 at Nokia's R&D unit near Tokyo in Japan.

On December 4, 1998, ETSI SMG, T1P1, ARIB, TTC, and TTA created 3GPP in Copenhagen, Denmark, and the first meeting of the 3GPP Technical Specification Groups was held in Sophia Antipolis, France, on December 7 and 8, 1998.

On April 27 and 28, 1999, Lucent Technologies, Ericsson, and NEC announced that they were selected by Nippon Telegraph and Telephone (NTT) DoCoMo to supply WCDMA equipment for NTT DoCoMo's next generation wireless commercial network in Japan. This was the first announced WCDMA 3G infrastructure deal.

3GPP approved the UMTS Release 4 specification in March 2001 in a meeting that took place in Palm Springs.

NTT DoCoMo launched a trial 3G service, an area-specific information service for i-mode on June 28, 2001. On September 25, 2001, NTT DoCoMo announced that three 3G phone models were commercially available. NTT DoCoMo launched the first commercial WCDMA 3G mobile network on October 1, 2001.

On March 14, 2002, UMTS Release 5 was issued.[11] UMTS Release 6 was issued on December 16, 2004, which was delayed from its initial target date of June 2003.

Ericsson demonstrates 9 Mbps with WCDMA, *High Speed Downlink Packet Access* (HSDPA) phase 2, on February 14, 2005. Ericsson and several operators in three Scandinavian countries demonstrated the 1.5 Mbps enhanced uplink in the live WCDMA system on May 10, 2005.[12] In fact, the peak data rate for HSDPA can reach up to 8–10 Mbps (and 20 Mbps for MIMO systems) over a 5 MHz bandwidth in WCDMA downlink. HSDPA implementations include Adaptive Modulation and Coding (AMC), Multiple-Input Multiple-Output (MIMO), Hybrid-Automatic Request (HARQ), fast cell search, and advanced receiver design. In the 3rd generation partnership project (3GPP) standards, Release 4 specifications provide efficient IP support, enabling the provision of services through an all-IP CN and Release 5 specifications focus on the HSDPA to provide data rates up to approximately

---

[11] Although the initial target date was December 2001, the launch was delayed by almost four months.

[12] HSDPA is a new 3GPP standard to facilitate the increase of the downlink throughput by changing the radio modulation (QPSK to 16QAM). 3GPP HSUPA for uplink will also be available.

Table 3.17   3G networks, number of licences and deployment requirements in different countries

Country	3G network	No. of lic	Government requirements
Australia	CDMA2000 & WCDMA	6	No coverage obligations
Austria	WCDMA	4	25% coverage by end-2003; 50% by end-2005
Belgium	WCDMA	3	30% coverage in 3yrs, 40% in 4yrs, 50% in 5 yrs; 85% by end-2006. In 2/02 delayed launch from 9/02 to 9/03
Brazil	CDMA2000 & WCDMA	NA	No special 3G requirements/policies announced
Canada	CDMA2000 & WCDMA	NA	No special 3G licence requirements. Operators use regular spectrum licences
China	CDMA2000 & WCDMA	NA	No special 3G requirements/policies announced
Denmark	WCDMA	4	30% of population by end-2004; 80% be end-2008; then sharing allowed for next 20%
Finland	WCDMA	4	No coverage req, but ministry may ensure implementation
France	WCDMA	2 (+1 pending)	25% voice coverage and 20% data 2 yrs after launch; 80% voice and 60% data 8 yrs after launch
Germany	WCDMA	6	25% of population covered by end-2003, 50% by end-2005, does not allow mergers of 3G licence holders
Greece	WCDMA	3	25% population coverage by 12/03, Olympic Games facilities 02/04, 50% population by 12/06
Hong Kong	WCDMA	4	50% coverage by end-2006; keep 30% available for MVNOs
Ireland	WCDMA	4	Licence A (more spectrum): 80% coverage by 2008 Licence B: cover 5 major cities (58% population) by 2008
Italy	WCDMA	5	Regional capitals covered within 30 mos, provincial capitals within 60 mos

10 Mbps to support packet-based multimedia services. MIMO systems are the work item in Release 6 specifications, which will support even higher data transmission rates up to 20 Mbps. HSDPA is evolved from and backward compatible with Release 99 WCDMA systems.

The comparison between 3GPP HSDPA and 3GPP2 1xEV-DV is made in Table 3.18.

The milestones of the development of UMTS are summarized as:

Table 3.17 (*continued*)

Country	3G network	No. of lic	Government requirements
Japan	CDMA2000 & WCDMA	3	Licences have temporary status, awarded permanent licences when ministry is satisfied with 3G status of each operator
The Netherlands	WCDMA	5	5 licences with specific coverage requirements. Infr sharing allowed in 9/01, but service separately
Norway	WCDMA	4	3 licences (4th given back): 90% coverage to largest cities within 5 yrs from launch; may fine if buildout not on track
Portugal	WCDMA	4	20% coverage in 1 yr, 40% in 3 yrs, 60% in 5 yrs; each lic holder committed $ 768.4 mil to infrastructure
Singapore	NA	3	Provisional deadline of 12/31/04 for nationwide network
South Korea	CDMA2000 & WCDMA	2 (+1 pending)	Government warned operators in 02/02 not to switch 3G techs
Spain	WCDMA	4	Launch by mid-2003, previously covered 23 cities by 06/02 (postponed from 08/01); pre-postponement required 90% coverage by 2005
Sweden	WCDMA	4	Access by 1/1/02; licences pay 0.15% of revs; 99.% overall coverage by end-2003. Telia suing for licence
Switzerland	WCDMA	4	50% coverage by end-2004; launch 12/31/02. Government said in 8/01 that they were willing to push back launch
United Kingdom	WCDMA	5	80% coverage by end-2007
United States of America	CDMA2000 & WCDMA	NA	No special 3G licence required, can use regular spectrum; doubts of enough spectrum being available for WCDMA

- Feb. 1992 (Malaga) ITU-R WRC identifies IMT2000 frequency bands.

- Jan. 1998 (Paris) ETSI selects WCDMA for paired (FDD) and TD-CDMA for unpaired (TDD) UMTS operation out of five competing modes.

- Nov. 1999 (Helsinki) ITU approves IMT-2000 Radio Interface specifications including FDD and TDD modes approved in ITU meeting (M.1457).

Table 3.18   The comparison of major technical features between 3GPP HSDPA and 3GPP2 1xEV-DV

Feature	HSDPA	1xEV-DV
Downlink frame size	2 ms TTI (3 slots)	1.25, 2.5, 5, 10 ms variable frame size (1.25 ms slot size)
Channel feedback	Channel quality reported at 2 ms rate or 500 Hz	C/I feedback at 800 Hz (every 1.25 ms)
Data user multiplexing	TDM/CDM	TDM/CDM (variable frame)
Adaptive modulation and coding	QPSK & 16-QAM mandatory	QPSK, 8-PSK & 16-QAM
Hybrid-ARQ	Chase or incremental redundancy (IR)	Async. Incremental redundancy (IR)
Spreading factor	SF = 16 using UTRA OVSF codes	Walsh code length 32
Control channel approach	Dedicated channel pointing to shared channel	Common control channel
Peak data rate	8–10 Mbps (20 Mbps with MIMO)	3.1 Mbps

- Dec. 1999 (Nice) 3GPP approves UMTS Release 99 specifications both for FDD and TDD.

- Mar. 2001 (Palm Springs) 3GPP approves UMTS Release 4 specifications both for FDD and TDD.

To better comprehend where the UMTS standard stands in the ITU IMT-2000 proposals, we provide Figure 3.18, where we have plotted all major ITU endorsed IMT-2000 candidate proposals which are later called *3G standards*. From among all the proposals or standards that were listed, we classified them into (1) TWO core technologies (TDMA and CDMA); (2) THREE systems (UMTS, CDMA2000 and UWC-136 or EDGE); and (3) FIVE radio interfaces, which include (a) IMT-DS (Direct Spread), used in UTRA-FDD; (b) IMT-MC (Multi-Carrier), used in the CDMA2000 system; (c) IMT-TC (Time Code), used in UTRA-TDD and TD-SCDMA; (d) IMT-SC (Single Carrier), used in UWC-136 or EDGE technology; and (e) IMT-FT (Frequency Time), used in the DECT system.

3.2.2   ETSI UMTS versus ARIB WCDMA

In this section, we focus our discussions on the ETSI UMTS WCDMA [425] technology due to the reason that it was a standard release issued by 3GPP. All 3GPP parties should make their best effort to commit to full compatibility to the 3GPP releases, whose major versions are listed in Table 3.19. On the other hand, the Japanese version of WCDMA launched by NTT DoCoMo in October 2001, is also called the *ARIB WCDMA system* [431] or the Freedom of Mobile Multimedia Access (FOMA) service.[13] It has some technical differences in comparison to UMTS standard, and we offer some explanations in this subsection.

Members of the 3GPP include organizations such as ETSI of Europe and ARIB of Japan, and so on, and individual members and market representatives, such as the GSM Association, and the like.

---

[13]FOMA is short for *Freedom of Mobile Multimedia Access*, which has been used to name the 3G handphones developed by NTT DoCoMo.

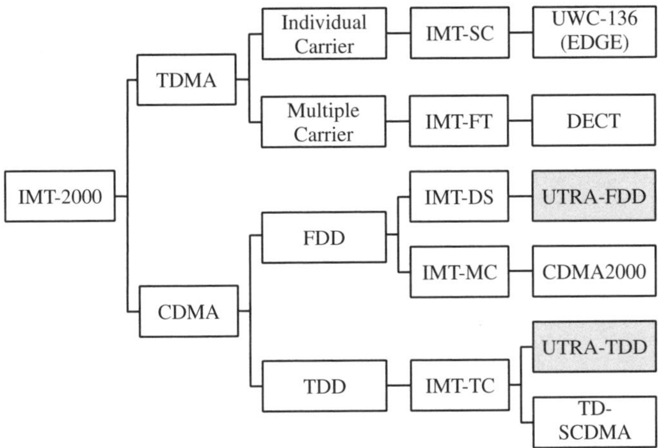

Figure 3.18 Family tree of all major ITU IMT-2000 candidate proposals.

Virtually every major OEM is included with the organizational and market representation partners. The 3GPP, similar to 3GPP2, is divided into several technical standards for their respective areas. Once these standards are written, the 3GPP endorses the standards and submits them to the ITU.

One aspect of 3G standard development that is often misunderstood by the public is the concept of releases, a system that also applies to 2G and 2.5G networks. 3G, in this case UMTS, does not consist only of one release, but a series of releases that build upon the previous releases. Initially, releases were noted by the year. For instance, Release'99, Release'00, and so on. However, later releases are no longer tied up to the year in which they are finalized. Instead, the 3GPP has defined the requirements in Release 4, has practically finalized Release 5, and has recently begun working on Release 6, all of which are subsequently the releases of the UMTS Release'99 standard.

What makes it even more complex is that within each release (e.g., Release'99) there are multiple versions. For instance, Release'99 began with the March 2000 version of the Release'99 standard, and has since evolved every three months since that time in conjunction with the quarterly 3GPP plenary meetings. Although the basic functionality of Release'99 does not change every quarter, the technical definition of how the functionality is implemented does change. Specifically, 3GPP members submit *Change Requests* (CRs), which identify changes to the baseline documentation. CRs can include anything from typographical or grammatical errors to additional/changed text that is inserted/replaced to clear up an ambiguity or correct an error, both of which could prevent a successful launch, in the documentation.

An interesting issue on the evolutional path of WCDMA technology is the compatibility between ARIB WCDMA technology [431] that was developed by NTT DoCoMo and UMTS-FDD [425] and proposed by ETSI.

In October 2001, NTT DoCoMo launched commercial *FOMA* services. Much has been reported on the launch, but we still believe that there are some widespread misunderstandings about what happened in Japan in comparison to European activities on UMTS-FDD.

FOMA networks use WCDMA, like the UMTS standard being deployed in Europe. However, FOMA uses an earlier version of the UMTS release. Without going into details, NTT DoCoMo got tired of waiting for Release'99 to become a standard. Instead, it elected to pursue 3G on its own, based upon the prefinalized Release'99 to meet its own particular technical requirements and subsequently required its suppliers to provide equipment (infrastructure and handsets) that met those requirements.

Table 3.19   Key Aspects of major 3GPP UMTS standard Releases

Release	Key Aspects
Release-99* (March 2000)	(1) New radio interface (UTRAN) (2) SMS, EMS, and MMS (3) FDD and TDD at 3.84 Mcps (4) Handover (5) CAMEL Phase 2 and 3 used for prepaid services and access charge in GSM and GPRS networks (6) EDGE (7) GSM-UMTS interworking (8) Call forwarding enhancement
Release-4 (March 2001)	(1) New TDD mode at 1.28 Mcps (2) Data synchronization (SyncML) (3) Evolution of UTRA to support IP (4) SMS and EMS enhancements (5) MMS Release 4 (6) Evolution of core transport to IP (7) UTRAN repeater
Release-5 (March 2002)	(1) Multirate speech codec (2) Provision of IP-based multimedia services (3) Wideband AMR (4) Security enhancements (5) IP multimedia services (6) MMS Release 5 (7) CAMEL Phase 4-optimal routing of mobile–mobile calls (8) Evolution of UTRAN transport to IP

*Release 99 of the UMTS standards is the first version to be deployed. Release 4 and Release 5 specify enhancements and optional features. A key philosophy of the 3GPP standards is that the first release specifies all mandatory features, while later releases add optional features only.

However, NTT DoCoMo publicly committed to bring its FOMA in line with the UMTS Release'99 in mid-2003.

FOMA has two operational modes: a dedicated 64 kbps circuit connection and a 384 kbps downlink, and a 64 kbps uplink best-effort connection. The dedicated 64 kbps circuit is currently intended for real-time services like video conference. It is important to note that UMTS and CDMA2000 do not have this 64 kbps circuit-switched mode. Thus, subscribers should not expect real-time high-data-rate services when the launches first occur. UMTS carriers will begin providing real-time services when they deploy a later release of the UMTS standard. For CDMA2000 carriers, they will not provide real-time services until they deploy 1xEV-DO (which is a data only scheme) or 1xE-DV (see Sections 3.1.13 and 3.1.12 for the details), which support higher peak data rates and have an all-IP core.

FOMA is a hybrid version of Release'99, but it will evolve to a higher compatibility with the UMTS standard. Before we move on to the details of the UMTS standard, we need to clear up any misunderstandings one may have about NTT DoCoMo's FOMA service and its relationship with the UMTS Release'99. NTT DoCoMo is a member of 3GPP and it is still involved in the 3GPP process. However, it was decided to deploy 3G services before the Release'99 standard was frozen for its own commercial consideration. Its FOMA service, therefore, was based on a prerelease version of the Release'99 standard. Since DoCoMo went at it alone, its 3G solution has evolved, and was not fully compatible with Release'99. However, DoCoMo has promised to make its FOMA service fully compatible with the UMTS standard in the next two years after its launch in October 2001.

3.2.3 UMTS Cell and Network Structure

UMTS can offer a different coverage or scale to different users. There are four different UMTS hierarchical cell structures altogether which include (1) picocell, which covers only a small area such as one office room; (2) microcell, which can cover a vicinity of several buildings to provide local UMTS services; (3) macrocell, which will span an area as large as a few kilometers in radius as a regional service provider; and finally (4) global cell, which will be covered by satellites and will be available to any place around the world. Under such a hierarchical cell structure, UMTS can provide services to the users located in various geographical regions on the earth. It is to be noted that the formation of a global cell needs to use technology other than UTRAN due to the nature of a long propagation delay in a satellite air-link sector.

Figure 3.19 shows a conceptual diagram of the UMTS hierarchical cell structure, which include all the four different cells.

A very basic UMTS network structure consists of three fundamental components: (1) Access Network, in which base stations play a key role in managing the air interface access between the UMTS network and *UE*; (2) CN, also called *Fixed Network*, which is responsible for handling all internal connections; and (3) Intelligent Network (IN), which is in charge of billing, subscriber location registration, roaming, handover, and so on. Figure 3.20 shows the UMTS basic network structure and the UMTS general reference architecture in the UMTS network. It can be seen from the figure that a *UMTS Terrestrial Radio Access Network* (UTRAN) contains several radio subsystems, so called *Radio Network Subsystems* (RNS) and contains functions for mobility management (MM). RNS controls the handover whenever a mobile changes cell, implements functions for encoding and administrates the resources of the UMTS radio interface. The U_u interface connects UTRAN with mobile end devices, or UE, and is comparable with U_m in a GSM network. UTRAN is connected over the I_u interface

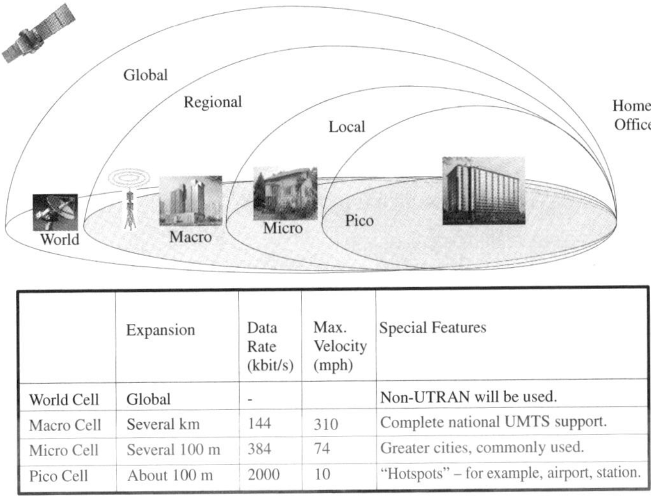

	Expansion	Data Rate (kbit/s)	Max. Velocity (mph)	Special Features
World Cell	Global	-		Non-UTRAN will be used.
Macro Cell	Several km	144	310	Complete national UMTS support.
Micro Cell	Several 100 m	384	74	Greater cities, commonly used.
Pico Cell	About 100 m	2000	10	"Hotspots" – for example, airport, station.

Figure 3.19 UMTS hierarchical cell structure, which includes four different cells: picocell, microcell, macrocell and global cell.

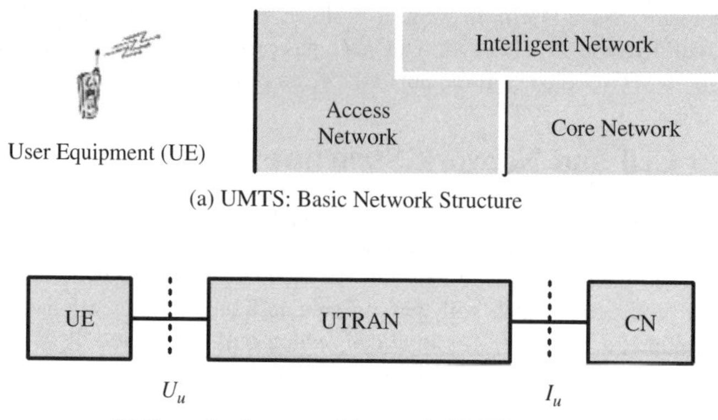

(a) UMTS: Basic Network Structure

(b) General reference architecture in UMTS network

Figure 3.20 (a) UMTS basic network structure, and (b) UMTS general reference architecture.

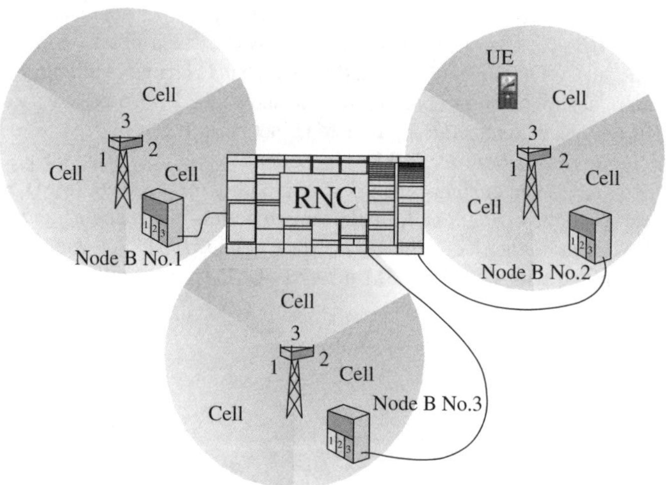

Figure 3.21 A typical UMTS network architecture, which consists of a radio network controller, cells, Node Bs, and UEs.

with the CN, comparable with the A interface in GSM between BSC and MSC. CN contains the interfaces to other networks and mechanisms for a connection handover to other systems.

A UMTS network can also be explained using a way commonly referred to in the literature, as shown in Figure 3.21, where there are four basic components, explained as follows:

- Cell: which specifies a basic coverage area. Hardware associated with the cell includes an antenna system, a high power amplifier (HPA), a transmitter, a receiver, and so on. The cell in UMTS is equivalent to a sector in GSM or CDMA2000.

- Node B: which is a common equipment at a cell site to control the cells, and is thus equivalent to RBS, BTS, or Base Station in GSM or CDMA2000.

- RNC: which is an equipment to control the Node Bs and interface them to the CN. This is equivalent to BSC in GSM or CDMA2000.

- UE: which is a subscriber equipment and equivalent to a mobile station in GSM or CDMA2000.

All of these naming conventions or acronyms will be extensively used in this section and followed whenever UMTS standard will be discussed.

3.2.4   UMTS Radio Interface

As mentioned earlier, the radio interface technology used in UMTS is called *UMTS Terrestrial Radio Access* (UTRA), in which two operating modes have been defined: UTRA *frequency division duplex* (FDD) and UTRA *time-division duplex* (TDD) modes.

In general, the UTRA-FDD operation mode is mainly suitable for suburban areas where a symmetrical transmission of speech and video is required. The data transmission rate in UTRA-FDD mode can go up to 384 kbps. The UTRA-FDD mode can also work for circuit- and packet-switched services in urban areas. On the other hand, the UTRA-TDD operation mode works mainly in households and other restricted areas, such as a company premises, and so on, similar to DECT. The UTRA-TDD is particularly suitable for the broadcast of speech and video, in both symmetrical (up to 384 kbps) and asymmetrical (up to 2 Mbps) ways.

Figure 3.22(a) draws a simple diagram to illustrate how the UTRA-FDD operation mode works in terms of its carrier frequency allocation. Similarly, Figure 3.22(b) depicts the operation principle for the UTRA-TDD scheme in the time and frequency domains.

It is seen from Figure 3.22 that UTRA-FDD places wideband CDMA (WCDMA) along with DSSS as a bandwidth expansion technique. All the channels are separated by carrier frequencies, spreading codes, and phase positions (only for uplink). There are 250 channels created for user data transmission in total, whose rates can go up to 384 kbps for mobile terminals. It has to be admitted that UTRA-FDD needs a relatively complex performance control mechanism due to the nature of FDD, which is particularly suitable for coverage driven roll-out.

UTRA-TDD makes use of wideband TDMA/CDMA techniques along with the DSSS scheme. Data signals can be sent and received on the same carrier due to the use of TDD to separate uplink and downlink transmissions. There are 120 channels created for user data transactions in total, whose rates can go up to 2 Mbps, which is higher than the UTRA-FDD scheme. Channel separation is implemented by using different spreading codes and time slots, and thus a lower spreading factor is required than that in the UTRA-FDD scheme. However, a UTRA-TDD operation needs cellwise precise synchronization in order to keep the same reference clock among different UEs. UTRA-TDD technology is suitable for small cells with more asymmetric traffic, as well as for unlicensed cordless and public wireless local loops. TDD technology is best suited for indoor use where interference from base stations is manageable and the lower range does not matter.

It is noted that FDD and TDD Node Bs can operate at the same RNC, as shown in Figure 3.23. The differences between UTRA-FDD and UTRA-TDD are listed in Table 3.20. Figure 3.24 depicts the frame structure used in UTRA-TDD.

To understand the UTRAN specifications better, one can find much more information from the 3GPP RAN documentations, which have been given in [309, 313]. One can easily download all of those specifications free of charge, with the need to register only once.

The general description of UTRAN specifications of 3GPP RAN are given in the following documentations:

- 3GPP TS 25.301: Radio Interface Protocol Architecture;
- 3GPP TS 25.302: Services provided by the physical layer;
- 3GPP TS 25.304: UE Procedures in Idle mode and Procedures for Cell Reselection in Connected Mode.

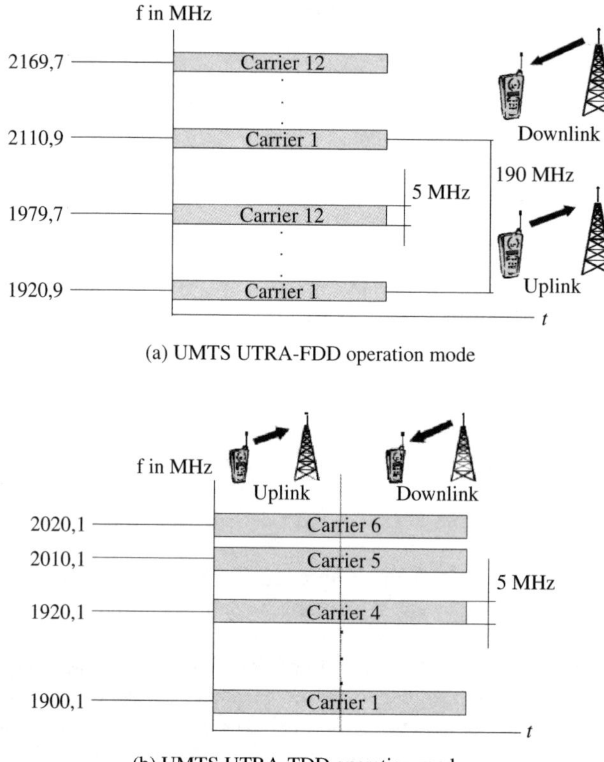

Figure 3.22 (a) UMTS UTRA-FDD operation mode, and (b) UMTS UTRA-TDD operation mode.

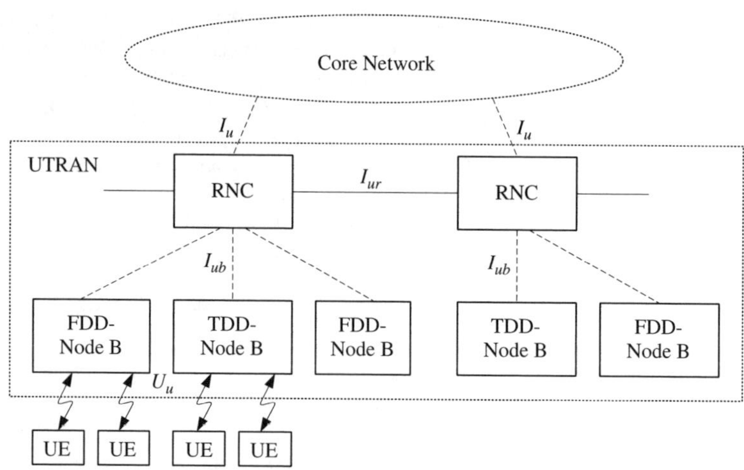

Figure 3.23 Illustration of coexistence of FDD and TDD Node Bs under the same UTRAN RNC in a UMTS system.

Table 3.20   Different parameters used in UTRA-FDD and UTRA-TDD modes

	UTRA-FDD	UTRA-TDD
Duplex method	FDD	TDD
Channel spacing (MHz)	5	5
Carrier chip rate (Mcps)	7.68 (HCR)	3.84 (LCR)
Time slot structure (slots/frame)	15	15
Frame length (ms)	10	10
Modulation	QPSK	QPSK
Detection	Based on pilot symbols	Based on midamble
Intrafrequency Handover	Soft	Hard (cell reselection)
Interfrequency Handover	Hard	Hard
Spreading factors	4...512	1...16

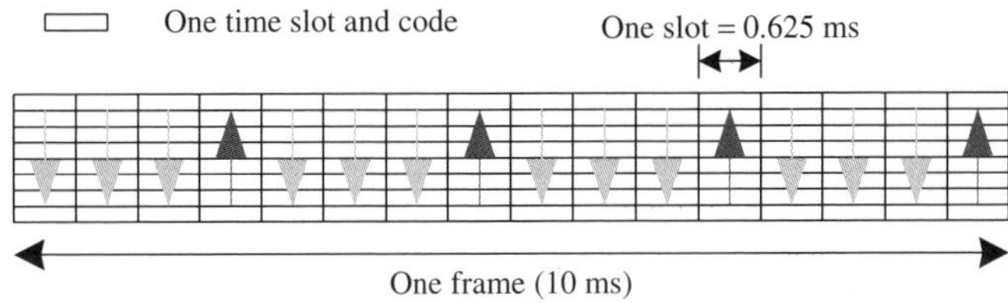

Figure 3.24   Illustration of UMTS UTRA-TDD frame structure.

The Layer 3 (RRC) protocols for both FDD and TDD operation modes are discussed in
- 3GPP TS 25.331: Description of the RRC Protocol.

The Layer 2 (MAC/RLC) specifications for both FDD and TDD operation modes are given in
- 3GPP TS 25.321: MAC Protocol Specification;
- 3GPP TS 25.322: Description of the RLC protocol.

The Layer 1 (physical layer) specifications for FDD operation mode are given in the documentations:
- 3GPP TS 25.211: Transport Channels and Physical Channels (FDD);
- 3GPP TS 25.212: Multiplexing and Channel Coding (FDD);
- 3GPP TS 25.213: Spreading and Modulation (FDD).

The Layer 1 (physical layer) specifications for TDD operation mode are given in the documentations:
- 3GPP TS 25.221: Transport Channels and Physical Channels (TDD);
- 3GPP TS 25.222: Multiplexing and Channel Coding (TDD);
- 3GPP TS 25.223: Spreading and Modulation (TDD).

3GPP UTRA specifications cover both the FDD and TDD operation modes. However, due to the limited space in this book, we shall not discuss them all in a very detailed manner. Therefore, in the discussions given in this book we focus on the FDD operation mode of UTRA. We discuss the TDD mode only if it is absolutely necessary, or if we need to compare the FDD mode with the TDD mode in terms of their performance and complexity, and so on.

3.2.5   UMTS Protocol Stack

The UMTS protocol stack architecture is shown in Figure 3.25, in which three main layers, the physical layer (L1), the MAC sublayer (L2), the Radio Link Control (RLC) sublayer (L2), and the Radio Resource Control (RRC) layer (L3) are illustrated. The *Control Plane* and *User Plane* in the UMTS protocol stacks are also shown.

Layer 1: Physical layer

Of course, the forward most part of the protocol stack in UMTS layered architecture is the physical layer, which offers information transfer services to the MAC layer. These services are denoted as Transport channels. There are also Physical channels, which is comprised of the following major functions:

- Various handover functions;
- Error detection and report to higher layers;
- Multiplexing of transport channels;
- Mapping of transport channels to physical channels;
- Fast Close loop Power control;
- Frequency and Time Synchronization;
- Other responsibilities associated with transmitting and receiving signals over the radio media.

The complete list and explanations on all physical channels will be given in the subsection followed.

Layer 2-1: MAC sublayer

The MAC sublayer offers data transfer to RLC and higher layers. The MAC sublayer is comprised of the following functions:

- The selection of appropriate TF (basically bit rate), within a predefined set, per information unit delivered to the physical layer;

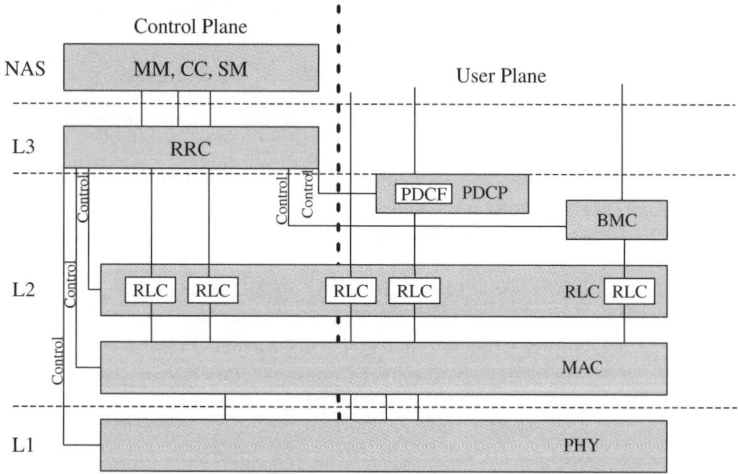

Figure 3.25   The UMTS protocol stack architecture (MM: mobility management; CC: call control; SM: service management)

- Service multiplexing on RACH, FACH, and dedicated channels;
- Priority handling between "data flows" of one user as well as between data flows from several users, the latter being achieved by means of dynamic scheduling;
- Access control on RACH;
- Address control on RACH and FACH;
- Contention resolution on RACH.

It is to be noted that the term "sublayer" is used here to distinguish the full layers, such as "physical layer," and so on, as MAC sublayer itself does not form a full layer. Instead, it is only a part of Layer 2 in the UTRA protocol stack structure. The MAC sublayer along with the RLC sublayer will form Layer 2.

Layer 2-2: RLC sublayer

The Radio Link Control (RLC) sublayer offers the following services to the higher layers:
- Layer 2 connection establishment/release;
- Transparent data transfer, that is, no protocol overhead is appended to the information unit received from the higher layer;
- Assured and unassured data transfer.

The RLC sublayer is comprised of the following functions:
- Segmentation and assembly;
- Transfer of user data;
- Error correction by means of retransmission optimized for the WCDMA physical layer;
- Sequence integrity (used by the control plane at the very least);
- Duplicate detection;
- Flow control;
- Ciphering.

Layer 3: RRC layer

The *Radio Resource Control* (RRC) layer is also called *Layer 3* in the UMTS protocol stack and offers the CN the following services:
- General control service, which is used as an information broadcast service;
- Notification service, which is used for the paging and notification of a selected UEs;
- Dedicated control service, which is used for the establishment/release of a connection and transfer of messages using the connection.

The RRC layer is comprised of the following functions:
- Broadcasting information from the network to all UEs;
- Radio resource handling (e.g. code allocation, handover, admission control, and measurement reporting/control);
- QoS Control;
- UE measurement reporting and control of the reporting;
- Power Control, Encryption, and Integrity protection.

3.2.6 UTRA Channels

The UTRA-FDD radio interface has logical channels, which are mapped to transport channels. The transport channels are further mapped to physical channels. Logical to Transport channel conversion happens in the *MAC* layer, which is a lower sublayer in the *Data Link Layer* (Layer 2).

There are six different *Logical Channels* in total, which are listed as follows:

- Broadcast Control Channel (BCCH), (DL);
- Paging Control Channel (PCCH), (DL);
- Dedicated Control Channel (DCCH), (UL/DL);
- Common Control Channel (CCCH), (UL/DL);
- Dedicated Traffic Channel (DTCH), (UL/DL);
- Common Traffic Channel (CTCH), (broadcasting).

where the abbreviations of "DL" and "UL" stand for downlink and uplink, respectively.

There are seven different *Transport Channels* in total:

- Dedicated Transport Channel (DCH), UL/DL, mapped to DCCH and DTCH;
- Broadcast Channel (BCH), DL, mapped to BCCH;
- Forward Access Channel (FACH), DL, mapped to BCCH, CCCH, CTCH, DCCH, and DTCH;
- Paging Channel (PCH), DL, mapped to PCCH;
- Random Access Channel (RACH), UL, mapped to CCCH, DCCH, and DTCH;
- Uplink Common Packet Channel (CPCH), UL, mapped to DCCH and DTCH;
- Downlink Shared Channel (DSCH), DL, mapped to DCCH and DTCH.

There are 13 different *Physical Channels* in total:

- Primary Common Control Physical Channel (P-CCPCH), mapped to BCH;
- Secondary Common Control Physical Channel (SCCPCH), mapped to FACH, PCH;
- Physical Random Access Channel (PRACH), mapped to RACH;
- Dedicated Physical Data Channel (DPDCH), mapped to DCH;
- Dedicated Physical Control Channel (DPCCH), mapped to DCH;
- Physical Downlink Shared Channel (PDSCH), mapped to DSCH;
- Physical Common Packet Channel (PCPCH), mapped to CPCH;
- Synchronization Channel (SCH);
- Common Pilot Channel (CPICH);
- Acquisition Indicator Channel (AICH);
- Paging Indication Channel (PICH);
- CPCH Status Indication Channel (CSICH);
- Collision Detection/Channel Assignment Indication Channel (CD/CA-ICH).

Figure 3.26 shows all the physical, transport, and logical channels, as well as their mapping relations among them in UMTS UTRA. It is noted that the mapping from logical channels to transport channels happens on the MAC layer (L2); while the mapping from transport channels to physical channels happen on the physical layer (L1). Figure 3.27 shows the general architecture of Layers 1, 2, and 3 in 3GPP UTRA standard. It is also clearly shown that the mapping from the logical channels to transport channels happens on the MAC layer.

Transport channels

More detailed information about the transport channels are given in the subsequent text. As shown in Figure 3.27, all the transport channels carry services offered by Layer 1 to the higher layers. A transport channel is defined by how and with what characteristics data is transferred over-the-air-interface.

There are two groups of transport channels: *Dedicated Transport Channels*, and Common Transport Channels (CTC). Only one Dedicated Transport Channel exists, that is, *DCH*, which is a downlink or an uplink transport channel, and is transmitted over the entire cell or only over a part of the cell using, for example, beam-forming antennas. DCH carries both the service data, such as speech frames

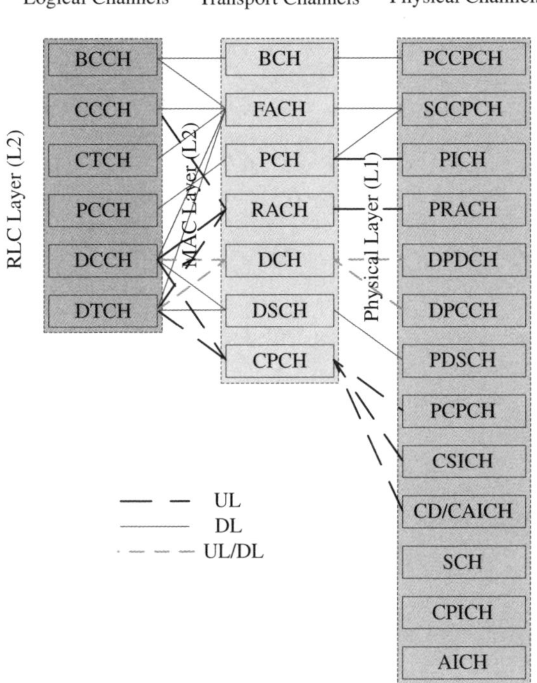

Figure 3.26 Physical, transport, and logical channels in the UMTS standard and their mapping relation among them.

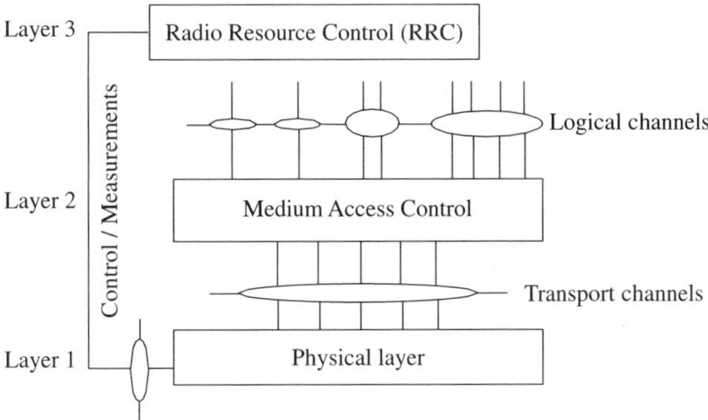

Figure 3.27 Illustration of Layer 1, 2, and 3 architecture in 3GPP UTRA specifications.

and higher-layer control information, such as handover commands or measurement reports from the terminal.

The content of the information carried on the DCH is not visible to the physical layer, and thus higher-layer control information and user data are treated in the same way. The physical layer parameters set by UTRAN may vary between control and data. DCH supports possible fast rate change (every 10 ms), fast power control, as well as soft handover.

However, there are six *CTC*, which are divided between all or a group of users in a cell. It is noted that they do not support soft handover, but some of them can support fast power control. They include: (1) BCH: Broadcast Channel; (2) FACH: Forward Access Channel; (3) PCH: Paging Channel; (4) RACH: Random Access Channel; (5) CPCH: Common Packet Channel; and (6) DSCH: DL Shared Channel.

BCH is a downlink transport channel that is used to broadcast system and cell-specific information. BCH is always transmitted over the entire cell. The most typical data needed in every network is the available random access codes and access slots in the cell, or the types of transmit diversity. BCH is transmitted with relatively high power. Single transport format offers a low and fixed data rate for the UTRA BCH to support low-end terminals.

PCH is also a downlink transport channel. PCH is always transmitted over the entire cell. PCH carries data relevant to the paging procedure, that is, when the network wants to initiate communication with the terminal. The identical paging message can be transmitted in a single cell or in up to a few hundreds of cells, depending on the system configuration.

Random Access Channel (RACH) is an uplink transport channel. RACH is intended to be used to carry control information from the terminals, such as requests to set up a connection. RACH can also be used to send small amounts of packet data from the terminal to the network. The RACH is always received from the entire cell. The RACH is characterized by collision risk. RACH is transmitted using open-loop power control.

Forward Access Channel (FACH) is a downlink transport channel. FACH is transmitted over the entire cell or over only a part of the cell using beam-forming antennas. FACH can carry control information, for example, after a random access message has been received by the base station. FACH can also transmit packet data. FACH does not use fast power control. FACH can be transmitted using slow power control. There can be more than one FACH in a cell. The messages transmitted need to include in-band identification information.

CPCH is an optional uplink transport channel. CPCH is an extension to the RACH channel that is intended to carry packet-based user data. CPCH is associated with a dedicated channel on the downlink that provides power control and CPCH Control Commands (e.g. Emergency Stop) for the uplink CPCH. The CPCH is characterized by initial collision risk and by using inner-loop power control. The CPCH may last several frames.

DSCH is an optional downlink transport channel shared by several UEs to carry dedicated user data and/or control information. The DSCH is always associated with one or several downlink DCH. The DSCH is transmitted over the entire cell or over only a part of the cell using beam-forming antennas. DSCH supports fast power control and variable bit rate on a frame-by-frame basis.

Physical channels

The mapping between the transport channels and physical channels happens in the UTRA physical layer, as shown in Figure 3.26. We would like to discuss uplink physical channels, followed by the downlink physical channels.

There are two Dedicated Uplink Physical Channels: *Uplink Dedicated Physical Data Channel* (UL DPDCH) and *Uplink Dedicated Physical Control Channel* (UL DPCCH). Also, there are two *Common Uplink Physical Channels*, which are *Physical Random Access Channel* (PRACH) and *Physical Common Packet Channel* (PCPCH).

The UL DPDCH carries the DCH transport channel (generated at Layer 2 and above). There may be zero, one, or several uplink DPDCHs on each radio link. The UL DPCCH carries control information generated at Layer 1. Only one UL DPCCH exists on each radio link.

The PRACH is used to carry the RACH. The random access transmission is based on a slotted ALOHA approach with fast acquisition indication. The UE can start the random-access transmission at the beginning of a number of well-defined time intervals, denoted as access slots. There are 15 access slots per two frames and they are spaced 5120 chips apart. Information on what access slots are available for random-access transmission is given from higher layers.

The PCPCH is used to carry the CPCH, whose transmission is based on Digital Sense Multiple Access – Collision Detection (DSMA-CD) approach[14] with fast acquisition indication. The UE can start transmission at the beginning of a number of well-defined time intervals.

There is only one type of downlink Dedicated Physical Channel (DPCH), that is *Downlink DPCH* (DL DPCH). Within one downlink DPCH, dedicated data generated at Layer 2 and above, that is, the dedicated transport channel (DCH), is transmitted in the time-multiplex with control information generated at Layer 1 (known pilot bits, Transmit Power Control (TPC) commands, and an optional TFCI).

Common Pilot Channel (CPICH) is a fixed rate (30 kbps, SF = 256) downlink physical channel that carries a predefined bit/symbol sequence. In case transmit diversity (open or closed loop) is used on any downlink channel in the cell, the CPICH should be transmitted from both antennas using the same channelization and scrambling code. There are two types of Common pilot channels: The Primary CPICH and the Secondary CPICH.

The *Primary Common Pilot Channel* (P-CPICH) has the following characteristics: The same channelization code is used for the P-CPICH; The P-CPICH is scrambled by the primary scrambling code; There is one and only one P-CPICH per cell; The P-CPICH is broadcast over the entire cell. The Primary CPICH is a phase reference for the following downlink channels: SCH, Primary CCPCH, AICH, PICH AP-AICH, CD/CA-ICH, CSICH, DL-DPCCH for CPCH and the SCCPCH. By default, the Primary CPICH is also a phase reference for downlink DPCH and any associated PDSCH. The Primary CPICH is always a phase reference for a downlink physical channel using closed-loop transmit diversity.

A *Secondary Common Pilot Channel* (S-CPICH) has the following characteristics: An arbitrary channelization code of SF = 256 is used for the S-CPICH; A S-CPICH is scrambled by either the primary or a secondary scrambling code; There may be zero, one, or several S-CPICHs per cell; A S-CPICH may be transmitted over the entire cell or only over a part of the cell; A Secondary CPICH may be a phase reference for a downlink DPCH. The Secondary CPICH can be a phase reference for a downlink physical channel using open-loop transmit diversity, instead of the Primary CPICH being a phase reference.

PCCPCH bears the following characteristics: it has a fixed rate: 30 kbps, SF = 256, and is used to carry the BCH transport channel. Neither TPC commands, nor TFCI nor pilot bits will be sent in P-CCPCH *Secondary Common Control Physical Channel* (SCCPCH) is used to carry the FACH and PCH. Two types of SCCPCHs exist: those that include TFCI and those that do not include TFCI. It is the UTRAN that determines if a TFCI should be transmitted, hence making it mandatory for all UEs to support the use of TFCI.

Synchronization Channel (SCH) is a downlink signal used for cell search. The SCH consists of the Primary and Secondary SCH. The 10 ms radio frames of the Primary and Secondary SCH are divided into 15 slots, each having a length of 2560 chips.

---

[14]DSMA/CD was initially used in Cellular Digital Packet Data (CDPD) networks as an access protocol. DSMA/CD is quite similar to the nonpersistent CSMA/CD. The difference is that the status of the shared channel cannot be sensed directly in CDPD networks. The status of the reverse channel is observed using special flags on the forward channel, because these flags are carried on the forward channel.

Physical Downlink Shared Channel (PDSCH) is used to carry the *DSCH*. A PDSCH corresponds to a channelization code below or at a PDSCH root channelization code. A PDSCH is allocated on a radio frame basis to a UE. Within one radio frame, UTRAN may allocate different PDSCHs under the same PDSCH root channelization code to different UEs based on code multiplexing. Within the same radio frame, multiple parallel PDSCHs, with the same spreading factor, may be allocated to a single UE. All the PDSCHs are operated with radio frame synchronization.

The *Acquisition Indicator Channel* (AICH) is a fixed rate (SF = 256) physical channel used to carry *Acquisition Indicators* (AI), which correspond to signatures on the PRACH.

CPCH Access Preamble Acquisition Indicator Channel (AP-AICH) is a fixed rate (SF = 256) physical channel used to carry the AP acquisition indicators (API) of CPCH. AP AI APIs correspond to the AP signature transmitted by UE.

CPCH Collision Detection/Channel Assignment Indicator Channel (CD/CA-ICH) is a fixed rate (SF = 256) physical channel used to carry the CD Indicator (CDI) only if the CA is not active, or to carry CD Indicator/CA Indicator (CDI/CAI) at the same time if the CA is active.

Paging Indicator Channel (PICH) is to provide terminals with efficient sleep mode operation. In order to detect the PICH, the terminal needs to obtain the phase reference from the CPICH, and as with the AICH, the PICH needs to be heard by all the terminals in the cell and thus needs to be sent at a high power level without power control. The PICH is a fixed rate (SF = 256) physical channel used to carry the paging indicators. The PICH is always associated with an SCCPCH to which a PCH transport channel is mapped.

CPCH Status Indicator Channel (CSICH) is a fixed rate (SF = 256) physical channel used to carry CPCH status information. The CSICH bits indicate the availability of each physical CPCH channel and are used to tell the terminal to only initiate access on a free channel, but, on the other hand, to accept a channel assignment command to an unused channel. A CSICH is always associated with a physical channel used for the transmission of CPCH AP-AICH and uses the same channelization and scrambling codes.

3.2.7   UTRA Multiplexing and Frame Structure

The frame structure is associated with specific transport or physical channels. In UTRA specifications, different transport channels are generally given distinct frame structures to fit particular requirements in terms of their contents, data rates, multiple access schemes, duplex techniques (FDD or TDD), downlink or uplink, and so on. Therefore, providing detailed information about the frame structures for all the different transport and physical channels will take up too much room, and thus we are not allowed to do so in this section. Instead, we should only present some discussions on the generic UTRA multiplexing and frame structures here, along with some examples.

As shown in Figure 3.20, the detailed operational parameters used in both FDD and TDD modes are given and then compared. In January 1998, ETSI has decided that the UMTS should be given an option to operate on two different duplex modes, FDD and TDD modes. The WCDMA technology was chosen for wide-area services and will use paired FDD bands: 1920–1980 MHz for uplink and 2110–2170 MHz for downlink. On the other hand, TD/CDMA was chosen for private, indoor services in unpaired TDD bands, that is, 1900–1920 MHz and 2010–2025 MHz.

UTRA-FDD

The UTRA-FDD operation mode is the multiplexing scheme, which separates downlink and uplink transmissions in different carriers, to implement full-duplex operation in radio links. Several salient features exist in the UTRA-FDD operation mode as listed below:
- It uses wideband DS-CDMA technology;
- It uses 4.096 Mcps chip rate, which is expandable to 8.192/16.384 Mcps;

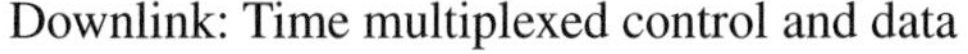

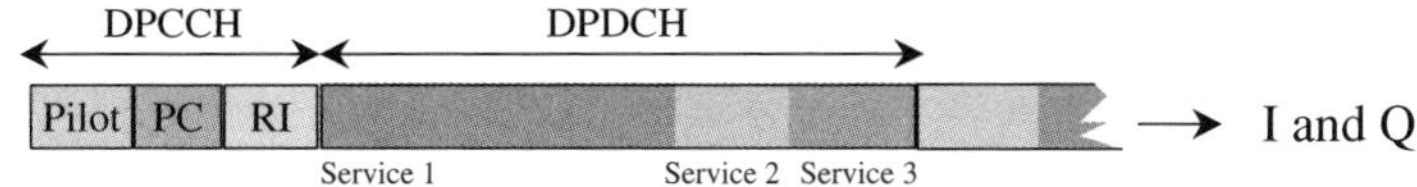

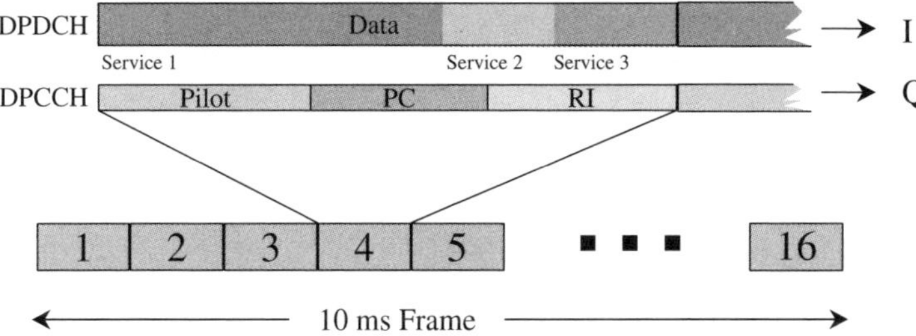

Figure 3.28 UTRA-FDD downlink and uplink time multiplexing schemes.

- It supports asynchronous base stations;
- It uses variable spreading and multicode operation;
- It enables coherent detection in both up- and downlinks;
- It offers optimized packet access on common or dedicated channels.

Figure 3.28 illustrates the frame structure used for the UTRA-FDD operation mode. It is seen from the figure that the frame length is 10 ms within total 16 slots. The downlink and uplink will use different multiplexing schemes. The downlink uses time multiplexed control and data frames, which will be transmitted via I and Q channels in quadrature digital modem. On the other hand, the uplink will use I/Q code multiplexed control and data frames, in which data signals will be sent in the *Dedicated Physical Data Channel* (DPDCH) channel, and control signals will be sent in the *Dedicated Physical Control Channel* (DPCCH) channel.

UTRA-FDD can select various ways to map transport channels into physical channels, depending on the different operational requirements, as shown in Figure 3.29. Figure 3.30 shows the super frame, frame and slot structures for the UTRA-FDD physical channels, the DPCCH, and the DPDCH.

UTRA-TDD

Time Division Duplex (TDD) mode should use harmonized parameters to UTRA-FDD, such as chip rate, frame length and slot size, modulation, and so on. Because of the relatively low average transmission power, UTRA-TDD is intended first and foremost for private, uncoordinated systems, which can be deployed in unpaired UMTS bands. This is of extreme importance especially for those countries where the spectral allocation has become very difficult.

Each 0.625 ms slot in UTRA-TDD can be allocated to either uplink or downlink transmissions, as shown in Figure 3.24. However, at least one slot should be assigned to downlink (BCCH) and one to uplink (RACH). The same asymmetry and frame synchronization is needed within continuous areas in coordinated systems. Up to eight codes are used for multiple access in each slot. It allows multicode transmission. Different users can share the same time slot. Since only a few codes are used in each time slot, joint detection is supported easily in the UTRA-TDD operation mode.

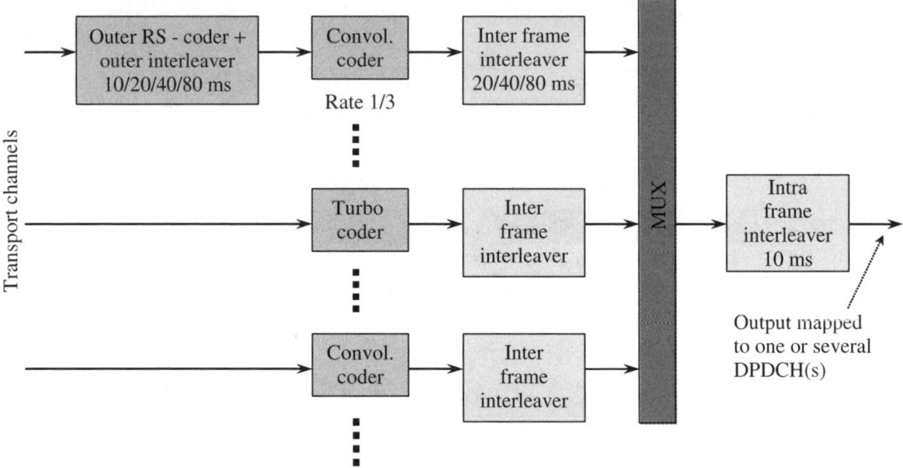

Figure 3.29 Mapping from transport channels to physical channels using deferent coding and multiplexing schemes in the UTRA-FDD.

3.2.8   Spreading and Carrier Modulations

Like any CDMA system, UMTS also needs spreading and carrier modulations. The carrier modulation is for sending baseband signals into the air through the radio frequency carrier. The spreading modulation functions as a vehicle to span the spectrum and implement multiple access.

OVSF codes

Both UTRA-FDD and UTRA-TDD use *Orthogonal Variable Spreading Factor* (OVSF) codes for spreading modulation. The chip duration of the OVSF codes is T_c. The OVSF codes can be generated from its unique tree structure. The OVSF codes in UTRA systems perform the following functions:
- Widen the band from $1/T_b$ to $1/T_c$;
- Characterize the users and user services in downlink;
- Characterize the user services in uplink.

The tree structure of OVSF codes is shown in Figure 3.31. There are several important characteristic features for the OVSF codes:
- They can maintain orthogonality among all the leaves (which is defined as the "end" of each branch);
- The number of available OVSF codes is exactly equal to the spreading factor (SF);
- They can be completely orthogonal if they operate in exactly synchronized channels.

In the discussions given in this section, we will concentrate on the UTRA-FDD operation mode due to the constraint on space.

Uplink spreading and modulation

Figure 3.32 illustrates the spreading and modulation for a single uplink DPDCH. Data modulation is carried out by dual-channel QPSK, where the uplink DPDCH and DPCCH are mapped to the I and Q branches, which are then spread to the chip rate with two different channelization codes c_D/c_C and subsequently complex scrambled by a mobile station–specific complex scrambling code c_{scramb}.

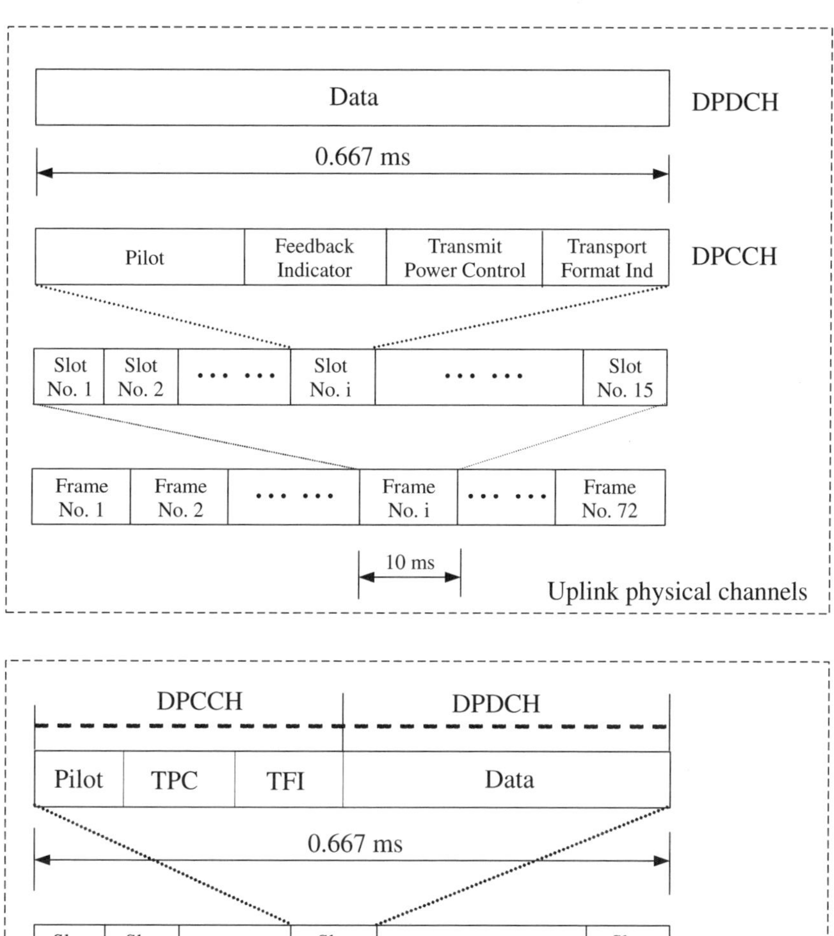

Figure 3.30 Super frame, frame, and slot structures for the UTRA-FDD physical channels, the DPCCH, and the DPDCH.

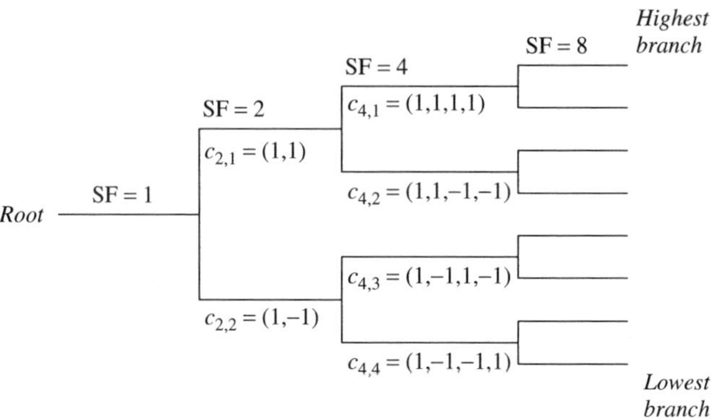

Figure 3.31 OVSF code generation tree in UTRA.

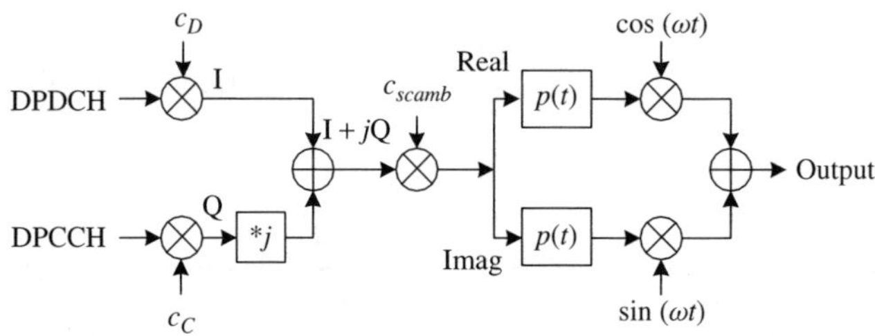

Figure 3.32 Spreading/modulation for uplink DPDCH/DPCCH in UTRA ("*j" denotes multiplication of imaginary part symbol).

For multicode transmission, each additional uplink DPDCH may be transmitted on either the I or the Q branch. For each branch, each additional uplink DPDCH should be assigned its own channelization code. Uplink DPDCHs on different branches may share a common channelization code.

The spreading and modulation of the message part of the random-access burst is basically the same as for the uplink dedicated physical channels, as shown in Figure 3.32, where the uplink DPDCH and uplink DPCCH are replaced by the data part and the control part, respectively. The scrambling code for the message part is chosen based on the base station-specific preamble code, the randomly chosen preamble sequence, and the randomly chosen access slot (random-access time-offset). This guarantees that two simultaneous random-access attempts that use different preamble codes and/or different preamble sequences will not collide during the data part of the random-access bursts.

The channelization codes of Figure 3.32 are the same type of OVSF codes as used in the downlink, as shown in Figure 3.33. Each connection is allocated at least one uplink channelization code, to be used for the uplink DPCCH. In most cases, at least one additional uplink channelization code is allocated for a uplink DPDCH. Further uplink channelization codes may be allocated if more than one uplink DPDCH is required. As different mobile stations use different uplink scrambling codes, the uplink channelization codes may be allocated with no coordination between different connections. Therefore, the uplink channelization codes are always allocated in a predefined order. The mobile station and network only need to agree on the number and length (spreading factor) of the uplink channelization codes.

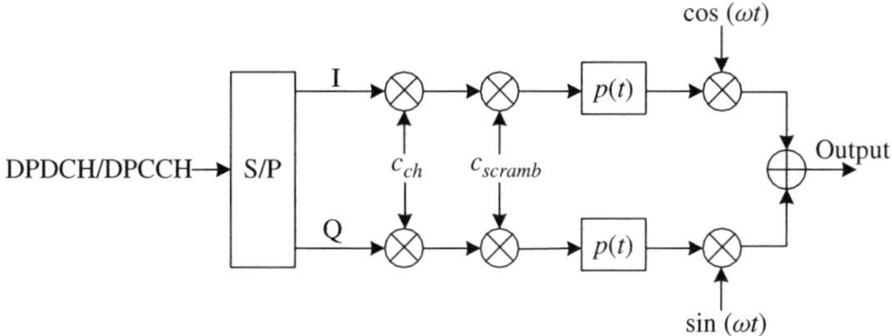

Figure 3.33 Spreading/modulation for downlink DPDCH/DPCCH in UTRA, where c_{ch}: channelization code; c_{scramb}: scrambling code; and $p(t)$: pulse-shaping filter (root-raised cosine, roll-off rate of 0.22).

Either short or long scrambling codes should be used on uplink. The short scrambling code is a complex code $c_{scramb} = c_I + jc_Q$, where c_I and c_Q are two different codes from the extended Very Large Kasami set of length 256. The network decides the uplink short scrambling code. The mobile station is informed as regards the type of short scrambling code to use in the downlink Access Grant Message that is the base-station response to an uplink Random Access Request. The short scrambling code may, in rare cases, be changed during a connection.

The long uplink scrambling code is typically used in cells without multiuser detection in the base station. The mobile station is informed if a long scrambling code should be used in the Access Grant Message following a Random-Access Request and in the handover message. The type of long scrambling code to be used is given directly by the short scrambling code. No explicit allocation of the long scrambling code is thus needed. The scrambling code sequences are constructed as the position wise modulo-2 sum of 40,960 chip segments of two binary m-sequences generated by means of two generator polynomials of degree 41. Let x and y be the two m-sequences respectively. The x sequence is constructed using the primitive (over GF(2)) polynomial $1 + X^3 + X^{41}$. The y sequence is constructed using the polynomial $1 + X^{20} + X^{41}$. The resulting sequences thus constitute segments of a set of Gold sequences. The scrambling code for the quadrature component is a 1024-chip shifted version of the in-phase scrambling code. The uplink scrambling code word has a period of one radio frame of 10 ms.

For random access channels, the spreading code for the preamble part is cell-specific and is broadcast by the base station. More than one preamble code can be used in a base station if the traffic load is high. The preamble codes must be code planned, since two neighboring cells should not use the same preamble code. The code used is a real-valued 256 chip Gold code. All 256 codes are used in the system. The preamble codes are generated in the same way as the codes used for the downlink synchronization channel.

The modulating chip rate is 4.096 Mcps. This basic chip rate can be extended to 8.192 or 16.384 Mcps. The pulse-shaping filters are *root-raised cosine* (RRC) with a roll-off factor of $\alpha = 0.22$ in the frequency domain. QPSK modulation is used.

Downlink spreading and modulation

Figure 3.33 illustrates the spreading and modulation for the downlink DPCH. Data modulation is QPSK, in which each pair of two bits are serial-to-parallel converted and mapped to the I and Q branches, respectively. The I and Q branches are then spread to the chip rate with the same channelization code c_{ch} (real spreading) and subsequently scrambled by the same cell specific scrambling code c_{scramb} (real scrambling).

For multicode transmission, each additional downlink DPCH should also be spread/modulated according to Figure 3.33. Each additional downlink DPCH should be assigned its own channelization code.

The channelization codes of Figure 3.33 are OVSF codes that preserve the orthogonality between downlink channels of different rates and spreading factors. The OVSF codes have been defined in Figure 3.31. Each level in the code tree defines channelization codes of length SF, corresponding to a spreading factor of SF in Figure 3.33. All codes within the code tree cannot be used simultaneously within one cell. A code can be used in a cell if and only if no other code on the path from the specific code to the root of the tree or in the subtree below the specific code is used in the same cell. This means that the number of available channelization codes is not fixed but depends on the rate and SF of each physical channel. The channelization code for the BCCH is a predefined code that is the same for all cells within the system. The channelization code(s) used for the *SCCPCH* is broadcast on the BCCH. The channelization codes for the downlink dedicated physical channels are decided by the network. The mobile station is informed about the type of downlink channelization codes to receive in the downlink Access Grant Message that is the base-station response to an uplink Random Access Request. The set of channelization codes may be changed during a connection, typically as a result of a change of service or an intercell handover. A change of downlink channelization codes is negotiated over a DCH.

The total number of available scrambling codes is 512, divided into 32 code groups with 16 codes in each group. The grouping of the downlink codes is done in order to facilitate a fast cell search. The downlink scrambling code is assigned to the cell (sector) at the initial deployment. The mobile station learns about the downlink scrambling code during the cell search process. The scrambling code sequences are constructed as the position wise modulo-2 sum of 40,960 chip segments of two binary m-sequences generated by means of two generator polynomials of degree 18. Let x and y be the two sequences respectively. The x sequence is constructed using the primitive (over GF(2)) polynomial $1 + X^7 + X^{18}$. The y sequence is constructed using the polynomial $1 + X^5 + X^7 + X^{10} + X^{18}$. The resulting sequences thus constitute segments of a set of Gold sequences. The scrambling codes are repeated for every 10 ms radio frame.

The modulating chip rate is 4.096 Mcps. This basic chip rate can be extended to 8.192 or 16.384 Mcps. The pulse-shaping filters are *root-raised cosine* (RRC) with a roll-off factor of $\alpha = 0.22$ in the frequency domain.

3.2.9 Packet Data

In the UMTS WCDMA systems, data applications are expected to dominate the overall traffic volume. The salient feature of 3G wireless is its capability to support packet-switched[15] traffic if compared to the 2G systems discussed earlier. The need for all-IP wireless services makes this feature of UMTS WCDMA system even more important.

In the packet-switched operation mode, the traffic generated by data applications is inherently bursty and asymmetric by nature, with higher data rates in the downlink than those in the uplink. In addition, the reverse transmissions of all users in one cell share the same set of OVSF channelization codes. Therefore, optimal resource utilization is essential in the downlink.

In UTRA-FDD WCDMA, there are three types of downlink transport channels that can be used to transmit bursty packet data: common, dedicated, and shared transport channels. Among these, the dedicated and shared channels are suited for the transmission of a medium to large amount of data, while the common channels are suited for the transmission of a small amount of data; such as signaling data or small IP packets. Consequently, we have to choose between the dedicated and shared channels for the transmission of bursty packet data.

---

[15]The difference between packet-switched and circuit-switched techniques is explained in Section 2.6.

The *downlink dedicated channel* (DCH) has a fixed SF that does not vary on a frame-by-frame basis and is determined by the highest transmission rate of the source. The variable data rate transmission may be implemented with *discontinuous transmission* (DTX) by gating the transmission on and off or by the use of flexible positions. If flexible positions are used, the transmission is continuous and the DTX is implemented by repeating the transmitted bits. Therefore, the use of the DCH channel for the transmission of bursty data is not efficient, as it results in low OVSF code utilization and decreased system capacity.

On the other hand, shared channels are made for the transmission of bursty data as they allow a single OVSF channelization code to be shared among several users. However, the use of the shared channels has some restrictions. The most important restrictions are: (1) The OVSF codes must be allocated from the same branch of the code tree. The root code of the DSCH subtree defines the SF for maximum data rate transmission, while the rest of the codes are used when lower rates are needed. (2) The DSCH does not support SF = 512 (the highest SF of the OVSF code tree). (3) HS-DSCH has a fixed SF.

A comparison among the DCH, DSCH, HS-DSCH and FACH channels in terms of their packet data operation parameters is shown in Table 3.21. Figure 3.34 shows the time diagrams for small/infrequent and large/frequent packet access in UTRA-FDD operation mode.

3.2.10 Power Control

Unlike the power control schemes used in IS-95, UMTS defines three main dissimilar power control mechanisms, that is, (1) Open-loop power control; (2) Inner-loop power control; and (3) Outer-loop power control, which will be introduced here.

Open-loop power control

In the UMTS standard, *Open-Loop Power Control* is defined as the ability of the UE transmitter to set its output power to a specific value. It is used for setting initial uplink and downlink transmission powers when a UE is accessing the network. The open-loop power control tolerance is ±9 dB (under normal conditions) and ±12 dB (under extreme conditions), respectively. Figure 3.35 shows the major functional blocks involved with power control mechanism in a UTRA transceiver.

Inner-loop power control

In UMTS, *Inner-Loop Power Control*, also called *Fast Closed-Loop Power Control*, in the uplink is defined as the ability of the UE transmitter to adjust its output power in accordance with one or more

Table 3.21   A comparison on packet data operation among the DCH, DSCH, HS-DSCH and FACH channels

Channel	HS-DSCH	DSCH	DCH (DL)	FACH
Spreading factor	Fixed 16	Variable (4–256) frame-by-frame	Fixed (4–512)	Fixed (4–256)
Power control	Fixed/slow power setting	Fast, based on associated DCH	Fast at 1500 kHz	Fixed/slow power setting
Interleaving	2 ms	10–80 ms	10–80 ms	10–80 ms
Soft handover	For associated DCH	For associated DCH	Yes	No
3GPP Release Version	Release 5	Release'99	Release'99	Release'99

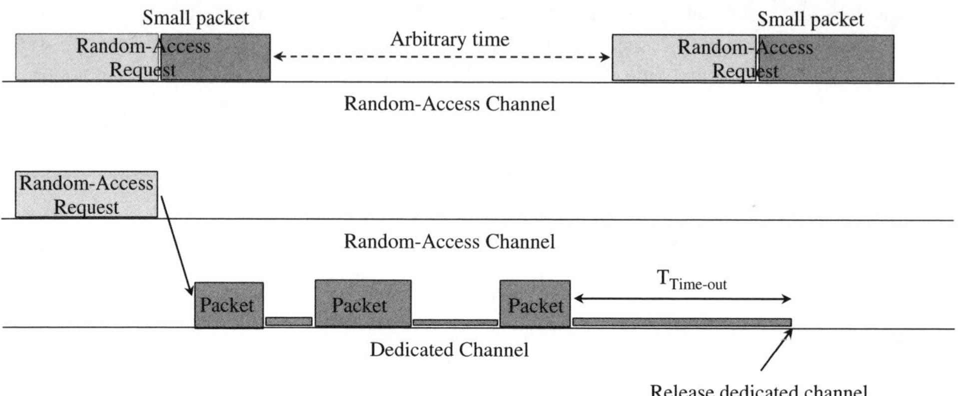

Figure 3.34 UTRA-FDD packet access with small infrequent and large frequent packet transactions.

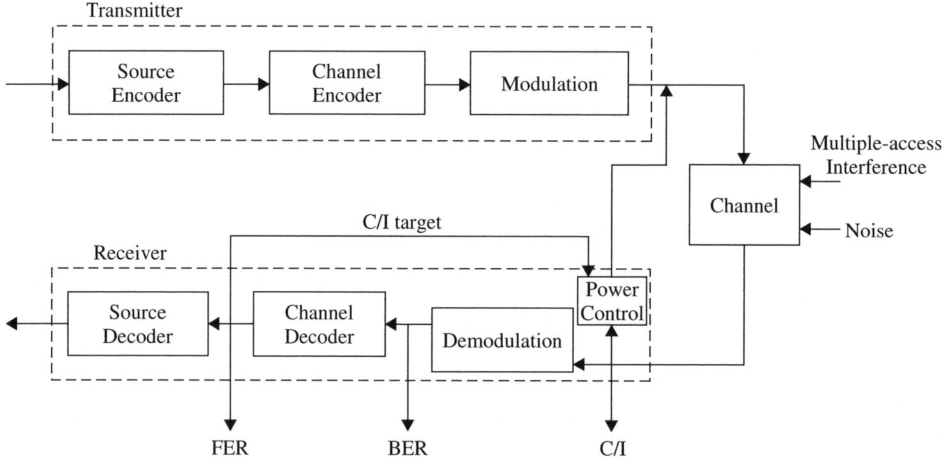

Figure 3.35 Major functional blocks involved with power control mechanism in a UTRA transceiver.

TPC commands received in the downlink, in order to keep the received uplink *Signal-to-Interference Ratio* (SIR) at a given SIR target. The UE transmitter is capable of changing the output power with a step size of 1, 2, and 3 dB respectively, in the slot immediately after the TPC_cmd can be derived. Inner-loop power control frequency is 1500 Hz. The serving cells estimate the SIR of the received uplink DPCH, generate TPC commands (TPC_cmd), and transmit the commands once per slot according to the following rule: if $SIR_{est} > SIR_{target}$ then the TPC command to transmit is "0," while if $SIR_{est} < SIR_{target}$ then the TPC command to transmit is "1."

Upon reception of one or more TPC commands in a slot, the UE derives a single TPC command for each slot, combining multiple TPC commands if more than one is received in a slot. Two algorithms can be used by the UE for deriving a TPC_cmd. Which of these two algorithms is used should be determined by a UE-specific higher-layer parameter, or "Power_Control_Algorithm." More specifically, the two algorithms used in the *Inner-Loop Power Control* can be explained as follows:

Algorithm 1: The power control step is the change in the UE transmitter output power in response to a single TPC command.

Algorithm 2: If all five estimated TPC commands are "down" the transmit power is reduced by 1 dB; If all five estimated TPC command are "up" the transmit power is increased by 1 dB; otherwise the transmit power is not changed.

As a matter of fact, the transmit power of the downlink channels is determined by the network. The power control step size can take four different values: 0.5, 1, 1.5 or 2 dB. It is specified in the UMTS standard that it is mandatory for UTRAN to support the step size of 1 dB; while the support of other step sizes is optional. The UE generates TPC commands to control the network transmit power and send them in the TPC field of the uplink DPCCH. Upon receiving the TPC commands, UTRAN adjusts its downlink DPCCH/DPDCH power accordingly.

To summarize, the *Inner-Loop Power Control* fulfils the following three functions: (1) It can mitigate fast fading effects at a rate of 1.5 kbps; (2) It functions in both down- and uplink; (3) It works based on a fixed quality target set in MS or BS, depending on down- or uplink.

Outer-loop power control

Outer-Loop Power Control is used to maintain the quality of communication at the level of bearer service quality requirement, while using as low a power as possible. The uplink outer-loop power control is responsible for setting a target SIR in the Node B for each individual uplink inner loop power control. This target SIR is updated for each UE according to the estimated uplink quality (e.g. Block Error Ratio, Bit Error Ratio, etc.) for each *Radio Resource Control* connection. The downlink outer-loop power control is the ability of the UE receiver to converge to requisite link quality (BLER) set by the network RNC in downlink.

It is to be noted that the power control in UMTS standard in the downlink common channels is determined by the network. In general, the ratio of the transmit power between different downlink channels is not specified in 3GPP specifications and may change with time, even dynamically. Additional special situations of power control are power control in compressed mode and downlink power during handover, and so on.

To summarize, the *Outer-Loop Power Control* works for the following four functionalities: (1) It can compensate changes in the environment; (2) It can adjust the SIR target to achieve the required FER/BER/BLER; (3) It depends on MS mobility and multipath diversity; and (4) In the case of soft handover it comes after frame selection.

3.2.11 Handovers

The handover defined in the UMTS standard is always associated with the cell reselection process, which occurs when a UE moves away from a cell under the control of one Node B to another. Defined in the UMTS standard, *Cell Reselection* is the process of selecting a new cell when the UE is not in traffic (e.g. Idle, cell FACH, cell PCH, cell URA). Although it is autonomously carried out by the UE, a number of system parameters carried in the *System Information Block* (System Information Block (SIB)) of types 3 and 11 influence the procedure.

There are three types of handovers specified in UMTS. (1) Intrafrequency handovers; (2) Interfrequency handovers; and (3) InterRAT handovers, where RAT stands for *Radio Access Technologies*.

Intrafrequency handover occurs between the cells on the same UMTS radio frequency. The UE can measure the signal strength of other cells without interrupting connectivity with the current cell. *Interfrequency handover* occurs between cells on different UMTS radio frequencies. To measure the signal strength of an interfrequency neighbor cell, the UE must tune away from the serving cell's frequency and tune to the neighbor cell's frequency without losing data. *InterRAT cell handover* occurs between cells on different *RAT*. For example, handover to a GSM, CDMA2000, or UMTS

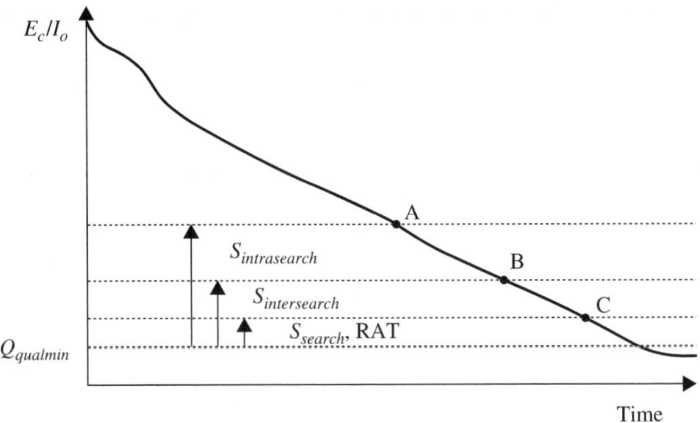

Figure 3.36 Cell reselection signal thresholds used for three different types of handovers in UMTS: intrafrequency handovers, interfrequency handovers and InterRAT handovers.

TDD is considered an interRAT handover (Release 99 considers only GSM). This requires significant reconfiguration of hardware and software in the UE. Measurements must be taken when it is possible to tune the radio away from the serving cell without losing data. For obvious reasons, the InterRAT cell handover is very important, especially during the initial deployment of UMTS in a region where GSM or any other mobile cellular systems are possibly operating, and thus the dual-mode UE will be dominant in the 3G handset market.

Figure 3.36 shows cell reselection signal thresholds used for three different types of handovers defined in UMTS, that is, intrafrequency handovers, interfrequency handovers and InterRAT handovers.

When an interfrequency cell reselection is in progress, it is recommended that the condition $S_{intersearch} < S_{intrasearch}$ must be satisfied. In this case, the interfrequency searching will only begin when the quality of the pilot signal is worse than the value at which intrafrequency searching begins. When an interRAT cell reselection is carried out, it is recommended that the requirement $S_{search,RAT} = 0$ dB must be satisfied. In this case, the interRAT reselection will be considered only when the quality of the pilot signal is below the level of $Q_{qualmin}$, thereby maximizing WCDMA coverage. The goal in setting these parameters is to ensure that intrafrequency neighbors are first considered for reselection, followed by the interfrequency neighbors. These parameters attempt to maximize the WCDMA usage, only considering other technologies when WCDMA is unsuitable. $Q_{qualmin}$ is defined as the minimum CPICH E_c/N_o for the cell to be suitable, where E_c stands for chip energy, N_o is the two-sided power spectral density of noise.

There are totally six reported events as a result of interfrequency measurement, which are listed below.

- Event 2a: Change of best frequency.

- Event 2b: The estimated quality of the currently used frequency is below a certain threshold and the estimated quality of a nonused frequency is above a certain threshold.

- Event 2c: The estimated quality of a nonused frequency is above a certain threshold.

- Event 2d: The estimated quality of the currently used frequency is below a certain threshold.

- Event 2e: The estimated quality of a nonused frequency is below a certain threshold.

- Event 2f: The estimated quality of the currently used frequency is above a certain threshold.

It is noted that the frequency quality estimate used in all aforementioned events is defined as:

$$Q_{frequency\ j} = 10W_j \log \left(\sum_{i=1}^{N_{A\ j}} M_{i,j} \right) + 10(1 - W_j) \log M_{Best\ j} \tag{3.1}$$

where $Q_{frequency\ j}$ is the estimated quality of the active or virtual Active Set on frequency j. $M_{i,j}$ is a measurement result of cell i in the active or virtual Active Set on frequency j. $N_{A\ j}$ is the number of cells in the active or virtual Active Set on frequency j. $M_{Best\ j}$ is the measurement result of the cell in the active or virtual Active Set on frequency j with the highest measurement result. W_j is a parameter sent from UTRAN to UE and used for frequency j.

Similarly, we can also have four different events reported from interRAT measurements:

- Event 3a: The estimated quality of the currently used UTRAN frequency is below a certain threshold and the estimated quality of the other system's frequency is above a certain threshold.

- Event 3b: The estimated quality of the other system's frequency is below a certain threshold.

- Event 3c: The estimated quality of the other system's frequency is above a certain threshold.

- Event 3d: Change of the best cell in the other system.

The frequency quality estimate for the serving UTRAN frequency used in Event 3a is the same as that used for interfrequency events. The triggering conditions for Event 3a are defined as:

$$\begin{cases} Q_{used} \le T_{used} - \frac{H_{3a}}{2} \\ M_{other\ RAT} + CIO_{other\ RAT} \ge T_{other\ RAT} + \frac{H_{3a}}{2} \end{cases} \tag{3.2}$$

where Q_{used} is the quality measurement of the serving UTRAN frequency. T_{used} is the absolute threshold that applies for the UTRAN system in that measurement. $M_{other\ RAT}$ is the measurement quantity for the cell of the other system. $CIO_{other\ RAT}$ is the cell individual offset for the cell of the other system. $T_{other\ RAT}$ is the absolute threshold that applies to the other system in that measurement. H_{3a} is the hysteresis parameter for Event 3a. Similar equations can be defined for all the other events.

3.3 TD-SCDMA

Vastly different from other 3G standards, such as CDMA2000 and WCDMA, which have been well covered in the literature, TD-SCDMA is a much less well-known 3G standard proposed by China as a possible IMT-2000 candidate submitted to ITU in 1998. This was done at almost the same time as other 3G standards were submitted.

The fact that TD-SCDMA has not been made widely known probably has two-fold implications. First, the development of TD-SCDMA technology lags behind the other main stream 3G systems, such as CDMA2000 and WCDMA, which have been deployed by many regions/countries in the world today. This has obviously meant that the TD-SCDMA technology has not been as mature as the other 3G technologies, and TD-SCDMA has simply not been ready to compete with other 3G standards in the world mobile communication market. Second, the slow development of the TD-SCDMA system is due partly to the fact that China had considered the TD-SCDMA as a national standard to replace imported technologies and to safeguard its national interest. China had extensively imported a variety of telecommunication equipments; consequently spending huge amounts of funds to buy that equipment. As a result, China has found that its IPRs are always in the hands of foreign vendors. This is the situation that China wants to eagerly change, and it has been trying hard to make the TD-SCDMA a home-made product, as it did with its nuclear and space technologies. Clearly, this position made it difficult for many foreign manufacturers to be involved with the TD-SCDMA system development.

However, China has recently changed its strategy drastically by offering a lot of opportunities to develop TD-SCDMA jointly with foreign vendors, such as Siemens and Ericsson, and so on. With the provision of China's 3G licences issuing policies, many foreign telecommunication firms have been showing great interest to join the Chinese team to make prototyping TD-SCDMA systems, including chip sets for handsets and base stations. Some analysts believe that TD-SCDMA, along with WCDMA and CDMA2000, will become a major player in China's huge 3G market, which will be as lucrative as its 2G mobile cellular market. Therefore, the incentive for developing the TD-SCDMA system has been there, and it is the time for TD-SCDMA technology, along with other 3G standards, to become a reality in the world 3G market.

3.3.1 Historical Background

The effort to propose the TD-SCDMA standard can be traced back to more than 10 years ago when CATT worked in TD-based CDMA architecture for a possible mobile communication standard for China. CWTS Group accepted the proposal from CATT and submitted the first version of TD-SCDMA to ITU in 1998 as one of the numerous other candidate proposals for IMT-2000 (such as UMTS WCDMA and CDMA2000, etc.), which was subsequently approved by the ITU as a 3G standard in May 2000. TD-SCDMA has been accepted by 3GPP in March 2001. Table 3.22 shows the milestones of the development of the TD-SCDMA standard.

China, as the largest developing country in the world, most likely enjoys the fastest economical growth rate in the world today. However, its telecommunication research and development capability still lags behind the United States, Europe, Japan, and Korea, although it has made impressive progress in some areas. Its unique economical and political structure further complicates the mobile market perspectives of China. Several problems/dilemmas exist in China in terms of its mobile communication perspective as detailed below:

- Increasing demand on mobile communication services versus China's inability to develop home-made mobile communication technology on its own;

Table 3.22   The milestones of development of the TD-SCDMA standard

Year	Milestones of TD-SCDMA
1998.06	Submission of TD-SCDMA standard by CATT of MII
1999.04	Agreed support by NTT DoCoMo, Panasonic, ARIB, Ericsson, Nokia, Siemens & CATT
2000.05	Official approval of TD-SCDMA as 3G standard by ITU
2000.10	3-party MOU signed by Siemens, CATT & Huawei
2000.12	TD-SCDMA Forum established
2001.03	The first call in Siemens 3G Lab.
2001.03	Official acceptance of TD-SCDMA by 3GPP
2001.04	The first call in Beijing by CATT, China
2001.05	The first mobile-to-mobile call in Beijing by CATT, China

CATT: China Academy of Telecommunication Technology
MII: Ministry of Information Industry
Huawei: A telecommunications equipment manufacturer in China

- High penetration of the foreign technologies and IPRs versus China's deep concern on possible boycott in case of conflicts in flashing spots, such as Taiwan and East Sea oil exploration with Japan;

- The post WTO-entry widely opened the mobile market versus the quasi-monopoly reality by China Mobile and China Unicom;

- The rule of free economy in mobile service providers versus national interest in its huge telecommunication market.

To solve these problems, China is seeking various solutions, one of which is to strengthen support to the locally grown telecommunication industry with various development programs, such as its *Digital Mobile Communication Production Localization Program*, which was initiated in 1999 in an effort to realize the goal that half of the total mobile handsets sold should be made in China in 2005. Such a measure is to help the local telecommunication manufacturer sector stand firm even after China's admission to the WTO. Another measure is to establish comprehensive regulations on the telecommunication services and the investment activities of foreign companies in China to consolidate national telecommunication business footage before its admission to the WTO. One of such regulations specifies that the ceiling for foreign investment in China's telecommunication manufacturer/services should be less than 50% even after the entry of the WTO, as shown in Table 3.23. Yet another important effort is to propose and develop its own mobile communication standard, with or without the partnership of overseas companies, so that it can control the most of related IPRs even for possible future technology transfers to other countries. The most important activity is the making of the TD-SCDMA standard.

China Academy of Telecommunication Technology (CATT) has been playing an important role in the TD-SCDMA proposal, and later in system architecture design and development. The first TD-SCDMA mobile-to-mobile call was successfully made in Beijing by CATT in May 2001. The time schedule for its formal service launch is very tight, which was initially planned between 2002 to 2003, but delayed many times until the present. At the moment, CATT is seeking foreign partners to develop TD-SCDMA systems jointly, including mobile handsets, base stations and networking solutions. The calls have attracted very positive responses and many foreign firms have expressed a strong interest in being involved with the *TD-SCDMA Forum*, which consists of about 210 members worldwide, including many major telecommunication end equipment suppliers, such as Nokia, Ericsson, Motorola, and so on, as shown in Figure 3.37. In particular, Siemens has been the first company to be

Table 3.23 Timetable for China to open its mobile and fixed-line services market to the United States, which was initially proposed during Sino-US bilateral negotiation for China's entry into the WTO

Service category	Open time	Foreign investment Ceiling	Open regions
Value-added & Paging	Immediately	30%	Beijing, Shanghai & Gungzhou
Value-added & Paging	2001	49%	14 major cities
Value-added & Paging	2002	50%	The whole country
Mobile	2001	25%	Beijing, Shanghai & Gungzhou
Mobile	2003	35%	14 major cities
Mobile	2005	49%	The whole country
Fixed-line	2003	25%	Beijing, Shanghai & Gungzhou
Fixed-line	2005	35%	14 major cities
Fixed-line	2006	49%	The whole country

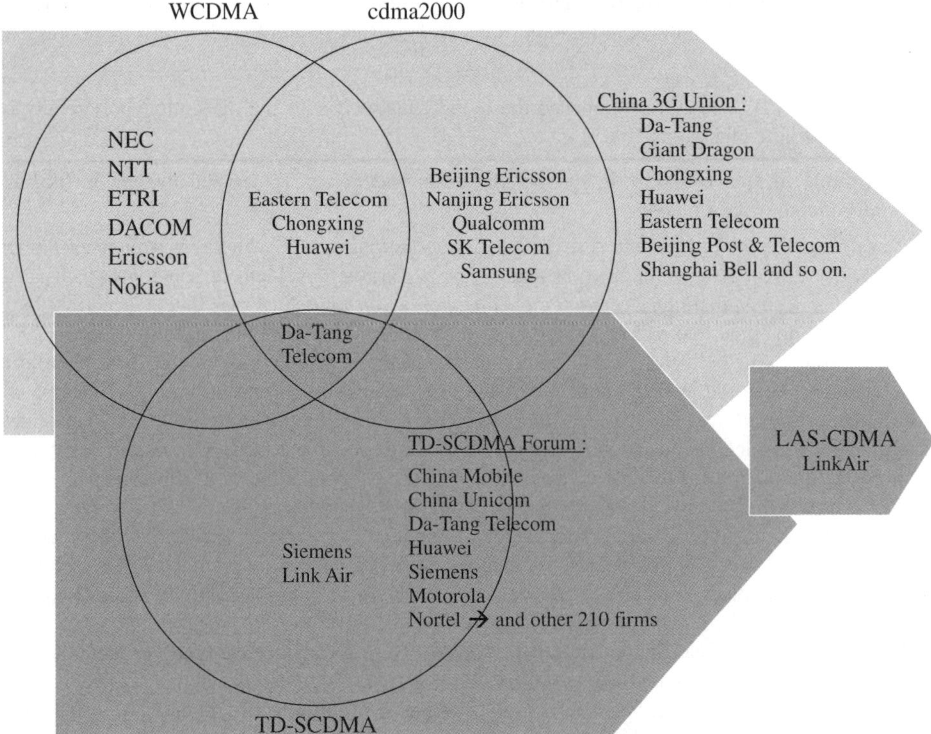

Figure 3.37 The major companies/research groups involved in the activities to develop 3- and 4G mobile communication systems in China.

seriously involved with the TD-SCDMA platform development. It is clear that the company considers TD-SCDMA technology to be a vital 3G solution with great opportunity for success. Siemens has noticeably lead other foreign competitors in TD-SCDMA system development. Currently, Siemens has invested a large amount in TD-SCDMA R&D facilities in China, where it has recruited several hundred research personnel working in the TD-SCDMA system. Several Korean companies and institutions, such as Samsung and ETRI, have also expressed a keen interest in TD-SCDMA systems development. In 2001, CATT also sent a large delegation to Taiwan to seek possible collaboration with Taiwanese companies in chip set design, silicon wafer fabrication support, and so on.

Since China has the largest number of GSM subscribers in the world, the technical similarity (especially in mobile CNs) between TD-SCDMA and GSM gives an advantage to those GSM operators who upgraded their networks into TD-SCDMA at a relatively low cost, in comparison with opting for other 3G standards. CATT estimates that the saving in the upgrading cost can be as much as 30%. Currently, both CATT and Siemens are developing dual-mode and dual-band terminals for use in GSM and TD-SCDMA networks to suit the great needs in the transition period from 2- to 3G systems in China, as well as other regions, where the TD-SCDMA will be selected as a 3G solution for the replacement of its legacy GSM networks.

3.3.2 Overview of TD-SCDMA

As its name suggested, the TD-SCDMA standard carries two important characteristic features: one is to adopt the *Time Division Duplex* (TDD) mode for uplink and downlink traffic separation. The other is to use synchronous CDMA technology, as the character "S" in front of "CDMA" implies.

The use of TDD in the TD-SCDMA standard offers several attractions. First, the agility in spectrum allocation for mobile services is a great advantage for the TDD operation mode, in comparison with FDD, which requires pair-wise spectrum allocation for uplink and downlink, causing a big burden for the countries where spectrum resources have already become very tight, such as the United States and Japan. Second, the use of the same carrier in both up- and downlinks helps with the implementation of smart antenna and other technologies that rely on identical propagation characteristics in both up- and downlinks. Third, TD-SCDMA facilitates asymmetric traffic support in up- and downlinks, associated with the increasing popularity of Internet services. The transmission rates in the two links can be dynamically adjusted according to specific traffic requirements, so that the overall bandwidth utilization efficiency can be maximized. Fourth, the TDD technology used in TD-SCDMA is attributed to the lower implementation cost of RF transceivers, which do not require a high isolation for the transmission and the reception of multiplexing as needed in an FDD transceiver; therefore an entire TD-SCDMA RF transceiver can be integrated into a single IC chip. On the contrary, an FDD transceiver requires two independent sets of RF electronics for uplink and downlink signal loops. The cost saving can be as much as 20–50% if compared with FDD solutions. Because of the aforementioned merits, some people expected the TDD technology to be a vital solution for 4G mobile communications, especially for the small coverage areas.

However, it is to be also noted that the use of the TDD operation in TD-SCDMA bears some technical limitations, if compared to the FDD mode. The relatively high peak-to-average power (PTAP) ratio is one problem. Because a CDMA transceiver is required to work in a good linearity, a relatively high PTAP ratio will limit the effective transmission range and consequently, the coverage area of a cell. Nevertheless, the TD-SCDMA's PTAP ratio is 10 dB less than that of the UTRA-TDD WCDMA proposal. Also, the discontinuity of slotted signal transmissions in the TDD mode also reduces its capability to mitigate fast fading and the Doppler effect in mobile channels, thus limiting the highest terminal mobility supported by the TDD systems. Fortunately, the highest mobility supportable by TD-SCDMA can be increased to 250 km/h with the help of antenna beam-forming and joint detection algorithms, which is comparable to the specification of the WCDMA standard, which is less than 300 km/h. It was recently revealed in a simulation report released by CATT that the smart antenna base station can adopt an 8-element circular array with a single-antenna mobile unit. The results showed a satisfactory performance for a vehicle mobility as high as 250 km/h.

The comparison of fundamental operational parameters of CATT TD-SCDMA, UMTS WCDMA, and TIA CDMA2000 standards is given in Table 3.24. We also provide a comparison between the ETSI UTRA-TDD system and the TD-SCDMA in Table 3.25, where the similarities and differences between the two can be seen. Because of the limits to the space in this book, we should mainly concern ourselves with the physical layer architecture of TD-SCDMA and we will not address the upper layer issues of the standard.

3.3.3 Frame Structure

TD-SCDMA combines both TDMA and CDMA techniques in one system, and the channelization in TD-SCDMA is performed by both time slots and signature codes to differentiate mobile terminals in a cell. The frame structure of TD-SCDMA is shown in Figure 3.38, where the hierarchy of four

Table 3.24 The comparison of the physical layer major operational parameters of TD-SCDMA, WCDMA, and cdma2000 standards

	cdma2000	WCDMA	TD-SCDMA
Multiple access	DS-CDMA/MC-CDMA	DS-CDMA	TDMA/DS-CDMA
CLPC	800 Hz	1600 Hz	200 Hz
PCSS	0.25…1.5 dB	0.25, 0.5, 1.0 dB	1, 2, 3 dB
Channel coding	Conv./Turbo	Conv./RS/Turbo	Conv./Turbo
Spreading code	DL: Walsh, UL: M-ary Walsh mapping	OVSF	OVSF
VSF	4…256	4…256	1…16
Carrier	2 GHz	2 GHz	2 GHz
Modulation	DL: QPSK, UL: BPSK	DL: QPSK, UL: BPSK	QPSK, 8PSK(2 Mbps)
Bandwidth	1.25*2/3.75*2 MHz	5*2 MHz	1.6 MHz
UL-DL spectrum	paired	paired	unpaired
Chip rate	1.2288/3.6864 Mcps	3.84 Mcps	1.28 Mcps
Frame length	20 ms, 5 ms	10 ms	10 ms
Interleaving periods	5/20/40/80 ms	10/20/40/80 ms	10/20/40/80 ms
Maximum data rate	2.4 Mbps	2 Mbps (low mobility)	2 Mbps
Pilot structure	DL: CCMP, UL: DTMP	DL: DTMP, UL: DTMP	CCMP
Detection	PSBC	PCBC	PSBC
Inter-BS timing	Sync.	Async./Sync.	Sync.

CCMP: common channel multiplexing pilot	DTMP: dedicated time multiplexing pilot	VSF: Variable spreading factor
CLPC: Close-loop power control	PCSS: Power control step size	DL: downlink
PSBC: Pilot symbol-based coherent	PCBC: Pilot channel-based coherent	UL: upper-link

Table 3.25 The comparison of the physical layer major operational parameters between TD-SCDMA and UTRA-TDD

	UTRA-TDD	TD-SCDMA
Bandwidth	5 MHz	1.6 MHz
Chip rate per carrier	3.84 Mcps	1.28 Mcps
Spreading	DS, SF = 1/2/4/8/16	DS, SF = 1/2/4/8/16
Channel coding	Convol. or Turbo coding	Convol. or Turbo coding
No. of time slots/subframe	15*2	7*2
Burst structure	Midamble	Midamble
Frame length	Super frame = 720 ms/Radio frame = 10 ms	Super frame = 720 ms/Radio frame − 10 ms
No. of channels/time slot	8	16
No. of channels/Carrier	8 * 7 = 56	16 * 3 = 48
Spectral efficiency	0.662 Mcps/MHz	1.232 Mcps/MHz

different layers of the frame structure, superframe, radio frame, subframe and time slot, are depicted. A subframe (5 ms) consists of seven normal time slots and three special time slots, where TS0 is reserved for downlink and TS1 is for uplink only; whereas the remaining time slots (TS2 to TS6) should form two groups; the first group (whose size can vary from 0 to 5) is for uplink and the second group (whose size can vary from 5 to 0) is for downlink. The size ratio of the two groups can take 0/5, 1/4, 2/3, 3/2, 4/1 and 5/0 to suit a particular traffic requirement. The agility in the support of asymmetric traffic is a very attractive feature of TD-SCDMA, which is of particular importance for the Internet and multimedia services required in 3G applications. The other three special time slots are the downlink pilot (DwPTS), guard period (GP) and uplink pilot (UpPTS) respectively. DwPTS and UpPTS are used as SCH (Synchronization Channel) for downlink and uplink respectively, which should be encoded by different PN codes to distinguish different base stations and mobiles respectively.

A time slot can exactly fit a burst, which consists of two data parts separated by a midamble part and followed by a guard period, as shown in Figure 3.38. Multiple bursts can be sent in the same time slot, where the data parts of those bursts should be encoded by up to 16 different OVSF channelization codes, whose *spreading factor* (SF) is fixed at 16 for downlinks and can vary from 1 to 16 for uplinks. However, each mobile can send up to two OVSF channelization codes in the same slot to form multicode transmission. The data parts of the burst should always be spread by using OVSF codes and scrambling codes, combined to distinguish the mobile and base station respectively. The information about the OVSF codes can be found in Subsection 3.2.8.

A TD-SCDMA physical channel is uniquely defined by frequency, channelization code, time slot, and radio frame allocation jointly.

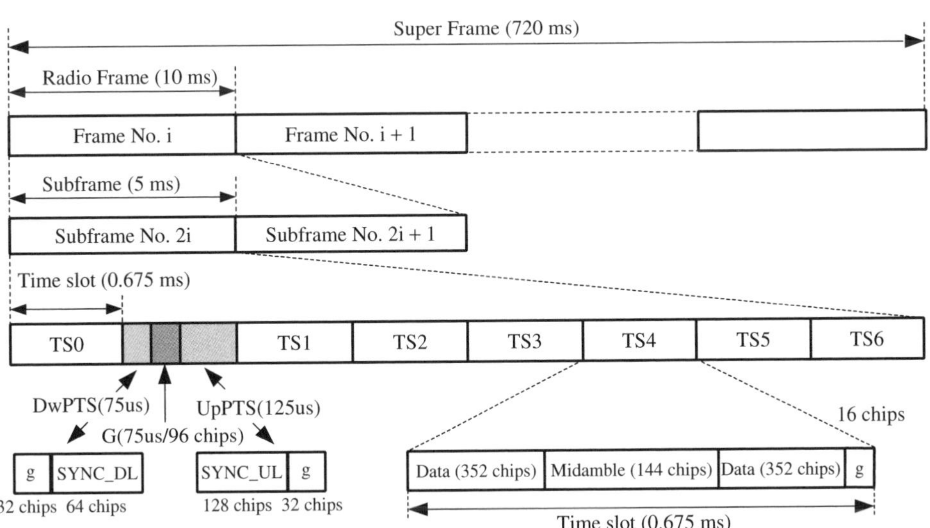

Figure 3.38 The four-layered frame hierarchy in TD-SCDMA standard. TS: time slot; DwPTS: downlink pilot time slot; UpPTS: uplink pilot time slot; G/g: guard period. TS0 is reserved for downlink and TS1 is for uplink only; while the remaining time slots (TS2 to TS6) can form two groups, the first group (which can consist of 0 slot) is for uplink and the second group is for downlink in order to suit a particular traffic requirement.

3.3.4 Smart Antenna

Smart antenna techniques have been integrated into the TD-SCDMA standard as they are an indispensable part of the standard. A smart antenna system is composed of an array of multiple antenna elements and coherent transceivers with an advanced digital signal processing unit. Instead of a single fixed beam pattern from a traditional antenna, the smart antenna can dynamically generate multiple beam patterns, each of which is pointed to a particular mobile; such beam patterns can adapt to follow any mobile adaptively. As a result, cochannel interference can be greatly reduced to enhance reception sensitivity, and therefore the capacity of the whole system. It can also effectively incorporate multipath components to combat multipath fading. The 5 ms subframe structure in TD-SCDMA is designed for the application of the smart antenna. More specifically, it implements fast beam-forming to follow the time variation of mobile channels. The 5 ms subframe length is a compromise by taking into account both the number of time slots and switching speed of the RF components used in a transmitter. It was reported that an 8-element circular array antenna with a diameter of 25 cm has been considered for use in TD-SCDMA base stations. If compared to an omni-directional antenna, there is an 8 dB gain obtainable by using such a circular array antenna. The TDD operation in TD-SCDMA ensures an ideally symmetric beam pattern for both the transmission of and the reception at the same base station, which improves channel estimation and beam-forming accuracy due to the same propagation characteristic in the uplink and downlink channels.

As mentioned above, a burst contains a 144-chip midamble, which functions as a training sequence for beam-forming carried out in the smart antenna system. The midamble is encoded by *basic midamble codes*. There are totally 128 different basic midamble codes of length 128 for the whole system, which are allocated into 32 code groups with four codes in each code group. The choice of code group is determined by base stations, such that four basic midamble codes are known to base stations and mobiles. The midambles of different users active in the same cell and the same time slot are cyclically shifted versions of one single basic midamble code.

Because of the provision for the use of transmit diversity, TD-SCDMA can also take full advantage of space-time coded signaling to further enhance the capacity of the system.

3.3.5 Adaptive Beam Patterns

There are two categories of transport channels in TD-SCDMA, which are *Dedicated Transport Channels* (DTC) and *CTCs*. The DTC is further divided into *DCH* and *ODMA Dedicated Transport Channels* (ODCH); the CTC is divided into six subtypes, as shown in Table 3.26.

It is specified in TD-SCDMA downlink transmissions from a base station that all CTCs (such as SCH, Pilot, BCH, PCH etc.) which usually carry the shared information of the network use omni-directional beam patterns to send their signals; all DTCs, which carry dedicated user or control signals, use directional beam patterns with the help of smart antenna technology. On the other hand, all the receiving channels in a base station should also use directional beam patterns to suppress the interferences from other unwanted transmissions. The use of different beam patterns for different transport channels in the TD-SCDMA system can effectively increase the utilization efficiency of transmission power from base stations and reduce cochannel interference in the cell, which contributes to the increase of cell capacity.

The introduction of beam-forming in all receiving channels can also facilitate mobile location positioning, based on the numerous new services (otherwise impossible) that can be added in a mobile cellular system.

3.3.6 Up-Link Synchronization Control

Another critical technique used in the TD-SCDMA is the synchronous CDMA transmission in downlink and uplink, both of which use OVSF codes for channelization due to its ideal orthogonality.

Table 3.26   Two types of transport channels in TD-SCDMA

Common Transport Channels (CTC)*	Dedicated Transport Channels (DTC)**
Broadcast Channel (BCH)	Dedicated Channels (DCH)
Paging Channel (PCH)	ODMA Dedicated Transport Channels (ODCH)
Forward Access Channel (FACH)	
Random Access Channel (RACH)	
Uplink Shared Channel (USCH)	
Downlink Shared Channel (DSCH)	

* CTC carries shared information of network
** DTC carries dedicated user/control signals between UE & network

In order to achieve the synchronization in the uplink, the TD-SCDMA introduces open-loop and close-loop synchronization control in its signaling design.

To pave the way for the successful application of orthogonal codes in asynchronous uplink channels, *uplink synchronization control*, which has been considered an option in the UMTS UTRA [425] and WCDMA [431] standards is necessary. However, real workable schemes have been solely implemented in the TD-SCDMA standard [432, 433] as an important part of the system architecture. Similar to the power control algorithm, there are two sectors of uplink synchronization control: the open-loop sector and the closed-loop sector, which ought to work jointly to achieve an accurate synchronization, up to 1/8 chip, as specified in the TD-SCDMA standard [432, 433]. With the help of such an accurate uplink synchronization control algorithm, the transmission channels in the uplink have been converted into quasi-synchronous ones, effectively enhancing the detection efficiency in the uplink channel of a CDMA system, which is often a bottleneck in the whole air-link section.

During a call set-up procedure, a mobile should first establish downlink synchronization with the base station by looking for DwPTS, after which it will initiate the uplink synchronization procedure. In the beginning, a mobile can estimate the propagation delay from a base by the received power level of DwPTS. Its first transmission in uplink is performed in the UpPTS time slot to reduce interference in the normal time slots. The timing used for the SYNC_UL burst is set according to the received power level of DwPTS. This executes the open-loop synchronization. At the detection of the SYNC_UL burst, the base station will evaluate the received power level and timing, and reply by sending the adjustment information to the mobile in order to modify its uplink transmission timing and power level in the next transmission.

To maintain the uplink synchronization, the midamble field of each uplink burst will be used. In each uplink time slot, the midamble from each mobile in the cell is distinct. The base station can estimate the power level and timing by measuring the midamble field from each mobile in the same time slot. In the next available downlink time slot, the base station will signal the *Synchronization Shift* (SS) and the *Power Control* (PC) commands, which occupy part of the midamble field, to enable the mobile to properly adjust its transmission timing and power level, respectively. The uplink synchronization can be checked once per TDD subframe and the step size in the uplink synchronization can be adapted from 1/8 chip to 1 chip duration, which is sufficiently accurate in order to maintain the orthogonality of OVSF codes from different mobiles. Figure 3.39 shows the flow-chart of the open/close-loop synchronization algorithm used by TD-SCDMA.

The detailed procedure of the uplink synchronization control algorithm can be explained as follows. During the cell search procedure in a TD-SCDMA system, a mobile will capture the information in downlink broadcasting slots to know the power level of a transmitted signal from a BS, based on which the mobile can roughly estimate the distance from the BS using a simple free-space propagation law to complete the open-loop uplink synchronous control stage. With this knowledge, the

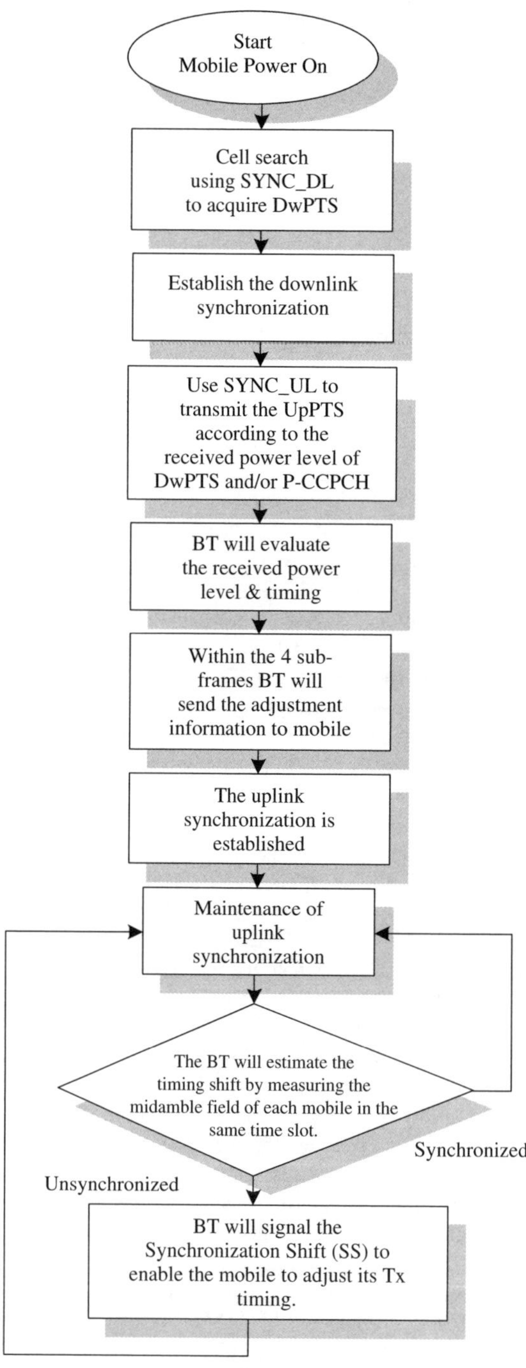

Figure 3.39 The flow-chart diagram of closed and open loops synchronization control used by TD-SCDMA for both uplink and downlink, from which it is seen that the downlink synchronization is established before the uplink synchronization.

mobile will send a testing burst in a special slot dedicated only for uplink testing bursts, called an *UpPTS slot*. If this testing burst has fallen within the *search-window* at the BS receiver, the testing burst will be successfully received and the BS will know if the timing for the mobile to send its burst is correct or not. If not, the BS should send *SS* instructions in the next downlink slots to ask the mobile to adjust its transmission timing to complete the closed-loop uplink synchronization control cycle. It is specified in the TD-SCDMA standard that the initial uplink synchronization procedure has to be finished within four subframes, followed by the uplink synchronization tracking process. A detailed illustration of both the open-loop and closed-loop uplink synchronization control algorithm implemented by TD-SCDMA is shown in Figure 3.40, where a scenario with three mobiles communicating with a BS is illustrated with UE3 being the mobile of interest, which wants to proceed with the uplink synchronization with the BS; furthermore, UE1 and UE2 are the mobiles that have already established communication links with the BS.

Obviously, the need for uplink synchronization control in the TD-SCDMA system is because of its use of OVSF codes, which are orthogonal codes, and perform poorly in asynchronous uplink channels due to the fact that the characteristics of their ACFs and CCFs in an asynchronous channel are very bad. However, it is still natural for us to question the justification of introducing such a complicated uplink synchronization control system simply for the application of orthogonal OVSF codes in uplink channels. Why do we not think about other better solutions, such as using some new spreading codes with an inherent isotropic or symmetrical performance? This indeed opens an interesting issue, which should be discussed in Chapter 7.

3.3.7 Intercell Synchronization

The TD-SCDMA standard adopts a technique used to achieve synchronization among neighboring base stations in order to optimize system capacity and to perform cell search in a handover procedure. A typical example for such a need is a scenario for coordinated operations with overlapping coverage areas of the cells, or there is contiguous coverage for a certain area. In fact, a TDD system requires such intercell synchronization, especially in the handover procedure, where a mobile will communicate with two or three base stations simultaneously. In such a scenario, a common clock source is needed to maintain the intercell synchronization. The synchronization between base stations and between cells is very important for the TDD mode to avoid interferences from nearby cells.

In the TD-SCDMA standard there are several possible ways to achieve the synchronous transmission among neighboring cells. The first way is to achieve the synchronization via the air interface, in which a special burst, *Network Synchronous Burst*, is employed. This burst should be sent on a predetermined time slot at regular intervals. The base stations involved should adjust their respective downlink signals timing in accordance with the network synchronous bursts. The second alternative way is to use other cell's DwPTS as a timing basis for the synchronous transmissions of base stations involved. Yet another way is to simply use a GPS as a common clock to synchronize the base stations. It is likely that the first generation TD-SCDMA network will work on a GPS in order to achieve the intercell synchronization to let the base stations have the same timing reference for transmitting and receiving. The accuracy for such intercell synchronization is required at about 5 μs.

With the intercell synchronization, the transmission time for each cell can be determined in network planing and controlled by the TD-SCDMA CN. The time offset in nearby cells is separated by at least one fixed time delay, which should be approximately 80% of the transmission time between two neighboring cells.

3.3.8 Baton Handover

Baton Handover is another salient feature offered by the TD-SCDMA standard, which is used to take advantage of both hard handoff and SHO and is particularly suited for the TDD mode operation.

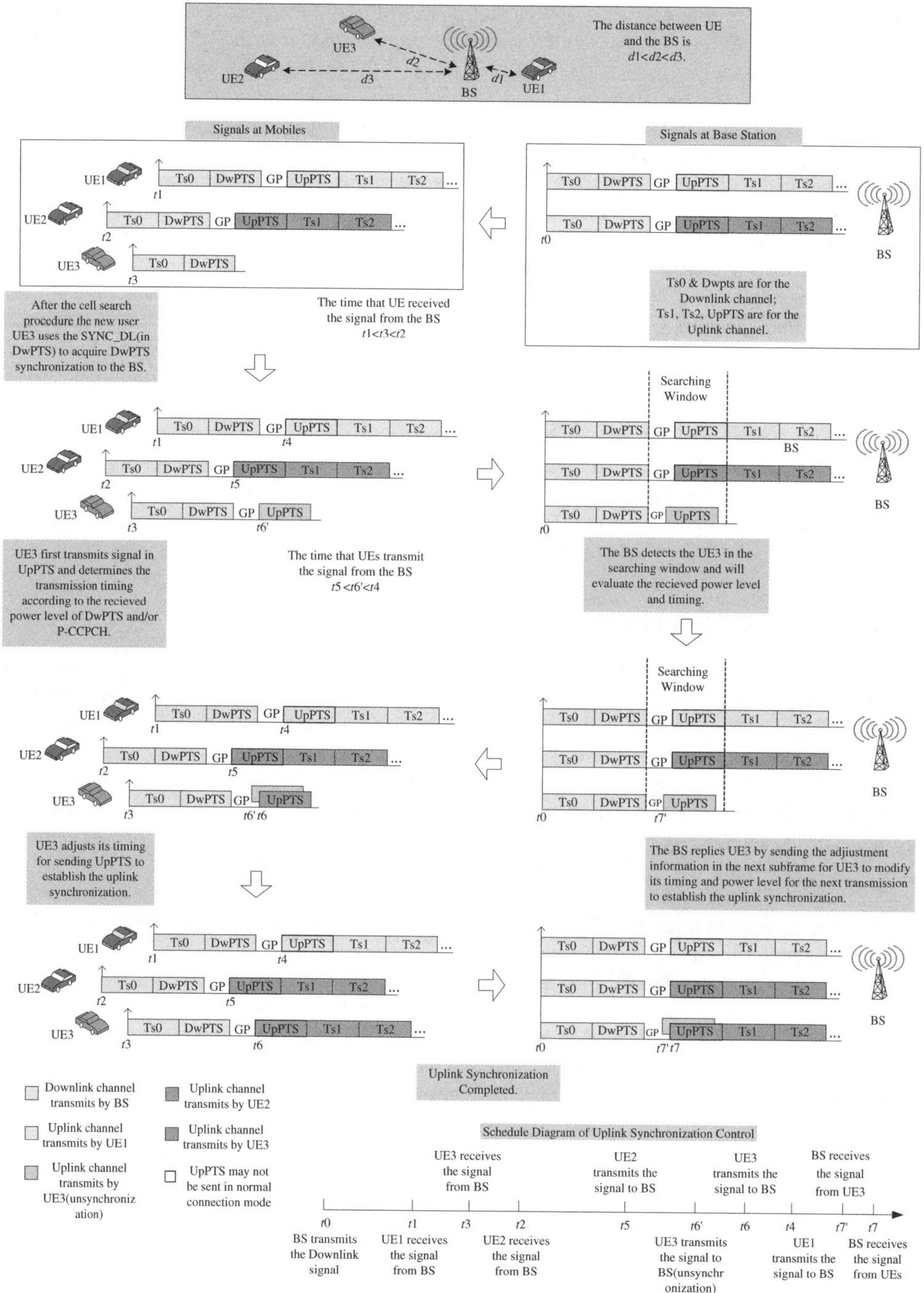

Figure 3.40 Illustration of open-loop and close-loop uplink synchronization control algorithm specified by the TD-SCDMA standard.

The baton handover, similar to the procedure as the handover of a baton is in relay, is based mainly on the user positioning capability provided by TD-SCDMA base stations using smart antenna technology.

In an urban pedestrian environment, it may obtain wrong information of the position for a mobile by use of a single base station because of serious multipath. Therefore, it has to be aided by cell search, based on the report from the mobile to make a decision on which the target base station is. The successful operation of baton handover is based on the fact that:

- the system knows the position of all mobiles;
- the system knows and determines the target cell for handover;
- the system informs the mobile about the base station in neighboring cells;
- the mobile measurement helps the system to make the final decision;
- after the cell search procedure, the mobile has already established synchronization to the base station in the target cell.

The procedure of the baton handover supported in TD-SCDMA can be explained as follows. Assume that BTS0 is the base station the mobile connects to earlier and BTS1 is the base station the mobile wants to handover. First, the mobile should listen to the broadcasted information from BTS0, which includes the data related to nearby cells including position, the operation carrier frequency, the Tx time offset, the short code distributed, and so on. The mobile will search the nearby cells based on the above received information. With that information the mobile is able to send relevant information to BTS1 via some common transport channel so that BTS1 can also measure the location of the mobile by the burst exchange between them. The handover procedure can be initiated by either a mobile or a BTS, but the network will decide when to execute the handover. Therefore, the baton handover is different from the soft handover that has been applied in IS-95, which makes use of macrodiversity.

By using the baton handover concept, the system will support both intrafrequency and interfrequency (in the TD-SCDMA system) handovers, and give higher accuracy and a shorter handover time period for handovers inside the TD-SCDMA system and between different systems. There are several different handover procedures defined in TD-SCDMA, which include intrasystem and intersystem handovers. The intersystem handover can be further divided into the TD-SCDMA/GSM handover and the TD-SCDMA/UTRA-FDD handover in order to provide future cooperation among different networks, which is extremely important especially in the initial period of TD-SCDMA network deployment when TD-SCDMA may coexist with GSM and other possible 3G systems such as UTRA-TDD, and so on.

3.3.9   Intercell Dynamic Channel Allocation

Channel allocation in TD-SCDMA can be made very flexible due to the use of synchronous TDD technology. It is possible that each TD-SCDMA base station can make use of three different carriers to occupy about 5 MHz bandwidth (each takes 1.6 MHz), which is the same as the bandwidth required by one carrier in UTRA-TDD. On the other hand, TD-SCDMA can also operate in a mode that each cell uses only one 1.6 MHz bandwidth and three neighboring cells can use three different carriers. On the other hand, each TD-SCDMA time slot can support 16 simultaneous code channels and each subframe has seven normal time slots, which can be made symmetric or asymmetric for downlink and uplink traffic. Therefore, the physical channels in TD-SCDMA can be viewed as a "pool," each element of which can be uniquely determined by three indices: carrier frequency, OVSF code and time slot. In this way, the channel allocation for each cell can be made a dynamic way in terms of three neighboring cells to further increase the bandwidth utilization efficiency of the overall system.

3.3.10 Flexibility in Network Deployment

TD-SCDMA carries many similar technical features as GSM and UTRA-TDD standards, which makes it possible for TD-SCDMA network to be deployed in an evolutionary, rather than a revolutionary way. It has been suggested that the TD-SCDMA network can be implemented via two phases, taking into account the currently operating networks in many countries around the world. The initial phase can only implement TD-SCDMA physical layer functionalities, with only some necessary modifications to the existing GSM second and third layer core networks to make them compatible with the TD-SCDMA upper layers requirements. Such an initial TD-SCDMA deployment can offer a maximum of 284 kbps data transmission rate services, which is comparable to 2.5G mobile communication system. If compared to the upgrade from GSM to WCDMA network, such an initial deployment of TD-SCDMA can save up to 50–70% cost, as estimated by some analysts. The saving in the initial deployment phases is significant in terms of view of business, because it greatly reduces the risk of the investment of service providers and paves the way for future network evolution toward full-functional 3G network. The second phase involves using full-functional TD-SCDMA physical layers and the second and third layers should use 3GPP compatible upper layers standard to meet the full functions required by IMT-2000. The maximum transmission rate can reach 2 Mbps, which is compatible with 3G requirement.

On the other hand, TD-SCDMA can also support the coexisting operation of different mobile networks, such as GSM and UTRA-TDD standards, which has been discussed in aforementioned sections on handover procedures across different mobile networks. Therefore, TD-SCDMA is particularly attractive for homogenous evolution from existing 2G toward 3G mobile networks at a relatively low upgrading cost and investment risk.

3.3.11 Technical Limitations of TD-SCDMA

There are several technical limitations in TD-SCDMA. Some of them stem from the TD-SCDMA system itself, and the other from TDD systems in general.

It is to be noted from Tables 3.24 and 3.25 that TD-SCDMA uses SF = 1 at a data rate of 2 Mbps, implying that no processing gain will be available in the highest transmission rate scenario. In such a case, multipath diversity gain will not be available, and the system should rely on other techniques to enhance the detection efficiency.

The use of OVSF codes in TD-SCDMA poses another problem for low-efficient and complex rate-matching algorithm for multimedia applications. The change of SF in OVSF codes must be made multiples of two, and as a result it is impossible to support arbitrary transmission rates to fit a particular data rate.

The application of uplink synchronization control also increases the complexity of the system, in both handsets and base stations. The success of the Baton Handover relies heavily on the accuracy in mobile positioning techniques provided by smart antenna, making it necessary to handle all handovers in a centralized way to increase overall networking traffic.

3.3.12 Global Impact of TD-SCDMA

At the time this book is written, China has not yet formally decided what standard it will adopt as a major 3G technology. However, there have been some signs that China is likely to support its own 3G standard and encourage its services providers to adopt them. If so, there will be some foreseeable impact to the world mobile communication market due to its sheer market size. The foreign mobile manufacturers should be very careful with China's 3G licensing process, which has not yet been decided. Table 3.27 shows the different natures of telecommunication markets in the United States, Europe, Japan, and China.

Table 3.27 Driving forces behind mobile communication technology development in the United States, Europe, Japan, and China

Region	Driven mainly by
United States	Market
Europe	Technology
Japan	Mobile Operators/Market
China	Government/Market

Technically speaking, TD-SCDMA is probably one of the most cost-effective solutions for the upgradation of existing GSM networks to 3G systems due to its unique technical feature. In this sense, the possible market for the TD-SCDMA system exists, simply because of the great success of GSM networks in the world. Therefore, the TD-SCDMA standard is in principle suitable not only for China, but also for any other regions where GSM is operating. Thus, the possible market competition with WCDMA (for both its TDD and FDD schemes) can be expected.

Since the submission of the TD-SCDMA proposal to ITU in 1998, China has taken a critical path in developing its own national 3G standard, which can be ready within years. China has become the largest single mobile communications market in the world and its great potential for 3G wireless applications has attracted all the major telecommunication companies in the world, especially after China's entry into the WTO. China's market is now open to foreign investment in terms of mobile communication equipments and services and is ready to market its own 3G technology to the world. To deal with the emergence of ever severe competition, China wants to promote its own 3G standard to save the cost for purchasing foreign IPRs and technologies and to eventually access the world-wide mobile market. The TD-SCDMA standard adopts numerous advanced technologies and offers a relatively cost-effective way to upgrade existing GSM networks to 3G CNs. Therefore, it is an attractive 3G technology, not only for China but also for the world. It can substantially reduce the investment risk, which is the most serious concern to almost all the existing 2G service providers with 3G licences in their hands. The impact of TD-SCDMA should never be under-estimated.

More information about the TD-SCDMA can be found in [432–439].

4

Wireless Data Networks

Why create a wireless network? The best-selling feature of most wireless technology is portability [453]. If every device in a network is joined wirelessly, then users benefit not only from the mobility of their telephones and notebook computers: They can interface a camera with a PC from the couch instead of sitting at their desks, where their cameras are connected to their PCs by some sort of cable or plug, and they can rearrange office equipment by moving devices, like printers or scanners, anywhere within range, without stringing new wires (and drilling new holes in the walls).

4.1 IEEE 802.11 Standards for Wireless Networks

The Institute of Electrical and Electronics Engineers (IEEE) develops and maintains technological standards based on the recommendations of individuals with expertise in the technology being standardized. Scientists, manufacturers, and end-users provide input to the institute, which comes to a consensus about the standards suitable for a particular technology. Use of an IEEE Standard is wholly voluntary and the existence of an IEEE Standard does not imply that there are no other ways to produce, test, measure, purchase, market, or provide other goods and services related to the scope of the IEEE Standard [452]. Research scientists, manufacturers, and end-users all benefit from the shared specifications contained in the standards. When everyone uses the standard, customers can use equipment from different manufacturers with no incompatibilities.

The IEEE 802 set of standards has to do with the physical layer (PHY) and data link layers of local and metropolitan area networks (LANs and MANs). These are the bottom two layers in the ISO/OSI networking model, far removed from the application layer, and are concerned with data transmission (and reception) between computers in LANs and MANs. The IEEE has split the data link layer into two different sublayers: logical link control (LLC) and media access control (MAC) (see Figure 4.1). The IEEE LLC protocol concerns the logical address, control information, and data portions of an HDLC (high-level data link control) frame, while the MAC protocols deal with synchronization, error control (EC), and physical addresses. MAC protocols are specific to the LAN using them (Ethernet, Token Ring, Token Bus, etc.) [455].

The IEEE 802.3 standards are concerned with Ethernet (wired) communications. Originally, they supported 10-Mbps data rates, but as network terminals became faster and thus capable of running multimedia applications, and as the need to share high-speed servers among LANs became widespread, faster data rates were included in the standards. They were updated in the mid-1990s to include "fast Ethernet" transmission rates of 100 Mbps, and in the late 1990s the Gigabit Ethernet was standardized

Next Generation Wireless Systems and Networks   Hsiao-Hwa Chen and Mohsen Guizani
© 2006 John Wiley & Sons, Ltd

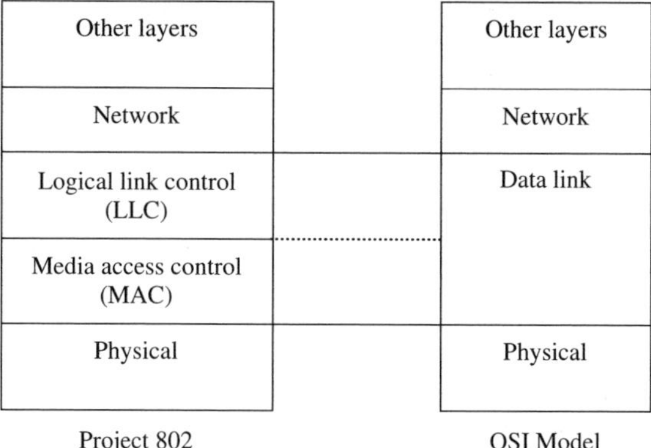

Figure 4.1 MAC and LLC split [455].

under 802.3 [454]. Experts attest that the two major driving forces of this industry have always been the ease of installation and increase of data rate, the two important characteristics of Fast Ethernet and Gigabit Ethernet. Thus, Ethernet dominated over other 802.3 LAN IEEE standards (the so-called Token Ring and Token Bus).

The 802.4 and 802.5 standards concern the PHY and MAC layers for *Token Bus* and *Token Ring* topologies, respectively. IEEE's 802.6 standards address the needs of MANs [454]. The 802.11 family of standards is devoted to the requirements of the bottom two ISO layers in wireless networks (wireless local-area networks (WLANs)). A complete list of the rest of the standards is given in Table 4.1.

When developing the standards for wireless networks, the IEEE observed the radio frequency regulations of the US Federal Communications Commission (FCC), since radio waves were the transmission medium of choice for wireless networking. In 1985, the FCC designated certain portions of the radio frequency spectrum for industrial, scientific, and medical use, and these became known as the *ISM bands*; they are: (1) 902–928 MHz, a bandwidth of 26 MHz; (2) 2.4–2.4835 GHz, a bandwidth of 83.5 MHz, commonly called the *2.4-GHz band*; and (3) 5.725–5.850 GHz, a bandwidth of 125 MHz, commonly called the *5-GHz band*.

Within certain guidelines, the FCC's regulations allow users to operate radios inside these bands without an FCC licence, an obvious boon for the developers of wireless network technology (and for the users who do not have to obtain a licence to operate their cell phones) [453].

The 802.11 standards have evolved over time, and presently six methods for wireless data transmission are defined in the 802.11 standards. Each means of transmission represents its own PHY within 802.11. The first IEEE 802.11 standards were completed in 1997, and defined three of these PHY for 1- and 2-Mbps data rates. An overview of these PHY is provided in Table 4.2 and also explained as follows:

- The Direct-Sequence Spread Spectrum (DSSS)[1] PHY uses the 2.4-GHz band and can transmit data at 1 or 2 Mbps. It was first used for military communications. To prevent jamming, and, to a lesser extent, eavesdropping, radios that use DSSS transmit their signals across the entire available ISM band at very low power. This prevents interference from narrowband signals (jammers or others) and lessens the likelihood of transmission errors. Eavesdroppers may interpret these signals as background noise [452, 453].

- The Frequency Hopping Spread Spectrum (FHSS) PHY also uses the 2.4-GHz band for transmission at 1 or 2 Mbps, and also originated in military applications. Two communicating radios

---

[1]More detailed discussions on DS and other station services (SS) techniques can be found in Section 2.2.

Table 4.1   802.11 standards list [486]

802.1	Higher-layer LAN protocols
802.2	Logical link control
802.3	Ethernet (wired)
802.4	Token Bus
802.5	Token Ring
802.6	MAN
802.7	Broadband
802.8	Fiber optic
802.9	Isochronous LAN
802.10	LAN/MAN Security
802.11a	Wireless LAN: 5-GHz band
802.11b	Wireless LAN: 2.4-GHz band
802.11c	Wireless LAN: higher layers
802.11d	Wireless LAN: MAC
802.11e	Wireless LAN: MAC
802.11f	Higher layers
802.11g	Wireless LAN: higher rate 2.4-GHz band
802.11h	Wireless LAN: MAC
802.11i	Wireless LAN: MAC
802.12	Demand priority
802.13	Not used
802.14	Cable modem
802.15	Wireless PAN
802.16	Broadband wireless access
802.17	Resilient packet ring
802.18	Radio regulations
802.19	Coexistence
802.20	Mobile broadband wireless access

Table 4.2   802.11 PHY layers

DSSS	2.4 GHz	1 or 2 Mbps
FHSS	2.4 GHz	1 or 2 Mbps
DFIR	850 to 950 nm (infrared)	None implemented
COFDM	5 GHz	54 Mbps
HR/DSSS	2.4 GHz	5.5 or 11 Mbps
OFDM	2.4 GHz	54 Mbps

using FHSS change frequencies according to a predetermined pseudorandom pattern, and only remain on a given frequency for a split second (FCC regulations require the frequency hops to take place in 400 ms or less). This technique minimizes the chances that more than one radio device will be transmitting on the same frequency at the same time. If a sender happens to detect interference from another radio at a particular frequency, it retransmits its data after the next hop to a new frequency [453]. FHSS was phased out of 802.11 in the 802.11b standards.

- The Diffused Infrared (DFIR) PHY uses near-visible light in the 850-nm to 950-nm range for signaling [452]. However, unlike infrared (IR) TV remote controls that need a line of sight to work, devices that follow the 802.11 DFIR standards do not need to be aimed at one another, permitting the construction of a true LAN [452]. But, there are no wireless networking products currently available that implement this PHY [453]. One potential source of interference when using this technology would be a human being walking between a PC and its printer when they were trying to communicate.

- A fourth 802.11 PHY is defined by IEEE's 802.11a standards: The Coded Orthogonal Frequency Division Multiplexing (COFDM) layer is capable of transmitting data at 54 Mbps by using the broader 5-GHz band. However, FCC regulations limit the transmission power used at these higher frequencies, and thus it reduces the distance higher-frequency transmissions can travel. For these reasons, radios that use COFDM technology must be closer together than those using the other PHY introduced above. The obvious benefit of COFDM is speed. The IEEE 802.11a standards are further discussed in Section 4.2.

- The IEEE 802.11b standards cover the fifth PHY, the High-Rate Direct-Sequence Spread Spectrum (HR/DSSS) layer. Using this layer, data can be transmitted at 5.5 or 11 Mbps, rivaling the standard Ethernet rate of 10 Mbps, and it has become the most widely used IEEE 802.11 PHY despite its recent entry onto the scene in 1999. HR/DSSS technology is an extension of DSSS technology and is designed to be backward compatible with its predecessor (both operate in the 2.4-MHz band) [453]. Further discussion on the 802.11b standards is presented in Section 4.1.7.

- The sixth 802.11 PHY is detailed in the IEEE 802.11g standards and is backward compatible with 802.11b. The Orthogonal Frequency Division Multiplexing (OFDM) PHY allows 54 Mbps data rates in the 2.4-MHz band. The speed of transmission under OFDM and COFDM is sufficient to carry voice and image data fast enough for most users. More on the IEEE 802.11g standards is given in Section 4.1.8.

4.1.1   Fundamentals of IEEE 802.11 Standards

Wireless LAN systems [472, 473, 481, 489] are different from wired LANs for a variety of reasons. The addressing schemes (and hence the contents of frames) must take into account the mobility of the network nodes, the PHY have to cope with the lower range and reliability of wireless media (WM), and the MAC sublayers have to ensure that these adjustments are presented to every higher layer (from the logical link layer on up) as a "generic" 802.11 LAN would. While one can easily draw the architecture of a wired LAN, for wireless PHYs, well-defined coverage areas simply do not exist. Propagation characteristics are dynamic and unpredictable (see Figure 4.2). Small changes in position or direction may result in dramatic differences in signal strength. Similar effects occur whether a station (STA) is stationary or mobile (as moving objects may impact station-to-station propagation). The shapes used in IEEE WLAN architecture drawings are there as a matter of convenience. In reality, the boundaries of WLANs are not well-defined from one moment to the next, mostly due to the mobility of the nodes (the addressable units of the WLAN).

In IEEE 802.11, the addressable unit is a STA. The STA is a message destination, but not (in general) a fixed location, as would be the case in a wired LAN. MAC frames are adjusted to take this change into account. The IEEE makes these observations about 802.11 PHYs, noting that they (a) Use a medium that has neither absolute nor readily observable boundaries outside of which stations with conformal PHY transceivers are known to be unable to receive network frames; (b) Are unprotected from outside signals; (c) Communicate over a medium significantly less reliable than wired PHYs; (d) Have dynamic topologies; (e) Lack full connectivity, and therefore the assumption normally made that every STA can hear every other STA is invalid (i.e., STAs may be "hidden" from each other); (f) Have time-varying and asymmetric propagation properties [452].

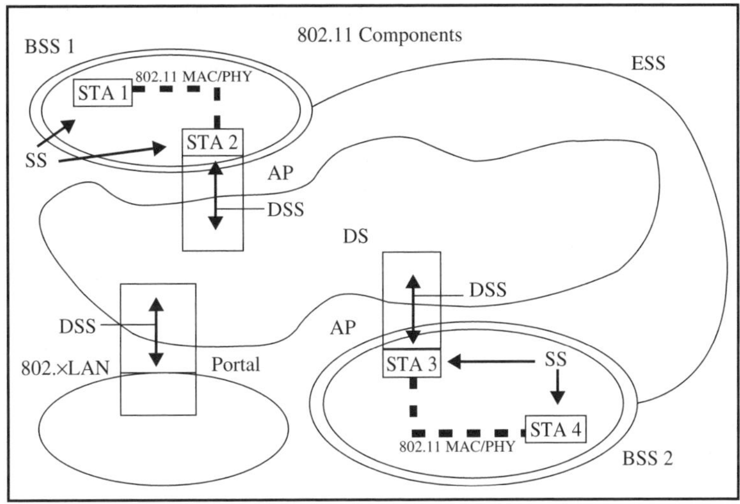

Figure 4.2 802.11 WLAN components [452].

Additionally, the specifications for the 802.11 PHYs must allow for both portable and mobile stations. Portable stations may change location from one access time to another, but mobile stations access the network while they are moving. Furthermore, the design of the PHYs recognizes that there is no guarantee that a particular station will be powered up at any particular time [452].

The architectural components of an 802.11 network include STA, basic service sets (BSSs), distribution systems (DS), WM, distribution system media (DSM), access points (AP) (also known as *base stations*), extended service sets (ESS), and portals, as shown in Figure 4.2 [452].

Stations are addressable units in a network and can be clients or servers. While it is possible for two personal computers to communicate with one another directly via a wireless connection, in a wireless LAN a personal computer is more likely to connect with a base station (or AP) for access to the rest of the network. Personal computers and personal digital assistants (PDAs) are the most common types of stations in a WLAN [453].

A BSS is the fundamental set of devices in a WLAN, and can comprise as few as two stations. The IEEE 802.11 (1999) documentation also uses the term BSS loosely to mean the coverage area within which the member stations of the BSS may remain in communication, allowing for the notion that a station can move "out" of its BSS, where it can no longer directly communicate with other members of the BSS [452]. An independent basic service set (IBSS) is possible if stations can communicate directly with one another. When an IBSS is created dynamically, for temporary use, it is referred to as an *ad hoc network*. If a station is a member of the infrastructure of a BSS, it is "associated" with the BSS by means of a distribution system service (DSS), which is discussed next. The associations are permitted to be dynamic, since stations come into and move out of range of the BSS, and can be turned off and back on [452].

A DS (not to be confused with a DSS) is the architectural element used to connect BSSs with one another. The DS maps addresses to actual destinations for mobile devices in multiple BSSs. In this type of architecture, the BSSs are not independent, but are components in a larger, extended network. The DS uses DSM, while the BSSs use what is referred to as *WM*. The terms are kept distinct because DSM and WM perform different jobs in the logical view of WLAN architecture. However, there is no IEEE "rule" that says the media used must be different if employed as DSM or WM. That is to say, one can use the same medium to perform both logical jobs (but, to allow for flexibility, one does not have to). The documentation expressly states that the IEEE 802.11 LAN architecture is specified independently of the physical characteristics of any specific implementation.

APs are stations that provide DS services. Since they are stations, they are addressable. APs connect STAs with their LAN. Administrators set parameters for APs, including the name of the wireless network, the channel used by the AP, and which Wired Equivalent Privacy (WEP) key is employed by the network for security [453]. Wireless networks use encryption to protect transmitted data from eavesdroppers – the data is usually sent over open airwaves – and WEP keys are one way to facilitate encryption and decryption. (As discussed in Section 4.3, WEP technology is vulnerable to crackers.) In short, data moves from STAs in a BSS, via an AP, to the DS, and vice versa.

When you use an AP to combine a DS, one or more BSSs, and potentially one or more LANs, the resulting network is called an *ESS* [453]. The IEEE 802.11 DS and BSSs allow IEEE 802.11 to create a wireless network of arbitrary size and complexity. The key concept is that the ESS network appears the same to an LLC layer as an IBSS network, and mobile stations may move from one BSS to another (within the same ESS) transparent to the LLC [452].

In an ESS, all of the following are possible. (a) The BSSs may partially overlap. This is commonly used to arrange contiguous coverage within a physical volume. (b) The BSSs could be physically disjointed. Logically there is no limit to the distance between BSSs. (c) The BSSs may be physically collocated. This may be done to provide redundancy. (d) One (or more) IBSS or ESS networks may be physically present in the same space as one (or more) ESS network(s). This may occur for a number of reasons. Two of the most common are when an ad hoc network is operating in a location that also has an ESS network, and when physically overlapping IEEE 802.11 networks have been set up by different organizations [452].

The last of the logical architectural units in an IEEE WLAN is the *portal*, which connects a traditional wired LAN to the 802.11 WLAN. The device acting as a portal can also act as an AP [452]. In very simple terms, a portal is the point where a wire (or cable) from a wired LAN meets a device on the wireless LAN that can read from the portal wire and transmit to the WLAN via its radio (or its wireless medium of choice). Needless to say, if no device on the WLAN is connected by wire to a wired LAN, then communication between the two networks will not take place (see Figure 4.3).

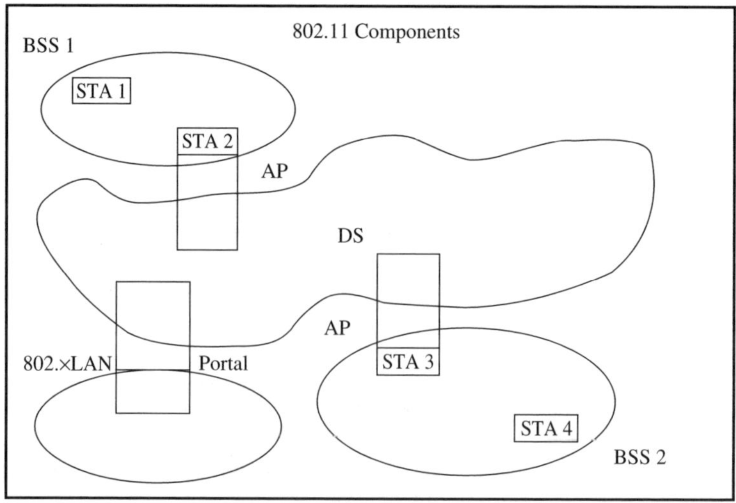

Figure 4.3 Portal connects wired LAN to WLAN [452].

Now that we are talking about joining wireless networks to other LANs, it is necessary to adopt the convention that IEEE uses to portray this concept in the 802.11 standards. In the first place, the DSS used by the joined networks do not have to be the same; in fact, IEEE 802.11 explicitly does not specify the details of DS implementations. Instead, IEEE 802.11 specifies services. The MAC sublayer of the WLAN utilizes these services while connecting the STAs on the network and to protect the data they wish to exchange. The services are divided into two categories: Services that are provided by every STA are called *station services* (SS), and services that are part of a DS are *DSS*, like the association of STAs to the infrastructure of a BSS mentioned above. The SSs are authentication (including preauthentication), deauthentication, privacy, and MAC service data unit (MSDU) delivery. Since APs are also STAs, APs provide SSs. APs also provide the DSSs; the DS accesses its DSSs from the APs. The DSSs are association, disassociation, distribution, integration, and reassociation (as shown in Figure 4.4). In the drawings included with the IEEE 802.11 documentation, DSSs are represented by arrows inside APs, and SSs are depicted as arrows between STAs [452].

IBSS networks do not have a physical DS and therefore must approach the provision of services different from the way in which ESSs do. Simply put, IBSS networks cannot provide the DSSs. The following descriptions of the SSs and DSSs assume a full-fledged ESS is in place.

Service 1: MSDU delivery: Networks are not much use without the ability to get the data to the recipient. Stations provide the MSDU delivery service, which is responsible for getting the data to the actual endpoint [456].

Service 2: Distribution: This is the primary service used by IEEE 802.11 STAs. It is conceptually invoked by every data message to or from an IEEE 802.11 STA operating in an ESS (when the frame is sent via the DS). Distribution is via a DSS [452]. When two BSSs are part of an ESS, STAs from the first BSS transmit messages to STAs in the second BSS via their respective APs, which communicate with each other via the DS. The IEEE 802.11 documentation refers to its Figure 7 and offers the

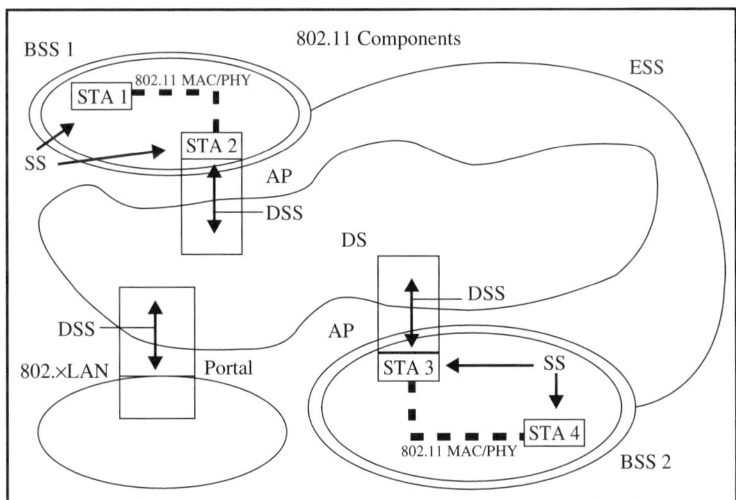

Figure 4.4 The Distribution service. STA 1, a unit in BSS 1, sends a transmission to STA 4 in BSS 2. The two BSSs contain APs that are connected by the DS of the overall ESS. When STA 1 sends its message, the data first travels to BSS 1's AP. The AP forwards the data to the distribution service of the distribution system (DS), and the distribution service maintained by the DS passes the data to the next appropriate recipient – in this case, BSS 2's AP. Once "inside" BSS 2, the data is forwarded to STA 4, their ultimate destination [452].

example of STA 1 in that drawing, a unit in BSS 1, sending a transmission to STA 4 in BSS 2. The two BSSs contain APs that are connected by the DS of the overall ESS. When STA 1 sends its message, the data first travels to BSS 1's AP. The AP forwards the data to the distribution service of the DS, and the distribution service maintained by the DS passes the data to the next appropriate recipient–in this case, BSS 2's AP. Once "inside" BSS 2, the data is forwarded to STA 4, their ultimate destination [452]. It must be stressed that any communication that uses an AP travels through the distribution service, including communications between two mobile stations associated with the same AP [456]. The DS makes use of its association-related services (the association, reassociation, and disassociation services) to gather the information necessary for the distribution system to locate the appropriate AP to receive a message being passed, as shown in Figure 4.4.

Service 3: Integration: If the distribution service finds that the appropriate next recipient of a message should be a portal, then the DS will activate the integration service. This service does whatever is needed to make the message compatible with the wire/cable/fiber that the portal will transmit on. The integration service is also called upon in the reverse situation – when a portal is passing a message to the DS – to make the message compatible with the wireless medium employed by the DS. This occurs before the message is handled by the distribution system. The IEEE 802.11 standard leaves the implementation of whatever is needed up to the DS implementers. (Implementation of the DS is outside the standards' scope.)

Service 4: Association: The association, reassociation, and disassociation services all ensure that the distribution service can do its job, which is to determine the next appropriate AP that a message needs to go to. These three services provide the DS with a mapping of the network's STAs to its APs. One STA can map to only one AP, but an AP may be mapped to several STAs. On a wired network this information can be keyed by an operator into a table and stored in a read-only format. On a wireless network, however, the mapping is dynamic because the STAs are mobile and the APs have limited ranges. The STAs are also fickle – they power down without bothering to inform the network's DS, or move out of range of the network entirely. A multitude of APs can improve the chance that a moving STA will remain within a network's transmission limits, but this scenario brings up another complication – how to maintain the DS's current "map" so that a STA is affiliated with only one of the network's APs (presumably the one with the strongest signal to the STA).

Before any STA can transmit messages on a network via a network AP, it must "join" the network. The term used by IEEE for this "joining" is association, and a STA that has "joined" a network has become associated with an AP on the network, in IEEE parlance. The actor in the network that accomplishes this joining is the DS's association service. It is invoked by an unassociated mobile STA when that STA requests association with an AP on the network (this is managed in the MAC sublayer). The DS stores the association – the STA-to-AP mapping – for use by the distribution service, and the STA is on the network.

Service 5: Reassociation: When an already-associated mobile STA moves and discovers the need to become associated with a different AP on the network, the reassociation service is invoked. Reassociations are initiated by mobile stations when signal conditions indicate that a different association would be beneficial. They are never initiated by the AP [456]. The reassociation service updates the DS's STA-to-AP map, and the distribution service has up-to-date information at its disposal.

Service 6: Disassociation: When a "polite" STA wishes to terminate its association, it calls upon the disassociation service, which removes data about the terminating association from the DS's map. "Impolite" STAs ignore this courtesy, abandon their APs, and the network relies on functions of the MAC sublayer to deal with the departed STAs' information. Disassociation can also be initiated by the partner AP (perhaps because the AP is leaving the network for maintenance service). Neither party can refuse disassociation – it is a notification, not a request.

Service 7: Authentication (and Preauthentication): IEEE 802.11 does not mandate the use of any particular authentication scheme, but it supports several authentication processes and allows the expansion of the supported authentication schemes. In both ESS and IBSS networks, before an association can be established, all STAs must confirm their identity. On a network with established associations, transmitting STAs must have authenticated themselves to the next logical destination STA – but a STA from which a message originates does not necessarily need to authenticate itself to the final destination STA. APs can be authenticated to numerous STAs at the same time.

Two authentication schemes are given in the 802.11 standards documentation: *Shared Key* and *Open System authentication*. On a Shared Key network, a secret encryption key is used for a STA to demonstrate that it has the right to be on the network. In this case the network must implement the optional WEP option. On an Open System network, any STA may become authenticated, but this may violate implicit assumptions made by higher network layers [452]. The authentication schemes are discussed in the Section 4.3.1, and WEP's vulnerability is covered in Section 4.3.

Preauthentication is a special case. It is also performed by the authentication service. Since STAs are mobile, they may need to reassociate with new APs at any moment, but they must be authenticated to the new AP before the new association is established, and authentication takes time. A STA can be preauthenticated with APs other than the one they are already associated with, to save time when they need to reassociate to another AP.

Service 8: Deauthentication: Deauthentication terminates an authenticated relationship. Because authentication is needed before network use is authorized, a side effect of deauthentication is the termination of any current association [456]. As with disassociation, deauthentication is not a request, it is a notification, and either partner in a mobile STA-AP relationship may call upon the deauthentication service – it is an SS. Deauthentication cannot be refused.

Service 9: Privacy: Even if an unauthenticated STA has no permission to send and receive messages on a network, if it is 802.11-compliant, it can hear them. For this reason, messages sent via the WM should be encrypted to be more secure. To this end, the optional WEP policy can be used by the privacy service for data encryption. Since the privacy service is an SS, all STA can invoke it. If, for some reason, unencrypted data frames arrive at a station configured to expect encrypted data, those frames are discarded and the LLC is not informed. They are acknowledged, however, to save the bandwidth that would be used to send duplicate frames in a Negative ACK (NACK) situation. The same is true when encrypted data arrive at a STA that does not have the appropriate key to decrypt them [452].

Again, it should be noted that WEP is not ironclad security – in fact, it has been proven recently that breaking WEP is easily within the capabilities of any laptop [456]. More details will be given in Section 4.3.

Before turning to address the way that *ad hoc networks* provide these services, some characterization of the 802.11 frame types is discussed. Frames are categorized as Class 1, Class 2, and Class 3 frames, and STAs are restricted as to which frame type they can send, on the basis of their authentication/association status. A STA has the status "State 1" if it is unauthenticated and unassociated with the network. A "State 2" STA is authenticated, but not associated, and a "State 3" STA is both authenticated and associated. A State 1 STA can send Class 1 frames, State 2 STAs can send Class 1 and 2 frames, and State 3 STAs can send any type of frame. The states are summarized in the 802.11 documentation's Figure 8 and shown in the Figure 4.5.

The 802.11 definitions of which kinds of frames (data, management, etc.) are considered to be of Class 1, 2, or 3, are listed in Tables 4.3, 4.4, and 4.5, respectively.

If STA A receives a Class 2 frame with a unicast address in the Address 1 field from STA B that is not authenticated with STA A, STA A should send a deauthentication frame to STA B.

If STA A receives a Class 3 frame with a unicast address in the Address 1 field from STA B that is authenticated but not associated with STA A, STA A should send a disassociation frame to STA B. This is an AP (STA A) receiving an illegal frame from a mobile, unassociated STA (STA B). The AP in this situation explicitly informs the mobile STA that it is not associated, and only has permission to send class 1 and 2 frames. In effect, the mobile STA is told that its status is presently State 2 [452].

If STA A receives a Class 3 frame with a unicast address in the Address 1 field from STA B that is not authenticated with STA A, STA A should send a deauthentication frame to STA B [452]. In this case, the AP receives an illegal frame from a STA that is not even authenticated, and tells the STA that its status is State 1 [456].

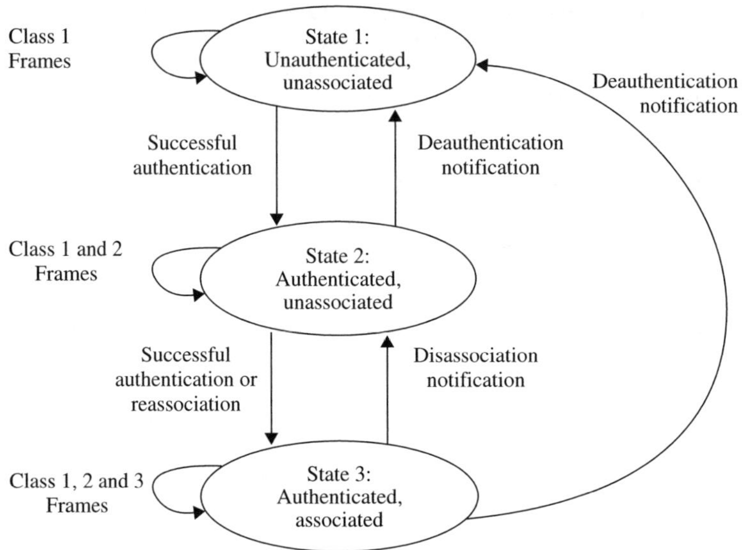

Figure 4.5 Classes of frames allowed to the three STA states [452].

Table 4.3   Class 1 frames (permitted from within States 1, 2, and 3)

(1) Control frames	(i) Request to send (RTS)
	(ii) Clear to send (CTS)
	(iii) Acknowledgment (ACK)
	(iv) Contention-Free (CF)- End+ACK
	(v) CF-End
(2) Management frames	(i) Probe request/response
	(ii) Beacon
	(iii) Authentication: Successful authentication enables a station to exchange Class 2 frames. Unsuccessful authentication leaves the STA in State 1.
	(iv) Deauthentication: Deauthentication notification when in State 2 or State 3 changes the STA's state to State 1. The STA should become authenticated again prior to sending Class 2 frames.
	(v) Announcement traffic indication message (ATIM)
(3) Data frames	(i) Data: Data frames with frame control (FC) bits "To DS" and "From DS" both set to false.

Table 4.4   Class 2 frames (if and only if authenticated; allowed from within States 2 and 3 only)

Management frames	(i) Association request/response —Successful association enables Class 3 frames. —Unsuccessful association leaves STA in State 2. (ii) Reassociation request/response —Successful reassociation enables Class 3 frames. —Unsuccessful reassociation leaves the STA in State 2 (with respect to the STA that was sent the reassociation message). Reassociation frames should only be sent if the sending STA is already associated in the same ESS. (iii) Disassociation —Disassociation notification when in State 3 changes a station's state to State 2. This station should become associated again if it wishes to utilize the DS.

Table 4.5   Class 3 frames (if and only if associated; allowed only from within State 3)

(1) Data frames	—Data subtypes: Data frames allowed. That is, either the "To DS" or "From DS" FC bits may be set to true to utilize DSSS.
(2) Management frames	—Deauthentication: Deauthentication notification when in State 3 implies disassociation as well, changing the STA's state from 3 to 1. The station should become authenticated again prior to another association.
(3) Control frames	—PS-Poll

The descriptions of the services (SS and DSS) presented above assumed that the network using them was an infrastructure ESS, with APs to provide the DSSs and a physical DS. IBSS networks do not have a DS and cannot support the DSSs, and in an IBSS, only frames of classes 1 and 2 are allowed [452].

4.1.2   Architecture and Functionality of a MAC Sublayer

Recall that the IEEE 802 family of standards has split the ISO/OSI data link layer into two parts: The upper sublayer is the LLC sublayer, and the lower is the MAC sublayer (just above the PHY) (as shown in Figure 4.1). This is in order to distinguish between medium access functionality and other data link issues. Each IEEE 802 PHY standard (Ethernet, Token Ring, Token Bus, and so on) specifies both the PHY aspects of the protocol as well as how medium access is to take place (as shown in Table 4.6). For example, the IEEE 802.3 standard (Ethernet) specifies the media types that can be used – a PHY issue – and specifies the use of the Carrier Sense Multiple Access/Collision Detection (CSMA/CD) medium access protocol – a data link layer and MAC sublayer issue [453]. In contrast, the LLC sublayer manages to provide a single interface to the network layer for the

Table 4.6   802 standards and medium access protocols

Standard	Medium access protocols
802.3	CSMA/CD
802.4	Token bus access
802.5	Token ring access
802.11	FHSS, DSSS, Infrared
802.11a	OFDM
802.11b	DSSS
802.11g	OFDM

numerous physical-layer topologies. This includes controlling the connection between sending and receiving computers, and seeing that frames are transferred without errors [453].

One of the MAC services, the asynchronous data transfer service, manages the exchange of data packets called *MSDUs* between devices (recall that every STA supports the MSDU delivery SS). Technically, MSDUs themselves are not passed from device to device. The MSDU is the packet of data going between the host computer's software and the wireless LAN MAC [457]. An MSDU is typically broken into smaller parts, each with a MAC header added, before encryption and transmission. This process is known as *fragmentation* (discussed at the end of this section). These pieces of the original MSDU are known as *MAC Protocol Data Units* (MPDUs). MPDUs are packets of data going between the MAC and the antenna. For transmissions, MSDUs are sent by the operating system (OS) to the MAC layer and are converted to MPDUs ready to be sent over the radio. For receptions, MPDUs arrive via the antenna and are converted to MSDUs prior to being delivered to the OS [457]. If an MPDU is lost in transmission, it can be resent instead of resending an entire MSDU.

All MAC frames share the same basic features: a MAC header for frame control, duration, address, and sequence control information, a frame body (which varies by frame type), and a frame check sequence (FCS) holding an IEEE 32-bit cyclic redundancy code (CRC). The FC field contains protocol version, type, subtype, to DS, from DS, more fragments, retry, power management, more data, WEP, and order subfields.

The 802.11 MAC supports CSMA/CA,[2] implemented in all STAs, as its fundamental distributed coordination function (DCF). This is almost the same DCF used in the IEEE 802.3 Ethernet LANs – CSMA/CD (CSMA with collision detection). CSMA is a "listen-before-talk" protocol: STAs "listen" to the transmission medium before sending a message. If the medium is in use, they use a back-off algorithm to reschedule their transmission for a later time, when the medium could potentially be free. Not all collisions are prevented with this scheme. If STA A sends a message, it will take time (the propagation delay) before it reaches STA B. In the meantime, STA B may sense the medium, not hear STA A's message yet, deduce that the medium is free, and send a message that collides with the first one. (On a LAN with an unusually long propagation period, or on a WAN, the propagation time between stations may be too long for carrier sensing to do much good.) Additionally, there is the "hidden terminal problem." On a wireless network, STA C may be physically prevented from ever hearing that STA A is transmitting, and constantly infer that it is safe to transmit to STA B, initiating collision after collision. In a wired LAN, collisions are detected to make sure messages involved in collisions are not lost for good, but time is lost and the medium is unnecessarily tied up. Wired LANs can easily detect collisions by listening for voltage spikes on the transmission medium. Wireless STAs cannot use this method because the signal of a transmitting STA dominates over all other nearby signals, and competing signals may not be detected. One solution would be to

---

[2]More discussions on random multiple access protocols can be found in Section 2.3.4.

deploy expensive directional antennas and front-end amplifiers at each STA, with one set for transmitting and one for receiving, to ensure that a STA could tell its transmitting antenna pattern from
its receiving antenna pattern. Arranging this situation is not convenient in radio terminals due to the
expense and the extra hardware encumbrance [454]. The collision avoidance (CA) method was developed to serve alongside CSMA in wireless networks and is the basic access method adopted by the
802.11 standards. Under CSMA/CA, STAs monitor the transmission medium by both virtual and physical means. The virtual monitor, the network allocation vector (NAV), is implemented in the MAC.
The NAV maintains a prediction of future traffic on the medium based on duration information that
is announced in RTS/CTS frames prior to the actual exchange of data. The duration information is
also available in the MAC headers of all frames sent during the CP other than the PS-Poll Control
frames. The physical monitor must be able to detect signals of certain types with certain degrees of
success [452].

Figure 4.6 provides an example of the operation of the CSMA/CA mechanism used in the IEEE
802.11 standard. Stations A, B, C, D, and E are engaged in contention for transmission of their
packet frames. Station A has a frame in the air when Stations B, C, and D sense the channel and
find it busy. Each of the three stations will run its random number generator to get a back-off time
at random. Station C followed by D and B draws the smallest number. All three terminals persist in
sensing the channel and defer their transmission until the transmission of the frame from terminal
A is completed. After completion, all three terminals wait for the interframe space (IFS) period and
start their counters immediately after completion of this period. As soon as the first terminal, Station
C in this example, finishes counting its waiting time, it starts transmission of its frame. The other
two terminals, B and D, sense C's transmission and freeze their counters to the value that they have
reached at the start of transmission for terminal C. During transmission of the frame from Station C,
Station E senses the channel, runs its own random number generator that in this case ends up with

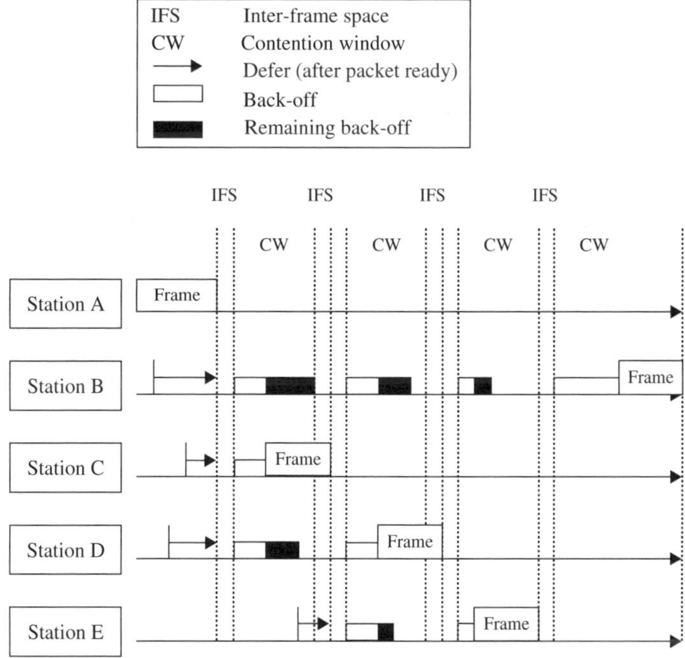

Figure 4.6 Illustration of CSMA/CA MAC layer protocol [454].

a number larger than the remainder of D's back-off time and smaller than the remainder of B's, and defers its transmission (until) after the completion of Station C's frame. In the same manner as the previous instance, all terminals wait for the IFS period and start their counters. Station D runs out of its random waiting time earlier and transmits its own packet. Stations B and E sense D's transmission, freeze their counters and wait for the completion of the frame transmission from terminal D and the IFS period after that before they start running down their counters. The counter for terminal E runs down to zero earlier, and this terminal sends its frame while B freezes its counter. After the IFS period following completion of the frame from Station E, the counter in Station B counts down to zero and B sends its own frame [454].

The advantage of this back-off strategy over the exponential back-off used in IEEE 802.3 is that the collision detection procedure is eliminated and the waiting time is fairly distributed in a way that on average enforces a first-come, first-served policy [454]. Under CSMA/CA, priority levels can be assigned to transmissions by shortening or lengthening the IFS time. Furthermore, all directed traffic uses immediate positive acknowledgment (ACK frame) where retransmission is scheduled by the sender if no ACK is received. CSMA/CA can be further refined by introducing RTS and CTS frames after determining that the medium is idle and after any deferrals or backoffs under certain circumstances (no RTSs are allowed prior to multicast and broadcast transmissions, for example, because multiple, colliding CTSs would be the result).

After the DCF, the next important MAC function is also an access method: the point coordination function (PCF). It is optional under the 802.11, but, if used, must coexist with the DCF. This access method uses a point coordinator (PC), which should operate at the AP of the BSS, to determine which STA currently has the right to transmit. The operation is essentially that of polling, with the PC performing the role of the polling master. The PCF should distribute information within Beacon management frames to gain control of the medium by setting the NAV in STAs. In addition, all frame transmissions under the PCF may use an IFS that is smaller than the IFS for frames transmitted via the DCF. The use of a smaller IFS implies that point-coordinated traffic should have priority access to the medium over STAs in overlapping BSSs operating under the DCF access method.

The access priority provided by a PCF may be utilized to create a CF access method. The PC controls the frame transmissions of the STAs so as to eliminate contention for a limited period of time. When both the DCF and the PCF are in operation on the same BSS, they alternate, and CF periods provided by the PCF alternate with contention periods under the DCF.

Another functionality of the MAC sublayer was brought up in the discussion of the asynchronous data transfer service, one of the MAC services. This is the fragmentation of large MAC frames into smaller MPDUs. Two types of MAC frames can be fragmented: MSDUs, which arrive in the MAC via the LLC, and MAC management protocol data units (MMPDUs), which originate in the MAC sublayer management entity (MLME) (discussed in the following text). When such frames are about to be sent, they are fragmented if they are larger than a specified limit (stipulated in a variable called a *Fragmentation Threshold*). Fragmentation creates MPDUs smaller than the original MSDU or MMPDU length to increase reliability, by increasing the probability of successful transmission of the MSDU or MMPDU in cases where channel characteristics limit reception reliability for longer frames [452]. Also, since each individual MPDU is acknowledged, the smaller MPDUs are all that needs to be resent if one is lost. Only unicast frames can be fragmented. The transmitting STA itself carries out the fragmentation process, and each receiving STA reassembles the MSDUs/MMPDUs. For historical reasons, the name for this recovery process is defragmentation.

The MAC data service translates LLC-style service messages into MAC-usable signals, and vice versa. The 802.11 MAC sublayer is managed by a logical abstraction called the *MLME* (the PHY is managed by an analogous unit called the *PHY layer management entity* (PLME)). The MLME provides an interface for invoking MAC layer management functions. Furthermore, a station management entity (SME) is resident in each STA. This is a layer-independent entity that may be viewed as residing in a separate management plane or as residing "off to the side."

These entities interact with one another via logical service access points (SAPs). There is a SME-MLME SAP, a SME-PLME SAP, and an MLME-PLME SAP. The 802.11 standards do not specify the exact functions of SMEs. However, the services provided by the MLME SAP to SMEs are defined. These include mundane information transfers revolving around requests by STAs to become associated (disassociated, reassociated) and authenticated (unauthenticated) with the network, STA power management and status, timer synchronization, and so on. Exchange of information concerning physical matters (getting hardware devices to cooperate with each other) is facilitated via the PLME SAPs [452].

4.1.3 IEEE 802.11 Frequency Hopping Spread Spectrum

Originally, 802.11 accommodated two transmission strategies: *frequency hopping spread spectrum* (FHSS) and *direct-sequence spread spectrum* (DSSS).

The 802.11 FHSS PHY was meant to operate in the 2.4-GHz band at speeds of 1 or 2 Mbps. The FHSS system was discarded by the 802.11b standards after it was found that having two transmission techniques for one standard meant that two kinds of (incompatible) equipment were necessary to implement the standard, and DSSS turned out to be the more reliable technique [465]. Legacy 802.11 FHSS equipment is not compatible with the current standards. Nevertheless, a discussion of FHSS is in order here, if only to see why DHSS came to be preferred for 802.11 networks.

When using the FHSS method, a transmitter shifts the center frequency of a signal several times per second, and both transmitter and receiver remain synchronized because the "hops" take place according to a pseudorandom pattern, which each device knows. In the United States, the FCC stipulates that at least 75 discrete frequencies must be employed for each transmission channel, and that a signal cannot remain on any particular frequency for more than 400 ms. In the 802.11, the maximum length of a packet is around 30 ms, and the hops are 1 MHz apart from one another [454]. FHSS can be employed for both analog and digital communications, but is currently implemented primarily for digital transmissions [454]. If 75 contiguous frequencies are used, then the bandwidth required for a transmission is 75 times larger than when only one frequency is used – the spectrum is spread over a larger portion of the transmission band (hence "frequency hopping spread spectrum"). The original motivation for developing this technique was a desire to avoid hostile jamming of a radio signal. If a transmission hops to a jammed frequency, the data sent in vain on that frequency are resent after the next hop. For wireless networks, FHSS has a desirable "side effect:" It minimizes the chances that different transmitters on the network will encounter interference from one another; otherwise the network could potentially disable itself.

The (original) 802.11 standards called for two-level *Gaussian frequency shift key* (GFSK) modulation for transmissions at 1 Mbps and four-level GFSK for 2 Mbps [453].

4.1.4 IEEE 802.11 Direct-Sequence Spread Spectrum

The DSSS[3] modulation spreads a signal across a wideband at very low power. The original 802.11 standards supported the DSSS data rates of 1 and 2 Mbps in the 2.4-GHz band. The widely spread signal can be recovered by a compliant receiver despite narrowband interference (and/or jamming) within the spectrum, and eavesdroppers may interpret any of the weak narrowband signals generated by DSSS that they discover as background noise [453]. Under the 802.11 standard, DHSS transmitters spread each data bit into 11 smaller pulses, called *chips*, and these chips are transmitted, spread over an extended (11 times wider) spectrum, for recovery and "despreading" by the DHSS receiver [454]. This chipping process increases the likelihood that the receiver can recover the original data on the first try. If some portion of chip is lost, the receiver can use statistical techniques to determine what the original was, without retransmitting the chip. The signal is effectively "louder" than if the data were transmitted raw [453]. In other words, this transmission method is robust.

---

[3]Section 2.2 has more discussions on the DSSS technique.

The 802.11 DHSS modulation splits the 2.4-GHz band into 14 five-MHz channels, 11 of which are available for use in the United States and Canada (not all are available elsewhere) [453]. Because of the signal spreading, the DHSS channels that are within 30 MHz of each other may interfere with one another. This means that only three WLANs should operate concurrently in the same area to assure unthreatened reliability [453].

The original 802.11 standards called for *differential binary phase-shift keying* (DBPSK) modulation for transmissions at 1 Mbps and differential quadrature phase-shift keying (DQPSK) for 2 Mbps [453].

4.1.5 The Reason DSSS Won

While the FHSS transmits over a wide spectrum on numerous narrowband frequencies, the DSSS's bandwidth is always wide. FHSS therefore needs a much slower sampling rate in its implementation and can accordingly be implemented with less expense. The decisive factor, however, is that DSSS systems provide a robust signal with better coverage area than the FHSS [454]. These amplifications are exactly what WLANs require.

4.1.6 IEEE 802.11 Infrared Specifications

A good deal of thought and work went into the 802.11 standards regarding the use of IR signals; unluckily, to little avail. Constructing a WLAN using IR technology would require that devices be placed closer together than when using radio signals, and this may have prevented implementers, striving for maximal computing mobility, from devoting much attention to the 802.11 IR. Plus, although the 802.11 standards allow for IR that is not line-of-sight, there is a common perception that IR devices must be "aimed at" each other, like a TV and its remote control. Another drawback may be that the 802.11 infrared-based LANs are only meant to operate indoors, constrained by exterior walls [452]. In any event, there are no wireless networking products currently available that implement this IR PHY [453].

The 802.11 infrared (IR) PHY stipulates that light in the 850- to 950-nm range is used for signals – which do not have to be directed – which permits the construction of a true LAN system. The range of such signals, with sensitive receivers allowed for under the 802.11, is about 20 m (not big enough for a regular classroom). However, in an environment lacking in reflective surfaces, the range would be reduced. On the other hand, an IR LAN in one room would not be susceptible to eavesdropping by (or interference from) a station in an adjacent room, because IR signals do not pass through walls. And the standards contain a cure for offending IR devices in the same room: If such a device does interfere, by transmitting continuously and with a very strong signal, it can be physically isolated (placing it in a different room) from the IEEE 802.11 LAN. The 802.11 standards designate temperatures from 0 to 40 °C as the range for full operation compliance with the IR PHY. When the standards were prepared, the only regulations on IR transmissions concerned safety issues, and there were no frequency allocation or bandwidth allocation restrictions on IR emissions worldwide [452].

4.1.7 IEEE 802.11b Supplement to 802.11 Standards

The IEEE's 802.11b supplement to the original 802.11 standards defined today's most common wireless standard, Wireless Fidelity (Wi-Fi). It is also known as *802.11 High Rate*, and supports speeds of up to 11 Mbps (comparable to the 10-Mbps speed of the original 802.3 Ethernet standard) [466]. The 802.11b operates in the same 2.4-GHz band as 802.11. Until the 802.11b, IEEE 802.11 networks operated at speeds of only 1 or 2 Mbps. The 802.11b's rapid rise in popularity is basically due to its faster data transmission rate: No longer was wireless the slower technology to standard Ethernet. Work could be done approximately as quickly on a wireless network as was being done on a wired standard Ethernet. Since the 802.11b is more widely embraced by commercial enterprisers, it became affordable for home networks.

It is important to realize that 802.11b only defines a new kind of PHY and MAC sublayer for 802.11, not an entirely new approach to wireless communications. The LLC sublayer (of the OSI data link layer) does not need to be changed by 802.11b (this is consistent with the IEEE 802 reference model). However, 802.11b does introduce two new sublayers into the PHY: the PHY convergence procedure (PLCP) sublayer, and the physical medium dependent (PMD) sublayer. The PMD sublayer provides a means and method of transmitting and receiving data through a wireless medium between two or more STAs, each using the high-rate (11 Mbps) system [467]. The PLCP sublayer allows the MAC to operate with minimum dependence on the PMD sublayer by facilitating the provision of MAC services (such as the asynchronous data transfer service, which handles the transport of MSDUs) [467].

Another prominent refinement introduced by 802.11b concerns data transmission procedures. The original 1997 802.11 standards support two entirely different methods of encoding – FHSS and DSSS – in order to grant some flexibility to implementers. As it turns out, this led to confusion and incompatibility between equipment. The 802.11b upgrade dropped FHSS in favor of DSSS. The DSSS has proven to be more reliable than the FHSS, and settling on one method of encoding eliminates the problem of having a single standard that includes two kinds of equipment that are not compatible with each other. Turning away from the FHSS, of course, means that the 802.11b devices are not backward compatible with the 802.11 devices using the FHSS. They are backward compatible with the 802.11 devices using the DSSS, and because nothing has changed in the LLC sublayer, they can operate in harmony with the standard wired Ethernet technology. The 802.11b looks like Ethernet to user applications [465]. IEEE gives a special name to the 802.11b DSSS: high-rate direct-sequence spread spectrum (HR/DSSS).

4.1.8   IEEE 802.11g Standard

The IEEE 802.11g standards amend the original 802.11 standards to allow for 54-Mbps data rates transmitted in the 2.4-MHz band. The 802.11g standards are designed to be compatible with the 802.11b standards – both share the same ISM band. The *extended rate physical* (ERP) layer is introduced to enable the faster data transmission rate.

The backward compatibility with the 802.11b means that the 802.11b STAs can connect to the 802.11g APs, albeit at only the 11-Mbps rate, and that the 802.11g STAs can connect with the 802.11b APs, again at the slower rate [453]. This means that users can deploy the new 802.11g equipment one piece at a time rather than all at once, without losing any of the functionality of their networks.

The 802.11g uses the OFDM, like the 802.11a, as well as the DSSS, like the 802.11b. All of the data rates supported in the 802.11a and the 802.11b are supported in the 802.11g. Table 4.7 lists the 802.11g data rates, transmission types and modulation schemes.

The 802.11b devices may not correctly detect that the medium they wish to use is busy when the 802.11g devices are using it for transmission. Therefore, the 802.11g standards provide protection mechanisms for the 802.11g STAs operating in a mixed 802.11b/802.11g environment, including RTS and CTS messages. The 802.11g throughput is greater than the 802.11b throughput at the same distance [480], with any of the protection mechanisms. There are also protection mechanisms for the 802.11g networks that sense that a 802.11b network is operating nearby, which could potentially cause interference [478].

4.2   IEEE 802.11a Supplement to 802.11 Standards

The IEEE's 802.11a supplement to the original 802.11 standards defines a new PHY for transmissions of up to 54 Mbps in the 5-GHz band using COFDM. The "Coded" in COFDM refers to error-control

Table 4.7   802.11g Data rates, transmission types, and modulation schemes [479]

Data rate (Mbps)	Transmission type	Modulation scheme
54	OFDM	64 QAM
48	OFDM	64 QAM
36	OFDM	16 QAM
24	OFDM	16 QAM
18	OFDM	QPSK1
12	OFDM	QPSK
11	DSSS	CCK2
9	OFDM	BPSK3
6	OFDM	BPSK
5.5	DSSS	CCK
2	DSSS	QPSK
1	DSSS	BPSK

codes. The MAC sublayer is unchanged from the original 802.11 version [454]. The 5-GHz band is also known as the *unlicensed national information infrastructure* (U-NII) band [452]. The European Telecommunications Standards Institute (ETSI) HIPERLAN-2 WLANs also employ the OFDM, as it is the most popular modulation technique for high-speed indoor WLANs [454].

The COFDM protocol is defined in the 802.11a standards. Using higher frequencies than the HR/DSSS (the 802.11b protocol) and several modulation schemes (see following text), COFDM delivers data rates of 6, 9, 12, 18, 24, 36, 48, and 54 Mbps. The 802.11a devices attempt to communicate at the highest rate and drop to the next-highest rate if they encounter too many transmission errors, dropping further still if necessary. The closer the devices are to one another, the faster they will be able to communicate, owing to higher signal strength [453]. The shorter data rates can cover distances of up to 100 m [454].

The COFDM's advantages over the HR/DSSS (802.11b) include higher transmission rates, roughly four times as many available channels – giving nearly eight times the transmission capacity, less risk of interference from Bluetooth devices and portable phones operating in the same ISM band, and up to five times the throughput in an office setting. An 802.11a administrator can achieve the throughput gain by deploying APs at the same cost as an 802.11b network, or can keep the throughput at the 802.11b level with lower AP deployment expenses [453].

OFDM is a variant of frequency division multiplexing (FDM). Both split bandwidth into smaller "subcarriers" and use the subcarriers as data transmission channels. FDM was used in first-generation mobile phones but wasted bandwidth by leaving an unused channel between used subcarriers to guard against interference from one phone to the next. In contrast, OFDM selects channels that can overlap without interfering with each other, conserving bandwidth [471][4].

OFDM encodes a single transmission into multiple subcarriers, unlike another emerging encoding technique, the code division multiple access (CDMA), which uses mathematical constructs more complicated than the OFDM's to send multiple transmissions on one carrier [471]. The advantage of the OFDM's less complicated mathematics is a savings on algorithm processing when transmissions are decoded at the receiver [454]. Under the OFDM, a wide frequency channel is divided into subchannels that each carry data and the subchannels are multiplexed into a single, fast channel for transmission [471].

---

[4]Section 7.5 offers more detailed discussions on OFDM technology.

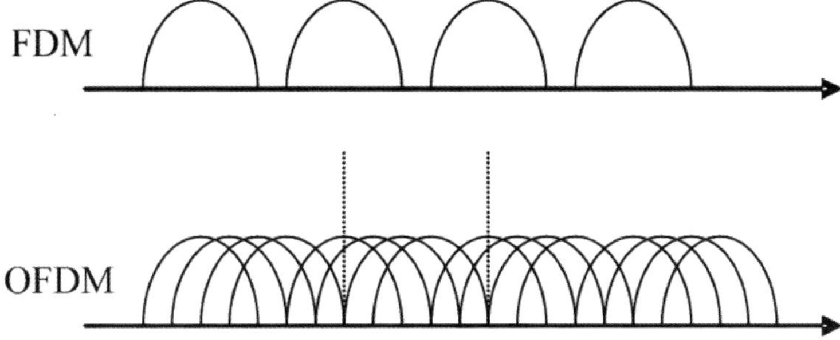

Figure 4.7 FDM versus OFDM [471].

OFDM achieves a gain in throughput over FDM by exploiting mathematical orthogonality. In essence, overlapping subcarriers peak in the frequency domain when their neighboring subcarriers have zero amplitude (see Figure 4.7). OFDM takes the coded signal for each subchannel and uses the inverse fast Fourier transform (IFFT) to create a composite waveform from the strength of each subchannel. OFDM receivers can then apply the FFT to a received waveform to extract the amplitude of each component subcarrier [471].

According to the 802.11a standards, the OFDM PHY consists of two protocol functions:

- A PHY convergence function, which adapts the capabilities of the PMD system to the PHY service. This function is supported by the PLCP, which defines a method of mapping the IEEE 802.11 PHY sublayer service data units (PSDUs) into a framing format suitable for sending and receiving user data and management information between two or more stations using the associated PMD system.

- A PMD system whose function defines the characteristics and method of transmitting and receiving data through a wireless medium between two or more stations, each using the OFDM system [452].

These two protocol functions are very much like the original 802.11 standards, but with changes to support the use of OFDM and the faster 5-GHz band.

The modulation schemes used by the 802.11a change when the supported transmission speeds rise. At the 6- and 9-Mbps rates, binary phase-shift keying (BPSK) modulation is employed, while the 12- and 18-Mbps rates use quadrature phase-shift keying (QPSK). The 24- and 36-Mbps rates use quadrature amplitude modulation (16-QAM), and the 48- and 54-Mbps rates employ 24-QAM [452].

4.3 IEEE 802.11 Security

In today's world, it is mandatory for the administrator of any computer network to make network security a top priority. The threat is not only that an intruder will type his way into his school records and change his grades: Computer networks are also under full-time assault from people with varying relationships to the enterprisers running them. Attacks come from anonymous virus distributors who have utterly no relationship at all with the enterpriser, from anarchic crackers who bear a grudge against particular computer or software manufacturers, or who delight in discovering vulnerabilities in OSs, from thrill-seeking "script kiddies" looking for some illegal amusement and peer recognition, from thieves searching for personal information for use in identity thefts or as blackmail material, from unscrupulous competitors looking for trade data

and secrets, and even from disgruntled or corrupted insiders that work at any level of the belea-
guered enterprisers themselves. Administrators of wired networks have to maintain solid firewalls,
keep and review network traffic logs, update software with anticracker patches, nag their users
to use strong passwords, scour their machines for "spyware" and purge it, make sure their virus
protection is up-to-date, and defend against physical break-ins. Every attack prepared for launch
against wired networks is automatically readymade for use against wireless networks as well. Wireless network administrators must defend against the same onslaught of adversaries that wired networks face, but in an environment that is much more convenient for attackers: WLANs operate within hard-to-control physical boundaries. Anyone who can receive their signals has already
accessed their network. From there, with enough determination and patience, it is only a matter
of time.

Administrators can make that time too long to be acceptable to attackers, say, with encryption.
But encryption comes with a cost of its own in quality of service, and a sufficient level of encryption
may come with too high a price to be acceptable to users. A 128-bit encryption steals nearly 2 Mbps
of bandwidth from every connection [458]. If throughput is already downgraded because of weak
signal strength, these 2 Mbps take on added significance: the 802.11b claims a speed of 11 Mbps.
A more realistic estimate of actual throughput is about 3.5 to 4.5 Mbps without WEP enabled and
2.5 to 3.5 Mbps with WEP enabled [465]. And some widely used encryption schemes have already
been broken, by techniques that grow in duration only linearly when 128-bit encryption is introduced,
rather than exponentially. Today's prevalent encryption schemes are under constant investigation. It
is still only a matter of time.

Unfortunately, all of the security mechanisms defined in the 1999 version of the IEEE 802.11
standard have been proven ineffective. The problems range all the way from issues with the lowest
primitives up to the high-level protocols used [457]. Already in 2003, the exhausting work of layering
security on top of the 802.11 was taking its toll, and some vendors were considering migrating to
a Bluetooth infrastructure to cut their losses [459]. Securing wireless networks [470] is a persistent,
dynamic struggle.

Therefore, WLAN administrators must be both wary and proactive, and just as determined as
their adversaries.

4.3.1 Authentication

The two IEEE 802.11 authentication schemes are *Open System authentication* and *Shared Key
authentication*. With either scheme, authentication takes place between two STAs, and authentication management frames may only be unicast. However, deauthentication frames may be sent to
groups. Authentication must be used between STAs and APs in an infrastructure BSS, and may be
used between STAs in an IBSS [452].

Open System authentication is the simplest of the available authentication algorithms. Essentially
it is a null authentication algorithm [452]. It simply provides a way for two parties to agree to exchange
data and provides no security benefits [456]. Any STA can (potentially) become authenticated to
a STA configured for Open System authentication. However, a STA configured for Open System
authentication may refuse to authenticate with another STA. APs can be set up to filter out the MAC
addresses of particular STAs – perhaps, past offenders [452, 458]. In open system authentication,
one party sends a MAC control frame, known as an *authentication frame*, to the other party. The
frame indicates that this is an open system authentication type. The other party responds with its
own authentication frame and the process is complete [456]. Obviously, a network using Open
System authentication must have other, higher-level security protocols in place if it is to maintain
any semblance of control over its data.

Shared Key authentication requires that the WEP option be implemented. With this scheme, STAs
can only authenticate to one another if they both share a common encryption key, previously delivered
in some secure way left unspecified by the IEEE 802.11 standards. The plan was fine for a while,

but WEP is now thoroughly compromised. It has been concluded that "WEP is not a well-designed cryptographic system" [456].

The following is a description of the four-step Shared Key authentication process:

- A station requesting 802.11 service sends an authentication frame to another station.

- When a station receives the initial authentication frame, the station replies with an authentication frame containing 40/128 octets of challenge text.

- The requesting station copies the challenge text into an authentication frame, encrypts it with a shared key using the WEP service, and sends the frame to the responding station.

- The receiving station decrypts the challenge text using the same shared key and compares it to the challenge text sent earlier. If they match, the receiving station replies with an authentication acknowledgement. If not, the station sends a negative authentication notice [456].

4.3.2   WEP

In this section, we discuss in the WEP theory of operation described in the IEEE 802.11 standards.

The WEP algorithm is a form of electronic code book in which a block of plaintext is bitwise XORed with a pseudorandom key sequence of equal length. The key sequence is generated by the WEP algorithm. WEP is a symmetric algorithm in which the same key is used for encipherment and decipherment. The secret key is concatenated with an initialization vector (IV) and the resulting seed is input to a pseudorandom number generator (PRNG). The PRNG outputs a key sequence k of pseudorandom octets equal in length to the number of data octets that are to be transmitted in the expanded MPDU plus 4 (since the key sequence is used to protect the integrity check value (ICV) as well as the data). Two processes are applied to the plaintext MPDU. To protect against unauthorized data modification, an integrity algorithm operates on P (the plaintext) to produce an ICV. Encipherment is then accomplished by mathematically combining the key sequence with the plaintext concatenated with the ICV. The output of the process is a message containing the IV and ciphertext.

As the IEEE 802.11 WEP theory continues, note the discrepancy between the asserted value of the IV and the way it is transmitted.

The WEP PRNG is the critical component of this process, since it transforms a relatively short secret key into an arbitrarily long key sequence. This greatly simplifies the task of key distribution, as only the secret key needs to be communicated between STAs. The IV extends the useful lifetime of the secret key and provides the self-synchronous property of the algorithm. The secret key remains constant while the IV changes periodically. Each new IV results in a new seed and key sequence, thus there is a one-to-one correspondence between the IV and k. The IV may be changed as frequently as every MPDU and, since it travels with the message, the receiver will always be able to decipher any message. The IV is transmitted in the clear since it does not provide an attacker with any information about the secret key, and since its value must be known by the recipient in order to perform the decryption.

IEEE recognized the vulnerability of WEP if implementers do not change the IV after each MPDU is generated: The contents of some fields in higher-layer protocol headers, as well as certain other higher-layer information, is constant or highly predictable. When such information is transmitted while encrypting with a particular key and IV, an eavesdropper can readily determine portions of the key sequence generated by that (key, IV) pair. If the same (key, IV) pair is used for successive MPDUs, this effect may substantially reduce the degree of privacy conferred by the WEP algorithm, allowing an eavesdropper to recover a subset of the user data without any knowledge of the secret key [452]. Here the standards caution implementers to change the IV for each MPDU.

The WEP algorithm is applied to the frame body of an MPDU. The (IV, frame body, ICV) triplet forms the actual data to be sent in the data frame. Decipherment begins with the arrival of a message. The IV of the incoming message should be used to generate the key sequence necessary to decipher the incoming message. Combining the ciphertext with the proper key sequence yields the original plaintext and ICV. Correct decipherment should be verified by performing the integrity check algorithm on the recovered plaintext and comparing the output ICV' to the ICV transmitted with the message. If ICV' is not equal to ICV, the received MPDU is in error [452].

Here is an expert's analysis of the WEP theory of operation.

To protect traffic from brute-force decryption attacks, WEP uses a set of up to four default keys, and it may also employ pairwise keys, called *mapped keys*, when allowed. Default keys are shared among all stations in a service set. Once a station has obtained the default keys for its service set, it may communicate using WEP.

Key reuse is often a weakness of cryptographic protocols. For this reason, WEP has a second class of keys used for pairwise communications. These keys are shared only between the two stations communicating. The two stations sharing a key have a key mapping relationship.

Reuse of the keystream is the major weakness in any stream cipher-based cryptosystem. When frames are encrypted with the same RC4 keystream, the XOR of the two encrypted packets is equivalent to the XOR of the two plaintext packets. By analyzing differences between the two streams in conjunction with the structure of the frame body, attackers can learn about the contents of the plaintext frames themselves. To help prevent the reuse of the keystream, WEP uses the IV to encrypt different packets with different RC4 keys. However, the IV is part of the packet header and is not encrypted, so eavesdroppers are tipped off to packets that are encrypted with the same RC4 key. Implementation problems can contribute to the lack of security. The 802.11 admits that using the same IV for a large number of frames is insecure and should be avoided. The standard allows for using a different IV for each frame, but it is not required.

That is why WEP is not a well-designed cryptographic system, and the extra bits in the key will help very little. The best publicly disclosed attack against WEP can recover the key in seconds, no matter what its length is [456]. To make matters worse, some "helpful" person invested their time into generating a cracker tool. Publicizing the threat was a service to everyone, but we leave it as an exercise for the readers to determine what satisfaction is obtained by the author of tools that turn threat into reality and waste millions of dollars of investment. However, the tool was published; it is available on the Internet, and attackers can use it to crack WEP systems open at will [457]. Such a "helpful" tool also enables countless others who would otherwise be unable to mount a successful attack on WEP. Researchers have discovered other problems with the WEP algorithm and the way it uses RC4. The first step an 802.11 security gives today should be "use something besides WEP."

WEP should not be relied on in 802.11 networks. Here is one expert's description of the WEP algorithm:

- It is assumed that the secret key has been distributed to both the transmitting and receiving stations by some secure means.

- On the transmitting station, the 40-bit secret key is concatenated with a 24-bit IV to produce a seed for input into the WEP PRNG.

- The seed is passed into the PRNG to produce a stream (keystream) of pseudorandom octets.

- The plaintext PDU (protocol data unit) is then XORed with the pseudorandom keystream to produce the ciphertext PDU.

- This ciphertext PDU is then concatenated with the IV and transmitted on the WM.

- The receiving station reads the IV and concatenates it with the secret key, producing the seed that it passes to its own PRNG.

- The receiver's PRNG should produce the identical keystream used by the transmitting station, so that when XORed with the ciphertext, the original plaintext PDU is produced [460].

Several researchers began publishing papers on WEP imperfections in 2001. The first such paper points out that since the IV used by WEP is transmitted in the clear, it is possible to gain enough information from transmissions with identical IVs to decrypt WEP-encrypted data without knowledge of the secret key [474]. When you XOR two ciphertexts that have the same IV, the keystream is cancelled out. The result is the XOR of the two original plaintexts. If one of the plaintexts is known, the other can now be obtained, as can the keystream used to generate both. A dictionary can then be created that specifies the keystream used for each IV. In this way, an attacker can eventually decrypt all transmissions on the WM without ever actually obtaining the secret key [460].

Here's how it works: K is for key, Cx is for ciphertext x, and Px is for plaintext x. Given two ciphertexts produced with the same <IV, K> pair:

C1 = RC4(IV, K) XOR P1

C2 = RC4(IV, K) XOR P2

XORing the two ciphertexts together removes the pseudorandom stream generated by RC4 and produces the XOR of the two plaintexts:

C1 XOR C2 = (RC4(IV, K) XOR P1) XOR (RC4(IV, K) XOR P2)

C1 XOR C2 = P1 XOR P2

The XOR of the two plaintexts makes it significantly easier to recover the two plaintexts because of their well-known structure [457].

A 128-bit encryption does not improve the situation. It is the reuse of IVs that creates the vulnerability, and this is nearly impossible to avoid [460]. Unfortunately, the IV in the IEEE 802.11 WEP is only 24 bits long. A 24-bit number has values from 0 to 16,777,216 – so there are about 17 million IV values possible. A busy access point at 11 Mbps is capable of transmitting/receiving about 700 average-sized packets a second. If a different IV value were used for every packet, all the values would be used up in less than seven hours. IV values are bound to be reused [457].

To make matters worse, many devices reset their IV to the same "startup" value when rebooted, and the same pseudorandom sequence of IVs results as after every previous reboot. "If there are 20 mobile devices turned on in the morning, and they all start with the same IV value and follow the same sequence, then the same IV value will appear 20 times for each value in the sequence" [457]. Collecting transmissions that have the same IV is child's play.

Another 2001 paper found a bias in the pseudorandom encryption stream produced by RC4, the PRNG algorithm used by the 802.11 [475]. The algorithm was reverse-engineered and made public in 1994. It uses a 256-byte array containing a permutation of the numbers 0–255 [457]. It was found that the second word of the pseudorandom stream is zero twice as often as should be expected (1 in 128 instead of 1 in 256) [475]. There are two consequences of this seemingly unimportant defect: it is easy to distinguish RC4 ciphertext from other algorithms' ciphertexts, and it sets the stage for the next paper's key-discovery attack [457].

Again in 2001, a "weak" class of keys used in RC4 was discovered. These "weak" keys expose information about the secret key, and, armed with this revelation, an attack was developed that recovers the entire secret key in WEP relatively quickly [476]. The researchers estimated that this could be accomplished after collecting approximately 4 million packets, although they did not test the attack on an actual WEP system. Investigators at AT&T did put the attack to the test, and achieved success with about 5 million packets – about three hours on a partially loaded network [460]. In a classic stroke of bad luck, the set of weak keys discovered were exactly those used by WEP [457]. They are formed when certain values of the IV are attached to the secret key and processed by RC4.

The attack, remarkably, only needs two pieces of information to work: the IV and the first encrypted byte. The former is transmitted in the clear, and the latter in most, if not all, WEP encrypted transmissions is the 802.2 LLC header that contains 0xAA [460]. The 128-bit encryption only makes the key recovery slightly more difficult than the 40-bit encryption [457]. Lengthening the IV (which would require a rewrite of the 802.11 standards) makes the time involved in this attack grow only linearly [460].

Yet another 2001 paper describes vulnerabilities in 802.11 beyond WEP. Its authors zero in on problems with the authentication and access control protocols, and show that the challenge-response authentication protocol for shared key authentication has a weakness that allows an attacker to determine the keystream used to encrypt a response, and then use this keystream to gain authentication to the network [477]. In addition, the service set identifier (SSID) used by each network is worthless as a security mechanism, because in many 802.11 management frames, it is transmitted in the clear [460]. Another problem identified in this paper has to do with hardware. Many 802.11 adapters allow their MAC address to be set by the user, and therefore it is a relatively simple procedure to sniff the WLAN for MAC addresses that are permitted access, change the MAC address of the (attacker's) 802.11 adapter, and gain access to the WLAN [460].

While 802.11 WLANs have come under increasing attack as the technology has become more popular, some denial of service (DoS) attacks can be entirely unintended: 802.11, 11b, and 11g all use the 2.4 to 2.5 GHz ISM band, which is extremely crowded at the moment. Cordless phones, baby monitors, X10 cameras, and a host of other devices (garage-door openers, microwave ovens, medical devices, and other wireless networks themselves [465]) operate in this band and can cause packet loss or outright disruption of service in 802.11 networks [459]. For safety's sake we turn off our cell phones when we enter a hospital, and turn off all wireless devices on a plane that is taking off or landing.

Another hardware-based DoS attack is not an accident. Any attacker with a bigger amplifier, antenna, or using more power, can deny service to an individual or group at the RF level and because our equipment is readily available to our attackers attempts to prevent such jamming in the consumer realm are futile [457].

DoS attacks can be launched against WLANs in the software realm, too. Because 802.11 management frames that control association services are unauthenticated, it is easy to forge a packet that purports to be from an AP and tells all the STAs associated with it to disassociate. There is nothing that can be done to prevent this and an attacker could keep the station(s) off the network indefinitely (or at least until someone turned off the computer sending the forged packets) [459]. Is there a cracker's tool available on the Internet that does this? You bet.

When cracker's tools are freely available, attackers don't have to be particularly computer-savvy. Anyone can implement the above attack, which is otherwise fairly complicated (and tedious – who wants to sit down and formulate the string of ones and zeros it would take to fabricate a packet?).

The 802.11 wireless networks are particularly vulnerable to what are known as *man-in-the-middle* (MiM) attacks, because "there are no integrity guarantees provided at the link layer" and it is simple to spoof MAC addresses [457]. MiMs can be established regardless of any protections (Wi-Fi Protected Access (WPA), Robust Security Network (RSN), Virtual Private Network (VPN), and so on) that you might be using but do not necessarily pose a threat if the (higher-level) security protocol is strong [457]. The attacker picks as his target a STA that is already associated with a legitimate AP, and sets up a false AP with the same SSID and MAC address as a genuine AP on the network within range of the victim STA (but not the one currently associated with the target). The next step is to forge a deauthentication message that appears to be from the associated AP and send it to the target STA. The STA has no choice but to drop its association with its current AP and attempt to reassociate (with any AP in range). The original AP denies it service because of the counterfeit deauthentication message, and the victim STA associates with the attacker's false AP. The false AP immediately associates with a valid AP, forwarding all traffic, and proceeds to authenticate. Now all messages sent between the real, hoodwinked, AP and the victim STA are controlled by the attacker, and these messages can be analyzed and then modified for use in further attacks. If DoS is the object, nothing prevents the attacker from "replaying" old messages and flooding the network [457].

Another MiM that "has been a plague on wired networks for some time" is known as *address resolution protocol* (ARP) spoofing, and it can be used against a WLAN if the encryption scheme has been broken, where it can still succeed more often than not [457]. ARP is used to learn the MAC address of a known IP address, and ARP packets do not have any integrity protection [457]. When a STA wants to communicate with a particular IP address it broadcasts an ARP-Request to acquire the corresponding MAC address. An attacker simply says "it's me" and the requesting STA puts an incorrect entry into its ARP cache. From then until the entry times out, the attacker receives all the data intended for the spoofed machine from the victim STA [457].

Additionally, when a STA has authenticated with an AP, an attacker can simply hijack the session by sending the STA phony disassociation frames and continuing to send forged management frames to keep the STA from reassociating. The AP has no idea this is happening; it believes an authenticated session is still in place. The attacker simply masquerades as the disassociated STA, taking over the session [457]. This technique can be used to gain unpaid access at a public hotspot, at least for a short time.

The most widespread attacks come by way of cracker's tools that are targeted specifically at 802.11 WLANs. There are tools to facilitate "war-driving" (identifying wireless networks, users, and authentication procedures, which literally involves driving around), tools for network mapping, tools for sniffing and monitoring wireless networks, and tools for gaining unauthorized access to WLANs [459]. Some of these tools only work in conjunction with a global positioning system (GPS) unit, and, of course, specialized antennas. Do not let that fool you into thinking that this is not an extensive problem: One antenna distributor is famous (and cracker-endorsed) for its "war-driving bundles" [459]. Considering the potential payoffs that can be reaped from a compromised network, intruders will not be dissuaded by the initial cost of a little apparatus. And some of the network-diagramming activities may not even be criminal (although they clearly involve an indisputable invasion of privacy).

War-driving (also known as *wilding*) is done from a vehicle containing a computer and whatever else is appropriate for the cracker's tool in use. The attacker simply drives around the likely location of a WLAN, or sits in an enterprise's parking lot, and the software does the real work. War-driving tools will find a network's SSID, MAC addresses, IP addresses, WEP status (on or off), and the length of the WEP key. Wireless mapping tools will map this information with GPS accuracy, giving, of course, the locations of the all-important APs, but also the locations of the GPS coordinates with the highest signal strength found for the AP (a tool does this!). In fact, a vehicle is not required if the attacker can walk through an enterprise's buildings: There are war-driving tools designed to run on Pocket-PCs. Some war-driving tools broadcast the 802.11 Probe Requests asking for network information, and can be blinded by administrators who do not allow Probe Responses to be sent to unknown machines. If the attacking machine is masquerading as a legitimate machine, this measure is ineffective. Other war-driving tools are entirely passive, listening only, and gleaning information from the network's transmissions. Some tools are specialized to collect only WEP-encrypted data for use in breaking the WEP key [457, 459].

Wireless sniffers are available that specifically decompose 802.11 packets and enable WEP cracking. Wireless monitoring tools will perform network analysis and filter out information the attacker has no interest in. In addition, they will identify the protocols being used by the target machines and unearth details about AP-to-client STA relationships [459].

Tools for gaining access to the 802.11 WLANs include a number of those that crack WEP encryption. In addition, there are tools that facilitate MiM attacks on 802.11 networks, and others that manage DoS attacks [457, 459]. Anyone can use them.

There is a special 802.11 implementation that is of concern, not only to administrators, but to casual laptop owners as well. When using a public hotspot, special care must be taken to protect the laptop from malicious hotspot users. On a corporate network, users are authorized to use the network because they are trusted, and presumably have no reason to attack each other. At a public hotspot, users are authorized to use the network because they paid a fee, not because they are trusted [457]. A measure of vigilance is in order. The threat is that your laptop is authenticated to a network on which potential attackers are now legitimate users, with all the rights and privileges that pertain thereto. If an attacker is present, he is already way too close (were you on your corporate network, he would already be past your corporate firewall). Data you send and receive can be intercepted, and your computer's files can be copied, modified, mutilated, or deleted. At a public hotspot, the prospect of someone accessing your computer should be taken very seriously [457]. If your computer is configured for file sharing among your colleagues when on your corporate network, and you forget to disable this feature before joining a hotspot network, there is a real danger it will be noticed by a stranger and investigated [457]. Worse, someone may already be interested in your computer and transmissions and "stalk" you until you enter a hotspot. They may want your company's plans, or they may want your personal information. Furthermore, since they are already on your network, attackers can easily slip a Trojan virus on your computer, and if your antivirus software does not detect it, your computer from then on, no matter where you are, will send out its location to the attacker and leave a back door open for unrestricted access by the attacker [457]. Public hotspots are attractive because they are convenient and the atmosphere is relaxed. Attackers can find them convenient, too, and their relaxed security environment attractive.

In 1999, 802.11b still relied on WEP for data encryption. In August, 2001 the 802.11 working group began work to remedy WEP's shortcomings [465], and in October 2002, the Wi-Fi Alliance announced a security solution that supercedes WEP called *WPA*. This standard was formerly known as *Safe Secure Network* (SSN). WPA is designed to work with existing 802.11-based products and offers forward compatibility with the 802.11i. All of the known shortcomings of WEP are addressed by WPA which features packet key mixing, a message integrity check, an extended initialization vector, and a rekeying mechanism [466].

These features of WPA are embedded in a new protocol called the *Temporal Key Integrity Protocol* (TKIP), which was developed specifically in order to facilitate security upgrades in WEP-enabled networks. TKIP is now a fundamental part of WPA and serves as a means to address the WEP weaknesses mentioned above. WEP's weaknesses can be summarized as: (1) The IV value is too short and is reused. (2) "Weak" keys are occasionally created and easy to crack. (3) Messages are not effectively checked for tampering. (4) The master key is applied in a perilous way and need not be updated. (5) Messages can be replayed with impunity during DoS attacks [457].

Under WPA, Per-Packet Key Mixing addresses the second and fourth vulnerabilities by changing the encryption key for every frame sent and constructing IVs in a way that avoids the creation of weak keys. A Message Integrity protocol to prevent tampering is in place to handle the third WEP weakness. The size of the IV is increased and the way in which it is chosen is changed in order to cover the first weakness. The fourth weakness is further addressed by adding a resource for changing the keys in use. Finally, a TKIP Sequence Counter is introduced to take care of the fifth insecurity [457].

The TKIP mechanism for per-packet key mixing is a significant improvement over the simple WEP key. The WPA preshared key differs dramatically from the WEP key. Under WPA, the preshared key is used only in the initial setup of the dynamic TKIP key exchange. This base key is never sent over the air or used to directly encrypt the data stream [468]. In TKIP, there are multiple levels of keys derived from a single master key. Session keys are derived from the master key. These keys are then split into pieces for various uses, one of which is encryption. What per-packet key mixing does is to further derive a key specifically for each and every packet sent. In other words, at the level of RC4 (802.11's choice of encryption algorithms), every packet uses a different, and apparently unrelated, key [457]. Note that it is not the session and master keys that change for every packet, but rather the derived, "mixed" key used for RC4 encryption.

The key mixing process makes use of the extended, 48-bit IV and adds the use of the MAC address of the sending machine to prevent two STAs from producing identical mixed keys. The problem is that the computation to derive the key can be processing intensive. There is not a lot of computing power in the MAC chip of most WEP-based Wi-Fi cards. So the calculation was divided into two phases. Phase 1 involves all the data that is relatively static, such as the secret session key, the high order 32 bits of the IV, and the MAC address. Phase 2 is a quicker computation and includes the only item that changes every packet – the low order 16 bits of the IV [457]. The innovation is that since the next IV value is known, the receiver can compute mixed keys in advance, in anticipation of the arrival of packets that were encrypted with them.

The generally employed methods for ensuring message integrity create a check value known as the *message authentication code* (MAC). The 802.11 standards call this the *message integrity code* (MIC) to avoid confusion with the medium access control sublayer (MAC). Several secure methods of computing a MIC have been devised for general use. However, they all require so much processing power (from the Wi-Fi MAC chip) that, if used in a WLAN, they would drop 11 Mbps throughput to less than 1 Mbps [457]. Since many APs do not have processing power equal to that of modern PCs, conducting these computations at the software level is out of question for contemporary WLANs. A means of creating MICs that would not degrade existing performance levels was needed. TKIP's message integrity check is implemented in a method called *Michael*, which solves the computation problem with a simple set of mathematical operations. This leaves Michael open to brute force attacks, but the solution to that problem can be found in defensive countermeasures. Essentially, Michael forces the abandonment of current keys when an attack on them is detected. New keys are established and the attack, as they say, is history. Michael employs a "blackout" rule that prohibits generation of new keys for 60 sec after an attack is detected, bringing the work of the compromised machines to a temporary halt, but also limiting an attacker to one attempt per minute [457].

In TKIP, IV length is doubled to 48 bits. The advantages of the extended-length IV are startling. Suppose you have a device sending 10,000 packets per second. This is feasible using 64-byte packets at 11 Mbps, for example. The 24-bit IV would roll over (begin generating already-used IV values) in less than half an hour, while the 48-bit IV would not roll over for over 900 years [457]. In addition, since the longer IV is used in per-packet key mixing as well, the value of the key used in RC encryption is different for every IV value [457].

The rekeying mechanism used in TKIP generates a new encryption key every 10,000 packets [469]. In addition, dynamic per-session and per-packet encryption keys prevent key reuse [468].

When all is said and done, WPA resolves the security problems associated with WEP, for the present. In the future, 802.11 standards (802.11i) will incorporate a RSN protocol, designed to be backward compatible with WPA.

4.4   IEEE 802.15 WPAN Standards

The 802.15 set of standards define the PHY and MAC layers for small, short-distance wireless networks. The term Personal Operating Space (POS) is introduced to refer to the small area covered by a wireless personal area network (WPAN). A POS is the space about a person or object that typically extends up to 10 m in all directions and envelops the person whether stationary or in motion. This standard has been developed to ensure coexistence with all IEEE 802.11 networks [482]. The 802.15 WPANs operate in the 2.4-GHz ISM band [482].

Much of this standard is derived from Bluetooth core, profiles, and test specifications [482]. The intent is to make the 802.15 WPANs capable of employing Bluetooth devices, if desired. The terms Bluetooth WPAN and IEEE 802.15.1 WPAN refer to the specific, single example of a WPAN presented in this standard [482].

According to the standards, a WPAN can be viewed as a personal communications bubble around a person. Within this bubble, which moves as a person moves around, personal devices can connect with one another. These devices may be under the control of a single individual or several people's devices may interact with each other [482].

Unlike the "traditional" devices used in WLANs, WPAN devices often include digital cameras, PDAs, and GPS units – devices that run on batteries – which are generally more mobile than the servers, client computers, and printers found on a WLAN. Therefore, the 802.15 standards define smaller power levels and area coverage for WPANs (around 10 m) than for WLANs (around 100 m).

In addition, while WLANs generally have an "unending" life span, the life span of a WPAN is for as long as a master device participates in the WPAN [482].

The 802.15 standards permit WPANs to connect to 802.11 WLANs or 802 (wired) LANs, if desired [482].

4.4.1   IEEE 802.15.3a Standard

The IEEE 802.15.3 standard defines low cost–and low power–use protocols for high-speed (up to 500 Mbps) WPANs. Ultra-wideband (UWB) technology is supported, including ad hoc piconet formation and disassembly. The speed is meant to satisfy user multimedia needs [483].

Low power use is ensured with protocols that either allow the WPAN to establish a (low) power setting for its devices, or allow communicating devices with a "good" connection to lower their power as long as the connection remains good [483]. More detailed discussions on UWB technology can be found in Section 7.6

4.4.2   IEEE 802.15.4 Standard

The 802.15.4 standard provides for ultralow complexity, ultralow cost, ultralow power consumption, and low data rate wireless connectivity among inexpensive devices. The raw data rate will be high enough (maximum of 250 kbps) to satisfy a set of simple needs such as interactive toys, but scalable down to the needs of sensor and automation needs (20 kbps or below) for wireless communications [484].

The 802.15.4 standard favors battery-powered devices. However, in certain applications some of these devices could potentially be mains-powered. Battery-powered devices will require duty-cycling to reduce power consumption. These devices will spend most of their operational life in a sleep state; however, each device should periodically listen to the RF channel in order to determine whether a message is pending. This mechanism allows the application designer to decide on the balance between battery consumption and message latency. Mains-powered devices have the option of listening to the RF channel continuously [484].

4.5   IEEE 802.16 WMAN Standards

The 802.16 standards describe protocols for a fast, potentially large, and cost-conscious wireless network. Raw data rates can be up to 120 Mbps, and power is conserved through transmission protocols that enable services to be tailored to the delay and bandwidth requirements of each user application [485].

4.6   ETSI HIPERLAN and ETSI HIPERLAN/2 Standards

The ETSI developed the HIPERLAN (High-Performance LAN) and HIPERLAN/2 (for WLANs) technologies concurrently with IEEE's work on 802.11. HIPERLAN achieves a 54-Mbps data transmission rate, made possible by the use of a modularization method called *OFDM* [463].[5] In HIPERLAN/2 networks, mobile terminals wirelessly connect to APs that are wired to a fixed, wired LAN. When the mobile terminals move out of range of their current AP, they are handed off to the closest AP.

HIPERLAN/2 connections can be assigned a priority level for quality of service purposes, depending on an application's bandwidth needs and jitter tolerance [463]. The HIPERLAN/2 APs have a built-in support for automatic transmission frequency allocation within the APs coverage area. This is performed by the Dynamic Frequency Selection (DFS) function. An appropriate radio channel is selected on the basis of what radio channels are already in use by other APs and to minimize interference with the environment [463].

---

[5]Refer to Section 7.5 for more detailed discussions on OFDM.

Mobile terminals and APs must authenticate to each other; each node has a HIPERLAN ID (HID) and a Node ID (NID) which uniquely identify the station. Encryption is used in communications to guard against eavesdroppers and MiM attacks [463].

The HIPERLAN/2 protocol model contains a PHY, a datal link control (DLC) layer, and a convergence layer (CL). The DLC consists of several sublayers: a MAC protocol sublayer, a radio link control (RLC) sublayer, and an EC sublayer. There is a radio link control protocol (RCP) for the following tasks: DLC connection control (DCC), radio resource control (RRC), and association control (AC). There are two models for the CL: cell based and packet based. The CL manages service requests from higher layers to the DLC [463].

4.7 MMAC by Japan

An MMAC is a Multimedia Mobile Access Communications system for connecting wireless devices to optical-fiber networks. The desire to obtain high-speed wireless transmission rates is the driving force behind MMAC designers, because MMACs are intended (as their name implies) for multimedia applications, especially business-related ones, including voice, music, and video teleconferencing transmissions [464]. The concept was originated in Japan.

The MMAC system divides the concept into four categories, based on target applications:

- High-Speed Wireless Access: a mobile communications system for mobile video telephone conversations. Meant for indoor or outdoor use with a transmission rate of 30 Mbps.

- Ultra High-Speed Wireless LAN: an indoor system for high-quality TV conferences, transmitting at 156 Mbps.

- 5-GHz Band Mobile Access: for multimedia transmissions indoors or outdoors, at 20 to 25 Mbps.

- Wireless Home-Link: for indoor use to link PCs and audiovisual equipment, with a 100-Mbps transmission rate.

4.8 Bluetooth Technologies

Bluetooth technology is based on a short-range radio specification defining transmission protocols between computers and other devices like cell phones and printers [462]. It was initially invented in 1994 by the Swedish L. M. Ericsson Company (who named it after the tenth-century Danish King Harald Blaatand "Bluetooth" II). In 1998 the Bluetooth Special Interest Group, Inc. (SIG) was founded by Ericsson, IBM, Intel, Nokia, and Toshiba. Their specifications for wireless connectivity were published in 1999. The Bluetooth SIG now consists of nearly 2000 companies, including Microsoft, Lucent, Motorola, and 3COM [463].

Bluetooth and 802.11b Wi-Fi can be thought of in some sense competitors, but in the real world they are complements of one another, or cousins. Bluetooth-compliant devices are easily upgradable to work with 802.15 WPAN-compliant devices [463]. Bluetooth's basic function is to provide a standard wireless technology to replace the multitude of propriety cables currently linking computing devices [462]. "The technology is designed to be low cost and low power to preserve the pocketbook and conserve battery life" [463].

These are the features of the Bluetooth technology:

- It separates the frequency band into hops. This spread spectrum is used to hop from one channel to another, which adds a strong layer of security.

- Up to eight devices can be networked in a piconet (the Bluetooth and 802.15 designation for a special personal area network (PAN)).

- Signals can be transmitted through walls and briefcases, thus eliminating the need for line-of-sight.

- Devices do not need to be pointed at each other, as signals are omni-directional.

- Both synchronous and asynchronous applications are supported, making it easy to implement on a variety of devices and for a variety of services, such as voice and Internet.

- Governments worldwide regulate it, so it is possible to utilize the same standard wherever one travels [462].

The way Bluetooth devices communicate is similar in concept to the IEEE 802.11b ad hoc mode. A Bluetooth device automatically and spontaneously forms informal PANs, called *piconets*, with other Bluetooth devices. The connection and disconnection of these devices is almost without any user command or interaction – a capability called *unconscious connectivity*. A particular Bluetooth device can be a member of any number of piconets at any moment in time. Each piconet has one master, usually the device that first initiates the connection. Other participants in a piconet are called *slaves* [453]. When only data is being communicated, a master can handle up to seven slaves in asynchronous connections When only voice is being communicated, a master can handle up to three slaves in synchronous connections. When both data and voice are being communicated, the piconet can contain only two devices; the voice connection is synchronous, and the data connection is asynchronous – taken together, the connection is isochronous. Voice transmission is accomplished via a 64-kbps Synchronous Connection-Oriented (SCO) link. Data transmission is via 1-Mbps Asynchronous Connectionless links (ACLs) (in actuality the transmission rate is lower than 1 Mbps) [453].

4.8.1 Bluetooth Protocol Stack

One of the distinct features of Bluetooth is that it provides a complete protocol stack that allows different applications to communicate over a variety of devices [454]. The Bluetooth protocol stack, from bottom (physical) to top (applications) includes a radio frequency layer (RF), a baseband layer, a link management protocol (LMP) layer, a LLC and adaptation protocol layer (L2CAP), a layer of three side-by-side protocols: the telephony control protocol (TCP) (represented, inexplicably, by "TCS" and "TSC" in the figures), the RFCOMM protocol, and the service discovery protocol (SDP), and the applications layer. The RF layer uses the 2.4-GHz ISM band for communications between devices within 10 m of one another. Bluetooth uses a special FHSS protocol to fully utilize the bandwidth and reduce interference (each piconet is assigned its own frequency hopping pattern [454]) [453]. The baseband layer specifies coding for the frequency hopping and packet assembly [454]. It also manages the RF channels, performs error correction and authentication, regulates the SCO and ACL links, and watches for inquiries from other Bluetooth devices in the vicinity [453]. The LMP layer keeps track of the status of the devices in the piconet and schedules traffic [454]. L2CAP allows applications to demand quality of service in terms of bandwidth, latency, and delay variation [453]. The TCP protocol works with cordless telephones and can interface with legacy telecommunication devices. The RFCOMM protocol provides wireless emulation of RS-232 signal control technology ("cable replacement"). The SDP protocol determines the characteristics of piconet devices to support printing, faxing, and teleconferencing [454]. The bottom three layers (RF, baseband, and LMP) are usually implemented in hardware or firmware. Software is used for the other layers [453].

The LMP and L2CAP layers take care of link setup, authentication, and configuration. In an 802.11 network the terminals can be master (M) or slave (S) terminals. In a Bluetooth network terminals can also be in standby (SB) mode or parked (P), and S terminals can join more than one piconet. An important issue in a truly ad hoc network is how to establish and maintain all the connections in a network whose elements appear and disappear in an ad hoc manner, and there is no central unit transmitting signals to coordinate these terminals. The Bluetooth specification achieves initiation of the network through a unique inquiry and page algorithm [454]. Initially, all terminals are in SB mode. The first device to initiate an inquiry becomes the M terminal. The inquiry process registers the SB terminals as S terminals. The M terminal sends timing information to the S terminals in a page message. The S terminals can return to power-saving modes: SB, hold, park, or sniff, depending on the device type. Parked devices do not send transmissions, they only listen to M messages to resynchronize. Sniffing devices monitor the piconet's traffic at intervals. Devices on hold status can join other piconets [454].

There can be some interference between Bluetooth's fast FHSS and 802.11's slow FHSS and DSSS. Students at Worcester Polytechnic Institute tested for interference on voice and data channels. They found that the closer a Bluetooth device is to an 802.11b device, the greater the packet loss rate for both networks.

4.8.2 Bluetooth Security

Wireless LANs running either 802.11b or Bluetooth have the advantage of being able to work from virtually anywhere; however, both these technologies depend on open communication from point to point. Unless access point devices are configured with some level of security, virtually anyone can connect [458]. Bluetooth uses two secret keys, a 128-bit key for authentication and an 8 to 128-bit key for encryption, along with a 128-bit random number and 48-bit MAC addresses. The encryption key length can be selected so that Bluetooth technology can be used in countries where encryption strength is regulated. A session key (called a *link key*) is also used between communicating devices, based on an initialization key, MAC addresses, and a PIN number. This protocol has been shown to have several vulnerabilities by which a malicious entity could obtain the PIN numbers and keys depending on how the session initialization of the communication protocol is performed [454].

5

All-IP Wireless Networking

The proliferation of competing wireless technologies has led the wireless communications industry to move toward mobile networks that follow an all-Internet protocol. Voice-over-IP (VoIP) telephone communications and videoconferencing have already become a reality, proving that all-IP technology can produce successful applications. It will take several years to finalize standards for all-IP networks, and many more for these networks to be deployed with great density. Therefore, there will be a long migration from current circuit-based networks and services to the all-IP environment [487]. The Internet Protocol, or IP, already a universal network-layer protocol for packet networks, is rapidly becoming a promising universal network-layer protocol for all wireless systems. The reason is that an IP terminal with multimode radio or software radio can possibly roam between different wireless systems if they all support IP as a common network layer. Unlike today's Radio Access Networks (RANs) which are mostly proprietary, IP provides an open interface (and promotes an open market). Distributed (autonomous) IP-based base stations will make the RANs more robust, scalable, and cost effective. In addition, deploying IP to wireless networks will enable IP-based applications to run over wireless networks, which will change mobile communications significantly [490]. In all-IP settings, mobile and wireless networks will be integrated together under an all-IP core network to support global roaming and services. Issues such as mobile security, quality-of-service (QoS), and mobility management are now being investigated to ensure that all-IP designs will meet the needs of the users of the new technology. For *mobile ad hoc and sensor networks*, critical issues include power saving, routing, sensing, media access control (MAC), and integration with other mobile networks [488]. Other challenges to realizing distributed all-IP wireless networks include technology independent IP signaling on all wireless and wired networks, IP-based mobility management supporting fast handoffs and universal roaming, real-time and non-real-time applications with guaranteed QoS, design, and realization of IP-based base station, authentication, authorization, and accounting (AAA), and smooth and seamless interworking with today's public switched telephone networks (PSTN) and first-generation/second-generation (1G/2G) wireless telephony [490]. Running an IP-based telephony system is much cheaper than running a traditional circuit-switch based system. VoIP has become the most important and the best accepted IP-based service in wireless networks. Naturally, the QoS is critical to VoIP and other real-time services, like video conferencing. Figure 5.1 shows the variety of services that can be interconnected using all-IP technology. At the bottom of the figure are the users of voice, data, and VoIP. In the middle are various access networks and technologies used to access public telecommunication services, and at the top are examples of the public telecommunication services: PSTN and IP networks.

Next Generation Wireless Systems and Networks   Hsiao-Hwa Chen and Mohsen Guizani
© 2006 John Wiley & Sons, Ltd

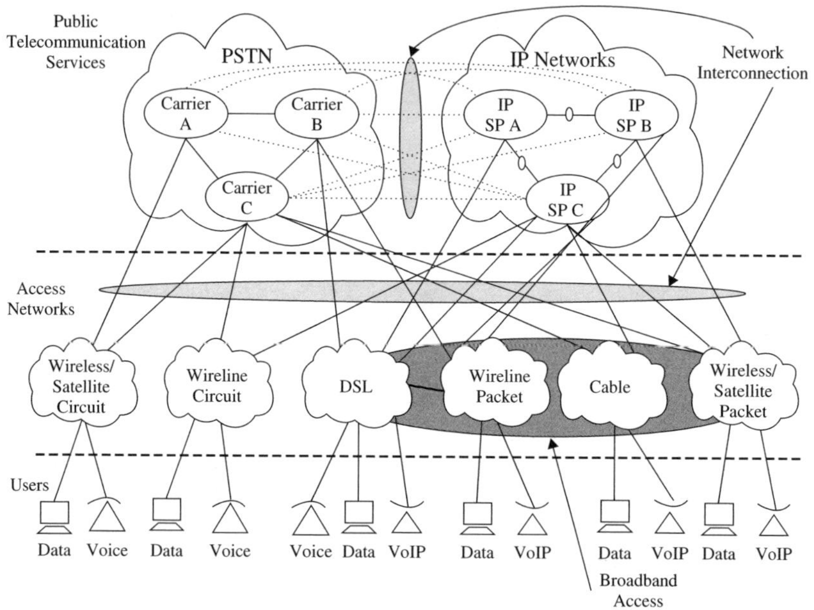

Figure 5.1 The variety of services that can be interconnected with all-IP technology [491].

5.1 Some Notes on 1G/2G/3G/4G Terminology

First generation (1G) wireless telecommunications – the bricklike analog phones that are now collector's items – introduced the cellular architecture that is being offered by most wireless companies today. Second generation (2G) wireless supported more users within a cell by using digital technology, which allowed many callers to use the same multiplexed channel. But 2G was still primarily meant for voice communications, not data, except some very low data-rate features, like short messaging service (SMS). So-called 2.5G allowed carriers to increase data rates with a software upgrade at the base transceivers stations (BTS), as long as consumers purchased new phones, too. Third generation (3G) wireless offers greater bandwidth to users, which allows them to send and receive more information.

All of these architectures, however, are still cellular. The cellular architecture is sometimes referred to as a *star architecture* or *star topology* or *spoke and hub*, because users within that cell access a common, centralized BTS. The advantage is that given enough time and money, carriers can build nationwide networks, which most of the big carriers have done. Some of the disadvantages include a singular point of failure, no load balancing, and spectral inefficiencies. The single biggest disadvantage in cellular networks is that as data rates increase, output power will have to increase, or the size of the cells will have to decrease to support these higher data rates. Since significant increases in output power scare both consumers and regulators, it is far more likely that we will see significantly smaller cells. This reduces the return on investment in 3G business plans.

Fourth generation (4G) wireless was originally conceived by the Defense Advanced Research Projects Agency (DARPA), the same organization that developed the wired Internet. It is not surprising, then, that DARPA chose the same distributed architecture for the wireless Internet that had proven so successful in the wired Internet. Although experts and policymakers are yet to agree on all the aspects of 4G wireless, two characteristics have emerged as all but certain components of 4G: end-to-end Internet Protocol (IP), and peer-to-peer networking. An all IP network makes sense because consumers will want to use the same data applications they are used to in wired networks. Peer-to-peer networks, where every device is both a transceiver and a router/repeater for other devices in the network, eliminate this spoke-and-hub weakness of cellular architectures, because the elimination of a single node does not disable the network. The final definition of "4G" will have to include something as simple as this: if a consumer can access the network at home or in the office while wired to the Internet, that consumer must be able to access it wirelessly in a fully mobile environment [492].

5.2 Mobile IP

The Internet Protocol (IP) is the most successful network layer protocol in computing due to its many strengths, but it also has some weaknesses, most of which have become more important as networks have evolved over time. Technologies like classless addressing and Network Address Translation combat the exhaustion of the IPv4 address space, while IPSec provides it with the secure communications it originally lacked. The TCP/IP suite of protocols work well as long as all of the nodes in a network stand still. Today, a conspicuous weakness of IP is that it was not designed with mobile computers in mind. IP's hierarchical addressing scheme assumes that once a node appears at a particular place on the network, it remains there for good. While mobile devices can certainly use IP, the way that devices are addressed and data is routed causes a problem when devices move from one network to another, making it impossible for unmodified IP to allow a portable computer to maintain an "always on" condition. At the time IP was developed, computers were large and rarely moved. Today, there are millions of notebook computers and smaller devices, some of which even use wireless networking to connect to the wired network. The importance of providing full IP capabilities for these mobile devices has grown dramatically. To support IP in a mobile environment, a new protocol called *IP Mobility Support*, or more simply, Mobile IP, was developed [496, 499, 500, 502, 515, 516].

Mobile computing and networking should not be confused with the portable computing and networking in use today. In mobile networking, computing activities are not disrupted when the user changes the computer's point of attachment to the Internet. Instead, all the needed reconnection occurs automatically, without interaction by the user. This means that users will have access to the Internet at any time, anywhere, and not be bound to the locations of their offices and studies.

There are some technical obstacles that must be overcome before mobile networking can become widespread. The most fundamental is the way the *Internet Protocol* routes packets to their destinations according to IP addresses. These addresses are associated with a fixed network location much as a nonmobile phone number is associated with a physical jack in a wall. When the packet's destination is a mobile node, this means that each new point of attachment made by the node is associated with a new network number and, hence, a new IP address, making transparent mobility impossible [493].

The basic concept behind mobile IP is simple. A mobile device's IP address must change as it moves from network to network, and mobile IP allows it to do so. Applications require a constant IP address, so it allows that too. The apparent conflict is resolved by maintaining two separate addresses for each device [497].

Mobile IP enables the routing of IP data to mobile nodes. The mobile node's home address (assigned by Dynamic Host Configuration Protocol (DHCP) or some other mechanism) always identifies the mobile node, regardless of its current point of attachment to the Internet or an organization's network. When away from home, a *care-of address* associates the mobile node with its home address by providing information about the mobile node's current point of attachment to the Internet or an organization's network. Mobile IP uses a registration mechanism to register the care-of address with a home agent (HA). The HA redirects data from the home network to the care-of address by constructing a new IP header that contains the mobile node's care-of address as the destination IP address. This new header then encapsulates the original IP datagram, causing the mobile node's home address to have no effect on the encapsulated datagram's routing until it arrives at the care-of address. This type of encapsulation is also called *tunneling*. After arriving at the care-of address, each datagram is de-encapsulated and then delivered to the mobile node [494].

A mobile node discovers its foreign and HAs via a process called *agent discovery*. During the agent discovery phase, the home agent and foreign agent advertise their services on the network by using the ICMP Router Discovery Protocol (IRDP). The mobile node listens to these advertisements to determine if it is connected to its home network or a foreign network. The mobile node then registers its current location with the foreign agent and HA during the registration process. The mobile node is configured with the IP address and mobility security association (which includes the shared key) of its HA. Thereafter, the mobile node sends packets using its home IP address, effectively maintaining the appearance that it is always on its home network. Even while the mobile node is roaming on foreign networks, its movements are transparent to correspondent nodes – this is sometimes called *tunneling*.

Mobile IP uses a strong authentication scheme for security purposes. All registration messages between a mobile node and HA are required to contain the Mobile-Home Authentication Extension (MHAE). The integrity of the registration messages is protected by a preshared 128-bit key between a mobile node and HA [495].

The real charm of the *Mobile IP* solution is that most of the elements of the Internet do not need to change. The server with which a mobile device is communicating does not need to do anything special. Most of the protocol stack on the device itself can be blissfully unaware that the device is moving, with the exception of the piece that negotiates with the foreign agent to establish the care-of address.

This plan works well in the mainstream of IP addressing, but complications are introduced by a number of other protocols that have become commonplace on the Internet. DHCP, in particular, is an example of an Internet protocol that needs modification to accommodate Mobile IP protocols. Right now it is possible for a DHCP client to obtain an IP address and information about DNS servers, gateway addresses, and resources on the local network. This capability is easily extended to support dynamic discovery of available HAs. Foreign agents, however, must be discovered through different

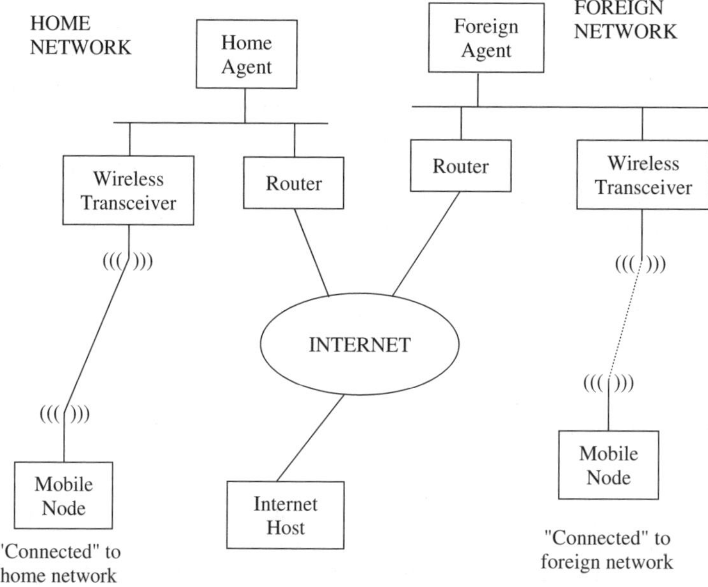

Figure 5.2 Principles of Mobile IP networking [501].

protocols, since the traveling device does not fully join the local network. A new option in DHCP will do this.

Another Internet component that complicates Mobile IP is the network firewall. Mobile IP elegantly routes packets from a mobile node directly to their destination, but a firewall that detects a packet originating from within its network that has a return address from somewhere else may become suspicious. In fact, many firewalls and border routers implement a feature called *ingress filtering*, which blocks any packet that exhibits just these characteristics. Mobile IP can be modified to encapsulate traffic from a mobile node and sent, as an intermediate step, to the HA, which can then forward the data to the intended destination. However, this procedure puts a heavier load on the network, and significantly increases the computational requirements of the HA. At this point in the development of Mobile IP, either the new protocols or the standing firewalls will have to give in [497].

Figure 5.2 shows the principles of Mobile IP networking, illustrating a mobile node that is connected to its home network, and hence has only a home address, and a mobile node that is connected to a foreign network, with both a home and care-of address.

5.3 IPv6 versus IPv4

IPv6 is a major revamp to the IPv4 architecture. The upgrade is largely seen as driven by a shortage of IP addresses, but it is more correct to say that it was motivated by fragmentation of the address space. However, the designers of IPv6 used the opportunity presented by this upgrade to address a number of lingering problems in IPv4. As a result, it is a major upgrade [497].

IPv6 increases the IP address size from 32 bits to 128 bits to support more levels of addressing hierarchy. IPv6 provides a much greater number of addressable nodes and employs a simpler automatic configuration of addresses. In addition, IPv6 defines a new type of address that is called an *anycast* address. An anycast address identifies sets of nodes. A packet that is sent to an anycast address is delivered to one of the nodes. The use of anycast addresses in the IPv6 source route allows nodes to control the path over which their traffic flows. Furthermore, some IPv4 header fields have been dropped or have been made optional. This change reduces the common-case processing cost of packet handling, and keeps the bandwidth cost of the IPv6 header as low as possible, despite the increased size of the addresses. Even though the IPv6 addresses are four times longer than the IPv4 addresses, the IPv6 header is only twice the size of the IPv4 header [498].

It would be difficult to predict what kind of new requirements the future will impose on Internet architecture. An extensible protocol like IPv6 has more prospects to meet the unforeseen needs than its rigid IPv4 counterpart. IPv6's flexibility is made possible by extension headers and options built into its design. While IPv6 enjoys clean extensibility in its architectural design, IPv4 is limited to a slow, costly, and limited patching process that further upsets its original design principles [504].

5.4 Mobile IPv6

Today's versions of Internet protocols assume implicitly that any node always has the same point of attachment to the Internet. Mobile IPv6 allows a host to leave its home subnet while transparently maintaining all of its current connections and remaining reachable by the rest of the Internet. Already, there are third-generation cellular phones that are packet-switched instead of circuit-switched, offering IP services, and the number of mobile Internet devices will continue to rise [505].

The current state of IPv4 architecture does not adequately serve the contemporary Internet in terms of security, mobility, extensibility, and dynamic reconfigurability. IPv6 is the proposed solution to these problems, and offers a larger IP address space and a number of fundamental features to enable wireless networking and mobility. IPv6 is designed to provide sufficient IP addresses for a multitude of mobile nodes to be deployed in a mobile network, without extensive software additions to keep the nodes in communication, and without the security vulnerabilities that lurk in every new layer of additional networking software. Mobility is built into the IPv6 protocol: a mobile IPv6 node can use mobility protocols wherever it can get ordinary IPv6 service. The mobile IPv6 protocol does not require or even define foreign agents, leading to scalable Internet-wide mobility management.

In addition, route optimization signaling enables a mobile IPv6 node to inform its correspondent node when it acquires a new care-of address. This allows both the mobile node and the correspondent node to send and receive packets using the shortest path between the two. One useful by-product of this feature is location-based services. The mobile IPv6 location update signaling can be used by a correspondent node to infer the geographic location of a mobile node, and hence provide customized service or content. This optional protocol signaling can be turned off if the mobile node's location privacy is an issue [504].

The term "binding" refers to the association of a mobile node's home address with its care-of address. When a new message is sent in IPv6, it carries with it the IPv6 Destination Options – additional information that only the destination node needs to examine. There are four Destination Options available: *Binding Update, Binding Acknowledgement, Binding Request*, and *Home Address*. A mobile node uses the Binding Update option to inform its HA and its correspondent nodes about its current care-of address. A Binding Acknowledgement acknowledges a Binding Update. A Binding Request asks a mobile node to send a Binding Update. The Home Address option is sent by a mobile node to convey its home address to another node.

IPv6 nodes maintain a *Binding Cache*, where they store the bindings for other nodes. Information from Binding Updates is sent to the Binding Cache, and when a node sends a packet, the cache is searched for an entry containing the destination node's care-of address. A node also maintains a Binding Update List containing information about every Binding Update that it has sent to its home agent and all correspondent nodes. In addition, nodes that serve as HAs maintain a *HAs List* where they keep track of information about all of the other HAs on their subnet.

A routing anomaly can occur with IPv6, called *triangle routing*: when a correspondent node wants to send a message to a mobile node, the message is routed to the mobile node's HA, and from there to the mobile node itself. However, when the mobile node wants to send a message to its correspondent node, the message is routed directly from the mobile node to the correspondent node. In fact, mobile nodes set the source address of their transmissions to their care-of address, and utilize the Home Address Destination Option in such transmissions.

To avoid triangle routing (and optimize routing), mobile nodes must send Binding Updates to any (mobile or stationary) correspondent node they communicate with. The correspondent nodes cache the current care-of address, and send messages directly to the mobile nodes on their list.

Route optimization is an integral feature of the IPv6, rather than an additional functionality as in the IPv4. In addition, the IPv6 does not need to rely on foreign agents (in foreign subnets) to configure the care-of addresses of mobile nodes – IPv6 uses a neighbor discovery mechanism and address auto-configuration to handle this task [505].

Implementation of the *Mobile IPv6* in 2 and 3G mobile networks primarily requires user plane (application layer) IPv6 support from the network, installing a *HA* router in the home network, and the use of mobile terminals supporting Mobile IPv6 and implementing *IP Security* (IPSec) infrastructure, because Mobile IPv6 uses IPSec for all its security requirements.

Mobile IPv6 allows for efficient roaming from a visited network to home network services and seamless roaming between different access technologies (WLAN, Bluetooth, etc.) using one address for a mobile node. In addition, mobile nodes can run peer-to-peer services directly without explicit support from their home network [506].

The number of wireless computing devices (such as mobile phones, laptops, and palmtops) in use on the Internet reached one billion sometime in 2005. The huge address space of IPv6 will meet the addressing requirements of the rapidly developing Internet. Since real-time applications like VoIP in mobile networks depend on smooth handoffs when mobile nodes transition between network links, a system of buffering packets during transition has been proposed to minimize packet loss.

In addition, the IPv6 introduces a packet option called *QoS Object* that can trigger QoS procedures when QoS-sensitive applications are in use [503].

5.5   Wireless Application Protocol (WAP)

The Wireless Application Protocol (WAP) [507, 512, 513] was originally conceived by four companies (Ericsson, Motorola, Nokia, and Unwired Planet – now called *Phone.com*) as a set of standards that would make it simple to access online services from mobile devices (primarily handheld devices, like mobile phones). An industry association named the *WAP Forum* was formed to promote WAP. The WAP Forum was consolidated into the *Open Mobile Alliance* (OMA) and no longer exists as an independent organization. The specification work currently being done by the relocated WAP Forum is accessible on the OMA web site (at http://www.openmobilealliance.org/tech/affiliates/wap/ wapindex.html). The OMA, formed in June, 2002 by nearly 200 companies, aims to consolidate into one organization with all the specification work devoted to the development of mobile service facilitators [514].

During its development, WAP has suffered some setbacks. Some of the original thoughts behind WAP were found to be erroneous, such as the idea that handheld terminals would be limited in their processing, memory and display abilities, that the communication channel would be too expensive to make full use of, that the human operator should be involved with the details of interacting with services, and that a special protocol stack would need to be introduced to find the optimal solution to these problems. Thus, in its first incarnation, WAP did not provide a truly end-to-end service, since the operator was kept in the middle of transactions with services. This meant that much of the content in a WAP transaction was in clear text, introducing serious security flaws. Most operators used SMS for WAP communications, which turned out to be prone to delays and expensive. Also, content had to be encoded more than once to accommodate a variety of terminal sizes and resolutions. To overcome these shortcomings, WAP has embraced the idea of an IP-based stack, with direct communications between servers and mobile devices, and support for HTTP and HTTPS being mandatory in the protocol. In its new form WAP no longer requires operators to manage WAP connections and transactions.

However, the WAP protocol still supports applications that use the original WAP protocol stack. It reflects the layered OSI stack, and like the OSI stack some layers are not mandatory. At the lowest level, the WAP Datagram Protocol (WDP) moves information from receiver to sender and is modeled after User Datagram Protocol (UDP) as a least effort method of delivery. On top of WDP sits the *WAP Transaction Protocol* (WTP), which ensures that the data fragments sent over the line are actually received, through typical ACK/NACK communication. Higher up, the WAP Session Protocol (WSP) handles the session between communicating systems. All sessions are granted unique IDs and are started/stopped/terminated accordingly (see Figure 5.3) [508].

Here is a description of the various layers in the WAP protocol stack.

WAE – The Wireless Application Environment holds the tools that wireless Internet content developers use. These include WML and WMLScript, which is a scripting language used in conjunction with WML. It functions much like *JavaScript*. WML is a markup language (*Wireless Markup Language*) developed for use with WAP-compliant devices.

WSP – The *Wireless Session Protocol* determines whether a session between the device and the network will be connection-oriented or connectionless. What this is basically talking about is whether the device needs to talk back and forth with the network during a session. In a connection-oriented session, data is passed both ways between the device and the network; WSP then sends the packet to the *Wireless Transaction Protocol* layer (see below). If the session is connectionless, commonly used when information is being broadcast or streamed from the network to the device, then WSP redirects the packet to the *Wireless Datagram Protocol* layer.

WTP – The *Wireless Transaction Protocol* acts like a traffic cop, keeping the data flowing in a logical and smooth manner. It also determines how to classify each transaction request: unreliable with

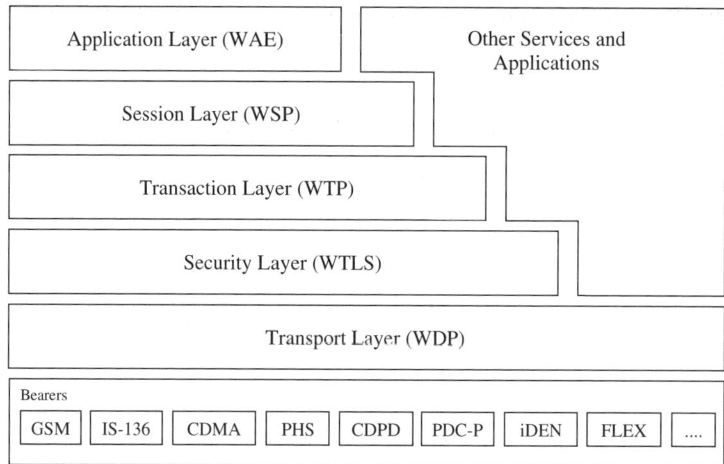

Figure 5.3 WAP architecture and reference model [508].

no result message, reliable with no result message, and reliable with one reliable result message. The WSP and WTP layers correspond to Hypertext Transfer Protocol (HTTP) in the TCP/IP protocol suite.

WTLS – Wireless Transport Layer Security provides many of the same security features found in the Transport Layer Security (TLS) part of TCP/IP. It checks data integrity, provides encryption and performs client and server authentication.

WDP – The Wireless Datagram Protocol works in conjunction with the network carrier layer. The WDP makes it easy to adapt the WAP to a variety of bearers because all that needs to change is the information maintained at this level.

Network Carriers – Also called *bearers*, these can be any of the existing technologies that wireless providers use, as long as information is provided at the WDP level to interface WAP with the bearer [509].

Some of the services now offered by the WAP include an External Functionality Interface (EFI) for access to external devices (like digital cameras and GPS units), a User Agent Profile (UAProf) to convey to an application server the preferences of a device's user and that device's inherent capabilities. A special set of rules supports WAP is "Push," which allows data to be sent ("pushed") to mobile devices for the enhancement of real-time applications. A *Persistent Storage Interface* standardizes the services that mobile devices use to organize and access data, and the Multimedia Messaging Service (MMS) makes the delivery of a variety of types of content to mobile devices possible. Pictograms – tiny images that can convey a message in a small space – have been integrated into WAP services [511].

The WAP is very similar to the combination of HTML and HTTP except that it is optimized for low-bandwidth, low-memory, and low-display capability environments, such as PDA (Personal Digital Assistant), wireless phones, and pagers [510].

5.6   IP on Mobile Ad Hoc Networks

A Mobile Ad Hoc Network (MANET) [518] consists of autonomous mobile users and their communications devices (PDAs, for example), which all act as wireless network nodes. When users activate their devices, the network self-organizes and the nodes find one another automatically. Once the network topology is discovered, the nodes collaborate to establish a stream of communication. In that stream, each node can act as a source, relay point, or destination. The communication flow starts with the source node and, in the case of out-of-range nodes, may hop across a number of intermediary nodes before reaching the destination node. These multiple hops use less power, cause less interference and utilize available frequencies better than direct links, and may enable more traffic to be carried on the MANET.

In addition, there is no single point of failure on a MANET, as there could be on a WLAN (access points) or cellular network (base stations). If a MANET node joins or leaves the network, the MANET can reconfigure itself appropriately [520].

IP-based technologies can be advantageously applied to MANETs. The protocols employed by such MANETs are standards-based and enjoy routing flexibility, efficiency, and robustness. Their interoperability with the Internet is greatly enhanced, and many QoS questions are taken care of by IP standardization [524].

When the MANET nodes utilize IP, they are assigned unique IP addresses. It is not necessary for all nodes to be in range of all the others – two nodes that are communicating and in the range of each other at one point in time might find themselves still communicating (via intermediary nodes), but out of range (due to their mobility) at a later time. One concern regarding MANETs is whether the nodes should keep track of routes to all possible destinations on the network, or only keep track of destinations that are of immediate use. There are trade-offs to consider with either approach. Keeping track of all possible routes means that initial latency is minimized, but additional control traffic needs to be constantly exchanged, lowering network efficiency and raising battery use. If routes are only discovered as needed, initial communication delays will be high, but power consumption and control traffic are kept low [523].

Some of the challenges faced by the developers of MANET technology and protocols stem directly from Internet connectivity. How many of the nodes in an ad hoc network should be allowed to directly connect to the Internet? Mobile IP protocol assigns a mobile node a care-of address along with a HA, effectively adding a new IP address to the mobile node. Decisions about which nodes in a MANET can function as Internet gateways and what to do when one of them leaves the network are still being deliberated. MANET routing becomes complicated when packets are routed across the MANET's boundary, and routing protocols for MANETs are still evolving [521].

One of the problems associated with MANETs stems from the lack of any centralized authority in an ad hoc network and the need for all the nodes to collaborate in order to perform infrastructural tasks like routing and forwarding: nodes need to cooperate in a "disinterested" manner to keep the network up and running. The fear is that, in the absence of an authority figure, some nodes may begin to function in a self-interested way, refusing to expend its resources for the good of the network. This may occur because of a particular device's internal set of battery conservation rules, or because a device may be programmed to "hoard" available bandwidth rather than relay packets for other nodes, for instance. Worse, a device may fail to abide by the network's back-off protocol or contention resolution rules. Current protocol proposals require that all nodes cooperate to correct route failures when a node leaves the network. This, in turn, requires that nodes transmit route failure messages to a sender "disinterestedly." If they fail to do so, the sender will erroneously interpret the lack of acknowledgements as a congestion situation and take inappropriate action. Research is under way to modify ad hoc network protocols to account for these possibilities [522].

Research is also under way to ensure the security of MANETs and put intrusion detection systems in place, especially for MANETs that arise when first responders (police, fire, and health officials) arrive on the scene of a public safety incident. The first responders' PDAs and laptops could quickly establish a network to work together, and researchers are developing secure routing protocols that do not rely on preexisting trust associations between nodes or the availability of an online service to establish trust associations. Intrusion detection is of obvious importance in such situations, first to maintain the privacy of affected individuals and second to prevent malicious nodes from entering and disrupting the network [519].

Because of their dynamic topology and variable link capacity, MANETs require special attention to QoS issues. The current model in existence relies on "best effort" routing and queuing mechanisms, but better methods are under research. This will become increasingly important as services such as streaming video are implemented in MANET devices [517].

5.7   All-IP Routing Protocols

Many of the fundamental characteristics of wired routing protocols can be found in all-IP routing protocols as well: they use routing tables and metrics to determine optimal paths for packets to travel, strive for simplicity and low overhead costs, endeavor to be robust and stable, and have some built-in flexibility for reacting to network changes and problems. However, wireless routing protocols must also take into consideration certain concerns that are specific to a wireless environment: they must be even more adaptable to changes in the network topology (moving nodes can find that their shortest paths to other moving nodes change dramatically), strive even harder to maximize throughput and minimize delay, and keep the power consumption level of the network as low as possible (since mobile nodes are typically run off battery power) [525].

Two well-known wired routing protocols are the Routing Information Protocol (RIP) and the Open Shortest Path First protocol (OSPF). Each has corresponding wireless counterparts: Ad hoc On-demand Distance Vector (AODV) routing can be thought of as RIP for wireless networks, and both Dynamic Source Routing (DSR) and the Zone Routing Protocol (ZRP) are roughly analogous to the OSPF. All of the ideas that have been proposed for wireless routing protocols can be found within AODV, DSR, and ZRP (when taken as a whole). The Distance-Vector family of protocols (which includes the Destination-Sequenced Distance Vector Routing protocol) is proactive. AODV and DSR are reactive protocols, whereas ZRP takes a hybrid approach.

AODV can handle both *unicast* and *multicast* routing. As its name implies, it was designed for use in ad hoc mobile networks and is an on-demand protocol that only constructs routes from source to destination at the request of a transmitting node. This is done using route request queries and route reply responses. When a transmitting node does not already have a route to a particular destination, it broadcasts a route request (RREQ) across the network. When nodes receive this request they update their information about the transmitting node, create backwards pointers to it in their route tables, and, if they are not the destination node and have not already established a route to the destination, rebroadcast the RREQ. If a node is the destination or has already established a route to the destination, it sends a route reply (RREP) back to the transmitting source node – via any intermediary node that had forwarded the RREQ. As the RREP returns to the source, the intermediary nodes create forward pointers to the destination node. When the source node receives the RREP it can begin to transmit data to the destination node. Such routes are maintained as long as they are "active," that is, as long as data packets are using the route within a set timeout period. If the route times out or a link in the route breaks, the sending node can reinitiate route discovery. Breaks in routes are reported to the source node in route error (RERR) messages when intermediary nodes perceive them [526].

AODV is the on-demand counterpart to table-based Dynamic State Routing DSDV wireless routing [526].

DSR is also an on-demand routing protocol, but, unlike the AODV, it does not use hop-by-hop routing. Instead, it employs packet headers that carry an ordered list of the nodes that constitute the route from source to destination. With DSR, intermediary nodes do not need to maintain route information about the various routes that they are a part of (although they do store the routes that they themselves have established when acting as a transmitting source). To discover a needed route, a transmitting source node broadcasts a ROUTE_REQUEST packet to neighboring nodes. Only nodes that have not yet seen this ROUTE_REQUEST forward it, and when they do so, they update the header with their own address (in the proper sequence). When either the destination node or a node which has already established a route to the destination receives the packet, it responds with a ROUTE_REPLY with the sequence of nodes in the route taken from the ROUTE_REQUEST header. If a route breaks and the source node learns that its messages are not reaching their destination, route discovery is reinitiated. DSR does not make use of periodic transmissions of routing information and therefore nodes consume less power than in other protocols. However, the large headers employed by DSR make it most efficient in networks of small diameter [525].

ZRP combines the advantages of the proactive (table-driven) protocols like OSPF and the reactive (on-demand) protocols like DSR and AODV into a hybrid routing protocol for ad hoc wireless networks. Purely proactive routing works best for networks with a high call rate, and purely reactive routing works best for networks with high node mobility. The hybrid ZRP is designed to work well in a network with both of these characteristics; that is, in a network with mobile nodes that frequently transmit data [528].

ZRP divides a network's map into zones, roughly centered on individual nodes or small clusters of nodes. These zones may overlap. The zone radius is an important property for the performance of ZRP. If a zone radius of one hop is used, routing is purely reactive and broadcasting degenerates into flood searching. If the radius approaches infinity, routing is reactive. The selection of radius is a trade-off between the routing efficiency of proactive routing and the increasing traffic for maintaining the view of the zone [529]. The design of ZRP assumes that the largest part of the traffic is directed to nearby nodes in an ad hoc network. Therefore, ZRP reduces the proactive scope to a zone centered on each node. In a limited zone, the proactive maintenance of routing information is easier. Further, the amount of routing information that is never used is minimized. Still, nodes farther away can be reached with reactive routing. Since all nodes proactively store local routing information, RREQs can be more efficiently performed without querying all the network nodes. ZRP refers to the locally proactive routing component as the Intrazone Routing Protocol (IARP). The globally reactive routing component is named the *Interzone Routing Protocol* (IERP) [529]. These are not specific, rigidly defined protocols because ZRP provides only a framework within which any of a number of well-defined protocols can be implemented, depending on the circumstances. In order to learn about its direct neighbors, a node may use the *MAC* protocols directly. Alternatively, it may require a Neighbor Discovery Protocol (NDP). Such a NDP typically relies on the transmission of "hello" beacons by each node. If a node receives a response to such a message, it may note that it has a direct point-to-point connection with this neighbor. The NDP is free to select nodes on various criteria, such as signal strength or frequency/delay of beacons. Once the local routing information has been collected, the node periodically broadcasts discovery messages in order to keep its map of neighbors up to date.

Communication between the different zones is controlled by the IERP and provides routing capabilities among peripheral nodes (nodes on the periphery of a zone) only. If a node encounters a packet with a destination outside its own zone, that is, it does not have a valid route for this packet, it forwards it to its peripheral nodes, which maintain routing information for the neighboring zones, so that they can make a decision of where to forward the packet. Through the use of a *bordercast* algorithm rather than flooding all peripheral nodes, these queries become more efficient [527].

Instead of broadcasting packets, ZRP uses a concept called bordercasting, which utilizes the topology information provided by IARP to direct query request to the border of the zone. The bordercast packet delivery service is provided by the Bordercast Resolution Protocol (BRP). BRP uses a map of an extended routing zone to construct bordercast trees for the query packets. Figure 5.4 shows the relationships between the various protocols of ZRP.

Route maintenance is especially important in ad hoc networks, where links are broken and established as nodes with limited radio coverage move. In purely reactive routing protocols, when routes containing broken links fail, a new route discovery or route repair must be performed. Until the new route is available, packets are dropped or delayed. In ZRP, the knowledge of the local topology can be used for route maintenance. Link failures and suboptimal route segments within one zone can be bypassed. Incoming packets can be directed around the broken link through an active multihop

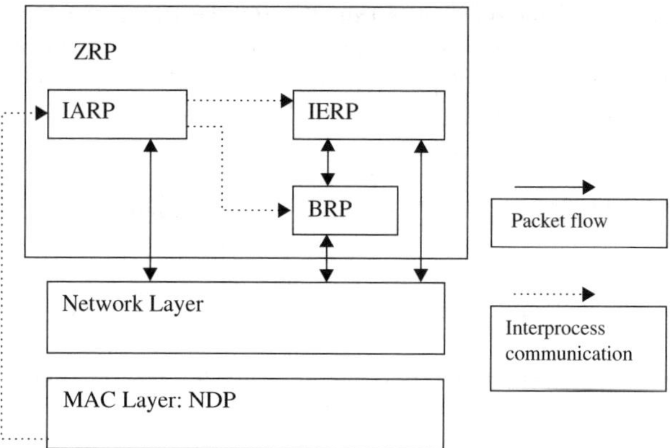

Figure 5.4 The components of ZRP [529].

path. Similarly, the topology can be used to shorten routes; for example, when two nodes have moved within each other's radio coverage. For source-routed packets, a relaying node can determine the closest route to the destination that is also a neighbor. Sometimes, a multihop segment can be replaced by a single hop. If next-hop forwarding is used, the nodes can make locally optimal decisions by selecting a shorter path [529].

6

Architecture of B3G Wireless Systems

The first cellular phone systems (the first wireless networks) were introduced in the late 1970s. They were modeled after wired phone systems and used transmitted analog data across a mobile network. They were called *first generation* (1G) wireless systems when the next generation of cellular networks was deployed in the 1990s. These "second generation" (2G) networks transmitted digital voice data on mobile networks. Their accompanying wireless e-mail and Internet applications are often referred to as *2.5G* technologies. The third generation (3G) of wireless technology is currently in use. It is designed for high-speed multimedia applications with data rates from 128 kbps to approximately 10 Mbps, and upgrades to around 100 Mbps in WLANs. Research and development efforts are now focused on the next generation of wireless technology – referred to as *4G* or *B3G* (for *beyond 3G*). These systems may deliver 1 Gbps transmission rates, with bandwidth up to 100 MHz. The year 2010 is often set as a rough target date for implementing B3G systems (but some applications will probably be deployed in 2006–2007). B3G technology will make it possible to watch movies and television on a (moving) cell phone. For this to happen, more of new technology must be put in place, involving upgrades of ad hoc mobile networking, satellite systems, spectrum allocation, and higher wireless data speeds. The proposed IEEE 802.20 standards will coordinate B3G design efforts. One important aspect of the standardization process will be to provide for ubiquitous access to the wide variety of wireless networks already in place (802.11 and HiperLAN/2 WLANs, 802.15 and Bluetooth Personal Area Networks (PAN)s, 802.16 MANs (Metropolitan Area Networks), and existing 3G networks) [531], which each have their own range, data rate, and mobility limits.

Many useful and interesting services and applications can be developed, assuming that ubiquitous and high-speed B3G wireless access is available ("always connected, everywhere" access). One of the main forces behind B3G development is the demand for higher data throughputs in a variety of scenarios. The planners of B3G include terminal and infrastructure equipment manufacturers, academics, operators, service providers, regulatory bodies, and governmental agencies. It should not be surprising to learn that finding a universal definition of B3G/4G is a very elusive task, even after several years of activity and numerous attempts in the literature.

B3G designers are aiming for the following technical targets: (1) data rates of 100 Mbps in wide coverage, and 1 Gbps in a local area; (2) all-IP networking; (3) ubiquitous, mobile, seamless communications; (4) shorter latency; (5) connection delays of less than 500 ms; (6) transmission delays of less than 50 ms; (7) costs per bit significantly lower, perhaps 1/10th to 1/100th lower than that

Next Generation Wireless Systems and Networks   Hsiao-Hwa Chen and Mohsen Guizani
© 2006 John Wiley & Sons, Ltd

Table 6.1   The goals of B3G planners

Data rates	100 Mbps in wide coverage, 1 Gbps in a local area
Networking	All-IP
Communications	Ubiquitous, mobile, seamless
Latency	Shorter than that of 3G
Connection delays	Less than 500 ms
Transmission delays	Less than 50 ms
Costs per bit	1/10 to 1/100 lower than that of 3G
Infrastructure cost	Lower, perhaps 1/10 lower than that of 3G

of 3G; and (8) lower infrastructure cost, perhaps 1/10 lower than that of 3G. The same is shown in Table 6.1.

It is envisioned that this type of technology will enable enhanced e-commerce, add to work productivity, and make available ways to improve personal free time. B3G technology may one day be found in vehicles, public places, health care, education, and in the entertainment industry. "Personal managers" may keep a user informed about personal finances, health, security, and local news and weather. "Home managers" may help manage comfort, security, and maintenance. B3G will likely facilitate mobile shopping, tourism, and mobile gaming scenarios [532].

6.1   Spectrum Allocation and Wireless Transmission Issues

B3G technology requires high bandwidth in order to provide multimedia services at a lower cost than is presently the case. In the United States, B3G systems will likely migrate to the 5.2–5.9 GHz range (assuming regulatory approval). It must be stressed, however, that there are serious spectrum allocation issues associated with B3G technology, simply because today unallocated spectrum either does not exist in some countries or is in short supply. Long-term planning is necessary to make spectrum available for B3G applications [530]. In addition, worldwide standardization of spectrum allocation for B3G systems would be desirable for maintaining connections when moving anyplace in the world – the "always connected, anywhere" philosophy.

In the United States, there is a bright spot in the spectrum allocation arena. The FCC has noted that there are large portions of allotted spectrum that are unused, and this is true both spatially and temporally. In other words, there are portions of assigned spectrum that are used only in certain geographical areas and there are some portions of assigned spectrum that are used only for brief periods of time. Studies have shown that even a straightforward reuse of such "wasted" spectrum can provide an order of magnitude improvement in available capacity. Thus, the issue is not that spectrum is scarce – the issue is that we do not currently have the technology to effectively manage access to it in a manner that would satisfy the concerns of the current licensed spectrum users.

The *Defense Advanced Research Projects Agency* (DARPA) is developing a new generation of spectrum access technology that is not only ostensibly oriented toward military applications, but also applicable to advanced spectrum management for communication services. The DARPA program is pursuing an approach wherein static allotment of spectrum is complemented by the opportunistic use of unused spectrum on an "instant-by-instant" basis in a manner that limits interference to primary users. This approach is called *opportunistic spectrum access* spectrum management and the basic parts of this approach are as follows: (1) sense the spectrum in which you want to transmit; (2) look for spectrum holes in time and frequency; and (3) transmit so that you do not interfere with licencees.

There are a number of research challenges to this adaptive spectrum management, including (1) wideband sensing; (2) opportunity identification; (3) network aspects of spectrum coordination when using adaptive spectrum management; (4) the need for a new regulatory policy framework; (5) traceability so that sources can be identified in the event that interference does occur; and (6) verification and accreditation.

The *National Science Foundation* (NSF) has a research program entitled *Programmable Wireless Networking* (NeTS-ProWiN). This research program addresses issues that result from the fact that wireless systems today are characterized by wasteful static spectrum allocations, fixed radio functions, and limited network and systems coordination. This has led to a proliferation of standards that provide similar functions – wireless LAN standards (e.g., Wi-Fi/802.11, Bluetooth) and cellular standards (e.g., 3G, 4G, CDMA, and GSM) – which in turn has encouraged hybrid architectures and services and has discouraged innovation and growth. Emerging programmable wireless systems can overcome these constraints as well as address urgent issues such as the increasing interference in unlicensed frequency bands and low overall spectrum utilization. The NSF research is based on the concept of programmable radios. Programmable radio systems offer the opportunity to use dynamic spectrum management techniques to help lower interference, adapt to time-varying local situations, provide greater QoS, deploy networks and create services rapidly, enhance interoperability, and in general enable innovative and open network architectures through flexible and dynamic connectivity [533].

Some of the proposed technologies for wireless transmissions in a B3G environment are detailed in the subsequent text. Each has its own implications for spectrum allocation concerns.

6.1.1 Modulation Access Techniques: OFDM and Beyond

Multi-carrier modulation has been identified as a key technology for B3G, and *Orthogonal Frequency Division Multiplexing* (OFDM) is the main technique under proposal. It is already present in IEEE 802.11a WLANs. OFDM was originally proposed for single users but extensions to multiusers, for example OFDMA,[1] support multiple access. Usually OFDM is combined with other access techniques, typically CDMA and TDMA, to allow more flexibility in multiuser scenarios. Multi-carrier Code Division Multiple Access (MC-CDMA) is another access technique with great potential. OFDM and CDMA are robust against multipath fading, which is a primary requirement for high data rate wireless access techniques. With overlapping orthogonal carriers, OFDM results in a spectrally efficient technique. Each carrier conveys lower data rate bits of a high-rate information stream; hence it can cope better with the *intersymbol interference* (ISI) problem encountered in multipath channels. The delay-spread tolerance and good utilization of the spectrum has put OFDM techniques in a rather dominant position among future communication technologies. OFDM, on the other hand, has strict time and frequency synchronization requirements and is prone to the *peak-to-average power ratio* (PAPR) problem.

6.1.2 Nonconventional Access Architectures

Wide coverage and local coverage are the two most distinctive B3G access components. It is expected that the requirement for higher data throughput and support for a great number of users will result in a shift to higher and less-congested frequency bands, for example the 5-GHz band, and wider bandwidths (20–100 MHz). In cellular access, this would mean that the link budget would be seriously degraded and unreasonable high power would have to be used to compensate for the higher attenuation occurring in this frequency band. This could easily exceed the regulation for power emission from base stations, and also it could dramatically reduce (the already challenged) battery life in terminals.

---

[1]OFDMA is short for orthogonal frequency division multiple access, which provides multiple access scheme for a multiuser communication system. On the other hand, OFDM is only a multiplexing scheme for a single user. More treatments on OFDM are given in Section 7.5.

Therefore, nonconventional access architectures for wide-area access are being considered to cope with this problem. Multi-hop cellular, and particularly two-hop, approaches appear to be an effective solution to the problem of achieving wide coverage and high data throughput. By using relaying (repeating) stations, the equivalent distance between base station and mobile station can be reduced. Efficient use of radio resources can also be attained since some resources can be reused in different hops. In principle, the relay stations can be fixed (called *infrastructure-based relaying*) or mobile (*ad hoc relaying*). In the distributed radio access approach, a base station has under its control a number of remote access sites, each with its own antenna(s) and covering a small area. The small-sized cells covering a large cell reduce the distance between the mobile terminal and its most suitable/closest access point. The base station is connected to the remote radio access sites by using optical fiber or radio links. Distributed radio access is a cost-effective approach to scalable networks. In local-area access, several architectures can be used in addition to the single-hop cellular access approach. Several ad hoc access concepts have shown their potential for short-range communications, including multi-hop, peer-to-peer, and cooperative communications. Collaboration among users (or nodes) aims to benefit either a single user or several (or all) collaborating users. Through cooperation (at intra- and/or interlayer level), the data throughput can be increased and signal quality can be enhanced. Moreover, power efficiency can be boosted, which equates to an increased battery life in terminals.

6.1.3 Multiantenna Techniques

Multiantenna techniques[2] are regarded as among the most important enabling technologies for B3G technology. In principle, no technique other than the use of multiple antennas will easily permit a high spectral efficiency. By exploiting these techniques, data throughput can be increased, link quality improved, cell coverage extended, and network capacity enlarged. Three approaches can be used, namely, diversity, beam-forming (smart antennas), and spatial multiplexing. Diversity techniques require widely separated antenna elements (several wavelengths at least). Actual separation depends on the type of channel. Directional channels (narrow angular spread) require large separation and vice versa. Diversity techniques exploit the fact that the associated channels fade independently, while diversity domains can be space, time, frequency, and polarization. Diversity gain will improve the average signal-to-noise ratio. In beam-forming, signals are coherently combined (either in reception or transmission) so as to enhance the array response in preferred directions. Nulls can also be spatially controlled. Beam-forming allows the establishment of directional links. In beam-forming, it is assumed that the channel or direction of arrival is known to the transmitter/receiver. Unlike with diversity, by using beam-forming, the variability of the signal (e.g., fading statistics) is not affected. The array gain is proportional to the number of elements of the array. Spatial multiplexing offers a linear increase in capacity by exploiting the parallel transmission of different information from different antennas. This is essential for attaining the high spectral efficiencies required by B3G. For the receiver to separate and decode the parallel streams, it is assumed that the signal propagates in a rich scattering channel and the number of reception antennas is at least equal to number of transmission antennas. The term *MIMO* refers in principle to any technique exploiting multiple antennas at the receiver and transmitter.

6.1.4 Adaptive Modulation and Coding

Adaptive Modulation and Coding (AMC) is a form of *link adaptation* that is used in response to the changing characteristics of a radio channel. AMC jointly selects the most appropriate modulation and coding scheme according to channel conditions. The better the radio conditions, the higher the modulation rate and code rate combination, and vice versa. Clearly, AMC is more effective in packet

---

[2]Section 8 has more discussions on multiantenna techniques, also called multiple-input-multiple-output (MIMO) systems.

networks – the networks envisioned for B3G. Conventional wireless services have mostly been designed for constant rate applications, such as voice transmission. To combat channel fading, communication systems have usually been designed to maximize time diversity with a combination of interleaving and coding for better bit error rate performance. B3G wireless systems must target packet data, and thus are usually designed to maximize throughput for a given battery energy budget while allowing a certain delay.

6.1.5 Software Defined Radio

Since different wireless interfaces will be used in B3G, *Software Defined Radio* (SDR) appears to be a cost-effective solution to implement several access approaches in one terminal. SDR uses a flexible architecture that allows the wireless interface to be reconfigured. This allows multistandard wireless interface operation with a common hardware platform, opening the door for forward compatibility. Furthermore, SDR is an enabler for cooperative networks. SDR allows dynamic modifications of the radio frequency, baseband processing, and even the MAC layer of the terminal (which can utilize a particular wireless interface by reconfiguring the system). The degree of flexibility brought by realtime reconfigurability opens up a new world of possibilities for users, operators, services providers, and terminal manufacturers. Users can establish connection to any network, allowing simple local and global roaming. Users can also benefit from the low-cost terminals that this technology can entails. Hardware and software updates can easily and wirelessly be carried out by users or operators. Manufacturers can also take advantage of SDR as large volumes of terminals with identical hardware (and fewer components) are produced. Even upgrades or changes in the terminals can be easily effected. In addition, service providers can exploit this flexibility to match their operation and services to user demands better [532].

The shift in B3G toward IP-based, high-speed multimedia wireless traffic demands a high spectral efficiency. A natural corollary to this is a need for cooperation across subnetworks and the use of multi-hop relaying. Regulatory reforms could free up bandwidth currently used for analog broadcasting – high-frequency bands – for B3G systems [534]. The more efficient modulation schemes discussed above cannot be retrofitted into 3G architecture, which is one of the reasons B3G research is being conducted before 3G systems are fully implemented (another reason is that 3G performance may not be sufficient for future high-performance applications like full-motion video and wireless teleconferencing). Spectrum regulation bodies must get involved in guiding the researchers by indicating which frequency bands might be used for B3G. Along with regulatory reforms, a number of spectrum allocation decisions, spectrum standardization decisions, spectrum availability decisions, technology innovations, component development, signal processing, and switching enhancements, plus intervendor cooperation have to take place before the vision of B3G will materialize. Standardization of wireless networks in terms of modulation techniques, switching schemes, and roaming is an absolute necessity for B3G technology. However, B3G is not an independent replacement architecture for existing systems. Network architects must base their vision of B3G architecture on hybrid network concepts that integrate wireless WANs, wireless LANs (IEEE 802.11a, IEEE 802.11b, IEEE 802.11g, IEEE 802.15, and IEEE 802.16), Bluetooth technology, and fiber-based backbones with broadband wireless (B3G) networks. Moreover, B3G planning must allow for a smooth transition from the current state of existing networks to their coexistence with B3G systems [535].

6.2 Integration of WMAN/WLAN/WPAN and Mobile Cellular

As mentioned above, B3G systems will need to assimilate and integrate existing technologies, rather than supplant them. It is envisioned that present mobile cellular systems will be "blended into" B3G technology, which will enable mobile cellular devices to roam seamlessly from Wireless Metropolitan Area Network (WMAN) to Wireless Local Area Network (WLAN) to Wireless Personal Area Network (WPAN) and vice versa without difficulty. Various WPAN technologies have emerged, and Bluetooth is well on its way to becoming the most widely deployed WPAN technology in handsets and other devices – with projections of nearly 300 million Bluetooth-enabled devices in the marketplace by 2007. In addition, a number of other wireless technologies are being tested and/or deployed. For example, Global Positioning System (GPS) is slated to ship in over 10 million phones this year, and several major device manufacturers are already shipping products with TV and/or radio receivers. Several operators and *original equipment manufacturers* (OEMs) are also experimenting with the inclusion of *digital video broadcast* (DVB) receivers in handsets, in some cases with general packet radio service (GPRS), which is a radio technology for GSM networks that adds packet switching protocols, shorter set-up time for ISP connections, and offers the possibility of charging by the amount of data sent rather than the connect time. To prepare for a future in which there are no barriers to access using a handheld device, engineers are investigating what measures are needed to create a "universal communicator," a device that is capable of communicating regardless of the connection options available to the user.

A B3G network, a "heterogeneously networked environment," will require that handheld devices evolve considerably, from the limited (often fixed function and fixed network) devices that predominate today, to powerful, flexible devices that can intelligently interact with multiple, heterogeneous networks and services. A universal communicator-class device is projected to be a flexible, powerful personal communication appliance that provides users with transparent access to any available network, at any time, including the ability to seamlessly roam across those networks. Such a device must also provide support for key usage models that are made possible by a mixed-network environment. These usage models include the following: (1) infofueling (smart data transfers using best available/most appropriate network); (2) simultaneous voice and data sessions; (3) rich media that scales across networks (for example, video quality increases in a higher-bandwidth environment); (4) cross-network voice, including support for seamless handoff; and (5) location-based services.

Enabling such ubiquitously connected devices poses numerous difficult technological challenges. These include the following:

- Multiple radio integration and coordination: the device integrates multiple radios.

- Intelligent networking – seamless roaming and handoff: users can expect to roam within and between networks like they do with today's cell phones.

- Power management: future handsets and other devices will run richer applications, and power management will become an even greater challenge.

- Support for cross-network identity and authentication: providing a trusted and efficient means of establishing identity is one of the key issues in cross-network connectivity.

- Support for rich media types: the addition of a high-bandwidth broadband wireless connection, such as a WLAN, will open up new opportunities for the delivery of rich media to handheld devices.

- Flexible, powerful computing platform: the foundation of a universal communicator-class device must be a flexible, powerful, general-purpose processing platform.

- Overall device usability: meeting all these challenges must not render the device "user-unfriendly" [536].

The plethora of network models that will be connected by B3G technology is shown in Table 6.2. Because the Internet and cellular systems were designed and implemented by people with different

Table 6.2 Wireless technologies [537]

	Standard	Usage	Throughput	Range	Frequency
UWB	802.15.3a	WPAN	110–480 Mbps	Up to 30 ft	7.5 GHz
Bluetooth	802.15.1	WPAN	Up to 720 kbps	Up to 30 ft	2.4 GHz
Wi-Fi	802.11a	WLAN	Up to 54 Mbps	Up to 300 ft	5 GHz
Wi-Fi	802.11b	WLAN	Up to 11 Mbps	Up to 300 ft	2.4 GHz
Wi-Fi	802.11g	WLAN	Up to 54 Mbps	Up to 300 ft	2.4 GHz
WIMAX	802.16d	WMAN Fixed	Up to 75 Mbps (20 MHz BW)	Typical 4–6 miles	<11 GHz
WIMAX	802.16e	WMAN Mobile	Up to 30 Mbps (10 MHz BW)	Typical 1–3 miles	2.6 GHz
EDGE	2.5G	WWAN	Up to 384 kbps	Typical 1–5 miles	1900 MHz
CDMA2000 /1xEVDO	3G	WWAN	Up to 2.4 Mbps (typical 300–600 kbps)	Typical 1–5 miles	400, 800, 900, 1700, 1800, 1900, 2100 MHz
WCDMA /UMTS	3G	WWAN	Up to 2 Mbps (Up to 10 Mbps with HSDPA Technology)	Typical 1–5 miles	1800, 1900, 2100 MHz

backgrounds in computers and communications, respectively, their integration will not be a simple task. Such integration, however, can be considered to be a first step toward B3G networks, where heterogeneous networks must work together in order to provide differentiated services to users in a seamless and transparent manner [538].

6.3 High-Speed Data

The introduction of multimedia services into mobile communications will require mobile transmission speeds of up to 100 Mbps. Therefore, a wider frequency band than that in 3G will have to be assigned to B3G mobile communication systems. Generally speaking, mobile communication systems should be assigned the lowest available frequency band when taking into account path loss in radio channels. However, it will probably be impossible to assign a lower frequency band than that of 3G to B3G systems because of the fact that these bands are already regulated and in use. Therefore, techniques that enable high-speed data transmission within a limited frequency band will become important in B3G systems. Simply put, techniques for increasing the efficiency of frequency utilization will play a great role in B3G systems. In addition, it is indispensable to combat severe selective fading in a mobile communication environment where such a high-speed data will be transmitted. While some techniques that satisfy the above requirement have been proposed and verified, the spatial signal processing technique has been recognized as one that can potentially increase the efficiency of frequency utilization and system capacity. Among the techniques, MIMO systems have attracted signal processing researchers since MIMO raises the possibility of increasing system capacity in proportion to the number of antennas installed in a transmitter and receiver, using spatial multiplexing. For instance, *Bell Labs Layered Space Time* (BLAST) Code has been experimentally verified to achieve high-capacity transmission rates in indoor scenarios. On the other hand, *Space Division Multiple Access*

(SDMA), utilizing spatial multiplexing as well as MIMO systems, is also considered a promising technique for improving system capacity. In these techniques, the orthogonalization of channels plays an important role in attaining high capacity. In contrast with these, *Multiuser Detection* (MUD) separates a user's signals, which are superposed at the top of a receiver. Therefore, MUD makes it possible to improve frequency utilization efficiency in cellular mobile communication systems. However, MUD with multiple antennas is considered to be a type of MIMO system without channel knowledge at a transmitter. Therefore, MUD also shows promise for improving channel capacity. While many types of MUDs have been investigated in CDMA, MUD is possible to implement in other systems, such as single-carrier systems. For instance, a MIMO turbo equalizer has been proposed that deploys a linear equalizer with iterative decoding (*Turbo decoding*) in addition to array signal processing. Besides, a *multibeam interference canceler* (MIC) has been proposed that deploys both MLSE (Maximum Likelihood Sequence Estimation) and an array antenna. MIC was shown to achieve the optimum transmission performance without any assistance from coding. Although MIC achieves excellent performance even in fading channels, it has a drawback in high hardware complexity, which grows exponentially as the number of the beams increases [540].

It will be technically challenging to enable high-speed data transfers in B3G mobile networks precisely because B3G systems will really be a means to integrate a variety of technologies, including cellular, cordless, WLAN, WMAN, and wired networks, with seamless global access among them. Planners aspire to achieve higher bit rates, higher spectral efficiency, and lower costs per bit than in 3G systems – all with lower power usage. Proposed B3G transmission protocols include OFDM, Wideband Orthogonal Frequency Division Multiplexing (W-OFDM), MC-CDMA, and Large-Area-Synchronized Code Division Multiple Access (LAS-CDMA). OFDM is good for high-bandwidth data transmission; it multiplexes thousands of orthogonal waves in one time waveform. W-OFDM enables data to be encoded on multiple high-speed radio frequencies concurrently. This allows for greater security, increased amounts of data transmission, and the industry's most efficient use of bandwidth. W-OFDM permits the implementation of low power multipoint RF networks that minimize interference with adjacent networks. This allows independent channels to operate within the same band, enabling multipoint networks and point-to-point backbone systems to be overlaid in the same frequency band. MC-CDMA is actually OFDM with a CDMA overlay. Like single-carrier CDMA systems, the users are multiplexed with orthogonal codes to distinguish users in (multi-carrier) MC-CDMA. However in MC-CDMA, each user can be allocated several codes, where the data is spread in time or in frequency. LinkAir Communications is the developer of LAS-CDMA, a patented B3G wireless technology. LAS-CDMA enables high-speed data transmission and increases voice capacity, using SDRs, and is advertised as the most spectrally efficient, high-capacity duplexing system available today [539].

6.4 Multimode and Reconfigurable Platforms

A major contributor toward the convergence of platforms in the B3G era is reconfigurability, which provides technologies (SDRs) that enable terminals and network segments to dynamically adapt to the set of radio access technologies (RATs) that are most appropriate for the conditions encountered in specific service area regions and at specific times of the day. RAT selection is not restricted to those preinstalled in the elements. On the contrary, the missing components can be dynamically downloaded, installed, and validated. Reconfigurability poses requirements on the functionality of wireless networks. Some of the challenges that have to be met to realize the reconfigurability concept are given below.

First, three families of scenarios that must be taken into account when designing the reconfigurability technology have been identified: the promises of ubiquitous access, pervasive services, and dynamic resources provisioning. Ubiquitous access is mainly targeted at increasing the worldwide access to services. It relates to the support of users who turn on a device in a wireless environment to which it has not been previously connected. Roaming is another example of this scenario, and the reconfigurability concept must increase roaming possibilities for users. The concept of pervasive services stresses the need for reconfigurability when several radio access technologies are present in a given wireless environment. Indeed, the proper use of these different access technologies and reconfigurable equipment needs many capabilities like system discovery, protocol reconfiguration, and a method of vertical handover. Dynamic resources provisioning involves a dynamic reconfiguration of the terminal and network elements to improve the bandwidth for users with better adapted radio interfaces as well as additional spectrum. In this case, the protocol stack must be updated in the terminal and in the network. Consequently, the different communication systems covering such areas must be able to adapt to load and services variations.

Second, reconfigurability research has identified the concept of a Management and Control System that enables network elements to operate in an end-to-end reconfigurability context. The main idea of this concept is a clear separation of network management and control functions. Reconfigurable components, like programmable processors and reconfigurable logic, are envisioned for reconfigurable equipment. B3G architecture needs to support the dynamic insertion and configuration of different protocol modules as devices join and leave the given wireless environment. Furthermore, the reconfigurability of SDR equipment is widely seen as one of the enabling technologies for communication systems beyond 3G.

Third, the full benefits of SDR show up only if the network infrastructure takes into account the specifics of a particular terminal and provides support for it. Network support for reconfigurable entities requires the definition of appropriate functions in existing network elements or separate reconfiguration entities (for example, reconfiguration proxies). The definition of reconfiguration signaling between reconfiguration functions and reconfigurable entities is another key point. On the basis of the network architectures derived, and the reconfiguration signaling between entities for installation, deinstalling and verification must also be researched. Intelligent and self-learning protocols dependent on the reconfiguration context will have to be evaluated. Reconfiguration security for secure download, installation, verification, and fault management must be addressed to ensure a reliable operation and to satisfy regulatory demands for radio software. A framework for secure access to reconfiguration functionality by operators, manufacturers, and third parties must be developed. Furthermore, the active network environment for the management of reconfiguration needs to be studied.

Lastly, efficient spectrum management (initially discussed in Section 6.1) is of prime importance for reconfigurability to be realized. In discussions on reconfigurability, efficient spectrum management is one of the components of radio resource management (RRM), which also includes a joint management of radio resources belonging to different (2G and 3G) RATs with fixed spectrum allocation, *cognitive radio*, and a progressive network planning process. RRM is a complex process, but is necessary for the deployment of B3G networks. It consists of dynamically managing a spectrum as well as allocating traffic dynamically to the RATs participating in a heterogeneous, wireless access infrastructure. The coexistence of diverse technologies that form part of a heterogeneous infrastructure has brought about the idea of flexibly managing the spectrum. This implies that fixed frequency bands are no more guaranteed for RATs, but through an intelligent management mechanism, bands are allocated to RATs dynamically in a way that ensures that the capacity of each RAT is maximized and interference is minimized. Furthermore, there is a tight relationship between spectrum management and cognitive radio. Cognitive radio will provide the technical means for determining in real time the best band and the best frequency to provide the services desired by a user. Additionally, the growing demand for high-speed access to all kinds of telecommunication systems has made the reconsideration of traditional network planning methods necessary. Taking into account the fact that the advent of composite reconfigurable networks has become an inseparable part of almost every communications conference and journal, dynamic network planning (DNP) is essential in order to handle the alternations that take place in frequent time periods, with respect to the demand pattern in a specific geographical area. So, the goal of DNP is to reduce the cost of network deployment by the selection of the appropriate RATs for operation at different times and in different regions [541]. More research needs to be done on RRM to make reconfigurability a reality in B3G systems.

Summarizing, reconfigurability requires enhanced functionality for both terminals and networks. Researchers are developing a system for the management and control of terminal and network equipment. Special attention is required for the interface between (the separated) network management and control functions. One preliminary concept would allow (re)configuration of all affected layers through a unified, generic interface. A more detailed specification of these services and functions is needed [542].

6.5 Ad Hoc Mobile Networking

Several challenges have to be met for an ad hoc network to be possible in a B3G mobile networking environment. However, most of the challenges that apply to B3G systems in general also apply to B3G ad hoc networking: spectrum allocation issues, the integration of WMAN/WLAN and cellular networks, the need for high-speed data in a heterogeneous environment, and the issue of reconfigurability. The concept of an ad hoc B3G network is somewhat hazy – if a B3G device can always connect to the seamless, ubiquitous, global network, when would the need for an ad hoc network ever arise? The answer is, in response to an extraordinary event. The first responders to accidents may use ad hoc networks for secure and congestion-free communications around the scene. Military uses for ad hoc networks are well defined. Spectrum allocation takes care of emergency and military frequency issues. Anyone else who might desire privacy from the ubiquitous, global net (for instance, for teleconferencing) could establish a private MANET – if they could gain access to some unused frequency band and keep it "private" for a period of time. Regulators will need to address this issue for B3G MANETs to thrive.

In general, devices intended for use in a B3G environment in general should be able to scan in a specific environment to discover candidate available for access networks and register some policy issues. Devices intended for some use in a B3G MANET should be able to scan for and identify the frequency band(s) available for temporary, "private" use. In the seamless, global B3G network, the authentication and authorization mechanisms for access to different networks could be connected to allow a user/device to move between different access networks without the need to log on multiple times. In a B3G MANET, the trick would be to keep a device from inadvertently leaving the MANET and joining the global one. A worse case would involve a MANET node authenticating an undesired device to the MANET (the undesired device would be scanning for its best connection at all times) [543]. B3G ad hoc networks should be able to robustly adapt to changing network conditions and topologies, having the capability to grow, fragment, and reorganize in the absence of centralized, hierarchical infrastructures [544].

Some issues for B3G MANETs have been identified:

Routing: for different ad hoc scenarios, the routing protocol differs dramatically. While the routing protocol for an "eHome" scenario can assume fixed wireless terminals (leading to a small dynamic for the routing), the terminals in a fire-fighting scenario are highly mobile (leading to a high dynamic for the routing). This leads to the assumption that different routing strategies have to be applied.

Auto-Configuration: If we focus on Internet Protocol (IP) services over ad hoc networks, we have to support the assignment of IP addresses. Protocols such as dynamic host configuration protocol (DHCP) will not work in an ad hoc environment.

Device Classes: The routing process depends on the device class of a wireless terminal. Device classes are based on power, range, air interface, costs, and so on. Terminals with batteries are not well suited for multi-hop routing since they tend to consume more resources [545].

Table 6.3 shows the characteristics of a variety of MANET technologies.

The specific characteristics of MANETs impose many challenges on network protocol designs on all layers of the protocol stack. The physical layer must deal with rapid changes in link characteristics.

Table 6.3   Mobile ad hoc network enabling technologies [546]

Technology	Theoretical bit rate	Frequency	Range	Power consumption
IEEE 802.11b	1, 2, 5.5, and 11 Mbps	2.4 GHz	25–100 m (indoor) 100–500 m (outdoor)	~30 mW
IEEE 802.11g	Up to 54 Mbps	2.4 GHz	20–50 m (indoor)	~79 mW
IEEE 802.11a	6, 9, 12, 24, 36, 49, and 54 Mbps	5 GHz	10–40 m (indoor)	40 mW, 250 mW or 1 W
Bluetooth (IEEE 802.15.1)	1 Mbps (v1.1)	2.4 GHz	10 m (up to 100 m)	1 mW (up to 100 mW)
UWB (IEEE 802.15.3)	110–480 Mbps	Mostly 3–10 GHz	~10 m	100 mW 250 mW
IEEE 802.15.4 (i.e., Zigbee)	20, 40, or 250 kbps	868 MHz, 915 MHz, or 2.4 GHz	10–100 m	1 mW
HiperLAN2	Up to 54 Mbps	5 GHz	30–150 m	200 mW or 1 W
IrDA	Up to 4 Mbps	Infrared (850 nm)	~10 m (line of sight)	Distance based
HomeRF	1 Mbps (v1.0) 10 Mbps (v2.0)	2.4 GHz	~50 m	100 mW
IEEE 802.16	32–134 Mbps	10–66 GHz	2–5 km	Complex power control
IEEE 802.16a	up to 75 Mbps	<11 GHz	7–10 km	
IEEE 802.16e (Broadband Wireless)	up to 15 Mbps	<6 GHz	(max 50 km) 2.5 km	

The MAC layer needs to allow fair channel access, minimize packet collisions, and deal with hidden and exposed terminals. At the network layer, nodes need to cooperate to calculate paths. The transport layer must be capable of handling packet loss and delay characteristics that are very different from wired networks. Applications should be able to handle possible disconnections and reconnections. Furthermore, all network protocol developments need to integrate smoothly with traditional networks and take into account possible security problems. The technological challenges that B3G ad hoc network protocol designers and network developers are faced with include routing, service and resource discovery, Internet connectivity, billing, and security.

As MANETs are characterized by a multi-hop network topology that can change frequently because of mobility, efficient routing protocols are needed to establish communication paths between nodes, without causing excessive control traffic overhead or computational burden on the power-constrained devices. Combinations of proactive and reactive protocols, where nearby routes (for example, maximum two hops) are kept up to date proactively, while faraway routes are set up reactively, are possible and fall in the category of hybrid routing protocols. A completely different approach is taken by the location-based routing protocols, where packet forwarding is based on the location of a node's communication partner. Location information services provide nodes with the location of the others, so packets can be forwarded in the direction of the destination. Simulation studies have revealed that the performance of routing protocols in terms of throughput, packet loss, delay, and control overhead strongly depends on the network conditions such as traffic load, mobility, density and, the number of nodes. Ongoing research is investigating the possibility of developing protocols capable of dynamically adapting to the network.

MANET nodes may have little or no knowledge about the capabilities of, or services offered by, each other. Therefore, service and resource discovery mechanisms, which allow devices to automatically locate network services and advertise their own capabilities to the rest of the network, are an important aspect of self-configurable networks. The possible services or resources include storage, access to databases or files, printers, computing power, and Internet access. "Directoryless" service and resource discovery mechanisms, in which nodes reactively request services when needed and/or nodes proactively announce their services to others, seem an attractive approach for infrastructureless networks. The alternative scheme is directory-based and involves directory agents where services are registered and service requests are handled. This implies that this functionality should be statically or dynamically assigned to a subset of the nodes and kept up to date. Existing directory-based services and resource discovery mechanisms are unable to deal with the dynamics in ad hoc networks. Currently, no mature solution exists, but it is clear that the design of these protocols should be done in close cooperation with the routing protocols and should include context awareness (location, neighborhood, user profile) to improve performance. Also, when ad hoc networks are connected to a fixed infrastructure (for example, the Internet or a cellular network), protocols and methods are needed to inject the available external services offered by the service and content providers into the ad hoc network.

To enable communication between nodes within the ad hoc network, each node needs an address. In stand-alone ad hoc networks, the use of IP addresses is not obligatory, as unique MAC addresses could be used to address nodes. However, all the current applications are based on transmission control protocol (TCP)/IP or user datagram protocol (UDP)/IP. In addition, as B3G mobile ad hoc networks will interact with IP-based networks and will run applications that use existing IPs such as TCP and UDP, the use of IP addresses is inevitable. Unfortunately, an internal address organization with prefixes and ranges as in the fixed Internet is hard to maintain in mobile ad hoc networks owing to node mobility and overhead reasons, and other solutions for address assignment are thus needed. One solution is based on the assumption (and restriction) that all MANET nodes already have a static, globally unique and preassigned IPv4 or IPv6 address. This solves the whole issue of assigning addresses, but introduces new problems when cooperating with fixed networks. Connections coming from and going to the fixed network can be handled using *mobile IP*, where the preassigned IP address serves as the mobile node's home address. All traffic sent to this IP address will arrive at the node's home agent. When the node in the ad hoc network advertises to its home agent the IP address of the Internet gateway as its *care of address*, the home agent can tunnel all traffic to the ad hoc network on which it is delivered to the mobile node using an ad hoc routing protocol. For outgoing connections, the mobile node has to route traffic to an Internet gateway, and for internal traffic an ad hoc routing protocol can be used. The main problem with this approach is that a MANET node needs an efficient way of figuring out if a certain address is present in the MANET or if it is necessary to use an Internet gateway, without flooding the entire network. Another solution is the assignment of random, internally unique addresses. This can be obtained by having each node pick a more or less random address from a very large address space, followed by duplicate address detection (DAD) techniques in order to impose address uniqueness within the MANET. Strong DAD techniques will always detect duplicates, but are difficult to scale in large networks. Weak DAD approaches can tolerate duplicates as long as they do not interfere with each other, that is, if packets always arrive at the intended destination. If interconnection to the Internet is desirable, outgoing connections could be realized using network address translation (NAT), but incoming connections still remain a problem if random, not globally routable, addresses are used. Also, the use of NAT remains problematic when multiple Internet gateways are present. If a MANET node switches to another gateway, a new IP address is used and ongoing TCP connections will break. Another possible approach is the assignment of unique addresses that all lie within one subnet (comparable to the addresses assigned by a DHCP server). When attached to the Internet, the ad hoc network can be seen as a separate routable subnet – probably the norm in a B3G environment. This simplifies the decision of whether a node is inside or outside the ad hoc network. However, no efficient solutions exist for choosing dynamically an appropriate, externally routable, and unique network prefix (for example, special MANET prefixes assigned to Internet gateways), handling the merging or splitting of ad hoc networks or handling multiple points of attachment to the Internet. It is clear that, although many solutions are being investigated, no common adopted solution for addressing and Internet connectivity is available yet. New approaches using host identities, where the role of IP is limited to routing and not addressing, combined with dynamic name spaces, could offer a potential solution, but may be problematic in a B3G environment.

The wireless mobile ad hoc nature of MANETs brings new security challenges to network design. Because the wireless medium is vulnerable to eavesdropping and ad hoc network functionality is established through node cooperation, mobile ad hoc networks are intrinsically exposed to numerous security attacks. During passive attacks, an attacker just listens to the channel in order to discover valuable information. This type of attack is usually impossible to detect, as it does not produce any new traffic in the network. On the other hand, during active attacks, an attacker actively participates in disrupting the normal operation of the network. This type of attack involves deletion, modification, replication, redirection, and fabrication of protocol control packets or data packets. Securing ad hoc networks against malicious attacks is difficult to achieve. Preventive mechanisms include authentication of message sources, data integrity, and protection of message sequencing, and are typically based on key-based cryptography.

Incorporating cryptographic mechanisms is challenging, as there is no centralized key distribution center or trusted certification authority at present. These preventative mechanisms need to be sustained by detection techniques that can discover attempts to penetrate or attack the network. Moreover, not all security problems in ad hoc networks can be attributed to malicious nodes that intentionally damage or compromise network functionality. Selfish nodes, which use the network but do not cooperate in routing or packet forwarding for others in order to conserve battery life or retain network bandwidth, constitute an important problem as network functioning entirely relies on the cooperation between nodes and their contribution to basic network functions. To deal with these problems, the self-organizing network concept must be based on an incentive for users to collaborate, thereby avoiding selfish behavior. Existing solutions aim at detecting and isolating selfish nodes using watchdog mechanisms, which identify misbehaving nodes, and reputation systems, which allow nodes to isolate selfish nodes. Another promising approach is the introduction of a billing system into the network based on economical models to enforce cooperation. Using virtual currencies or micropayments, nodes pay for using other nodes' forwarding capabilities or services and are remunerated for making theirs available. This approach certainly has potential in scenarios in which a part of the ad hoc network and services is deployed by companies or service providers (for example, location- or context-aware services, a sports stadium, or a taxi cab network). Also, when ad hoc networks are interconnected to fixed infrastructures by gateway nodes, which are billed by a telecom operator, billing mechanisms are needed to remunerate these nodes for making these services available. Questions about who is billing whom, and for what, need to be answered and may lead to complex business models. Further research into security mechanisms, mechanisms to enforce cooperation between nodes, and billing methods are needed for B3G MANETs to function [546].

6.6   Networking Plan Issues

DNP is considered as a subset of a more general framework: *Dynamic Network Planning and Management* (DNPM) – a framework dealing with planning and managing a reconfigurable network [541].

Reconfigurability and spectrum issues are changing the way wireless networks are planned. Planners are mindful of QoS constraints and the need to reduce infrastructure costs in the B3G era.

Traditionally, mobile operators have designed and deployed the radio access networks to cover the traffic demand of the planned services in a static approach, considering the busy hour traffic in a given geographical zone. This means that the operator installs as many base stations as needed to attend to the traffic foreseen in each zone. In doing so, the conventional network planning methods consist of some predefined phases, namely, the initial dimensioning and the detailed planning with the help of an appropriate planning tool, and such methods can be applied only prior to the network deployment.

The current planning process follows several steps to obtain the site locations and configurations that satisfy the network planning requirements of coverage, capacity, and QoS in a geographical area. An initial number of sites and configurations can be obtained as a preliminary dimensioning exercise, based on the network data obtained by the operator in this first phase. According to the estimated number of sites in the dimensioning phase, sites are selected in the desired geographical area in the second phase. This selection could become a complicated task. Although some algorithms can be used in the planning process to assist the planners in the selection of sites, and this task can be carried out by automatic tools, the restrictions to the problem sometimes make the effort of using these algorithms not worthwhile. These restrictions in the selection of sites are due to the difficulty of the operators in choosing the desirable positions for the sites. Increasingly, people and governments are more concerned about mobile telephony and antennas on the roofs of the city, and it is very complicated for the operators to acquire new sites. NPs must often restrict themselves to the set of sites they have from earlier network deployments. Once the sites are selected and placed on the scenario, the radio network deployment should be analyzed to check that the initial requirements of coverage, capacity, and quality of service are satisfied. This evaluation can be performed by means of a radio network planning tool.

However, reconfigurable networks are continuously transforming, according to time- and space-variant demand. More specifically, the distribution pattern of subscribers, user-related information (profiles), and available terminal types are different from those of conventional networks. This means that the reconfiguration mechanisms for the base stations of a particular RAT can control the change-able parameters and operational modes, targeting optimal network configurations. Moreover, software download support must be integrated into network infrastructure. A flexible management covers electric tilting of antenna angles, frequency settings, the maximum size of the active/candidate cell for Mobile Terminals (MT), power allocation for high-speed data services, which has adaptive modulation and code schemes implemented, and complete reconfiguration between RATs for a common platform. According to the temporal-spatial changing traffic, some of these parameters are subject to change. Therefore, the busy-hour traffic for some particular hotspots in the conventional network planning paradigm is not the only criteria for planning anymore. Moreover, there will be no exact separation between planning and management, but DNPM has to be applied to reconfigurable contexts.

Consequently, while considering the gains and characteristics exclusively offered by the flexibility of the reconfigurable system, the suitable planning methods and the affecting factors need to be studied; innovative engineering mechanisms need to be defined, in order to guarantee for the best possible planning design, not only before network deployment but also during network operation.

In the reconfigurability context, DNPM is a complete framework that cooperates with other mechanisms such as the Joint Radio Resource management (JRRM) and Dynamic Spectrum Management (DSM), for efficient network deployment. During network planning, modeling of network performance, taking into consideration a given traffic distribution and network deployment cost, is needed. The measurements of network performance should not only be based on the carrier strength that a MT can receive but also on the performance improvement given by other resource management mechanisms. In the optimization phase, algorithms like "Greedy," "Taboo Search," and "Simulated

Annealing" are considered in an approach involving combinations of snapshot simulations. In the management phase of DNPM, radio network elements and some key resource management related parameters are subject to reconfiguration. Reconfiguration is triggered by the management entities like the network element manager so that self-tuning of a radio network targeting optimal parameter settings can be carried out. Typical examples are the vertical antenna tilting, power adjustment, spectrum management, and multistandard base station reconfiguration. For an on-the-fly reconfiguration, a faster heuristic search, rather than the classic algorithms, needs to be used.

Early research in the field of reconfigurable networks shows significant dependencies between network planning and network management resulting from the time- and space-variant conditions that render initial planning insufficient. The assumption is that the transceivers within the service area are reconfigurable. The situation that arises owing to the changes requires reallocation of RATs to the transceivers of the "target" region. The problem tackled is called the *RDQ-A problem* because its solution aims at new assignments of RATs to transceivers, demand to transceiver/RATs, and applications to QoS levels.

The RDQ-A problem can be generally described from a certain input and a certain objective (output). The input to this problem provides information on the service area and demand, as well as on the system. The service area is divided into a set of area portions, called *pixels*. What is of interest are the applications (services) offered in the service area, the quality levels (QoS levels) through which each service can be offered, the RATs through which each service can be offered and the expected demand per service and pixel. Moreover, the additional requirements are the utility volume and the resource consumption, when a service is offered at a certain quality level, through a certain RAT. The aspects of the system that need to be taken into account are the set of sites that cover the service area region that needs reconfiguration, and their locations (pixels), the set of transceivers per site, the set of RATs that can be used per transceiver, and the coverage and the anticipated capacity, when a certain RAT is used by a certain transceiver, taking into account intra- and inter-RAT interference. The objective (output) of the RDQ-A problem is to determine new configurations, for example, new allocations of RATs to transceivers, demand to transceiver/RATs, and applications to QoS levels. The three allocations should optimize a utility-based objective function, which is associated with the resulting QoS levels. Moreover, the allocations should respect constraints. The demand in the service area should be satisfied. Applications should be assigned to acceptable QoS levels. Permissible RATs should be assigned to transceivers. The allocations of RATs to transceivers should provide adequate capacity and coverage levels.

Initially, the overall RDQ-A problem is split into a number of subproblems, depending on the corresponding number of available transceivers and RATs, that have to be solved in parallel. In each of the resultant subproblems, the transceivers are assigned with a specific RAT. The second phase includes the solution of these subproblems, which can be done in parallel. Each subproblem aims at allocating the demand to the available transceivers. For this procedure, it is assumed that the lowest QoS levels are assigned to the offered services. In the third phase, called *improvement phase*, the QoS levels to be assigned are gradually augmented in a greedy fashion. Finally, the fourth phase summarizes the three past phases and selects the best combination of allocations that maximizes an objective function associated with the utility, by means of the resulting QoS levels. In the sequel, there are some indicative results from the application of the aforementioned algorithm to a simulated network that deploys reconfigurable transceivers working at multiple RATs [547].

The network planning problem can be solved with the utilization of the appropriate optimization functionality. This refers mainly to the respective midterm algorithms, necessary for dynamic network planning issues. Simulations for dynamic networks taking into account multistandard radio network elements must be performed and the requisite recommendations for network planning must be deduced. Automatic network planning is another use-case for reconfigurable, multistandard network elements, for example, the autonomous selection of carrier frequencies [548].

6.7   Satellite Systems in B3G Wireless

There was a notion that satellites are an expensive way to deliver services and cannot compete in terms of QoS with terrestrial service providers. The satellite and terrestrial communities are changing their minds about competing with one another, and in B3G systems they will strive to collaborate, cooperate, and find ways to integrate their devices into the ubiquitous net.

In satellite communications, a division is made between Fixed satellite service (FSS), Broadcast satellite service (BSS) and Mobile satellite service (MSS) delivery. In FSS, satellites operate mainly in a point-to-point mode as part of the core network and provide high-capacity links in telecommunications and ISPs. On a point-to-point basis, IPv6 operations pose no problems and are currently in operation over many satellite links. Within the FSS/BSS domain, satellites are used extensively to deliver video services to cable heads or direct to home (DTH). Most of these services have now been transferred to MPEG-DVB-S packetized transmission. Interactive television is available in this digitized service through low rate channels with proprietary protocols via, mostly at the moment, landline. The *DVB-S standard* is becoming an industry standard for the delivery of IP via satellite, although it was not primarily designed for this purpose and is not optimal.

For the mobile domains, the success has been with delivery to a niche market to ships, planes, and land vehicles. The service is now digital and packet based, using again a proprietary protocol, and with the introduction of the latest satellites will be capable of delivering 3G-like services (however, it is still TDMA). The Regional GEO system *Thuraya* seems to be taking off well with a good customer basis and has wide coverage over Asia and Europe, providing 2G+ services in areas not well served by terrestrial infrastructure. All of these systems are capable of extensions to 3G-like services and are especially suited to those services in which location data and communications together are key.

So far mobile satellites have either selected niche areas or tried to compete in the mass market with cellular services. In the long run, such competition will not be fruitful, but collaboration with cellular services in the access networks will be beneficial. This is true mainly in two areas. The first is in the coverage of remote areas that would be too expensive to be served by cellular services. Providing such services would be more expensive but could form a value-added offering for mobile service providers. An adjunct to this would be the provision of disaster services to back up cellular services. The second, and perhaps more interesting, is the delivery of multimedia broadcast and multicast service (MBMS) services to the mass market. Within 3G networks and also beyond, these services are very difficult and expensive to be delivered in a cellular format. However, they are ideally matched to the attributes of a satellite network in terms of the broadcast coverage and thus we have a win–win situation. The delivery of multimedia content in MBMS to a large customer base via satellite can reduce the cost by orders of magnitude over cellular networks. Moreover, with sufficient large storage capacity in the user terminals, unidirectional point-to-multipoint services are able to provide on-demand and interactive applications because push and store mechanisms make the point-to-multipoint relationship transparent to users.

In a truly integrated satellite/cellular system to be used by mobile operators, satellites will be complemented by terrestrial repeaters known as *intermediate module repeaters* (IMRs) to provide cost-effective services. This collaboration is what is envisioned in a B3G environment [549].

6.8   Other Challenging Issues

Several other technologies can be thought of as essential for B3G systems and require more research.

- Ultra-wideband (UWB) techniques[3] for short-range communications.

- Optical wireless techniques for short-range communications.

---

[3]More discussions on UWB technologies are given in Section 7.6.

- Techniques for seamless vertical and horizontal handovers.

- Cross-layer design and optimization.

- Advanced RRM, with multidimensional. scheduling (time, frequency, space) and intelligent radio technology[4].

- Techniques for reducing the PAPR problem typical of multi-carrier systems.

- Advanced channel coding techniques (turbo codes, LDPC codes,[5] etc.).

- Sensor networks.

- Mesh networks.

- Network security.

- Battery technology.

Introducing a new system always involves risks, and B3G is not an exception, even when considering that it will also integrate legacy systems. Integration in B3G means having different networks, different terminals, and different services working together seamlessly. It is precisely the integrative capability of B3G that is one of the crucial challenges, as access solutions for different B3G scenarios are being developed independently by different parties. Seamless operation is one of the pillars of B3G, implying transparent intra- and internetwork (horizontal/vertical) handovers. Turning a very heterogeneous network into a single, simple, and monolithic network (in the eyes of the user) could entail a colossal task, in particular, if the integration aspects are left for the final phase of B3G development after different access techniques are developed. The risk is not only in the integration of access technologies but also in their adoption. Indeed, not all proponent solutions being currently developed are complementary; many of them would compete with each other [532].

Here is another list of challenges facing the developers of B3G systems:

- Lower price points only slightly higher than alternatives.

- More coordination among spectrum regulators around the world.

- More academic research.

- Standardization of wireless networks in terms of modulation techniques, switching schemes, and roaming is an absolute necessity for B3G.

- Justification for voice- independent business.

- Integration across different network topologies.

- Nondisruptive implementation: B3G must allow us to move from 3G to B3G [550].

---

[4]One of the most prominent issues in this topic is cognitive radio, which is discussed extensively in Chapter 9.
[5]"LDPC" code stands for low density parity check code, which is an emerging effective channel coding scheme.

7

Multiple Access Technologies for B3G Wireless

Some general discussions on multiple access technologies were given in some of the earlier chapters, such as Chapters 2 and 6. In this chapter, the multiple access technologies suitable for beyond 3G wireless communications are discussed. Before proceeding to the discussions, we would like to take a brief look at the history of wireless communications under the context of the multiple access technologies.

Multiple access is always an important issue that should be addressed carefully in the design of any wireless communication systems. Many peculiar properties of a wireless transmission medium, as discussed in Chapter 2, make it critical to choose appropriate multiple access technologies, ensuring an efficient and yet fair sharing of precious radio spectrum resources in a wireless communication system.

In Chapter 3, we discussed a variety of 3G standards for mobile cellular communication systems. It was seen from the discussions that the evolution of multiple access technologies has been driven by the need to deliver increasingly high-data-rate and multimedia services. The first-generation mobile cellular systems operated mainly on analog electronics based on the Frequency-Division Multiple Access (FDMA) technology. At that time, the mobile cellular systems were voice application oriented. The ultimate requirements of the systems then were to provide customers with a satisfactory voice quality at a reasonable cost. The simple idea of separating users via different frequency channels could hardly offer a very high capacity to bring down the operation cost of mobile cellular systems. Analog radio technology was not concerned with the issue of *data transmission rate*, and thus only voice channels per MHz were an important merit parameter of the systems.

It so happened that the demand for mobile voice communications effectively saturated the capacity of the entire analog cellular networks. This was a strong push to search for a new and more effective multiple access technology to replace the legacy FDMA technology, as an effort to support more users within a limited radio spectrum bandwidth. Time-Division Multiple Access (TDMA) was put forward as the right choice to meet the needs of the second-generation mobile cellular systems, which were proposed as an effort to make international roaming possible. The TDMA scheme works on digital mobile cellular architectures, which also become much more complex than FDMA-based systems. Unlike FDMA, the TDMA technology works on the idea that the transmission time in a cell is divided into many frames, each of which consists of many short time slots. The signal transmissions from all users need to be synchronized in time, and every active user is assigned a particular time slot in the

Next Generation Wireless Systems and Networks   Hsiao-Hwa Chen and Mohsen Guizani
© 2006 John Wiley & Sons, Ltd

frame for its transmission. A specific user will transmit only at a time slot assigned to it, and should refrain from transmission until the same slot appears in the next frame, and so on. Therefore, the number of users a cell can support is exactly equal to the number of time slots available in a complete signal frame. The on-and-off transmission nature in each user makes it easier for a TDMA system to adopt digital transmission technologies. The Global System for Mobile (GSM) communication and IS-54B (and later IS-136) standards were proposed on the basis of TDMA technologies.

It is noted that the IS-136 system was proposed at almost the same time as the IS-95A, which took a very different path from that of GSM and IS-136 systems. The IS-95A standard adopted *Code Division Multiple Access* (CDMA) technology as its multiple access scheme. CDMA technology was developed from *Spread Spectrum* transmission technology, which had been used mainly in military applications for a long time before the 1970s. As discussed in Subsection 2.3.3, CDMA makes use of the orthogonality or the quasi-orthogonality of signature codes to divide users in a cell. Thus, different users should be assigned different codes, which should maintain acceptably low cross-correlation functions (CCFs) between any two codes. Like TDMA technology, CDMA should also be implemented by digital technology, and should provide many unique desirable features that are otherwise impossible when using other multiple access technologies, as discussed in Subsection 2.3.3.

CDMA technology has become a prime multiple access technology in third-generation wireless communication systems. Almost all 3G standards submitted to ITU as candidate proposals of the IMT-2000 system chose CDMA as their multiple access technology. Three major 3G standards, CDMA2000, WCDMA, and TD-SCDMA have been discussed in Sections 3.1, 3.2, and 3.3 respectively. It is to be noted that the CDMA technologies used in all the 3G standards share almost the same core technologies as those introduced by IS-95A. There was no technological revolution in them. It is regretful to say so here, but it is true. For instance, the channelization codes used in WCDMA and TD-SCDMA standards are Orthogonal Variable Spreading Factor (OVSF) codes, which is, in fact, exactly the same as the Walsh-Hadamard codes used by the IS-95A. Therefore, many people agree that the proposals for 3G standards were made in too short a time frame to choose technologically right solutions. In other words, it has been suggested by many people that the development of the 3G standards were somehow driven by politics rather than by technologies. The triggering factor was the competition between two Asian countries, Japan and Korea, who have a history of enmity with each other. Japan worried about the fact that Korea was one step ahead in developing CDMA-based technologies, and thus pushed hard on Europe to jointly propose a WCDMA standard in a hurry. The worldwide 3G standardization activities might be a different story if Korea had not decided to purchase Qualcomm CDMA technologies in the early 1990s.

After having reviewed the 1- to 3G mobile cellular standards from a perspective of multiple access technologies, we would like to talk about the scenarios beyond 3G wireless applications. In fact, the services for all 3G systems have been very different from those offered in the 2G systems, which concentrate on voice-centric applications. 3G networks should carry a lot of multimedia traffic or contents, such as videoconferencing, digital TV broadcast, DVD quality interactive gaming, and so on. Therefore, it is foreseeable that a direct impact of multiple access technologies on the B3G systems is the capability of ensuring efficient and fair sharing of a limited radio spectra among concurrent wideband transactions. Therefore, it is very challenging to design a multiple access scheme for future B3G wireless applications. More discussions on B3G multiple access technologies for wireless communications can also be found in [551–561, 707, 764–766, 769–771].

7.1   What does B3G Wireless Need?

Beyond 3G (B3G) wireless systems should deliver higher data transmission rates and more diverse services than current 2- to 3G systems can. The all-IP wireless architecture has emerged as the most preferred platform for B3G wireless communications. Therefore, the design of a future wireless air interface has to take into account the fact that the dominant load in B3G wireless channels will be high-speed burst-type traffic. The necessity to support such high-capacity bursty traffic in extremely unpredictable wireless channels has already posed a great challenge to all existing air link technologies based on TDMA or CDMA alike. Many research initiatives have been underway to investigate the type of multiple access technologies that could be the most suitable for B3G wireless applications. Some suggested that the current CDMA technologies, all based on direct-sequence (DS) CDMA, are only suited for slow-speed continuous transmission applications such as voice services, but may not be a good choice for high-speed burst type traffic in future all-IP B3G wireless systems. Therefore, a new wave of worldwide research is underway to search for next generation multiple access technologies, which should effectively address all the constraints and problems existing in current TDMA and CDMA technologies, such as poor bandwidth efficiency, strictly interference-limited capacity, difficulties in performing rate-matching algorithms, and complexity in implementing fast adaptive equalizers. The study of next-generation multiple access technologies involves many cutting-edge research topics, such as novel CDMA code design, time-frequency adaptive equalization, interference-free CDMA architecture [781, 782], high-data-rate TDMA, Orthogonal Frequency-Division Multiplexing (OFDM) techniques, and Multiple-Input Multiple-Output (MIMO) algorithms.

7.2 A Feature Topic on B3G Wireless

The authors of this book were the Guest Editors for a Feature Topic on *Multiple Access Technologies for B3G Wireless Communications* in the IEEE Communications Magazine, which was published in the 2005 February issue [562]. That feature topic was published to serve as a stimulus to accelerate the technological evolution of multiple access technologies for B3G wireless applications. Several important issues on multiple access technologies that are suitable for B3G wireless systems have been addressed in the feature topic. It should be mentioned that the call for papers for the feature topic received an overwhelming response from the research community. More than forty high-quality submissions were received from both the academia and the industry of different regions around the world. This was a very positive sign, which showed that people around the world have been aware of the importance of the research issues related to the feature topic. Unfortunately, because of limited space and volume, only eight papers were accepted in the feature topic. In the following text, we give a brief introduction to the eight articles published in that feature topic, as they addressed relevant issues to what the B3G wireless needs in the perspective of multiple access technologies.

The first article was written by H. Wei, L-L. Yang, and L. Hanzo, titled *Interference-Free Broadband Single- and Multi-carrier DS-CDMA* [563]. The article addressed a very interesting issue: the choice of the DS spreading code for a DS-CDMA system. It was demonstrated in the article that the family of codes exhibiting an interference-free window (IFW) outperforms classic spreading codes, provided that the interfering multiuser and multipath components arrive within this IFW, which may be ensured with the aid of quasi-synchronous adaptive timing advance control. The article further showed that the IFW duration may be extended with the advent of multicarrier DS-CDMA proportionate to the number of subcarriers. Hence, the resultant MC DS-CDMA system is capable of exhibiting near-single-user performance without employing a multiuser detector. The authors also addressed the limitations of the system, such as the number of spreading codes exhibiting a certain IFW being limited, although this problem may be mitigated with the aid of novel code design principles. This contribution was interesting because of the fact that all existing CDMA systems fail to offer satisfactory performance and capacity, which is usually far less than half of the Processing Gain (PG) of CDMA systems. Therefore, the work presented in the article can be a direction finder for further research on the design of next generation CDMA systems, whose performance should not be interference-limited.

The second article [564] in the feature topic was written by William C. Y. Lee, who proposed a new up and down link duplexing scheme called *code division duplexing* (CDD), whose physical layer scheme can work harmoniously with a 4G wireless architecture called *code spreading* (CS) orthogonal frequency-division multiple access (OFDMA). The CS technique used a set of smart codes with its meritorious properties, which can eliminate both multipath and multiuser interferences. The article showed that the same number of smart codes applied repeatedly in each subcarrier of OFDMA could be used to increase the number of traffic channels. The author further illustrated that the scheme could take advantage of both the attributes of CDMA and OFDMA toward a simpler and more spectrally efficient system as applied in a CDD system.

The third article, *Ultra-wideband for Multiple Access Communications* [565], was contributed by Robert C. Qiu, Huaping Liu, and Sherman Shen. This article offered a comprehensive review of UWB multiple access and modulation schemes. It also included a comparison with narrowband radios. The authors outlined other issues with Ultra-wideband (UWB) signal reception and detection, and explored various suboptimal low complexity receiving schemes. The focus of the article was on balancing the treatment of theoretical and practical designs. They also mixed the needs of two major applications (IEEE 802.15.3a and 4a). They believed that UWB holds a promising future for wireless multiple access, simply because of the unprecedented huge chunk of the unlicensed spectrum. The article then pointed out that in the long run, UWB will change some of the basic ideas in the standard wireless textbooks, and will enable the concept of gigabit wireless communications.

Next, Farooq Khan offered his opinion on a time-orthogonal CDMA high-speed up link data transmission scheme, suitable for B3G wireless systems. In this article [566], the author presented a new time-orthogonal CDMA approach called *high-speed up link data burst transmission mode*. The concept was based on slot-synchronized slot-orthogonal transmissions, whereby high-speed data transmissions take place in slots orthogonal to the slots used for physical layer control signaling and low-data-rate transmission such as resource requests. Using this approach, very high data rates and capacity are achieved during data burst transmission because of the availability of high signal-to-interference-plus-noise ratio (SINR) resulting from the orthogonality of the transmissions. The simulation results given in this article showed that the up link spectral efficiency of the proposed scheme was approximately four times better than the up link spectral efficiency achieved with existing 3G systems.

The fifth article [567], written by Romano Fantacci *et al.*, gave a review of the implementation of future CDMA communication systems. More specifically, the authors highlighted the issues regarding coping with the challenging requirements future systems ought to face. In the article, particular attention was given to addressing some inherent weak points of current CDMA systems, and then to suitable detection techniques to overcome major impairments in the channels. The article went on to further point out that high-data-rate transmissions in B3G wireless require proper link adaptation techniques. The article concluded by focusing on the design of suitable protocol strategies for new heterogenous multimedia packet services, which are constrained by strict quality of service (QoS) requirements.

The sixth article [568] discussed high-performance MIMO OFDM wireless LAN systems. The article was jointly authored by S. Nanda *et al.*, from Qualcomm, Inc. The article started by enumerating valuable lessons to be learned from the design of wireless local-area networks (WLANs) that provide data rates in excess of hundreds of megabits per second. In particular, they presented a MIMO WLAN design and prototype that exploits these attributes to provide data rates in excess of 200 Mbps above the medium access control (MAC). Although the design and prototype given in the article were only focused on WLAN, it was demonstrated that very high data rates over wireless are feasible. The article concluded with a summary of several key attributes of next generation high-performance wireless networks.

The seventh article [569] offered a tutorial on multiple access technologies for beyond 3G mobile networks, written by Abbas Jamalipour *et al.* In this article, the state-of-the-art technologies for multiple access schemes that have been adopted in third-generation wireless cellular systems were reviewed, and a path for the development of appropriate multiple access technologies for B3G mobile networks was suggested. The authors provided an expression that could interconnect all those multiple access schemes that are usually separated under time-division, frequency-division, and code-division techniques. Several combinations of these multiple access schemes were discussed, and their application in different cellular mobile standards was addressed. The authors suggested that next generation networks will be developed through the good management and combination of advanced multiple access technologies, rather than the development of new schemes.

Finally, Markku Juntti *et al.*, [570] proposed a MIMO system based on space–time turbo coded modulation and layered spatial multiplexing architectures for cellular multicarrier (MC) CDMA systems. The authors discussed the issues related to the design of appropriate receiver algorithms, and compared the performance of the new system with competing schemes in a single cell basis. The performance evaluation of the proposed scheme was carried out for a seven-cell system with universal frequency reuse. It was shown that the proposed MIMO scheme could significantly improve the throughput compared to the corresponding single-antenna communication systems even in the presence of spatial correlation.

It was hoped that by highlighting some of the current research work on multiple access technologies for B3G Wireless Communications in the feature topic [562], the researchers would be encouraged to consider some specific research issues raised therein. We believe that this feature topic has triggered further interest in the research areas on multiple access technologies for B3G wireless communications.

7.3   Next-Generation CDMA Technologies

As mentioned at the beginning of this chapter, the CDMA technologies used in all 3G wireless communication systems share the same core IPRs as proposed by the IS-95A standard, which was initially designed by Qualcomm Inc. in the early 1990s. Those core IPRs include closed and open-loop power control, RAKE receiver, DS spreading, soft-handover technique, and so on. Therefore, it is sad to say that CDMA technology has not been innovated in 3G wireless communications. For this reason, we call the CDMA technology adopted in 2- to 3G wireless systems *The First-Generation CDMA technology*, or simply 1G-CDMA technology.

1G-CDMA technology has to undergo a thorough technological innovation to suit the needs of B3G wireless communications. Several technical aspects that limit further performance enhancement in the 1G-CDMA system should be addressed. In particular, a CDMA system should not always be considered as an interference-limited system, whose capacity can never reach half of the PG. To design the next-generation CDMA technology, we should bring several fundamental changes to the existing 1G-CDMA on CDMA sequence generation, spreading and carrier modulation schemes, and overall system architecture designs, and so on. In the following text, we discuss these issues based on our previous research experience.

7.3.1   Importance of Using Good CDMA Codes

As mentioned in Section 2.3.3, the characteristic features of a CDMA system can be basically determined by the spreading codes used by the CDMA system. Thus, it is of ultimate importance to use the right codes to make sure the performance of the overall CDMA system will not be limited by the inherent defects of the codes. For instance, the use of the Walsh-Hadamard sequences in the IS-95A/B and the CDMA One systems for down link channelization gives a very poor performance when Multipath Interference (MI) is present. Even the use of a RAKE receiver will not help much in such a scenario, as the output from each finger of the RAKE consists not only of the useful signal components, but also of plenty of unwanted interference components caused by multipath returns. In this case, the choice of the Walsh-Hadamard sequences has predetermined the basic performance of the IS-95A/B system, whose cell-wise capacity never exceeds one third of the Walsh-Hadamard code length of 64/3.

It is also noted that the need for many complex subsystems required by two to three wireless systems based on 1G-CDMA technology are due to the poor spreading codes. These complex subsystems include closed- and open-loop power control, CDMA multiuser detection, smart antenna with adaptive beam-forming, RAKE receiver, and so on. The closed- and open-loop power control is required because of the near–far effect caused by nontrivial CCFs of the CDMA codes. The multiuser detection should be applied to all 1G-CDMA-based wireless systems because of the strong correlation existing in user signature codes due to their imperfect correlation properties. The reason that we need to use smart antenna and adaptive beam-forming techniques in a cell site is to suppress strong cochannel interference produced by poor CCFs among user signature codes. A RAKE receiver is used in all 1G-CDMA systems to overcome MI, which will never be harmful if the signature codes offer ideal correlation properties (zero autocorrelation side lobes for any code and zero CCFs for any pair of codes). Therefore, the use of more desirable spreading codes in a CDMA system will not only improve the overall performance, but also greatly simplify the hardware complexity, without the need for all those complex subsystems designed in particular for 1G-CDMA technology.

7.3.2 System Model and Assumptions

Several assumptions should be made to facilitate our discussions carried out in this section.

- First, we consider a generic CDMA system that uses short codes (with chip width being T_c) to spread data bits. The code length is exactly equal to the data bit duration (T_b).

- Second, the wireless system under consideration consists of mobile terminals and a base station (BS). We will focus on the intracell physical layer architecture of a new CDMA system and will not address any upper layer issues of a mobile network, nor those involving different cells in a mobile network.

- Third, in the discussions given in this section all CDMA codes are classified into two categories, *unitary codes* and *complementary codes*. The former includes almost all traditional spreading codes, such as Gold, Kasami, Walsh-Hadamard, and OVSF codes, and so on, which work on a one-code-per-user basis. The latter forms another group of CDMA codes working on a flock of codes basis. Each user in such complementary code–based CDMA systems should be assigned a flock of M element codes, which ought to be sent via different channels (either in *frequency* or *time*) to a specific receiver for complementary auto- and CCF reconstruction. Obviously, the unitary codes are only a special case of the complementary codes with $M = 1$.

- Fourth, our discussion on the CDMA codes will not be limited to any specific chip value, being either complex, real, or binary, to make the discussions as general as possible.

Let us consider a generic K-user CDMA system, where each user is assigned one unique flock of M codes $(\mathbf{c}_{k,1}, \mathbf{c}_{k,2}, \ldots, \mathbf{c}_{k,M})$ for CDMA purpose. Each code $\mathbf{c}_{k,m}$ consists of N chips, where $1 \leq k \leq K$ and $1 \leq m \leq M$. Assume that the signal of interest is from the user 1. If $M = 1$, then the system model is equivalent to a traditional unitary code based CDMA system; otherwise, if $M > 1$ it makes a complementary code–based CDMA system. Therefore, the discussions given in this section will make sense in general for new CDMA systems using either unitary or complementary codes.

7.3.3 Spreading and Carrier Modulations

The most important role of spreading modulation is to achieve a PG for some operational advantages over a non-spread-spectrum communication system. On the other hand, carrier modulation functions as a vehicle to send user data to a receiver through RF transmission. They work closely together and should thus be jointly considered in a CDMA system design. The major concern in the design of spreading and carrier modulations for the next generation CDMA, similar to all traditional CDMA, is centered on bandwidth efficiency and power efficiency. Obviously, they form a dual, which often work in a counteractive way. Therefore, to achieve a good trade-off between the two becomes extremely important. In this subsection, we discuss two different spreading modulation schemes, DS spreading, and offset-stacking (OS) spreading, both of which could possibly be used in the next generation CDMA systems, and their impact on the carrier modulations.

DS spreading versus OS spreading

Both DS and OS spreading modulation schemes can be applied to a CDMA system based on complementary codes (here, any unitary code is treated as a special case only), resulting in either a DS-CDMA or an OS-CDMA scheme. The DS spreading modulation has been widely used in 2- to 3G CDMA mobile cellular standards, whereas the OS spreading scheme was only introduced recently [210]. The basic idea behind the OS spreading is that a new bit will start immediately after n-chip shift relative to its previous bit, and thus the consecutive bits are stacked over one another with n relative offset chips, where n can take any integer from 1 to N (N is the element code length). For more detailed information of the OS spreading technique, the readers may refer to [210], in which, however, only the case with $n = 1$ is discussed. Clearly, if we allow arbitrary relative offset chips n, where $1 \leq n \leq N$, between two consecutive data bits in an OS spreading modulator, a DS-CDMA system becomes only a special case of the OS-CDMA scheme with its relative offset chips being equal to the element code length or N chips. Thus, the study on an OS-CDMA scheme makes sense in general. The use of more relative offset chips between two consecutive bits will result in a slower transmission rate. Figure 7.1 illustrates the variable numbers of relative offset chips between two consecutive bits in an OS-CDMA system, where only two short element codes, $(+ + + -)$ and $(+ - + +)$, which are assigned to the same user and sent via different carriers (f_1 and f_2), are shown in Figure 7.1 for the simplicity of illustration.

Therefore, we can introduce a merit figure of spreading efficiency (SE) to describe how many data bits can be conveyed in one chip duration. If a fixed chip width is considered, the SE figure only gives the bandwidth efficiency of a CDMA system. The greater the SE is, the higher the bandwidth efficiency of a CDMA system will be. Obviously, we have the relation between the number of relative offset chips n and SE as $SE = \frac{1}{n}$, or $\frac{1}{N} \leq SE \leq 1$. Therefore, the use of $n = 1$ relative offset chip in an OS-CDMA scheme leads to the highest SE figure equal to one, which is exactly N times higher than that of a traditional DS-CDMA system, and thus a substantial gain in bandwidth efficiency can be achieved. Unfortunately, all unitary codes, such as Gold, Kasami, Walsh-Hadamard, OVSF codes, and so on, are not suitable for the OS spreading due to the excessively high CCF between any two codes if modulated by an OS spreader. Only orthogonal complementary (OC) codes can be successfully applied to an OS-CDMA system, giving a multiple access interference (MAI) free operation, because of the ideal CCFs between any two OC codes sent in either synchronous or asynchronous channels. We name this superior property of the OC codes as *isotropic* MAI-free operation. Therefore, an OC code–based OS-CDMA system has a great advantage over its counterparts in terms of cell-wise average capacity, in addition to its extremely high-bandwidth efficiency.

Binary versus M-ary carrier modulations

If a transmitter is implemented with a spreading modulator followed by a carrier modulator, the choice of the latter greatly depends on the former's operation mode (i.e., either DS or OS spreading). For instance, if DS spreading is considered, the simplest form of the carrier modem can be BPSK (although other more complex modems, such as QPSK, quadrature amplitude modulation (QAM), and so on, can also be applied). However, if using an OS spreader, we have to deal with a multileveled

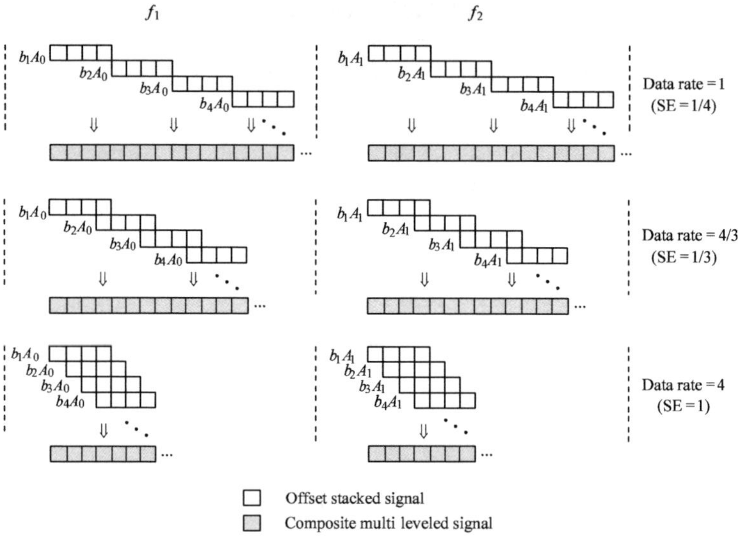

Figure 7.1 Variable SE figures and agility in changing transmission rate in an OCC/OS-CDMA system, where only two short element codes are shown and sent via two different carriers f_1 and f_2. The top layer of the figure considers DS spreading scheme or a special case of OS spreading with the number of relative offset chips being $N = 4$. The composite multileveled signals (shaded blocks) are also shown, along with the offset stacked binary signal (transparent blocks).

baseband signal, as shown by the shaded blocks in Figure 7.1. The dynamic range of the composite multileveled signal depends on relative offset chips n between two neighboring bits and the element code length N. In general, if each chip takes only binary values, the dynamic range, D_{OS}, of the output signal from an OS spreader can be $D_{OS} = \lceil \frac{N}{n} \rceil + 1$, where $1 \leq n \leq N$ and the notation $\lceil x \rceil$ stands for the smallest integer greater than x. The output signal from an OS spreader has to be modulated by a multileveled or M-ary digital modem, such as M-QAM, M-PSK or M-PAM, and so on, where $M \geq D_{OS}$ and D_{OS} is always an odd number, as shown in Figure 7.1. Also note that the appearance frequencies of different levels in the composite multileveled signal are different, with zero appearing most frequently if $n < N$. The greater the absolute value of a level is, the less frequently it appears. Thus, the histogram of the different levels gives a Gaussian-shaped envelope with its center or the mean value being zero. If an M-QAM modem should be used to map each level to a particular constellation point in its two-dimensional constellation plan, then there is an interesting design problem as to how to minimize the average symbol or the bit error rate (BER) with respect to the ways to map all the D_{OS} levels onto the same number of constellation points, which can be placed onto any points in an X-Y plan.

Therefore, the use of OS spreading can enhance bandwidth efficiency due to its very high SE, approaching to one for $n = 1$. However, the realization of this high-bandwidth efficiency is conditioned on the relatively low-power efficiency of an M-ary modem. If the latter fails to perform satisfactorily in a channel where many impairing factors such as MI and noise may exist, its bandwidth efficiency will be a question. On the other hand, DS spreading can never be comparable to the OS spreading in terms of bandwidth efficiency with its SE being only $\frac{1}{N}$, where N is the element code length. However, the output signal from DS spreading forms a binary bit stream, which can be modulated via an extremely power efficient BPSK or QPSK scheme. Therefore, the overall performance difference between a DS spreader cum BPSK/QPSK modem and an OS spreader cum M-QAM modem in varying circumstances is an interesting topic of study. Therefore, the final selection for either DS spreading or OS spreading should be exercised very carefully, depending on the nature of the operation environment.

7.3.4   Why the REAL Approach?

As mentioned earlier, the characteristic features of a CDMA code set govern the performance and the limitations of a CDMA system. For instance, the use of OVSF codes in UMTS-UTRA and WCDMA standards requires that a dedicated rate-matching algorithm has to be used whenever the transmission rate changes to match a specific spreading factor, or the system wants to admit as many users as possible in a cell. In addition, the rate change in UMTS-UTRA and WCDMA can be made only in multiples of two, meaning that continuous rate change is impossible. This requirement is a direct consequence of the OVSF code generation tree, where the codes in upper layers bear lower spreading factors; whereas those in lower layers offer higher spreading factors. Therefore, the occupancy of any node in the upper layers effectively blocks all nodes in the lower layers such that fewer users can be accommodated in a cell. The rate-matching algorithms also consume a great amount of hardware and software resources and affect the overall performance, such as increased battery consumption and processing time latency, and so on. Therefore, the choice of CDMA codes is extremely important and should be exercised very carefully at the early stage of a CDMA system design; otherwise the shortcomings of the system will carry on forever.

The search for promising CDMA codes or sequences used to be a very active research topic. Numerous candidates were proposed and their performance and possible applications in a CDMA system were investigated extensively in the literature. Almost all of those popular CDMA codes involved in the current 2- to 3G systems are unitary codes, meaning that they work on a single-code basis. All those unitary codes can be further classified into two subgroups, one being quasi-orthogonal codes (such as m-sequences, Gold codes, Kasami codes, etc.) and the other being orthogonal codes (such as Walsh-Hadamard sequences, OVSF codes, etc.). In addition to these commonly used unitary CDMA codes, there are many other less widely quoted ones, such as *GMW* codes [191–193], *No* codes [194], *Bent* codes [204], and so on.

Another group of CDMA codes is the complementary codes. They were first studied by Golay and Turyn [206, 207] for their applications in radar systems. Later on, there was some sporadic research on the complementary codes but no serious attention was given to them, mainly due to their implementation complexity and relatively small set sizes [208, 209]. It was found in our earlier study [210] that the joint application of the OC codes and OS spreading could greatly improve the bandwidth efficiency of a CDMA system, along with several other desirable properties which include isotropic MAI-free operation, ability to implement rate-matching, and suitability for burst traffic applications. Unfortunately, all of the previous studies on complementary codes could never explain why they could offer an isotropic MAI-free and MI-free (only for DS spreading and those OS spreading schemes with n being larger than the delay spread) operation, whereas a unitary code set cannot. Some generation method was reported for generating only a small subgroup of these, such as *complete complementary codes* [208, 209], and so on, which could hardly be applied to a real system due to the very small set sizes: only $\sqrt[3]{L}$ users can be supported for a code set with its PG value being L.

Traditionally, the merits of a CDMA code set are always judged according to its autocorrelation function (ACF) and CCFs. In most cases, only even periodic ACFs and CCFs are considered. Therefore, the code design approach, adopted in all traditional unitary codes, completely ignored the real application environment, where many external impairing factors exist, such as asynchronous transmission and MI. To make a substantial improvement on traditional CDMA systems, the CDMA code design approach should be innovated. In particular, we need a new code design approach that could take the real working environment into account. Our objective is to find new generation

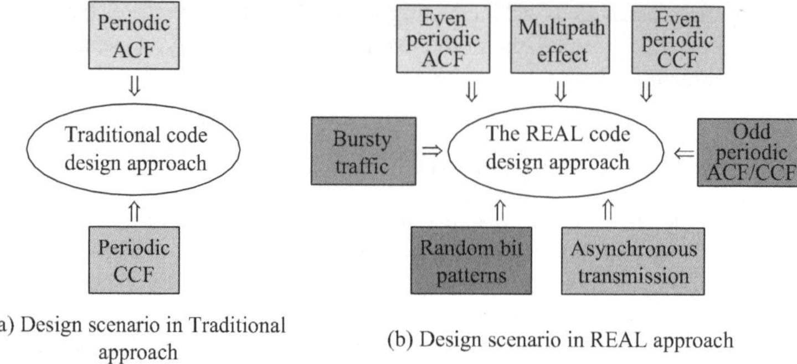

(a) Design scenario in Traditional approach   (b) Design scenario in REAL approach

Figure 7.2 (a) Factors taken into account in the traditional code design approaches and (b) Factors taken into account in the REAL approach.

CDMA codes, which should offer an inherent immunity against major impairing factors, such as MAI and MI, without the help of external subsystems, such as multiuser detection, power control, and so on.

Here we would like to introduce a new method, called the *Real Environment Adapted Linearization* (REAL) approach, to design these desirable CDMA code sets. The word *linearization* reflects the fact that it converts a nonlinear problem into a linear one by making use of some carefully selected *seed code*, as explained in the following text. Figure 7.2 illustrates the differences between traditional code design approach and the REAL approach. It is seen from the figure that the new approach considers many real working conditions that the traditional methods never did. To formulate a design problem as general as possible, we start with a generic complementary code set with its set size, flock size, and element code length being K, M and N, respectively. It is noted that any unitary code only becomes a special case with $M = 1$. Here, we do not impose any constraints on the values each chip may take, which can be either binary, real, or complex. If we let all chips in a code set be independent variables, we will have KMN unknown variables to solve in total in order to obtain a wanted new code set.

In the REAL approach we should consider all the impairing factors shown in Figure 7.2(b). Its design procedure is explained below for two different scenarios: (1) DS spreading CDMA and (2) OS spreading CDMA.

7.3.5 REAL Approach for DS-CDMA

Let us again consider a generic CDMA system, where K mobiles are communicating with a BS both through asynchronous up link and synchronous down link channels under MI. First, assume only one code $\mathbf{x}$ is present in the system, where $\mathbf{x} = \{\mathbf{x}_1, \mathbf{x}_2, \ldots, \mathbf{x}_M\}$, $\mathbf{x}_m = \{x_{m,1}, x_{m,2}, \ldots, x_{m,N}\}$ and $1 \leq m \leq M$. Observing at a correlator tuned to $\mathbf{x}$, we can write down all the possible output signal patterns, which represent either EPACFs or OPACFs with all the possibly delayed versions due to different multipath returns, as shown in the lower two shaded portions of Figure 7.3. The first (transparent) row in Figure 7.3 represents the local correlator bank of the receiver. It is to be noted that different element codes should be sent via different carriers, and combined only after individual matched filtering at a receiver. Obviously, the EPACFs (light shaded rows, caused by the same signed consecutive bits) are the results from down link synchronous transmissions; while the OPACFs (dark shaded rows, caused by different signed consecutive bits) are from up link asynchronous transmissions. In the REAL approach, both the EPACFs and the OPACFs should be taken into account at the

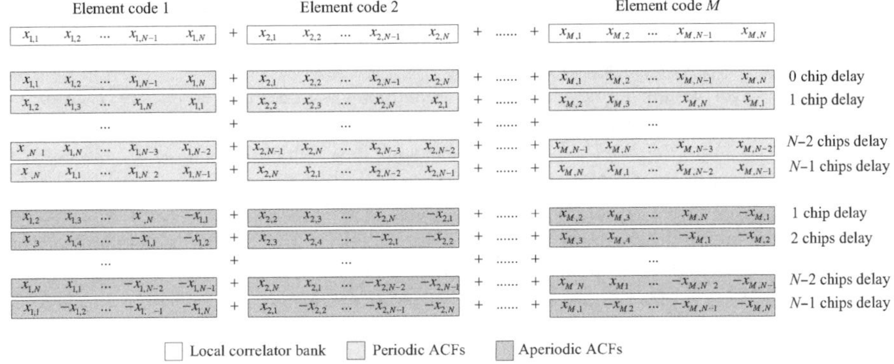

Figure 7.3 All possible patterns of EPACFs and OPACFs of a generic complementary code. The set size, flock size, and element code length are K, M, and N, respectively. All possible multipath returns, asynchronous transmission mode, and random bit patterns have been taken into account.

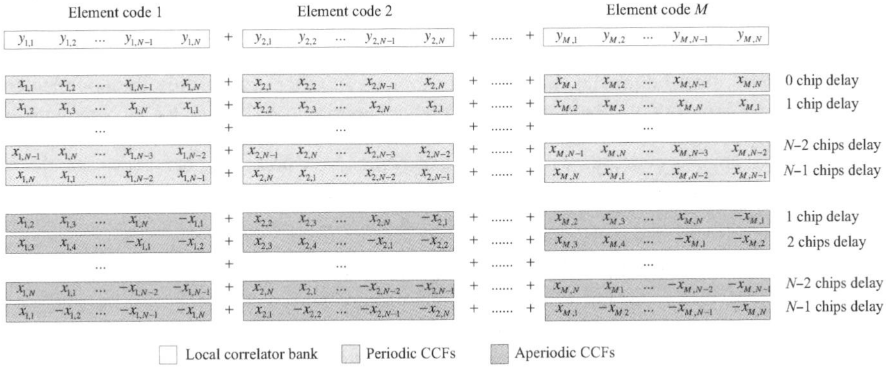

Figure 7.4 All possible patterns of EPCCFs and OPCCFs of a generic complementary code. The set size, flock size, and element code length are K, M, and N, respectively. All possible multipath returns, asynchronous transmission mode, and random bit patterns have been taken into account.

same time to ensure that they should work properly under any working condition. Thus, letting the correlation function between the local correlator and the first row indicated by "0 chip delay" be NM (which in fact yields the autocorrelation peak), and the correlation functions between the local correlator and all the rest of the rows be zero, we obtain a nonlinear equation set with totally $2N - 1$ nonlinear equations and NM unknown chips or variables to determine.

Now we introduce the second code $\mathbf{y}$ to the system. Thus, observing at a receiver tuned to $\mathbf{y}$, we can establish $2N - 1$ nonlinear equations from Figure 7.4 due to the incoming code $\mathbf{x}$, where all CCFs (both EPCCFs and OPCCFs) caused by all possible multipath returns and transmission modes, either up link or down link, have been considered. Of course, code $\mathbf{y}$ itself should also satisfy the conditions for an ideal ACF (for both EPACFs and OPACFs), similar to what is shown in Figure 7.3, but with $\mathbf{x}$ being replaced by $\mathbf{y}$. It should be noted that the nonlinear equations established by EPACFs and OPACFs in fact specify the conditions for *isotropic MI-free operation*, and those established by

EPCCFs and OPCCFs specify the conditions for *isotropic MAI-free operation*. Thus, the solutions, if available to all these nonlinear equations will yield the code sets that guarantee *isotropic MI-free and MAI-free operation* of a CDMA system.

Similarly, we can establish more nonlinear equations for the third code, say $\mathbf{z}$, and so on. The procedure can be continued until all K codes of interest have been introduced, resulting in $K \left(\lfloor \frac{N}{2} \rfloor + \lfloor \frac{N-1}{2} \rfloor + 1 \right) + (2N - 1) \frac{K(K-1)}{2}$ nonlinear equations, which contain MNK unknown variables for a code set size of K. It is possible that these nonlinear equations might be solvable if the inequality $K \left(\lfloor \frac{N}{2} \rfloor + \lfloor \frac{N-1}{2} \rfloor + 1 \right) + (2N - 1) \frac{K(K-1)}{2} \geq MNK$ is satisfied, which is only a necessary, rather than a sufficient, condition. Obviously, the solutions to a nonlinear equation set are not guaranteed even if the above inequality is held. Fortunately, all the nonlinear equations established with respect to CCFs between $\mathbf{x}$ and $\mathbf{y}$ can in fact be transformed into a homogenous linear equation set if $\mathbf{x}$ is already known. We will take this known first code $\mathbf{x}$ as a *seed code*, which should itself satisfy all the conditions for an MI-free operation or both periodic and aperiodic out-of-phase ACF of the code should be zero. If so, we readily have a linear equation set from Figure 7.4 to solve the second code $\mathbf{y}$ jointly with those equations specifying its own MI-free condition. The solution to these equation sets or the code $\mathbf{y}$ must have already satisfied the MAI-free and MI-free conditions. The same procedure should be repeated until all the codes in the set are determined. It is noted here that an original nonlinear equation set has been converted into a linear one with the help of the seed code. That is the reason that we call the algorithm the REAL approach.

Two important conclusions can be made from the REAL approach (for DS spreading only):

- To ensure an MAI-free and MI-free operation in a DS-CDMA system, the flock size M of the CDMA codes must be greater than one (or $M > 1$). In other words, only the complementary code sets can possibly achieve perfect periodic and aperiodic ACFs and CCFs, simultaneously.

- To generate an OC code set with its set size being K, whose member codes should retain perfect periodic and aperiodic ACFs and CCFs, the flock size M of the codes must not be less than the set size K, or $M \geq K$. Usually, $M = K$ suffices.

From the second conclusion, a CDMA code set obtained from the REAL approach can support as many as $K = M$ users. As the PG of a complementary code set is MN, we have $K = M = PG$ if $N = 1$. It means that an OC code set (with isotropic MI-free and MAI-free properties) so generated can support as many as $K = M = PG$ users (if $N = 1$), which is much larger than any other complementary code reported in the literature [208, 209]. Also, to achieve a cell-wise capacity of $K = M = PG$ users, all of which can operate in an interference-free mode, each user should use M element codes sent via different carriers, resulting in a multi-carrier CDMA system with M subcarriers. To avoid using M coherent oscillators in each transceiver, OFDM is a natural choice to simplify the overall implementation hardware of a complementary code based CDMA.

Table 7.1 summarizes the major steps in the REAL approach to generate an OC code set for an interference-free CDMA system. Tables 7.2 and 7.3 give two examples of the OC code sets obtained using the REAL approach. Figures 2.33 and 2.34, given in Section 2.3.3, show how the ideal PACF and PCCF of two OC codes (their information is given in Table 7.3) are achieved in a *complementary* way. It should be stressed that, although Figures 2.33 and 2.34 only show the PACFs and periodic cross-correlation functions (PCCFs) of the two OC codes, their aperiodic autocorrelation functions (AACFs), and aperiodic cross-correlation functions (ACCFs) will give similar results, implying that the OC code set can provide an *isotropic* (or two-directional) MI-free and MAI-free operation. More OC codes are listed in Appendix A.

Table 7.1 Procedure to generate orthogonal complementary codes in the REAL approach

Step	Operation
1	Specify K, M, and N, where the conditions for $M > 1$ and $M = K$ must be satisfied.
2	Generate a *seed code* $\mathbf{x}$ using conditions for $\mathbf{x}$ to maintain ideal PACFs[a] and AACFs[b].
3	Proceed to search for the second code $\mathbf{y}$ using homogenous linear equation set specifying perfect PCCFs[c] and ACCFs[d] between $\mathbf{x}$ and $\mathbf{y}$. If $M > 1$ and $M = K$, the homogenous linear equation set should give some suitable solutions, which take the form of expressions with some undetermined variables.
4	Valid solutions can be obtained by solving the undetermined variables using the equations specifying ideal PACF and AACF for $\mathbf{y}$ itself.
5	Repeat steps 1–4 above to generate all K codes in the set.

[a]PACFs: Periodic autocorrelation functions.
[b]AACFs: Aperiodic autocorrelation functions.
[c]PCCFs: Periodic cross-correlation functions.
[d]ACCFs: Aperiodic cross-correlation functions.

Table 7.2 The OC code set with $K = 8$, $M = 8$ and $N = 4$. (Note: k is the flock index and m is the element index.)

Flock\Element	$m = 1$	$m = 2$	$m = 3$	$m = 4$	$m = 5$	$m = 6$	$m = 7$	$m = 8$
$k = 1$	$+++-$	$+-++$	$++-+$	$+---$	$+++-$	$+-++$	$--+-$	$-+++$
$k = 2$	$+++-$	$-+--$	$++-+$	$-+++$	$+++-$	$-+--$	$--+-$	$+---$
$k = 3$	$+-++$	$+++-$	$+---$	$++-+$	$+-++$	$+++-$	$-+++$	$--+-$
$k = 4$	$+-++$	$---+$	$+---$	$--+-$	$+-++$	$---+$	$-+++$	$++-+$
$k = 5$	$+++-$	$+-++$	$++-+$	$+---$	$---+$	$-+--$	$++-+$	$+---$
$k = 6$	$+++-$	$-+--$	$++-+$	$-+++$	$---+$	$+-++$	$++-+$	$-+++$
$k = 7$	$+-++$	$+++-$	$+---$	$++-+$	$-+--$	$---+$	$+---$	$++-+$
$k = 8$	$+-++$	$---+$	$+---$	$--+-$	$-+--$	$+++-$	$+---$	$--+-$

Table 7.3 The OC code set with $K = 4$, $M = 4$ and $N = 8$. (Note: k is the flock index and m is the element index.)

Flock\Element	$m = 1$	$m = 2$	$m = 3$	$m = 4$
$k = 1$	$+++-++-+$	$+++---+-$	$+++-++-+$	$---+++-+$
$k = 2$	$+-+++---$	$+-++-+++$	$+-+++---$	$-+--+---$
$k = 3$	$+++-++-+$	$+++---+-$	$---+--+-$	$+++---+-$
$k = 4$	$+-+++---$	$+-++-+++$	$-+---+++$	$+-++-+++$

7.3.6   REAL Approach for OS-CDMA

Following similar procedures, we can use the REAL approach also to generate CDMA codes for the OS spreading scheme. However, the signal of interest now, as considered in Figures 7.3 and 7.4, should be replaced by the composite multileveled signal shown by the shaded parts in Figure 7.1. Therefore, if we take all chips as unknown variables again, we can obtain very similar nonlinear equation sets as those obtained for the DS spreading scheme discussed in the previous subsection. The OS spreading process is only a linear operation and cannot change the highest order (being only two) of the nonlinear equation sets obtained. In this way, we can again establish these nonlinear equations specifying ideal EPACFs and OPACFs, which are needed for *isotropic MI-free operation*, and the nonlinear equations specifying ideal EPCCFs and OPCCFs, which are required to ensure *isotropic MAI-free operation*. Similar steps can be followed to generate the CDMA codes suitable for OS spreading modulation.

The following conclusions can also be made for the CDMA codes suitable for the OS spreading modulation:

- Any unitary code set can offer neither isotropic MAI-free nor isotropic MI-free operation in an OS-CDMA system. Therefore, only complementary codes can be possibly used to implement an OS-CDMA system as long as the number of relative offset chips is less than N, or $1 \leq n \leq N$.

- Any OC code set, obtained from the REAL approach to offer isotropic MI-free and isotropic MAI-free operation in a DS-CDMA system (as discussed in the previous subsection), can always be used in an OS-CDMA system to provide isotropic MAI-free operation.

- Define autocorrelation interference-free-window (ACIFW) as the widest time opening, in which all autocorrelation side lobes for any code are zero, and cross-correlation interference-free-window (CCIFW) as the widest time opening, in which all CCFs for any pair of codes are zero. Thus, OC code sets obtained from the REAL approach for DS spreading always have their ACIFW and CCIFW with the largest possible value, or $\text{ACIFW} = N - 2$ and $\text{CCIFW} = N$, if they are used in a DS spreading system. On the other hand, OC code sets obtained from the REAL approach for DS spreading usually have their $\text{ACIFW} < N - 2$ and $\text{CCIFW} = N$ if they are applied to an OS-CDMA system, implying that they could offer an MAI-free operation in an OS-CDMA system, but not always an MI-free operation in an OS-CDMA system in multipath channels with an arbitrary delay spread. However, if the ACIFW of an OC code set can be made larger than the delay spread of the multipath channel, an MI-free operation can still be made possible.

Figure 7.5 shows the openings of ACIFW and CCIFW for an OC code set working in an OS-CDMA system. It is seen from Figure 7.5 that the ACIFW is equal to $n - 2$ chips if the number of relative offset chips is n, implying that we have to slowdown the transmission rate if we want to empower an OS-CDMA with a greater immunity against MI. This leads to another interesting conclusion with the OS-CDMA system.

The immunity against MI for an OS-CDMA system depends on the highest transmission rate or the SE figure (with a fixed chip width), which represents also the bandwidth efficiency. The higher the transmission rate or the SE figure is, the poorer its immunity against MI. In general, an OS-CDMA system can work in an interference-free mode if the number of relative offset chips n can be made at least two chips larger than the delay spread of a multipath channel.

7.3.7   Implementation and Performance Issues

In the previous section, we discussed issues on how to design isotropic MAI-free and MI-free CDMA codes suitable for next-generation CDMA technology. In the following text, we are going to address the issues on the implementation and performance evaluation of a next-generation CDMA system.

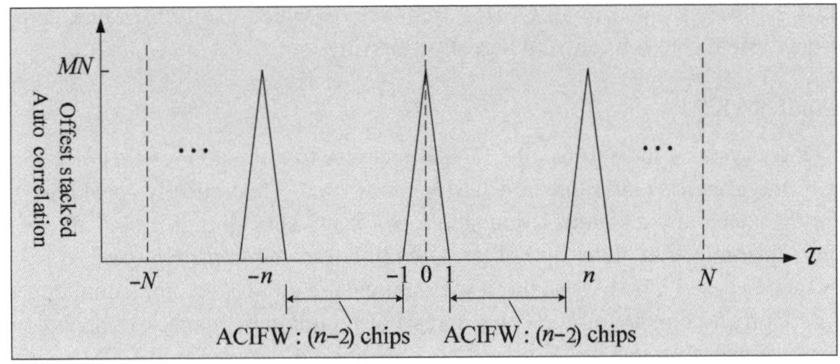

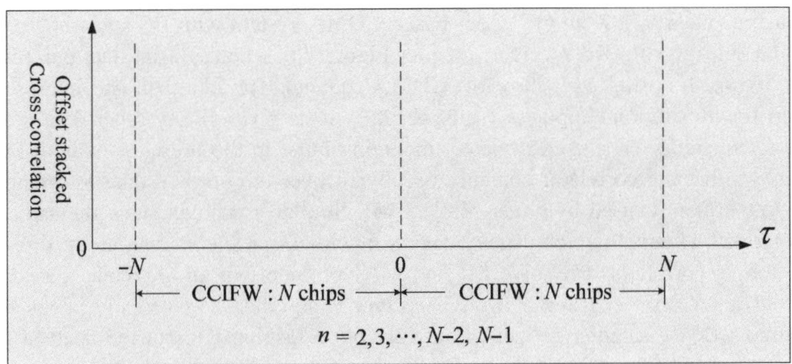

Figure 7.5 The widths of ACIFW and CCIFW in an OCC/OS-CDMA system versus n, where n is the relative offset chips between two consecutive bits, as shown in Figure 7.1. Obviously, if $n = N$, an OCC/OS-CDMA system will become an OCC/DS-CDMA system.

FDM versus TDM for element code division

To implement a CDMA system based on OC codes, an important requirement is to send M element codes, which are assigned to a specific user, via different channels to an intended receiver, where each element code is despread *separately* and their outputs are combined to form a decision variable. The most straightforward way to implement element code division in such a CDMA system is to use the frequency-division multiplex (FDM) approach, in which M element codes are sent through M different carriers f_m, where $1 \leq m \leq M$. In this way, the proposed CDMA system looks just like a MC-CDMA scheme. However, the major difference between an OCC-CDMA/FDM system and a traditional MC-CDMA lies in the fact that the former uses different carriers to convey different information without providing any diversity in frequencies, while the latter usually does.

Obviously, another scheme to implement element code division in an OC code–based CDMA is to send element codes in different time slots or simply via the time-division multiplex (TDM) method. The FDM and TDM implementations of element code division offer distinct system operational advantages. One of the benefits of using FDM is to allow the new CDMA system to work harmonically with the frequency division duplex (FDD) operation mode used in most mobile cellular standards, such as WCDMA, CDMA2000, and so on. On the other hand, the TDM implementation can fit the TDD mode naturally, but the TDD operation is suited only for covering a relatively small cell size.

Another salient feature of the FDM option is the reduction of the overall hardware complexity with the help of the OFDM technology. Like any MC-CDMA system, an OC code–based CDMA/FDM system can also be implemented in an OFDM architecture, which can transform a complicated MC RF transmission system into a base band signal processing unit.

With or without RAKE?

All current CDMA systems have to use the RAKE receiver to mitigate the otherwise formidable MI, which is due to the imperfect ACF of the CDMA codes used. Theoretically speaking, the OC codes virtually do not produce any autocorrelation side lobes if DS spreading is assumed, and thus an MI-free operation is guaranteed in either up link or down link transmissions. However, if OS spreading is used in an OC code–based CDMA system, it may not ensure an MI-free operation, depending on the values of relative offset chips n and the delay spread of a multipath channel, as shown in Figure 7.5. Another alternative way to mitigate MI in an OS spreading OC code–based CDMA system is to use a recursive filter with the help of adaptive channel estimation and the pilot signal, as suggested in [210]. To illustrate clearly how an OC code–based CDMA system with DS spreading can overcome MI without the help of a RAKE receiver, we plot Figure 7.6, where a 3-bit data burst is sent into a two-user and two-path up link asynchronous CDMA channel. The interpath and interuser delays are only one chip for illustration simplicity (same result will be given if any other delays are applied). A simple correlator receiver is used to detect incoming burst in the presence of both MAI and MI. The figure shows that the correlator can successfully recover the original data information $(+ - +)$ without any impairment caused by either MAI or MI. Similarly, we can show the same result for a down link multipath channel, which usually causes much less problems than an up link channel.

Thus, a simple correlator can solve MI and MAI problems in an OC code–based DS-CDMA system. The advantage of not using a RAKE receiver is significant as it really paves a way for an OC code–based CDMA receiver to work in a truly blind fashion without the need of any *a priori* channel information. On the other hand, a RAKE has to acquire virtually all channel information, such as delays and amplitudes of all multipath returns, for its maximal-ratio-combining (MRC) operation, whose impact on implementation complexity in a mobile handset can never be underestimated. Obviously, in an OC code–based DS-CDMA, we can also use a RAKE receiver to further boost up the signal-to-interference ratio. The use of RAKE here can provide a much higher multipath-diversity gain than achievable in a conventional CDMA system due to the fact that the output signal from each finger now contains only the useful autocorrelation peak, as shown explicitly in Figure 7.6. On the other hand, the output from a finger in a conventional CDMA RAKE receiver contains both useful and unwanted signals caused by nontrivial autocorrelation side lobes.

Furthermore, if a RAKE has to be used in an OC code–based DS-CDMA receiver, an equal gain combining (EGC) is preferred to yield a satisfactory detection efficiency. The advantage of using EGC rather than MRC in a RAKE is to save a complicated multipath amplitude estimation unit.

Multiuser detection

As shown in Figure 7.6, the isotropic MAI-free property in an OC code–based CDMA based on either DS spreading or OS spreading makes it unnecessary to use multiuser detection to decorrelate transmission signals from different users, due to the fact that the transmissions from different users in the OC code–based CDMA system are already predecorrelated at the transmitter side because of its MAI-free signaling structure.

Is power control necessary?

Owing to the isotropic MAI-free and MI-free properties, near–far effect will cause no harm to the signal detection process at a correlator in an OC code–based DS-CDMA system, as long as bit-synchronization can be achieved prior to the data detection process. In other words, the OC code–based DS-CDMA is a system with an excellent near–far resistance, as shown in Figure 7.7. Therefore, complicated open-loop and closed-loop power control is no longer a necessity. More

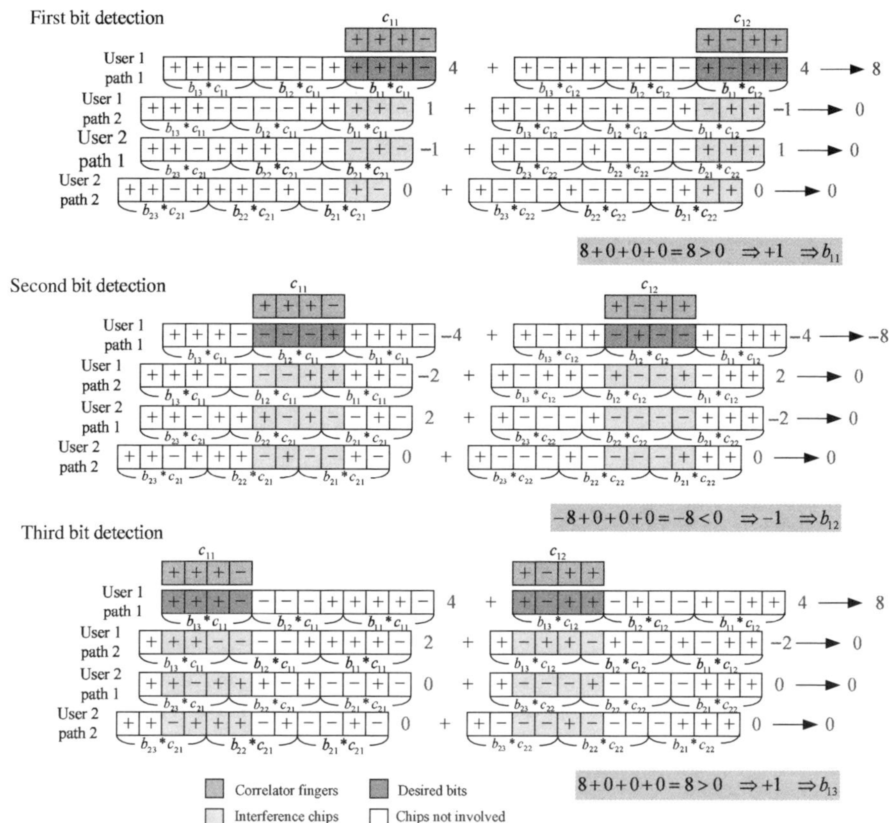

Figure 7.6 Illustration of MI-free and MAI-free operation for a two-user OC code–based DS-CDMA system in asynchronous up link channels, where a two-ray multipath channel is considered with both interpath delay and interuser delay being one chip for illustration simplicity. The parameters of the OC codes used in this figure are $K = M = 2$ and $N = 4$. The 3-bit data information sent is $\{+ - +\}$.

precisely, the power control in an OC code–based DS-CDMA system is used merely to reduce unnecessary power emission at the terminals, whose requirements on power control response time and accuracy can be made much more relaxed than necessary in a conventional DS-CDMA system. For an OC code–based OS-CDMA system, a similar conclusion can be drawn with respect to the power control requirements, due to its ideal MAI-free property. However, the OS spreading scheme is not always MI-free (dependent on the value of n) and thus the requirements on the power control can be a bit higher than that in an OC code–based DS-CDMA system.

Suitability for bursty traffic

As an important requirement for a CDMA technology suitable for future Gigabit all-IP wireless communications, we should pay sufficient attention to the detection efficiency at the edges of a short packet or burst. In this sense, an OC code–based CDMA scheme is particularly well suited, as shown in the upper parts of Figure 7.6, which illustrates the signal detection process for the first (or the rightmost) bit of a packet. In detecting these bits at the edges of a packet, partial ACFs and partial

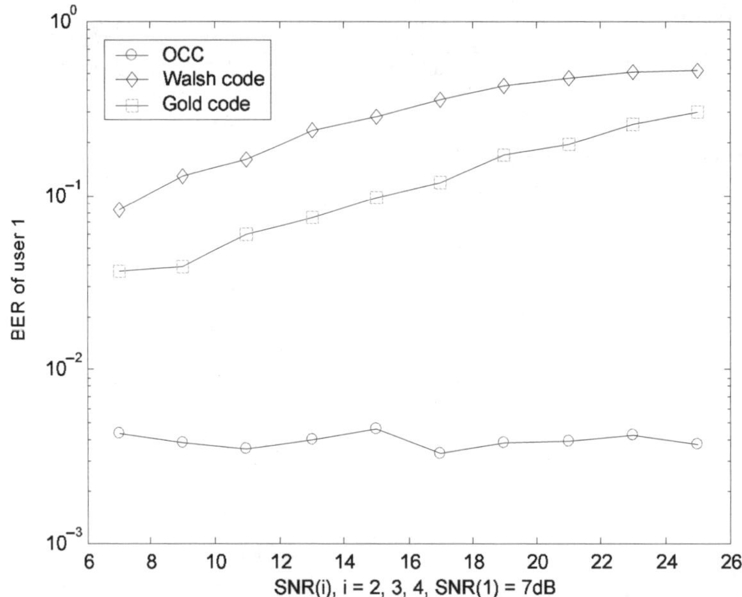

Figure 7.7 Illustration of inherent near–far resistance for an OC code–based CDMA system, where the performance of a conventional DS-CDMA system based on Gold and Walsh-Hadamard codes under the near–far effect is also plotted. Four users are present in the system and the signal of interest is from user 1. An up link asynchronous multipath channel with its path gain being [0.8557, 0.5134, 0.0642] is considered. Interpath and interuser delays are one and three chips, respectively. The parameters of the OC code are $K = M = 4$, $N = 4$, and thus the PG is $MN = 16$. The length of the Gold and Walsh-Hadamard codes is 15 and 16, respectively.

CCFs of the codes will become extremely important. It is seen from the figure that the OC code–based DS-CDMA system yields zero partial CCFs, which ensure an ideal performance for signal detection even at the edges of a frame or packet. This observation is significant owing to the domination of burst traffic in future wireless systems.

Easy rate-matching

One of the most celebrated merits of an OC code–based OS-CDMA system is its continuously adjustable transmission rates without using a complex rate-matching algorithm that is required in the WCDMA standard due to the treelike code generation process of the OVSF codes. The unique OS spreading technique adopted by the OC code–based OS-CDMA system can easily slow down the transmission rate by simply shifting more than one chip (at most N chips) between two consecutive OS bits. If N chips are shifted between two consecutive bits, the OS spreading reduces to a DS spreading, yielding the lowest data rate. On the other hand, the highest data rate can be achieved if only one chip is shifted between two neighboring OS bits. In doing so, the highest spreading efficiency equal to one can be obtained, implying that every chip can carry one bit information. Figure 7.1 shows how the OS spreading can offer an agile rate-matching algorithm for various data transmission rates.

The OS spreading also helps support asymmetrical transmissions in the up link and down link channels, pertaining to thriving Internet applications (such as web browsing, etc.). The data rates in up link and down link can be made truly scalable, such that the *rate on demand* is achieved by simply adjusting relative offset chips between two neighboring spreading modulated bits. It should be stressed that such an agile rate-matching property for an OC code–based OS-CDMA system can never be made possible by using other traditional unitary codes, such as Gold codes, OVSF codes, Walsh-Hadamard codes, and so on.

Yet another advantage of the rate change scheme implemented in OC code–based OS-CDMA is that the same PG will apply to all different transmission rates. However, the rate-matching used in the UMTS-UTRA and WCDMA standards (as discussed in Chapter 3.2) is PG-dependent; the slower the transmission rate is, the higher the PG or spreading factor, if transmission bandwidth is kept constant. To maintain an even detection efficiency for different rates at a receiver, the transmitter has to adjust its transmitting power for different rate services.

The issues to be explored further

As mentioned earlier, a CDMA system based on OC codes needs different channels to send different element codes assigned to the same user. In general, either frequency, or time can be used to separate the transmissions of element codes, resulting in either an FDM or TDM scheme. In an OC code–based CDMA system based on the FDM scheme, we should be very careful as to how to minimize the impact of frequency selectivity or variable propagation losses in different carriers that send different element codes. A receiver may fail to reconstruct ideal correlation properties if no other effective measure is taken to mitigate the frequency dispersion problem. In [210], using a pilot-aided approach to overcome the problem with the help of adaptive channel estimation techniques was suggested, although other techniques can also be used.

On the other hand, if TDM is used for element code separation, an OC code–based CDMA system may operate sensitively to time-selective fading as long as channel coherent time becomes comparable to a symbol duration, which spans at least over a length of M element codes. The use of a high data rate may help alleviate the problem, but at a price of worsened inter-element-code-interference, which may impair the performance dramatically. Therefore, the TDM scheme is not preferred to be used in an OC code–based CDMA system unless some more cost-effective measures can be found.

7.4 Multicarrier CDMA Techniques

As discussed in the previous section, a multi-carrier scheme is used to implement an OC code–based CDMA system, if the FDM technique is used to send different element codes.

There is big difference between an OC code–based CDMA system implemented by a multi-carrier scheme and a general multi-carrier CDMA system, which is the focal point in this section. An OC code–based CDMA system uses a multi-carrier architecture to send different element codes of a complementary code assigned to a specific user. Thus, different frequency channels convey totally different information, each piece of which is indispensable to the successful reconstruction of ACFs or CCFs at a receiver. In this way, there is no *frequency diversity* in an OC code–based CDMA system. On the other hand, a conventional multi-carrier CDMA system uses different carrier frequencies to multiplex the same input wideband data stream, such that each frequency carries only a narrowband bit stream, whose transmission rate is only $\frac{1}{M}$-th of that of the input wideband stream if M subcarriers are used. Each narrowband subcarrier channel is less vulnerable to frequency-selective fading if the coherent bandwidth of the channel is wider than the bandwidth occupied by each subcarrier channel. Even if some subcarrier channels unfortunately fail into the nulls of deep frequency fades, the joint use of interleaving and error-correction coding can effectively recover the information corrupted by the deep fade nulls in those subcarrier channels.

Therefore, multicarrier modulation (MCM) is the principle of transmitting data by dividing the input wideband stream into several parallel narrowband bit streams, each of which has a much lower bit rate, and using these substreams to modulate different carriers. On the basis of this principle, several derivative MC-CDMA schemes have been introduced and they have been studied extensively in [571–584]. On the basis of the division on duplication or multiplex schemes, there are two MC-CDMA forms, that is, (1) duplicated MC-CDMA (2) multiplexed MC-CDMA; on the basis of which domain the DS spreading takes place, there are also two alternative schemes, which are (1) time-spreading MC-CDMA, and (2) frequency-spreading MC-CDMA. Therefore, by mixing all of them in various combinations, we will have four different MC-CDMA systems as follows:

- Duplicated time-spreading MC-CDMA;

- Duplicated frequency-spreading MC-CDMA;

- Multiplexed time-spreading MC-CDMA;

- Multiplexed frequency-spreading MC-CDMA.

The term *duplicated* or *multiplexed* means that the data streams in different subcarrier branches in a transmitter are the "duplicated" or the "multiplexed" version of the input data stream, respectively. We give a brief introduction to them in this section.

7.4.1 Duplicated Time-Spreading MC-CDMA

A duplicated time-spreading MC-CDMA scheme, in which M subcarriers are used, is illustrated in Figure 7.8. The input data stream is duplicated in M subcarrier branches, which undergo spreading modulation using the same signature code assigned to the k-th user. Only one transmitter for the k-th user is shown in the figure for simplicity; while the signal of interest to the receiver is the k-th

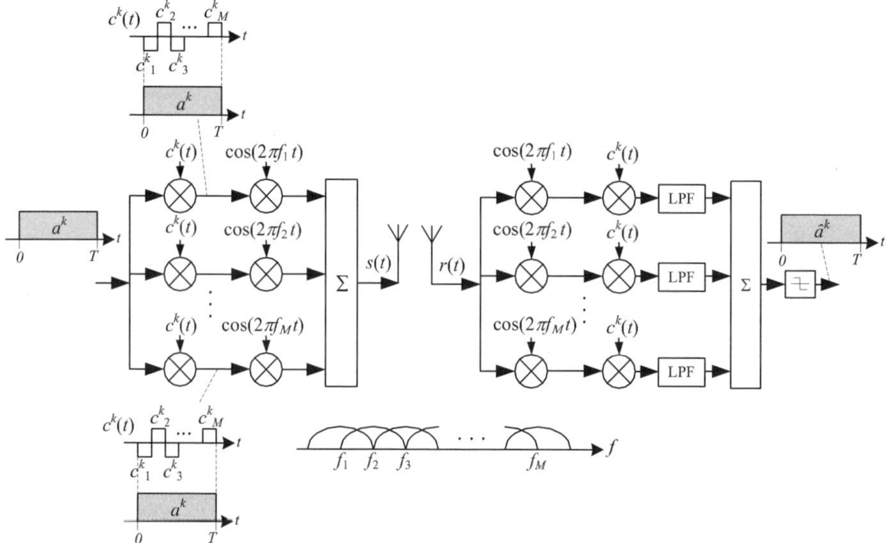

Figure 7.8 Transmitter and receiver of duplicated time-spreading MC-CDMA scheme, where M subcarriers are used.

transmission. Time-domain spreading takes place in this scheme, and thus the input data stream will be converted into a wideband subcarrier-modulated signal before it is sent into the channel.

Obviously, this scheme provides M duplicated SS signals in all M subcarriers. In other words, the signals conveyed in M different subcarrier channels are exactly the same replicas of the original input data. Therefore, there is a great deal of redundancy in the transmission, which can be explored to offer sufficient frequency diversity for signal detection. In this way, the scheme is very robust in terms of its immunity against frequency selectivity of the channel. Of course, the price paid for this strong immunity is the consumption of a great amount of bandwidth resource in sending all M identical data streams in different subcarriers. Thus, the bandwidth efficiency of the duplicated time-spreading MC-CDMA scheme is relatively low. For this reason, this scheme is not a popular option of MC-CDMA systems.

7.4.2 Duplicated Frequency-Spreading MC-CDMA

Figure 7.9 depicts a conceptual block diagram of a duplicated frequency-spreading MC-CDMA, where M subcarriers, or $\{f_1, f_2, \ldots, f_M\}$, are used. Only the k-th transmitter is shown in the figure and the receiver is dedicated for the k-th user's transmission. The k-th user is assigned a signature code $\{c_1^k, c_2^k, \ldots, c_M^k\}$.

It is noted that DS spreading takes place in the frequency domain, and thus a matched filtering is applied to the receiver at the outputs from all M low-pass filters to yield a decision variable. All subcarriers are orthogonal with each other, with each overlapped with its neighboring subchannels by half, to improve the overall bandwidth efficiency.

The multi-carrier modulated signals are not spread in the time domain in this scheme. The PG of this system will be achieved in the frequency domain, where MAI will be rejected only at the last stage of the receiver, that is, the frequency-domain matched filter. In this duplicated frequency-spreading MC-CDMA scheme, the PG value is exactly equal to the number of subcarriers or M, which becomes a very important system parameter of the scheme.

In the duplicated frequency-spreading MC-CDMA scheme, all subcarriers convey the same data, but are spread modulated by different chips. Therefore, this scheme provides redundancy in the frequency domain, but it will not help improve the system's robustness against the frequency-selective fading caused by the multipath propagation effect. The reason is because spreading over the different

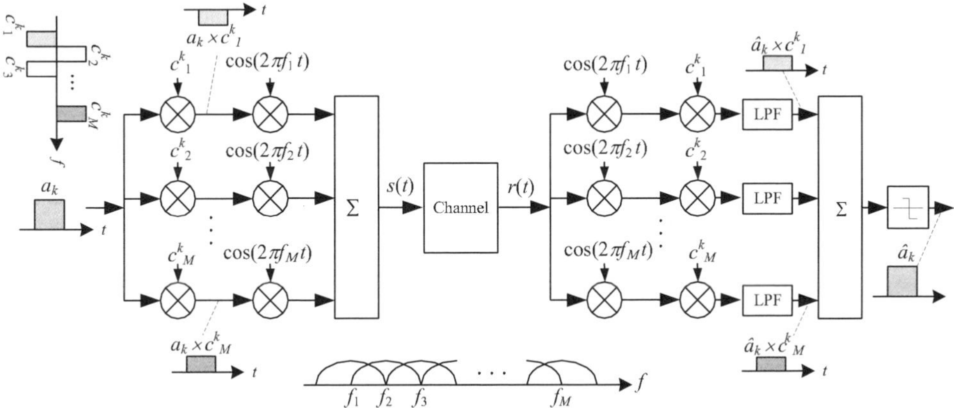

Figure 7.9 Transmitter and receiver of the duplicated frequency-spreading MC-CDMA scheme, where M subcarriers are used.

subcarrier frequencies, $\{f_1, f_2, \ldots, f_M\}$, requires that each subcarrier channel should give the identical gain; otherwise it will cause a serious problem in the reconstruction of ideal ACFs and CCFs at the *matched filter* in a receiver, thus introducing serious MAI.

Figure 7.9 does not include the interleaving and error-correction coding blocks for simplicity of illustration. However, a practical duplicated frequency-spreading MC-CDMA system should always work with them to achieve a better performance against MI.

7.4.3 Multiplexed Time-Spreading MC-CDMA

Figure 7.10 illustrates a conceptual block diagram of a multiplexed time-spreading MC-CDMA system, where the k-th transmitter and a receiver tuned to the k-th transmission are shown. It is seen from the figure that the input data stream $\{a_1^k, a_2^k, \ldots, a_M^k\}$ with a bit rate of $\frac{1}{T}$ is first multiplexed into M substreams, each of which has a rate of $\frac{1}{MT}$ and will be spread modulated by the same signature code $c^k(t)$ assigned to the k-th user, followed by carrier modulation using subcarrier f_i, where $i = 1, 2, \ldots, M$. The spectra of all subcarriers overlap with one another with a spectral offset of $\frac{1}{2MT}$, thus forming an orthogonal multi-carrier CDMA signaling.

The received signal in this MC-CDMA scheme will be fed into a receiver. First, it undergoes subcarrier demodulation using all M different subcarriers, followed by spreading demodulation using the signature code $c^k(t)$, which is assigned to the k-th transmitter. The use of this signature code in the receiver allows CDMA in this MC-CDMA system. The signal in each branch of the MC-CDMA receiver will go through low-pass filtering, the decision device, and then the de-multiplexing unit to yield a recovered wideband data stream, $\{\hat{a}_1^k, \hat{a}_2^k, \ldots, \hat{a}_M^k\}$.

In the multiplexed time-spreading MC-CDMA scheme, each subcarrier branch carries different information due to the use of the multiplexing unit or the *serial-to-parallel* unit at the transmitter. Thus, no redundance exists in different subcarriers. However, each subcarrier will convey only a narrowband signal and it will only experience a flat fading, instead of a frequency-selective fading. With the help of interleaving cum error-correction algorithms, this MC-CDMA scheme can work satisfactorily in a channel with MI. In addition, a possible multipath diversity can be achieved if a

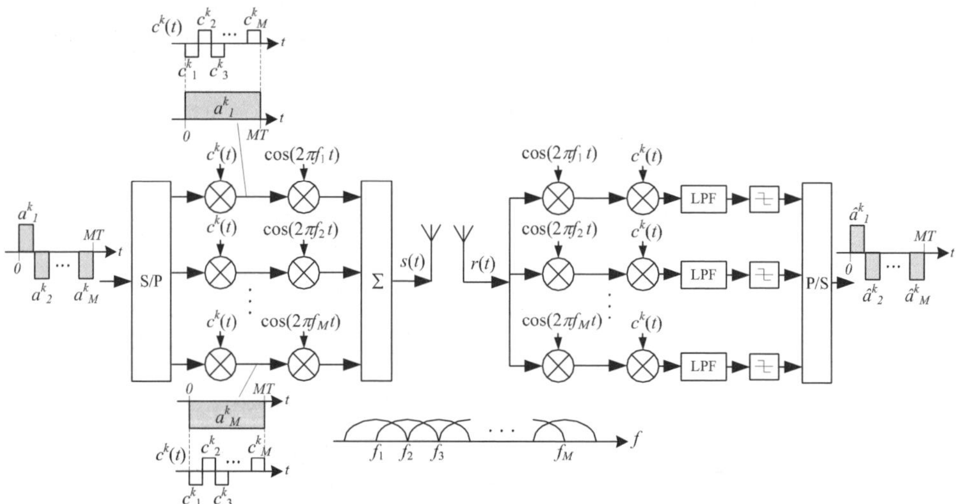

Figure 7.10 Transmitter and receiver of the multiplexed time-spreading MC-CDMA scheme, where M subcarriers are used.

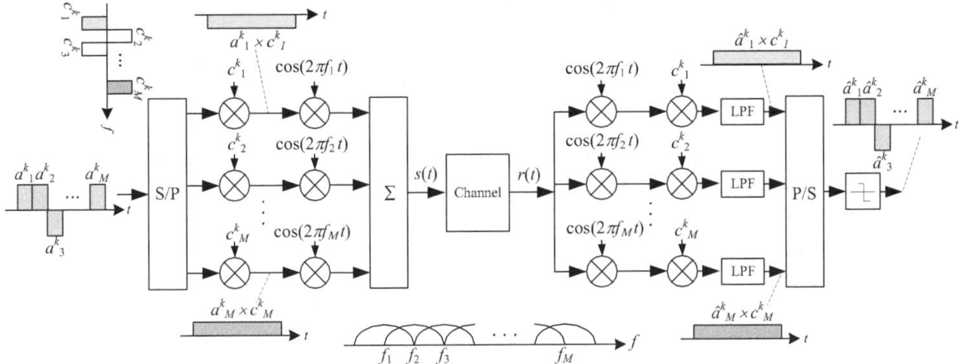

Figure 7.11 Transmitter and receiver of multiplexed frequency-spreading MC-CDMA scheme, where M subcarriers are used.

RAKE receiver can be applied to replace a conventional matched filter, as used in Figure 7.10, that is, the despreading unit and low-pass filter (LPF) in each subcarrier branch in the receiver.

7.4.4 Multiplexed Frequency-Spreading MC-CDMA

Finally, we would like to introduce the last MC-CDMA scheme called *Multiplexed frequency-spreading MC-CDMA*, whose block diagram is shown in Figure 7.11.

In this MC-CDMA scheme, the wideband input data stream is multiplexed into M narrowband substreams, each of which should be carrier modulated by M different subcarriers $\{f_1, f_2, \ldots, f_M\}$. Frequency-domain spreading takes place across all subcarriers in this scheme. Therefore, MAI suppression is ensured using some spreading codes with satisfactory CCF between any pair of them. Thus, there is no redundancy in the frequency domain and different subcarriers deliver totally different information. Also, because of the use of frequency-domain spreading, the transmitting signal will not be spread in time. Obviously, this scheme will offer the highest bandwidth efficiency of all the four MC-CDMA schemes discussed in this section. However, this scheme does not provide any frequency diversity and time diversity, and its performance cannot be comparable to any of the other MC-CDMA schemes.

Having introduced the four different MC-CDMA schemes, namely, duplicated time-spreading MC-CDMA, duplicated frequency-spreading MC-CDMA, multiplexed time-spreading MC-CDMA, and multiplexed frequency-spreading MC-CDMA, we can compare them in terms of different performance parameters, as given in Table 7.4.

7.5 OFDM Techniques

OFDM technology was developed from multi-carrier CDMA techniques. Although bulk research on OFDM has not been seen until the late 1990s, a primitive concept of OFDM first appeared in the open literature in the mid-1960s, and many discussions on OFDM can be found from the discussions given in [585–593, 814].

The major difference between OFDM and traditional FDM lies in the fact that OFDM uses multiple carriers that are orthogonal to one another, such that the frequency space between two neighboring subcarriers is equal to Δf, where $\Delta f = \frac{1}{T_s}$ and T_s is the symbol duration. Two of the most important

Table 7.4   Performance   comparison   of   four   different   MC-CDMA schemes

	FD	TD	BE	DE	IC
DTS-MC-CDMA	Good	Good	Poor	Excellent	Normal
DFS-MC-CDMA	Good	Poor	Poor	Normal	Normal
MTS-MC-CDMA	Normal	Good	Good	Good	High
MFS-MC-CDMA	Poor	Poor	Excellent	Poor	High

FD: Frequency diversity
TD: Time diversity
BE: Bandwidth efficiency
DE: Detection efficiency
IC: Implementation complexity
DTS-MC-CDMA: Duplicated time-spreading MC-CDMA
DFS-MC-CDMA: Duplicated frequency-spreading MC-CDMA
MTS-MC-CDMA: Multiplexed time-spreading MC-CDMA
MFS-MC-CDMA: Multiplexed frequency-spreading MC-CDMA

modules used in any OFDM system are the inverse fast fourier transform (IFFT) and fast fourier transform (FFT) algorithms, the former used in transmitters and the latter used in receivers. We will show later in this section that IFFT and FFT can be used to replace multi-carrier modulation/demodulation units in a multi-carrier system.

When compared with FDMA, TDMA, and CDMA, OFDM represents a different system design approach. It can be thought of as a combination of modulation and multiple access techniques, which divide a communication channel in such a way that multiple users can share it without much mutual interference. FDMA, TDMA, and CDMA schemes segment the channel according to times, frequencies, and codes, respectively. OFDM also segments the channel according to the frequency/tone. In more specific words, it divides the spectrum into a number of equally spaced tones and carries a portion of a user's information on each tone. A tone can be thought of as a frequency, much in the same way that each key on a piano represents a unique frequency. OFDM can be viewed as a form of FDM. However, OFDM has an important special property that each tone is orthogonal with every other tone. FDM typically requires some frequency guard bands between the frequencies so that they do not interfere with each other. OFDM allows the spectrum of each tone to overlap because they are orthogonal, and thus they do not interfere with each other. By allowing the tones to overlap, the overall amount of spectrum required is reduced, as shown in Figure 7.12.

OFDM is a unique modulation technique that enables user data to be modulated onto the tones. The information is modulated onto a tone by adjusting the tone's phase, amplitude, or both. In the most basic form, a tone may be present or disabled to indicate a one or zero bit of information. Either phase-shift keying (PSK) or QAM is typically employed. An OFDM system takes a data stream and splits it into M parallel data streams, each at a rate $1/M$ of the original rate. Each stream is then mapped to a tone at a unique frequency and combined together using the IFFT to yield the time-domain waveform to be transmitted. For example, if a 100-tone system were used, a single data stream with a rate of 1 Mbps would be converted into 100 streams of 10 kbps. By creating slower parallel data streams, the bandwidth of the modulation symbol is effectively decreased by a factor of 100, or equivalently, the duration of the modulation symbol is increased by a factor of 100. Proper selection of system parameters, such as the number of tones and tone spacing, can greatly reduce, or even eliminate, ISI, because typical multipath delay spread represents a much smaller proportion of the lengthened symbol time. Viewed in another way, the acceptable coherence bandwidth of the channel can be much smaller, because the symbol bandwidth has been reduced.

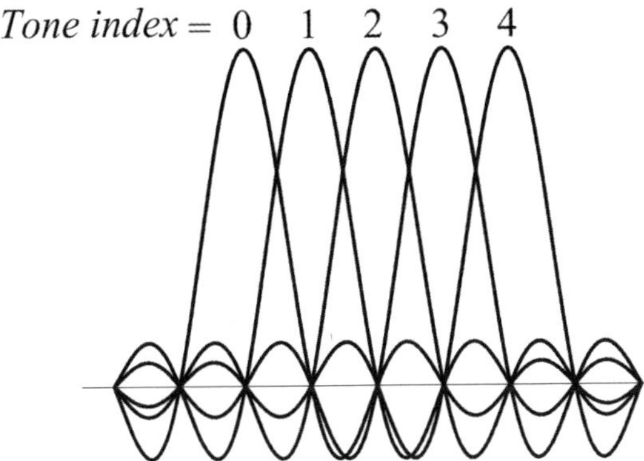

Figure 7.12 Orthogonality in overlapped OFDM multitones in the frequency domain, where the peak of each tone corresponds to a zero level, or null, of every other tone.

OFDM can also be considered as a multiple access technique, because an individual tone or groups of tones can be assigned to different users. Multiple users share a given bandwidth in this manner, yielding the system called *OFDMA*. Each user can be assigned a predetermined number of tones, or alternatively a user can be assigned a variable number of tones based on the amount of information that they have to send. The assignments are controlled by the MAC layer, which schedules the resource assignments based on user demand.

OFDM can be combined with *frequency hopping* to create an SS system, realizing the benefits of frequency diversity and interference averaging previously used in CDMA, as discussed in Section 2.2.2. In a FHSS system, each user's set of tones change after each time period (usually corresponding to a modulation symbol). By switching frequencies after each symbol time, the losses due to frequency-selective fading are minimized.

7.5.1 From Multicarrier System to OFDM

Let us consider a simple multi-carrier system with a transmitter and a receiver, as shown in Figure 7.13, where M subcarriers $\{f_1, f_2, \ldots, f_M\}$ are shown. Each subcarrier signal can be written as

$$s_m(t) = x_m(t)e^{j2\pi f_m t} = x_m(t)e^{j\omega_m t}, \qquad (m = 0, 2, \ldots, M-1) \tag{7.1}$$

where $x_m(t)$ is a complex data symbol in the m-th subcarrier branch. If the input data stream is first cut into frames, each of which has a length of M symbols, we will have the output signal from this simple multi-carrier transmitter as follows:

$$s(t) = \frac{1}{M} \sum_{m=0}^{M-1} x_m(t)e^{j\omega_m t}, \tag{7.2}$$

where $\omega_m = 2\pi f_m$, $\omega_m = \omega_0 + m\Delta\omega(m = 0, 2, \ldots, M-1)$ and $\frac{1}{M}$ is the normalized factor. Of course, we can also let $x_m(t) = a_m(t)e^{j\phi_m(t)}$, such that

$$s(t) = \frac{1}{M} \sum_{m=0}^{M-1} a_m(t)e^{j[\omega_m t + \phi_m(t)]}, \tag{7.3}$$

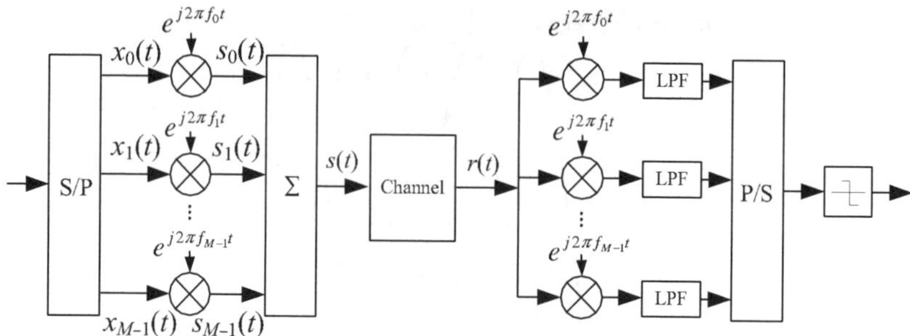

Figure 7.13 A simple multi-carrier system consisting of a transmitter and a receiver, where M complex subcarriers are present.

where $a_m(t)$, $\phi_m(t)$, and ω_m are amplitude, initial phase shift, and carrier frequency of the m-th subcarrier, respectively.

Now we sample the above equation in the time domain with a sampling frequency of $\frac{1}{\Delta t}$ to have $a_m(t) = a_m$, $\phi_m(t) = \phi_m$ and the output multi-carrier signal as

$$s(n\Delta t) = \frac{1}{M} \sum_{m=0}^{M-1} a_m e^{j\phi_m} e^{j2\pi(m\Delta f)(n\Delta t)}, \qquad (n = 1, 2, \ldots, M-1) \tag{7.4}$$

where we have assumed the carrier frequency $\omega_0 = 0$, if only a baseband equivalent model is considered here; otherwise a complex carrier term $e^{j\omega_0 t}$ should be multiplied to both the sides of (7.4). On the other hand, the M-point inverse discrete fourier transform (IDFT) between a time domain discrete signal $f(n\Delta t)$ and its frequency domain discrete signal representation $F(m\Delta f)$ is defined by

$$f(n\Delta t) = \frac{1}{M} \sum_{m=0}^{M-1} F(m\Delta f) e^{j\frac{2\pi nm}{M}}, \qquad (n = 1, 2, \ldots, M-1) \tag{7.5}$$

which will be equal to (7.4) if and only if $\Delta f = \frac{1}{M\Delta t}$. Under this condition, Δf is the frequency spacing between two consecutive subcarriers, $M\Delta t$ is the frame length, and Δt is the symbol duration. We also have $s(n\Delta t) = f(n\Delta t)$ (which is a time-domain signal) and $a_m e^{j\phi_m} = F(m\Delta f)$ (which is a frequency-domain signal). Also note that the IFFT algorithm is only a fast computation method for the IDFT. Therefore, a multi-carrier transmitter can be effectively implemented by an IFFT unit cascaded by a complex carrier modulator (if $\omega_0 \neq 0$), as shown in Figure 7.14. A reverse process can be used to show that a FFT algorithm can be used at an OFDM receiver to implement a multi-carrier receiver, as shown in Figure 7.13.

A MC-CDMA system can thus be implemented by an OFDM architecture. Figure 7.14 shows an OFDM system, which functions as a multiplexed time-spreading MC-CDMA system shown in Figure 7.10. Similarly, we can use some other suitable OFDM structures to implement other MC-CDMA schemes, which have been discussed in the previous section.

One of the most attractive features of an OFDM system is that it can use more flexible and powerful baseband digital technologies to replace otherwise rigid and complicated analog multiple carrier modulation units in a multi-carrier system. FFT and IFFT can also be implemented readily by software, and thus a digital signal processor in an OFDM system is enough to replace all those analog circuitries in a multi-carrier scheme. The complexity of implementing these carrier oscillators is formidable, especially if a large number of subcarriers is needed.

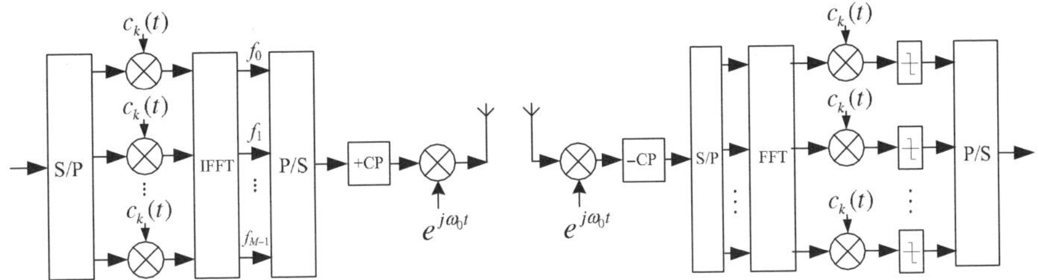

Figure 7.14 An OFDM implementation for a multiplexed time-spreading MC-CDMA system (as shown in Figure 7.10), where $c_k(t)$ is the signature code assigned to the k-th user and M subcarriers are used.

However, nothing is free. The simplicity of the baseband digital processing algorithms in an OFDM scheme also produces some problems pertaining to their unique characteristics, such as *peak-to-average-power-ratio*, and so on, which will be discussed in the following subsections.

7.5.2 Cyclic Prefix

The sinusoidal waveforms making up the tones in OFDM have the very special property of being the only Eigen-functions of a linear channel. This special property prevents adjacent tones in OFDM systems from interfering with one another, in much the same way as the human ear can clearly distinguish between each of the tones created by the adjacent keys of a piano. This property, and the incorporation of a small amount of *guard time* to each symbol, enables the orthogonality between tones to be preserved in the presence of MI. This is what enables OFDM to avoid the multiple access interference that is present in CDMA systems.

Figure 7.15 illustrates how an OFDM transceiver works in terms of time and frequency domain signaling, where M subcarriers are present and no cyclic prefix is added for simplicity of illustration. The left side of Figure 7.15 shows the input and output signals of the IFFT unit at a transmitter; while the right side shows those signals of the FFT unit at a receiver. The signal waveforms shown in the input side of IFFT unit are $sinc(f)$ functions because of the square window truncated sinusoidal waveforms at the output side of the IFFT unit.

The frequency-domain representation of a number of tones, shown in Figure 7.12 or Figure 7.15, highlights the orthogonal nature of the tones used in the OFDM system. Notice that the peak of each tone corresponds to a zero level, or null, of every other tone. The result of this is that there is no interference between the tones. When the receiver samples at the center frequency of each tone, the only energy present is that of the desired signal, plus whatever other noise happens to be in the channel.

To maintain orthogonality between tones in an OFDM system, it is important to ensure that the symbol time contains one or multiple cycles of each sinusoidal tone waveform. This is normally the case, because the system numerology is constructed such that tone frequencies are integer multiples of the symbol period, as is subsequently highlighted, where the tone spacing is $1/T$. Viewed as sinusoids, Figure 7.16 shows three tones over a single symbol period, where each tone has an integer number of cycles during the symbol period.

In absolute terms, to generate a pure sinusoidal tone requires that the signal starts at time minus infinity. This is important, because tones are the only waveform that can ensure orthogonality. Fortunately, the channel response can be treated as finite, because multipath components decay over time and the channel is effectively band-limited. By adding a guard time, called a *cyclic prefix* (CP), the channel can be made to behave as if the transmitted waveforms were from time minus infinite, and

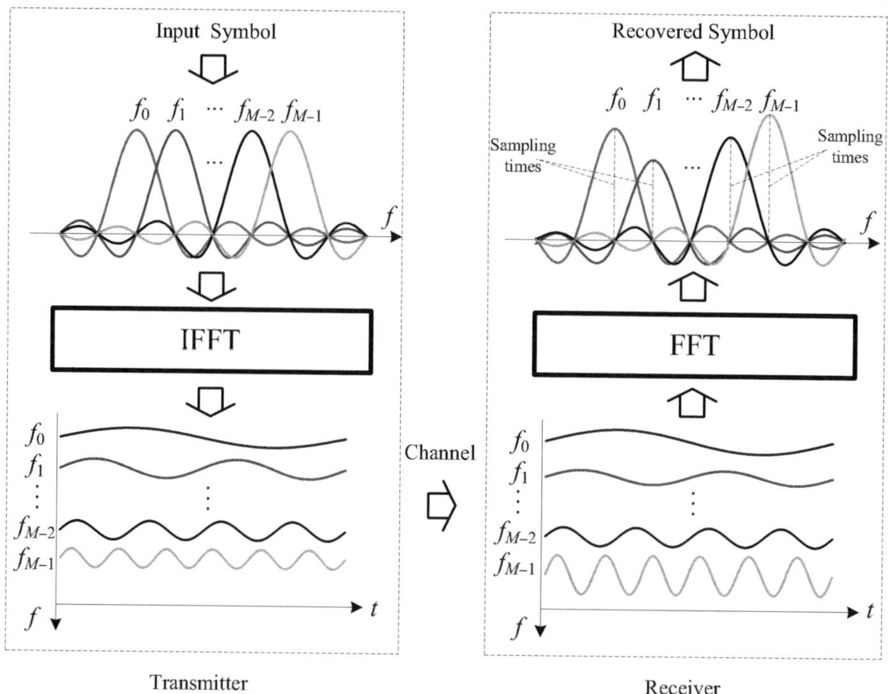

Figure 7.15 The OFDM signaling illustration in both the transmitter and receiver based on IFFT and FFT units, where the cyclic prefix is not added.

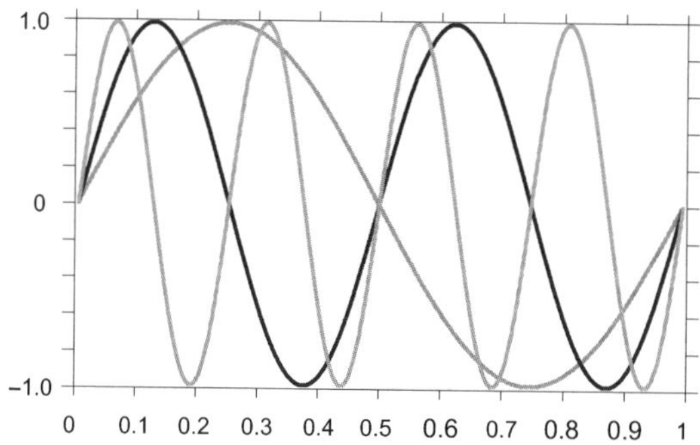

Figure 7.16 The three tones in a complete symbol duration, in which each tone occupies an integer number of cycles.

thus ensure orthogonality, which essentially prevents one subcarrier from interfering with another (called *intercarrier interference*, or (ICI)).

The CP is actually a copy of the last portion of the data symbol appended to the front of the symbol during the guard interval, as shown in Figure 7.17. Multipath causes tones and delayed replicas of tones to arrive at the receiver with some delay spread. This leads to misalignment between sinusoids,

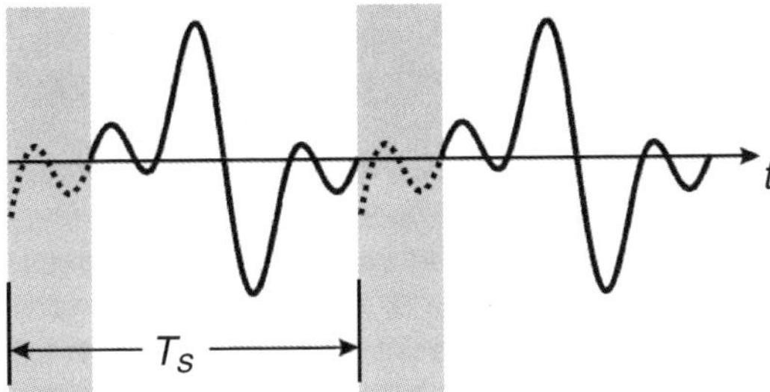

Figure 7.17 The cyclic prefix (shown as the shaded parts) used in an OFDM system as a guard time, which should be longer than the delay spread of the channel. Usually, the CP spans about one fourth of the symbol duration.

which need to be aligned as in Figure 7.17 to be orthogonal. The CP allows the tones to be realigned at the receiver, thus regaining orthogonality.

The CP is sized appropriately to serve as a guard time to eliminate ISI. This is accomplished because the amount of time dispersion from the channel is smaller than the duration of the CP. A fundamental trade-off is that the CP must be long enough to account for the anticipated multipath delay spread experienced by the system. However, the amount of overhead increases as the CP gets longer. The sizing of the CP forces a trade-off between the amount of delay spread that is acceptable and the amount of Doppler shift that is acceptable. In most practical application scenarios, the length of the CP is made about one fourth of the symbol duration.

7.5.3 PAPR Issues

As shown in Figure 7.14, an OFDM transmitter should finish all baseband signal processing before the carrier modulation, which is a simple amplitude modulation (AM). Thus, all important data information in an OFDM transmitter is carried on the amplitude of a carrier signal. An RF power amplifier usually works in a saturated status to achieve a relatively high power efficiency, and thus it will behave like a hard-limiter, which will cut off all useful data information if the dynamic range, also called *Peak-to-Average Power Ratio* (PAPR), of the input signal exceeds a certain level. The PAPR is a very important issue for any OFDM system.

It is seen from Figure 7.15 that the output from an IFFT unit is in fact the sum of all tones corresponding to the presence of the subcarriers in the input side of the IFFT unit. In other words, if only one of the M subcarriers exists at the input of the IFFT unit (or the input data symbol pattern is all "0" but one "1"), then the output will consist of only single tone. On the other extreme, if the all-one pattern appears at the input symbol, the output signal will be the sum of all tones, resulting in a very high PAPR, which requires that an RF power amplifier with a very big dynamic range can be used.

Therefore, it is in our best interest to reduce the PAPR to avoid the appearance of all-one or any other symbol patterns (which will generate high PAPR levels) at the input side of the IFFT unit. Therefore, many real OFDM systems always use an interleaver at the input side of the IFFT unit to make the symbol patterns appear more balanced in terms of zeros and ones.

7.5.4 OFDMA Technologies

To address the unique demands posed by mobile users of high-speed data applications, new air interfaces must be designed and optimized across all the layers of the protocol stack, including the networking layers. A prime example of this kind of optimization is found in OFDMA technology.

As its name suggests, the system is based on OFDM, however, OFDMA is much more than just a physical layer solution. It is a cross-layer-optimized technology that exploits the unique physical properties of OFDM, enabling significant higher layer advantages that contribute to very efficient packet data transmission in a cellular network.

Packet-switched air interface

The telephone network, designed basically for voice, is an example of circuit-switched systems. Circuit-switched systems exist only at the physical layer that uses the channel resource to create an end-to-end bit pipe. They are conceptually simple as the bit pipe is a dedicated resource, and the pipe does not need to be controlled once it is created (some control may be required in setting up or tearing down the pipe). Circuit-switched systems, however, are very inefficient for burst data traffic. Packet-switched systems, on the other hand, are very efficient for data traffic but require that the upper layers be controlled in addition to the physical layer that creates the bit pipe. The MAC layer is required for the many data users to share the bit pipe. The data link layer is needed to take the error-prone pipe and create a reliable link for the network layers to pass packet data flows over. The Internet is the best example of a packet-switched network. Because all conventional cellular wireless systems, including 3G, were fundamentally designed for circuit-switched voice, they were designed and optimized primarily at the physical layer. Some people suggested that the choice of CDMA as the physical layer multiple access technology was also dictated by voice requirements. OFDMA, on the other hand, is a packet-switched scheme designed for data and is optimized across the physical, MAC, data link, and network layers. The choice of OFDM as the multiple access technology is based not only on physical layer consideration, but also on the MAC layer, data link layer, and network layer requirements.

Physical layer advantages: OFDMA

As discussed earlier, most of the physical layer advantages of OFDM are well understood. Most notably, OFDM creates a robust multiple access technology to deal with the impairments of the wireless channel, such as multipath fading, delay spread, and Doppler shifts. Advanced OFDM-based data systems typically divide the available spectrum into a number of equally spaced tones. For each OFDM symbol duration, information carrying symbols (based on modulation such as QPSK, QAM, etc.) are loaded on each tone.

The OFDMA can also use fast hopping across all tones in a predetermined pseudorandom pattern, making it an SS technology. With fast hopping, a user that is assigned one tone does not transmit every symbol on the same tone, but uses a hopping pattern to jump to a different tone for every symbol. Different BSs use different hopping patterns, and each uses the entire available spectrum (thus to realize frequency reuse of 1). In cellular deployment, this adds to the advantages of CDMA systems, including frequency diversity and out of cell (intercell) interference averaging spectral efficiency benefit that narrowband systems such as conventional TDMA do not have.

As discussed earlier, different users within the same cell use different resources (tones) and hence do not interfere with each other. This is similar to TDMA, where different users in a cell transmit at different time slots and do not interfere with one another. In contrast, CDMA users in a cell do interfere with each other, increasing the total interference in the system. OFDMA therefore has the physical layer benefits of both CDMA and TDMA and is at least three times (3*times*) more efficient than CDMA. In other words, at the physical layer, OFDMA creates the biggest pipe of all cellular technologies. Even though the 3*times* advantage at the physical layer is a huge advantage, the most significant advantage of OFDMA for data is at the MAC and link layers.

MAC and link layer advantages

OFDMA exploits the granular nature of resources in OFDM to come up with extremely efficient control layers. In OFDM, when designed appropriately, it is possible to send a very small amount (as little as one bit) of information from the transmitter to the receiver with virtually no overhead. Therefore, a transmitter that is previously not transmitting can start transmitting as little as one bit of information, and then stop, without causing any resource overhead. This is unlike CDMA or TDMA, in which the granularity is much coarser, and merely initiating a transmission wastes a significant resource. Hence, in TDMA, for example, there is a frame structure, and whenever a transmission is initiated, a minimum of one frame (a few hundred bits) of information is transmitted. The frame structure does not cause any significant inefficiency in user data transmission, as data traffic typically consists of a large number of bits. However, for the transmission of control-layer information, the frame structure is extremely inefficient, as the control information typically consists of one or two bits but requires a whole frame. Not having a granular technology can therefore be very detrimental from a MAC layer and link layer point of view.

OFDMA takes advantage of the granularity of OFDM in its control-layer design, enabling the MAC layer to perform efficient packet switching over the air and at the same time provide all the hooks to handle QoS. It also supports a data link layer that uses local (as opposed to end-to-end) feedback to create a very reliable link from an unreliable wireless channel, with very low delays. The network layer's traffic therefore experiences small delays and no significant delay jitter. Hence, interactive applications such as (packet) voice can be supported. Moreover, Internet protocols such as TCP/IP run smoothly and efficiently over an OFDMA air link. As discussed in Chapter 3, TCP/IP performance on 3G networks is very inefficient because the data link layer introduces significant delay jitter so that channel errors are misinterpreted by TCP as network congestion and TCP responds by backing off to the lowest rate.

Packet switching leads to efficient statistical multiplexing of data users and helps the wireless operators to support a much greater number of users for a given user experience. This desirable feature in OFDMA, together with QoS support and a three *times* bigger pipe, allows the operators to profitably scale their wireless networks to meet the burgeoning data traffic demand in an all-you-can-eat pricing environment.

7.6 Ultra-Wideband Technologies

As mentioned in Section 2.2.3, the UWB technology can be viewed as a derivative from the spreading spectrum technology, in particular, the time hopping spread spectrum (THSS) technique, which is also considered as a multiple access technology, being particularly suited for extreme narrow pulse transmissions. Before discussing the technical details about the UWB technologies, we would like to review briefly the history as well as the recent research activities carried out in this area.

Since the introduction of UWB technology to commercial applications in the early 1990s [674], much of its initial research has been focused on the application of the THSS [675], where several pulses in each symbol duration are sent with a particular time offset pattern determined by a unique signature code for multiple access. The implementation of a THSS-UWB system requires a precise network-wise synchronization clock. This inevitably increases overall hardware complexity at a transceiver, which used to be a major concern in realizing a feasible UWB system at its early stage. On the other hand, DS techniques can also work jointly with UWB systems to provide multiple access among different users within the same wireless personal area network (WPAN). The operation of a DS-UWB system does not need an accurate synchronization clock and the use of antipodal pulses in DS modulation can boost up effective transmission power, which is very important to improve the detection efficiency of a UWB receiver, due to the severe emission constraints imposed on the power spectral mask specified in the FCC Part15.209, in which the maximal transmitting power for a UWB transmitter should be lower than -41.3 dBm within the bandwidth from 3.1 to 10.6 GHz.

The UWB technologies have been standardized in IEEE 802.15.3a as a technology for WPANs. Figure 7.18 shows all IEEE 802 standards, including those for WLANs as IEEE 802.11 standards,

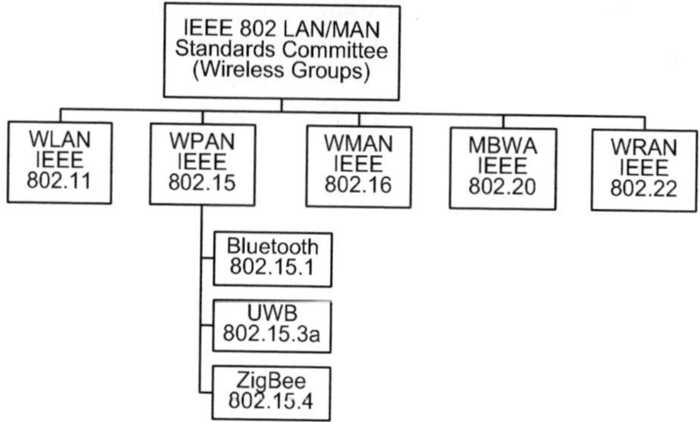

Figure 7.18 Various IEEE 802 standards, in which UWB technologies have been covered in IEEE 802.15.3a standard for WPAN applications.

wireless metropolitan area networks (WMANs) as IEEE 802.16 standards, WPANs as IEEE 802.15 standards, and so on. It is noted that IEEE 802.15.4.a is emerging as the standard for low-data-rate transmission.

The FCC issued a notice of inquiry (NOI) in September 1998 and within a year the Time Domain Corporation, US Radar, and Zircon Corporation had received waivers from the FCC to allow limited deployment of a small number of UWB devices to support continued development of the technology, and the University of Southern California (USC) UltRa Lab had an experimental licence to study UWB radio transmissions. A notice of proposed rule making (NPRM) was issued in May 2000. In April 2002, after extensive commentary from the industry, the FCC issued its first report and order on UWB technology, thereby providing regulations to support deployment of UWB radio systems. This FCC action was a major change in the approach to the regulation of RF emissions, allowing a significant portion of the RF spectrum, originally allocated in many smaller bands exclusively for specific uses, to be effectively shared with low-power UWB radios.

The FCC regulations classify UWB applications into several categories (see Table 7.5) with different emission regulations in each case. Maximum emissions in the prescribed bands are at an effective

Table 7.5 The application categories specified by FCC UWB regulations

Application	Frequency band for operation at Part 1 limit	User limitations
Communications and measurement systems	3.1 to 10.6 GHz (different out-of-band emission limits for indoor and outdoor devices)	No
Imaging: ground penetrating radar, wall, medical imaging	<960 MHz or 3.1 to 10.6 GHz	Yes
Imaging: through wall	<960 MHz or 1.99 to 10.6 GHz	Yes
Imaging: surveillance	1.99 to 10.6 GHz	Yes
Vehicular	24 to 29 GHz	No

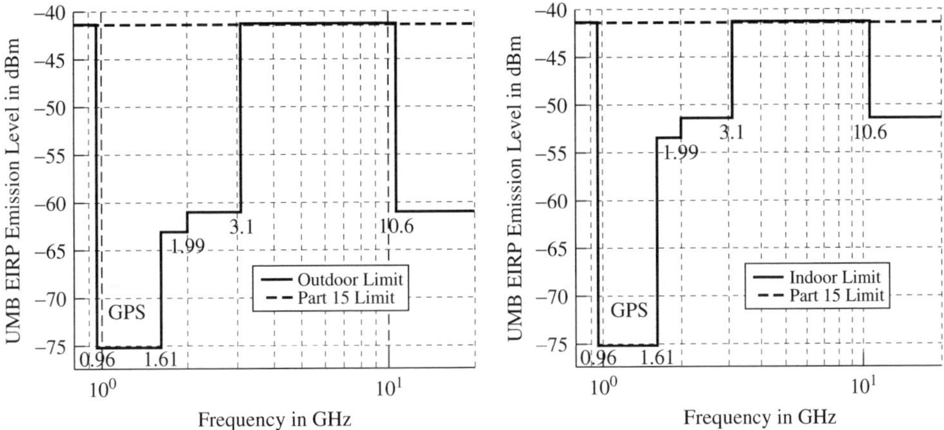

Figure 7.19 FCC regulated spectral masks regarding the indoor and outdoor UWB communications applications.

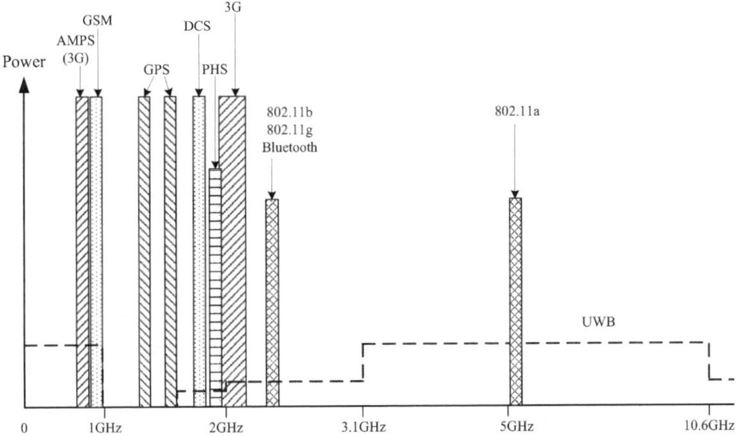

Figure 7.20 Other communications applications in the vicinity of UWB operating bands.

isotropic radiated power (EIRP) of -41.3 dBm per MHz, and the -10 dB level of the emissions must fall within the prescribed band, as shown in Figure 7.19, which should be compared with Figure 7.20 to know other communication applications in the vicinity of the UWB operating bands.

7.6.1 Major UWB Technologies

There are four major UWB technologies that have been proposed in the literature. The first type is Time Hopped (TH) UWB or Time-modulated (TM) UWB,[1] which is a traditional UWB scheme

---

[1] The traditional impulse radio technology can be called as either time hopped (TH) UWB or time modulated (TM) UWB. It should be noted that both names are used very often.

and is often called *impulse radio* (IR) UWB. The TH-UWB is by far the earliest version of UWB technology and remains an important solution even today. The TH-UWB can be further divided into two subcategories, that is, analog impulse radio multiple access (AIRMA) and digital impulse radio multiple access (DIRMA), which were suggested and studied in [613, 624, 637]. The second UWB technology is called *direct-sequence CDMA-based UWB* and can be implemented with a multi-carrier CDMA architecture. The DS-CDMA UWB scheme will be discussed in detail in Subsections 7.6.1, 7.6.2, 7.6.3, 7.6.4, and 7.6.5. Another UWB scheme that has gained much popularity is based on OFDM technology, namely, OFDM-UWB, which can be implemented on a multiband (MB) OFDM scheme. The MB-OFDM UWB technology is particularly useful when cognitive radio technology is used, as discussed in Chapter 9. In addition, some people also proposed frequency-modulation (FM) based UWB systems, which can be implemented by swept frequency technology. Figure 7.21 shows a family tree for all possible UWB technologies that have been proposed so far. Because of limited space, we will only focus on the discussions on TH-UWB (or TM-UWB) and DS-CDMA UWB in this subsection.

TH-UWB technology

The basic concept of a TH-UWB system is shown in Figure 7.22, where the system consists of four major parts, namely, modulator[2], delay unit, transmission time controller, and a pseudorandom sequences generator. Obviously, in such a TH-UWB system, the data is sent in bursts and transmission time is controlled by the pseudorandom sequences generator.

Understandably, the bandwidth of such a TM-UWB system is determined by the width and shape of impulses, which usually takes some special waveforms, such as "monocycle." The design of the monocycles suitable for IR applications is a very interesting research topic in that the shape of the monocycles should provide a very good time ACF for a better detection efficiency and fit FCC spectral mask as illustrated in Figure 7.19. There are many pulse waveforms that have been proposed, such as Gaussian pulse and its derivative functions, Hermite pulse and its modified versions, prolate spheroidal waveforms, Laplacian monocycle, cubic monocycle, wavelets, and so on. For more information on these popular impulses suitable for UWB applications, please refer to the large number of references given at the end of this book [604–691].

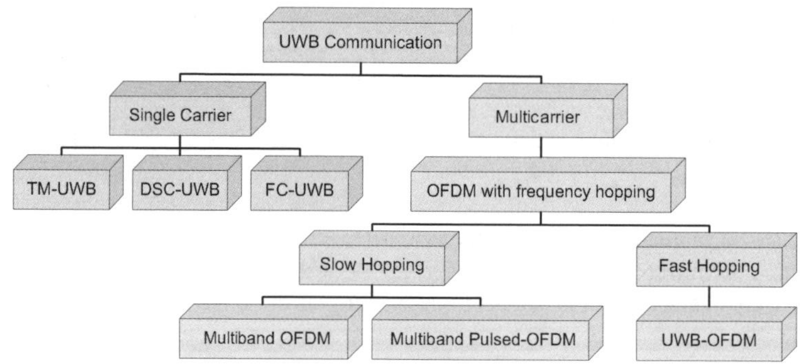

Figure 7.21   Family tree for various UWB technologies proposed so far.

---

[2]The most commonly used modulator scheme in an IR (or time hopping UWB) is pulse position modulation (PPM), although many other modulation schemes can also be used, such as pulse amplitude modulation (PAM), on-off-keying (OOK), pulse shape modulation (PSM), and so on.

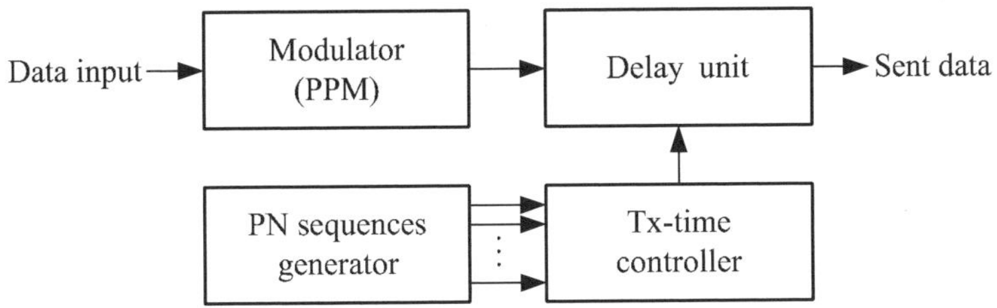

Figure 7.22 Block diagram for a TH-UWB IR transmitter.

The data signal should be sent out from an IR system, as shown in Figure 7.22, using carrier-less transmission. The base band signal can be converted directly from the received signal and no intermediate frequency unit is required, thus reducing the implementation complexity. The TM-UWB scheme can provide a relatively large PG due to the fact that it has a very narrow impulse (whose width is of the order of a nanosecond). This large PG also entails several other operational advantages, which are explained as follows. First of all, it offers an excellent multipath immunity because of its very high so time resolution that almost all multipath components can be separated and combined coherently at a receiver. If the time between two pulses is longer than the channel delay spread, there will be no ISI between two consecutive pulses, nor between two symbols.[3] Second, it gives a good resistance to external interference based on the same reasons as any SS system. The big PG also ensures a relatively low-power spectral density, which helps in not causing interference to other existing wireless applications, as shown in Figure 7.20.

It is to be noted that the data-carrying modulation in an IR-UWB system is usually PPM, which controls the appearance position of a pulse in a certain duration to represent different data-information. On the other hand, the multiple access capability of an IR-UWB system is implemented through time hopping schemes, as briefly discussed in Subsection 2.2.3. Different users in a pico-cell can be assigned different PN sequences that control the timing of pulses, as shown in Figure 7.23, where only three users are present for simplicity of illustration and 13 hopping slots are shown in one symbol duration. In this case, there is no overlapping in the hopping slots among the three users, implying that there will be no MAI.

A TH-UWB can offer a very good time diversity gain if multiple hopping patterns can be assigned to a single user. Therefore, it is intuitively true that it can be made very robust against time-selective fading, especially suitable for the applications where fast mobility is present.[4]

DS-CDMA UWB technology

The direct-sequence CDMA UWB scheme is the focus of discussion in this subsection. The analysis of a DS-CDMA UWB system is given in the following subsections. A DS-CDMA UWB scheme works like a conventional DS-CDMA system. The pulse trains are used to perform DS modulation to spread the signal. A PN code is assigned to a particular user and will be used to spread its data bit into multiple chips. In the same way as in IR, various data modulation schemes, such as PAM, OOK, PSM, and so on, can also be used in the DS-CDMA UWB system. Figure 7.24 shows an example of the PAM-modulated DS-CDMA UWB scheme.

---

[3]This is particularly true if a UWB system is operating in an indoor environment where the delay spread is relatively small.

[4]Because of the fact that most UWB systems are operated in an indoor environment, this advantage may not be well exploited.

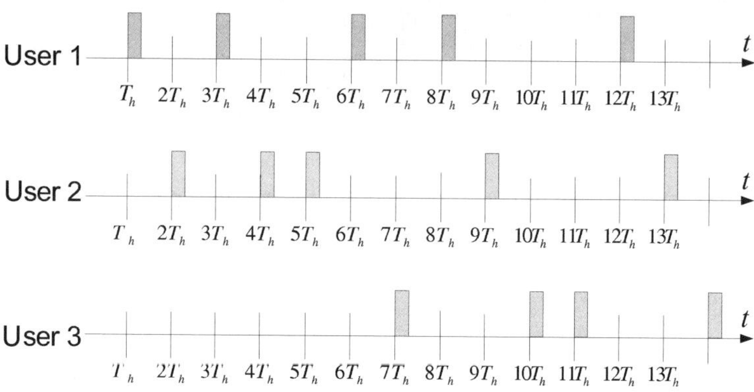

Figure 7.23 Multiple access capability provided by a TH-UWB IR system.

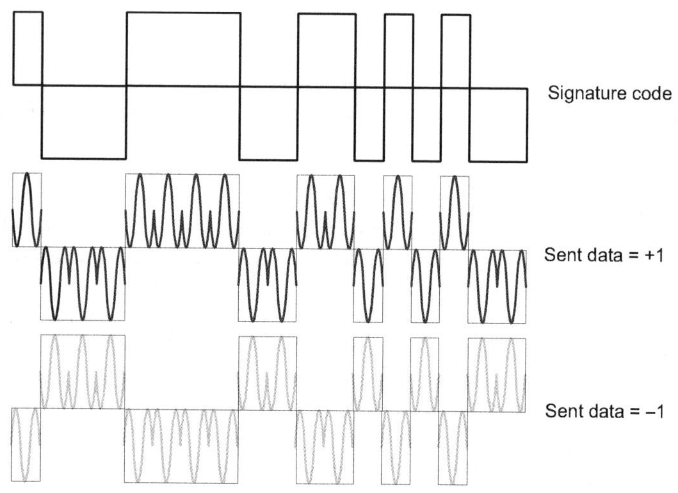

Figure 7.24 Conceptual diagram of a DS-CDMA UWB system with PAM.

Many results have been reported on the performance of the DS-CDMA-based UWB systems, as shown in [594–673]. Srinivasa [677] presents a comparison between a TH-PPM UWB and a TH DS spreading with antipodal signaling (TH/DS-BPSK) in terms of their multiple access performance, where the study was limited to an AWGN channel only. Foester [678] characterized the performance of a direct sequence UWB system in the presence of multipath and narrowband interferences. It was shown that the code design that tries to minimize sequence autocorrelation sidelobes as well as cross correlation among spreading codes is critical for a good performance under multipath, multiuser, and narrowband interferences at the same time.

A comprehensive review on almost all possible multiple access techniques suitable for UWB-based WPANs or piconet was given in [679]. It was suggested that, among all multiple access schemes (i.e. FDMA, TDMA, and CDMA), CDMA is the most suitable for UWB applications. The use of CDMA allows multiple piconets to be relatively independent, and it is able to produce the highest aggregate data rate. It was also pointed out that CDMA is completely compatible with high QoS, video streaming capable MAC layer protocols, such as the TDMA-based IEEE 802.15.3. On the implementation side, to map to high-speed low-voltage low-power IC technologies, UWB systems must use low peak-to-average pulse trains with a relatively high chip rate. These high chip rates are perfectly suited for building UWB CDMA systems.

Qinghua Li and Rusch [680, 681] studied the effectiveness of an adaptive MMSE multiuser detection for a DS-CDMA-based UWB system, particularly under the interference of an IEEE 802.11a OFDM transmitter, as shown in Figure 7.20. Extensive simulations were performed using channel sounding techniques in the 2- to 8-GHz band in a residential environment, which was characterized by a high level of multipath fragmentation. It was demonstrated that the adaptive MMSE is able to reject intersymbol and interchip interference for those channels much more effectively than by using a RAKE receiver with four to eight fingers. It was also shown that the same receiver setting can reject a narrowband interferer generated from an adjacent IEEE 802.11a transmitter. The majority of the work was carried out on the basis of computer simulations.

Sadler and Swami [682] investigated a DS-UWB system with so-called *episodic* transmission, that is, the system should send *n* pulses per information bit and allow for off time separation between pulses. Several issues on the design of a DS-UWB system, such as PG, jamming margin, coding gain, and multiple access interference, power control, and so on, were investigated. The BER performance was studied using a *Chernoff bound* and considering a single-user matched-filter receiver in an AWGN channel scenario.

The comparison between two UWB techniques for implementing multiple access communications, specifically TH-PPM and DS-BPSK schemes, was made by Canadeo *et al.* [683]. They carried out a spreading-code-dependent study on both UWB schemes. A generic channel model based on a very simple delay tapped line was used. The coefficients in this multipath channel model were constants, implying that no fading was considered.

Boudaker and Letaief [684] outlined the attractive features of DS-UWB multiple access systems employing antipodal signaling and compared it with the TH scheme. An appropriate DS-UWB transmitter and receiver were designed, and the system signal processing formulation was investigated. The performance of such communication systems in an AWGN channel in terms of multiple access capability, error rate performance, and achievable transmission rate were evaluated without MI. Only a single matched-filter detector was considered.

An interesting method for implementing a DS-UWB system based on a new multi-carrier pulse waveform was proposed in [685]. A unique frequency domain processing technique was used at the receiver side to exploit diversity in the frequency domain and provide resistance against intersymbol interference and multiple access interference. The performance of such a frequency domain processing DS-UWB scheme was compared with a DS-UWB system using traditional time-domain processing techniques.

An UWB system with PPM for data modulation and DS spreading for multiple access in an indoor fading environment was considered in [686]. A RAKE receiver was used to combine a subset of the resolvable multipath components using MRC technique. In the following subsections, we will consider a multipath environment, modeled by a discrete-time linear filter with an impulse response whose coefficients are lognormally distributed random variables.

Runkle *et al.* [688] compared a multi-carrier UWB with a DS-CDMA UWB. The results illustrated that a significant advantage can be obtained if a UWB system is implemented by DS-CDMA techniques. The multi-carrier UWB was implemented by a MB OFDM architecture. The authors explained how the DS-CDMA UWB architecture could support robust and flexible multiuser capabilities, protect against in-band interference, and provide high resolution ranging capability for safety-of-life applications.

A comparison of the average BER and outage probability performance of the three UWB multiple access and modulation combinations for a single-user UWB radio was reported by [689]. The three schemes are TH with bit flipping modulation, TH-PPM, and DS with bit flipping modulation. The authors used the channel models recommended for use in the IEEE 802.15.3a evaluation. The results showed that direct sequence multiple access coding was more likely to achieve the lowest BER for a fixed channel.

Unfortunately, most of the currently reported researches on UWB have separated the issues on pulse waveform design from system-level performance, such as bit error probability, and so on. In other words, the previous system-level analysis on BER performance seldom considered the characteristic features of UWB pulses used in the system, as seen from all the papers referred in the preceding text [675–689]. On the other hand, most of the current research on UWB pulse waveforms was focused on the requirements concerning their spectral shapes and has little to do with the overall system BER performance. In the following subsections, we demonstrate a BER performance analysis that is associated with the characteristic feature of UWB pulse waveforms. We give a unified approach to derive a closed form BER expression by taking into account major factors of a UWB system, such as noninteger chip asynchronous transmission of the signals, multiple access interference, MI, and so on, as well as their impact on the BER performance. In particular, we introduce a merit parameter, namely, *normalized mean squared autocorrelation function* (NMSACF) of the pulse waveform denoted by $\sigma^2_{mp\_normal}$. It will be used to characterize different UWB pulses in terms of their ACF. In fact, $\sigma^2_{mp\_normal}$ measures the average interchip interference level associated with the autocorrelation side lobes of the pulse waveforms. We will illustrate from the analysis that $\sigma^2_{mp\_normal}$ should be made as small as possible to ensure a desirable BER performance.

7.6.2 DS-CDMA UWB System Model

Let us consider a DS-CDMA UWB radio system with K users. An ultranarrow pulse waveform $g(t)$ defined over $(0,\ T_c)$ is used to directly modulate the binary data stream $\{b_j^{(k)}\}_{j=-\infty,\dots,\infty}^{k=1,\dots,K}$ without using a sinusoidal carrier. The k-th user is assigned a signature sequence $\{a_n^{(k)}\}_{n=0,\dots,N-1}^{k=1,\dots,K}$ to modulate antipodal pulses. Presumably, a pulse covers just a chip duration T_c, and a signature code has N chips such that $T_b = NT_c$, where N is the PG.

The block diagram of this generic DS-CDMA UWB transceiver is shown in Figure 7.25, where each transmitted signal will experience fading in the channel with its impulse response being $h_k(t)$ for the k-th user. The receiver model is tuned to the first user's transmitted signal with its signature code being $\{a_n^{(1)}\}_{n=0}^{N-1}$. The received signal will be processed by signature code matched filtering as well as pulse waveform–correlation before making a decision for the j-th bit, or at time $t = (j+1)T_b$.

The transmitted signal from the k-th user can be written as

$$s_k(t) = \sum_{j=-\infty}^{\infty} \sum_{n=0}^{N-1} b_j^{(k)} a_n^{(k)} g(t - jT_b - nT_c) \tag{7.6}$$

where $g(t)$ is defined as

$$\begin{cases} g(t) \neq 0, & 0 \leq t \leq T_c \\ g(t) = 0, & t < 0,\ t > T_c \\ \max\left[g(t)\right] = 1, & 0 \leq t \leq T_c \end{cases} \tag{7.7}$$

We are considering an asynchronous DS-CDMA UWB system and its pulse waveform–dependent bit error performance analysis. The k-th user's channel impulse response is $h_k(t) = \alpha\delta(t - \tau_k)$, where α is a fading coefficient, which may obey any distribution dependent on a particular environment, and $\delta(t - \tau_k)$ is an impulse function being unit at $t = \tau_k$ and zero elsewhere. The received signal can be expressed as

$$r(t) = \sum_{k=1}^{K} s_k(t) \otimes h_k(t) + n(t) = \sum_{k=1}^{K} \alpha s_k(t - \tau_k) + n(t) \tag{7.8}$$

where symbol $\otimes$ denotes the convolution operation, $\{\tau_k\}_{k=1}^{K}$ is the delay of the k-th user, $n(t)$ obeys Gaussian Distribution $\mathcal{N}(0, \sigma_n^2)$ or can simply be denoted as $n(t) \sim \mathcal{N}(0, \sigma_n^2)$, which specifies a

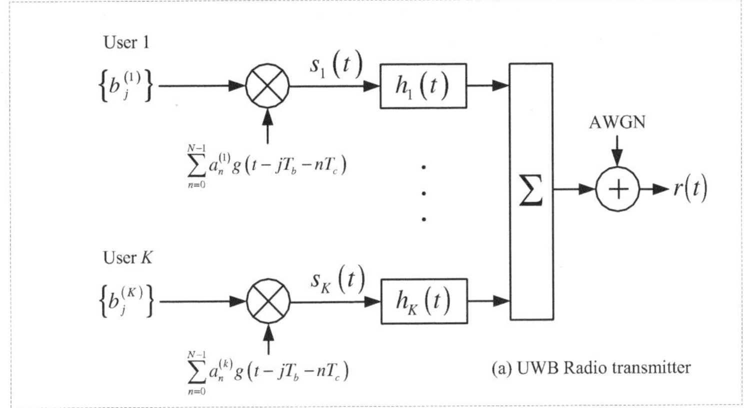

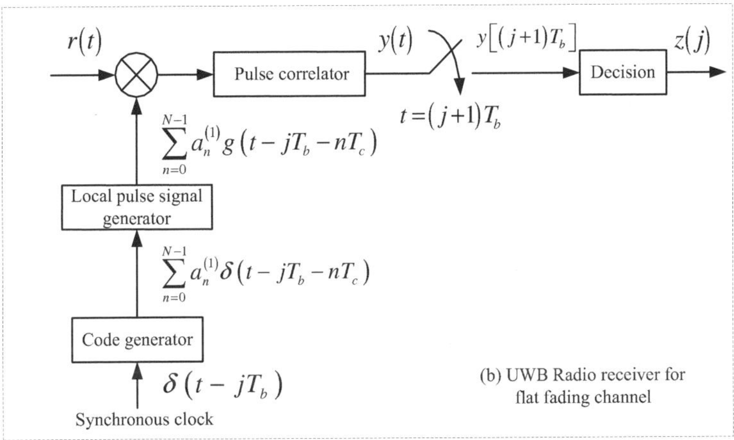

Figure 7.25 A block diagram of a DS-CDMA UWB transceiver. (a) Transmitter model; (b) Receiver model, where the receiver is intended for user k and a flat fading channel is used.

relationship between $n(t)$ and a Gaussian distribution with zero mean and variance σ_n^2. Here, the receiver intends to detect the first user's transmission. Without loss of generality, let $\tau_1 = 0$ and τ_k be the relative delay between the first and k-th users' transmissions. Inserting $s_k(t)$ and $h_k(t)$ into Equation (7.8), we obtain

$$r(t) = \sum_{j=-\infty}^{\infty} \sum_{n=0}^{N-1} \alpha b_j^{(1)} a_n^{(1)} g(t - jT_b - nT_c)$$

$$+ \sum_{k=2}^{K} \sum_{j=-\infty}^{\infty} \sum_{n=0}^{N-1} \alpha b_j^{(k)} a_n^{(k)} g(t - \tau_k - jT_b - nT_c) + n(t) \qquad (7.9)$$

The decision variable at the receiver can be written as

$$y[(j+1)T_b] = \int_{jT_b}^{(j+1)T_b} r(t) \sum_{n=0}^{N-1} a_n^{(1)} g(t - jT_b - nT_c)\, dt = S + I + \eta \qquad (7.10)$$

where decision variable $y\left[(j+1)\,T_b\right]$ has been decomposed into three components, useful signal S, multiple access interference I, and noise η. The useful signal component is written as

$$
S = \int_{jT_b}^{(j+1)T_b} \sum_{j=-\infty}^{\infty} \sum_{n=0}^{N-1} \alpha b_j^{(1)} a_n^{(1)} g(t - jT_b - nT_c) \sum_{n=0}^{N-1} a_n^{(1)} g(t - jT_b - nT_c)\, dt
$$

$$
= \alpha N b_j^{(1)} E_{mp} \tag{7.11}
$$

where $E_{mp} = \int_0^{T_c} g^2(t)\, dt$ gives the energy of a single pulse. The MAI component can be expressed by

$$
I = \sum_{k=2}^{K} I_k = \sum_{k=2}^{K} \int_{jT_b}^{(j+1)T_b} \sum_{j=-\infty}^{\infty} \sum_{n=0}^{N-1} \alpha b_j^{(k)} a_n^{(k)} g(t - \tau_k - jT_b - nT_c)
$$

$$
\times \sum_{n=0}^{N-1} a_n^{(1)} g(t - jT_b - nT_c)\, dt \tag{7.12}
$$

The noise term becomes

$$
\eta = \int_{jT_b}^{(j+1)T_b} n(t) \sum_{n=0}^{N-1} a_n^{(1)} g(t - jT_b - nT_c)\, dt \tag{7.13}
$$

As $n(t) \sim \mathcal{N}\left(0, \sigma_n^2\right)$, we have $E(\eta) = 0$ and the variance of η becomes

$$
\sigma_\eta^2 = E\left(\eta^2\right) = E\left[\int_{jT_b}^{(j+1)T_b} n(t) \sum_{n=0}^{N-1} a_n^{(1)} g(t - jT_b - nT_c)\, dt \right]^2 \tag{7.14}
$$

Letting $q(t) = \sum_{n=0}^{N-1} a_n^{(1)} g(t - jT_b - nT_c)$, we obtain

$$
\sigma_\eta^2 = E\left(\eta^2\right) = E\left[\int_{jT_b}^{(j+1)T_b} n(t) q(t)\, dt \right]^2
$$

$$
= \int_{jT_b}^{(j+1)T_b} \int_{jT_b}^{(j+1)T_b} E[n(t)n(\zeta)] q(t)\, q(\zeta)\, dt\, d\zeta
$$

$$
= \sigma_n^2 \int_{jT_b}^{(j+1)T_b} \sum_{n=0}^{N-1} \left(a_n^{(1)}\right)^2 g_1^2(t - jT_b - nT_c)\, dt = \sigma_n^2 N E_{mp} \tag{7.15}
$$

Therefore, we have $\eta \sim \mathcal{N}\left(0, \sigma_n^2 N E_{mp}\right)$.

7.6.3 Flat Fading Channel

In this subsection, we will proceed to determine the MAI term in a flat fading channel. In general, the calculation of the variance of multiple access interference component I involves ACF of pulse waveforms on a chip-by-chip basis, which is to be explained in the sequel.

Chip-wise pulse autocorrelation function

Assume that the signal of interest is the first user's transmission. Let us first consider the interference component caused by the k-th transmission, which can be written as

$$I_k = \int_{jT_b}^{(j+1)T_b} \sum_{j=-\infty}^{\infty} \sum_{n=0}^{N-1} \alpha b_j^{(k)} a_n^{(k)} g(t - \tau_k - jT_b - nT_c) \sum_{n=0}^{N-1} a_n^{(1)} g(t - jT_b - nT_c)\, dt \qquad (7.16)$$

The relative delay between the first and k-th users is $\tau_k \in (0, T_b)$, and two consecutive interfering bits with respect to $b_j^{(1)}$ are $b_{j-1}^{(k)}$ and $b_j^{(k)}$. We obtain

$$I_k = \alpha \int_{jT_b}^{(j+1)T_b} \left[\sum_{j=-\infty}^{\infty} \sum_{n=0}^{N-1} b_j^{(k)} a_n^{(k)} g(t - \tau_k - jT_b - nT_c) \right]$$

$$\times \sum_{n=0}^{N-1} a_n^{(1)} g(t - jT_b - nT_c)\, dt$$

$$= \alpha \left[b_{j-1}^{(k)} \int_0^{\tau_k} \sum_{n=0}^{N-1} a_n^{(k)} g(t - \tau_k + T_b - nT_c) \sum_{n=0}^{N-1} a_n^{(1)} g(t - nT_c)\, dt \right.$$

$$\left. + b_j^{(k)} \int_{\tau_k}^{T_b} \sum_{n=0}^{N-1} a_n^{(k)} g(t - \tau_k - nT_c) \sum_{n=0}^{N-1} a_n^{(1)} g(t - nT_c)\, dt \right] \qquad (7.17)$$

As shown in Appendix A, the k-th interfering component can be determined by

$$I_k = \int_{jT_b}^{(j+1)T_b} \sum_{j=-\infty}^{\infty} \sum_{n=0}^{N-1} \alpha b_j^{(k)} a_n^{(k)} g(t - \tau_k - jT_b - nT_c) \sum_{n=0}^{N-1} a_n^{(1)} g(t - jT_b - nT_c)\, dt$$

$$= \alpha \left\{ b_{j-1}^{(k)} \left[C_{k,1}(i_k - N + 1) \int_0^{\gamma_k T_c} g(t) g(t + (1 - \gamma_k) T_c)\, dt \right. \right.$$

$$\left. + C_{k,1}(i_k - N) \int_{\gamma_k T_c}^{T_c} g(t) g(t - \gamma_k T_c)\, dt \right] + b_j^{(k)} \left[C_{k,1}(i_k) \int_{\gamma_k T_c}^{T_c} g(t) g(t - \gamma_k T_c)\, dt \right.$$

$$\left. \left. + C_{k,1}(i_k + 1) \int_0^{\gamma_k T_c} g(t) g(t + (1 - \gamma_k) T_c)\, dt \right] \right\} \qquad (7.18)$$

where γ_k is the fractional-chip relative delay between the first and k-th users (as shown in Figure 7.26), and $C_{k,1}(i)$ is the discrete aperiodic partial cross-correlation between signature codes of the first and k-th users, which is defined as

$$C_{k,1}(i) = \begin{cases} \displaystyle\sum_{j=0}^{N-1-i} a_j^{(k)} a_{j+i}^{(1)}, & 0 \le i \le N - 1 \\[2.5em] \displaystyle\sum_{j=0}^{N-1+i} a_{j-i}^{(k)} a_j^{(1)}, & -(N-1) \le i \le 0 \\[2em] 0, & \text{otherwise} \end{cases} \qquad (7.19)$$

Let us define *chip-wise pulse waveform autocorrelation function* as

$$R_p(\tau) = \int_{-T_c}^{T_c} g(t) g(t - \tau)\, dt \qquad (7.20)$$

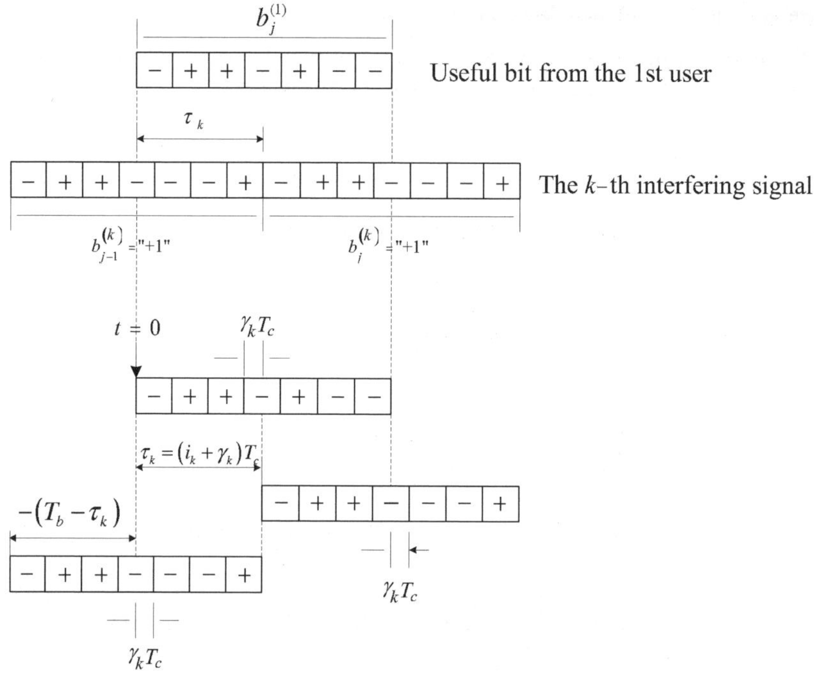

Figure 7.26 Calculation of asynchronous cross-correlation function with fractional-chip delay, that is, $\tau_k = (j + \gamma_k) T_c$, between transmitting signals from the first and k-th users, where γ_k takes a real value such that $0 \le \gamma_k \le 1$.

Thus, we have

$$R_p (\gamma_k T_c) = \int_{\gamma_k T_c}^{T_c} g(t) g(t - \gamma_k T_c) \, dt \tag{7.21}$$

$$R_p (- (1 - \gamma_k) T_c) = \int_0^{\gamma_k T_c} g(t) g(t + T_c - \gamma_k T_c) \, dt \tag{7.22}$$

The illustrations of $R_p (\gamma_k T_c)$ and $R_p (- (1 - \gamma_k) T_c)$ have been given in Figure 7.27. Therefore, the k-th interfering component can be written as

$$I_k = \alpha \left\{ b_{j-1}^{(k)} \left[C_{k,1} (i_k - N + 1) R_p (- (1 - \gamma_k) T_c) + C_{k,1} (i_k - N) R_p (\gamma_k T_c) \right] \right.$$

$$\left. + b_j^{(k)} \left[C_{k,1} (i_k) R_p (\gamma_k T_c) + C_{k,1} (i_k + 1) R_p (- (1 - \gamma_k) T_c) \right] \right\} \tag{7.23}$$

which contains four random variables. $b_{j-1}^{(k)}$ and $b_j^{(k)}$ are two consecutive bits of the k-th user, γ_k is the fractional-chip relative delay between the first and k-th users, and α is the flat fading coefficient of the channel. Theoretically speaking, if we know the distributions of all four random variables, it is possible to obtain the distribution of I_k from Equation (7.23), and thus the distribution of $I = \sum_{k=2}^{K} I_k$ by convolution of $f_I(x) = f_{I_2} \otimes f_{I_3} \otimes \ldots \otimes f_{I_K}$. Unfortunately, the intricate relationship between the four random variables in Equation (7.23) makes it almost impossible to calculate explicitly the distribution of I_k. Therefore, we will use the Gaussian approximation to obtain the close form of BER.

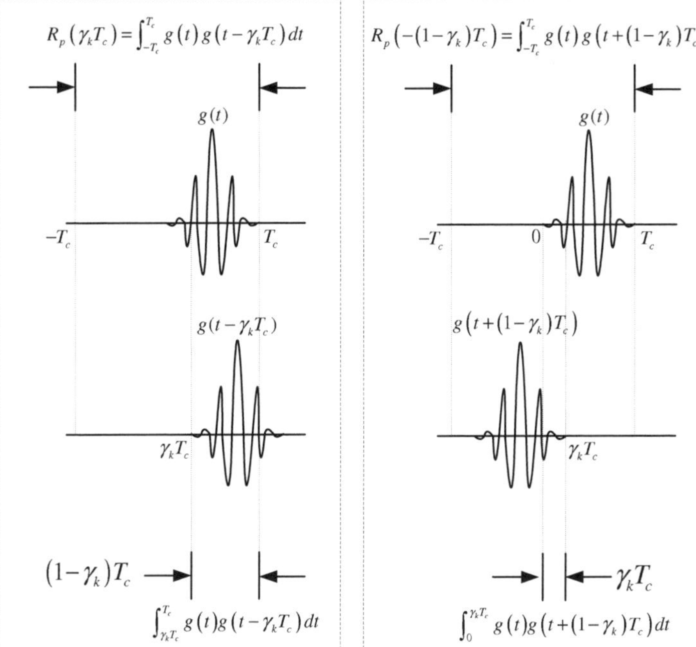

Figure 7.27 This figure illustrates how to calculate chip-wise pulse autocorrelation functions $R_p(\gamma_k T_c)$ and $R_p(-(1-\gamma_k)T_c)$.

MAI statistics in a flat fading channel

If K is sufficiently large, we can approximate the multiple access interference term, that is, I, as a Gaussian random variable. Therefore, the decision variable $y[(j+1)T_b]$ defined in Equation (7.10) will also be Gaussian. It is assumed that the appearance frequency of two consecutive bits for the k-th user is independently equiprobable; that is, $P(b_j^{(k)} = 1) = P(b_j^{(k)} = -1) = \frac{1}{2}$. Besides, we also have

$$E\left[(b_{j-1}^{(k)})^2\right] = E\left[(b_j^{(k)})^2\right] = 1; \quad E\left[b_{j-1}^{(k)} b_j^{(k)}\right] = E\left[b_{j-1}^{(k)}\right] E(b_j^{(k)}) = 0 \cdot 0 = 0 \tag{7.24}$$

Therefore, the conditional mean of I_k can be calculated as

$$\begin{aligned}
E(I_k | i_k, \gamma_k) &= \tfrac{1}{2}\alpha \left\{ C_{k,1}(i_k - N + 1) R_p\left[-(1-\gamma_k)T_c\right] + C_{k,1}(i_k - N) R_p(\gamma_k T_c) \right. \\
&\quad \left. + C_{k,1}(i_k) R_p(\gamma_k T_c) + C_{k,1}(i_k + 1) R_p\left[-(1-\gamma_k)T_c\right]\right) \\
&\quad + \tfrac{1}{2}\alpha_c \left[C_{k,1}(i_k - N + 1) R_p\left[-(1-\gamma_k)T_c\right] + C_{k,1}(i_k - N) R_p(\gamma_k T_c) \right. \\
&\quad \left. - C_{k,1}(i_k) R_p(\gamma_k T_c) - C_{k,1}(i_k + 1) R_p\left[-(1-\gamma_k)T_c\right]\right\} \\
&= \alpha \left\{ C_{k,1}(i_k - N + 1) R_p\left[-(1-\gamma_k)T_c\right] + C_{k,1}(i_k - N) R_p(\gamma_k T_c) \right\} \tag{7.25}
\end{aligned}$$

Using the results given in Equation (7.24), we obtain the conditional second order moment of I_k as

$$\begin{aligned}
E(I_k^2 | i_k, \gamma_k) &= \alpha^2 \left\{ \left[C_{k,1}(i_k - N + 1) R_p(-(1-\gamma_k)T_c) + C_{k,1}(i_k - N) R_p(\gamma_k T_c)\right]^2 \right. \\
&\quad \left. + \left[C_{k,1}(i_k) R_p(\gamma_k T_c) + C_{k,1}(i_k + 1) R_p(-(1-\gamma_k)T_c)\right]^2 \right\} \tag{7.26}
\end{aligned}$$

Therefore, the conditional variance of I_k becomes

$$\begin{aligned}
Var\,(I_k|i_k,\gamma_k) &= E\left[I_k^2|i_k,\gamma_k\right] - \left[E\,(I_k|i_k,\gamma_k)\right]^2 \\
&= \alpha^2\left\{C_{k,1}\,(i_k)\,R_p\,(\gamma_k T_c) + C_{k,1}\,(i_k+1)\,R_p\left[-(1-\gamma_k)\,T_c\right]\right\}^2
\end{aligned} \tag{7.27}$$

Because of $I = \sum_{k=2}^{K} I_k$, we have the conditional mean of I as

$$\begin{aligned}
E\,(I|i,\gamma) &= \sum_{k=2}^{K} E\,(I_k|i_k,\gamma_k) \\
&= \sum_{k=2}^{K}\alpha\left\{C_{k,1}\,(i_k-N+1)\,R_p\left[-(1-\gamma_k)\,T_c\right] + C_{k,1}\,(i_k-N)\,R_p\,(\gamma_k T_c)\right\}
\end{aligned}$$

$$\tag{7.28}$$

We also have

$$I^2 = \sum_{k=2}^{K} I_k^2 + \sum_{x=2}^{K}\sum_{x\neq y,y=2}^{K} I_x I_y \tag{7.29}$$

If all interference components, that is, $I_2,\ldots,I_K$, are mutually independent, the conditional variance can be written as

$$\begin{aligned}
Var\,(I|i,\gamma) &= E\left[I^2|i_k,\gamma_k\right] - \left[E\,(I|i_k,\gamma_k)\right]^2 \\
&= \sum_{k=2}^{K} E\,(I_k^2|i_k,\gamma_k) + \sum_{x=2}^{K}\sum_{x\neq y,y=2}^{K} E\,(I_x|i_x,\gamma_x)\,E\,(I_y|i_y,\gamma_y) \\
&\quad - \sum_{k=2}^{K} E^2\,(I_k|i_k,\gamma_k) - \sum_{x=2}^{K}\sum_{x\neq y,y=2}^{K} E\,(I_x|i_x,\gamma_x)\,E\,(I_y|i_y,\gamma_y) \\
&= \sum_{k=2}^{K} E\,(I_k^2|i_k,\gamma_k) - \sum_{k=2}^{K} E^2\,(I_k|i_k,\gamma_k)
\end{aligned} \tag{7.30}$$

Therefore we obtain

$$\begin{aligned}
Var\,(I|i,\gamma) &= \sum_{k=2}^{K} E\,(I_k^2|i_k,\gamma_k) - \sum_{k=2}^{K} E^2\,(I_k|i_k,\gamma_k) \\
&= \sum_{k=2}^{K}\alpha^2\left\{\left\{C_{k,1}\,(i_k-N+1)\,R_p\left[-(1-\gamma_k)\,T_c\right] + C_{k,1}\,(i_k-N)\,R_p\,(\gamma_k T_c)\right\}^2\right. \\
&\quad \left. + \left\{C_{k,1}\,(i_k)\,R_p\,(\gamma_k T_c) + C_{k,1}\,(i_k+1)\,R_p\left[-(1-\gamma_k)\,T_c\right]\right\}^2\right\} \\
&\quad - \sum_{k=2}^{K}\alpha^2\left\{C_{k,1}\,(i_k-N+1)\,R_p\left[-(1-\gamma_k)\,T_c\right] + C_{k,1}\,(i_k-N)\,R_p\,(\gamma_k T_c)\right\}^2 \\
&= \sum_{k=2}^{K}\alpha^2\left\{C_{k,1}\,(i_k)\,R_p\,(\gamma_k T_c) + C_{k,1}\,(i_k+1)\,R_p\left[-(1-\gamma_k)\,T_c\right]\right\}^2
\end{aligned} \tag{7.31}$$

from which we observe that the variance of the combined interference component is closely related to three parameters; one being the square of the fading coefficient α, the other being the partial CCF of the signature codes of the first and k-th users, and the last being the ACF of UWB pulse waveform $R_p(\gamma_k T_c)$ and $R_p\left[-(1-\gamma_k)T_c\right]$. In fact, both partial cross-correlation function of the signature codes and ACF of the pulse waveform can be calculated explicitly if we have the knowledge of user signature codes and pulse waveforms, and thus the variance of MAI can also be determined.

Random sequences

In this subsection, purely random sequences will be used as spreading sequences, whose chips $\{a_n^{(k)}\}_{n=0,\ldots,N-1}^{k=1,\ldots,K}$ will take "-1" and "$+1$" equally likely. In addition, $a_i^{(k)}$ and $a_j^{(k)}$ should be independent if $i \neq j$. The integer-chip relative delay between the first and k-th users' transmissions, $\{i_k\}_{k=1}^{K}$, is uniformly distributed over $(0, N-1)$; and the fractional-chip relative delay between the first and k-th users' transmissions, $\{\gamma_k\}_{k=1}^{K}$, is uniformly distributed over $(0, 1)$. On the basis of these assumptions, we can calculate the unconditional expectation of discrete partial cross-correlation functions of the first and k-th spreading sequences as follows:

$$\begin{cases} E\left[C_{k,1}(i)\,|i\right] = E\left(\sum_{j=0}^{N-1-i} a_j^{(k)} a_{j+i}^{(1)}\right) = 0, & 0 \leq i \leq N-1 \\ E\left[C_{k,1}(i)\,|i\right] = E\left(\sum_{j=0}^{N-1+i} a_{j-i}^{(k)} a_j^{(1)}\right) = 0, & -(N-1) \leq i \leq 0 \end{cases} \tag{7.32}$$

The unconditional second order moment of discrete partial CCFs can be calculated section by section as

$$\begin{aligned} E\left[(C_{k,1}(i))^2\,|i\right] &= E\left(\sum_{l=0}^{N-1-i}\sum_{j=0}^{N-1-i} a_j^{(k)} a_{j+i}^{(1)} a_l^{(k)} a_{l+i}^{(1)}\right) \\ &= \sum_{l=0}^{N-1-i}\sum_{j=0}^{N-1-i} E\left(a_j^{(k)} a_{j+i}^{(1)} a_l^{(k)} a_{l+i}^{(1)}\right) \\ &= \sum_{l=0}^{N-1-i}\sum_{j=0}^{N-1-i} E\left(a_{j+i}^{(1)} a_{l+i}^{(1)}\right) E\left(a_j^{(k)} a_l^{(k)}\right) \\ &= \sum_{j=0}^{N-1-i} E\left[\left(a_{j+i}^{(1)}\right)^2\right] E\left[\left(a_j^{(k)}\right)^2\right] = N-i, \quad (0 \leq i \leq N-1) \end{aligned} \tag{7.33}$$

and

$$\begin{aligned} E\left[(C_{k,1}(i))^2\,|i\right] &= E\left(\sum_{l=0}^{N-1+i}\sum_{j=0}^{N-1+i} a_{j-i}^{(k)} a_j^{(1)} a_{l-i}^{(k)} a_l^{(1)}\right) \\ &= \sum_{l=0}^{N-1+i}\sum_{j=0}^{N-1+i} E\left(a_{j-i}^{(k)} a_j^{(1)} a_{l-i}^{(k)} a_l^{(1)}\right) \end{aligned}$$

$$= \sum_{l=0}^{N-1+i} \sum_{j=0}^{N-1+i} E\left(a_{j-i}^{(k)} a_{l-i}^{(k)}\right) E\left(a_j^{(1)} a_l^{(1)}\right)$$

$$= \sum_{j=0}^{N-1+i} E\left[\left(a_{j-i}^{(k)}\right)^2\right] E\left[\left(a_j^{(1)}\right)^2\right] = N+i, \quad [-(N-1) \leq i \leq 0] \quad (7.34)$$

Inserting the above results into Equation (7.28), we obtain

$$E\left(I|\gamma\right) = E\left[E\left(I|i,\gamma\right)\right]$$

$$= \sum_{k=2}^{K} \sum_{i_k=0}^{N-1} \frac{\alpha}{N} \left\{ E\left[C_{k,1}\left(i_k - N + 1\right)\right] R_p\left(-(1-\gamma_k)T_c\right) + E\left[C_{k,1}\left(i_k - N\right)\right] R_p\left(\gamma_k T_c\right) \right\}$$

$$= 0 \tag{7.35}$$

Applying Equations (7.33) and (7.34) to Equation (7.31), we have

$$Var\left(I|\gamma\right) = E\left[Var\left(I|i,\gamma\right)\right]$$

$$= \sum_{k=2}^{K} \sum_{i_k=0}^{N-1} \frac{1}{N} \alpha^2 E\left\{\left[C_{k,1}\left(i_k\right) R_p\left(\gamma_k T_c\right) + C_{k,1}\left(i_k + 1\right) R_p\left(-(1-\gamma_k)T_c\right)\right]^2\right\}$$

$$= \sum_{k=2}^{K} \sum_{i_k=0}^{N-1} \frac{1}{N} \alpha^2 \left\{(N - i_k)\left[R_p\left(\gamma_k T_c\right)\right]^2 + (N - i_k - 1)\left[R_p\left(-(1-\gamma_k)T_c\right)\right]^2\right\}$$

$$= \sum_{k=2}^{K} \alpha^2 \left\{\frac{(N+1)}{2}\left[R_p\left(\gamma_k T_c\right)\right]^2 + \frac{(N-1)}{2}\left[R_p\left(-(1-\gamma_k)T_c\right)\right]^2\right\} \tag{7.36}$$

Let us define parameter σ_{mp}^2 as

$$\sigma_{mp}^2 = E\left[R_p^2\left(\tau\right)\right] = \frac{1}{2T_c} \int_{-T_c}^{T_c} \left(\int_{-T_c}^{T_c} g(t-\tau)g(t)\,dt\right)^2 d\tau, \tag{7.37}$$

then we have

$$Var\left(I\right) = E\left[Var\left(I|\gamma\right)\right]$$

$$= \sum_{k=2}^{K} \alpha^2 \left\{\frac{(N+1)}{2} E\left(\left[R_p\left(\gamma_k T_c\right)\right]^2\right) + \frac{(N-1)}{2} E\left(\left[R_p\left(-(1-\gamma_k)T_c\right)\right]^2\right)\right\}$$

$$= (K-1)\alpha^2 N \sigma_{mp}^2 \tag{7.38}$$

Now that we have identified all deterministic and random components in the decision variable as

$$\begin{cases} S = \alpha N b_j^{(1)} E_{mp} \\ \eta \sim \mathcal{N}\left(0, \sigma_n^2 N E_{mp}\right) \\ I \sim \mathcal{N}\left(0, (K-1)\alpha^2 N \sigma_{mp}^2\right), \end{cases} \tag{7.39}$$

together with the decision rule of

$$z\left(j\right) = \begin{cases} +1, & y\left((j+1)T_b\right) > 0 \\ -1, & y\left((j+1)T_b\right) \leq 0, \end{cases} \tag{7.40}$$

we can proceed to derive the BER expression as

$$
\begin{aligned}
P_e &= \int_{-\infty}^{\infty} Q \left\{ \frac{\alpha N E_{mp}}{\sqrt{\sigma_n^2 N E_{mp} + (K-1)\,\alpha^2 N \sigma_{mp}^2}} \right\} f_\alpha(\alpha)\, d\alpha \\
&= \int_{-\infty}^{\infty} Q \left\{ \frac{1}{\sqrt{\frac{1}{\alpha^2} \frac{\sigma_n^2}{N E_{mp}} + (K-1)\,\frac{\sigma_{mp}^2}{N E_{mp}^2}}} \right\} f_\alpha(\alpha)\, d\alpha \\
&= \int_{-\infty}^{\infty} Q \left\{ \frac{1}{\sqrt{\frac{1}{\alpha^2}\frac{1}{SNR} + (K-1)\,\frac{\sigma_{mp\_normal}^2}{N E_{mp}}}} \right\} f_\alpha(\alpha)\, d\alpha
\end{aligned}
\tag{7.41}
$$

where $\sigma_{mp\_normal}^2$ has been defined as

$$
\sigma_{mp\_normal}^2 = \frac{\sigma_{mp}^2}{E_{mp}}
\tag{7.42}
$$

and $f_\alpha(\alpha)$ is the probability density function of fading coefficient α.

7.6.4 Frequency-Selective Fading Channels

In this subsection, we analyze pulse waveform–dependent BER performance of a DS-CDMA UWB radio under a frequency-selective fading environment. The multipath fading channel is modeled by a modified Saleh–Valenzuela (S–V) indoor channel model, which was initially proposed by A. Saleh and R. Valenzuela [690] and was modified by J. Foerster and Q. Li [691].

Modified S–V channel model

The basic idea of the modified S–V channel model can be summarized as follows [691].

- The signal arrivals from an indoor channel can be decomposed into several clusters, each of which consists of several multipath rays. Different clusters are formed because of the building's structure, such as different storeys; while different rays in the same cluster are formed owing to different reflecting objects in the propagation path of the cluster.

- The amplitude of the rays attenuates according to a Lognormal distribution (instead of a Rayleigh Distribution as suggested in the original S–V model [690]). In addition, the variance of amplitude attenuation decays exponentially with the delays of different clusters as well as different rays in the same cluster.

- The arrival processes of both clusters and rays obey Poisson distributions and thus their interarrival times are exponentially (instead of uniform, as specified in the original S–V model [690]) distributed.

The modified S–V channel model can be expressed mathematically by

$$
h(t) = \sum_{q=1}^{Q} \sum_{l=1}^{L} w_{q,l}\alpha_{q,l}\delta\left(t - T_q - \tau_{q,l}\right)
\tag{7.43}
$$

where $q \in (1, Q)$ and $l \in (1, L)$ stand for cluster and ray indices, respectively; $\{w_{q,l}\}_{l=1,\ldots,L}^{q=1,\ldots,Q}$ takes either $+1$ or -1 to denote either positive or negative path return; $\{T_q\}_{q=1}^{Q}$ is the delay of the first ray in the q-th cluster; $\{\tau_{q,l}\}_{l=1,\ldots,L}^{q=1,\ldots,Q}$ is the relative delay between the first and l-th rays in the q-th cluster with $T_1 = 0$ and $\{\tau_{q,1}\}_{q=1}^{Q} = 0$, without losing generality. To model the exponential distributions of interarrival times for both clusters and rays, we need to define two parameters, Λ and λ, which represent cluster arrival rate and ray arrival rate, respectively, with their distributions being

$$\begin{cases} p\left(T_q | T_{q-1}\right) = \Lambda \exp\left[-\Lambda \left(T_q - T_{q-1}\right)\right], & q > 0 \\ p\left(\tau_{q,l} | \tau_{q,l-1}\right) = \lambda \exp\left[-\lambda \left(\tau_{q,l} - \tau_{q,l-1}\right)\right], & l > 0 \end{cases} \tag{7.44}$$

Because of the properties of exponentially distributed interarrival times for both clusters and rays, we have

$$\begin{cases} T_q - T_1 = \left(T_q - T_{q-1}\right) + \left(T_{q-1} - T_{q-2}\right) + \cdots + \left(T_2 - T_1\right) \\ \tau_{q,l} - \tau_{q,1} = \left(\tau_{q,l} - \tau_{q,l-1}\right) + \left(\tau_{q,l-1} - \tau_{q,l-2}\right) + \cdots + \left(\tau_{q,2} - \tau_{q,1}\right) \end{cases} \tag{7.45}$$

and thus the mean interarrival times for both clusters and rays can be expressed by

$$\begin{cases} E\left(T_q\right) = (q - 1) \frac{1}{\Lambda} \\ E\left(\tau_{q,l}\right) = (l - 1) \frac{1}{\lambda} \end{cases} \tag{7.46}$$

The mean relative delay between the l-th ray in the q-th cluster and the first ray in the first cluster is

$$E\left(T_q + \tau_{q,l}\right) = (q - 1) \frac{1}{\Lambda} + (l - 1) \frac{1}{\lambda} \tag{7.47}$$

The channel gain $\alpha_{q,l}$ is a Lognormal random variable, whose relation can be written as

$$20 \log_{10}\left(\alpha_{q,l}\right) \sim N\left(\mu_q, \sigma^2\right) \tag{7.48}$$

Also, $\alpha_{q,l}$ in Equation (7.43) is always positive and its second order moment obeys a dual-exponential distribution as

$$E\left(\alpha_{q,l}^2\right) = \Omega_1 e^{-T_q/\Gamma} e^{-\tau_{q,l}/\gamma} \tag{7.49}$$

where Γ and γ are the attenuation coefficients for the clusters and rays, respectively; Ω_1 is the mean power of the first ray in the first cluster. Therefore, the mean of $\alpha_{q,l}$ in Equation (7.48) can be written as

$$\mu_q = \frac{10 \ln\left(\Omega_1\right) - 10 T_q/\Gamma - 10 \tau_{q,l}/\gamma}{\ln(10)} - \frac{\sigma^2 \ln(10)}{20} \tag{7.50}$$

Received signal in modified S–V channel

The transmitter block diagram for a DS-CDMA UWB radio under frequency-selective fading channels is almost exactly the same as Figure 7.25(a). The only difference is that the channel impulse responses $\{h_k(t)\}_{k=1}^{K}$ in Figure 7.25(a) should be replaced by the modified S–V model defined in Equation (7.43), or

$$h_k(t) = \sum_{q=1}^{Q_k} \sum_{l=1}^{L_k} w_{k,q,l} \alpha_{k,q,l} \delta\left(t - T_{k,q} - \tau_{k,q,l} - \tau_k\right); \tag{7.51}$$

where $\{\tau_k\}_{k=1}^{K}$ is the relative delay between the first and k-th users' transmissions and $\tau_1 = 0$. The receiver model is shown in Figure 7.28, where a RAKE receiver is used for the reception of the signal from user 1. The transmitted signal from the k-th user can be written as

$$s_k(t) = \sum_{j=-\infty}^{\infty} \sum_{n=0}^{N-1} b_j^{(k)} a_n^{(k)} g(t - jT_b - nT_c) \tag{7.52}$$

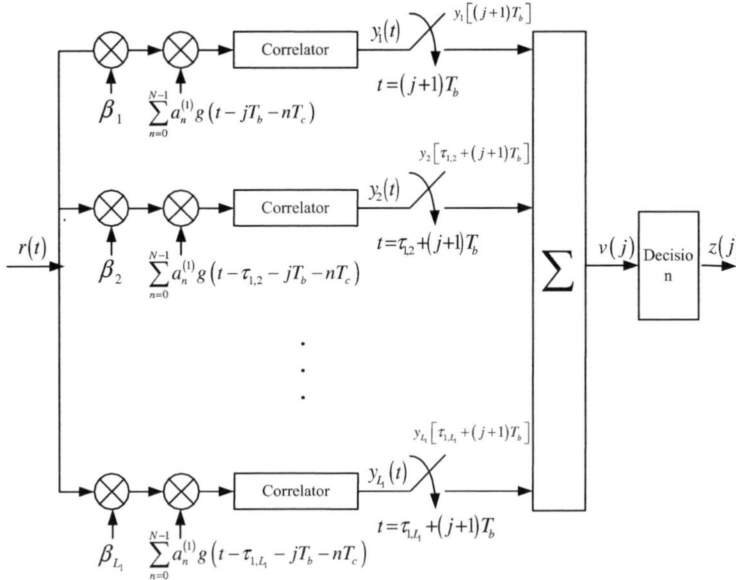

Figure 7.28 The RAKE receiver for reception of the signal from the first user using the modified S–V channel model, where there are totally L_1 fingers, the combining coefficients are $\{\beta_p\}_{p=1}^{L_1}$, and $\tau_{1,1} = 0$ without losing generality.

where $\{b_j^{(k)}\}_{k=1}^K$ is the input binary information from the k-th user, T_b is the bit duration, $\{a_n^{(k)}\}_{n=0}^{N-1}$ is the spreading sequence for the k-th user with its length being N, T_c is the chip duration with $T_b = NT_c$, and $g(t)$ denotes the pulse waveform, which was defined in Equation (7.7).

The received signal can be written as

$$r(t) = \sum_{k=1}^{K} s_k(t) \otimes h_k(t) + n(t)$$

$$= \sum_{k=1}^{K} \sum_{q=1}^{Q_k} \sum_{l=1}^{L_k} w_{k,q,l}\alpha_{k,q,l} s_k\left(t - T_{k,q} - \tau_{k,q,l} - \tau_k\right) + n(t) \tag{7.53}$$

where $\otimes$ denotes convolution operation; $n(t)$ is the AWGN component added in the channel with its mean and variance being zero and σ_n^2, respectively, or simply represented by $n(t) \sim \mathcal{N}(0, \sigma_n^2)$. After the convolution, the received signal can be rewritten as

$$r(t) = \sum_{q=1}^{Q_1} \sum_{l=1}^{L_1} \sum_{j=-\infty}^{\infty} \sum_{n=0}^{N-1} w_{q,l}\alpha_{q,l} b_j^{(1)} a_n^{(1)} g(t - T_q - \tau_{q,l} - jT_b - nT_c)$$

$$+ \sum_{k=2}^{K} \sum_{q=1}^{Q_k} \sum_{l=1}^{L_k} \sum_{j=-\infty}^{\infty} \sum_{n=0}^{N-1} w_{k,q,l}\alpha_{k,q,l} b_j^{(k)} a_n^{(k)} g(t - T_{k,q} - \tau_{k,q,l} - \tau_k - jT_b - nT_c)$$

$$+ n(t) \tag{7.54}$$

It is assumed that an L_1-finger RAKE receiver will capture the L_1 rays in the first cluster multipath returns. The output from the p-th finger is

$$y_p\left[\tau_{1,p} + (j+1)\,T_b\right] = \int_{\tau_{1,p}+jT_b}^{\tau_{1,p}+(j+1)T_b} r(t)\beta_p \sum_{n=0}^{N-1} a_n^{(1)} g(t - \tau_{1,p} - jT_b - nT_c)\,dt \tag{7.55}$$

Then, the decision variable for the j-th bit, $v(j)$, becomes

$$v(j) = \sum_{p=1}^{L_1} y_l\left[\tau_{1,p} + (j+1)\,T_b\right]$$

$$= \sum_{p=1}^{L_1} \int_{\tau_{1,p}+jT_b}^{\tau_{1,p}+(j+1)T_b} r(t)\beta_p \sum_{n=0}^{N-1} a_n^{(1)} g(t - \tau_{1,p} - jT_b - nT_c)\,dt \tag{7.56}$$

which can be decomposed into four terms, that is, useful signal, MI, MAI, and noise, as

$$v(j) = S + I_L + I_K + \eta = \sum_{p=1}^{L_1} \left(S_p + I_{L,p} + I_{K,p} + \eta_p\right) \tag{7.57}$$

where S_p, $I_{L,p}$, $I_{K,p}$, and η_p are the useful signal, MI, MAI, and noise terms generated from the p-th finger, respectively. The useful signal component can be written as

$$S_p = \int_{\tau_{1,p}+jT_b}^{\tau_{1,p}+(j+1)T_b} \beta_p \sum_{j=-\infty}^{\infty} \sum_{n=0}^{N-1} w_{1,p}\alpha_{1,p} b_j^{(1)} a_n^{(1)} g(t - \tau_{1,p} - jT_b - nT_c)$$

$$\times \sum_{n=0}^{N-1} a_n^{(1)} g(t - \tau_{1,p} - jT_b - nT_c)\,dt$$

$$= \beta_p w_{1,p}\alpha_{1,p} N b_j^{(1)} E_{mp} \tag{7.58}$$

where $E_{mp} = \int_0^{T_c} g^2(t)\,dt$ denotes the energy of a single pulse. The weighted sum of the useful signal component yields

$$S = \sum_{p=1}^{L_1} S_p = N b_j^{(1)} E_{mp} \sum_{p=1}^{L_1} \beta_p w_{1,p}\alpha_{1,p} \tag{7.59}$$

AWGN statistics

The noise component from the first finger can be written as

$$\eta_1 = \int_{jT_b}^{(j+1)T_b} \beta_1 n(t) \sum_{n=0}^{N-1} a_n^{(1)} g(t - jT_b - nT_c)\,dt \tag{7.60}$$

Obviously, since $n(t) \sim \mathcal{N}\left(0, \sigma_n^2\right)$ we readily get $E\left(\eta_1\right) = 0$ from Equation (7.60). The variance of the noise component from the first finger can be calculated as

$$\sigma_{\eta_1}^2 = E\left(\eta_1^2\right) = E\left[\int_{jT_b}^{(j+1)T_b} \beta_1 n(t) \sum_{n=0}^{N-1} a_n^{(1)} g(t - jT_b - nT_c)\,dt\right]^2 \tag{7.61}$$

Let $q(t) = \beta_1 \sum_{n=0}^{N-1} a_n^{(1)} g(t - jT_b - nT_c)$. Equation (7.61) can be written as

$$
\begin{aligned}
\sigma_{\eta_1}^2 = E\left(\eta_1^2\right) &= E\left[\int_{jT_b}^{(j+1)T_b} n(t) q(t)\, dt\right]^2 \\
&= \int_{jT_b}^{(j+1)T_b} \int_{jT_b}^{(j+1)T_b} E[n(t)n(\zeta)] q(t)\, q(\zeta)\, dt\, d\zeta \\
&= \sigma_n^2 \int_{jT_b}^{(j+1)T_b} q^2(t)\, dt = \sigma_n^2 \int_{jT_b}^{(j+1)T_b} \beta_1^2 \sum_{n=0}^{N-1} \left[a_n^{(1)}\right]^2 g_1^2(t - jT_b - nT_c)\, dt \\
&= \sigma_n^2 \beta_1^2 N E_{mp}
\end{aligned}
\tag{7.62}
$$

Therefore, the noise term in the output from the first finger is still Gaussian as $\eta_1 \sim \mathcal{N}\left(0, \sigma_n^2 \beta_1^2 N E_{mp}\right)$. The weighted sum of all noise terms in the output of the RAKE receiver will be Gaussian too, or

$$
\eta \sim \mathcal{N}\left(0, \sigma_n^2 N E_{mp} \sum_{p=1}^{L_1} \beta_p^2\right)
\tag{7.63}
$$

MI statistics in modified S–V channel

From Equation (7.57) we have the MI component in the decision variable $v(j)$ as $I_L = \sum_{p=1}^{L_1} I_{L,p}$. It is known from Equations (7.54) and (7.56) that the MI term obtained from the output of the first finger is

$$
\begin{aligned}
I_{L,1} &= \sum_{l=2}^{L_1} I_{1,l,1} + \sum_{q=2}^{Q_1} \sum_{l=1}^{L_1} I_{q,l,1} \\
&= \sum_{l=2}^{L_1} \int_{jT_b}^{(j+1)T_b} \beta_1 w_{1,l} \alpha_{1,l} \sum_{j=-\infty}^{\infty} \sum_{n=0}^{N-1} b_j^{(1)} a_n^{(1)} g(t - \tau_{1,l} - jT_b - nT_c) \\
&\quad \times \sum_{n=0}^{N-1} a_n^{(1)} g(t - jT_b - nT_c)\, dt \\
&\quad + \sum_{q=2}^{Q_1} \sum_{l=1}^{L_1} \int_{jT_b}^{(j+1)T_b} \beta_1 w_{q,l} \alpha_{q,l} \sum_{j=-\infty}^{\infty} \sum_{n=0}^{N-1} b_j^{(1)} a_n^{(1)} g(t - T_q - \tau_{q,l} - jT_b - nT_c) \\
&\quad \times \sum_{n=0}^{N-1} a_n^{(1)} g(t - jT_b - nT_c)\, dt
\end{aligned}
\tag{7.64}
$$

Using the approach as shown in Appendix B, we can obtain

$$
\begin{aligned}
I_{L,1} &= \sum_{l=2}^{L_1} I_{1,l,1} + \sum_{q=2}^{Q_1} \sum_{l=1}^{L_1} I_{q,l,1} \\
&= \sum_{l=2}^{L_1} \beta_1 w_{1,l} \alpha_{1,l} \left\{ b_{j-1}^{(1)} \left[C_1 \left(i_{1,l} - N + 1\right) R_p \left(-(1 - \gamma_{1,l}) T_c\right) \right. \right. \\
&\quad \left. \left. + C_1 \left(i_{1,l} - N\right) R_p \left(\gamma_{1,l} T_c\right) \right]
\end{aligned}
$$

$$+ b_j^{(1)} \big[C_1 \left(i_{1,l} \right) R_p \left(\gamma_{1,l} T_c \right) + C_1 \left(i_{1,l} + 1 \right) R_p \left(- \left(1 - \gamma_{1,l} \right) T_c \right) \big] \Big\}$$

$$+ \sum_{q=2}^{Q_1} \sum_{l=1}^{L_1} \beta_1 w_{q,l} \alpha_{q,l} \Big\{ b_{j-1}^{(1)} \big[C_1 \left(i_{q,l} - N + 1 \right) R_p \left(- \left(1 - \gamma_{q,l} \right) T_c \right)$$

$$+ C_1 \left(i_{q,l} - N \right) R_p \left(\gamma_{q,l} T_c \right) \big]$$

$$+ b_j^{(1)} \big[C_1 \left(i_{q,l} \right) R_p \left(\gamma_{q,l} T_c \right) + C_1 \left(i_{q,l} + 1 \right) R_p \left(- \left(1 - \gamma_{q,l} \right) T_c \right) \big] \Big\} \qquad (7.65)$$

Similarly, we can also have

$$I_{L,p} = \sum_{l=1, l \neq p}^{L_1} I_{1,l,p} + \sum_{q=2}^{Q_1} \sum_{l=1}^{L_1} I_{q,l,p}$$

$$= \sum_{l=1, l \neq p}^{L_1} \beta_p w_{1,l} \alpha_{1,l} \Big\{ b_{j-1}^{(1)} \big[C_1 \left(i_{1,l,p} - N + 1 \right) R_p \left(- \left(1 - \gamma_{1,l,p} \right) T_c \right)$$

$$+ C_1 (i_{1,l,p} - N) R_p (\gamma_{1,l,p} T_c) \big]$$

$$+ b_j^{(1)} \big[C_1 \left(i_{1,l,p} \right) R_p \left(\gamma_{1,l,p} T_c \right) + C_1 \left(i_{1,l,p} + 1 \right) R_p \left(- \left(1 - \gamma_{1,l,p} \right) T_c \right) \big] \Big\}$$

$$+ \sum_{q=2}^{Q_1} \sum_{l=1}^{L_1} \beta_p w_{q,l} \alpha_{q,l} \Big\{ b_{j-1}^{(1)} \big[C_1 \left(i_{q,l,p} - N + 1 \right) R_p \left(- \left(1 - \gamma_{q,l,p} \right) T_c \right)$$

$$+ C_1 (i_{q,l,p} - N) R_p (\gamma_{q,l,p} T_c) \big]$$

$$+ b_j^{(1)} \big[C_1 \left(i_{q,l,p} \right) R_p \left(\gamma_{q,l,p} T_c \right) + C_1 \left(i_{q,l,p} + 1 \right) R_p \left(- \left(1 - \gamma_{q,l,p} \right) T_c \right) \big] \Big\} \qquad (7.66)$$

Assume that the first user sends its binary bit stream with "+1" and "−1" appearing equal likely or $P(b_j^{(1)} = 1) = P(b_j^{(1)} = -1) = \frac{1}{2}$, and thus $E\{[b_{j-1}^{(1)}]^2\} = E\{[b_j^{(1)}]^2\} = 1$ and $E[b_{j-1}^{(1)} b_j^{(1)}] = E[b_{j-1}^{(1)}] E[b_j^{(1)}] = 0 \cdot 0 = 0$. From Equation (7.66) we can proceed to derive the conditional mean and variance of $I_{1,l,1}$, which corresponds to the signal captured from the l-th ray in the first cluster of the first user, as

$$E \left(I_{1,l,1} | i_{1,l,1}, \gamma_{1,l,1}, w_{1,l}, \alpha_{1,l}, \beta_1 \right)$$

$$= \beta_1 w_{1,l} \alpha_{1,l} \big[C_1 \left(i_{1,l,1} - N + 1 \right) R_p \left(- \left(1 - \gamma_{1,l,1} \right) T_c \right) + C_1 \left(i_{1,l,1} - N \right) R_p \left(\gamma_{1,l,1} T_c \right) \big]$$

$$Var \left(I_{1,l,1} | i_{1,l,1}, \gamma_{1,l,1}, w_{1,l}, \alpha_{1,l}, \beta_1 \right)$$

$$= \beta_1^2 w_{1,l}^2 \alpha_{1,l}^2 \big[C_1 \left(i_{1,l,1} \right) R_p \left(\gamma_{1,l,1} T_c \right) + C_1 \left(i_{1,l,1} + 1 \right) R_p \left(- \left(1 - \gamma_{1,l,1} \right) T_c \right) \big]^2$$

Similarly, we can derive the conditional mean and variance for the signal captured from the l-th ray in the q-th cluster of the first user as

$$E \left(I_{q,l,1} | i_{q,l,1}, \gamma_{q,l,1}, w_{q,l}, \alpha_{q,l}, \beta_1 \right)$$

$$= \beta_1 w_{q,l} \alpha_{q,l} \big[C_1 \left(i_{q,l,1} - N + 1 \right) R_p \left(- \left(1 - \gamma_{q,l,1} \right) T_c \right) + C_1 \left(i_{q,l,1} - N \right) R_p \left(\gamma_{q,l,1} T_c \right) \big]$$

$$Var \left(I_{q,l,1} | i_{q,l,1}, \gamma_{q,l,1}, w_{q,l}, \alpha_{q,l}, \beta_1 \right)$$

$$= \beta_1^2 w_{q,l}^2 \alpha_{q,l}^2 \big[C_1 \left(i_{q,l,1} \right) R_p \left(\gamma_{q,l,1} T_c \right) + C_1 \left(i_{q,l,1} + 1 \right) R_p \left(- \left(1 - \gamma_{q,l,1} \right) T_c \right) \big]^2$$

Therefore, the conditional mean and variance of the MI term from the first finger become

$$
E\left(I_{L,1}|i_{q,l,1}, \gamma_{q,l,1}, w_{q,l}, \alpha_{q,l}, \beta_1\right)
$$

$$
= \sum_{l=2}^{L_1} E\left(I_{1,l,1}|i_{1,l,1}, \gamma_{1,l,1}, w_{1,l}, \alpha_{1,l}, \beta_1\right)
$$

$$
+ \sum_{q=2}^{Q_1} \sum_{l=1}^{L_1} E\left(I_{q,l,1}|i_{q,l,1}, \gamma_{q,l,1}, w_{q,l}, \alpha_{q,l}, \beta_1\right)
$$

$$
= \sum_{l=2}^{L_1} \beta_1 w_{1,l}\alpha_{1,l}\left[C_1\left(i_{1,l,1} - N + 1\right) R_p\left(-\left(1 - \gamma_{1,l,1}\right) T_c\right)\right.
$$

$$
\left. + C_1\left(i_{1,l,1} - N\right) R_p\left(\gamma_{1,l,1}T_c\right)\right]
$$

$$
+ \sum_{q=2}^{Q_1} \sum_{l=1}^{L_1} \beta_1 w_{q,l}\alpha_{q,l}\left[C_1\left(i_{q,l,1} - N + 1\right) R_p\left(-\left(1 - \gamma_{q,l,1}\right) T_c\right)\right.
$$

$$
\left. + C_1\left(i_{q,l,1} - N\right) R_p\left(\gamma_{q,l,1}T_c\right)\right] \tag{7.67}
$$

$$
Var\left(I_{L,1}|i_{q,l,1}, \gamma_{q,l,1}, w_{q,l}, \alpha_{q,l}, \beta_1\right)
$$

$$
= \sum_{l=2}^{L_1} Var\left(I_{1,l,1}|i_{1,l,1}, \gamma_{1,l,1}, w_{1,l}, \alpha_{1,l}, \beta_1\right)
$$

$$
+ \sum_{q=2}^{Q_1} \sum_{l=1}^{L_1} Var\left(I_{q,l,1}|i_{q,l,1}, \gamma_{q,l,1}, w_{q,l}, \alpha_{q,l}, \beta_1\right)
$$

$$
= \sum_{l=2}^{L_1} \beta_1^2 w_{1,l}^2 \alpha_{1,l}^2\left[C_1\left(i_{1,l,1}\right) R_p\left(\gamma_{1,l,1}T_c\right)\right.
$$

$$
\left. + C_1\left(i_{1,l,1} + 1\right) R_p\left(-\left(1 - \gamma_{1,l,1}\right) T_c\right)\right]^2
$$

$$
+ \sum_{q=2}^{Q_1} \sum_{l=1}^{L_1} \beta_1^2 w_{q,l}^2 \alpha_{q,l}^2\left[C_1\left(i_{q,l,1}\right) R_p\left(\gamma_{q,l,1}T_c\right)\right.
$$

$$
\left. + C_1\left(i_{q,l,1} + 1\right) R_p\left(-\left(1 - \gamma_{q,l,1}\right) T_c\right)\right]^2 \tag{7.68}
$$

Here, again we take purely random sequences as the spreading codes to derive a closed form of the BER expression because in such a case the partial ACF, as defined in (C.4), can be quantified. It is shown in Appendix B that the conditional mean and variance of the MI component seen from the output of the first finger can be written as

$$
E\left(I_{L,1}|\gamma_{q,l,1}, w_{q,l}, \alpha_{q,l}, \beta_1\right) = E\left[E\left(I_{L,1}|i_{q,l,1}, \gamma_{q,l,1}, w_{q,l}, \alpha_{q,l}, \beta_1\right)\right]
$$

$$
= \sum_{l=2}^{L_1} \beta_1 w_{1,l}\alpha_{1,l}\left\{E\left[C_1(i_{1,l,1} - N + 1)\right] R_p\left[-(1 - \gamma_{1,l,1})T_c\right]\right.
$$

$$
\left. + E\left[C_1(i_{1,l,1} - N)\right] R_p(\gamma_{1,l,1}T_c)\right\}
$$

$$+ \sum_{q=2}^{Q_1} \sum_{l=1}^{L_1} \beta_1 w_{q,l} \alpha_{q,l} \left\{ E\left[C_1(i_{q,l,1} - N + 1)\right] R_p \left[-(1 - \gamma_{q,l,1})T_c\right] \right.$$

$$\left. + E\left[C_1(i_{q,l,1} - N)\right] R_p \left(\gamma_{q,l,1} T_c\right) \right\} \tag{7.69}$$

$$Var\left(I_{L,1} | w_{q,l}, \alpha_{q,l}, \beta_1\right)$$

$$= \sum_{l=2}^{L_1} \beta_1^2 w_{1,l}^2 \alpha_{1,l}^2 \left\{ \left(N - \frac{l-1}{\lambda T_c} + \frac{1}{2}\right) \sigma_{mp}^2 + \left(N - \frac{l-1}{\lambda T_c} - \frac{1}{2}\right) \sigma_{mp}^2 \right\}$$

$$+ \sum_{q=2}^{Q_1} \sum_{l=1}^{L_1} \beta_1^2 w_{q,l}^2 \alpha_{q,l}^2 \left\{ \left(N - \frac{q-1}{\Lambda T_c} - \frac{l-1}{\lambda T_c} + \frac{1}{2}\right) \sigma_{mp}^2 \right.$$

$$\left. + \left(N - \frac{q-1}{\Lambda T_c} - \frac{l-1}{\lambda T_c} - \frac{1}{2}\right) \sigma_{mp}^2 \right\}$$

$$= \sum_{l=2}^{L_1} \beta_1^2 w_{1,l}^2 \alpha_{1,l}^2 \left\{ \left(2N - \frac{2(l-1)}{\lambda T_c}\right) \sigma_{mp}^2 \right\}$$

$$+ \sum_{q=2}^{Q_1} \sum_{l=1}^{L_1} \beta_1^2 w_{q,l}^2 \alpha_{q,l}^2 \left\{ \left(2N - \frac{2(q-1)}{\Lambda T_c} - \frac{2(l-1)}{\lambda T_c}\right) \sigma_{mp}^2 \right\} \tag{7.70}$$

where σ_{mp}^2 is defined in Equation (7.37). It is to be noted that parameter σ_{mp}^2 is used here to characterize pulse waveforms used in a DS-CDMA UWB radio system and it plays an important role in determining overall BER performance of the UWB system. Finally, the conditional variance of the MI term generated by the RAKE receiver is

$$Var\left(I_L | w_{q,l}, \alpha_{q,l}, \beta\right) = \sum_{p=1}^{L_1} Var\left(I_{L,p} | w_{q,l}, \alpha_{q,l}, \beta_p\right)$$

$$= \sum_{p=1}^{L_1} \left[\sum_{l=1, l \neq p}^{L_1} \beta_p^2 w_{1,l}^2 \alpha_{1,l}^2 \left\{ \left(2N - \frac{2(l-1)}{\lambda T_c}\right) \sigma_{mp}^2 \right\} \right.$$

$$\left. + \sum_{q=2}^{Q_1} \sum_{l=1}^{L_1} \beta_p^2 w_{q,l}^2 \alpha_{q,l}^2 \left\{ \left(2N - \frac{2(q-1)}{\Lambda T_c} - \frac{2(l-1)}{\lambda T_c}\right) \sigma_{mp}^2 \right\} \right] \tag{7.71}$$

MAI statistics in modified S–V channel

From Equation (7.57), the MAI component from the first finger of the RAKE receiver, as shown in Figure 7.28, can be written as

$$I_{K,1} = \sum_{k=2}^{K} \sum_{q=1}^{Q_k} \sum_{l=1}^{L_k} I_{k,q,l,1}$$

$$= \sum_{k=2}^{K} \sum_{q=1}^{Q_k} \sum_{l=1}^{L_k} \int_{jT_b}^{(j+1)T_b} \beta_1 w_{k,q,l} \alpha_{k,q,l}$$

$$\times \sum_{j=-\infty}^{\infty} \sum_{n=0}^{N-1} b_j^{(k)} a_n^{(k)} g(t - T_{k,q} - \tau_{k,q,l} - \tau_k - jT_b - nT_c)$$

$$\times \sum_{n=0}^{N-1} a_n^{(1)} g(t - jT_b - nT_c)\, dt \tag{7.72}$$

The relation between the relative delay of clusters and rays can be expressed by $T_{k,q} + \tau_{k,q,l} + \tau_k = i_{k,q,l,1} T_c + \gamma_{k,q,l,1} T_c$. Similarly, we can calculate the conditional mean and variance of the MAI term generated from the first finger due to the l-th ray in the q-th cluster for the k-th user as

$$E\left(I_{k,q,l,1} | i_{k,q,l,1}, \gamma_{k,q,l,1}, w_{k,q,l}, \alpha_{k,q,l}, \beta_1\right)$$

$$= \beta_1 w_{k,q,l} \alpha_{k,q,l} \left[C_{k,1}\left(i_{k,q,l,1} - N + 1\right) R_p \left(-\left(1 - \gamma_{k,q,l,1}\right) T_c\right)\right.$$

$$\left. + C_{k,1}(i_{k,q,l,1} - N) R_p(\gamma_{k,q,l,1} T_c)\right] \tag{7.73}$$

$$Var\left(I_{k,q,l,1} | i_{k,q,l,1}, \gamma_{k,q,l,1}, w_{k,q,l}, \alpha_{k,q,l}, \beta_1\right)$$

$$= \beta_1^2 w_{k,q,l}^2 \alpha_{k,q,l}^2 \left[C_{k,1}\left(i_{k,q,l,1}\right) R_p \left(\gamma_{k,q,l,1} T_c\right)\right.$$

$$\left. + C_{k,1}(i_{k,q,l,1} + 1) R_p(-(1 - \gamma_{k,q,l,1}) T_c)\right]^2 \tag{7.74}$$

Thus, we obtain

$$E\left(I_{K,1} | i_{k,q,l,1}, \gamma_{k,q,l,1}, w_{k,q,l}, \alpha_{k,q,l}, \beta_1\right)$$

$$= \sum_{k=2}^{K} \sum_{q=1}^{Q_k} \sum_{l=1}^{L_k} \beta_1 w_{k,q,l} \alpha_{k,q,l} \left[C_{k,1}\left(i_{k,q,l,1} - N + 1\right) R_p \left(-\left(1 - \gamma_{k,q,l,1}\right) T_c\right)\right.$$

$$\left. + C_{k,1}\left(i_{k,q,l,1} - N\right) R_p \left(\gamma_{k,q,l,1} T_c\right)\right] \tag{7.75}$$

$$Var\left(I_{K,1} | i_{k,q,l,1}, \gamma_{k,q,l,1}, w_{k,q,l}, \alpha_{k,q,l}, \beta_1\right)$$

$$= \sum_{k=2}^{K} \sum_{q=1}^{Q_k} \sum_{l=1}^{L_k} \beta_1^2 w_{k,q,l}^2 \alpha_{k,q,l}^2 \left[C_{k,1}\left(i_{k,q,l,1}\right) R_p \left(\gamma_{k,q,l,1} T_c\right)\right.$$

$$\left. + C_{k,1}\left(i_{k,q,l,1} + 1\right) R_p \left(-\left(1 - \gamma_{k,q,l,1}\right) T_c\right)\right]^2 \tag{7.76}$$

Assume that the spreading codes are purely random sequences, fractional-chip delay $\gamma_{k,q,l,1}$ is a uniformly distributed random variable, denoted by $\gamma_{k,q,l,1} \sim \mathcal{U}(0, 1)$, and the relative delay between the k-th and first users, $\{\tau_k\}_{k=1}^{K}$, is a uniformly distributed random variable, defined as $\tau_k \sim \mathcal{U}(0, NT_c)$. Using the same approach as suggested in Appendix A, we can obtain

$$E\left(I_{K,1} | i_{k,q,l,1}, \gamma_{k,q,l,1}, w_{k,q,l}, \alpha_{k,q,l}, \beta_1\right) = 0 \tag{7.77}$$

$$Var\left(I_{K,1} | w_{k,q,l}, \alpha_{k,q,l}, \beta_1\right)$$

$$= \sum_{k=2}^{K} \sum_{q=1}^{Q_k} \sum_{l=1}^{L_k} \beta_1^2 w_{k,q,l}^2 \alpha_{k,q,l}^2 \left[\left(N - \frac{q-1}{\Lambda T_c} - \frac{l-1}{\lambda T_c} - \frac{NT_c}{2} + \frac{1}{2} \right) \sigma_{mp}^2 \right.$$

$$+ \left(N - \frac{q-1}{\Lambda T_c} - \frac{l-1}{\lambda T_c} - \frac{NT_c}{2} - \frac{1}{2} \right) \sigma_{mp}^2 \Bigg]$$

$$= \sum_{k=2}^{K} \sum_{q=1}^{Q_k} \sum_{l=1}^{L_k} \beta_1^2 w_{k,q,l}^2 \alpha_{k,q,l}^2 \left(2N - NT_c - \frac{2(q-1)}{\Lambda T_c} - \frac{2(l-1)}{\lambda T_c} \right) \sigma_{mp}^2 \qquad (7.78)$$

Thus, we obtain the conditional variance of the MAI term at the output of the RAKE receiver as

$$Var\left(I_K | w_{k,q,l}, \alpha_{k,q,l}, \beta \right)$$

$$- \sum_{p=1}^{L_1} Var\left(I_{K,p} | w_{k,q,l}, \alpha_{k,q,l}, \beta_p \right)$$

$$= \sum_{p=1}^{L_1} \sum_{k=2}^{K} \sum_{q=1}^{Q_k} \sum_{l=1}^{L_k} \beta_p^2 w_{k,q,l}^2 \alpha_{k,q,l}^2 \left(2N - NT_c - \frac{2(q-1)}{\Lambda T_c} - \frac{2(l-1)}{\lambda T_c} \right) \sigma_{mp}^2 \qquad (7.79)$$

Effect of lognormal fading

The conditional variances of MI and MAI terms from the RAKE receiver, as given in Equations (7.71) and (7.79), are still conditioned on variables $w_{k,q,l}$ and $\alpha_{k,q,l}$, where $w_{k,q,l}$ is a discrete random variable uniformly distributed over $(-1, 1)$ and $\alpha_{k,q,l}$ is a Lognormal random variable. To apply MRC to the RAKE receiver, we let

$$\beta_p = w_{1,p} \alpha_{1,p}, \qquad (7.80)$$

which could be inserted into Equations (7.59), (7.63), (7.71) and (7.79) to obtain the useful signal component. The distribution of the noise term, the variances of MI and MAI terms after the MRC-RAKE receiver, respectively, can be expressed as

$$S = N b_j^{(1)} E_{mp} \sum_{p=1}^{L_1} \alpha_{1,p}^2, \qquad (7.81)$$

$$\eta \sim \mathcal{N}\left(0, \sigma_n^2 N E_{mp} \sum_{p=1}^{L_1} \alpha_{1,p}^2 \right), \qquad (7.82)$$

$$Var\left(I_L | \alpha_{q,l} \right) = E\left[Var\left(I_L | w_{q,l}, \alpha_{q,l} \right) \right]$$

$$= \sum_{p=1}^{L_1} \alpha_{1,p}^2 \left[\sum_{l=1,l\neq p}^{L_1} \alpha_{1,l}^2 \left\{ \left(2N - \frac{2(l-1)}{\lambda T_c} \right) \sigma_{mp}^2 \right\} \right.$$

$$\left. + \sum_{q=2}^{Q_1} \sum_{l=1}^{L_1} \alpha_{q,l}^2 \left\{ \left(2N - \frac{2(q-1)}{\Lambda T_c} - \frac{2(l-1)}{\lambda T_c} \right) \sigma_{mp}^2 \right\} \right], \qquad (7.83)$$

$$Var\left(I_K | \alpha_{k,q,l} \right) = E\left[Var\left(I_K | w_{k,q,l}, \alpha_{k,q,l} \right) \right]$$

$$= \sum_{p=1}^{L_1} \sum_{k=2}^{K} \sum_{q=1}^{Q_k} \sum_{l=1}^{L_k} E\left(w_{1,p}^2 \right) E\left(w_{k,q,l}^2 \right) \alpha_{1,p}^2 \alpha_{k,q,l}^2$$

$$\times \left(2N - NT_c - \frac{2(q-1)}{\Lambda T_c} - \frac{2(l-1)}{\lambda T_c} \right) \sigma_{mp}^2$$

$$= \sum_{p=1}^{L_1} \alpha_{1,p}^2 \sum_{k=2}^{K} \sum_{q=1}^{Q_k} \sum_{l=1}^{L_k} \alpha_{k,q,l}^2 \left(2N - NT_c - \frac{2(q-1)}{\Lambda T_c} - \frac{2(l-1)}{\lambda T_c} \right) \sigma_{mp}^2 \quad (7.84)$$

where it should be noted that $w_{k,q,l}$ and $\alpha_{k,q,l}$ can be considered as two independent random variables [691], and thus we can first remove the condition on $w_{k,q,l}$. Also, as $w_{k,q,l}$ takes either "-1" or "$+1$," we always have $w_{k,q,l}^2 = 1$, assuming that fading on different rays happens independently. Applying Equation (7.49) we can remove the condition on $\alpha_{1,p}$ attached on Equation (7.83) as follows:

$$Var\left(I_L | \alpha_{1,p} \right)$$

$$\triangleq \sum_{p=1}^{L_1} \alpha_{1,p}^2 \Omega_1 \sigma_{mp}^2 \left[2Nq\left(L_1, \frac{1}{\lambda\gamma} \right) - \frac{2}{\lambda T_c} u\left(L_1, \frac{1}{\lambda\gamma} \right) \right.$$

$$+ 2Nq\left(L_1, \frac{1}{\lambda\gamma} \right) e^{-\frac{1}{\Lambda\Gamma}} q\left(Q_1 - 1, \frac{1}{\Lambda\Gamma} \right)$$

$$\left. - \frac{2}{\Lambda T_c} q\left(L_1, \frac{1}{\lambda\gamma} \right) u\left(Q_1, \frac{1}{\Lambda\Gamma} \right) - \frac{2}{\lambda T_c} e^{-\frac{1}{\Lambda\Gamma}} q\left(Q_1 - 1, \frac{1}{\Lambda\Gamma} \right) u\left(L_1, \frac{1}{\lambda\gamma} \right) \right]$$

$$= \sum_{p=1}^{L_1} \alpha_{1,p}^2 \Omega_1 \sigma_{mp}^2 \mathcal{R}\left(I_L \right) \quad (7.85)$$

where we have used

$$\mathcal{R}\left(I_L \right)$$

$$= 2Nq\left(L_1, \frac{1}{\lambda\gamma} \right) - \frac{2}{\lambda T_c} u\left(L_1, \frac{1}{\lambda\gamma} \right) + 2Nq\left(L_1, \frac{1}{\lambda\gamma} \right) e^{-\frac{1}{\Lambda\Gamma}} q\left(Q_1 - 1, \frac{1}{\Lambda\Gamma} \right)$$

$$- \frac{2}{\Lambda T_c} q\left(L_1, \frac{1}{\lambda\gamma} \right) u\left(Q_1, \frac{1}{\Lambda\Gamma} \right) - \frac{2}{\lambda T_c} e^{-\frac{1}{\Lambda\Gamma}} q\left(Q_1 - 1, \frac{1}{\Lambda\Gamma} \right) u\left(L_1, \frac{1}{\lambda\gamma} \right). \quad (7.86)$$

$$\begin{cases} q(L_1, \delta) = \sum_{p=1}^{L_1} e^{-(p-1)\delta} = \frac{1-e^{-\delta L_1}}{1-e^{-\delta}} \\ u(L_1, \alpha) = \sum_{l=1}^{L_1} (l-1) e^{-\alpha(l-1)} \end{cases} \quad (7.87)$$

Similarly, we can also remove the conditions on $\alpha_{k,q,l}$ in the MAI term after the RAKE receiver in Equation (7.84), by using Equation (7.49), to yield

$$Var\left(I_K | \alpha_{1,p} \right)$$

$$= \sum_{p=1}^{L_1} \alpha_{1,p}^2 (K-1) \Omega_1 \sigma_{mp}^2 \left[N(2-T_c) q\left(Q_k, \frac{1}{\Lambda\Gamma} \right) q\left(L_k, \frac{1}{\lambda\gamma} \right) \right.$$

$$\left. - \frac{2}{\Lambda T_c} u\left(Q_k, \frac{1}{\Lambda\Gamma} \right) q\left(L_k, \frac{1}{\lambda\gamma} \right) - \frac{2}{\lambda T_c} q\left(Q_k, \frac{1}{\Lambda\Gamma} \right) u\left(L_k, \frac{1}{\lambda\gamma} \right) \right]$$

$$= \sum_{p=1}^{L_1} \alpha_{1,p}^2 (K-1) \Omega_1 \sigma_{mp}^2 \mathcal{X}\left(I_K \right) \quad (7.88)$$

where

$$
\mathcal{X}(I_K) = N(2 - T_c) q\left(Q_k, \frac{1}{\Lambda\Gamma}\right) q\left(L_k, \frac{1}{\lambda\gamma}\right)
$$

$$
- \frac{2}{\Lambda T_c} u\left(Q_k, \frac{1}{\Lambda\Gamma}\right) q\left(L_k, \frac{1}{\lambda\gamma}\right) - \frac{2}{\lambda T_c} q\left(Q_k, \frac{1}{\Lambda\Gamma}\right) u\left(L_k, \frac{1}{\lambda\gamma}\right) \quad (7.89)
$$

The functions $q(L_1, \delta)$ and $u(L_1, \alpha)$ have been defined in Equation (7.87)

After having obtained all means and variances of the four variables in the decision variable, defined in Equation (7.57), or $v(j) = S + I_I + I_K + \eta$, as shown in Equations (7.81), (7.82), (7.85), and (7.88), respectively, we can use the Gaussian approximation method to derive a closed form BER expression for a DS-CDMA UWB system under a frequency-selective fading channel specified by the modified S–V channel model as

$$
P_b\left(E|\alpha_{1,p}\right)
$$

$$
= Q\left(\frac{N E_{mp} \sum_{p=1}^{L_1} \alpha_{1,p}^2}{\sqrt{\sum_{p=1}^{L_1} \alpha_{1,p}^2 \Omega_1 \sigma_{mp}^2 \mathcal{R}(I_L) + \sum_{p=1}^{L_1} \alpha_{1,p}^2 (K-1)\Omega_1\sigma_{mp}^2 \mathcal{X}(I_K) + \sigma_n^2 N E_{mp} \sum_{p=1}^{L_1} \alpha_{1,p}^2}} \right)
$$

$$
= Q\left(\xi \sqrt{\frac{1}{\Omega_1}\sum_{p=1}^{L_1} \alpha_{1,p}^2} \right) \quad (7.90)
$$

where

$$
\xi = \frac{N E_{mp}}{\sqrt{\sigma_{mp}^2 \mathcal{R}(I_L) + (K-1)\sigma_{mp}^2 \mathcal{X}(I_K) + \frac{\sigma_n^2 N E_{mp}}{\Omega_1}}}
$$

$$
= \frac{1}{\sqrt{\frac{\sigma_{mp\_normal}^2}{N^2 E_{mp}}\mathcal{R}(I_L) + \frac{(K-1)\sigma_{mp\_normal}^2}{N^2 E_{mp}}\mathcal{X}(I_K) + \frac{1}{\Omega_1}\frac{1}{SNR}}} \quad (7.91)
$$

$E_{mp} = \int_0^{T_c} g^2(t)\, dt$ denotes the energy contained in a single pulse, and $\sigma_{mp\_normal}^2$ is defined in Equations (7.37) and (7.42), representing normalized mean squared ACF of the pulse waveform used in the UWB system, which varies in $(0, 1)$. The parameter $\sigma_{mp\_normal}^2$ can sensitively affect the overall performance of a UWB radio system. The impact of the parameter $\sigma_{mp\_normal}^2$ will be quantified to offer us some useful guidance in the design of proper pulse waveforms for their application in a UWB radio.

Let $\zeta_p = \alpha_{1,p}^2$. On the basis of Equation (7.48), the probability density distribution of ζ_p can be written as

$$
f_{\zeta_p}(\zeta_p) = \frac{1}{2\sqrt{2\pi}\sigma\zeta_p} \exp\left(-\frac{(10\log_{10}\zeta_p - \mu_p)^2}{2\sigma^2} \right) \quad (7.92)
$$

where μ_p is given in Equation (7.50). Therefore, Equation (7.90) can be rewritten as

$$
P_b(E) = \int_{-\infty}^{\infty} Q\left(\xi\sqrt{\frac{1}{\Omega_1}\sum_{p=1}^{L_1}\zeta_p} \right) f_{\zeta_p}(\zeta_p)\, d\zeta_p \quad (7.93)
$$

7.6.5 DS-CDMA UWB System Performance

On the basis of the BER expressions derived in Equations (7.41) and (7.93), we can study the impact of various parameters on the overall system performance of a DS-CDMA UWB radio. First, let us examine the bit error performance of a DS-CDMA UWB radio under a flat fading channel. Figure 7.29 plots BER of a DS-CDMA UWB radio under a flat fading channel. The number of active users varies from 5, 10, 15 to 20. The length of the purely random spreading code is equal to 64. The normalized mean squared ACF of the pulse's waveform $\sigma^2_{mp\_normal} = 0.3$.

Figure 7.30 shows the BER performance versus the changing number of active users in a DS-CDMA UWB radio, with SNR (SNR = 5, 10, and 20 dB) and spreading code length ($N = 64, 128$) being its parameters. The normalized mean squared ACF of the pulse waveform $\sigma^2_{mp\_normal}$ is made equal to 0.1 to plot this figure. It is observed from the figure that the UWB radio gives a similar BER for different spreading code lengths when very few users are present. However, the BER gap between them enlarges as the number of active users increases, especially if SNR remains a relatively low value, for example, equal to 5 dB.

In Figure 7.31, BER is plotted versus SNR, the normalized mean squared ACF of the pulse waveform ($\sigma^2_{mp\_normal}$ takes 0.1 or 0.3), and the length of spreading codes (N is 64 or 128). Increase in the length of the spreading codes or decrease in the normalized mean squared ACF of the pulse waveform improves the BER performance, as shown in the figure. The improvement becomes more obvious when a relatively large SNR is applied.

The relation between BER and the normalized mean squared ACF of the pulse waveform $\sigma^2_{mp\_normal}$ has been illustrated in Figure 7.32, where $\sigma^2_{mp\_normal}$ changes from 0 to 1. The number of active users is used as a third parameter in this figure. Predictably, the BER deteriorates as the number of users increases. A very interesting result shown in this figure is the trend, at which the BER changes with the different values of the normalized mean squared ACF of the pulse's waveform

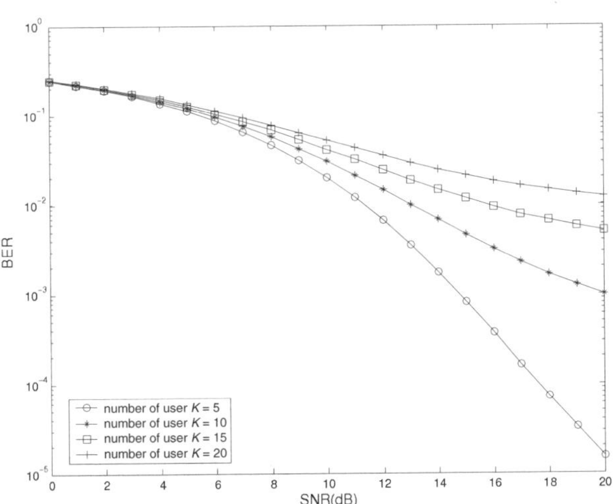

Figure 7.29 Bit error rate performance of a DS-CDMA UWB radio under a flat fading channel. The number of active users varies from 5, 10, 15 to 20. Purely random spreading code length is equal to 64. The normalized mean squared autocorrelation function of the pulse waveform $\sigma^2_{mp\_normal} = 0.3$.

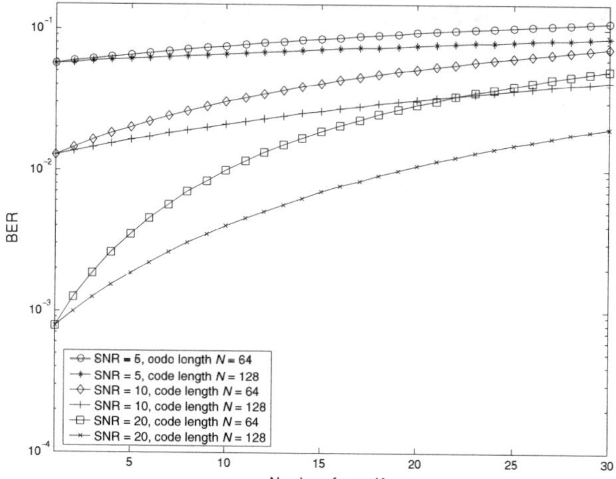

Figure 7.30 Bit error rate performance versus the changing number of active users in a DS-CDMA UWB radio under a flat fading channel. The number of active users varies from 0 to 30. The length of the purely random spreading code is equal to 64 and 128. The normalized mean squared ACF of the pulse waveform $\sigma^2_{mp\_normal} = 0.1$. The SNR takes three different values, that is, 5, 10, and 20 dB, respectively.

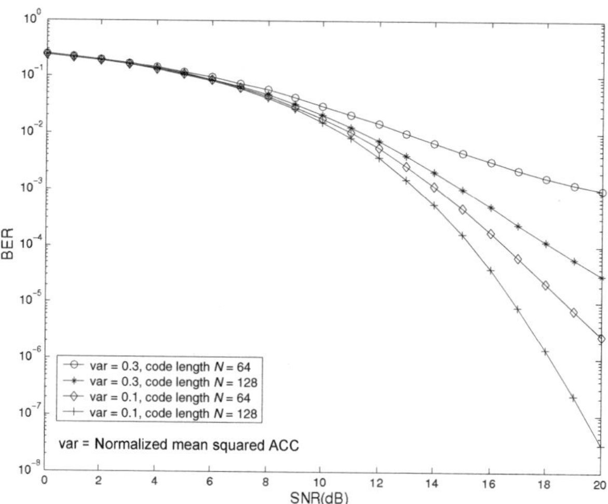

Figure 7.31 Bit error rate performance versus SNR in a DS-CDMA UWB radio under flat fading channel with the normalized mean squared ACF of the pulse waveform ($\sigma^2_{mp\_normal} = (0.1, 0.3)$) and the length of spreading code ($N = 64$ and 128) being its controlling parameters. The number of active users is equal to 10.

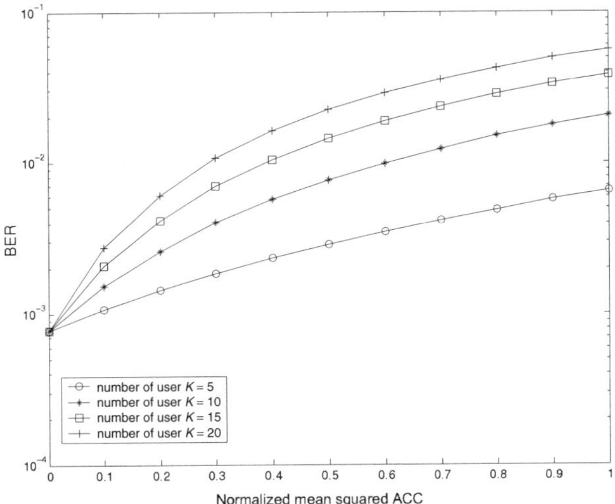

Figure 7.32 Bit error rate performance versus the normalized mean squared ACF of pulse waveform ($\sigma^2_{mp\_normal} \in (0, 1)$) in a DS-CDMA UWB radio under a flat fading channel with the number of active users ($K = 5, 10, 15$, and 20) being its parameter. The length of spreading code is 128 ($N = 128$). The SNR is equal to 20 dB.

$\sigma^2_{mp\_normal}$. $\sigma^2_{mp\_normal}$ was defined in Equations (7.37) and (7.42), which is rewritten as

$$\sigma^2_{mp\_normal} = \frac{\sigma^2_{mp}}{E_{mp}} = \frac{E\left[R_p^2(\tau)\right]}{\int_0^{T_c} g^2(t)\,dt} = \frac{\frac{1}{2T_c}\int_{-T_c}^{T_c}\left(\int_{-T_c}^{T_c} g(t-\tau)g(t)\,dt\right)^2 d\tau}{\int_0^{T_c} g^2(t)\,dt}, \qquad (7.94)$$

The mean squared ACF or $\sigma^2_{mp\_normal}$ gives the amount of average interchip interference associated with autocorrelation side lobes of the pulse waveform [785–788]. In general, the mean squared ACF of the pulse waveform should be made as relatively small as possible when compared with the energy of the pulse waveform, which can usually be normalized into unit without losing generality. To illustrate it clearly, let us consider an ideal case, in which $R_p(\tau)$ takes an impulse waveform such that the integral, $\frac{1}{2T_c}\int_{-T_c}^{T_c} R_p^2(\tau)\,d\tau$, will reach its minimal value or zero. In this case, the highest time resolution has been achieved because of its extremely narrow and infinitively high autocorrelation peak. At the same time, the small value generated from the integral $\frac{1}{2T_c}\int_{-T_c}^{T_c} R_p^2(\tau)\,d\tau$ gives very low interchip interference caused by ACF side lobes of the pulse waveform. Both factors can help in improving the overall performance of a UWB radio.

On the other hand, if the ACF takes a flat-looking shape to yield a relatively large value of the integral $\frac{1}{2T_c}\int_{-T_c}^{T_c} R_p^2(\tau)\,d\tau$ the poor time resolution of the ACF, which may induce a high average interchip interference, is responsible for an impaired BER performance. Although we cannot give an optimal pulse waveform from the simple discussion given here, it does make sense for us to pay enough attention to design suitable pulse waveforms for successful applications in a DS-CDMA-based UWB radio system by following the guidelines specified in Equation (7.94). The analysis of the pulse waveform–dependent BER conducted here can also be very useful to quantify how much the impact of the mean squared ACF of pulse waveforms will exert on the overall BER. For instance, the usefulness of the results given in Figure 7.32 lies in the fact that we can use the curves in the figure to predict the BER performance of a DS-CDMA UWB radio under a flat fading channel for a

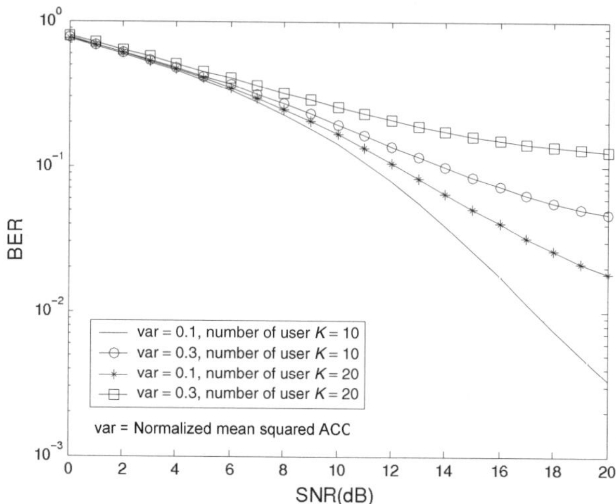

Figure 7.33 Bit error rate versus SNR with number of active users ($K = 10$ or 20) and normalized mean squared ACF of pulse waveform ($\sigma^2_{mp\_normal}$) being parameters. The spreading code length is 1024. The first cluster of the first user consists of 10 rays with totally 15 clusters. For all other users' signals, there are five clusters, each of which consists of 10 rays. A modified S–V channel model is considered, in which the arrival rates for the clusters and rays are 1/11 (1/ns) and 1/0.35 (1/ns), respectively. $\sigma = 4.8$ dB. The mean attenuation power of the first ray $\Omega_1 = 1$, and all other rays decay exponentially with its decay factor Γ being 16. $\gamma = 8.5$.

particular pulse waveform function $g(t)$, which is defined in Equation (7.7), before we actually apply it to a real system.

Figures 7.33, 7.34, and 7.35 show the BER performance of a DS-CDMA UWB radio under a frequency-selective fading channel, which is characterized by the modified S–V channel model. All results were obtained using an MRC-RAKE receiver.

Figure 7.33 shows BER performance versus SNR with the number of active users ($K = 10$ or 20) and the normalized mean squared ACF of pulse waveform $\sigma^2_{mp\_normal}$ being the parameters. The modified S–V channel model in this figure bears the following characteristics. The first cluster of the first user consists of 10 rays with altogether 15 clusters. In all other users' signals, there exist five clusters, each of which consists of 10 rays. The arrival rates for the clusters and rays are 1/11 (1/ns) and 1/0.35 (1/ns), respectively. Channel parameter σ is 4.8 dB. The average power of the first ray $\Omega_1 = 1$, and all other rays decay exponentially with their decay factor Γ being 16. The path attenuation factor γ is equal to 8.5. All the variables σ, Ω_1, Γ and γ are defined in Equations (7.49) and (7.50). Figure 7.33 shows that the BER of a DS-CDMA UWB radio under a frequency-selective channel varies again sensitively with respect to the normalized mean squared ACF of pulse waveform $\sigma^2_{mp\_normal}$ under the same number of active users in the system. The SNR gain obtained by replacing $\sigma^2_{mp\_normal} = 0.3$ with $\sigma^2_{mp\_normal} = 0.1$ at a fixed BER performance can be as large as 3 to 6 dB. This gain can be even more substantial than that achievable by reducing the number of users by half, that is, from $K = 20$ to $K = 10$. Therefore, the use of some properly designed pulse waveforms with relatively small $\sigma^2_{mp\_normal}$ is strongly recommended.

Figure 7.34 studies the effect of normalized mean squared ACF and spreading code length on the BER performance of a DS-CDMA UWB radio. The figure shows that the impact of the normalized

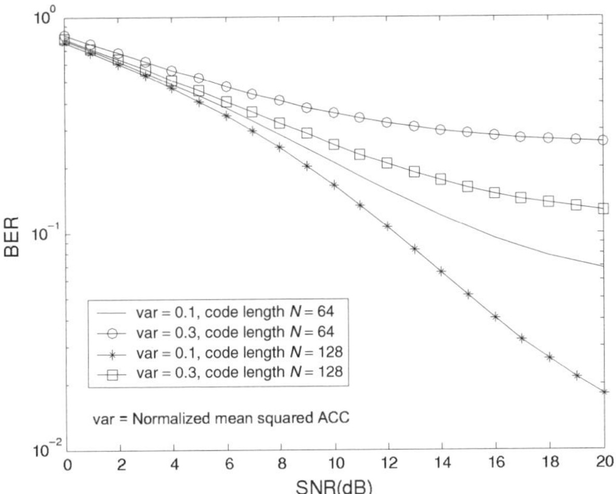

Figure 7.34 Bit error rate versus SNR in modified S–V multipath channel with spreading code length ($N = 64$ or 128) and normalized mean squared ACF of pulse waveform ($\sigma^2_{mp\_normal} = 0.1$ or 0.3) being parameters. The number of active users is equal to 20. The first cluster of the first user consists of 10 rays with a total of 15 clusters. For all other users' signals, there are five clusters, each of which consists of 10 rays. A modified S–V channel model is considered, in which the arrival rates for the clusters and rays are 1/11 (1/ns) and 1/0.35 (1/ns), respectively. Channel parameter $\sigma = 4.8$ dB. The mean attenuation power of the first ray $\Omega_1 = 1$, and all other rays decay exponentially with their decay factor Γ being 16. $\gamma = 8.5$.

mean squared ACF is much bigger than the spreading code length to the overall BER. The reduction of the normalized mean squared ACF from 0.3 to 0.1 can offer a SNR margin improvement of about 2 to 10 dB depending on the code length and the current SNR level. On the other hand, the double of the code length from 64 to 128 will only bring a limited SNR margin improvement, much less than that obtained from the reduction in normalized mean squared ACF. Therefore, the use of proper pulse waveforms with a small normalized mean squared ACF is of great importance.

Finally, Figure 7.35 considers two modified S–V channel models proposed by Intel in [691], where the channel parameters for these two modified S–V models are explained as follows. Both the models have line-of-sight propagation path. It is assumed that the first cluster of the first user consists of 10 rays with altogether 15 clusters. For all other users' signals there exist five clusters, each of which consists of 10 rays. Two modified S–V channel models are considered, in which the arrival rates for the clusters and rays are 1/22 (for Model 4), 1/60 (for Model 5) (1/ns) and 1/0.94 (for Model 4), 1/0.5 (for Model 5) (1/ns), respectively. Channel parameter σ is again 4.8 dB. The mean attenuation power of the first ray Ω_1 is 1 and all other rays decay exponentially with their decay factor Γ being 7.6 (for Model 4) and 16 (for Model 5), respectively. The path attenuation factor γ is equal to 0.94 (for Model 4) and 1.6 (for Model 5). Model 4 and Model 5 are two modified S–V channel models suggested by Intel for the environments with line-of-sight path [691].

Several observations can be made from Figure 7.35. First, the results show once again that the variation of normalized mean squared ACF is very sensitive to the BER of a UWB radio. Second, the improvement on BER error rate due to the use of pulse waveform with small normalized mean squared ACF will become significant only in the regions with a relatively large SNR, that is, greater

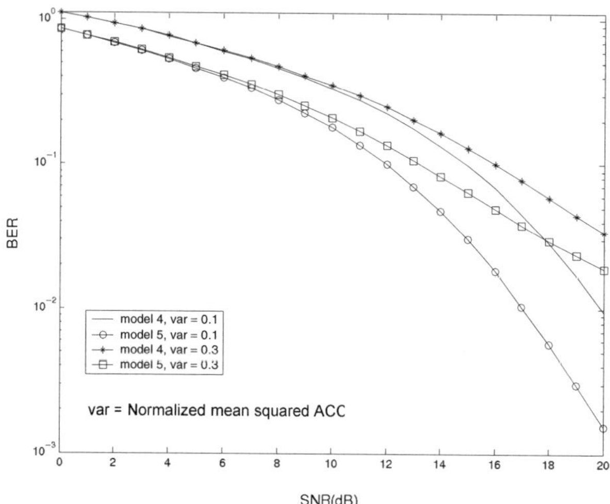

Figure 7.35 Bit error rate versus SNR in a modified S–V multipath channel with normalized mean squared ACF of pulse waveform ($\sigma^2_{mp\_normal} = 0.1$ or 0.3) being parameters. The number of active users is equal to 20. The spreading code length is 64. The first cluster of the first user consists of 10 rays with a total of 15 clusters. For all other users' signals, there are five clusters, each of which consists of 10 rays. A modified S–V channel model is considered, in which the arrival rates for the clusters and rays are 1/22 (Model 4), 1/60 (Model 5) (1/ns) and 1/0.94 (Model 4), 1/0.5 (Model 5) (1/ns), respectively. Channel parameter $\sigma = 4.8$ dB. The mean attenuation power of the first ray $\Omega_1 = 1$ and all other rays decay exponentially with their decay factor Γ being 7.6 (Model 4) and 16 (Model 5). The path attenuation factor γ is equal to 0.94 (Model 4) and 1.6 (Model 5). Model 4 and Model 5 are two modified S–V models suggested by Intel for channel environment with line-of-sight path [691].

than 10 dB. Third, the choice of channel models has a significant impact on the performance analysis, as shown in the curves of the figure. The difference in BER given by two different models can be as large as 3 dB, indicated by the curves for Models 4 and 5 with a fixed normalized mean squared ACF of 0.1 and a fixed BER of about 10^{-2}.

8

MIMO Systems

Since the late 90's, many space–time coding schemes have been proposed in the literature to achieve open loop transmitter diversity. Because of the limited space here, we can only review some of the examples. In 1998, Tarokh *et al.* in their paper [692] proposed a space–time trellis coding (STTC) scheme, which could maximize both coding and diversity gains. The complexity of the scheme increases with the trellis states in an exponential way, thus prohibiting its possible hardware realization with currently available digital technologies. Alamouti [693], almost at the same time, suggested a simple but extremely effective space–time block coding (STBC) algorithm, which can achieve a maximal space-diversity gain without introducing a complexity problem. One year after Alamouti published his simple STBC scheme, Tarokh *et al.* [694] in 1999 extended Alamouti's STBC algorithm into a general form. Many more research works on the MIMO systems have been published, as shown in [692–699].

In this chapter, we first introduce the concept of SIMO, MISO and MIMO systems, followed by several widely referred *space–time coding schemes*. However, the major content of the chapter focuses on the introduction and analysis of STBC-CDMA systems. To do so, we first introduce STBC-CDMA system models suitable for performance analysis with either unitary codes or orthogonal complementary codes. Then, we proceed to study a STBC-CDMA system based on conventional unitary codes, followed by that using orthogonal complementary codes. The results obtained from both the analysis and the simulation will be given and discussed. At the end of this book, Appendices D and F show the derivations of the two key equalities used in the main text, and Appendix E describes several important properties of orthogonal complementary codes dealt with in this chapter.

8.1 SIMO, MISO and MIMO Systems

The study on MIMO systems begins with the simplest *single input single output* (SISO) model, which is just an ordinary point-to-point radio link with a single transmitter antenna and a single receiver antenna, as shown in Figure 8.1(a).

The time-variant impulse response of the channel can be written as

$$h(\tau, t) = p(\tau, t) * g(\tau) \tag{8.1}$$

where $p(\tau, t)$ is the impulse response of the time-variant propagation channel, and $g(\tau)$ is the combined effect of the transmitter pulse-shaping and the receiver matched-filtering. Thus, the received

Next Generation Wireless Systems and Networks   Hsiao-Hwa Chen and Mohsen Guizani
© 2006 John Wiley & Sons, Ltd

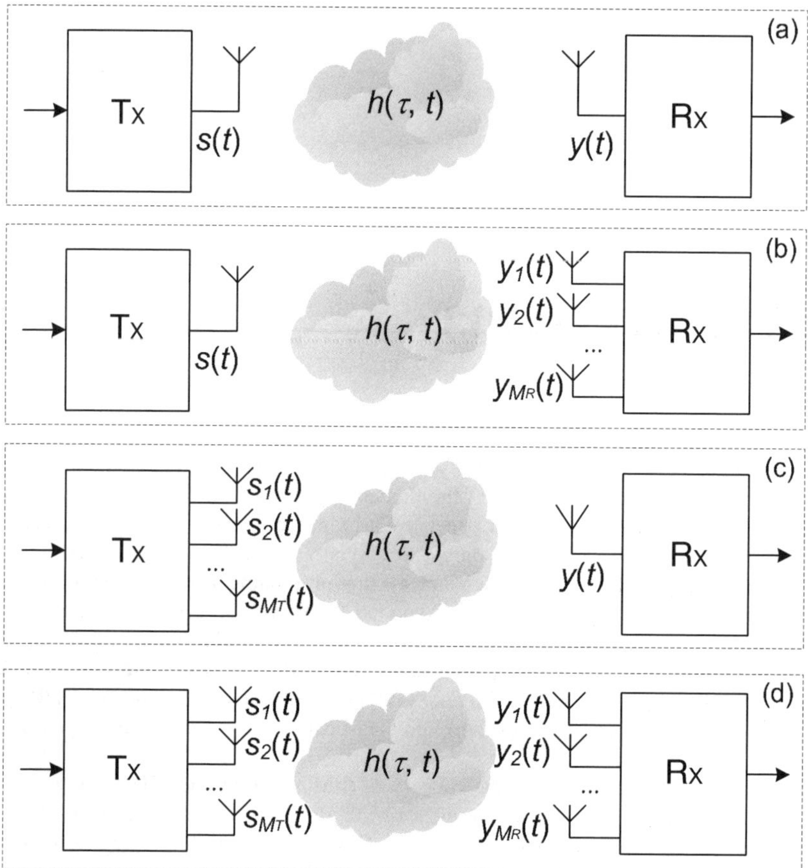

Figure 8.1 (a) The SISO model, where $h(\tau, t)$ is the time-variant impulse response of the channel, $s(t)$ and $y(t)$ are the input and output signals observed at the transmitter antenna and the receiver antenna. (b) The SIMO model, where $s(t)$ and $\{y_i(t)\}_{i=1}^{M_R}$ are the input and output signals observed at the transmitter and receiver antennae, respectively. (c) The MISO model, where $\{s_j(t)\}_{j=1}^{M_T}$ and $y(t)$ are the input and output signals observed at the transmitter and receiver antennae, respectively. (d) The MIMO model, where $\{s_j(t)\}_{j=1}^{M_T}$ and $\{y_i(t)\}_{i=1}^{M_R}$ are the input and output signals observed at the transmitter and receiver antennae, respectively.

signal at the receiver antenna side (omitting the noise) can be written as

$$y(t) = \int_0^{\tau_{max}} h(\tau, t)s(t - \tau)d\tau = h(\tau, t) * s(t) \tag{8.2}$$

We should be careful to identify the meaning of the two time variables appearing in the channel impulse response $h(\tau, t)$, where t is normal time index observed in one particular delay profile; while τ denotes the instant, at which one is observing the delay profile of the channel. Therefore, τ just denotes the time we take a snap-shot of the channel impulse response. The channel impulse response or the delay profile will look different if observed at different τ.

If linear modulation schemes are applied to the transmitter, the user data stream $s(t)$ is, in effect, a train of discrete amplitude pulses with symbol spacing T_s, or

$$s(t) = \sum_{m=-\infty}^{\infty} (a_m + jb_m)\delta(t - mT_s) \tag{8.3}$$

where $\{a_m\}_{m=-\infty}^{\infty}$ and $\{b_m\}_{m=-\infty}^{\infty}$ are real and imaginary parts of the symbols, and $\delta(t - mT_s)$ is the impulse function that is equal to nonzero value only at $t = \{mT_s\}_{m=-\infty}^{\infty}$.

If the transmitter still uses a single antenna while the receiver uses multiple M_R antennae, an SIMO system is formulated, as shown in Figure 8.1(b), where M_R receiver antennae are present.

The SIMO channel can be represented by a $M_R \times 1$ vector $\mathbf{h}(\tau, t)$, or

$$\mathbf{h}(\tau, t) = \left[h_1(\tau, t), h_2(\tau, t), \ldots, h_{M_R}(\tau, t) \right]^T \tag{8.4}$$

For the input signal $s(t)$, the output signal at the i-th receiver antenna is given by

$$y_i(t) = h_i(\tau, t) * s(t), \qquad i \in \{1, 2, \ldots, M_R\} \tag{8.5}$$

Defining $\mathbf{y}(t) = \left[y_1(t), y_2(t), \ldots, y_{M_R}(t), \right]^T$, we can write the output signal vector as

$$\mathbf{y}(t) = \mathbf{h}(\tau, t) * s(t) \tag{8.6}$$

where the notation "$*$" denotes the convolution operation.

Similarly, for a MISO system (as shown in Figure 8.1(c)) that has M_T transmitter antennae and single receiver antenna we can represent the MISO channel by a $1 \times M_T$ vector $\mathbf{h}(\tau, t)$, or

$$\mathbf{h}(\tau, t) = \left[h_1(\tau, t), h_2(\tau, t), \ldots, h_{M_T}(\tau, t) \right]^T \tag{8.7}$$

For input signal $\{s_j(t)\}_{j=1}^{M_T}$, the output signal is given by

$$y(t) = \sum_{j=1}^{M_T} h_j(\tau, t) * s_j(t) \tag{8.8}$$

Defining $\mathbf{s}(t) = \left[s_1(t), s_2(t), \ldots, s_{M_T}(t) \right]$, in vector representation we have

$$y(t) = \mathbf{h}(\tau, t) * \mathbf{s}(t) \tag{8.9}$$

After having discussed SIMO and MISO systems, we can readily introduce the MIMO system, as shown in Figure 8.1(d), where there are M_T transmitter antennae and M_R receiver antennae.

The channel impulse response between the j-th transmitter antenna ($j = 1, 2, \ldots, M_T$) and the i-th receiver antenna ($i = 1, 2, \ldots, M_R$) in a MIMO system can be represented by $\{h_{i,j}(\tau, t)\}_{i=1,\ldots,M_T}^{j=1,\ldots,M_R}$. Thus, a $M_R \times M_T$ MIMO channel matrix $\mathbf{H}(\tau, t)$ can be written as

$$\mathbf{H}(\tau, t) = \begin{bmatrix} h_{1,1}(\tau, t) & h_{1,2}(\tau, t) & \cdots & h_{1,M_T}(\tau, t) \\ h_{2,1}(\tau, t) & h_{2,2}(\tau, t) & \cdots & h_{2,M_T}(\tau, t) \\ \vdots & \vdots & \ddots & \vdots \\ h_{M_R,1}(\tau, t) & h_{M_R,2}(\tau, t) & \cdots & h_{M_R,M_T}(\tau, t) \end{bmatrix} \tag{8.10}$$

Therefore, we have the output signal vector as

$$\mathbf{y}(t) = \mathbf{H}(\tau, t) * \mathbf{s}(t) \tag{8.11}$$

where we have $\mathbf{y}(t) = \left[y_1(t), y_2(t), \ldots, y_{M_R}(t) \right]$ and $\mathbf{s}(t) = \left[s_1(t), s_2(t), \ldots, s_{M_T}(t) \right]$.

8.2   Spatial Diversity in MIMO Systems

As seen from the introduction to SIMO, MISO, and MIMO system models, the multiple antennae used in either transmitters or receivers will create more signal passage channels under the condition that they will be able to be separated at the receiver without mutual interference. Only in this way, the signal flows independence among different Tx–Rx channels can be exploited to achieve certain kinds of gains in "spatial diversity" or "multiplexing," depending on the applications.

8.2.1   Diversity Combining Methods

In order to achieve some kind of *diversity gain*, three important conditions have to be fulfilled, as listed below:

- Redundancy: different signal channels should carry the same information.

- Separation: the signals carried in the different channels can be separated at a receiver without much distortion.

- Independence: the channels carrying the same signal should be statistically independent from one another.

The above three conditions are necessary for achieving signal reception diversity. In other words, all three conditions have to be satisfied to achieve diversity; and if all three conditions are met, the diversity gain will be guaranteed if the received signal is processed in some appropriate way. The diversity gain provides a receiver with multiple (ideally, independent) observations of the same transmitted signal. Each observation is considered as a *diversity branch*. Receivers can exploit the diversity gain by combining these diversity branches by mainly using four different methods, namely (1) Random Hopping (RH), (2) Selection Combining (SC), (3) Equal Gain Combining (EGC), and (4) Maximal Ratio Combining (MRC).

The RH scheme selects a branch by randomly hopping from one branch to another. The SC scheme always chooses a branch with the maximal signal power level, and consequently the selected branch may not be fixed. The EGC scheme works in such a way that it combines the branches with equal weighting coefficients to all branches, and therefore, the decision variable always involves all branches. Finally, the MRC scheme uses different weighting coefficients for different branches according to their different signal power levels. In general, the MRC scheme offers the best performance at a price of complexity, as it needs to know the full channel or the "branch" statistics information. On the other hand, RH, SC, and EGC can work with none or only partial channel or "branch" statistics information.

8.2.2   Receiver Diversity

Channel-coding often works jointly with spatial diversity schemes. The effectiveness of channel-coding schemes is always measured by the so-called *coding gain*. It is to be noted that the coding gain shifts the *bit error rate* (BER) curve horizontally to the left in a log–log BER plot, as shown in Figure 8.2(a).

On the other hand, the *diversity gain* changes the slope of the BER versus the signal-to-noise-ratio (SNR) curve in a log–log plot, as shown in Figure 8.2(b). The asymptotic slope is also called *diversity order*.

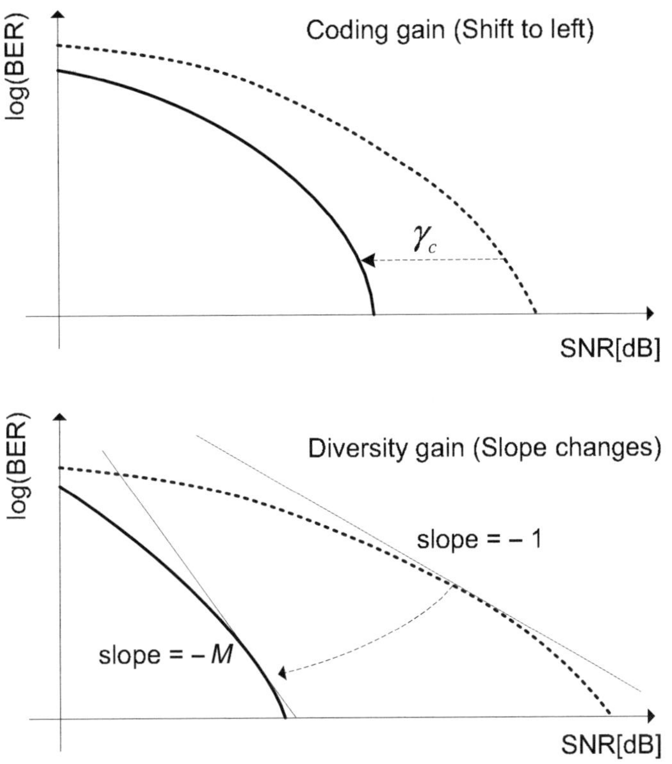

Figure 8.2 (a) Upper: Coding gain (γ_c) moves the log–log BER curve horizontally to the left. (b) Lower: Diversity gain (M) changes the slope of the log–log BER curve.

Both coding and diversity gains will improve the detection efficiency of a receiver. However, the diversity gain M will increase the SNR; whereas the SNR advantage due to the coding gain γ_c does not depend on SNR. The BER expression for a generic MIMO communication system can be written as

$$P_e \propto \frac{c}{(\gamma_c \rho)^M} \tag{8.12}$$

where γ_c, M, and ρ are coding gain, diversity gain, and SNR value, respectively. c is a coefficient. Let us consider a SIMO system with a single transmitter antenna and M_R receiver antennae, as shown in Figure 8.1(b). Assume a flat fading channel and the noise to be identically independently distributed (i.i.d) *spatially white*. Therefore, the received signal at the receiver can be written as

$$\mathbf{y} = \sqrt{E_s}\mathbf{h}s + \mathbf{n} \tag{8.13}$$

where $\mathbf{y} = \left[y_1, y_2, \ldots, y_{M_R}\right]^T$, $\mathbf{h} = \left[h_1, h_2, \ldots, h_{M_R}\right]^T$, and $\mathbf{n} = \left[n_1, n_2, \ldots, n_{M_R}\right]^T$. We have used discrete form to simplify the expression, and thus $s = s(n) = s(n\Delta t)$.

It is assumed that the receiver has complete knowledge of the channel $\mathbf{h}$. To maximize the received SNR, a receiver will use the MRC scheme to combine the different diversity branches.

Therefore, the decision variable z can be calculated from a linear combination of the diversity branches $[y_1, y_2, \ldots, y_{M_R}]^T$. The combining weights are chosen proportional to the channel gains, or

$$
\begin{aligned}
z &= \mathbf{h}^H \mathbf{y} \\
&= \sqrt{E_s} \mathbf{h}^H \mathbf{h} s + \mathbf{h}^H \mathbf{n} \\
&= \sqrt{E_s} \parallel \mathbf{h} \parallel^2 s + \mathbf{h}^H \mathbf{n}
\end{aligned}
\tag{8.14}
$$

where $\parallel \mathbf{h} \parallel$ is the norm of vector $\mathbf{h}$. The SNR η conditioned on the channel vector $\mathbf{h}$ at the receiver becomes

$$
\eta = \parallel \mathbf{h} \parallel^2 \rho
\tag{8.15}
$$

The channel gains can be considered to be i.i.d random if the M_R receiver antenna elements keep a relatively large spacing, usually being separated by more than the coherent distance (which is at least equal to half of the wavelength of the carrier frequency), and $\{h_i\}_{i=1}^{M_R}$ is a zero mean, complex circularly symmetric Gaussian variable. It can be shown that the average BER of such a SIMO system will be bounded by

$$
P_e \leq c \prod_{i=1}^{M_R} \frac{1}{1 + \frac{\rho d_{\min}^2}{4}}
\tag{8.16}
$$

where ρ, M_R, and c are the SNR, the number of receiver antennae, and a constant. If the SNR ρ is sufficiently large, the above expression will be simplified into

$$
P_e \leq c \left(\frac{\rho d_{\min}^2}{4} \right)^{-M_R}
\tag{8.17}
$$

which is the asymptotical BER when the SNR becomes very large. Therefore, the *diversity order* for this SIMO system is simply equal to the number of receiver antennae or M_R. In addition, since $E\left[\parallel \mathbf{h} \parallel^2 \right] = M_R$, the average SNR at the receiver can be written as

$$
\bar{\eta} = E\left[\eta \right] = \rho M_R
\tag{8.18}
$$

Hence, the average SNR at the receiver has been enhanced by a factor of M_R when compared to that in a 1×1 SISO link due to the *array gain*, which can be defined as $10\log_{10} M_R$ dB.

From the above analysis, we can obtain several useful conclusions. The receiver diversity in a SIMO system can be used to extract full diversity gain and full array gain. The performance improvement achieved in a SIMO system is proportional to the number of receiver antennae or M_R. To ensure the full diversity gain, it is important to know the channel information on a real-time basis. However, it should be noted that the receiver diversity achieved in a SIMO system may not be applicable in some applications, such as a mobile cellular downlink transmission, where a mobile can hardly be implemented with multiple receiver antennae in a tiny handset.

8.2.3 Transmitter Diversity

Next we discuss the transmitter diversity implemented in a MISO system, as shown in Figure 8.1(c). For analytical simplicity, we assume that the receiver has the full information about the channel in question. Let us consider a simple transmitter diversity scheme with two transmitter antennae and one receiver antenna, also called the *Alamouti scheme*, where the transmitter does not need to know the channel information, whereas the receiver does. Therefore, in this simple MISO scheme, as shown in Figure 8.1(c) with the number of transmitter antennae fixed at two, there are

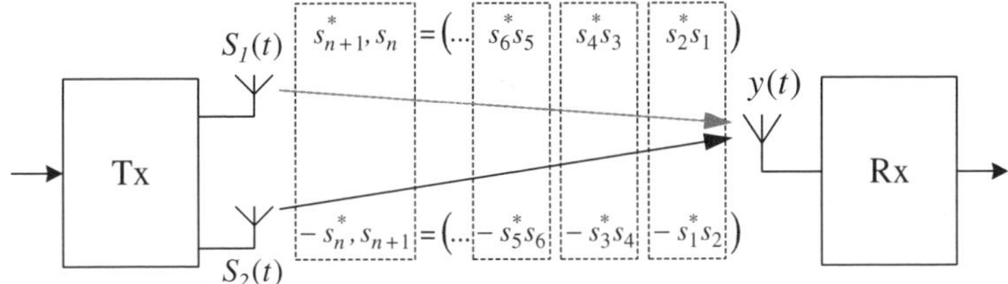

Figure 8.3 A simple block coding scheme used in the Alamouti scheme.

two channel impulse response functions or h_1 and h_2, which are known to the receiver but not the transmitter.

Now we introduce a simple block coding scheme to the signal transmitted from two antennae, as shown in Figure 8.3. This STBC scheme was first proposed by Alamouti [693], and it works to group all the two bits together to form a transmitting symbol. The two antennae will send the same two bits information, but will use the block coding scheme shown in Figure 8.3. These codes are transmitted using an orthogonal block structure which enables simple decoding at the receiver. The detection will take place at a receiver symbol by symbol. The basic ideal for using such a STBC scheme was to facilitate the separation of the signals sent via different antennae if the detection proceeds block-wise at a receiver. Therefore, it is easily seen that the working principle of the Alamouti STBC scheme fits the three conditions required to implement diversity perfectly well, as mentioned in subsection 8.2.1, or specifically: (1) Two antennae send the same data information; (2) The signals sent via two antennae are statistically independent (due to sufficient spacing between the two antennae); and (3) The signals sent via two antennae are separated (with the help of two-bit block coding scheme). The same principle can be applied to other MIMO systems to achieve spatial diversity gain.

Thus, the signals received at the receiver become

$$\begin{bmatrix} r_1 \\ r_2 \end{bmatrix} = \begin{bmatrix} s_1 & s_2 \\ s_2^* & -s_1^* \end{bmatrix} \begin{bmatrix} h_1 \\ h_2 \end{bmatrix} + \begin{bmatrix} n_1 \\ n_2 \end{bmatrix} \tag{8.19}$$

which can be written into vector representation as

$$\mathbf{r} = \mathbf{Sh} + \mathbf{n} \tag{8.20}$$

It is noted in the above equation that the 2×2 symbol block $\mathbf{s}$ has orthogonal columns. The Equation (8.19) can be rewritten as

$$\begin{bmatrix} r_1 \\ r_2^* \end{bmatrix} = \begin{bmatrix} h_1 & h_2 \\ -h_2^* & h_1^* \end{bmatrix} \begin{bmatrix} s_1 \\ s_2 \end{bmatrix} + \begin{bmatrix} n_1 \\ n_2^* \end{bmatrix} \tag{8.21}$$

or

$$\mathbf{y} = \mathbf{Hs} + \mathbf{n}' \tag{8.22}$$

where

$$\mathbf{H} = \begin{bmatrix} h_1 & h_2 \\ -h_2^* & h_1^* \end{bmatrix} \tag{8.23}$$

which is just the channel matrix as given in Equation (8.10) with $M_T = 2$ and $M_R = 1$. Also, the 2×2 channel matrix $\mathbf{H}$ has orthogonal columns. Applying Zero-Forcing (ZF) detection to the above equation, we obtain

$$\hat{\mathbf{s}} = \frac{\mathbf{H}^H \mathbf{y}}{\mid h_1 \mid^2 + \mid h_2 \mid^2} \tag{8.24}$$

If (n_1, n_2) are complex white Gaussian distributed variables, that is, $(n_1, n_2) \sim (\mathbf{0}, \sigma^2 \mathbf{I})$, so are (n_1, n_2^*) with its mean and variance being $(n_1, n_2^*) \sim (\mathbf{0}, \sigma^2 \mathbf{I})$, where $\mathbf{0}$ and $\mathbf{I}$ denote a 2×2 zero matrix and 2×2 unity matrix, respectively. The asymptotic BER performance for the Alamouti scheme with $M_T \times M_R$ being 1×2, 1×4, 2×1 and 2×2 is shown in Figure 8.4 [693], from which it is observed that the diversity gain changes the slope of BER curves, or the BER curves with the same $M_T \times M_R$ value has the same slope, as illustrated in Figure 8.2. The results shown in Figure 8.4 were obtained with the consideration of a flat Rayleigh fading channel using coherent BPSK with two-branch transmit diversity. The Alamouti's transmitter diversity scheme is also called *transmitter beam-forming*, which acts as a dual to receiver beam-forming in a SIMO system.

The *array gain* and *diversity order* are compared among various MIMO configurations in Table 8.1.

In addition to the STBC MIMO schemes, another commonly used S-T coding scheme is the STTC scheme, which works on convolutional codes extended to the cases of multiple transmitter

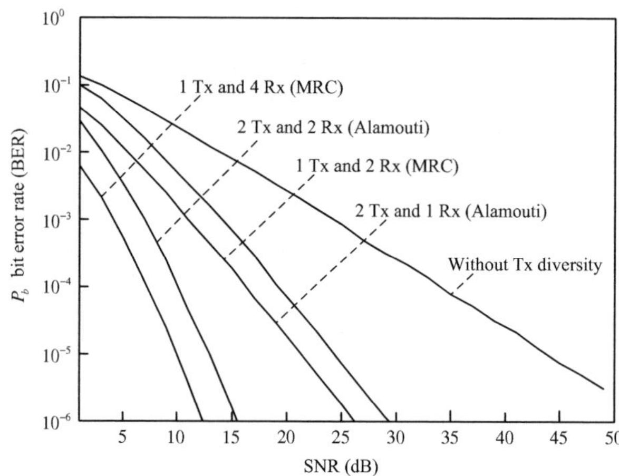

Figure 8.4 Asymptotic BER performance for Alamouti STBC schemes under flat Rayleigh fading, where coherent BPSK with two-branch transmit diversity is considered. The performance is compared to traditional receiver diversity schemes.

Table 8.1 Comparison of array gain and diversity order for different configurations of transmitter/receiver antennae

Configuration	Channel	Array gain	Diversity order
SIMO	Unknown at Tx	M_R	M_R
SIMO	Known at Tx	M_R	M_R
MISO	Unknown at Tx	1	M_T
MISO	Known at Tx	M_T	M_T
MIMO	Unknown at Tx	M_R	$M_R \times M_T$
MIMO	Known at Tx	$E\left[\lambda_{\max}\right]$	$M_R \times M_T$

and receiver antennae. We will not discuss them here in detail. In general, space–time trellis codes perform better than the Alamouti STBC scheme. In fact, the implementation of STTC will not be more complex than that of STBC. Yet another form of S-T coding is the differential STBC scheme, which works using differential coding to the STBC scheme, allowing a simplified decoding algorithm at a receiver. For more information on these schemes, please refer to [692–703].

8.3 Spatial Multiplexing in MIMO Systems

In the previous section, we have discussed the issues on how a MIMO system can improve the communication quality by exploiting the *diversity gain* provided by multiple independent samples of the same signal at the receiver. Now we are going to touch upon the other facets of MIMO systems due to the fact that MIMO techniques can also increase the capacity via its *spatial multiplexing* capability in different Tx–Rx channels formed by multiple antennae. It is noted that MIMO can enhance the capacity or data transmission rates at no extra bandwidth cost, which is an extremely attractive feature.

To implement spatial multiplexing in a MIMO system, multiple data streams are transmitted simultaneously and on the same frequency using a transmitter array to improve the system capacity or transmission rate. Different data substreams should be sent from different transmitter antennae, as shown in Figure 8.5, where $M_T = 3$ transmitter antennae are assumed. It should be noted that in such a transmitter multiplexing scheme, the transmitter usually need not know the channel state information. Another important advantage for spatial multiplexing is that the system does not require any fast feedback channels.

To detect the information sent via different transmitter antennae, several detection algorithms have been proposed. The *maximum likelihood* (ML) algorithm is an optimal solution but its implementation can be very complex as the number of multiplexing channels increases. The ML algorithm detects the information according to the following criteria

$$\hat{\mathbf{x}} = \arg \min_{\mathbf{x}_k \in \{\mathbf{x}_1, \dots, \mathbf{x}_{C^{M_T}}\}} \|\mathbf{r} - \mathbf{H}\mathbf{x}_k\|^2 \tag{8.25}$$

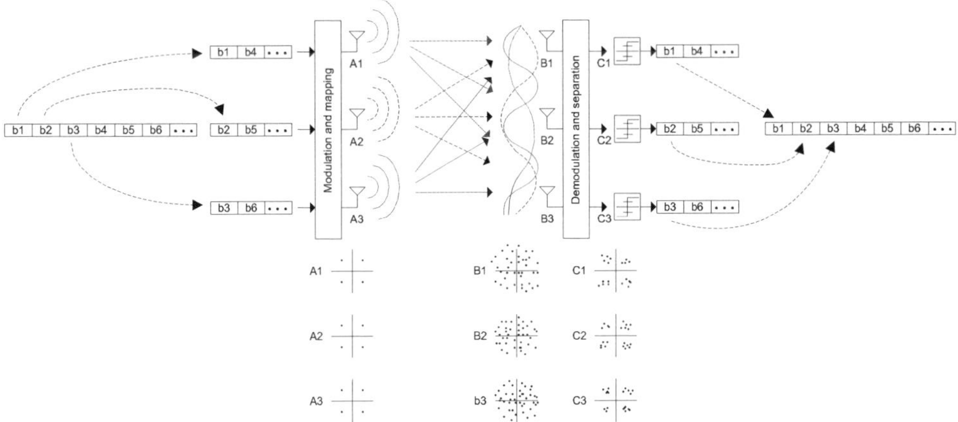

Figure 8.5 Conceptual diagram for using MIMO to implement spatial multiplexing.

where $\mathbf{r}$, $\mathbf{H}$, and $\mathbf{x}_k$ are the received signal vector, the channel matrix, and the sent kth symbol respectively. C is the constellation size and M_T is the number of transmitter antennae used in this MIMO multiplexing system.

The receiver in a spatial multiplexing system can also use linear detection algorithms, two of which are ZF and *Minimum Mean-Squared Error* (MMSE) detectors. The ZF algorithm performs pseudoinverse of the channel, so that we can obtain

$$\hat{\mathbf{x}} = \left(\mathbf{H}^H \mathbf{H}\right)^{-1} \mathbf{H}\mathbf{r} = \mathbf{H}^H \mathbf{r} \tag{8.26}$$

where $\mathbf{H}$ and $\mathbf{r}$ are the channel matrix and the received signal vector. $\mathbf{H}^H$ performs the Hermitian transform of matrix $\mathbf{H}$. The ZF detector has the simplest implementation complexity.

The MMSE algorithm, on the other hand, offers a good performance with an intermediate complexity if compared to the ZF detector. The MMSE detector performs the following detection algorithm:

$$\hat{\mathbf{x}} = \left(\mathbf{H}^H \mathbf{H} + \frac{\mathbf{I}_{M_R}}{SNR}\right)^{-1} \mathbf{H}^H \mathbf{r} \tag{8.27}$$

where $\mathbf{I}_{M_R}$ is a unit square matrix with its dimension being $M_R \times M_R$.

The typical example of a spatial multiplexing system is the V-BLAST scheme [701], which extracts multiplexed data streams by the ZF or the MMSE filter with ordered *successive interference cancelation* (SIC). The V-BLAST detection process includes five steps, which proceeds as follows:

- First step—Ordering: the receiver should order all incoming information channels to choose the best to proceed with the detection.

- Second step—Nulling: the scheme proceeds with the detection using either the ZF or the MMSE algorithm.

- Third step—Slicing: the system makes a decision symbol by symbol.

- Fourth step—Canceling: the receiver subtracts the detected symbol from the received signal.

- Fifth step—Iteration: the scheme goes back to the first step to detect the next symbol. The iteration continues until all the symbols have been detected.

We can make several important observations from the above discussions on spatial multiplexing using MIMO configurations. MIMO can achieve spatial multiplexing using multiple antennae at both ends (transmitter and receiver) of a radio link. It can increase the data rate substantially by transmitting independent information streams on different antennae. No channel state information is required at the transmitters. If scattering is rich enough (or, the channel matrix $\mathbf{H}$ has a relatively high rank), the multiplexing gain will be high. Spatial data pipes are created with the same bandwidth and the multiplexing gain can be achieved at no extra bandwidth or power.

To measure the efficiency of spatial multiplexing, the multiplexing gain can be defined intuitively by the number of parallel spatial data pipes in the same frequency band between the transmitter and the receiver. More specifically, the multiplexing gain m can be expressed as

$$m \overset{\Delta}{=} \lim_{\rho \to \infty} \frac{C(\rho)}{\log_2 \rho} = L \tag{8.28}$$

where $C(\rho)$ is the system capacity function with SNR ρ as its variable, and L is a constant.

It is noted that, as multiplexing gain and diversity gain are a dual, they work in a counteractive way with each other. The *multiplexing-diversity trade-off curve* can be used to illustrate how this

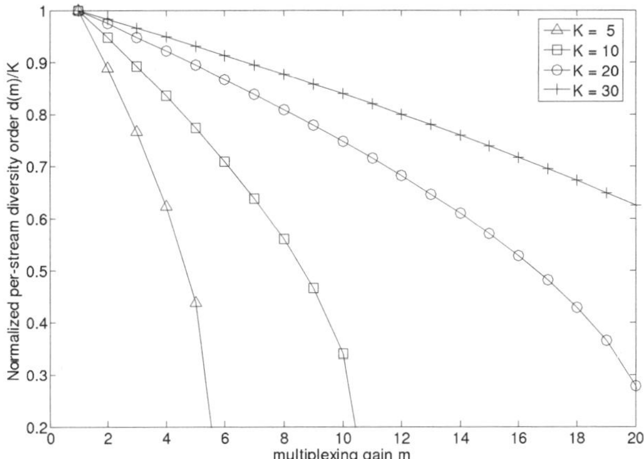

Figure 8.6   Multiplexing-diversity trade-off curves for $K=5$, 10, 20, and 30, where $K = max(M_T, M_R)$.

counteraction will take place in a particular MIMO system setup. The multiplexing-diversity trade-off curve tells us how much diversity each stream can achieve if a multitude of independent streams are spatially multiplexed. Therefore, the multiplexing-diversity trade-off curve is given by

$$d(m) = \frac{m}{\sum_{j=1}^{m} \frac{1}{K-j+1}}$$

(8.29)

where $K = max(M_T, M_R)$, and m is the multiplexing gain. Obviously, the multiplexing-diversity trade-off can be defined in different ways in terms of either outage capacity, or error probability of the MIMO system. Figure 8.6 shows a typical trend of multiplexing-diversity trade-off curve for $K = 5$, 10, 20, and 30.

To summarize, we can identify the four major improvements achievable in a wireless communication system with the help of MIMO technologies as follows:

- Spectral efficiency improvement due to multiplexing gain.

- Link reliability improvement due to diversity gain.

- Coverage improvement due to diversity gain and array gain.

- Capacity improvement due to cochannel interference suppression using both transmitter beam-forming and receiver beam-forming.

8.4   STBC-CDMA Systems

This section and the following ones will discuss the STBC-CDMA system, in particular, the usage of complementary codes.

The current 2- to 3G CDMA–based wireless systems use *unitary* signature codes, which provide channelization for a CDMA system on an one-code-per-channel basis. These unitary codes can also be further classified into two major subsets, that is, quasi-orthogonal codes such as Gold codes [195], Kasami codes [196–199], m-sequences [132–138], and so on, and orthogonal codes such as Walsh-Hadamard sequences [200–203], OVSF codes, and so on. One of the most problematic issues in these unitary code–based DS-CDMA systems is multiple access interference (MAI), which dominates the undesirable signals present at a receiver, making all the traditional CDMA systems strictly interference-limited. The maximum capacity of such a CDMA system is usually at only a fraction of the system processing gain (PG), more specifically being 0.5 to 0.7 of the PG, dependent on its operational environment. There are also many other inherent limitations to a traditional DS-CDMA system using unitary spreading codes, such as its great sensitivity to the near–far effect and multipath interference (MI), and so on. Obviously, most of these technical limitations in the current CDMA–based 2- to 3G systems stem from many far-from-ideal properties of those unitary codes, to be explained in the sequel.

All these unitary signature codes were designed and generated based only on their seemly acceptable periodic auto-correlation (PAC) and periodic cross-correlation (PCC) functions. Unfortunately, both PAC and PCC become completely irrelevant if a CDMA system works in an asynchronous transmission mode, let alone a working environment where MI is present. Many quasi-orthogonal codes, such as Gold codes, Kasami codes, M-sequences, and so on, have nontrivial cross-correlation functions, periodic and aperiodic alike, which cause serious MAI when more than one active user exists in the system. Even for those *orthogonal* codes, such as Walsh-Hadamard sequences and OVSF codes, the MAI can also be rampant when they are working as channelization codes in asynchronous channels, no matter whether the multipath effect is present or not. Thus, it is clear that the use of these traditional unitary codes on their own can never get rid of those annoying MAI due to their poor correlation properties. The same conclusion is also applicable to such a CDMA system working jointly with a space–time coding scheme.

Orthogonal complementary (OC) codes were proposed in [206–220] as CDMA signature codes because of their following salient features:

- The OC codes [206–220] retain *perfect* auto-correlation functions and cross-correlation functions in both synchronous and asynchronous transmission modes. Here, the word *perfect* means zero auto-correlation side lobes and zero cross-correlation levels for any relative shifts. This property enables desirable MAI-free operation in a CDMA system.

- The OC codes have inherent immunity against MI due to their perfect auto-correlation and perfect cross-correlation functions regardless of their operation modes, in either synchronous or asynchronous channels. This property ensures MI-free operation of a CDMA system.

- The OC codes offer perfect partial cross-correlation functions, which makes them particularly suitable for the applications where traffic is dominated by short bursts, such as future all-IP wireless networks.

- The OC codes work on a-flock-of-codes basis. Every user may use M different carriers to send its M element codes in parallel. Thus, the OFDM technology can be a suitable solution to implement such a multi-carrier CDMA in a reasonable hardware complexity.

Recently, Hochwald *et al.* [695] successfully applied the initial STBC scheme into a generic CDMA system to establish a theoretical framework to study an S-T coded CDMA system. In this chapter, we will use a similar analytical method to study a generic STBC-CDMA system. More specifically, we will study an STBC-CDMA system based on two major categories of spreading codes, either unitary codes or orthogonal complementary codes. In particular, the study on an STBC-CDMA system with complementary codes has not been reported in the literature. We will show that an STBC-CDMA system and OC codes can work jointly to achieve a full space-diversity gain, in addition to its superior MAI-free property inherited from OC code sets. The major part of this section focusses on the analytical treatment of code-dependent analysis of an STBC-CDMA system under MAI and flat Rayleigh fading. The comparison is also made among the STBC-CDMA systems using both analysis and simulation.

8.5   Generic STBC-CDMA System Model

Figure 8.7 illustrates a generic STBC-CDMA system model with K users, each of whom is assigned a flock of M element codes, $\{\mathbf{c}_{k,1}, \mathbf{c}_{k,2}, \ldots, \mathbf{c}_{k,M}\}$, $k \in (1, K)$. Each element code $\mathbf{c}_{k,m}$ consists of N chips with their waveforms being square pulses, where $k \in (1, K)$ and $m \in (1, M)$. If $M = 1$, the same system model can be used to describe an STBC-CDMA system based on unitary spreading codes; otherwise it is for an STBC-CDMA system using OC codes. In this model, it will be assumed that each transmitter will use only two antennae to achieve transmitter diversity. However, it will be generalized into n_t transmitter antennae in the analysis. The channel coefficients for different transmitters are assumed to be $h_{k,1}$ and $h_{k,2}$, respectively for two antennae where $k \in (1, K)$. It is also assumed that signals from different antennae in a transmitter experience independent Rayleigh fading and additive white Gaussian noise (AWGN). We do not consider multipath effect here to make the analysis tractable. It concerns a synchronous downlink channel here, which is relevant to most application scenarios with transmitter diversity.

Figure 8.8 illustrates a generic architecture of the k-th transmitter, where the original data stream should go through an STBC encoder, a DS spreading modulation, and a BPSK carrier modulation before being sent out via two antennae. It is noted that for the cases of OC codes, each transmitter uses M carriers to send M different element codes, which will undergo separate matched filtering and then

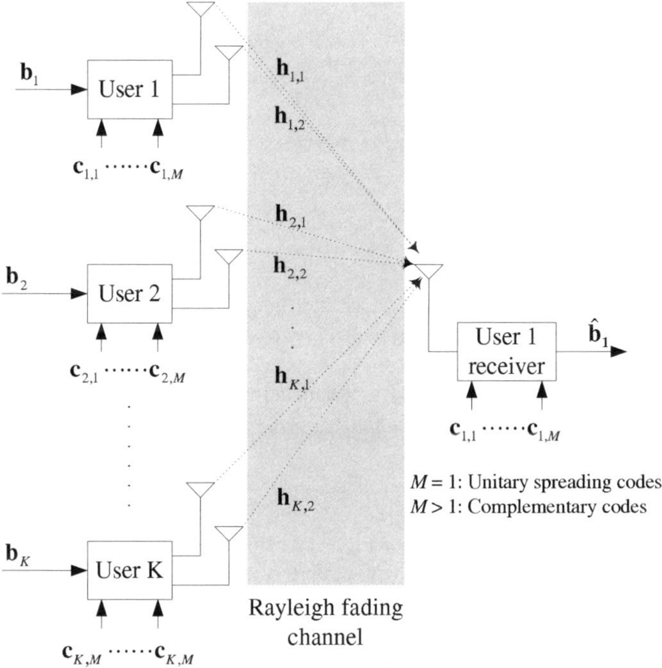

Figure 8.7 A generic downlink model for an STBC-CDMA system, where K users are present, each user is assigned a flock of M element codes, each element code has N chips, two (which can be further extended to a more general case of n_t in the analysis) transmitter antennae are assigned to each user, as is a flat Rayleigh fading channel, and the signal of interest to the receiver is user 1. If $M = 1$, this model represents an STBC-CDMA system based on unitary codes; otherwise it will be a model for an STBC-CDMA using complementary codes.

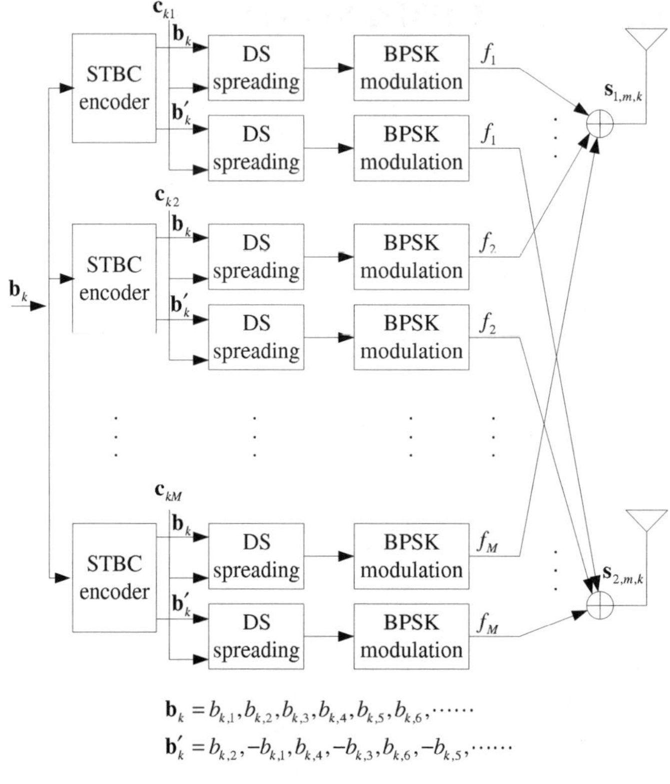

$$\mathbf{b}_k = b_{k,1}, b_{k,2}, b_{k,3}, b_{k,4}, b_{k,5}, b_{k,6}, \cdots\cdots$$

$$\mathbf{b}'_k = b_{k,2}, -b_{k,1}, b_{k,4}, -b_{k,3}, b_{k,6}, -b_{k,5}, \cdots\cdots$$

Figure 8.8 The block diagram for the k-th transmitter in a generic STBC-CDMA system, where the original data stream $\mathbf{b}_k$ is encoded into $\mathbf{b}_k$ and $\mathbf{b}'_k$ two substreams, and the DS spreading is then modulated by a flock of M element codes $\mathbf{c}_{k,1}, \mathbf{c}_{k,2}, \ldots, \mathbf{c}_{k,M}$, followed by BPSK carrier modulation with respect to M different carriers. If $M = 1$, the model represents an STBC-CDMA system based on unitary codes; otherwise it will be a model of an STBC-CDMA system using complementary codes.

be summed over to reconstruct the orthogonality of complementary codes at a receiver. Similarly, the model will suit an STBC-CDMA system with unitary codes if we let $M = 1$ in Figure 8.8.

Figure 8.9 shows a generic model of an STBC-CDMA receiver for both unitary codes ($M = 1$) and OC codes ($M > 1$). This model concerns an STBC decoder in the receiver intended for the k-th user ($k = 1$ is assumed in the analysis followed) in downlink channels, meaning that we only need to consider one complete data block consisting of two consecutive symbols, that is, $b_{k,o}$ and $b_{k,e}$, without losing the generality, where the subscripts "o" and "e" stand for *odd* and *even* symbols, respectively. To facilitate the analysis, two *extended element codes* for the m-th element code of user k or $\mathbf{c}_{k,m}$ are formulated as

$$\mathbf{c}_{o,k,m} = [\mathbf{c}_{k,m}, 0, 0, \ldots, 0] \tag{8.30}$$

$$\mathbf{c}_{e,k,m} = [0, 0, \ldots, 0, \mathbf{c}_{k,m}] \tag{8.31}$$

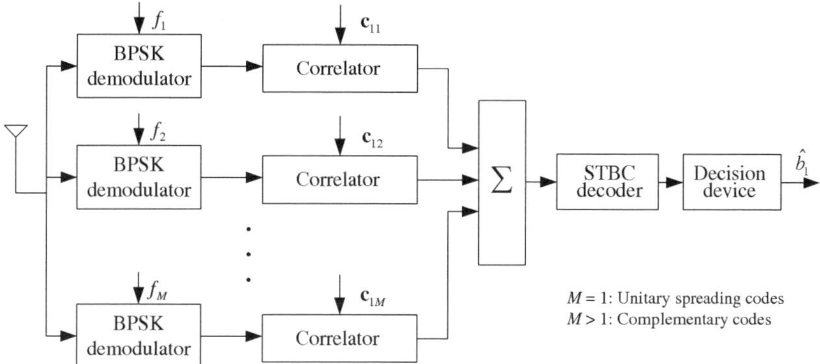

Figure 8.9 The block diagram for a generic STBC-CDMA receiver whose signal of interest is the k-th user ($k = 1$ is assumed here), where M BPSK demodulators are tuned to different carriers, $f_1, f_2, \ldots, f_M$. M element codes should undergo individual matched filtering and then be summed up to form a variable prior to STBC decoding. If $M = 1$, the model represents an STBC-CDMA system based on unitary codes; otherwise it will be a model for an STBC-CDMA system using complementary codes.

each of which has $2N$ chips in length if N is the length of an element code. The definitions given in Equations (8.30) and (8.31) are applicable to both unitary codes and complementary codes alike.

Two operators, $\otimes$ and $\oplus$, are introduced as *element-wise product* (EWP) and *half-length addition* (HLA) operations, respectively, to perform the calculations with respect to two vectors, $\mathbf{x}$ and $\mathbf{y}$:

$$\mathbf{x} = [\alpha_1, \alpha_2, \ldots, \alpha_N, \beta_1, \beta_2, \ldots, \beta_N] \tag{8.32}$$

$$\mathbf{y} = \left[\gamma_1, \gamma_2, \ldots, \gamma_N, \delta_1, \delta_2, \ldots, \delta_N\right] \tag{8.33}$$

as follows

$$\mathbf{x} \otimes \mathbf{y} = \left[\alpha_1\gamma_1, \alpha_2\gamma_2, \ldots, \alpha_N\gamma_N, \beta_1\delta_1, \beta_2\delta_2, \ldots, \beta_N\delta_N\right] \tag{8.34}$$

$$\mathbf{x} \oplus \mathbf{x} = [\alpha_1 + \alpha_2 + \cdots + \alpha_N, \beta_1 + \beta_2 + \cdots + \beta_N] \tag{8.35}$$

where it is noted that the result from the HLA operation is a two-element vector and both sides of the operator $\oplus$ must be the same; otherwise the result should be treated as a null vector, or

$$\mathbf{x} \oplus \mathbf{y} \equiv [0, 0], \quad \mathbf{x} \neq \mathbf{y}. \tag{8.36}$$

Therefore, the EWP operation does not change the dimension of the vectors involved; while the HLA operation reduces the dimension of the original vectors involved into two. If each block consists of two symbols, as required by the Alamouti block encoder [693], the operation of $(\mathbf{x} \otimes \mathbf{y}) \oplus (\mathbf{x} \otimes \mathbf{y})$ between any two extended element codes $\mathbf{x}$ and $\mathbf{y}$ gives rise to *block-wise correlation* (BWC), which generates a block-wise auto-correlation function if $\mathbf{x} = \mathbf{y}$ and a block-wise cross-correlation function if $\mathbf{x} \neq \mathbf{y}$. Therefore, we have

$$f_{bwc}(\mathbf{x}, \mathbf{x}) = (\mathbf{x} \otimes \mathbf{x}) \oplus (\mathbf{x} \otimes \mathbf{x})$$

$$= \left[\alpha_1^2 + \alpha_2^2 + \cdots + \alpha_N^2, \beta_1^2 + \beta_2^2 + \cdots + \beta_N^2\right] \tag{8.37}$$

$$f_{bwc}(\mathbf{x}, \mathbf{y}) = (\mathbf{x} \otimes \mathbf{y}) \oplus (\mathbf{x} \otimes \mathbf{y})$$

$$= \left[\alpha_1\gamma_1 + \alpha_2\gamma_2 + \cdots + \alpha_N\gamma_N, \beta_1\delta_1 + \beta_2\delta_2 + \cdots + \beta_N\delta_N\right] \tag{8.38}$$

It can be shown that both EWP and HLA operators are linear. So are $f_{bwc}(\mathbf{x}, \mathbf{x})$ and $f_{bwc}(\mathbf{x}, \mathbf{y})$. In particular, we have the following property for the HLA operator

$$(\mathbf{x} + \mathbf{y}) \oplus (\mathbf{x} + \mathbf{y}) = \mathbf{x} \oplus \mathbf{x} + \mathbf{y} \oplus \mathbf{y} + \mathbf{x} \oplus \mathbf{y} + \mathbf{y} \oplus \mathbf{x}$$

$$= (\mathbf{x} \oplus \mathbf{x}) + (\mathbf{y} \oplus \mathbf{y}) \qquad (8.39)$$

where either $\mathbf{x} \oplus \mathbf{y}$ or $\mathbf{y} \oplus \mathbf{x}$ has been defined to be $[0, 0]$, which is reasonable because the following two operations with respect to extended element codes $\mathbf{x}$ and $\mathbf{y}$ will give exactly the same result:

- First add the two vectors and then carry out the HLA operation.

- First perform the HLA operation for each vector individually, followed by a summation.

8.6   Unitary Code–Based STBC-CDMA System

If $M = 1$, Figures 8.7, 8.8, and 8.9 in fact show a downlink STBC-CDMA system model, a transmitter and a receiver model with unitary codes, respectively. In this case, the actual system architecture for a unitary code STBC-CDMA system becomes much simpler than what appears in Figures 8.7, 8.8, and 8.9. For instance, only a single-carrier frequency is necessary in Figure 8.8, and therefore, the overall complexity of a transmitter with unitary codes will only be about $\frac{1}{M}$ of that using complementary codes. Therefore, the even and odd extended element codes for the k-th user given in Equations (8.30) and (8.31) can be written as

$$\mathbf{c}_{o,k} = [\mathbf{c}_k, 0, 0, \ldots, 0] \qquad (8.40)$$

$$\mathbf{c}_{e,k} = [0, 0, \ldots, 0, \mathbf{c}_k] \qquad (8.41)$$

each being $2N$ chips long for $k \in (1, K)$. The equivalent baseband signal from the k-th transmitter can be written as

$$\mathbf{t}_k = (b_{k,o}\mathbf{c}_{o,k} + b_{k,e}\mathbf{c}_{e,k})h_{k,1} + (b_{k,e}\mathbf{c}_{o,k} - b_{k,o}\mathbf{c}_{e,k})h_{k,2} \qquad (8.42)$$

when two transmitting antennae are considered. Thus, the received signal at a receiver tuned to the signal from the first user can be represented by

$$\mathbf{r}_1 = \sum_{k=1}^{K} \mathbf{t}_k + \mathbf{n}_1 \qquad (8.43)$$

where $\mathbf{n}_1$ is an AWGN vector observed at the receiver, and K users are totally present in the system. Referring to the receiver model shown in Figure 8.9 (for $M = 1$), we can write a block of the output signal from the correlator as

$$[d_{1,1}, d_{1,2}] = \left[\left(\mathbf{r}_1 \otimes [\mathbf{c}_{o,1} + \mathbf{c}_{e,1}] \right) \oplus \left(\mathbf{r}_1 \otimes [\mathbf{c}_{o,1} + \mathbf{c}_{e,1}] \right) \right]$$

$$= [(h_{1,1}b_{1,o} + h_{1,2}b_{1,e} + \mathbf{c}_{o,1}^H \mathbf{n}_1), (-h_{1,2}b_{1,o} + h_{1,1}b_{1,e} + \mathbf{c}_{e,1}^H \mathbf{n}_1)]$$

$$+ \sum_{k=2}^{K} \left[\left(\mathbf{t}_k \otimes [\mathbf{c}_{o,1} + \mathbf{c}_{e,1}] \right) \oplus \left(\mathbf{t}_k \otimes [\mathbf{c}_{o,1} + \mathbf{c}_{e,1}] \right) \right] \qquad (8.44)$$

where $\mathbf{x}^H$ is the Hermitian form of matrix $\mathbf{x}$ and the detail derivation of Equation (8.44) is shown in Appendix D. The second summation term in Equation (8.44) represents the MAI caused by other unwanted transmissions, which can be expanded into

$$\mathbf{I}_1 = \sum_{k=2}^{K} \left[\left(\mathbf{t}_k \otimes \left[\mathbf{c}_{o,1} + \mathbf{c}_{e,1} \right] \right) \oplus \left(\mathbf{t}_k \otimes \left[\mathbf{c}_{o,1} + \mathbf{c}_{e,1} \right] \right) \right]$$

$$= \left[\begin{array}{c} b_{2,o}\mathbf{c}_{o,1}^H \mathbf{c}_{o,2}h_{2,1} + b_{2,e}\mathbf{c}_{o,1}^H \mathbf{c}_{o,2}h_{2,2} \\ b_{2,e}\mathbf{c}_{e,1}^H \mathbf{c}_{e,2}h_{2,1} - b_{2,o}\mathbf{c}_{e,1}^H \mathbf{c}_{e,2}h_{2,2} \end{array} \right]^T + \left[\begin{array}{c} b_{3,o}\mathbf{c}_{o,1}^H \mathbf{c}_{o,3}h_{3,1} + b_{3,e}\mathbf{c}_{o,1}^H \mathbf{c}_{o,3}h_{3,2} \\ b_{3,e}\mathbf{c}_{e,1}^H \mathbf{c}_{e,3}h_{3,1} - b_{3,o}\mathbf{c}_{e,1}^H \mathbf{c}_{e,3}h_{3,2} \end{array} \right]^T$$

$$+ \cdots + \left[\begin{array}{c} b_{K,o}\mathbf{c}_{o,1}^H \mathbf{c}_{o,K}h_{K,1} + b_{K,e}\mathbf{c}_{o,1}^H \mathbf{c}_{o,K}h_{K,2} \\ b_{K,e}\mathbf{c}_{e,1}^H \mathbf{c}_{e,K}h_{K,1} - b_{K,o}\mathbf{c}_{e,1}^H \mathbf{c}_{e,K}h_{K,2} \end{array} \right]^T$$

$$= \sum_{k=2}^{K} \left[\begin{array}{c} b_{k,o}\mathbf{c}_{o,1}^H \mathbf{c}_{o,k}h_{k,1} + b_{k,e}\mathbf{c}_{o,1}^H \mathbf{c}_{o,k}h_{k,2} \\ b_{k,e}\mathbf{c}_{e,1}^H \mathbf{c}_{e,k}h_{k,1} - b_{k,o}\mathbf{c}_{e,1}^H \mathbf{c}_{e,k}h_{k,2} \end{array} \right]^T$$

$$= \sum_{k=2}^{K} \xi_k \left[\begin{array}{c} b_{k,o}h_{k,1} + b_{k,e}h_{k,2} \\ b_{k,e}h_{k,1} - b_{k,o}h_{k,2} \end{array} \right]^T \tag{8.45}$$

where we have used a definition of ξ_k given by

$$\xi_k = \mathbf{c}_{o,1}^H \mathbf{c}_{o,k} = \mathbf{c}_{e,1}^H \mathbf{c}_{e,k} = \mathbf{c}_1^H \mathbf{c}_k \tag{8.46}$$

which stands simply for the inner product or the PCC of the two element codes of $\mathbf{c}_1$ and $\mathbf{c}_k$. Therefore, Equation (8.44) can also be expressed as

$$\left[\begin{array}{c} d_{1,1} \\ d_{1,2} \end{array} \right] = \left[\begin{array}{cc} h_{1,1} & h_{1,2} \\ -h_{1,2} & h_{1,1} \end{array} \right] \left[\begin{array}{c} b_{1,o} \\ b_{1,e} \end{array} \right] + \sum_{k=2}^{K} \xi_k \left[\begin{array}{c} b_{k,o}h_{k,1} + b_{k,e}h_{k,2} \\ b_{k,e}h_{k,1} - b_{k,o}h_{k,2} \end{array} \right] + \left[\begin{array}{c} \mathbf{c}_{o,1}^H \\ \mathbf{c}_{e,1}^H \end{array} \right] \mathbf{n}_1$$

$$= \left[\begin{array}{cc} h_{1,1} & h_{1,2} \\ -h_{1,2} & h_{1,1} \end{array} \right] \left[\begin{array}{c} b_{1,o} \\ b_{1,e} \end{array} \right] + \sum_{k=2}^{K} \xi_k \left[\begin{array}{cc} h_{k,1} & h_{k,2} \\ -h_{k,2} & h_{k,1} \end{array} \right] \left[\begin{array}{c} b_{k,o} \\ b_{k,e} \end{array} \right] + \left[\begin{array}{c} \mathbf{c}_{o,1}^H \\ \mathbf{c}_{e,1}^H \end{array} \right] \mathbf{n}_1 \tag{8.47}$$

which can be rewritten into a vector form as

$$\mathbf{d}_1 = \mathbf{H}_1 \mathbf{b}_1 + \sum_{k=2}^{K} \xi_k \mathbf{H}_k \mathbf{b}_k + \mathbf{z}_1 \tag{8.48}$$

where we have used the following definitions: $\mathbf{d}_1 = \left[\begin{array}{c} d_{1,1} \\ d_{1,2} \end{array} \right]$, $\mathbf{b}_1 = \left[\begin{array}{c} b_{1,o} \\ b_{1,e} \end{array} \right]$, $\mathbf{b}_k = \left[\begin{array}{c} b_{k,o} \\ b_{k,e} \end{array} \right]$,

$\mathbf{H}_1 = \left[\begin{array}{cc} h_{1,1} & h_{1,2} \\ -h_{1,2} & h_{1,1} \end{array} \right]$, $\mathbf{H}_k = \left[\begin{array}{cc} h_{k,1} & h_{k,2} \\ -h_{k,2} & h_{k,1} \end{array} \right]$ and $\mathbf{z}_1 = \left[\begin{array}{c} \mathbf{c}_{o,1}^H \\ \mathbf{c}_{e,1}^H \end{array} \right] \mathbf{n}_1$. Thus, after ST-decoding we obtain

$$\Re\left(\mathbf{H}_1^H \mathbf{d}_1 \right) = \Re\left(\mathbf{H}_1^H \mathbf{H}_1 \mathbf{b}_1 \right) + \sum_{k=2}^{K} \xi_k \Re\left(\mathbf{H}_1^H \mathbf{H}_k \mathbf{b}_k \right) + \Re\left(\mathbf{H}_1^H \mathbf{z}_1 \right)$$

$$= \left[\begin{array}{cc} h_{1,1}^2 + h_{1,2}^2 & 0 \\ 0 & h_{1,1}^2 + h_{1,2}^2 \end{array} \right] \left[\begin{array}{c} b_{1,o} \\ b_{1,e} \end{array} \right] + \sum_{k=2}^{K} \xi_k \Re\left(\mathbf{H}_1^H \mathbf{H}_k \mathbf{b}_k \right) + \Re\left(\mathbf{H}_1^H \mathbf{z}_1 \right)$$

$$= \left[\begin{array}{cc} h_{1,1}^2 + h_{1,2}^2 & 0 \\ 0 & h_{1,1}^2 + h_{1,2}^2 \end{array} \right] \left[\begin{array}{c} b_{1,o} \\ b_{1,e} \end{array} \right] + \mathbf{I}_1 + \bar{\mathbf{n}}_1 \tag{8.49}$$

where $\Re(\mathbf{x})$ is used to calculate the real part of $\mathbf{x}$, the first term in Equation (8.49) is the useful component that has achieved a full diversity gain, the second term, $\mathbf{I}_1$, is the MAI vector caused by other unwanted transmissions, and the last term, $\bar{\mathbf{n}}_1$, is because of noise. Therefore, the decision on the block can be made from Equation (8.49) corrupted by MAI and noise. Now let us first fix $\mathbf{H}_1$ as a constant matrix and treat $\mathbf{H}_k$ as a matrix with all its elements being Rayleigh distributed random variables. It can be shown that the variance of each element in the MAI vector or $\mathbf{I}_1$ is

$$\sigma_{MAI}^2 = \sigma_I^2 \sum_{j=1}^{n_t} h_{1,j}^2 \tag{8.50}$$

where

$$\sigma_I^2 = 2n_t \sum_{k=2}^{K} \xi_k^2 \sigma^2 \tag{8.51}$$

in which we have generalized the results to the cases with n_t transmitter antennae. Similarly, the variance for each element in the noise vector or $\bar{\mathbf{n}}_1$ in Equation (8.49) is

$$\sigma_{noise}^2 = N_o \sum_{j=1}^{n_t} h_{1,j}^2 \tag{8.52}$$

If a BPSK modem is used for carrier modulation and demodulation, we can immediately write down the BER for an unitary code–based STBC-CDMA system as

$$P_{unitary} = Q \left(\sqrt{\frac{2\alpha_M SNR}{n_t(1 + SNR \, \sigma_I^2/n_t)}} \right) \tag{8.53}$$

where $SNR = E_b/N_o$ and the random variable α_M is defined as

$$\alpha_M = \sum_{j=1}^{n_t} h_{1,j}^2 \tag{8.54}$$

which obeys the following distribution

$$f_{\alpha,n_t}(r) = \left(\frac{1}{2\sigma^2} \right)^{n_t} \frac{r^{n_t-1}}{(n_t - 1)!} \exp\left(-\frac{r}{2\sigma^2} \right), \qquad 0 \le r < \infty. \tag{8.55}$$

Therefore, the average BER for downlink transmissions in a unitary code STBC-CDMA system can be expressed by

$$\bar{P}_{unitary} = \int_0^{\infty} Q \left(\sqrt{\frac{2rSNR}{n_t(1 + SNR \, \sigma_I^2/n_t)}} \right) f_{\alpha,n_t}(r) \, dr$$

$$= \int_0^{\infty} Q \left(\sqrt{\frac{2rSNR}{n_t(1 + SNR \, \sigma_I^2/n_t)}} \right) \left(\frac{1}{2\sigma^2} \right)^{n_t} \frac{r^{n_t-1}}{(n_t - 1)!} e^{-\frac{r}{2\sigma^2}} \, dr \tag{8.56}$$

8.7 Complementary Coded STBC-CDMA System

As the analysis for an STBC-CDMA system with OC codes can be more complicated than that with the unitary codes, we address the issue in two separate steps: We first start the analysis with a relatively simple two-antenna system, and then extend the analysis to an OC code–based STBC-CDMA system with an arbitrary number of transmitter antennae.

8.7.1 Dual Transmitter Antennae

To study an STBC-CDMA system based on OC codes,[1] the assumption of $M > 1$ should be applied in the system models illustrated in Figures 8.7, 8.8, and 8.9.

On the basis of the Alamouti STBC algorithm [693], an encoded signal block (for the m-th element code) from two transmitter antennae of the k-th user in an OC code–based STBC-CDMA system can be written as

$$\mathbf{s}_{1,k,m} = (b_{1,o}\mathbf{c}_{o,k,m} + b_{1,e}\mathbf{c}_{e,k,m}) \tag{8.57}$$

$$\mathbf{s}_{2,k,m} = (b_{1,e}\mathbf{c}_{o,k,m} - b_{1,o}\mathbf{c}_{e,k,m}) \tag{8.58}$$

where $k \in (1, K)$ and $m \in (1, M)$. If a perfect coherent demodulation process is assumed, the received signal at a receiver tuned to user 1 in the m-th carrier frequency f_m becomes

$$\mathbf{r}_{1,m} = \sum_{k=1}^{K} \big[(b_{k,o}\mathbf{c}_{o,k,m} + b_{k,e}\mathbf{c}_{e,k,m})h_{k,1}$$
$$+ (b_{k,e}\mathbf{c}_{o,k,m} - b_{k,o}\mathbf{c}_{e,k,m})h_{k,2} + \mathbf{n}_{k,m}\big] \tag{8.59}$$

where $m \in (1, M)$, $h_{k,1}$ and $h_{k,2}$ are independent Rayleigh fading channel coefficients due to two sufficiently spaced antennae at transmitter 1, and $\mathbf{n}_{k,m}$ is an AWGN term with zero mean and variance being $N_o/(2N)$ observed in each chip interval.

As shown in Figure 8.9, the received signal should first undergo separate matched filtering for different element codes before summation. Let the first user's transmission be the signal of interest or $k = 1$. The received signal $\mathbf{r}_{1,m}$ from different carrier frequencies should be matched-filtered with respect to different extended element codes or $\mathbf{c}_{o,1,m} + \mathbf{c}_{e,1,m}$, $m \in (1, M)$. For analytical simplicity, we would like to carry out the EWP operation first, followed by the HLA operation, as shown in the sequel.

$$\begin{cases} \mathbf{w}_{1,1} = \mathbf{r}_{1,1} \otimes (\mathbf{c}_{o,1,1} + \mathbf{c}_{e,1,1}) \\ \mathbf{w}_{1,2} = \mathbf{r}_{1,2} \otimes (\mathbf{c}_{o,1,2} + \mathbf{c}_{e,1,2}) \\ \quad\vdots \\ \mathbf{w}_{1,M} = \mathbf{r}_{1,M} \otimes (\mathbf{c}_{o,1,M} + \mathbf{c}_{e,1,M}) \end{cases} \tag{8.60}$$

or

$$\begin{cases} \mathbf{w}_{1,1} = \displaystyle\sum_{k=1}^{K} \big[(b_{k,o}\mathbf{c}_{o,k,1} + b_{k,e}\mathbf{c}_{e,k,1})h_{k,1} + (b_{k,e}\mathbf{c}_{o,k,1} - b_{k,o}\mathbf{c}_{e,k,1})h_{k,2} + \mathbf{n}_{k,1}\big] \\ \quad \otimes (\mathbf{c}_{o,1,1} + \mathbf{c}_{e,1,1}) \\ \mathbf{w}_{1,2} = \displaystyle\sum_{k=1}^{K} \big[(b_{k,o}\mathbf{c}_{o,k,2} + b_{k,e}\mathbf{c}_{e,k,2})h_{k,3} + (b_{k,e}\mathbf{c}_{o,k,2} - b_{k,o}\mathbf{c}_{e,k,2})h_{k,4} + \mathbf{n}_{k,2}\big] \\ \quad \otimes (\mathbf{c}_{o,1,2} + \mathbf{c}_{e,1,2}) \\ \quad\vdots \\ \mathbf{w}_{1,M} = \displaystyle\sum_{k=1}^{K} \big[(b_{k,o}\mathbf{c}_{o,k,M} + b_{k,e}\mathbf{c}_{e,k,M})h_{k,(2M-1)} + (b_{k,e}\mathbf{c}_{o,k,M} - b_{k,o}\mathbf{c}_{e,k,M})h_{k,2M} + \mathbf{n}_{k,M}\big] \\ \quad \otimes (\mathbf{c}_{o,1,M} + \mathbf{c}_{e,1,M}) \end{cases} \tag{8.61}$$

---

[1]This new STBC scheme based on complementary codes is also called a *Space–Time Complementary Coding* (STCC) scheme.

which can be rewritten into

$$
\begin{cases}
\mathbf{w}_{1,1} = \left[(b_{1,o}\mathbf{c}_{o,1,1} + b_{1,e}\mathbf{c}_{e,1,1})h_{1,1} + (b_{1,e}\mathbf{c}_{o,1,1} - b_{1,o}\mathbf{c}_{e,1,1})h_{1,2} + \mathbf{I}_{1,1} + \mathbf{n}_{1,1}\right] \\
\quad \otimes(\mathbf{c}_{o,1,1} + \mathbf{c}_{e,1,1}) \\
\mathbf{w}_{1,2} = \left[(b_{1,o}\mathbf{c}_{o,1,2} + b_{1,e}\mathbf{c}_{e,1,2})h_{1,3} + (b_{1,e}\mathbf{c}_{o,1,2} - b_{1,o}\mathbf{c}_{e,1,2})h_{1,4} + \mathbf{I}_{1,2} + \mathbf{n}_{1,2}\right] \\
\quad \otimes(\mathbf{c}_{o,1,2} + \mathbf{c}_{e,1,2}) \\
\quad\vdots \\
\mathbf{w}_{1,M} = \left[(b_{1,o}\mathbf{c}_{o,1,M} + b_{1,e}\mathbf{c}_{e,1,M})h_{1,(2M-1)} + (b_{1,e}\mathbf{c}_{o,1,M} - b_{1,o}\mathbf{c}_{e,1,M})h_{1,2M} + \mathbf{I}_{1,M} + \mathbf{n}_{1,M}\right] \\
\quad \otimes(\mathbf{c}_{o,1,M} + \mathbf{c}_{e,1,M})
\end{cases}
$$

$$(8.62)$$

where $\mathbf{I}_{1,m}$, $m \in (1, M)$, is the interference term defined by

$$
\mathbf{I}_{1,m} = \sum_{k=2}^{K} \left[(b_{k,o}\mathbf{c}_{o,k,m} + b_{k,e}\mathbf{c}_{e,k,m})h_{k,(2m-1)} + (b_{k,e}\mathbf{c}_{o,k,m} - b_{k,o}\mathbf{c}_{e,k,m})h_{k,2m}\right] \qquad (8.63)
$$

To proceed with the correlation process, we need to sum up all the items given in Equation (8.62) to obtain

$$
\begin{aligned}
\mathbf{w} &= \sum_{m=1}^{M} \mathbf{w}_{1,m} \\
&= (h_{1,1} + h_{1,3} + \cdots + h_{1,2M-1})\left\{b_{1,o}[1, 1, \ldots, 1, 0, 0, \ldots, 0] + b_{1,e}[0, 0, \ldots, 0, 1, 1, \ldots, 1]\right\} \\
&\quad + (h_{1,2} + h_{1,4} + \cdots + h_{1,2M})\left\{b_{1,e}[1, 1, \ldots, 1, 0, 0, \ldots, 0] - b_{1,o}[0, 0, \ldots, 0, 1, 1, \ldots, 1]\right\} \\
&\quad + \sum_{m=1}^{M}(\mathbf{I}_{1,m} + \mathbf{n}_{1,m}) \otimes (\mathbf{c}_{o,1,m} + \mathbf{c}_{e,1,m})
\end{aligned}
$$

$$(8.64)$$

which results in a row vector. To complete the correlation process, we need the HLA operator that will generate the output from the matched filter as

$$
\begin{aligned}
[d_{1,1}, d_{1,2}] &= \mathbf{w} \oplus \mathbf{w} \\
&= \begin{bmatrix} (h_{1,1} + h_{1,3} + \cdots + h_{1,2M-1})b_{1,o} + (h_{1,2} + h_{1,4} + \cdots + h_{1,2M})b_{1,e} \\ -(h_{1,2} + h_{1,4} + \cdots + h_{1,2M})b_{1,o} + (h_{1,1} + h_{1,3} + \cdots + h_{1,2M-1})b_{1,e} \end{bmatrix}^{T} \\
&\quad + \left\{\sum_{m=1}^{M}(\mathbf{I}_{1,m} + \mathbf{n}_{1,m}) \otimes (\mathbf{c}_{o,1,m} + \mathbf{c}_{e,1,m})\right\} \\
&\quad \oplus \left\{\sum_{m=1}^{M}(\mathbf{I}_{1,m} + \mathbf{n}_{1,m}) \otimes (\mathbf{c}_{o,1,m} + \mathbf{c}_{e,1,m})\right\}
\end{aligned}
$$

$$(8.65)$$

In Appendix F, we show the validity of the following equation

$$
\left\{\sum_{m=1}^{M} \mathbf{I}_{1,m} \otimes (\mathbf{c}_{o,1,m} + \mathbf{c}_{e,1,m})\right\} \oplus \left\{\sum_{m=1}^{M} \mathbf{I}_{1,m} \otimes (\mathbf{c}_{o,1,m} + \mathbf{c}_{e,1,m})\right\} = [0, 0] \qquad (8.66)
$$

Define

$$[v_{1,1}, v_{1,2}] = \left\{ \sum_{m=1}^{M} \mathbf{n}_{1,m} \otimes (\mathbf{c}_{o,1,m} + \mathbf{c}_{e,1,m}) \right\}$$

$$\oplus \left\{ \sum_{m=1}^{M} \mathbf{n}_{1,m} \otimes (\mathbf{c}_{o,1,m} + \mathbf{c}_{e,1,m}) \right\}. \tag{8.67}$$

Equation (8.65) can be rewritten as

$$\begin{cases} d_{1,1} = (h_{1,1} + h_{1,3} + \cdots + h_{1,2M-1})b_{1,o} + (h_{1,2} + h_{1,4} + \cdots + h_{1,2M})b_{1,e} + v_{1,1} \\ d_{1,2} = -(h_{1,2} + h_{1,4} + \cdots + h_{1,2M})b_{1,o} + (h_{1,1} + h_{1,3} + \cdots + h_{1,2M-1})b_{1,e} + v_{1,2} \end{cases}$$

which can be further written into a matrix form as

$$\begin{bmatrix} d_{1,1} \\ d_{1,2} \end{bmatrix} = \begin{bmatrix} h_{1,1,sum} & h_{1,2,sum} \\ -h_{1,2,sum} & h_{1,1,sum} \end{bmatrix} \begin{bmatrix} b_{1,o} \\ b_{1,e} \end{bmatrix} + \begin{bmatrix} v_{1,1} \\ v_{1,2} \end{bmatrix} \tag{8.68}$$

where we have used the following equations

$$\begin{cases} h_{1,1,sum} = h_{1,1} + h_{1,3} + \cdots + h_{1,2M-1} \\ h_{1,2,sum} = h_{1,2} + h_{1,4} + \cdots + h_{1,2M} \end{cases} \tag{8.69}$$

Thus, we obtain

$$\mathbf{d}_{1,sum} = \mathbf{H}_{1,sum} \mathbf{b}_{1,1} + \mathbf{v}_{1,sum} \tag{8.70}$$

where we have used the following definitions:

$$\begin{cases} \mathbf{d}_{1,sum} = \begin{bmatrix} d_{1,1} \\ d_{1,2} \end{bmatrix} \\ \mathbf{H}_{1,sum} = \begin{bmatrix} h_{1,1,sum} & h_{1,2,sum} \\ -h_{1,2,sum} & h_{1,1,sum} \end{bmatrix} \\ \mathbf{b}_{1,1} = \begin{bmatrix} b_{1,o} \\ b_{1,e} \end{bmatrix} \\ \mathbf{v}_{1,sum} = \begin{bmatrix} v_{1,1} \\ v_{1,2} \end{bmatrix} \end{cases} \tag{8.71}$$

Next we can perform an STBC decoding by multiplying both the sides of Equation (8.70) with $\mathbf{H}_{1,sum}^{H}$ and retaining only the real part to get the decision variables as

$$\begin{bmatrix} g_{1,o} \\ g_{1,e} \end{bmatrix} = \Re\left(\mathbf{H}_{1,sum}^{H} \mathbf{d}_{1,sum}\right) \Re\left(\mathbf{H}_{1,sum}^{H} \mathbf{H}_{1,sum} \mathbf{b}_{1,1}\right) + \Re\left(\mathbf{H}_{1,sum}^{H} \mathbf{v}_{1,sum}\right)$$

$$= \begin{bmatrix} |h_{1,1,sum}|^2 + |h_{1,2,sum}|^2 & 0 \\ 0 & |h_{1,1,sum}|^2 + |h_{1,2,sum}|^2 \end{bmatrix} \begin{bmatrix} b_{1,o} \\ b_{1,e} \end{bmatrix}$$

$$+ \Re\left(\mathbf{H}_{1,sum}^{H} \mathbf{v}_{1,sum}\right)$$

$$\tag{8.72}$$

where the operators $\mathbf{x}^{H}$ and $\Re(\mathbf{x})$ are used to calculate the Hermitian form and to retain the real part of a complex vector $\mathbf{x}$, respectively. The significance of Equation (8.72) is to show that the output from an STBC decoder in an OC code STBC-CDMA system with two transmitter antennae can achieve a full diversity gain, in addition to the inherent MAI-free property of the system.

8.7.2　Arbitrary Number of Transmitter Antennae

Similarly, the above analysis for a two-antenna OC code–based STBC-CDMA system can be extended to the cases with n_t transmitter antennae at each user, while every receiver will still use a single antenna for signal reception.

It can be shown that the generalized form of Equation (8.72), which is the output from an STBC decoder or the decision variable vector, becomes

$$
\begin{cases}
\tilde{g}_{1,o} = \left(\mid h_{1,1,sum}\mid^2 + \mid h_{1,2,sum}\mid^2 + \cdots + \mid h_{1,n_t,sum}\mid^2\right) b_{1,o} + \displaystyle\sum_{j=1}^{n_t} h^*_{1,j,sum} v_{1,j} \\[2ex]
\tilde{g}_{1,e} = \left(\mid h_{1,1,sum}\mid^2 + \mid h_{1,2,sum}\mid^2 + \cdots + \mid h_{1,n_t,sum}\mid^2\right) b_{1,e} + \displaystyle\sum_{j=1}^{n_t} h_{1,j,sum} v^*_{1,j}
\end{cases}
\tag{8.73}
$$

Note that now $h_{1,j,sum}$, $j \in (1, n_t)$, results from the summation of M Rayleigh fading channel coefficients, or

$$
\begin{cases}
h_{1,1,sum} = h_{1,1} + h_{1,1+n_t} + \cdots + h_{1,n_t M-(n_t-1)} \\
h_{1,2,sum} = h_{1,2} + h_{1,2+n_t} + \cdots + h_{1,n_t M-(n_t-2)} \\
\quad\vdots \\
h_{1,n_t,sum} = h_{1,n_t} + h_{1,2n_t} + \cdots + h_{1,n_t M}
\end{cases}
\tag{8.74}
$$

Equation (8.74) will be reduced to Equation (8.69) if $n_t = 2$. The right-hand side of each equation in (8.74) is the summation of M terms, each of which is an identical and independent distributed (*i.i.d.*) Rayleigh random variable. Let $h_{1,i}$, $i \in (1, n_t M)$, be a generic term at the right-hand side of Equation (8.74), whose probability density function (pdf) is

$$
f_{h_{1,i}}(r) = \frac{r}{\sigma^2}\exp\left(-\frac{r^2}{2\sigma^2}\right), \qquad 0 \le r < \infty
\tag{8.75}
$$

with its variance being σ^2. Let $\beta = (h_{1,i})^2$, which obeys exponential distribution as

$$
f_\beta(r) = \frac{1}{2\sigma^2}\exp\left(-\frac{r}{2\sigma^2}\right), \qquad 0 \le r < \infty.
\tag{8.76}
$$

In this OC code–based STBC-CDMA system there are K users in total, each of which is assigned M element codes as its signature codes sent via M different carriers. Therefore, we have

$$
Var(h_{1,j,sum}) = M\sigma^2, \qquad j \in (1, n_t).
\tag{8.77}
$$

The BER of the system can be derived from Equation (8.73) due to the fact that either $\tilde{g}_{1,o}$ or $\tilde{g}_{1,e}$ is Gaussian under the condition of fixing all $h_{1,j,sum}$, $j \in (1, n_t)$. As shown in Figure 8.9, the BPSK modem here is used in the system. Therefore, the average BER of an OC code–based STBC-CDMA system can be obtained if we know the SNR at the input side of the decision unit in Figure 8.9. Define $\tilde{\alpha}_M$ as

$$
\tilde{\alpha}_M = \sum_{j=1}^{n_t} \mid h_{1,j,sum}\mid^2
\tag{8.78}
$$

From Equation (8.73), fixing $h_{1,j,sum}$ and thus $h_{1,j,sum}^*$ we obtain the variance of the noise terms as

$$\sigma_{n-total}^2 = Var\left(\sum_{j=1}^{n_t} h_{1,j,sum}^* v_{1,j}\right) = \sum_{j=1}^{n_t} \mid h_{1,j,sum} \mid^2 Var(v_1) = \tilde{\alpha}_M M N_o \tag{8.79}$$

where we have used $Var(v_1) = 2M\frac{N_o}{2}$ from Equation (8.67). Thus, the SNR at the output of an STBC decoder becomes

$$\frac{\tilde{\alpha}_M^2 E_b}{\sigma_{n-total}^2} = \frac{\tilde{\alpha}_M^2 E_b}{\tilde{\alpha}_M M N_o} = \frac{\tilde{\alpha}_M E_b}{M N_o} \tag{8.80}$$

Therefore, we have the average BER of an OC code–based STBC-CDMA system as

$$P_{OCC} = \int_0^\infty Q\left(\sqrt{\frac{2E_b r}{n_t M N_o}}\right) f_{\beta,n_t}(r)\, dr \tag{8.81}$$

where the factor n_t counts for the normalization of transmitting power for n_t antennae and $f_{\beta,n_t}(r)$ is the pdf function for α_M, which takes the form of

$$f_{\beta,n_t}(r) = \left(\frac{1}{2M\sigma^2}\right)^{n_t} \frac{r^{n_t-1}}{(n_t-1)!} \exp\left(-\frac{r}{2M\sigma^2}\right), \quad 0 \le r < \infty. \tag{8.82}$$

which differs from Equation (8.55) only on a factor of M multiplying with σ^2. Thus, the BER expression can be rewritten as

$$P_{OCC} = \int_0^\infty Q\left(\sqrt{\frac{2E_b r}{n_t M N_o}}\right) \left(\frac{1}{2M\sigma^2}\right)^{n_t} \frac{r^{n_t-1}}{(n_t-1)!} \exp\left(-\frac{r}{2M\sigma^2}\right) dr \tag{8.83}$$

If letting $z = \frac{E_b r}{n_t M N_o}$, we can simplify Equation (8.83) into

$$P_{OCC} = \int_0^\infty Q\left(\sqrt{2z}\right) \frac{(n_t N_o)^{n_t} z^{n_t-1}}{(2E_b\sigma^2)^{n_t}(n_t-1)!} \exp\left(-\frac{n_t N_o z}{2E_b\sigma^2}\right) dz$$

$$= \left(\frac{1-\mu}{2}\right)^{n_t} \sum_{n=1}^{n_t-1} \binom{n_t-1+n}{n} \left(\frac{1+\mu}{2}\right)^n \tag{8.84}$$

where μ has been defined as

$$\mu = \sqrt{\frac{\frac{2E_b\sigma^2}{n_t N_o}}{1 + \frac{2E_b\sigma^2}{n_t N_o}}} = \sqrt{\frac{\gamma}{1+\gamma}} \tag{8.85}$$

Here, we have used the expression

$$\gamma = \frac{2E_b\sigma^2}{n_t N_o} \tag{8.86}$$

as a normalized SNR with respect to the number of transmitter antennae or n_t. As long as $2\sigma^2 = 1$ or $\sigma^2 = 0.5$, we will have $\gamma = \frac{E_b}{n_t N_o}$, which just gives normalized SNR in an OC code–based STBC-CDMA system with n_t transmitter antennae. Therefore, we can see from Equations (8.84) to (8.86) that the average BER performance of an OC code STBC-CDMA system is under complete control by a single parameter n_t and has nothing to do with the other system variables, including K, M, N, and so on, implying that it is a noise-limited system with a full STBC diversity gain.

It is also in our interest to note that Equation (8.84) resembles the analytical BER results obtained in [699], which concerned a point-to-point Rayleigh fading downlink channel with a single transmitter antenna and n_t receiver antennae. However, the system in [699] was a non-CDMA digital communication system with ordinary BPSK modulation and coherent detection.

8.8   Discussion and Summary

On the basis of the analysis carried out in the above sections, we can evaluate the performance of an STBC-CDMA system using different signature codes, such as OC codes, Gold codes, and M-sequences. We take Gold codes and M-sequences as examples here for traditional unitary codes for the following reasons. Gold code has a relatively well-controlled three-level cross-correlation function, representing a good model of the unitary codes; on the other hand, M-sequence does not have regular cross-correlation functions, thus being a bad model of the unitary codes. With these two unitary codes we can make an objective and yet unbiased comparison with OC codes, which is the focal point here.

As a benchmark to the theoretical analysis, computer simulations have also been carried out and the results obtained from both will be compared with each other.

Figure 8.10 shows BER versus SNR for an STBC-CDMA system using OC codes with variable numbers of transmitter antennae, from 2 up to 32 antennae. It illustrates that a great advantage can be obtained by using a relatively large number of transmitter antennae. The results reveal that the BER performance for an OC code STBC-CDMA system under the Rayleigh fading channels can monotonously approach that of the single user noise only bound if a sufficiently large number of antennae can be made available. Figure 8.10 gives purely theoretical results and deals with only the OC codes.

The comparison between STBC-CDMA systems under flat Rayleigh fading with different codes is made in Figure 8.11, which shows the BER performance versus SNR for a system setup with two transmitter antennae and one receiver antenna. The PG values are 31 and 63 for Gold codes, but only 63 for M-sequence. In this figure, we do not give simulation results. It is seen from the figure that an STBC-CDMA with the OC codes perform much better than that with either Gold codes or M-sequences.

Figure 8.12 compares the BER results obtained from the theoretical analysis and computer simulations for an STBC-CDMA system with either OC codes or M-sequences. The number of users in the STBC-CDMA with M-sequences changes from 2, 4, and 8. It is not surprising that the STBC-CDMA

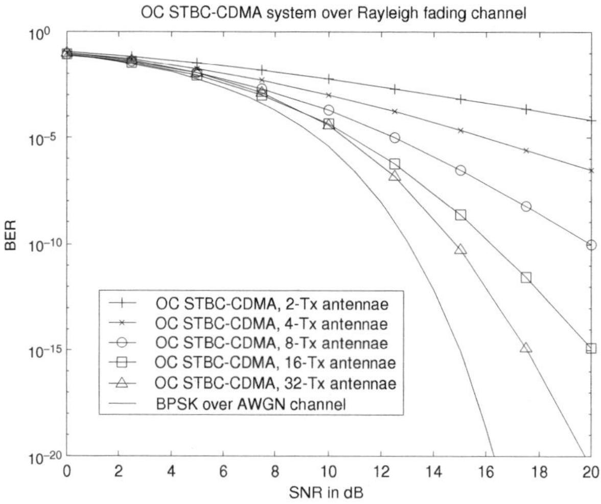

Figure 8.10   BER versus signal-to-noise-ratio for an STBC-CDMA system in Rayleigh fading channels with a variable number of transmitter antennae. The performance is independent of the number of users and PG values, showing the MAI-free property of the proposed system.

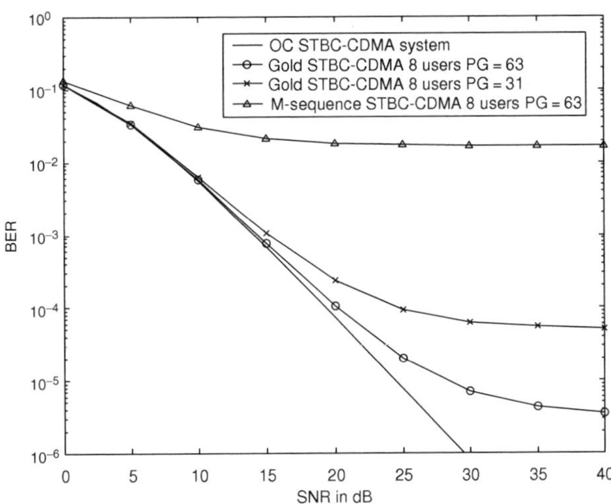

Figure 8.11 The BER performance comparison under the Rayleigh fading channel for an STBC-CDMA system using orthogonal complementary code, Gold codes (PG = 31 and 63) and M-sequence (PG = 63), where eight users are present in the system. Only theoretical results are shown. Two transmitter antennae and one receiver antenna are used.

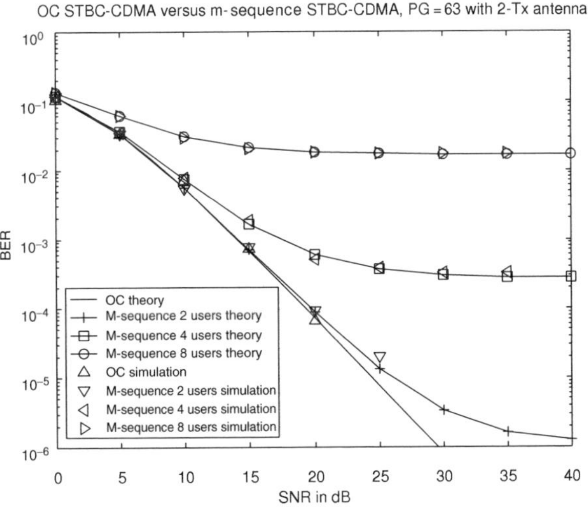

Figure 8.12 BER performance comparison for an STBC-CDMA system with orthogonal complementary codes and M-sequences (PG = 63) with both theoretical analysis and computer simulation. A flat Rayleigh fading channel is present and the number of users varies from 2, 4, and 8. Two transmitter antennae and one receiver antenna are used.

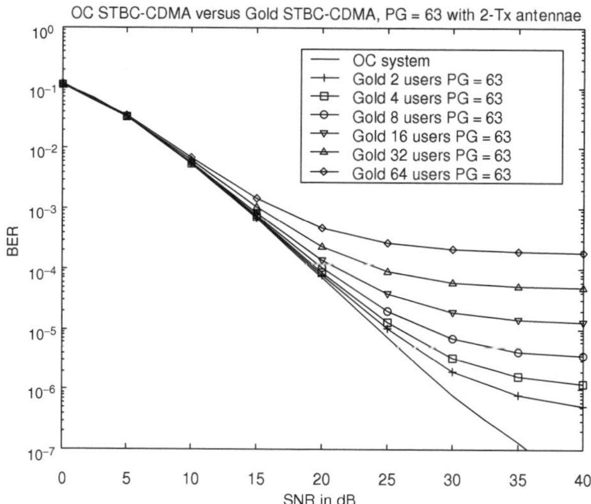

Figure 8.13 Comparison of BER versus signal-to-noise-ratio for an OC STBC-CDMA and an STBC Gold code DS-CDMA systems in Rayleigh fading channels with a variable number of users. The PG for OC codes and Gold codes is 64 and 63 respectively. Two transmitter antennae and a single receiver antenna are used.

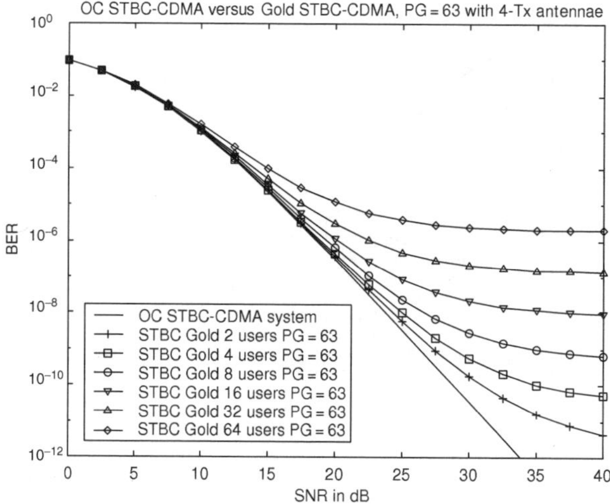

Figure 8.14 Comparison of BER versus signal-to-noise-ratio for STBC-CDMA and STBC Gold code DS-CDMA systems in Rayleigh fading channels with a variable number of users. The PG for OC codes and Gold codes is 64 and 63 respectively. Four transmitter antennae and a single receiver antenna are used.

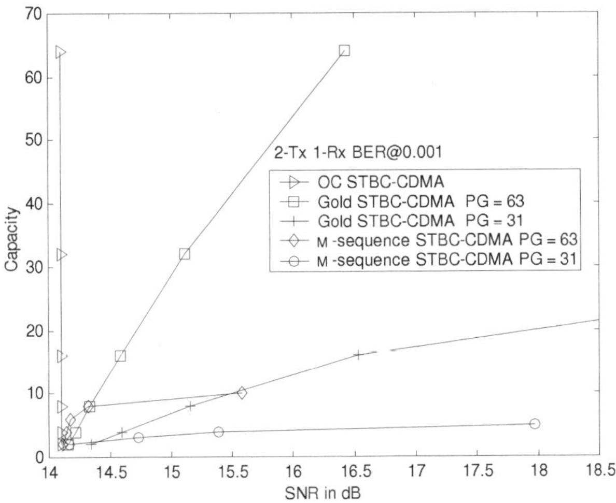

Figure 8.15 Capacity comparison for an STBC-CDMA system with orthogonal complementary codes, Gold codes (PG = 31 and 63) and M-sequences (PG = 31 and 63). Two transmitter antennae and one receiver antenna are used. The BER requirement is fixed at 0.001. A flat Rayleigh fading channel is considered.

system with the M-sequences is very sensitive to the change in user population; while the system with the OC codes offers a BER performance independent of user population, manifesting an MAI-free operation. A very good match between the results obtained from analysis and simulation is also shown in the figure. Similar conclusions can be made with respect to a system using other unitary codes.

Figures 8.13 and 8.14 compare the BER performance of an STBC-CDMA system with OC codes and Gold codes (PG = 63). The two figures are obtained by using a similar system setup, except for the difference in the number of transmitter antennae, being two in Figure 8.13 and four in Figure 8.14, respectively. The number of users in the system with Gold codes varies from 2 to 64, demonstrating how BER will change with the MAI level. It is clearly shown that the curve for the OC code behaves like a single user bound for the curves obtained for Gold codes. A similar observation can also be made from Figure 8.12, where an OC code is compared with M-sequences.

To explicitly show how much the difference in terms of capacity can be by using different codes, Figures 8.15 and 8.16 are given, which basically concern a similar working environment, except for the difference in the number of transmitter antennae, being two in Figure 8.15 and four in Figure 8.16, respectively. Both the figures were obtained by fixing the BER at 0.001. Three different codes are compared with one another, which are OC code, Gold codes with PG being 31 and 63, and M-sequences with PG being 31 and 63.

The capacity advantage for an STBC-CDMA system based on an OC code over its counterpart, either Gold codes or M-sequences, can be significant due to its interference-free operation. Assume, for instance, that the required BER is about 10^{-3} as specified in both the figures. It is observed from Figure 8.16 that an OC code–based STBC-CDMA system with four antennae can support as many as 64 users at SNR = 10.06, which is in fact limited only by the set size of the OC code set (PG = 64). However, either a Gold code (PG = 63) or an M-sequence (PG = 63) STBC-CDMA with four antennae can only support about 2 users, differing from that of the OC code STBC-CDMA system by as many as 62 users! Alternatively, in order to achieve the same capacity, an unitary code–based STBC-CDMA system has to use much more (which must be more than 32 antennae

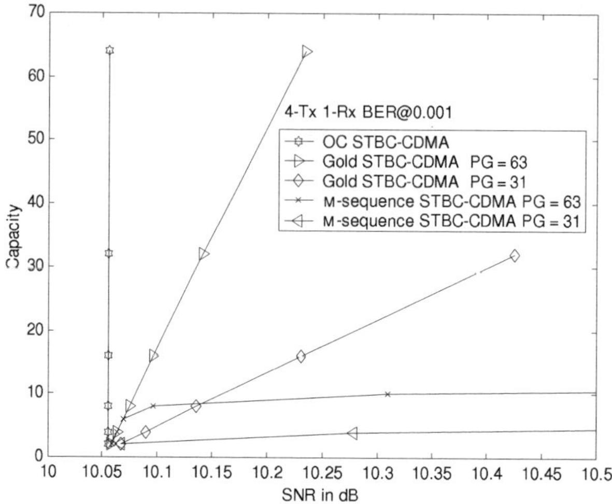

Figure 8.16 Capacity comparison for an STBC-CDMA system with orthogonal complementary codes, Gold codes (PG = 31 and 63) and M-sequences (PG = 31 and 63). The four transmitter antennae and one receiver antenna are concerned. The BER requirement is fixed at 0.001. A flat Rayleigh fading channel is considered.

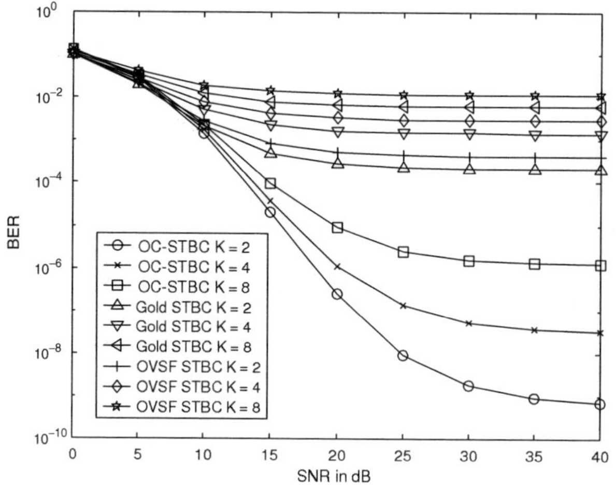

Figure 8.17 BER comparison versus the number of users for an STBC-CDMA system under a normalized three-path channel (whose delay profile is [$\sqrt{0.6}$, $\sqrt{0.3}$, $\sqrt{0.1}$]) with orthogonal complementary codes (PG = 8×8), Gold codes (PG = 63) and OVSF code (PG = 64). Two transmitter antennae and one receiver antenna are concerned.

from our study) transmitter antennae at the same BER performance (10^{-3}), thus definitely resulting in much greater complexity.

Figure 8.17 shows the performance of an STBC-CDMA scheme using different CDMA codes under a multipath channel with its delay profile being ($\sqrt{0.6}$, $\sqrt{0.3}$, $\sqrt{0.1}$). The three codes used here are Gold code, OVSF code, and OC code with their PG values being 63, 64 and 8×8, respectively. It is seen from the figure that the scheme with the OC code offers a superior BER performance under various scenarios of user population in the system. The OVSF code performs worst and the Gold code gives a slightly better BER than that of the OVSF code, but is never comparable to that of the OC code.

We can summarize the results obtained so far regarding the OC codes–based STBC-CDMA systems as follows. In these sections we have studied an STBC-CDMA system in downlink Rayleigh fading channels. A comprehensive analysis has been carried out to derive the BER performance expression of such a system under MAI and flat fading. It has been shown through the analysis that an OC code–based STBC-CDMA system can achieve an ideal MAI-free operation and a full diversity gain jointly under a single system framework. The results obtained from the theoretical study have also been compared with those generated from computer simulations, and they have been shown to match one another, for the most part. A unitary code–based STBC-CDMA still suffers serious MAI problems even with the help of the full diversity gain of the STBC scheme. On the other hand, an OC code STBC-CDMA system can offer a capacity limited only by noise and fading, and not by interference. The results obtained here concluded that the integration of an OC code–based CDMA and STBC system is technically feasible.

9

Cognitive Radio Technology

Cognitive radio (CR) is a newly emerging technology [789, 790], which has been recently proposed to implement some kind of intelligence to allow a radio terminal to automatically sense, recognize, and make wise use of any available radio frequency spectrum at a given time. The use of the available frequency spectrum is purely on an opportunity driven basis. In other words, it can utilize any idle spectrum sector for the exchange of information and stop using it the instant the primary user of the spectrum sector needs to use it. Thus, cognitive radio is also sometimes called *smart radio*, *frequency agile radio*, *police radio*, or *adaptive software radio*,[1] and so on. For the same reason, the cognitive radio techniques can, in many cases, exempt licensed use of the spectrum that is otherwise not in use or is lightly used; this is done without infringing upon the rights of licensed users or causing harmful interference to licensed operations.

9.1 Why Cognitive Radio?

The discussion on cognitive radio technology can best begin with the remark made by Ed Thomas, former Chief Engineer of the Federal Communication Commission (FCC). "If you look at the entire radio frequency (RF) up to 100 GHz, and take a snapshot at any given time, you'll see that only 5 to 10 % of it is being used. So there's 90 GHz of available bandwidth." This shows that the usage of the radio spectrum is severely inefficient, and therefore the cognitive radio can be extremely useful to exploit the unused spectrum from time to time, as long as the vacancy appears in the spectrum.

The radio spectrum, as regulated by the FCC in the United States (in a similar way in many other countries also), is divided into channels which are usually licensed by individuals, corporations, and municipalities as primary users. Most of these channels actively transmit only for the duration of a small fraction of time. This is an inefficient use of the available spectrum. Clearly, if all those unused spectra can be utilized, many more radio users can be accommodated without the need to create a new spectrum.

Radio spectrum is one of the most important natural resources in the world today, and it is necessary to build up a wireless information infrastructure. Insufficient radio spectrum has always been a serious bottleneck for the deployment of a wireless information superhighway in the world. For a long time, we have been resorting to three major strategies to accommodate growing radio/wireless

---

[1] There is a difference between Software Definable Radio (SDR) and cognitive radio, which has to be explained later. SDR has been discussed briefly in Section 6.1.5.

Next Generation Wireless Systems and Networks   Hsiao-Hwa Chen and Mohsen Guizani
© 2006 John Wiley & Sons, Ltd

based applications. First of all, we have been trying hard to persuade existing radio spectrum owners or licencees to vacate their legacy radio applications (for instance, the terrestrial microwave relay trunk systems) to make way for the deployment of newly emerging wireless services, such as mobile cellular networks, and so on. Nowadays, almost all (if not all) legacy radio users in most developed countries who can possibly be reallocated have been moved away from the prime spectrum sectors (roughly from 800 MHz to 5 GHz bands). Therefore, this strategy for spectrum clearance will be of little help in solving the problem with severe spectrum shortage. The second traditional way of accommodating new wireless applications is to move the carrier frequency to new high spectrum sectors, which have been occupied by very few radio applications. Those new high radio spectra includes millimeter waves from 10–30 GHz bandwidth. The positive aspect of using a relatively high frequency spectrum is the ease with which broadband applications where very high data rates can be implemented are supported. However, the shortcomings of using a very high carrier frequency are obvious. One of the most problematic issues is that radio propagation properties in very high frequency spectra are very sensitive to rain, dust, water vapor, and other small particles in the air. In other words, the radio transmission in very high frequency ranges will no longer be weather-proof. Therefore, the outage rate will become unacceptably high under rain, snow, and/or other weather conditions. Finally, the third approach used to support more radio applications in an already crowded spectrum is to overlay/underlay the new wireless applications on top of existing radio services. The second generation mobile cellular standard, IS-95A/B, which works on direct-sequence CDMA technology, was initially proposed for the work on the 900 MHz PCS spectrum in North America to overlay many existing radio applications. The success of the overlay operation is largely based on relatively low power spread spectrum transmissions from the CDMA technology. Another example of such overlay applications is the ultra-wideband (UWB) technology, whose bandwidth overlaps with those previously allocated for GPS, radar, and satellite services. Therefore, a strict low power spectral density (PSD) emission mask is necessary to control the UWB transmission power level below a certain threshold in order to not interfere with them.

It is obvious that all the above three major strategies to introduce new radio applications on top of an already very crowded spectrum chart cannot solve the problem. Therefore, the need to search for a more effective solution to solve the problems of severe spectrum shortage has become imperative. Cognitive radio technology was introduced for this purpose.

To have a real picture of the current radio spectrum allocation situation, the readers may refer to the US Frequency Allocation Chart [792] (in this chart, all radio spectrum allocations from 3 kHz up to 300 GHz are shown), which is available from the web site of US National Telecommunications and Information Administration. Similar situations can be found in many other developed countries, such as Japan and in Europe. The US Frequency Allocation Chart is shown in Figure 9.1, which is too large to show all spectrum allocation details clearly within a page. We use it here just to give readers an idea of what it looks like.

The justification to use cognitive radio technology on top of the existing spectrum licencees to provide various wireless applications on a licensed exempt basis can be summarized as follows.

Firstly and as discussed above, an unused spectrum is not desirable. Users should not be allowed to own a spectrum that they do not use. It is also recommended that allocated spectra should not be underutilized. Whatever the reasons for not fully using the allocated spectrum (economic, historic, or other systemic reasons), it does not represent the best and highest use of this valuable and scarce public resource.

Secondly, layering more licensed allocations on top of existing allocations as a solution to the underutilized spectrum does not, in many people's view, increase the economic incentives for new applications in these spectrum slots, since obtaining investor support required to build licensed services becomes problematic when the economic history of a particular allocation in a particular geographic area has shown little promise for significant profits. On the other hand, licence exempt use can support business models which do not require large capital investment to roll out services because of the low cost of unlicensed equipment and the lack of the high up-front costs of acquiring

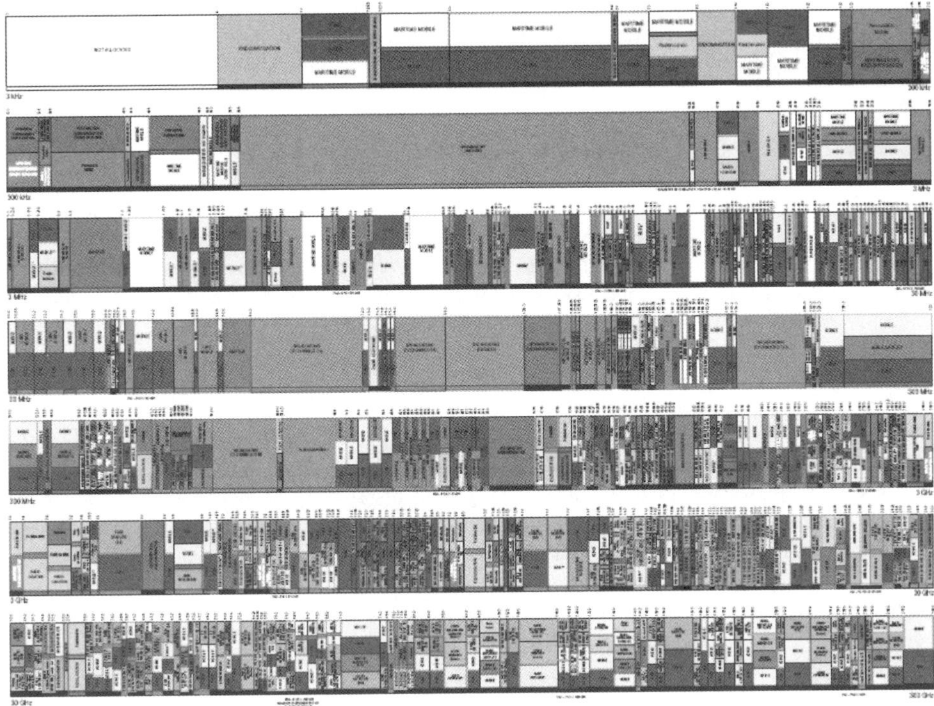

Figure 9.1 US Frequency Allocation Chart [792] from 3 kHz up to 300 GHz. It should be noted that a vacancy can only be found from 3 kHz to 9 kHz, which is shown in the left corner of the first row.

a spectrum at an auction (especially the case in the United States, Europe, and many other developed countries). As a result, rural and other low population density areas could obtain services which would otherwise be unavailable from the business entities which operate on licensed spectra and tend to focus their investments on the larger, more profitable, urban and suburban marketplaces. For similar reasons, community based networks and other not-for-profit groups could make use of otherwise unused spectra to offer their constituencies innovative services and applications that would otherwise be viewed as uneconomic, and, as a result, ignored by profit-oriented entities.

Thirdly, the assertions made by some people that licence exempt use interferes with business opportunities flies in the face of the clear evidence that a vast amount of spectrum remains unused because the high cost of rolling out licensed infrastructure is not justified on investment basis. Without the opportunity to reclaim this spectrum in the public interest using cognitive radio technology under licence exempt rules, this fallow spectrum would continue to be underutilized. In the broader context of licence exempt sharing of licensed spectrum, it is widely believed that opportunities exist to apply sophisticated cognitive radio technologies to recover otherwise underutilized spectrum for uses which have significant economic and societal benefits without harming the interests of licensed services.

Finally, the current state-of-the-art radio technology has made it possible to implement a practical cognitive radio in various wireless applications, such as wireless regional area networks (WRANs), wireless metropolitan area networks (WMANs), wireless local area networks (WLANs), and wireless personal area networks (WPANs), and so on, at a reasonable cost. Therefore, the radio terminals can be given some intelligence to work automatically on the available frequency spectrum at any given time.

In fact, a cognitive radio extends the functionality of a software-definable radio (SDR) to permit it to react and adapt intelligently to its environment. It provides a central nervous system to communications and computing platforms. This permits intelligent access and configuration by the radio devices.

9.2   History of Cognitive Radio

The cognitive radio is an emerging new technology, which is far from mature in terms of real applications in current wireless systems and networks. Today, to implement a practical cognitive radio, many hurdles should be overcome, and it is still too early to tell what a cognitive radio should look like for different wireless applications. Therefore, the history of cognitive radio technology is still relatively short.

Mitola's work

A comprehensive description of the term *cognitive radio* was first discussed in a paper written by J. Mitola III and Gerald Q. Maguire in 1999 [793]. In 2000, J. Mitola III wrote his PhD dissertation [794] on cognitive radio as a natural extension of the SDR concept. When addressing the broad issue of wireless personal digital assistants (PDAs) in his dissertation, Mitola mentioned that the term *cognitive radio* identifies the point at which wireless PDAs and the related networks are sufficiently computationally intelligent regarding radio resources and related computer-to-computer communications to (a) detect user communications needs as a function of use context, and (b) to provide radio resources and wireless services most appropriate to those needs.

FCC's initiatives

In 2002, the FCC's Spectrum Policy Task Force Report [797] identified that most spectra go unused most of the time, as shown in Figure 9.2. Consequently, it was then realised that spectrum scarcity is driven mainly by archaic systems for spectrum allocation and not by a fundamental lack of spectra.

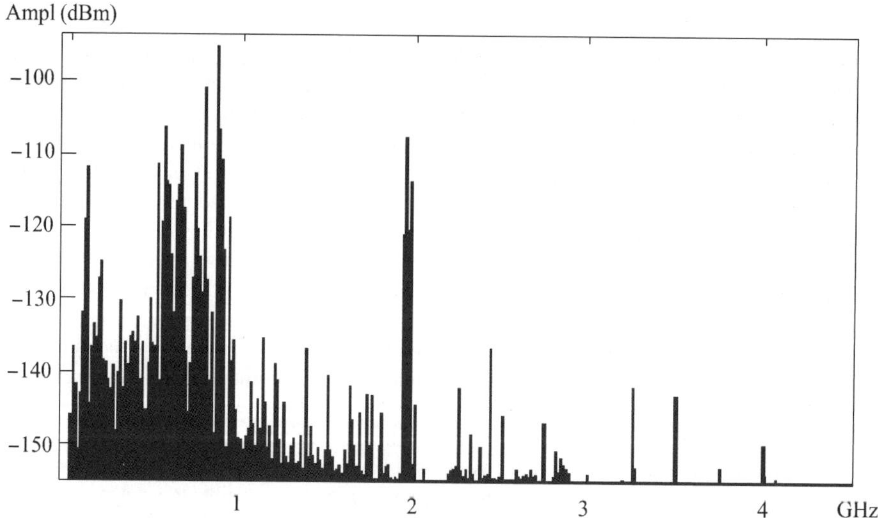

Figure 9.2 A sample of the snapshot of radio spectrum utilization up to 6 GHz. It is shown that most frequency bands were not used at the time when this snap shot was taken.

How to open up additional spectra, whether it should be licensed or unlicensed, and the economic implications of these decisions, have been topics of considerable debate [798]. Cognitive radio technology offers a possible solution based on a more sophisticated or intelligent system for allocating spectra that can dramatically increase the amount of spectra available to network operators and individual users. In particular, on December 20, 2002, it was stated in FCC's "Notice of Inquiry" (NOI) titled "Additional Spectrum for Unlicensed Devices Below 900 MHz and in the 3 GHz Band" (FCC-02-328) that it opens the question of using fallow TV band channels for unlicensed services on a noninterference basis. In the NOI, the FCC states that specifically, an unlicensed device should be able to identify unused frequency bands before it can transmit, that is, by using Dynamic Frequency Selection (DFS) and Incumbent Profile Detection (IPD) algorithms.

On November 13 of 2003, FCC issued NOI and "Notice of Proposed Rulemaking" (NPRM) titled "Establishment of an Interference Temperature Metric..." (FCC-03-289), in which it proposed an interference temperature model for quantifying and managing interference. The interference temperature is calculated by $T_{int} = \frac{N+I}{kB}$. It also stated that for an interference temperature limit to function effectively on an adaptive or real-time basis, a system (cognitive radio) would be needed to measure, and a response process would also be needed.

In another NPRM and order titled "Facilitating Opportunities for Flexible, Efficient, and Reliable Spectrum Use Employing Cognitive Radio Technologies" (FCC-03-322), issued by FCC on December 17, 2003, it was stated that a wide ranging NPRM exploring a broad range of issues related to cognitive radio technology will be required. It pointed out that the FCC wants to push for advances in technology which support more effective spectrum use. Among these advances are cognitive radio technologies that can possibly make more intensive and efficient spectrum use by licencees within their own networks, and by spectrum users sharing spectrum access on a negotiated or an opportunistic basis.

The FCC's action sparked a lot of response from both industry and academia, and some research activities on cognitive radio [798–801] in the last few years. However, the most important event in the development of cognitive radio happened in 2004, when the FCC issued yet another NPRM that raised the possibility of permitting unlicensed users to temporarily "borrow" spectrum from licensed holders as long as no excessive interference was seen by the primary user [795]. Devices that borrow spectrum on a temporary basis without generating harmful interference are commonly referred to as "*cognitive radios*" [796]. Basic cognitive radio techniques, such as DFS and transmit power control (TPC), already exist in many unlicensed devices. However, to make a practical cognitive radio terminal, we have to deal with many serious challenges.

The FCC is proposing specific rulemaking in the unlicensed arena related to cognitive technology as follows:

- Opening three new bands to unlicensed operation based on DFS and TPC protocols (interference temperature NPRM), which include 6525–6700 MHz (175 MHz), 12.75–13.15 GHz (400 MHz), and 13.2125–13.25 GHz (37.5 MHz);

- Allowing six times more transmitter power for cognitive radio devices (under Part 15.247 and Part 15.249) where the ISM band is lightly used (cognitive radio NPRM);

- DFS thresholds at which frequency change is required: For Tx power levels < 23 dBm: −62 dBm; For Tx power levels > 23 dBm: −64 dBm; DFS threshold averaging time varies with rule: unlicensed national information infrastructure (U-NII) is 1 μs, new interference temperature bands: 1 ms; DFS thresholds are referenced to the output of an omni-directional antenna.

- The definition of an unoccupied band: RSL < −83 dBm measured in a 1.25 MHz bandwidth using an omni-antenna.

- Minimum TPC backoff from maximum allowed Tx power: −6 dB, triggered by a vendor specific criterion for link quality.

Related IEEE standards

On the other hand, the standardization work done by the Institute of Electrical and Electronics Engineers (IEEE) has also been carried out parallel to the FCC's action. Recent IEEE 802 standards activity in cognitive radio includes a recently approved amendment to the IEEE 802.11 operation, or the IEEE 802.11h, which incorporates DFS and TPC protocols for 5-GHz operations under the IEEE 802.11a standard [449–451].

Because 802.11a wireless networks operate in the 5-GHz radio frequency band and support as many as 24 nonoverlapping channels, they are less susceptible to interference than their 802.11b/g counterparts. However, regulatory requirements governing the use of the 5-GHz band vary from country to country, hampering 802.11a deployment. In response, the International Telecommunication Union (ITU) recommended a harmonized set of rules for WLANs to share the 5-GHz spectrum with primary-use devices such as military radar systems. Approved in September 2004, the IEEE 802.11h standard defines mechanisms that 802.11a WLAN devices can use to comply with the ITU recommendations. These mechanisms are DFS and TPC. WLAN products supporting 802.11h have already been available in the second half of 2005. DFS detects other devices using the same radio channel, and it switches the WLAN operation to another channel if necessary. DFS is responsible for avoiding interference with other devices, such as radar systems and other WLAN segments, and for uniform utilization of channels.

Among other activities carried out by the IEEE is 802.18 SG1, which was established at the Albuquerque Plenary in November 2003, and focused on creating the following: (1) Recommendations for a rule making proposal to the FCC on TV band use by unlicensed devices. (2) A Project Authorization Request (PAR) and associated five Criteria documents to create a network standard aimed at unlicensed operation in the TV band. In "Reply to Comments of IEEE 802.18" prepared by Carl R. Stevenson (carl.stevenson@ieee.org) in May 2004, it indicated clearly that IEEE 802.18 supports the opportunistic use of fallow spectrum by licence exempt networks on a noninterfering basis with licensed services using cognitive radio techniques. IEEE 802.18 supports the FCC's approach to rural applications of cognitive radio technology as a means to increase the coverage area of wisps and other unlicensed services in the ISM bands.

Earlier similar works

It has to be noted that, although the terminology of "cognitive radio" was only proposed recently, the concept of intelligent radio is not completely new. Many previously carried out researches on wireless communications and networks bear some similarity to what a cognitive radio does. The first example of such research is the collision avoidance protocol used in IEEE 802.3 standard or Ethernet standard: carrier sense multiple access (CSMA).[2] The basic idea for CSMA is to sense before transmitting, which works in a very similar way to what a cognitive radio unerringly does. This polite radio transmission etiquette forms the core of today's cognitive radio technology.

Another example of similar research is the so-called "dynamic channel selection/allocation," which has been extensively used in user traffic channel assignment schemes in mobile cellular systems. A new mobile terminal will be assigned a traffic channel with an available idle channel from the traffic channel pool. Its utilization will be released back to the pool when its transmission ends, thus making it available to others' use. Naturally, the intelligence level possessed in a cognitive radio will be much higher than that available in all previous wireless applications.

9.3   What is Cognitive Radio?

In this section, we define cognitive radio and investigate the algorithms and types of technologies that already exist.

---

[2]CSMA has been discussed in Section 2.3.4 of this book.

9.3.1 Definitions of Cognitive Radio

As any newly emerging technology, the definition of "cognitive radio" can be seen in many different ways. In fact, the term *Cognitive Radio* means different things to different audiences. The earlier definition by Joseph Mitola in his dissertation titled "Cognitive Radio – An Integrated Agent Architecture for Software Defined Radio" [794], was given as follows. The cognitive radio identifies the point at which wireless PDAs and the related networks are sufficiently computationally intelligent on the subject of radio resources and related computer-to-computer communications to (a) detect user communications needs as a function of use context, and (b) to provide radio resources and wireless services most appropriate to those needs. Cognitive radio increases the awareness that computational entities in radios have of their locations, users, networks, and the larger environment. Mitola included the concept of machine learning as a property of cognitive radio. Mitola's definition on cognitive radio includes a high level of awareness and autonomy, in a sense that cognition tasks, that might be performed, range in difficulty from the goal driven choice of RF band, air interface, or protocol to higher-level tasks of planning, learning, and evolving new upper layer protocols.

The FCC gave the following definition on cognitive radio [795]. A cognitive radio is a radio that can change its transmitter parameters based on interaction with the environment in which it operates. At the same time, it should also note that FCC refers to a SDR as a transmitter in which the operating parameters . . . can be altered by making a change in software that controls the operation of the device without . . . changes in the hardware components that affect the radio frequency emissions. It went on to claim that the majority of cognitive radios will probably be SDRs, but neither having software nor being field reprogrammable are requirements of a cognitive radio.

To summarize from the aforementioned two versions of definitions on cognitive radio, we can see that Mitola emphasized the level of device/network intelligence which adapts to user activity; while the FCC seems primarily concerned with a regulatory friendly view, focused on transmitter behavior at the moment. Therefore, the relationship between the cognitive radio and SDR from the views of Mitola and the FCC can be seen in Figure 9.3, where cognitive radio adapts to the spectrum environment; while SDR adapts to the network environment. They partially overlap in their functionalities.

9.3.2 Basic Cognitive Algorithms

It is therefore not difficult to discern that a fully functional cognitive radio should have the ability to do the following works: (1) Tune to any available channel in the target band. (2) Establish network communications and operate in all or part of the channel. (3) Implement channel sharing and power

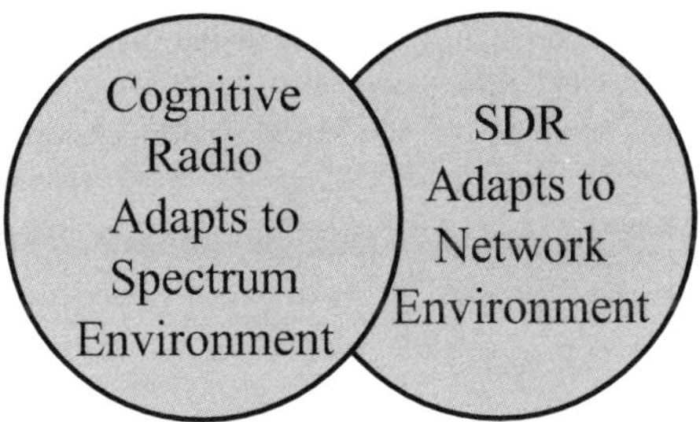

Figure 9.3 The cognitive radio adapts to the spectrum environment; while SDR adapts to the network environment. Their functionalities are partially overlapped.

control protocols which adapt to spectra occupied by multiple heterogeneous networks. (4) Implement adaptive transmission bandwidths, data rates, and error correction schemes to obtain the best throughput possible. (5) Implement adaptive antenna steering to focus transmitter power in the direction required to optimize received signal strength.

The core of a cognitive radio is its inherent intelligence, which makes it different from any normal wireless terminal available today, in either 2- or 3G systems. This intelligence will allow a cognitive radio to scan all possible frequency spectra before it makes an intelligent decision on how and when to make use of a particular sector of the spectrum for communications. Therefore, it is inevitable that a cognitive radio needs great signal processing power to deal with the vast amounts of data it captures from various radio channels. Thus, the capability to process all those enormous amounts of data on a real-time or quasi-real-time basis is a must for any cognitive radio.

It is still too early to specify exactly the algorithms that a cognitive radio should use at the moment of writing this book. However, we would like to provide some evidence as to how a primitive cognitive radio may behave. Obviously, any cognitive radio has to use the following two protocols for its very basic operation: (1) DFS, and (2) TPC.

The DFS was originally used to describe a technique to avoid radar signals by 802.11a networks which operate in the 5 GHz U-NII band. Now, it has been generalized to refer to an automatic frequency selection process intended to achieve some specific objective (like avoiding harmful interference to a radio system with a higher regulatory priority). On the other hand, TPC was originally a mechanism for 802.11a networks to lower aggregate transmit power by 3 dB from the maximum regulatory limit to protect Earth Exploration Satellite Systems (EESS) operations. Now it has been generalized to a mechanism that adaptively sets transmit power based on the spectrum or regulatory environment. These two protocols will become a must for all cognitive radios.

In addition, a cognitive radio should have IPD capability [799], which is another key cognitive radio behavior. The IPD is the ability to detect an incumbent user (one with regulatory priority) based on a specific spectrum signature. The operation of IPD bears the following characteristics: (1) DFS requires an IPD protocol to identify unoccupied, or lightly used frequencies. (2) IPD includes detection schemes focused on the characteristics of the specific incumbents in the band, or bands, that the cognitive radio is designed to support. (3) IPD eliminates the need for geo-location techniques (GPS, etc.) to determine the location of the radio and, using a database, identifies unused channels.

As both TPC and IPD algorithms are intuitive, as suggested by its name, we will only explain the implementation of the DFS cognitive algorithms in depth, in the following text.

The DFS algorithm was originally proposed in the ITU-R recommendation M.1461 [807] to avoid possible interference to existing radar operations in the vicinity. Many radar systems and unlicensed devices operating co-channels in proximity could produce a scenario where mutual interference is experienced. The DFS methodology is used to compute the received interference power levels at the radar and unlicensed device receivers. A DFS algorithm may provide a means of mitigating this interference by causing the unlicensed devices to migrate to another channel once a radar system has been detected on the currently active channel. This model first considers the interference caused by the radar to the unlicensed device at the output of the unlicensed device antenna. If the received interference power level at the output of the unlicensed device antenna exceeds the DFS detection threshold, the unlicensed device will cease transmissions and move to another channel. The algorithm then computes the aggregate interference to the radar from the remaining unlicensed devices. Each of the technical parameters used in the method and the radar interference criteria will also be described.

The received signal level from the radar at the output of the unlicensed device antenna can be evaluated by using the following equation:

$$I^U = P_{\text{Radar}} + G_{\text{Radar}} + G_U - L_{\text{Radar}} - L_U - L_P - L_L - FDR \tag{9.1}$$

where I^U is the received interference power at the output of the unlicensed device antenna in dBm, P_{Radar} is the peak power of the radar in dBm, G_{Radar} is the antenna gain of the radar in the direction of the unlicensed device in dBi, G_U is the antenna gain of the unlicensed device in the direction of the radar in dBi, L_{Radar} is the radar transmit insertion loss in dB, L_U is the unlicensed device receive insertion loss in dB, L_P is the propagation loss in dB, L_L is the building and nonspecific terrain losses in dB, and FDR is the frequency dependent rejection in dB.

Equation (9.1) is calculated for each unlicensed device in the distribution. The value obtained is then compared to the DFS detection threshold under investigation. Any unlicensed device for which the threshold has been exceeded will begin to move to another channel, and consequently is not considered in the calculation of interference to the radar, as given by

$$I^{\text{RADAR}} = P_U + G_U + G_{\text{Radar}} - L_U - L_{\text{Radar}} - L_P - L_L - FDR \tag{9.2}$$

where I^{RADAR} is the received interference power at the input of the radar receiver in dBm, P_U is the power of the unlicensed device in dBm, G_U is the antenna gain of the unlicensed device in the direction of the radar in dBi, G_{Radar} is the antenna gain of the radar in the direction of the unlicensed device in dBi, L_U is the unlicensed device transmit insertion loss in dB, L_{Radar} is the radar receive insertion loss in dB, L_P is the radio-wave propagation loss in dB, L_L is the building and nonspecific terrain losses in dB, and FDR is the frequency dependent rejection in dB.

With the help of equation (9.2), we can calculate each unlicensed device being considered in the analysis that has not detected energy from the radar in excess of the DFS detection threshold. These values are then used in the calculation of the aggregate interference to the radar by the unlicensed devices using the following equation:

$$I^{AGG} = \sum_{j=1}^{N} I_j^{\text{RADAR}} \tag{9.3}$$

where I^{AGG} is the aggregate interference to the radar from the unlicensed devices in Watts, N is the number of unlicensed devices remaining in the simulation, and I_{RADAR} is the interference caused to the radar from an individual unlicensed device in Watts.

It is necessary to convert the interference power calculated in Equation (9.2) from dBm to Watts before calculating the aggregate interference seen by the radar using Equation (9.3).

The parameters used in the above DFS algorithm can be explained as follows: To obtain "radar antenna gain" (G_{Radar}), we need to know the azimuth and elevation antenna pattern models for the radar considered. The models should provide the antenna gain as a function of an off-axis angle for a given main beam antenna gain. The unlicensed device power level (P_U) in this analysis is assumed to be 38 dBm and 6.6 dBm. The building and nonspecific terrain losses (L_L) include building blockage, terrain features, and multipath. In the above analysis, this loss has been treated as a uniformly distributed random variable between 1 and 10 dB for each radar unlicensed device path. When determining Radar and Unlicensed Device Transmit and Receive Insertion Losses (L_{Radar} and L_U), we have assumed that the analysis includes a nominal 2 dB for the insertion losses between the transmitter and receiver antenna and the transmitter and receiver inputs for the radar and the unlicensed device. Finally, to compute the radio-wave propagation loss (L_P), the NTIA Institute for Telecommunication Sciences Irregular Terrain Model (ITM) was used [808]. The ITM model computes radio-wave propagation based on the electromagnetic theory and on the statistical analysis of both terrain features and radio measurements to predict the median attenuation as a function of distance and variability of the signal in time and space.

9.3.3 Conceptual Classifications of Cognitive Radios

The characteristic features of a cognitive radio have a lot to do with the spectrum facts in different regions or countries. If we are only looking at the US market, we will see that a lot of spectra have been assigned for licensed use by the FCC. Actual spectrum use varies dramatically from region to region: spectrum is more congested in urban areas, and hardly used in rural areas. Some licensed services only operate in a few locations nationally (for example, Fixed Satellite Services). Even in urban areas, only a fraction of available spectra is in continuous use. We have to admit that, in terms of reclaiming fallow spectrum, a lot of low hanging fruit is available for harvest using cognitive techniques. Regulatory activity is just beginning to open up opportunities to reclaim lightly used spectra for new services.

Currently, there are two conceptual forms of cognitive radios. One is called *full cognitive radio*, in which every possible parameter observed by the wireless node and/or the network is taken into account while making a decision on the transmission and/or reception parameter change. The other is called *Spectrum Sensing Cognitive Radio*, which is a special case of Full Cognitive Radio in which only the RF spectrum is observed.

Also, depending on the parts of the spectrum available for cognitive radio, we can distinguish "Licensed Band Cognitive Radio" and "Unlicensed Band Cognitive Radio." When a cognitive radio is capable of using bands assigned to licensed users, apart from the utilization of unlicensed bands such as the U-NII band or the ISM band, it is called a *Licensed Band Cognitive Radio*. One of the Licensed Band Cognitive Radio-like systems is the IEEE 802.15 Task group 2 [802] specification. On the other hand, if a cognitive radio can only utilize the unlicensed parts of a RF spectrum, it is an Unlicensed Band Cognitive Radio. An example of an Unlicensed Band Cognitive Radio is IEEE 802.19 [803].

Although cognitive radio was initially thought of as an SDR extension (Full Cognitive Radio), most of the current research work is focused on Spectrum Sensing Cognitive Radio, particularly on the utilization of TV bands for communication. The essential problem of Spectrum Sensing Cognitive Radio is the design of high-quality spectrum sensing devices and algorithms for exchanging spectrum sensing data between different nodes in a cognitive radio network. It has been shown in [804] that a simple energy detector cannot guarantee the accurate detection of signal presence. This calls for more sophisticated spectrum sensing techniques and requires that information about spectrum sensing must be regularly exchanged between nodes. In [805], the authors showed that the increasing number of cooperating sensing nodes decreases the probability of false detection. To adaptively fill free RF bands, OFDM seems to be a perfect candidate. Indeed in [801] T. A. Weiss and F. K. Jondral from the University of Karlsruhe, Germany, proposed a Spectrum Pooling system in which free bands sensed by nodes were immediately filled by OFDM subbands. Some of the applications of Spectrum Sensing Cognitive Radio include emergency networks and WLAN higher throughput, and transmission distance extensions.

9.4 From SDR to Cognitive Radio

SDR has now been widely accepted as the implement of choice for a variety of platforms and applications. The success in harnessing the promised flexibility and incredible processing power of the SDR has led designers to consider implementing cognitive radios that adapt to their environment by analyzing the RF environment and adjusting the spectrum use appropriately. The key components for the successful implementation of cognitive radio are low latency and adaptability to the operating conditions. These are the essential characteristic features that are needed for the deployment of cognitive radios in multiservice scenarios such as communications, electronic warfare (EW), and radar. Cognitive radios thus represent a huge evolution of SDRs.

Therefore, the cognitive radio has a lot to do with SDR [789–791]. As a matter of fact, the cognitive radio works largely on the basis of many functionalities of SDR.[3] It is of imperative importance for us to understand how a software definable radio works in order to gain a better understanding of cognitive radio. The discussion on SDR is to be covered in the subsection that follows.

---

[3] A very brief introduction on SDR is also available in 6.1.5.

9.4.1 How Does SDR Work?

An SDR is a collection of hardware and software technologies that enable reconfigurable system architectures for wireless networks and user terminals. It provides an efficient and comparatively inexpensive solution to the problem of building multimode, multiband, and multifunction wireless devices that are able to work adaptively in a complex radio environment. In an SDR, all functions, operation modes, and applications can be configured and reconfigured by various software. If the configuration automation can be implemented in an SDR, a primitive cognitive radio will result.

The fundamental idea of SDR is to sample the received signal in the RF band right after the RF low noise amplifier. It is also noted that the most important part of an SDR is its receiver part, rather than its transmitter part. The reason is simple: the major difference between a conventional radio and an SDR lies mainly in their methods of recovering required signals. Therefore, in this subsection we will concentrate on the discussions on the SDR receiver.

The best way to describe what an SDR system looks like is to compare it with a traditional heterodyne radio, as shown in Figure 9.4, which consists of a bandpass filter (BPF), a low noise amplifier (LNA), a mixer, a frequency synthesizer, an intermediate frequency (IF) amplifier, an automatic gain controller (AGC), a demodulator, an analog to digital converter (ADC), and a digital signal processor (DSP), and so on. It is noted that filtering, amplification, and carrier down conversion are implemented by analogue circuits. There might be several stages of IF amplification, thus needing several IF filters, which makes it very difficult to miniaturize the terminal design due to their bulky sizes.

On the contrary, in an ideal SDR receiver, as shown in Figure 9.5, the signal captured from a wideband antenna will be directly sampled and analogue-to-digital converted; thus all postantenna signal processing will be carried out in the digital domain. Therefore, the physical layer (PHY) air interface signaling format will be determined "over the air," or controlled by either a network or a terminal operator. This feature is critical for the implementation of a cognitive radio. The only difference is that a cognitive radio needs to scan a wide range of frequency spectra before deciding which band to use, instead of a predefined one, as an SDR terminal does.

One of the most important characteristic features of an SDR terminal is that its signal is processed almost completely in the digital domain, needing very little analogue circuit. This brings a tremendous benefit to make the terminal very flexible (for a multimode terminal) and ultrasmall size with the help of state-of-the-art microelectronics technology.

To implement an SDR receiver as shown in Figure 9.5, we have to raise the sampling frequency up to at least twice as high as the carrier frequency seen from the antenna. For instance, if we are interested in receiving the signals in a 10 GHz band, an ADC with a sampling rate of at least 20 Giga samples per second has to be used. This will pose an even higher challenge if a cognitive radio needs to scan an entire frequency spectrum up to millimeter bands. A compromise is to retain the RF front-end

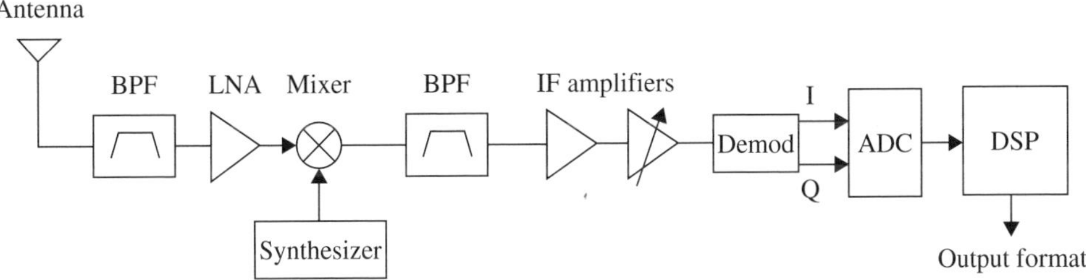

Figure 9.4 A traditional heterodyne radio receiver structure used in a GSM terminal.

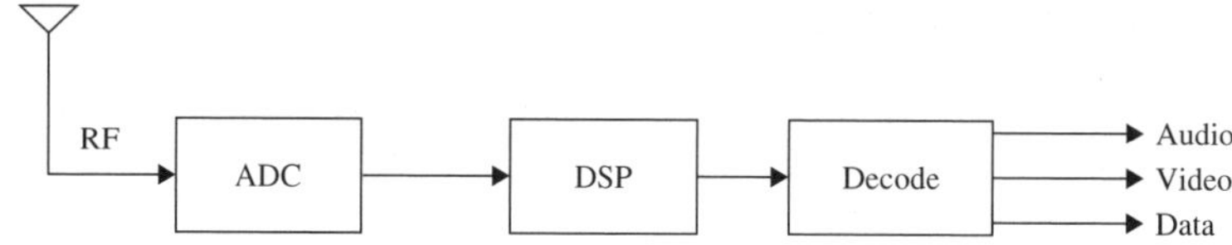

Figure 9.5 A generic SDR receiver structure, which directly samples signals in the RF band.

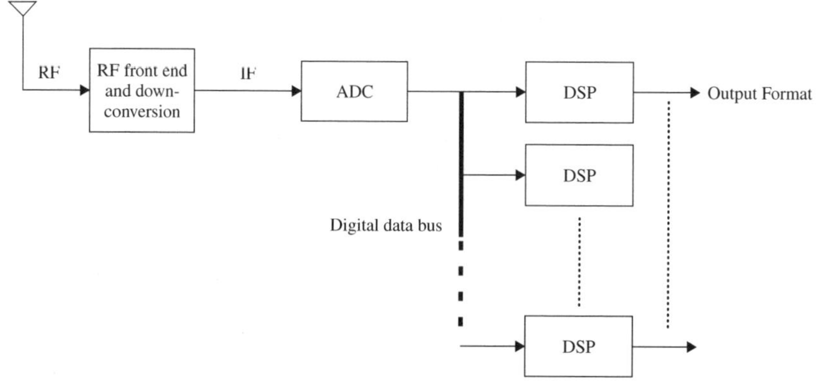

Figure 9.6 An SDR receiver structure with IF sampling implementation.

IF amplifier, and the signal will be sampled only at the IF bands, which will be much lower than RF bands of interest and an ADC with a fixed sampling rate can be applied to all RF signals if the IF is fixed. This can greatly simplify the architecture of an SDR receiver and lower the implementation cost. An SDR receiver with IF sampling is shown in Figure 9.6, where different DSP chips will be used for decoding different pay-loads carried in the RF signals.

9.4.2 Digital Down Converter (DDC)

An SDR terminal should be able to work under different air interface standards/modes. As mentioned earlier, this requires that the signal be digitized as early as possible at a receiver, preferably right after the antenna's front end. However, the complexity of implementing direct RF sampling can be formidable, so that the compromise that uses IF sampling is usually an attractive solution.

However, the use of the IF sampling technique gives rise to a new problem where the DSP bandwidth and processing speed sometimes do not match the output signal from the ADC placed after the IF amplifiers. Therefore, it is commonplace to use a digital down converter (DDC) to bridge the gap between the DSP and the ADC output signal. The block diagram for the DDC is shown in Figure 9.7, where signal processing algorithms can be explained by the following analysis.

First, the input wideband signal should be converted into complex baseband signal as

$$x[n] = r[n]\, e^{-j2\pi f_c nT_s} = r[n]\left\{\cos\left(2\pi f_c nT_s\right) - j\sin\left(2\pi f_c nT_s\right)\right\} \tag{9.4}$$

where $r[n]$ is the sampled IF signal, f_c is the carrier frequency, and T_s is the sampling interval. Now, this complex baseband signal is fed into an M-stage *finite impulse response* (FIR) filter, whose

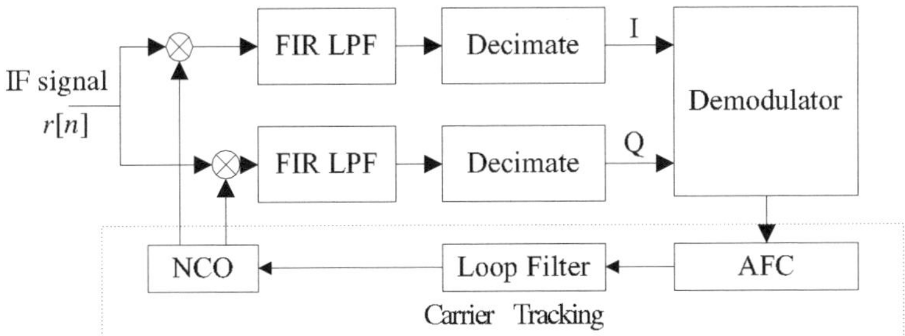

Figure 9.7 Block diagram of a digital down-converter used in SDR receiver.

impulse response is $h[m]$, to obtain

$$y[n] = \sum_{m=0}^{M} h[m]\, x[n-m] = \sum_{m=0}^{M} h[m]\, r[n-m]\, e^{-2\pi f_c (n-m) T_s} \tag{9.5}$$

which can be rearranged to yield

$$y[n] = e^{-j2\pi f_c n T_s} \sum_{m=0}^{M} h[m]\, r[n-m]\, e^{j2\pi f_c m T_s} = e^{-j2\pi f_c n T_s} \sum_{m=0}^{M} c[m]\, r[n-m] \tag{9.6}$$

where $c[m] = h[m]\, e^{j2\pi f_c m T_s}$ is the combined coefficient of the FIR filter. Figure 9.8 shows the frequency domain representation for the signals before and after DDC and filter-cum-decimation.

The use of the DDC in an SDR contributes to a great reduction of computation load in the following DSP chip. In addition, the programmability of the DDC unit makes it possible to select any portion of a signal as the one of interest. This brings about a great flexibility to an SDR. This characteristic feature will also be very useful for the implementation of a cognitive radio.

9.4.3 Analog to Digital Converter

Another important element in an SDR is ADC, which performs the functions to sample, quantize and encode continuous-time analog signals into a digital signal stream, suitable for digital signal processing in the DSP unit. Obviously, the performance of an ADC unit will affect the overall performance of the whole SDR system. In the following text, we would like to introduce several important merit parameters for the ADC unit, which will be used in an SDR terminal.

The first merit parameter we want to discuss is the quantization noise. There are two fundamental ways to perform quantization algorithms: uniform quantization and nonuniform quantization. The nonuniform quantization algorithms include A-law quantization, μ-law quantization, adaptive quantization, and differential quantization, and so on. In this book, we will only concern ourselves with uniform quantization algorithm, whose quantization noise can be expressed by

$$P_{qn} = \frac{q^2}{12R} \tag{9.7}$$

where q is the quantization step size, and R is the input impedance of the A-D converter.

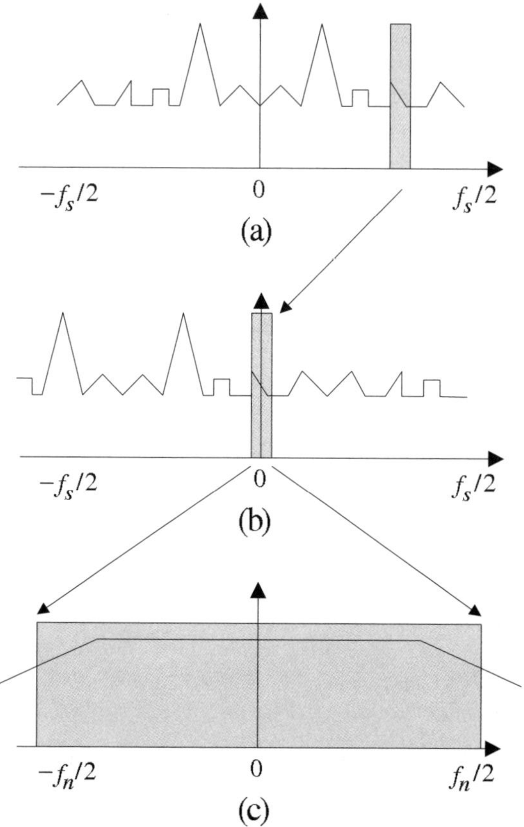

Figure 9.8 The frequency domain representation for the signals before and after digital down-converter. (a) Input wideband signal (the dark area is the bandpass signal); (b) Frequency shift to form the baseband signal; and (c) Signal after filtering and decimation.

The signal-to-noise ratio (SNR) derived from quantization noise and aperture jitter can be expressed respectively by

$$SNR = 6.02B + 1.76 + 10\log_{10}\left(\frac{f_s}{2f_{\max}}\right) \quad (\text{dB}) \tag{9.8}$$

and

$$SNR_{aj} = 20\log_{10}\left(\frac{1}{2\pi f_{\max}t_a}\right) \quad (\text{dB}) \tag{9.9}$$

where B is the resolution of the ADC in terms of the number of bits, f_s is the sampling frequency, $f_{\max}$ is the highest frequency in the input analog signal, and t_a is the aperture jitter of the ADC.

Other important parameters for an ADC are the signal to noise plus distortion (denoted by SINAD), and the effective number of bits (denoted by ENOB), whose theoretical value can be expressed by

$$SINAD = 6.02B + 1.76 + 10\log\left(\frac{f_s}{2 \cdot BW}\right) \tag{9.10}$$

and

$$ENOB = \frac{SINAD - 1.8}{6.02} \times B \tag{9.11}$$

where B is the resolution of an ADC in terms of the number of bits, f_s is the sampling frequency, and BW is the signal bandwidth. More merit parameters for an ADC are listed in Table 9.1.

Obviously, the dynamic range of an ADC is proportional to its resolution in the number of bits. Thus, the increase by one more bit in resolution results in a 6 dB increase in the dynamic range. The second order intermodulation distortion (IMD) is generated due to the nonlinearity of the ADC, producing an $f_1 \pm f_2$ interfering component. On the other hand, the third order IMD produces $2f_2 \pm f_1$ and $2f_1 \pm f_2$ interfering components.

9.4.4 A Generic SDR

In this subsection, we will take a look at a generic SDR, which will be used to detect a DS/SS-BPSK modulated signal. The block diagram for the SDR is shown in Figure 9.9, which consists of the following major elements, a down converter, a BPF, an ADC and a DSP unit.

The transmitting DS/SS-BPSK modulated signal can be written as

$$s(t) = Ab(t)c(t)\cos(2\pi f_c t + \theta) \tag{9.12}$$

where $b(t)$ is a binary data information carried in the received signal, A is the signal amplitude, $c(t)$ is the spreading sequence, f_c is the carrier frequency, and θ is the initial carrier phase. The received signal can be expressed by

$$r(t) = s(t) + n(t) \tag{9.13}$$

where $n(t)$ is zero-mean additive white Gaussian noise (AWGN), whose double-sided PSD is $\frac{N_0}{2}$. As shown in Figure 9.9, the received signal should first be fed into the mixer by multiplying a mixing

Table 9.1   Major merit parameters for analog to digital converters for SDRs

A-D converter merit parameters
Sample rate
SNR (signal-to-noise ratio)
SINAD (signal-to-noise plus distortion)
ENOB (effective number of bits)
THD (total harmonic distortion)
IMD (intermodulation distortion)
SFDR (spurious free dynamic range)
Settling
Quantization noise
Resolution
Dynamic range
Differential nonlinearity
Integral nonlinearity

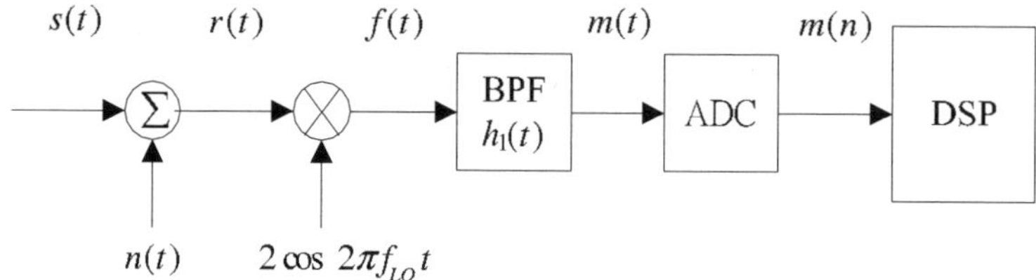

Figure 9.9 A generic SDR block diagram for the reception of DS/SS-BPSK signal.

signal $2\cos 2\pi f_{LO}t$, whose frequency is f_{LO}, where $f_{LO} < f_c$. The output can be written as

$$f(t) = r(t) \times 2\cos 2\pi f_{LO}t$$

$$= Ab(t)c(t)\big[\cos 2\pi(f_c + f_{LO})t + \cos 2\pi(f_c - f_{LO})t\big] + n_m(t) \qquad (9.14)$$

where we have $n_m(t) = n(t) \times 2\cos 2\pi f_{LO}t$. The output from the mixer will be fed into an ideal BPF, whose impulse response is $h_1(t)$. Removing the high frequency component $Ab(t)c(t)\cos 2\pi(f_c + f_{LO})t$, we can obtain the IF signal as

$$m(t) = f(t) \otimes h_1(t)$$

$$= Ab(t)c(t)\cos 2\pi(f_{IF}t + \theta) + n_f(t) \qquad (9.15)$$

where we have $f_{IF} = f_c - f_{LO}$ and $n_f(t) = n_m(t) \otimes h_1(t)$. $m(t)$ will be sent into the ADC, using a subsampling method to obtain

$$m(n) = m(iT_s) = \sum_{i=-\infty}^{\infty} m(t)\delta(t - iT_s)$$

$$= \sum_{i=-\infty}^{\infty} m(iT_s)\delta(t - iT_s) + n_c(iT_s) + n_q(iT_s) + n_j(iT_s) \qquad (9.16)$$

where $n_c(n) = n_f(iT_s)$ is the zero-mean AWGN term after sampling, $n_q(n)$ is the quantization noise generated in the sampling process, $n_j(n)$ is the noise term caused by the sampling clock jitter. It can be shown that the combination of $n_c(n) + n_q(n) + n_j(n)$ is still a white noise. Performing the Fourier transform against (9.16), we obtain

$$m_d(f) = \frac{1}{T_s}\sum_{n=-\infty}^{\infty} m(f - nf_s) \qquad (9.17)$$

which shows that the original signal spectrum will be reproduced in multiples of the sampling frequency f_s, as shown in Figure 9.10. If we let the sampling frequency equal the IF, we will obtain a complete wanted signal spectrum at zero frequency, which can be sent into DSP for low-pass filtering to remove other unwanted components before carrying out any other signal processing tasks in the DSP unit.

It can be shown, under the assumption, that a Gaussian approximation method can be used for performance analysis; the bit error rate (BER) performance for such a DS/SS-BPSK SDR receiver can be written into

$$P_e = Q\left(\sqrt{2 \times SNR}\right) \qquad (9.18)$$

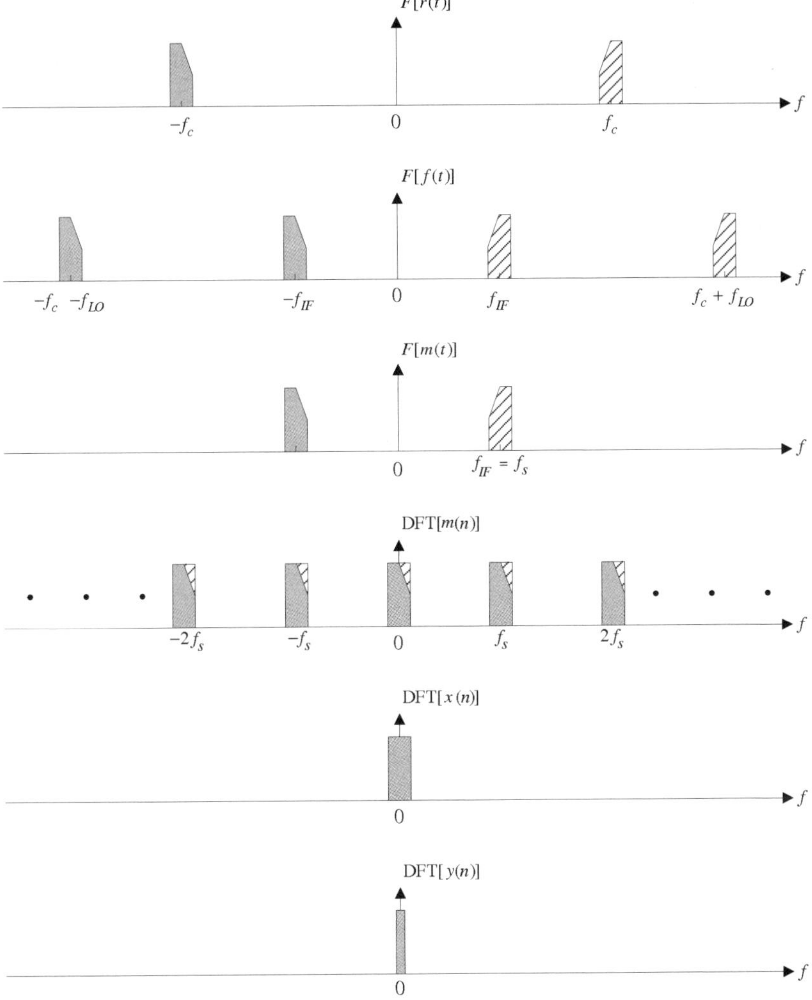

Figure 9.10 Frequency domain signal representation for a generic SDR as shown in Figure 9.9.

where $Q(z)$ is defined as and

$$Q(z) = \int_z^\infty \frac{1}{\sqrt{2\pi}} e^{-\frac{x^2}{2}} \, dx \qquad (9.19)$$

and

$$SNR = \frac{E_b}{N_0 + N_q + N_j}$$

$$= \frac{E_b}{N_0 + \frac{R^2}{3}\left(\frac{1}{2^{2Bq}}\right)\frac{1}{f_{IF}} + \frac{4R^2}{f_{IF}}\left[1 - \exp(-2\pi^2 f_{IF}^2 \sigma_j^2)\right]} \qquad (9.20)$$

9.4.5 Three SDR Schemes

Based on the discussions given in the above subsection, we can obtain three major SDR schemes, as shown in Figure 9.11.

Scheme A (as shown in Figure 9.11a) performs sampling in the IF signal, producing a complete signal spectrum copy near zero frequency. Then it uses DDC to down convert it into a baseband to perform the signal processing before the DSP.

Scheme B also performs undersampling in the IF band to generate a complete signal spectrum copy at baseband, which will be sent into the DSP for signal processing.

The simplest scheme is Scheme C, which performs undersampling directly in the RF band, to produce a complete signal spectrum copy at zero frequency, which is sent to the DSP for any other signal processing tasks.

9.4.6 Implement Cognitive Radio Based on SDR

As discussed in the previous subsections, we can clearly see that SDR forms the basis for implementing cognitive radio. The reason is simple: the SDR provides a very flexible radio platform, which can

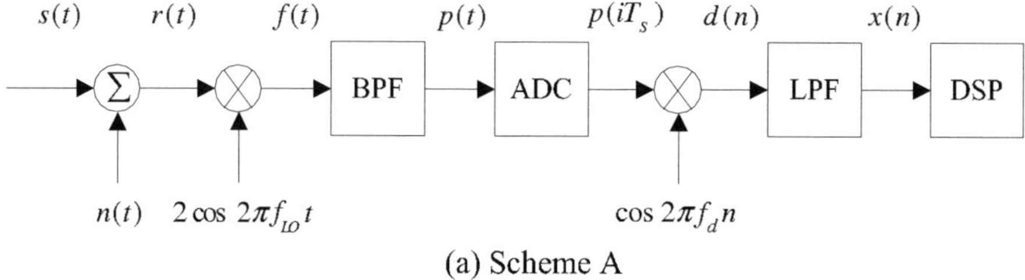

(a) Scheme A

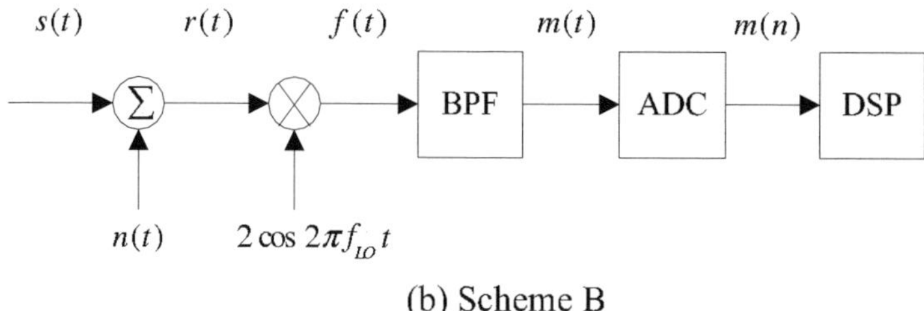

(b) Scheme B

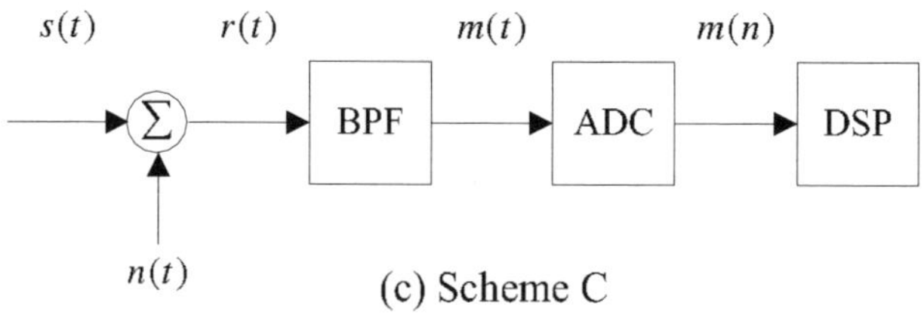

(c) Scheme C

Figure 9.11 Three major SDR implementation schemes.

be programmable and adaptively controlled by a central monitoring unit. The current state-of-the-art electronic technologies, including the ADC, the DDC, the high-speed frequency synthesizer, the ultra-accurate timing controller, the microprocessor, the microelectronics fabrication process, and so on, have made it possible to implement an SDR at a very reasonable cost and small size. The ready availability of SDRs can make the implementation of a cognitive radio a reality in the near future, although we fully understand that there are still many challenges awaiting us on the road to the implementation of practical cognitive radios for commercial applications.

A conceptual block diagram of a cognitive radio scheme based on SDR modules has been shown in Figure 9.12, in which there are two fundamentally important units, one being the receiver module, as illustrated within the upper dashed line block, and the other the transmitter module, as shown in the lower dashed line block in the figure.

Let us introduce the whole cognitive radio transceiver on a block by block basis as follows. The first block on the left-side of the figure is the "wideband antenna," which behaves like a gate to the cognitive radio and controls the bandwidth the cognitive radio will operate in terms of its RF frequency. As it has been acknowledged from the earlier discussions, a cognitive radio may need to scan a fairly wide bandwidth to respond to the changing environment, and thus the total bandwidth for a particular cognitive radio will depend on its applications and services. The initial interest for the FCC in its NPRM [795] was the TV broadcast bands, which are not necessarily used all the time. Therefore, the bandwidth of such a cognitive radio should cover all those TV bands in the design of the wideband antenna, and so on. It has to be noted that the total bandwidth for the "wideband antenna" is denoted by $\sum_{i=1}^{N} \Delta f_i$ in the figure. This total bandwidth should be divided into N sectors, each of which should be assigned to a particular SDR to work on.

Also, a multiple antenna array is preferred for a cognitive radio, such that spatial beam-forming (i.e., beam steering and null steering) and spatial diversity gain can be exploited to enhance the spatial resolution and detection efficiency in performing various cognitive algorithms. As a matter of fact, space, in addition to frequency and time, is another important domain a cognitive radio should take into account to realize spectral reuse.

Following the "wideband antenna," a "duplexer" will control the antenna sharing with receiving and transmitting signals to provide a sufficient isolation between the incoming and outgoing signals. The DFS, was originally introduced to avoid radar signals by IEEE 802.11a networks which operate in the 5 GHz U-NII band, and here refers to an automatic frequency selection process, intended to achieve some specific objective (such as avoiding harmful interference to a radio system with a higher regulatory priority) in the cognitive radio.

There are N SDR units working in parallel in the receiver module of Figure 9.12, and each will take care of a particular sector of the bandwidth of interest. The reason to use several parallel SDRs, instead of only one single SDR, is that it needs to process a huge amount of data in each bandwidth sector (Δf_i, where $i = 1, \ldots, N$) before any kind of "intelligent" decision can be made in the cognitive radio. On the other hand, we can also implement a cognitive radio using only a single SDR unit. However, in this case, a very powerful SDR will be required to finish all data processing within a reasonably short period of time. Unfortunately, it is still impossible to use currently available DSP to fulfill such a challenging computation load for cognitive radio applications.

All output data will then be fed into a unit, which is responsible for making intelligent decisions on them. Those decisions include the selection and combination of detected information, to obtain the information we really want as the output.

On the other hand, the transmitter module in the cognitive radio scheme, as shown in Figure 9.12, should carry out the tasks for information dispatch. The "adaptive synthesizer" works to generate the correct local carrier reference to perform the modulation process and the up-conversion operation. To do so, the transmitter module also needs useful information from the IPD unit, which provides the current carrier frequency allocation chart, spectrum licencees' program timetables and their transmitting power profiles, and so on. This information is vital to determine a correct transmitting power level, so that the transmission from the cognitive radio will not interfere with existing incumbent users. Similar functions will be performed in the "Timing Gate" unit, which controls the transmission time slots, so that the transmissions from the cognitive radio will happen only when the spectrum sector is free.

The generic layered architecture for the cognitive radio scheme shown in Figure 9.12 is illustrated in Figure 9.13, where only two layers (PHY and data link layer) are shown because all the other upper layers are applications dependent and thus not to our interest here.

The spectrum scanning is one of the most important PHY functions of a cognitive radio, scanning all spectrum sectors of interest over the entire operating bandwidth. The scanning should also be done to record the duty cycles of each carrier frequency, so that a cognitive radio will be able to find the right time slot in the right carrier frequency to send the data. This will require the ability to process a wide bandwidth of spectrum and then perform a wideband spectral, spatial, and temporal analysis. It will be necessary for the cognitive radio to exchange their local sensing information to optimally detect interfering incumbent users. This cooperation among different secondary users in the same communication group will be important to estimate accurate interference activities.

Channel measurement has to be used to determine the quality of scanned channels shared with incumbent users. The channel parameters (such as transmit power, bit rate, and so on.) have to be determined based on the channel measurement results.

A cognitive radio must have the ability to operate at variable data transmission rates, modulation formats, different channel-coding schemes, and to transmit power levels. A multiple-in multiple-out

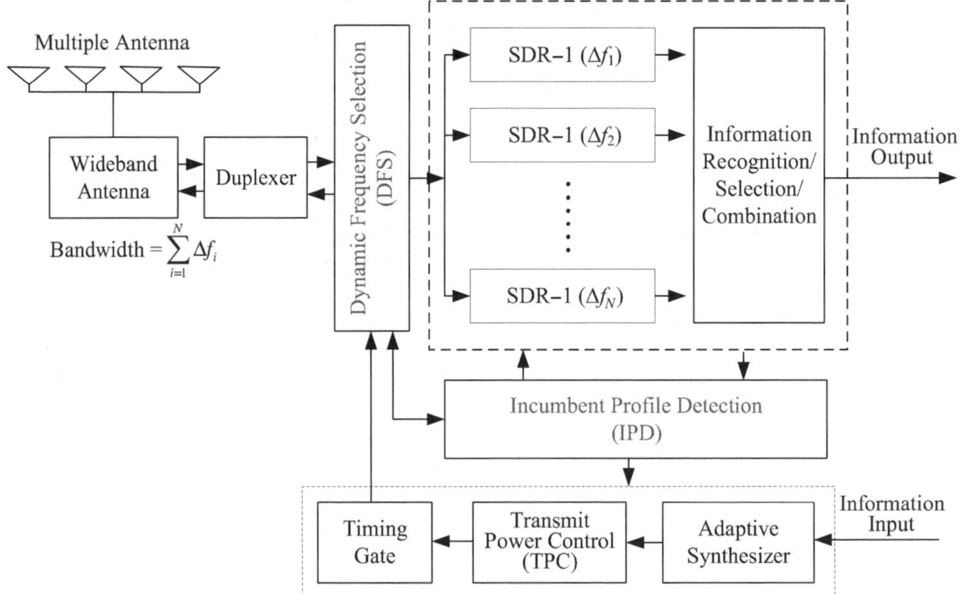

$$\text{Bandwidth} = \sum_{i=1}^{N} \Delta f_i$$

Figure 9.12 A conceptual block diagram for a cognitive radio implementation scheme based on software definable radio. There are two major parts in this cognitive radio configuration: the upper dashed line block represents the receiver module, and the lower dashed line block is the transmitter module. The MIMO system is used in a cognitive radio to enhance the spatial processing capability with the joint application of its array gain, diversity gain, and multiplexing capability.

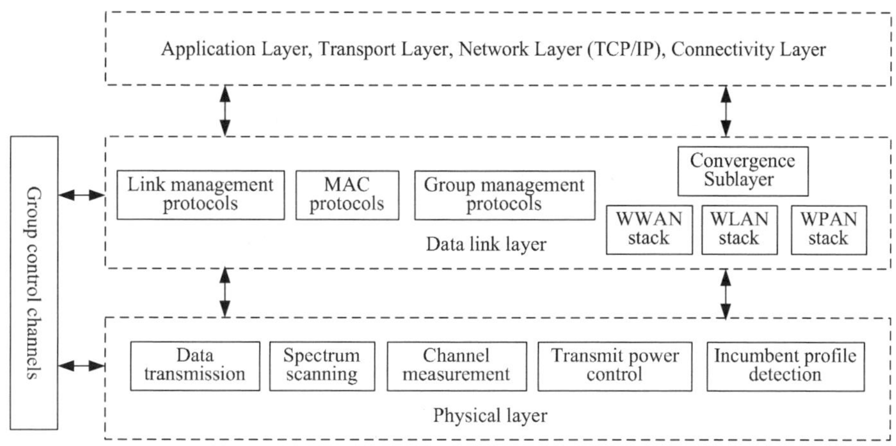

Figure 9.13 Generic layered architecture for a cognitive radio shown in Figure 9.12.

(MIMO) system also might be used to suppress interference spatially and increase throughput via multiplexing. The OFDM technique may also be used to further improve the bandwidth efficiency and detection efficiency. The data transmission block will take care of all the above mentioned tasks.

There are other PHY functions, such as TPC and IPD, and so on, whose functions have been discussed previously. On the other hand, the data link layer consists of three major blocks, including "group management protocols," "Medium Access Control (MAC) protocols" and "link management protocols."

It is assumed that any secondary user belongs to a secondary user group. The group management protocols will be used to coordinate all secondary users in the same group. Any new user can get all the necessary group information when joining a particular group. Link management protocols take care of the link setup to enable communication between two secondary users and maintain the link connection during the entire communication sessions. The data link protocol is used to select a suitable channel to create the communication link. This selection should be made based on the information obtained from spectrum scanning and IPD. Once the secondary link is established, the data link protocols are responsible for maintaining the link connection. The MAC protocols work on the basis of the information obtained from PHY, such as "spectrum scanning" and "IPD," and so on. MAC protocols will decide the ways of accessing the channels, depending on the patterns of shared channels operated by the primary users.

The convergence sublayer in the data link layer is to provide a coordinating mechanism in a cognitive radio to operate in different wireless environments, such as WWANs, WLANs, and WPANs, and so on (also possible for WMANs, although it is not shown in Figure 9.13).

It is admitted that much work needs to be done before a complete layered architecture can be designed. It should also be noted that the layered architecture has to be designed very carefully in order not to increase the process latency, which is a very critical parameter in a cognitive radio. This was the reason for suggestions that a cross-layer design approach [783, 784] may also be a choice for cognitive radio. A primitive cross-layer optimization design for spectrum sensing is shown in Figure 9.14.

We would like to use the following sentence to end this section. There are three primary signal domains we can maneuver: time, frequency, and space, each of which has opportunities for spectral reuse where a cognitive radio is concerned.

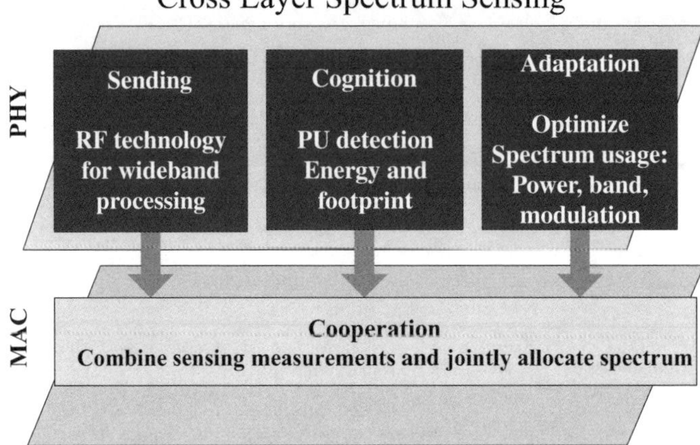

Figure 9.14 A cross-layer design for spectrum sensing algorithms used between the physical layer and the MAC layer in cognitive radio.

9.5 Cognitive Radio for WPANs

In this and the following sections, we will discuss the applications of cognitive radio technology in different wireless networks. In this section, we are going to discuss the issues on the possible application of cognitive radio technology in a WPAN environment, in which a UWB (IEEE 802.15.3a)[4] network operates, although the ideas presented here are equally applicable to other WPANs, such as Bluetooth (IEEE 802.15.1) [461], ZigBee (IEEE 802.15.4), and so on.

In December 2004 the FCC officially began the effort to extricate the concept from academia by issuing a NPRM [795] calling for input on how cognitive radio could be realized. In the response stage, the NPRM underscores the unprecedented willingness of the commission in recent years to explore innovative ways to open up new spectra to commercial unlicensed use. One of the examples includes the release of new spectra in the 5 GHz U-NII band in 2003, as well as the opening up of 7.5 GHz of bandwidth for UWB signaling in the region between 3.1 and 10.6 GHz. Though the power levels allowed for UWB were extremely low, with its roof being 41 dBm, the move marked the first time the FCC had allowed unlicensed use across otherwise licensed bands.

The UWB technology, as an important part of the WPANs, was one of the first beneficiary parties of the FCC's NPRM. In fact, UWB transmits signals that are already below the noise floor. With the help of cognitive radio technology, a UWB terminal can also operate by jumping out really fast from one channel to another if it detects incumbent users, and yet it can transmit at a very reasonably low power level.

As mentioned in Section 7.6, UWB technology has generated a great deal of interest recently as an attractive means to provide high-speed short-range communications. However, it has also generated a lot of controversy. As UWB signals are flat over a broad range of spectrum at a power level close to the noise floor, some people are concerned that UWB will artificially raise up the noise floor and degrade the performance of the existing primary users located at the same spectrum. After a lot of public debates, the FCC approved the first Report and Order on February 14, 2002, in which the FCC not only gave the definition of the UWB signals, but also defined a spectrum mask that

---

[4]More discussions on UWB are given in Section 7.6.

specifies the amount of power that can be sent out by any UWB system working in the band. The spectrum mask is shown in Figure 7.19. The FCC's definition is explained in the following text.

According to the FCC ruling that permitted the operation of UWB systems, an UWB signal is defined as a signal that must satisfy the criteria mentioned below.

Assume that f_H is the upper 10 dB cutoff frequency and f_L the lower 10 dB cutoff frequency. Also identify that the center frequency is $f_c = \frac{f_H + f_L}{2}$ and the fractional bandwidth is $F_p = s\frac{f_H - f_L}{f_H + f_L}$. Then, it is a UWB signal if the condition that either $F_P > 0.2$ or $f_H + f_L > 500$ MHz holds good.

Two types of interference problems

The FCC's spectral mask for UWB signals can help to control the interference to other users of the spectrum, such as GPS, radar and satellite systems, and so on. However, this mask also requires UWB to avoid sending signals in the given band even if it is not in use by any other devices within the spectrum. On the other hand, according to the power emission requirements as specified in IEEE 802.15.3a, the UWB signal is generally sent out at or close to the thermal noise floor. Therefore, it is essentially imperceptible in most cases by an incumbent user more than a few tens of meters away. In addition, most UWB devices will normally work indoors, and thus the possible interference to those incumbent users (i.e., GPS, radar and satellite systems, and so on.) is very unlikely. However, the outdoor operation of UWB systems will generate the problem of the interference to primary users operating in the same band.

A more serious concern of interference is due to the fact that IEEE 802.11 WLANs and IEEE 802.15.1 Bluetooth devices are also operating in the same 2.4 GHz ISM bands. Therefore, there are basically two major interference issues that we need to investigate in the WPAN working environment: one being the UWB's interference to the licensed users allocated in the same band (such as GPS, radar and satellite systems, and so on.), and the other being the mutual interference among unlicensed users (such as WLANs and Bluetooth devices, and so on.). The cognitive radio technology is well suited for the UWB applications to overcome the above two interference problems.

To overcome the first kind of interference problem in UWB applications, it is impossible to mandate a coordination mechanism for spectrum sharing, as non-UWB applications are primary users of the spectrum. Therefore, there will be no collaboration in this scheme and the use of the cognitive radio in the UWB devices will try to avoid using the band if it detects it is in use by some incumbent users, in order not to interfere with their normal operations. Clearly, in this case, the cognitive radio is used by UWB devices on a noncollaborative basis, and only part of cognitive radio functionalities can be used.

As an example to using cognitive radio in a UWB system to avoid interfering with other incumbent users, let us consider a multiband OFDM UWB system, whose signal can be shaped in the frequency domain to avoid generating interference in some particular parts of the spectrum. This particularly shaped signal spectrum can be based on *a priori* decision from the spectrum scanning, which is an important part of the cognitive radio's functionalities. Furthermore, the tone weighting coefficients can also be derived from the spectrum scanning process, which in fact measures the interference temperature of the working environment. In this way, a multiband OFDM UWB signal permits great flexibility in this kind of "spectrum sculpture," which is hardly possible with other UWB schemes, such as pulsed or direct-sequence UWB signals.

On the other hand, we can also consider a WPAN environment, in which other nonlicensed devices, such as Bluetooth and Wi-Fi networks, may exist. In order to achieve the optimal performance for all nonlicensed devices, cognitive radio technology can be applied to all the terminals which form a cognitive radio network, and can work jointly on a collaborative basis. Only in this case, can the benefit of a cognitive radio network be fully realized. Basically, there are four parameters that a cognitive radio network can try to optimize: (1) transmission data throughput, (2) error rate, (3) quality of service, and (4) cost of connection. The optimization of the aforementioned four parameters can be achieved by leveraging the following seven approaches: (1) power level control, (2) antenna beam steering, (3) carrier frequency, (4) channel-coding adaptation, (5) transmission time slot, (6) MAC protocol adaptation, and (7) using CDMA to manage interference.

Obviously, the set of MAC protocols is generally fixed to some extent, but the other parameters can normally be chosen independently from a fairly wide range of values. There is probably some fruitful research into methods for including protocol as part of the optimization; for example, a WLAN user can often choose to modify packet size and data rate, and so does a Bluetooth network. Optimal choices of these parameters in a WPAN using cognitive radio technology are an active research area.

PulseLINK cognitive radio UWB

Before ending this section, we would like to talk about one already-in-market UWB cognitive radio chipset developed by PulseLINK, Inc., which conducted private showings of the PulseLINK chip along with various demonstrations of its UWB technology at the ITU Global Conference on Ultra Wideband held on from June 9–18, 2004 in Boston, Massachusetts. PulseLINK, Inc. is a private Delaware Corporation founded in June 2000 and headquartered in Carlsbad, California. PulseLINK has over 190 issued and pending patents pertaining to UWB wired and wireless communications technology.

The company unveils an architecture that supports PulseLINK's UWB wireless, UWB power line (i.e., UWB across electric power lines), and UWB cable communications technologies (i.e., UWB for Cable Television Networks) simultaneously on the same chipset.

Combining these networking technologies on a single chipset allows consumer electronics and computing devices around the home to be seamlessly networked together wirelessly, through existing home electrical wiring or across CATV cabling as conditions and usage warrants. For Cable Television customers, those home networks can then be networked to the rest of the world through a massive new two-way UWB data pipe enabled through the existing CATV infrastructure, all on the same chipset.

PulseLINK's target data rates for its chipset are up to 1 Gbps for UWB wireless connectivity, up to 200 Mbps for UWB power line communications (electrical wiring in homes and small offices), and up to 1 Gbps of new downstream bandwidth across existing cable television networks in addition to hundreds of megabits of new bandwidth per node upstream.

The fundamental architecture of the chipset, implemented on Jazz Semiconductor's SiGe 120 process, is that of a Software-Defined Cognitive Radio. The hardware platform used in the chipset reduces the complexity and implementation of wireless and wired PHY to a software abstraction. Thus, future evolutions can be anticipated based on this same chip to be able to support and deliver narrowband carrier signals such as Wi-Fi, WiMAX, or IEEE 802.15.3a wireless UWB standard as well. Because it is driven by software, such evolution will not require any modifications to the chipset hardware architecture and will be purely a matter of software/firmware upgrades.

9.6 Cognitive Radio for WLANs

It is widely believed that the technical foundations established by WLANs provide a launching pad for cognitive radios. WLANs already incorporate essential cognitive radio features such as DFS and TPC. Also, while the RF front ends may require wideband receivers and transmitters, the hardware exists now, and the software only involves the generation of the software engineering to make functions like filtering, band selection, and interference mitigation available as plug-in software modules for the radios.

In the following text, we will discuss several study cases for WLAN implementation using cognitive radio technology.

Cognitive radio in IEEE 802.11h

The first case we would like to discuss here is the IEEE 802.11h standard, which was a modified version of IEEE 802.11a standard, as an effort to reduce possible interference to some existing users in the same RF band, such as radar applications, and so on. The two important elements, DFS and TPC, used in IEEE 802.11h standard, bear the important characteristics of a cognitive radio. Therefore, it is justified to say that the IEEE 802.11h standard incorporates some functions only used in cognitive radios.

As we have discussed in Section 4.2, the IEEE 802.11a wireless networks operate in the 5-GHz RF band and support as many as 24 nonoverlapping channels, which are less susceptible to interference than their IEEE 802.11b or IEEE 802.11g counterparts. However, regulatory requirements governing the use of the 5-GHz band vary from country to country, hampering 802.11a's rapid deployment in different regions of the world.

To overcome the problem, the ITU recommended a harmonized set of rules for IEEE 802.11a WLANs to share the 5-GHz spectrum with primary user devices such as military radar systems, and so on. Issued in 14 October 2003, the IEEE 802.11h standard [806] defines mechanisms that 802.11a WLAN devices can use to comply with the ITU recommendations. These mechanisms are DFS and TPC, which are two very important functionalities that every cognitive radio must provide. WLAN products supporting 802.11h are already available in the market.

The DFS detects other devices working in the same RF channel, and it switches WLAN operation to another channel whenever necessary. DFS is responsible for avoiding interference with other devices, such as radar systems and other WLAN segments, and for uniform utilization of channels.

An access point (AP) specifies that it uses DFS in the frames WLAN stations use to find APs. When a WLAN station associates or reassociates with an AP, the station reports a list of channels that it can support. When it is necessary to switch to a new channel, the AP uses this data to determine the best channel.

The AP initiates a channel switch by sending a frame to all stations associated with the AP that identifies the new channel number, the length of time until the channel switch takes effect, and whether or not transmission is allowed before the channel switch. Stations that receive the channel switch information from the AP change to the new channel after the elapsed time.

An AP measures channel activity to determine if there is other radio traffic in the channels being used for a WLAN or other applications, such as radar systems. The AP sends a measurement request to a station or group of stations identifying the channel where activity is to be measured, the start time of the measurement and the duration of the measurement. The station performs the requested measurement of channel activity and generates a report to the AP.

On the other hand, the TPC is intended to reduce interference from WLANs to satellite services by reducing the radio transmit power that WLAN devices use. The TPC can also be used to manage the power consumption of wireless devices and the range between APs and wireless devices.

An AP specifies telephony control protocol (TCP) support in the frames it generates to WLAN stations. These frames also specify the maximum transmit power allowed in the WLAN and the transmit power the AP is currently using. The transmit power used by stations associated with an AP cannot exceed the maximum limit that the AP sets.

When a WLAN station associates with an AP, the station indicates its transmit power capability. The AP uses this data about the stations associated with it to determine the maximum power for the WLAN segment. This means the radio power in a WLAN segment can be adjusted to reduce interference with other devices while still maintaining a sufficient link margin for the operation of the wireless network.

Frames are also sent between the AP and the stations to monitor the signal strength of the wireless network. The AP can dynamically adjust the radio signal strength, if necessary, to maintain wireless communications.

The original motivation for the DFS and TPC mechanisms defined in 802.11h ensure a standard method of operation under the regulatory requirements governing the 5-GHz band, which will spur the deployment of 802.11a wireless networks around the world, especially those places where a strict regulation is imposed. Along with meeting regulatory requirements, the DFS and TPC can be used to improve the management, deployment, and operation of WLANs.

Figure 9.15 illustrates the five-step working procedure specified in IEEE 802.11h protocol based on DFS and TPC etiquettes.

Cognitive radio to enhance WLANs security

Another application of cognitive radio to help WLANs is the security enhancement of existing WLANs standards, such as IEEE 802.11b/g/a, the explosive growth of which has created managerial challenges for those who monitor them. With wireless intrusion threatening security and RF interference impeding performance, WLAN managers need an up-to-date, detailed understanding of the RF environment to make informed decisions about how to solve these problems. Cognitive radio technology enables a WLAN device and its antenna to sense its RF environment and adapt its spectrum use as needed to avoid interference. Integrated software and silicon solutions enable cognitive radio to be built into enterprise-class WLAN APs to boost security and optimize performance. APs with this feature are expected to soon be available in the market.

Wireless intrusion occurs when unauthorized WLAN users gain access to a secured network. The causes of intrusion include hackers creating ad hoc networks with WLAN clients or a rogue AP connected to a wired network without proper security levels. In either case, the keys to prevention are quick identification, containment, and defensive action. The first step in preventing wireless intrusion is to identify the intrusion point. Because WLANs are fixed to a specific WLAN channel during operation, they cannot simultaneously detect intrusion points on other WLAN channels. Relying on a single-radio AP to provide access and security is far from sufficient. Network managers must be able to monitor the full WLAN frequency range to be able to reliably detect and identify intrusion points.

Cognitive radio for WLANs provides a means to observe the RF environment in both 2.4-GHz and 5-GHz frequency bands, on which IEEE 802.11 WLANs operate without disrupting normal wireless voice over IP (VoIP) and data traffic. Continuous scanning of both bands for IEEE 802.11 and non-802.11 devices allows for the timely detection of intruders. With detailed information from multiple cognitive radios within a WLAN, administrators can take preventive action.

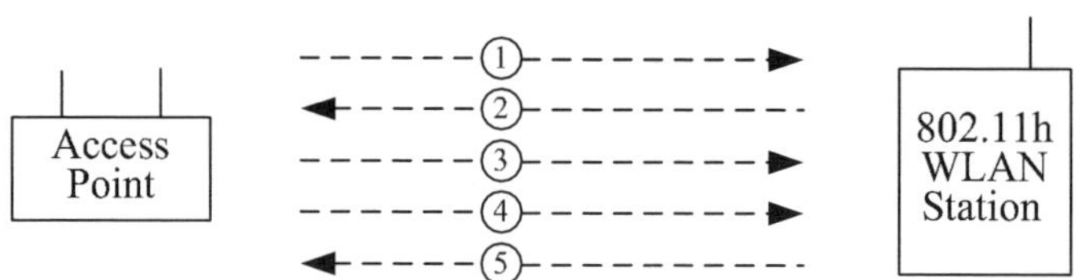

Figure 9.15   Five-step working procedure in IEEE 802.11h protocol based on dynamic frequency selection (DFS) and transmit power control (TPC) etiquettes. (1) An AP broadcasts that spectrum management is required in the frames that advertise the existence of the WLAN; (2) The WLAN station includes the operating channels it supports in the frames it sends to request an association with the AP; (3) The AP responds with a message completing the association process of the WLAN station and the AP; (4) After determining that it must change the operating channel for the WLAN segment, the AP sends a message to all associated WLAN stations, announcing that the WLAN segment will switch to a new channel, the time when the change will happen, and the information about the new channel; and (5) At the designated time, the WLAN stations switch to the new channel.

Cognitive radio can detect non-802.11 devices. This is important because interference, regardless of sources, lowers effective data throughput and overall network performance. IEEE 802.11b/g/a-based WLANs operate within the unlicensed radio spectra around 2.4 GHz (802.11b and 802.11g) and 5 GHz (802.11a). Other wireless connectivity standards, such as Bluetooth and HomeRF, operate in the same unlicensed radio spectra. Microwave ovens, cordless phones and industrial equipments can also generate noise in these bands.

This technology allows a network to detect, identify, and avoid these noise sources. Proactive behavior allows a network to be established on clearer channels during initial deployment. Ongoing monitoring allows network administrators to take rapid corrective action against RF noise. This affords the optimal network performance that WLAN users expect. Integrated software and silicon solutions enable APs to be developed to provide simultaneous WLAN access in the 2.4-GHz and 5-GHz bands while providing integrated cognitive radio functionality that ensures security and performance for enterprising wireless networks.

Figure 9.16 shows how a cognitive radio AP works to identify interference, rogue APs, and other wireless intruders to enhance the security level of WLANs.

More works on WLANs

IEEE is very active in promoting the applications of cognitive radios in all its already very popular IEEE 802.11 standards, as we have discussed in the previous text on IEEE 802.11h standard. In fact, there are at least two other IEEE 802.11 standards that have either been issued or are in their standard-making process, including IEEE 802.11e for the support of Quality of Service (QoS), and IEEE 802.11k for new types of radio resource measurements, which are two well suited candidate technologies for future cognitive radios.

Cognitive radios coordinate the usage of radio spectra without the involvement of restrictive radio regulation. They operate in spectra when it is not used by licensed radio systems, and therefore share spectra with the radio systems that have priority access. This is referred to as *vertical sharing*. An unused radio spectrum is called a *spectrum opportunity*. In the vertical sharing scenario, cognitive radios adapt their transmission schemes in such a way that they fit into the identified spectrum usage patterns of the incumbent radio systems.

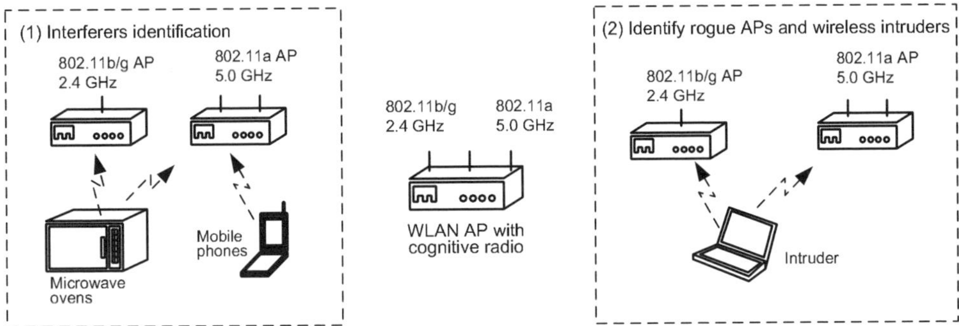

Figure 9.16 Cognitive radio can be used to enhance WLAN security. (1) Upon initial deployment of an AP, cognitive radio performs a full scan of the RF channels to detect both 802.11 and other interferers. The AP selects the WLAN channels with the least interference to maximize performance. (2) Continual scanning of all WLAN channels in both frequency bands enables the WLAN system to identify rogue APs and wireless intruders.

Note that cognitive radio does not only refer to a novel radio technology. It also requires a revolutionary change in how our spectrum will be regulated. However, with this change and the new cognitive radio approach for open spectrum sharing, it will be difficult to achieve complete fairness and efficient spectrum sharing. This is particularly challenging in ad hoc mesh networks, where cognitive radio systems rely on their own capabilities not only to maintain connectivity, but also to increase spectrum efficiency. This *horizontal sharing* problem in mesh networks is a challenge for cognitive radios, and a lot of work needs to be done in designing new spectrum etiquette rules, which are voluntary rules based on the mechanisms like DFS, TPC, adaptive duty cycles, or carrier sensing.

In addition to the design of spectrum etiquettes, we also need to provide insights on cognitive algorithms and reasoning based on machine-understandable languages and logics. The support of the quality of service in spectrum sharing scenarios is another challenging problem. Decentralized cognitive algorithms on the basis of spectrum observation for mutual coordination are required.

Here we will focus on a cognitive radio operating in the unlicensed radio band. Particularly, we focus on the challenges faced by IEEE 802 WLANs operating in the newly allocated 5.470–5.725 GHz band and the 3.650–3.700 GHz band currently under consideration.

In order to utilize these bands a cognitive radio must sense the presence of the incumbent spectrum owner and vacate within a very short period of time with a minimum number of transmissions. It must also perform communications with a minimum amount of transmit power. In the 5.470–5.725 GHz band the incumbent users are radio-location, radio navigation, and meteorological radars. In the 3.650–3.700 GHz band the incumbent user is the C-band fixed space-to-earth satellite services.

The transmission characteristics can vary significantly even within the same band. According to ITU-R M.1461 [807] pulse repetition rates for radars operating in the 5 GHz band can vary from 20 to 100,000 pulses per second and the 3-dB bandwidth of the transmitter varies from approximately 500 kHz to approximately 150 MHz.

Detection must be successful irrespective of the varied transmission characteristics of the incumbent users and irrespective of the instantaneous RF propagating conditions. In order to ensure that minimum power is transmitted, the device must determine the minimum power that will be necessary to maintain communications in a dynamic RF environment. Furthermore, in order to avoid unnecessarily declaring a channel of the network as occupied, the designer must trade off the stringent requirements of the probability of detection against an unnecessarily high probability of false alarm.

Usually 802.11 system designers perform this trade-off by combining an energy detect threshold that can be triggered by noise with a correlator which detects packet preambles. The initial energy detect can be used to adapt the gain in the front end of the receiver; while the correlator can be used to establish if a valid packet is truly present. Unfortunately, this is not applicable to radar detection due to the varied characteristics of the signals that the system is trying to detect.

Furthermore, using a static energy detect threshold is particularly problematic as the duration and bandwidth of the incumbent signals is variable. The system must have a spectrum analysis capability in order to effectively contend with this situation.

Another challenging issue in the implementation of cognitive radio for WLANs is that of moving the network and resuming communications with minimum disruption to the network. In the European Telecommunications Standards Institute (ETSI) regulatory domain a WLAN must vacate a channel in 10 s after the first radar pulse is detected with a channel closing time, that is, maximum transmission time during a channel move, of 260 ms. Furthermore, before a channel can be utilized it must be sensed for a minimum of 60 s. This presents some very significant challenges. One of the most important issues is how to detect the interferer while maintaining communications and simultaneously being prepared to move to a new channel without ceasing operations within 60 s.

9.7 Cognitive Radio for WMANs

As discussed in Section 4.5, a WMAN will cover an area much larger than a WLAN does. The radius of a WMAN can reach several kilometers. The most widely referred standard for WMANs is IEEE 802.16, also often called as *WiMAX standard*, or *broadband wireless access* (BWA) technology.

Different from WLAN technologies, WMANs usually support mobile terminals. For instance, the IEEE 802.16e standard can support a terminal moving at a vehicular speed. Like WLANs, a WMAN can also use cognitive radio to make it possible to underlay/overlay in those spectra that have been already occupied by incumbent users.

It is to be noted that WiMAX is flexible in its channel sizes and can use the 6 MHz width of the underused TV channels. For a WiMAX system using a bandwidth below 900 MHz, its coverage can be three times larger than that in 2.4 GHz, reducing the number of base stations required well below 3G's requirements, making mobile WiMAX clouds an even stronger proposition against cellular, both in licensed and unlicensed modes.

It is very interesting to note that, many technologies could use TV spectra, though the WiMAX community is keen to claim it for its own. In March 2004, when the IEEE 802.22[5] group was set up, the 802.16 Working Group was angered when its proposal that the cognitive radio work should be under its auspices, rather than in a separate group, was defeated. However, this has not halted its supporters, led by Intel, in their quest to turn the 802.22 efforts to their advantage.

The story behind the conflict between the IEEE 802.16 and the IEEE 802.22 groups highlights the importance of cognitive radio technology and its applications in either WMANs or WRANs (also called "*Wi-TV*" technology) operating in underused VHF/UHF TV bands.

The IEEE 802.22 working group insisted at the time the group began work that Wi-TV would work with existing IEEE 802 architectures, serving as a "regional area network" complementing both the Wi-Fi LAN and the WiMAX MAN. The IEEE 802.22 group also pointed out that WiMAX is not suitable for TV spectrum because it does not include cognitive radio functions.

On the other hand, IEEE 802.16 group denounced that claim, noting that WiMAX does in fact have a provision for cognitive radios (IEEE 802.16h) to avoid interference with other WiMAX devices at higher frequencies, and that it could easily be adapted for UHF frequencies.

In December 2004, the IEEE 802.22 group accused the 802.16 group of overstepping its scope and developing its cognitive radios for "coexistence with primary users," not just WiMAX users, and asked the WiMAX committee to reaffirm the limits of its scope. The WiMAX group declined and received support from another IEEE task group, 802.19, which saw no reason why both groups could not work on the problem separately.

The company that controls cognitive radio technology really will have its hand on the rudder of the creation of next-generation communications, and the commercial vendors behind the shadowy IEEE task groups know that.

As another development, very recently (according to a report written on February 28, 2005), telecom agencies in India and Canada are working together on a cognitive radio-based broadband wireless technology: The networks will operate in the 5 GHz spectrum (or possibly the licensed Multichannel Multipoint Distribution Service (MMDS) bands) and transmit as far as one to two kilometers. The base station would use as many as 48 antenna beams. The system would use cognitive radio technology to identify interference and poor links and then change its own signal transmission to improve the weak links. Ultimately, the agencies hope to develop this as a low-costs system that can be used in underdeveloped regions or areas with aging telecom infrastructure.

It is interesting to see that an Indian government body is involved in such a research project which appears to be primarily targeted at unlicensed bands, given that India only very recently de-licensed the 2.4 GHz band for Wi-Fi. The 5.1 GHz band is still only available for unlicensed use indoors.

---

[5]IEEE 802.22 is a new standard for WRANs using vacated VHF/UHF TV bands.

9.8 Cognitive Radio for WWANs

In this section, we take a look at the very basic functions of a cognitive radio in a WWAN environment.

Before beginning operations, cognitive radios must obtain an estimate of the PSD of the radio spectrum to determine which frequencies are used and which frequencies are unused. In order to measure the spectrum in a very accurate way, a highly sensitive radio receiver will be required to measure signals at a cell edge. Consider the example of digital TV which is also situated at the cell edge; the received signal will be barely above the sensitivity of the digital TV receiver. For a cognitive radio equipped with a mobile handset to be able to detect this signal, it needs to have a radio receiver that is considerably more sensitive than the digital TV receiver. If the cognitive radio is not capable of detecting the digital TV signal, then it will incorrectly determine that the spectrum is unused; thereby leading to potential interference if this radio spectrum is used, that is, the signal transmitted by the cognitive radio (with the mobile handset) will interfere with the signal that the digital TV is trying to decode. This situation is often referred to as the "*hidden node problem.*"

Another example of a hidden node is shown in Figure 9.17. In this example, transmissions between UE1 and UE2 cannot be detected by UE3 or UE4, even though UE3 and UE4 are within signaling distance of UE1 and UE2. Therefore, UE3 and UE4 may determine that the spectrum is unused and decide to send out a signal and potentially interfere with the signal reception at UE1 and UE2. These examples show the necessity for highly sensitive receivers for cognitive radios.

On the other hand, if the spectrum is occupied, the cognitive radio must also be able to estimate the *interference temperature* that the primary user can tolerate, that is, the transmit power level that a cognitive device can utilize without raising the noise floor of the primary user's device substantially beyond a specified level. In many cases, this specified level is of the order of 0.5 dB to 1.0 dB, but will depend heavily on the link margin available at the primary user's receiver. The interference temperature can be determined with at least two pieces of information: (1) an estimate of the signal bandwidth used by the primary user, and (2) the distance between the cognitive radio and the primary user's device. The signal bandwidth can be used to determine the noise floor for the primary user, while the distance can be used to determine the received signal strength seen by the primary user as a function of transmit power used at the cognitive radio. Assuming that the noise floor for the primary user is allowed to rise by a prespecified level, it is easy to calculate the maximum allowed transmit power for the cognitive radio. Of course, this analysis is very basic and should be refined if the cognitive radio can blindly classify the type of signal and corresponding data rates used by the primary user. This extra knowledge would determine the exact sensitivity requirements for the primary user.

Clearly, the previous discussion depends on an accurate estimate of the radio spectrum utilization conditions in a WWAN. As the spectrum use is constantly changing, the estimate should be updated in real time in order to ensure that the primary users are always being protected from interference. In addition, a simple view of the radio transmission path has been assumed in the aforementioned analysis. In a real application, however, the propagation path from the cognitive radio to the primary

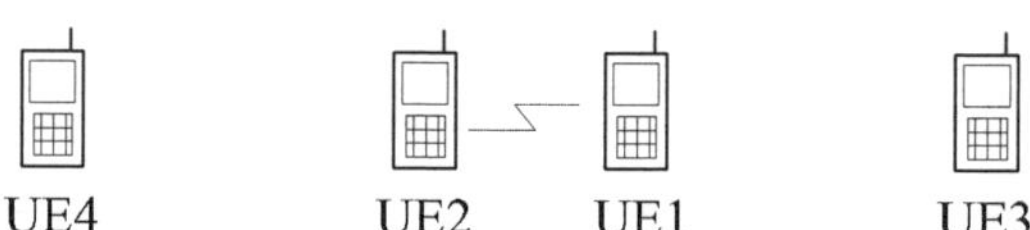

Figure 9.17 Hidden terminal problem, which may affect the performance of a WWAN using cognitive radio. In this example, transmissions between UE1 and UE2 will not be detected by UE3 or UE4, even though UE3 and UE4 are within signaling range of UE1 and UE2.

user might be very complicated. For example, obstacles between the two can attenuate the signal substantially, meaning that the cognitive radio can transmit at a much higher power level than would normally be assumed. The other possibility that the reflections of the transmitted signal may enhance interference at the primary receiver is also valid. Consequently, the lack of information about the transmission paths between the transmitters and receivers in a cognitive radio network can create a great design challenge. Until now, our discussion has been primarily focused on protecting the primary users from interference. However, a significant challenge also lies in the design of a cognitive radio receiver. Emissions from primary users will also result in interference to cognitive radios. As emissions from primary users can be very high, the cognitive radios must use sophisticated radio designs, such as a multiuser detection, a MIMO system, and so on, to deal with the interference and to ensure that the desirable signals can be reliably delivered.

Also, to take advantage of constantly changing spectrum conditions, cognitive radios need to be adaptive in the spectrum, power levels, modulation schemes, and MAC and other upper layer protocols that they use. This is especially challenging for mobile cellular environments in which channel conditions can change rapidly where mobility is concerned. Thus, considerable research still needs to be done in these areas in order to make cognitive radio a reality for WWANs.

9.9   Cognitive Radio for WRANs: IEEE 802.22

As mentioned earlier, the fight between IEEE 802.16 and IEEE 802.22 groups was about who should dominate in the technological development to reuse the UHF/UHF TV bands with the help of cognitive radio technology.

Initially, the FCC was looking at a means by which the 6-MHz wide licensed spectrum in the UHF band [795], currently assigned to TV broadcasters, can be reused in secondary markets as a path to last mile data access. This proposal would set power levels up to 8 dB higher than those currently allowed in those bands, thereby greatly increasing the range and coverage areas. In addition, the propagation characteristics at the lower UHF frequencies are particularly attractive, offering the possibility of longer distance and lower power. This new wireless service scheme is also called *WRANs*.

If a WRAN operates in 6-MHz channels, it will certainly be difficult to get the 54 Mbps as we get with IEEE 802.11g or a Wi-Fi. Nonetheless, data rates provided in a WRAN will be respectable, especially in rural areas where channels can be concatenated.

Now, many research projects are going on to grapple with how to identify channels and the so-called *white space* within the TV spectrum, as well as the uncertainty surrounding FCC certification rules for radios with no assigned frequency bands.

Work began very recently on one of the IEEE's most important projects to devise an intelligent air interface (cognitive radio), that can tap into unused television frequencies. This will be the standard for WRANs for fixed wireless systems that use cognitive radio techniques to automatically switch to a clear area of the band, and to avoid interfering with other occupying devices. The IEEE is particularly focused on systems for the underused US television spectrum between 54 MHz and 862 MHz, which is being vacated (though very reluctantly) as broadcasters move to digital. The FCC proposed to open up 300 MHz of this UHF/VHF spectrum as its first major test of cognitive radios. It would permit fixed access systems to transmit up to 1 W in power and portable devices up to 100 mW.

If successful, the FCC could push to open up other licensed bands to coexist with unlicensed devices, and encourage other regulators around the world to follow suit. For instance, Intel and others are lobbying to increase spectrum for Wi-Fi and WiMAX in order to stimulate their own sales potential.

The standardization for the WRANs is under way and its name is IEEE 802.22. The FCC database can only give IEEE 802.22 modest bandwidth, because not everything that has been licensed is actually used. On the other hand, smart IEEE 802.22 may take too much, because a device's measurements can only cover signal strength at a single point in space and time. The final IEEE 802.22 may need to use both methods.

This year, the IEEE 802.22 group was designated to develop a standard. It is interesting to note that the IEEE 802.16 standard group (for WPANs) argued that cognitive radio work should be under its tent, rather than in a separate group, but was defeated. So the quest to turn IEEE 802.22 into a real 700 MHz alternative is under way.

The first focus of the IEEE 802.22 working group is on rural fixed wireless access. This is the ideal spectrum for deploying regional networks to provide broadband services in sparsely populated areas. The ultimate goal is to equal or exceed the quality of xDSL or cable modem services, and to be able to provide that service in areas where wire-line service is economically infeasible, due to the distance between potential users. In fixed networks, IEEE 802.22-based technologies could achieve a 40 km range and complement local Wi-Fi and 802.16 backhaul. The 802.22 people, apparently, do not want their work too closely identified with 802.16 (WiMAX), which belongs to a WMAN technology.

Some of the major ideas of IEEE 802.22 standard [809] are excerpted in the following text.

Over-the-air broadcast TV channels are separated by unused frequencies. This *white space* in the broadcast spectrum varies with the channels present in a locale and creates opportunities for other applications. As a step in putting these unused TV channels to practical use, IEEE 802.22 standard is to be made to enable the deployment of WRANs using the unused TV channels, while not interfering with the licensed services now operating in the TV bands.

The IEEE P802.22 project, "Standard for Wireless Regional Area Networks (WRAN)-Specific requirements-Part 22: Cognitive Wireless RAN MAC and PHY Specifications: Policies and procedures for operation in the TV Bands," will specify a cognitive air interface for fixed, point-to-multipoint WRANs that operate on unused channels in the VHF/UHF TV bands between 54 and 862 MHz.

Signals at these frequencies in the TV bands can propagate 40 km or more from a well suited base station, depending on terrain. This is the ideal spectrum for deploying regional networks to provide broadband services in sparsely populated areas, where vacant channels are available. The IEEE 802.22 standard will enable the creation of interoperable IEEE 802.22 WRAN products. It has generated a great deal of interest in wireless internet service providers, community networking organizations, government bodies, and other parties.

Protocols in the IEEE 802.22 standard will ensure that this new service does not cause harmful interference to the licensed incumbent services in the TV broadcast bands. The standard will provide for broadband systems that choose portions of the spectrum by sensing what frequencies are unoccupied.

In the United States, the FCC has issued a NPRM to open the 54–698 MHz portion of the TV spectrum for unlicensed usage. IEEE 802.22 will enable compliance to these rules once they are finalized. The IEEE 802.22 standard, which will work with existing IEEE 802 architectures, will give IEEE 802.11 WLANs in outlying areas a fatter pipe for receiving and transmitting data. It will also complement IEEE 802.16 MANs, which do not include cognitive radio functions for sharing TV spectrum.

The concepts underlying the IEEE 802.22 standard are attractive to both developed and undeveloped countries that have little wire-line infrastructure. By extending out to 40 km or more, most regional area networks should have enough potential subscribers within their coverage areas to make them viable ventures.

The formation of the IEEE 802.22 working group has involved broad participation from those in the TV broadcast sector and the public safety community who are licensed users of the target spectrum, as well as from chip vendors, wireless equipment suppliers, and even other countries having large, relatively sparsely populated areas.

IEEE 802.22 is sponsored by the IEEE Computer Society and supported by IEEE 802 Local and Metropolitan Standards Committee.

9.10 Challenges to Implement Cognitive Radio

As mentioned earlier, the research on cognitive radio technology has a very short history, which spans less than 10 years. We have to get over many technical hurdles before cognitive radios can be deployed on a mass commercial application scale. Many challenges need to be overcome to implement a cognitive radio system for practical applications. Cognitive radio is a methodology for the opportunistic utilization of fallow spectrum. This technology can be categorized into two broad classes:

- Unlicensed cognitive radios operating in the unlicensed bands;

- Unlicensed cognitive radios operating in the licensed bands.

Each class has unique challenges to ensure its successful operation. The implementation of the second class of cognitive radios is in particular challenging since there are many parts of the radio spectra that are used by passive receivers such as radio astronomy where very weak distant objects are being observed. A typical signal power in radio astronomy is less than a trillionth of a watt. Detecting and avoiding these passive receivers is an extremely difficult issue and one method of solving this problem is to require any device operating in this band to be able to determine its location and avoid utilizing that part of the spectrum once in the proximity of this sensitive receiver.

Future research areas in cognitive radio include, but are not limited to: (1) New concepts and algorithms for agile radio and spectrum etiquette protocols; (2) Architecture and design of adaptive wireless networks based on cognitive radios; (3) Detailed evaluation of large-scale cognitive radio systems using alternative methods; (4) Spectrum measurement and field validation of proposed methods; and (5) Cognitive radio hardware and software platforms.

User-level field trials of emerging cognitive radios and related algorithms/protocols may also be useful in gaining experience, including: (1) controlled test bed experiments comparing different methods; (2) large-scale spectrum server trial for 802.11x coordination; (3) experimental deployments in the proposed US FCC cognitive radio band.

It has to be admitted that there exists a huge gap between what we expect a cognitive radio to do and what we can use to implement a prototyping cognitive radio. Without doubt, success with the development of cognitive radio technologies should lead to major improvements in spectrum efficiency, performance, and in the interoperability of different wireless networks as a whole. The wide application of cognitive radio technologies will also bring a fundamental change to the philosophy in global radio spectrum allocation and specification across different frequency bands.

9.11 Cognitive Radio Products and Applications

As the last section in this chapter, we would like to discuss a currently available cognitive radio product developed by Adapt4 XG1, where "XG" stands for "Next-Generation" Communications. The appearance of XG1 is shown in Figure 9.18.

The Adapt4 Inc. is a Florida based organization started in early 2003 as an affiliate of Data Flow Systems, Inc., which is the largest manufacturer of wireless turn-key Supervisory Control and Data Acquisition (SCADA) systems for water utilities in the Southeastern US. The XG1 is the world's first cognitive radio. The XG1 Cognitive Radio's technology allows it to adapt to its surroundings

Figure 9.18 The Adapt4 XG1 cognitive radio terminal.

and *time-share* existing licensed channels, but only when they are not in use. XG1 cognitive radios provide the opportunity to implement new networks or to expand existing ones that are currently restricted by the scarcity of licencable FCC frequencies. This is accomplished without resorting to unreliable, unlicensed alternatives. The XG1 operates within licensed frequency bands on a secondary user basis.

In fact, the XG1 project was largely funded by DARPA. Its aim is to develop technology that allows multiple users to share spectra in a way that coexists with and complements sharing protocols included in today's Wi-Fi technologies. The XG1 explores the idea in the most extreme way. The product can work on a *dc-to-daylight* box, covering a broad range of frequency bands up to the microwave area, and it can use any spectrum at any time and adapt accordingly.

The XG1 is also well suited for military applications. From a military perspective, the ability of the cognitive radio to handle functions that best serve its user translates to sufficient situational and mission awareness to help the soldiers reach an objective. It has to be admitted that the kind of data required for this is still in the labs. Academia are studying this and industry wants to work with them. Examples of data for a soldier could include local topography, mission objectives and timescales for those objectives, as well as the knowledge of, and access to, the radio networks in the area as well as the location of friendly and enemy positions and artillery.

Now, we would like to discuss the way the product works and its key technical features.

The XG1 uses its Automatic Spectrum Adaptation Protocol (ASAP) technology to manage time, space, frequency, and power to provide reliable communications without causing interference to other licensed users. A patent-pending feature in XG1 allows all XG1 Adaptive radios within a network to monitor activity by other users in a specified band and identify unused bandwidth. The network generates a set of parallel carriers and transmits on these channels when they are not in use. When another licensed user is sensed, the network stops using that frequency until it becomes dormant once more. Two additional features are used to further reduce the likelihood of interference to other users. A frequency hopping technique is used to minimize the amount of time that any single frequency utilizes and the radios' transmit power is dynamically regulated so that the minimum amount of power needed to establish solid communications is used. Three key technologies have been used in the XG1 terminal, and they are introduced in the following text.

Dynamic frequency selection and avoidance

The XG1 uses advanced signal processing techniques to identify unused bandwidth, normally licensed to others, and creates a selection of channels available for use by the XG1 network. Individual XG1 sites detect non-XG1 radio users and relay this information to the central radio. This information is processed by the central radio to create a *composite* usage map covering the entire XG1 network. Networks with XG1 adaptive radios can span a radius of over 50 miles.

This map is continuously updated and distributed throughout the network. Using a spectrum usage policy in accordance with FCC and network operator guidelines, the XG1 determines opportunities to use these channels on a shared, noninterfering basis. When any XG1 adaptive radio in the network recognizes the presence of another licensed user within the specified band, that frequency is immediately removed from the selection of available channels until the network determines that it is not being used once more. The network also employs a programmable *keep away* feature that allows the operator to determine the amount of *white space* buffer to maintain around other primary users.

Dynamic power management

The adaptive radios in an XG1 network regulate their power output to further decrease the chance of interference to licensed users. Each remote radio in the network uses the minimum amount of transmit power needed to maintain reliable communications with the central radio, dynamically adjusting as necessitated by the variables in the environment. The central radio maintains an output power equal to the level demanded by the network link with the highest wattage requirement.

Frequency hopping

In order to further reduce the likelihood of causing interference to primary channel users, the network employs a frequency hopping technique that minimizes the amount of time the network uses a particular frequency. Typically, these bursts use up to 40 channels simultaneously, but only spend about 10 continuous milliseconds on any one channel.

Additional features

The XG1 Adaptive Radio employs unique modulation techniques to achieve unequaled transmission efficiency. It uses multiple, narrow band carriers to maximize the use of available channels and throughput within the defined band. Networks support a variety of architectures such as mobile, point-to-multipoint and point-to-point with multihopping. The radio's software-defined features allow them to easily adapt to meet unique networking traffic requirements.

A feature-rich network management system is available for providing software downloads over the air, traffic monitoring, fault reporting, policy engine updates, and complete configuration. Link transmission parameters such as the TPC (up to FCC limits) transmitter disable and frequency blocking are provided.

User data can be supplied over the Ethernet port (10/100 MHz) or a serial RS-232 port. In addition to a DSP, the XG1 includes a Power PC microprocessor and software to support a wide variety of user data processing, including encryption, protocol emulation, and network management.

Adapt4 has made use of the fourth dimension, time, to provide highly reliable, interference-free wireless communications solutions for a nearly limitless variety of applications. Operating on FCC-licensed channels, Adapt4's XG1 cognitive radio effectively manages space, frequency, power, and time, to provide adjustable throughput rates approaching 200 kbps, even in the presence of other primary users. The XG1 provides protected QoS communications without interference to or from other users, whether licensed or not.

GENERAL

Frequency Band, licenced		217.0–220.0 MHz
		PLMRS Industrial Business
Data Rates	Configurable	9.6–192 kpbs[1]
Range[2]		
38.4 kbps[2]	Fixed Range (typical)	25 miles
	Mobile Range (parked)	20 miles
	Mobile Range (moving)	15 miles
128 kbps[2]	Fixed Range (typical)	17 miles
	Mobile Range (parked)	12 miles
	Mobile Range (moving)	8 miles
192 kbps[2]	Fixed Range (typical)	14 miles
	Mobile Range (parked)	9 miles
	Mobile Range (moving)	5 miles
Modulation		QPSK with MSFH[3]
Number of MSFH Carriers		1–45

RADIO, *COGNITIVE* SDR

Output Power, Maximum		2 watts
Instantaneous Bandwidth		3 MHz
Channel BW, occupied		$n \times 6.25$ kHz/MSFH Chan
Frequency Stability		± 1 ppm
Antenna Impedance		50 ohm
Duplexing Mode		TDD, Half-Duplex
Receiver Sensitivity	38 kbps with 10–6 BER[4]	−105 dBm
	128 kbps with 10–6 BER[4]	−99 dBm
	192 kbps with 10–6 BER[4]	−97 dBm
Transmitting Mode		Burst
Forward Error Correction		Encoding, Convolutional Decoding, Viterbi
FEC Rates		7/8, 3/4, 1/2, none

PHYSICAL INTERFACE

Ethernet (1)	10 BaseT, RJ45
Serial (2)	User:RS-232/V.24, DB-9, DCE
Antenna	BNC
LED	LAN, Power, Summary Fault
Power	2.5MM Contact, Locking

PROTOCOLS

MAC	XG+, Polled TDMA, Poll-Select
Ethernet	IEEE 802.3
	Transparent Bridge to (IP, IPX, DHCP, ICMP, UDP, ARP, others)
Serial	PPP[5], Transparent Async Encapsulation over IP (tunneling) for serial async multidrop protocols including Modbus, DNP.3, DFI, BSAP

ENVIRONMENTAL

Temperature	−30° to −50°C
Humidity	95%, noncondensing (outdoor pkg opt.)

POWER

Input Voltage	12±2 VDC
Current	1A, Receive
	1.5A Transmit

MANAGEMENT

From Hub	Element Management System EMS SNMPvl/v2/v3, MIB II, Enterprise MIB
From Internet	Network Management System Manage Multiple Networks, Northbound Interface
From Remote	Field Service Laptop

SECURITY

User Data	Approved Site List Frequency Hopping
Software Access	User ID/Password

NMS

	User ID
	Password Login /"Lockdown"
Hardware	Electronic Serial Number (ESN), Unalterable

Figure 9.19 The specifications of Adapt4 XG1 cognitive radio terminal.

Using Adapt4's ASAP, the XG1 offers a multitude of ideal communications solutions: (SCADA) communications for Critical Infrastructure; office-in-the-vehicle and Internet connectivity for Public Safety and Transportation needs; drop-and-insert and frame video capabilities are perfectly suited for Homeland Security, Disaster Response, and Military needs.

The detailed technical specifications for the XG1 adaptive radio are shown in Figure 9.19.

10

E-UTRAN: 3GPP's Evolutional Path to 4G

The third-generation (3G) mobile networks have been deployed in many countries in the world. However, even before its deployment, its enhancement activity had already started and has been actively pursued in the Third Generation Partnership Project (3GPP) [810, 811, 816]. Furthermore, in various fora and organizations for wireless research, such as International Telecommunication Union (ITU) [813] and the Wireless World Research Forum (WWRF) [812], there have been active discussions about 4G systems, which have been targeted for deployment around 2010. Intensive research and development activities have been carried out in both academia and industry around the world.

These R&D activities are quite appropriate if we recall that research in 3G systems began more than 10 years ago, and take a look at the explosive growth of the mobile/wireless communication market in these 10 years. According to the statistic given in [813], the number of mobile subscribers worldwide increased from 300 million in 1997 to about 1200 million in 2005. Based on this estimated growth trend, it is predicted that by 2010 there will be about 1800 million subscribers worldwide. Therefore, it is considered a timely strategy to start the R&D a couple of years before the 4G wireless is expected to be ready in or around 2010. A vision for 4G wireless communications in terms of mobility support and data transmission rate is shown in Figure 10.1.

In this chapter, we are going to focus on the R&D activities for 4G wireless communications carried out under the 3GPP, which is an international organization responsible for the development and harmonization of the UMTS UTRA series standards (such as the WCDMA standard and the TD-SCDMA standard, etc.), as discussed in Sections 3.2 and 3.3.

As discussed in Section 3.2, it is noted that High Speed Downlink Packet Access (HSDPA), as an extension of the UMTS 3G system, can already offer a data rate up to 10 Mbps on a downlink (DL) channel. HSDPA is an enhanced 3GPP-3G standard to increase the DL throughput by replacing QPSK in UMTS 3G by 16QAM in HSDPA. It works to offer a combination of channel bundling (TDMA), code division multiplex (CDM) and improved coding adaptive modulation and coding (AMC). It also introduces a separate control channel to facilitate the data transmission speed. Similar techniques will also be used in the uplink (UL) to yield High Speed Uplink Packet Access (HSUPA) standard. In order to make 3GPP UTRA/UTRAN technology even more competitive in the world, and chiefly to compete with the new technologies proposed by 3GPP2, 3GPP decided to go for an Evolved UTRA and UTRAN (also called *Super-3G standard*), which will work for the next 10 years and beyond, as a long-term evolution of the 3GPP radio-access technology.

Next Generation Wireless Systems and Networks   Hsiao-Hwa Chen and Mohsen Guizani
© 2006 John Wiley & Sons, Ltd

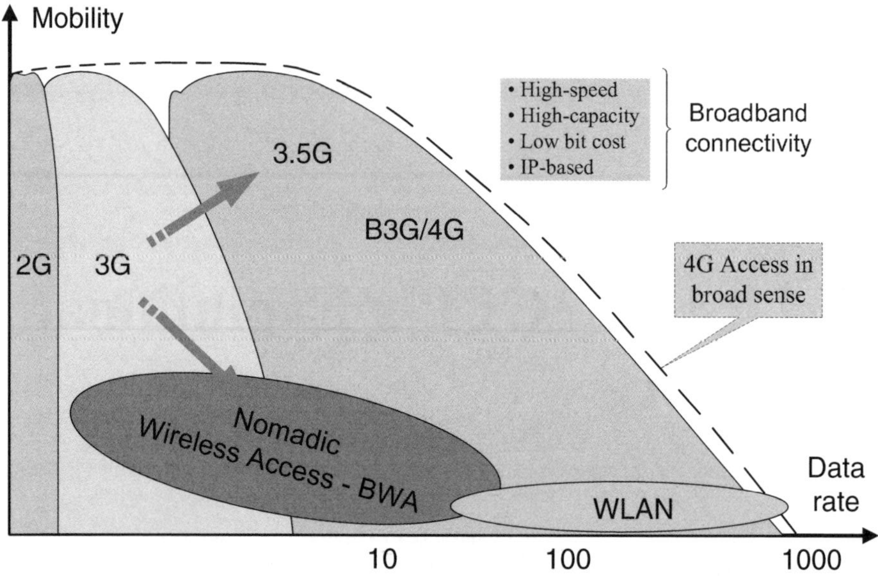

Figure 10.1 A vision for 4G wireless communications in terms of mobility support and data transmission rate.

It has to be noted that, due to the limitations of space in this book, we will not discuss the 4G R&D activities carried out by 3GPP2 in parallel to these discussions; 3GPP2 being the other camp formed under the umbrella of CDMA2000 series standards, as discussed in Section 3.1.

10.1 3GPP TSG for E-UTRAN

In this chapter, we will concentrate on the details regarding 3GPP's Evolved UTRAN standardization process, E-UTRAN, also called 3GPP Long-Term Evolution (3GPP-LTE) standardization, or sometimes called the Super-3G technology. However, before discussing the detailed aspects of the E-UTRAN, we feel that it is appropriate to introduce the 3GPP and its activities. The 3GPP is a collaborative agreement between the Standards Development Organizations (SDOs) and other related bodies for the production of a complete set of globally applicable Technical Specifications and Reports for a 3G System.

The membership for 3GPP is open to (1) all national/regional SDOs irrespective of their geographical location, Organizational Partners (OPs), (2) all organizations that can offer market advice and a consensus view of market requirements, Market Representation Partners (MRPs), (3) SDOs who have the qualifications to become future OPs (Observers).

While the OPs determine the general policy and strategy of 3GPP, the running of the Project is performed by the Project Coordination Group (PCG), under which lie four Technical Specification Groups (TSGs), that is, (1) Services and System Aspects (SA), (2) Core Network and Terminals (CT), (3) GSM EDGE Radio Access Network (GERAN), and (4) Radio Access Network (RAN). It

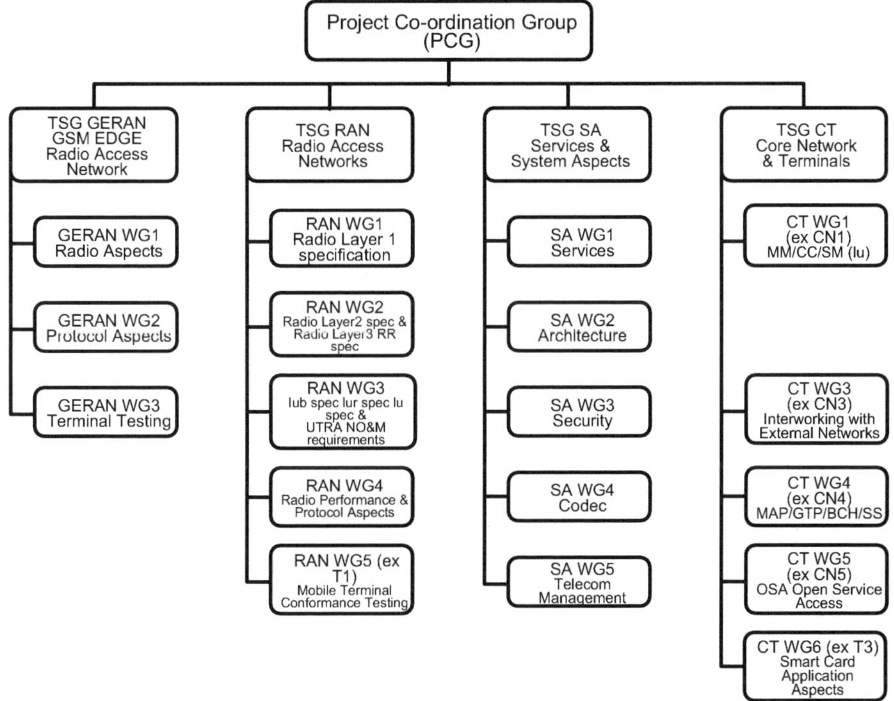

Figure 10.2 The organization of the 3GPP Project Coordination Group (PCG), under which lie four Technical Specification Groups (TSGs), that is, (1) Services and System Aspects (SA), (2) Core Network and Terminals (CT), (3) GSM EDGE Radio Access Network (GERAN), and (4) Radio Access Network (RAN).

is noted that each of these TSGs has a number of Working Groups (WGs) populated by individuals from companies that are members of one or more of the OPs, MRPs, or Observers. The detailed organization of the PCG is illustrated in Figure 10.2.

It is noted that 3GPP has no legal status, but the 3GPP IPRs are jointly owned by the OPs. It is the responsibility of the OPs to transpose the results into their own regional deliverables (e.g. standards).

The support function of the 3GPP is the Mobile Competence Center (MCC) located in Sophia Antipolis, France. In fact, the MCC provides support to the 3GPP, as well as to the European Telecommunications Standards Institute's (ETSI) Technical Committee MSG, and the ETSI Projects SCP and RT.

In this chapter, we will focus on the technical aspects proposed in the TSG RAN (new), shown in the second column in the PCG illustrated in Figure 10.2. The all TSG RAN meetings after the TSG RAN meeting #26 are denoted by (new) TSG RAN meetings. The TSG RAN is responsible for the definition of the functions, requirements, and interfaces of the UTRA network in its two modes, FDD and TDD. More precisely: radio performance, physical layer, layer 2 and layer 3 Radio Resource

(RR) specifications in the Universal Terrestrial Radio Access Network (UTRAN); the specification of the access network interfaces (Iu, Iub, and Iur); the definition of the Operations and Maintenance (O&M) requirements in UTRAN and conformance testing for the Base Stations.

10.2   Origin of E-UTRAN

At the 3GPP TSG RAN #26 meeting, the Study Item description on "Evolved UTRA and UTRAN" was approved [815]. It is noted that all 3GPP TSG RAN meetings after the #26 meeting have been called *3GPP TSG RAN (new) meetings.*

The justification of the Study Item was that with enhancements such as HSDPA and HSUPA, the 3GPP radio-access technology will be highly competitive for several years. However, to ensure competitiveness in an even longer time frame, that is, for the next 10 years and further, a long-term evolution of the 3GPP radio-access technology needs to be considered.

Important parts of such a long-term evolution include reduced latency, higher user data rates, improved system capacity and coverage, and reduced cost for the operator. In order to achieve this, an evolution of the radio interface as well as the radio network architecture should be considered.

Considering a desire for even higher data rates and also taking into account future additional 3G spectrum allocations, the long-term 3GPP evolution should include an evolution toward support for wider transmission bandwidth than 5 MHz. At the same time, support for transmission bandwidths of 5 MHz and less than 5 MHz should also be investigated in order to allow for more flexibility in whichever frequency bands the system may be deployed.

3GPP work on the Evolution of the 3G Mobile System started with the RAN Evolution Workshop, held from 2–3 November 2004 in Toronto, Canada. The Workshop was open to all interested organizations and members and nonmembers of 3GPP. Operators, manufacturers, and research institutes presented more than 40 contributions with views and proposals on the evolution of the UTRAN. A set of high-level requirements were identified in the Workshop including: (1) Reduced cost per bit, (2) Increased service provisioning – more services at a lower cost with better user experience, (3) Flexibility of use of existing and new frequency bands, (4) Simplified architecture, Open interfaces, and (5) Agreement toward reasonable terminal power consumption.

It was also recommended that the Evolved UTRAN should bring significant improvements to justify the standardization effort and it should avoid unnecessary options. In a certain light, the collaboration with 3GPP SA WGs was found a must with regards to the new split between the Access Network and the Core, and the characteristics of the throughput that new services would require.

With the conclusions of this Workshop and with broad support from 3GPP members, a feasibility study on the UTRA and UTRAN Long-Term Evolution was started in December 2004. The objective was to develop a framework for the evolution of the 3GPP radio-access technology toward a high-data–rate, low-latency and packet-optimized radio-access technology. The study should be completed by June 2006 (at the time when this book is finished, it seems that this deadline for final E-UTRAN standard is likely to be postponed), with the selection of a new air-interface and the layout of the new architecture. At that point, Work Items will be created to introduce the E-UTRAN in 3GPP Work Plan.

10.3   General Features of E-UTRAN

The study being carried out under the 3GPP Work Plan is focussing on supporting services provided by the packet-switched (PS) domain with activities in the following areas, at the very least. (1) services related to the radio-interface physical layer (DL and UL), for example, to support flexible transmission bandwidth up to 20 MHz, and new transmission schemes and advanced multiantenna technologies; (2) services related to the radio-interface layer 2 and 3: for example, signaling optimization; (3) services related to the UTRAN architecture: (a) identify the most optimum UTRAN network architecture and the functional split between RAN network nodes, and (b) RF-related issues. It is very important to note that the E-UTRAN scheme leaves open an option to operate at a bandwidth that is much wider than its predecessor, the WCDMA UTRA, which has a fixed signal bandwidth at 5 MHz; this paves the way for providing a much higher data rate transmission in the E-UTRAN than was possible in its 3G standard, WCDMA, as discussed in Section 3.2.

Also note that, as a packet-based data service in WCDMA DL with data transmission up to 8–10 Mbps (and 20 Mbps for MIMO systems), HSDPA also operates over a 5 MHz bandwidth in WCDMA DL. Unlike standard WCDMA, the HSDPA uses several advanced technologies in its implementations, including AMC, Multiple-Input Multiple-Output (MIMO), Hybrid Automatic Request (HARQ), fast cell search, and advanced receiver design. However, its fixed bandwidth operation limits its further enhancement in its data transmission rate. Therefore, in this sense, the E-UTRAN is a big step forward toward 4G wireless technology.

All RAN WGs will participate in the study on E-UTRAN, with collaboration from SA WG2 in the key area of the network architecture. The first part of the study was the agreement of the requirements for the E-UTRAN. Two joint meetings, with the participation of all RAN WGs, were held in 2005:

(1) RAN WGs on Long-Term Evolution, 7–8 March 2005, Tokyo, Japan;
(2) RAN WGs on Long-Term Evolution, 30–31 May, Quebec, Canada.

In the above two meetings, TR25.913 [817] was drafted and completed. This Technical Report (TR) contains detailed requirements or the following key parameters, which will be introduced individually in the sequel.

Peak Data Rate

E-UTRA should support significantly increased instantaneous peak data rates, which should scale according to different sizes of the spectrum allocation.

E-UTRAN should provide instantaneous DL peak data rate of 100 Mb/s within a 20 MHz DL spectrum allocation (5 bps/Hz), and instantaneous UL peak data rate of 50 Mb/s (2.5 bps/Hz) within a 20 MHz UL spectrum allocation. It is therefore noted that the occupied bandwidth for the E-UTRAN has been increased four times as wide as what its 3G system does.

Note that the peak data rates may depend on the numbers of transmit and receive antennae at the UE. The above targets for DL and UL peak data rates were specified in terms of a reference UE configuration comprising: (1) DL capability with two receive antennae at UE, (2) UL capability with one transmit antenna at UE. In case of spectra shared between DL and UL transmission, E-UTRA does not need to support the above instantaneous peak data rates simultaneously.

It is noted that the DL peak data rate supported by HSDPA (an enhanced 3GPP 3G version) is about 10 Mbps (as discussed in Section 3.2.1). Thus, the bandwidth efficiency required by E-UTRAN (assume that the 20 MHz bandwidth will be used) has been doubled if compared to that of the HSDPA, which uses 5 MHz bandwidth for its operation.

In the design of E-UTRAN architecture, emphasis has been laid on the increasing cell edge bit rate while maintaining the same site locations as deployed in UTRAN/GERAN today.

C-plane and U-plane latency

It is required that a significantly reduced Control-plane (C-plane) latency (e.g. including the possibility to exchange user-plane data starting from a camped state with a transition time of less than 100 ms, excluding DL paging delay) should be ensured.

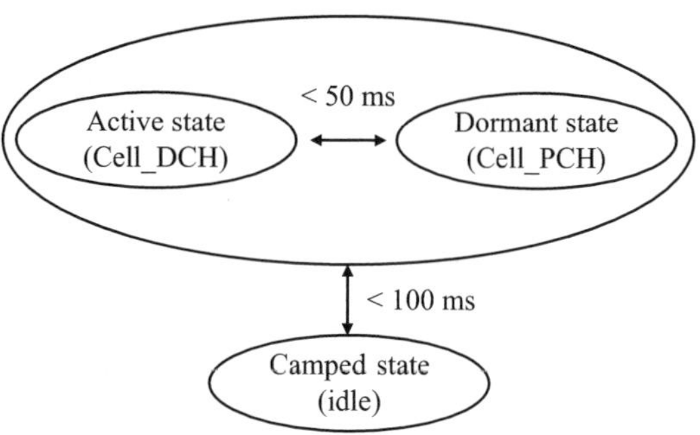

Figure 10.3 An example of state transition in E-UTRAN architecture.

E-UTRAN should have a transition time of less than 100 ms from a camped state, such as Release 6 Idle Mode, to an active state such as Release 6 CELL_DCH. It also needs to provide a transition time of less than 50 ms between a dormant state such as Release 6 CELL_PCH and an active state such as Release 6 CELL_DCH. An example of state transition in E-UTRAN is shown in Figure 10.3.

It is also required that the possibility for a RAN U-plane latency below 10 ms should be included. The U-plane delay is defined as the one-way transit time between a packet being available at the IP layer in either the UE/RAN edge node and the availability of this packet at the IP layer in the RAN edge node/UE. The RAN edge node is the node providing the RAN interface toward the core network. Specifications should enable an E-UTRA U-plane latency of less than 5 ms in unload conditions (i.e. a single user with a single data stream) for small IP packet, for example, zero byte payload plus IP headers. Obviously, E-UTRAN bandwidth allocation modes may impact the experienced latency substantially.

The protocol stacks for the C-plane and U-plane are shown in Figures 10.8 and 10.7, respectively.

Data throughput

The DL data throughput in E-UTRAN will be three to four times higher than that specified in the Release 6 HSDPA UL specifications in terms of an averaged user throughput per MHz. It is noted that the DL throughput performance concerned has assumed that the Release 6 reference performance is based on a single Tx antenna at the Node B with an enhanced performance type one receiver in the UE; while the E-UTRA may use a maximum of two Tx antennae at the Node B and two Rx antennae at the UE. Also, it is understandable that the supported user throughput should scale with the spectrum bandwidth allocation schemes.

On the other hand, the UL throughput in E-UTRAN will be two to three times higher than that given in the Release 6 Enhanced Uplink or the HSUPA in terms of averaged user throughput per MHz. It is assumed that the Release 6 Enhanced Uplink is deployed with a single Tx antenna at the UE and two Rx antennae at the Node B; and the E-UTRA uses a maximum of a single Tx antenna at the UE and two Rx antennae at the Node B. Of course, a greater user throughput should be achievable using more Tx antennae at the UE.

Spectrum efficiency

E-UTRA should deliver significantly improved spectrum efficiency and increased cell edge bit rate while maintaining the same site locations as UTRAN and GERAN deployed today.

In a loaded network, the spectrum efficiency in the DL channels in E-UTRAN should be three to four times higher than the Release 6 HSDPA if measured in bits/sec/Hz/site. This should be achieved assuming that the Release 6 reference performance is based on a single Tx antenna at the Node B with enhanced performance type 1 receiver in UE; while the E-UTRA may use a maximum of two Tx antennae at the Node B and two Rx antennae at the UE.

The spectrum efficiency in the UL channels in E-UTRAN should be two to three times higher than the Release 6 Enhanced Uplink deployed with a single Tx antenna at the UE and two Rx antennae at the Node B. This spectrum efficiency in the UL channels in E-UTRAN should be achievable by the E-UTRA using a maximum of a single Tx antenna at the UE and two Rx antennae at the Node B.

It should be noted that the discrepancy in the spectrum efficiency between the DL and UL channel underlines the different operational environments between the DL and UL. Usually, the UL is much more susceptive to channel impairments, such as multipath interference, and so on, and thus the cost to maintain a satisfactory detection efficiency in UL channels is higher than that in DL channels.

E-UTRAN should support a saleable bandwidth allocation scheme, that is, 5, 10, 20, and possibly 15 MHz. Support to scale the bandwidth in an increment factor of 1.25 or 2.5 MHz should also be considered to allow flexibility in narrow spectral allocations where the system may be deployed.

Mobility support

E-UTRAN should be optimized in terms of its performance for low mobile users at a speed from 0 to 15 km/h. Higher mobile users at a speed between 15 and 120 km/h should be supported with a satisfactorily high performance. Supportable mobility across the cellular networks should be maintained at speeds from 120 km/h to 350 km/h (or even up to 500 km/h depending on the frequency band allocated). The provision for mobility support up to 350 km/h is important to maintain an acceptable service quality to the users who need the services at high-speed railway systems, such as the Euro-Star trains running between the United Kingdom and France. In such a case, a special scenario applies for issues such as mobility solutions and channel models. For the physical layer parameterizations, E-UTRAN should be able to maintain the connection up to 350 km/h, or even up to 500 km/h depending on the frequency band.

The E-UTRAN should also support techniques and mechanisms to optimize delay and packet loss during intrasystem handovers. Voice and other real-time services supported in the Circuit Switched (CS) domain in R6 should be supported by E-UTRAN via the PS domain with a minimum of equal quality as supported by UTRAN (e.g. in terms of guaranteed bit rate) over the whole speed range. The impact of intra E-UTRA handovers on quality (e.g. interruption time) should be less than or equal to that provided by CS-domain handovers in GERAN.

Coverage

E-UTRA should be sufficiently flexible to support a variety of coverage scenarios for which the aforementioned performance targets should be met assuming the reuse of existing UTRAN sites and the same carrier frequency. For more accurate comparisons, reference scenarios should be defined that are representatives of the current UTRAN (WCDMA) deployments.

The throughput, spectrum efficiency, and mobility support mentioned above should be met for 5 km cells in radius, and with a slight degradation for 30 km cells in radius. A cell range of up to 100 km should not be precluded.

As mentioned earlier, E-UTRAN should operate in spectrum allocations of different bandwidths, such as 1.25 MHz, 2.5 MHz, 5 MHz, 10 MHz, 15 MHz, and 20 MHz, in both the UL and DL. Operations in paired and unpaired spectra should also be supported. Operation in paired and unpaired spectra should not be excluded.

The system should be able to support content delivery over an aggregation of resources, including Radio Band Resources (as well as power, adaptive scheduling, etc.) in same as well as different bands, in both UL and DL, and in both adjacent and nonadjacent channel arrangements. A "Radio Band Resource" is defined as an all spectrum available to an operator.

Enhanced MBMS

Multimedia Broadcast Multicast Service (MBMS), has been introduced in 3GPP UTRAN services. E-UTRA systems should support enhanced MBMS modes if compared to UTRA operation. For the unicast case, E-UTRA should be capable of achieving the target performance levels when operating from the same site locations as existing UTRA systems.

E-UTRA should provide enhanced support for MBMS services. Specifically, E-UTRA's support for MBMS should take the following requirements into account. (1) Physical Layer Component Reuse: in order to reduce E-UTRA terminal complexity, the same fundamental modulation, coding, and multiple access approaches used for unicast operations should apply to MBMS services, and the same UE bandwidth mode set supported for unicast operations should be applicable to the MBMS operation. (2) Voice and MBMS: the E-UTRA approach to MBMS should permit simultaneous, tightly integrated, and efficient provisioning of dedicated voice and MBMS services to the user. (3) Unpaired MBMS Operation: the deployment of E-UTRA carriers bearing MBMS services in unpaired spectrum arrangements should be supported.

Spectrum deployment

E-UTRA is required to work with the following spectrum deployment scenarios:

- Coexistence in the same geographical area and colocation with GERAN/UTRAN on adjacent channels.

- Coexistence in the same geographical area and colocation between operators on adjacent channels.

- Coexistence on overlapping and/or adjacent spectra at country borders.

- E-UTRA should possibly operate stand-alone, that is, there is no need for any other carrier to be available.

- All frequency bands should be allowed following release of independent frequency band principles.

It is noted that in case of border coordination requirements, other aspects such as possible scheduling solutions should be considered, along with other physical layer behaviors.

Coexistence and interworking with 3GPP RAT

E-UTRAN should support interworking with existing 3G systems and non-3GPP specified systems. E-UTRAN should provide a possibility for simplified coexistence between the operators in adjacent bands as well as cross-border coexistence.

Basically, all E-UTRAN terminals that are also supporting UTRAN and/or GERAN operations should be capable of supporting the measurement of, and the handover from and to, both 3GPP UTRA and 3GPP GERAN systems. In addition, E-UTRAN is required to efficiently support inter-RAT (Radio Access Technology) measurements with an acceptable impact on terminal complexity and network performance, for instance, by providing UEs with measurement opportunities through DL and UL scheduling.

Therefore, note that the question here is not about backward compatibility, but only about the support for handover mechanism between different 3GPP networks. Also note that HSPDA is still a 3G solution from 3GPP, and it is fully backward compatible to WCDMA networks. Backwards compatibility is highly desirable in E-UTRAN, but the trade-off versus performance and/or capability enhancements should be carefully considered. It is interesting to note that, like the compatible problems existing between UTRAN (based on WCDMA) and GERAN (based on GSM), the problem has surfaced again here between E-UTRAN and UTRAN.

Requirements that are applicable to interworking between E-UTRA and other 3GPP systems are listed below:

- The interruption time during a handover of real-time services between E-UTRAN and UTRAN should be less than 300 ms.

- The interruption time during a handover of non real-time services between E-UTRAN and UTRAN should be less than 500 ms.

- The interruption time during a handover of real-time services between E-UTRAN and GERAN is less than 300 ms.

- The interruption time during a handover of non real-time services between E-UTRAN and GERAN should be less than 500 ms.

- Nonactive terminals (such as the one in Release 6 idle mode or CELL_PCH) that support UTRAN and/or GERAN in addition to E-UTRAN should not need to monitor paging messages only from one of GERAN, UTRA or E-UTRA.

The above requirements are set for the cases where the UTRAN and/or GERAN networks provide support for E-UTRAN handovers. The interruption times required above are to be considered as maximum values, which may be subject to further modifications when the overall architecture and the E-UTRA physical layer has been defined in more detail.

Architecture and migration

A single E-UTRAN architecture should be agreed upon in TSG. The E-UTRAN architecture should be packet-based, although provisions should be made to support real-time and conversational class traffic. E-UTRAN architecture should simplify and minimize the number of interfaces where possible.

E-UTRAN should offer a cost-effective migration from Release 6 UTRA radio interface and architecture. The design of the E-UTRAN network should be under a single E-UTRAN architecture, which should be packet-based (thus, all IP wireless architecture will be dominant in the E-UTRAN networks), although provisions should be made to support real-time and conversational class traffic.

E-UTRAN architecture should minimize the presence of "single point of failures," and thus some backup measures should be considered. The E-UTRAN architecture should support an end-to-end Quality of Service (QoS) requirement. Also, backhaul communication protocols should be optimized in E-UTRAN. QoS mechanism(s) should take into account the various types of traffic that exist to provide efficient bandwidth utilization.

E-UTRAN should efficiently support various types of services, especially from the PS domain (e.g. Voice over IP, Presence). The E-UTRAN should be designed in such a way as to minimize the delay variation (jitter) for the TCP/IP packet communication.

Radio resource management

As mentioned earlier, the E-UTRAN RR management requires that: (1) an enhanced support for end-to-end QoS is in place; (2) efficient support for transmission of higher layers is needed; and (3) the support of load sharing and policy management across different Radio Access Technologies is necessary.

Complexity issues

E-UTRA and E-UTRAN should satisfy the required performance. Additionally, system complexity should be minimized in order to stabilize the system and interoperability in the earlier stages; it also serves to decrease the cost of terminal and UTRAN. To fulfill these requirements, the following points should be taken into account.

To reduce the implementation complexity in both hardware and software, the design of E-UTRAN networks should minimize the number of options, and also ensure the elimination of any redundant mandatory features. It is also important to reduce the number of necessary test cases, for example, to reduce the number of the states of protocols, minimize the number of procedures, appropriate parameter range, and granularity.

The proposed E-UTRA/E-UTRAN requirements should minimize the complexity of the E-UTRA UE in terms of size, weight, and battery life (standby and active), which should be consistent with the provision of the advanced services of the E-UTRA/UTRAN. To satisfy these requirements, the following factors should be taken into account:

- UE complexity in terms of its capability to support multi-RAT (GERAN/UTRA/E-UTRA) should be considered when considering the complexity of E-UTRA features.

- The mandatory features should be kept to the minimum.

- There should be no redundant or duplicate specifications of mandatory features, for accomplishing the same task.

- The number of options should be minimized. Sets of options should be realizable in terms of separate distinct UE types/capabilities. Different UE types/capabilities should be used to capture different complexity versus performance trade-offs, for instance, for the impact of multiple antennae.

- The number of necessary test cases should be minimized so it is feasible to complete the development of the test cases within a reasonable time frame after the Core Specifications are completed.

10.4   E-UTRAN Study Items

The E-UTRAN WGs have dedicated normal meeting times to the Evolution activity, as well as separate Ad Hoc meetings. RAN WG1 held one of these Ad Hoc meetings on June 20–21, 2005 (3GPP TSG RAN WG1 Ad Hoc on UTRA/UTRAN LT evolution, held in Sophia Antipolis, France), where it started looking at, and evaluating new air-interface schemes. A set of six basic layer 1 or physical layer proposals were then agreed for further study, which included the following:

- FDD UL based on SC-FDMA, FDD DL based on OFDMA

- FDD UL based on OFDMA, FDD DL based on OFDMA

- FDD UL/DL based on MC-WCDMA

- TDD UL/DL based on MC-TD-SCDMA

- TDD UL/DL based on OFDMA

- TDD UL based on SC-FDMA, TDD DL based on OFDMA.

The evaluations of these technologies against the requirements for the physical layer are collected in TR25.814 [818].

The TSG RAN WG2 has also organized the first meeting to propose and discuss the air-interface protocols of the Evolved UTRAN [819]. Although the details of these are very dependent on the solutions chosen for the physical layer, some assumptions, and agreements have been taken, which are summarized as follows:

- Simplification of the protocol architecture and the actual protocols is necessary.

- There should be no dedicated channels, and so they form a simplified Medium Access Control (MAC) layer (without MAC-d entity).

- A debate over Radio Resource Control (RRC) was held. It is generally supported that it should be simplified and have less states. The location of its functions is open.

- Currently, there are very similar functions in the Radio Network and the Core. This should be simplified.

- Other open issues include: (1) Macro diversity.[1] (2) Security and ciphering; (3) Handover support; and (4) Measurements.

The TSG RAN WG3 (as shown in Figure 10.2, the fifth layer from the top in the second column from the left) is working closely with SA WG2 (as shown in Figure 10.2, the fourth layer from the top in the third column from the left) in the definition of the new E-UTRAN architecture. SA WG2 has started its own study for the System Architecture Evolution whose objective is to develop a framework for an evolution or migration of the 3GPP system to a higher-data-rate, lower-latency, and packet-optimized system that supports multiple RATs. The focus of this work will be on the PS domain with the assumption that voice services are supported in this domain.

This study builds on the RAN Long-Term Evolution and on the All-IP Network work carried out in SA WG1, and a long list of open points that needed clarification were identified, which include the items stated below.

First, how will we achieve mobility within the Evolved Access System? This issue is closely associated with the ways to overcome serious Doppler spread problems in a fast fading channel environment. As the allowed mobility supported in E-UTRAN will be higher than the 3GPP 3G system, this problem is very critical to the overall success of the E-UTRAN project.

Then, is the Evolved Access System envisioned to work on new and/or existing frequency bands? As 3GPP UTRAN is working in 2 GHz carrier frequency bands with its bandwidth being 5 MHz, the E-UTRAN may not be suitable for its operation in the same 2 GHz band as WCDMA is. The main reason is that E-UTRAN can work on a much wider bandwidth (up to 20 MHz), and the existing bandwidth allocation at 2 GHz is already very crowded. The real situation could be different from country to country. As an example, the US radio spectrum allocation situation can be seen from Figure 9.1 [792].

The more frequently discussed issues in SA WG1 include the following:

- Is connecting the Evolved RAN to the legacy PS core necessary?

- How do we add support for non-3GPP Access Systems (ASs)?

- WLAN 3GPP IP AS might need some new functionalities for Intersystem Mobility with the Evolved AS.

- Clarify which interfaces are the roaming interfaces, and how roaming works in general.

- The issues on inter-AS mobility should be discussed.

- Possible difference between PCC functionalities, mainly stemming from the difference in how Inter-AS mobility is provided.

- How do UEs discover ASs and corresponding radio cells? The options include autonomous per AS versus the UEs scans/monitors of any supported AS to discover systems and cells. Or, do ASs advertise other ASs to support UEs in discovering alternative ASs? How is such advertising performed (e.g. system broadcast, requested by UE, etc.)? How do these procedures impact battery lifetime?

---

[1]General agreement is that it should be avoided in the DL design.

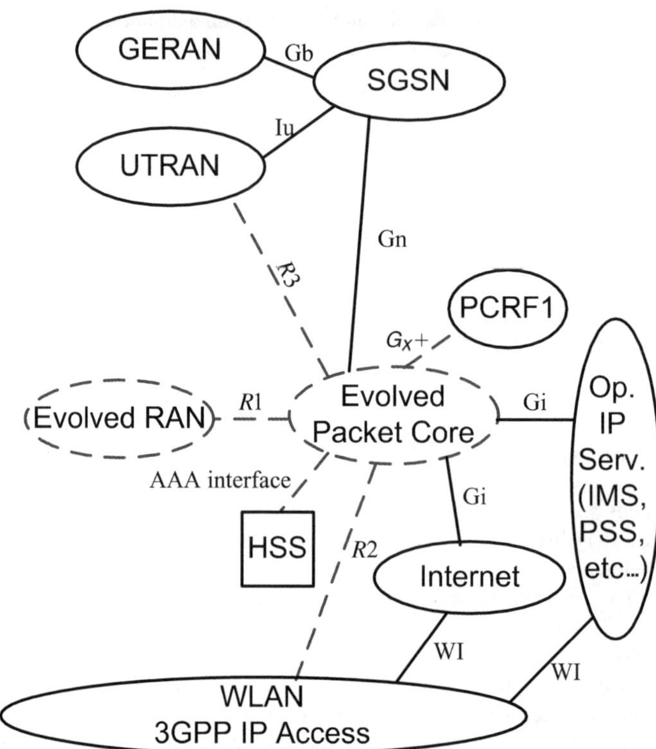

Figure 10.4 The E-UTRAN Model architecture B1 for non-roaming scenario, where $R1$, $R2$ and R are working names for reference points; G_x+ denotes evolved or extended G_x; PCRF1 represents evolved Policy and Charging Rules Function; the dash links and circles represent new functional elements/interfaces in E-UTRAN architecture.

- In the case of ASs advertising other ASs: will any AS provide seamless coverage (avoiding the loss of network/network search), or is a hierarchy of ASs needed to provide seamless coverage for continuous advertisement?

The two model architectures [820], which summarize the broad range of proposals that have been presented in several WG meetings, are shown in Figures 10.4 and 10.5. Note that the key difference in the two model E-UTRAN architectures lies in the way that intersystem mobility is achieved and managed, and thus the interactions among the E-UTRAN network and other 3GPP networks, such as UTRAN (based on WCDMA technology) and GERAN (based on GSM standard).

10.5 E-UTRAN TSG Work Plan

As mentioned earlier, the E-UTRAN standardization process is still going on. Only some very general technical aspects have been agreed upon in the TSG RAN meetings, and even for them the subsequent meetings can revise them from time to time. So far, a detailed work plan of the aforementioned Study Items has been made by 3GPP and can be summarized in terms of the milestones per TSG RAN

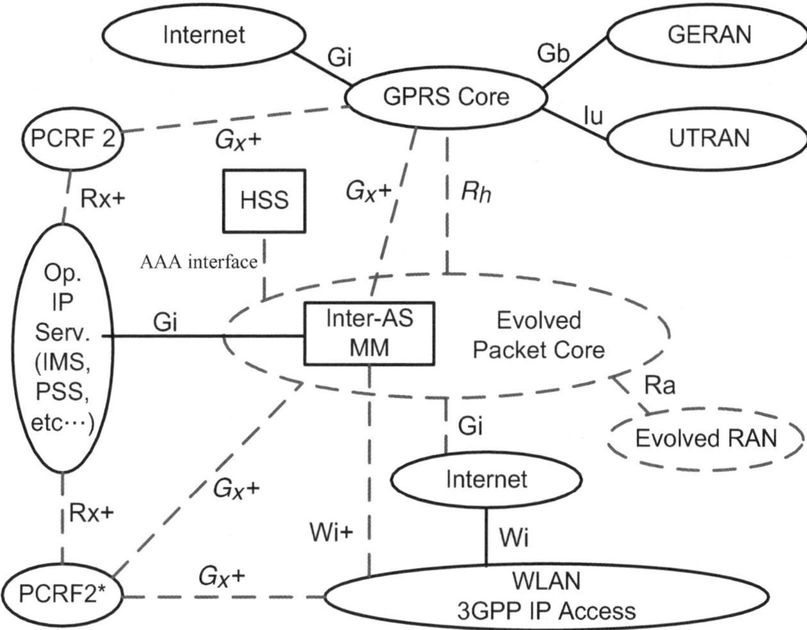

Figure 10.5 The E-UTRAN Model architecture B2, where R_h provides functionality to prepare handovers so that interruption time is reduced. It is intended that this interface should be generic enough to cope with other combinations of RATs, for which handover preparation is needed. G_x+ denotes G_x with added Inter-Access-System mobility support. W_x+ denotes W_x with added Inter-Access-System mobility support. Inter-AS MM denotes Inter-Access-System Mobility Management. PCRF2 elements are drawn twice only for figure topology reasons. PCRF2 represents the evolved Policy and Charging Rules Function. The dash links and circles represent new functional elements/interfaces in E-UTRAN architecture.

meetings. The work plan for SAE is included below by taking into account the time alignment between the LTE and SAE works.

TSG RAN #28 meeting (June 2005, Quebec)

- Revised Work plan;

- Requirement TR Approved: (1) Deployment Scenarios included; (2) Requirements on Migration Scenarios included.

TSG RAN #29 meeting (September 2005, Tallin)

Revised Work plan

TSG RAN #30 meeting (December 2005, MT)

- Revised Work plan;

- Physical Layer basics: (1) Multiple access scheme; (2) Macro diversity or not.

TSG RAN #31 meeting (March 8–10, 2006, China)

- More detailed L1 concepts to be used for evaluation, such as: (1) MIMO scheme; (2) Intercell interference mitigation scheme; (3) Scheduling and link adaptation principles; (4) Physical channel structure (including control signaling, reference signals).

- Simulation conditions and methodology for the system evaluation.

- RF Scenarios.

- Radio Interface Protocol Architecture: Functions of RRC, MAC, and so on.

- RAN Architecture including migration scenarios: (1) Radio interface protocol termination points: RRC, Outer ARQ termination points, and so on; (2) Security: User-plane and control-plane ciphering; Control-plane integrity protection.

- Core Network Architecture related to E-UTRA/UTRA/GSM: (1) Control-plane functional termination points; (2) User-plane functional termination points.

- Overall System Architecture: Nodes and interfaces related to E-UTRA/UTRA/GSM.

- States and state transitions: (1) Final state model; (2) State transition between E-UTRA and UTRA/GERA.

- Intra E-UTRA and E-UTRA-UTRA/GSM mobility in Active and Idle modes: (1) Mobility concept including measurements and signaling; (2) Interruption time node and interface budget.

- Service Requirements: (1) Are there any legacy service requirements that are obsolete? Or are they still very important? (2) Location Services.

- Legal intercept Requirements.

- Revised work plan.

TSG RAN #32 meeting (May 31–June 2, 2006, Poland)

- RAN TR25.912 ready for approval: (1) TR has its level of details at stage 2 and this is necessary for the smooth transition to Work Item phase; (2) The TR should include performance assessments, UE capabilities, and system and terminal complexities.

- Mobility between 3GPP and non-3GPP accesses.

- QoS concept.

- MBMS architecture.

- Documentation of overall system migration scenarios.

- Optimal routing and roaming including local breakout.

- Addressing/identification requirements and solutions.

- SA2 TR ready for approval: (1) Containing architecture diagram showing the main functional entities and interfaces; (2) Signaling flow diagram with delay estimations.

- Work Items created and their time plans agreed.

More information about the upcoming TSG RAN meetings for E-UTRAN architecture is shown in Figure 10.6.

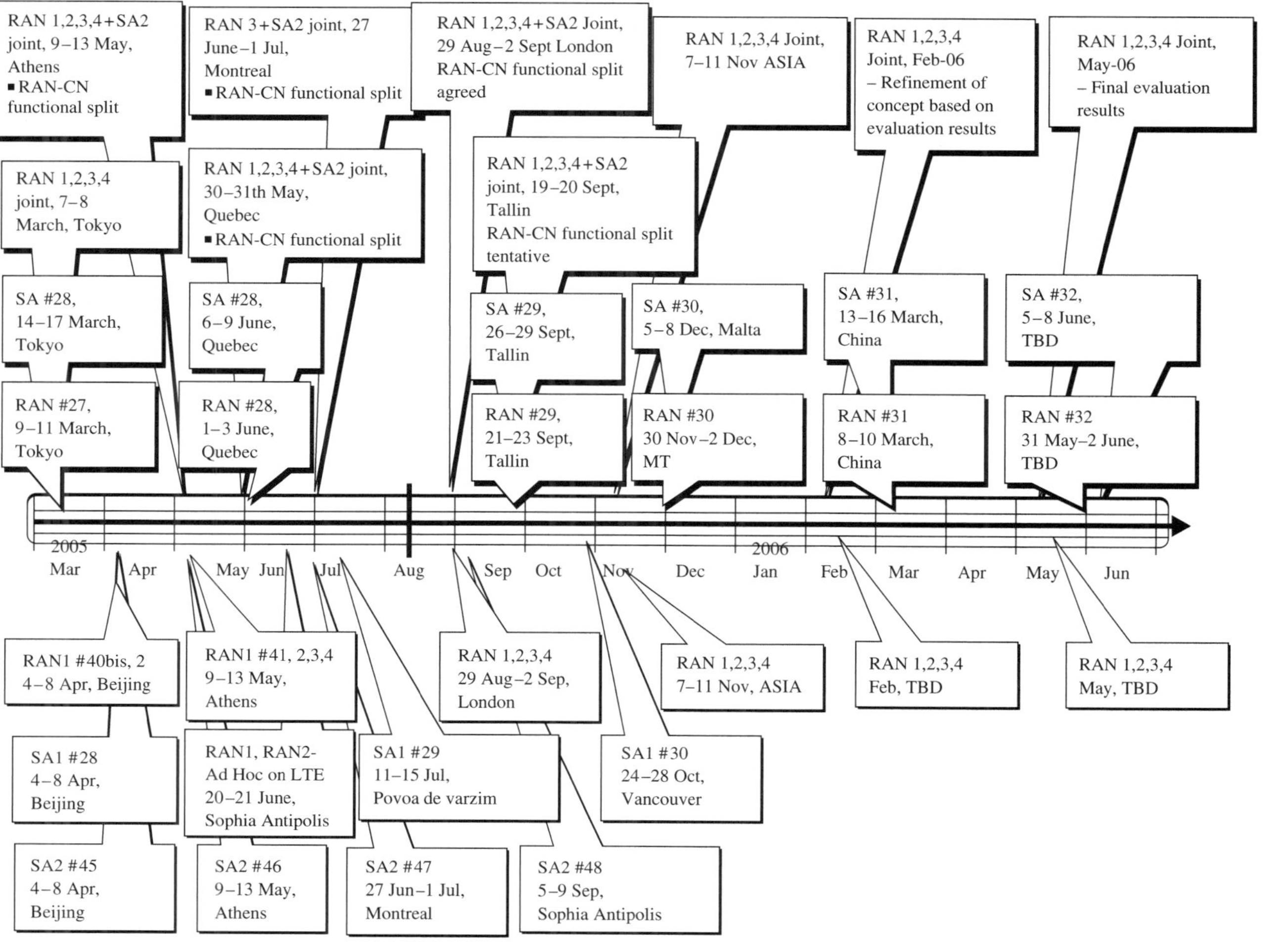

Figure 10.6 The tentative schedule for the upcoming 3GPP TSG RAN and SA LTE joint meetings schedule for the E-UTRAN architecture standard. Therefore, it is seen that the E-UTRAN standardization process has not yet been finalized.

10.6 E-UTRAN Radio Interface Protocols

This section is to describe the radio-interface protocol evolution for Evolved UTRA and Evolved UTRAN [821]. This activity involves the TSG RAN working group of the 3GPP studies for evolution and has impacts both on the UE and the Access Network of the 3GPP systems. It should be noted that the information provided in this section should not be considered as the final standard, but rather only the results from the discussions made in various 3GPP TSG WG meetings held previously up to the time when this book was written.

Before introducing the E-UTRAN Radio Interface Protocols, we would like to define various acronyms used in the discussions followed, as shown in Table 10.1.

10.6.1 E-UTRAN Protocol Architecture

The E-UTRAN protocol architecture bears a similar form as the one defined for the UTRAN. Two layered protocol stacks have been defined for the E-UTRAN, including the user-plane protocol stack and the control-plane protocol stack, as shown in Figures 10.7 and 10.8, respectively.

It is to be noted that in the E-UTRAN user-plane protocol stack, a MAC sublayer exists right above the physical layer (or Layer-1). The dashed line in Figure 10.7 means that the existence of a separate RLC layer is still open. The *Packet Data Convergence Protocol* (PDCP) will exist in the E-UTRAN protocol stack with its exact functionalities to be revisited in the future.

Table 10.1   Various acronyms used in the discussions on E-UTRAN Radio Interface Protocols

ARQ	Automatic Repeat Request
AS	Access Stratum
CN	Core Network
DL	Downlink
E-UTRAN	Evolved UMTS Terrestrial Radio Access Network
HARQ	Hybrid Automatic Repeat Request
HO	Handover
L1	Layer 1 (physical layer)
L2	Layer 2 (data link layer)
L3	Layer 3 (network layer)
MAC	Medium Access Control
NAS	Nonaccess Stratum
NW	Network
PDCP	Packet Data Convergence Protocol
PDU	Protocol Data Unit
RAN	Radio access network
RLC	Radio Link Control
RRC	Radio Resource Control
SDU	Service Data Unit
TCH	Traffic Channel
UE	User Equipment
UL	Uplink
UMTS	Universal Mobile Telecommunications System
UTRA	UMTS Terrestrial Radio Access
UTRAN	UMTS Terrestrial Radio Access Network

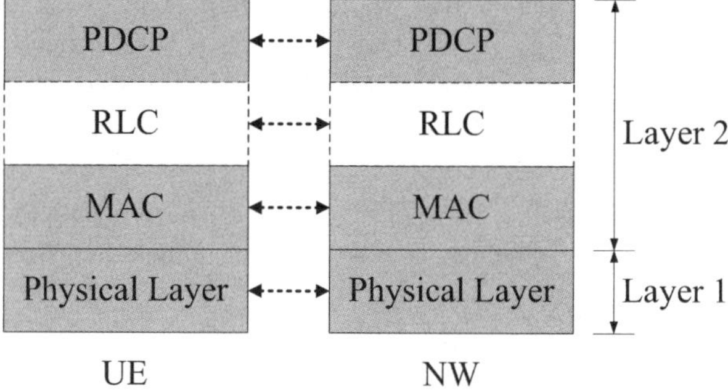

Figure 10.7 The E-UTRAN user-plane protocol stack.

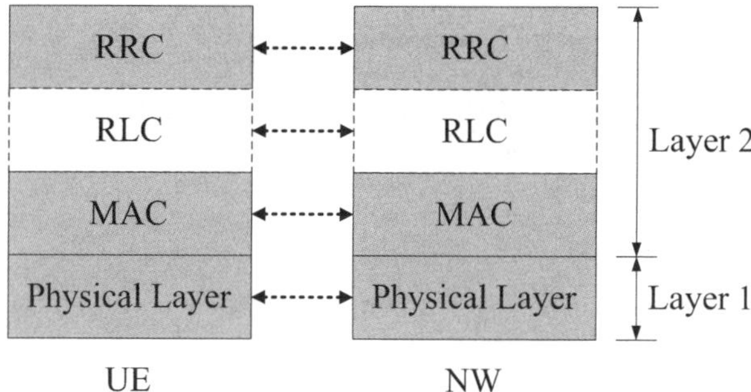

Figure 10.8 The E-UTRAN control-plane protocol stack.

On the other hand, the E-UTRAN control-plane protocol stack also has a MAC sublayer. Similar to the one in the user-plane, the existence of a separate RLC entity is still undetermined. A simplified RRC layer will be used in the E-UTRAN standard.

10.6.2 E-UTRAN Layer 1

Layer 1 in E-UTRAN is defined as exactly the same as what we often refer to in the Physical Layer. In this subsection, we will introduce the service, functions, and transport channels of Layer 1 (or the physical layer) in the E-UTRAN architecture.

The physical layer offers information transfer services to MAC and all other higher sublayers/layers, as shown in Figures 10.7 and 10.8. The physical layer transport services are described by how and with what characteristics data are transferred over the radio interface. An adequate term for this is "Transport Channel." It should be noted that, on the other hand, the classification of what is transported is what relates to the concept of logical channels at the MAC layer.

Downlink transport channels

There are three types of DL transport channels in total, which are explained as follows:

- Broadcast Channel (BCH): characterized by (1) low fixed bit rate; (2) requirement to be broadcast in the entire coverage area of the cell.

- Downlink Shared Channel (DL-SCH): characterized by: (1) the possibility of using HARQ; (2) the possibility of applying link adaptation by varying the modulation, coding, and transmit power; (3) the possibility to be broadcast in the entire cell; (4) the possibility to use beam-forming; (5) dynamic or semistatic resource allocation; (6) the possibility of channel-quality indication (CQI) reporting;[2] and (7) the support of UE power saving.[3]

- Paging Channel and Notification Channel (PCH and NCH): characterized by: (1) the support of UE power saving; (2) the requirement to be broadcast in the entire coverage area of the cell.

Uplink transport channels

There are two types of UL transport channels, which are explained as follows:

- Uplink Shared channel (UL-SCH):[4] characterized by: (1) the possibility to use beam-forming; (2) the possibility of applying link adaptation by varying the transmit power and potentially modulation and coding; (3) the possibility to use HARQ; (4) dynamic or semistatic resource allocation;[5] (5) the possibility of CQI reporting.[6]

- Random Access Channel(s) (RACH):[7] characterized by: (1) limited data field; (2) collision risk; and (3) the possibility of using HARQ.

10.6.3 E-UTRAN Layer 2

Layer 2 in the E-UTRAN protocol architecture consists of three sublayers, including the MAC sublayer, the RLC sublayer and the PDCP sublayer.

The MAC sublayer provides data transfer services on logical channels. A set of logical channel types is defined for different kinds of data transfer services as offered by MAC. Each logical channel type is defined by what type of information is transferred.

Some RR control (scheduling of user data, common channel transmissions, resource allocations, etc.) is also performed in MAC. MAC should: (1) be QoS aware; (2) assign resource blocks based on QoS attributes, buffer occupancy, and radio measurements; (3) include support of HARQ mechanism; (4) include segmentation/reassembly, if taken out of RLC and considered needed in L2. The possibility to cipher all flows in MAC exists.

A general classification of logical channels is divided into two groups: Control Channels (for the transfer of control-plane information) and Traffic Channels (for the transfer of user-plane information).

MAC Control Channels

Control channels are used for the transfer of control-plane information only. There are five different control channels offered by MAC.

- Broadcast Control Channel (BCCH): A DL channel for broadcasting system control information.

---

[2] Some new attributes should be discussed on whether there should be two types of DL-SCH.

[3] This function of DL-SCH is about an association with a physical layer signal, the Page Indicator, to support efficient sleep mode procedures.

[4] It is to be noted that the possibility of using UL synchronization and timing advance depends on the physical layer.

[5] Note: This is a new attribute for future studies on whether there should be two types of UL-SCH.

[6] It is also a study topic on whether a Random Access Channel is included. If yes, it will be characterized by the following attributes.

[7] The possibility to use open-loop power control depends on the physical layer solution.

- Paging Control Channel (PCCH) and Notification Control Channel (NCCH): A DL channel that transfers paging information (and notifications for MBMS). This channel is used when the network does not know the location cell of the UE.

- Common Control Channel (CCCH): This channel is used by the UEs having no RRC connection with the network.[8]

- Multicast Control Channel (MCCH):[9] a point-to-multipoint DL channel used for transmitting MBMS scheduling and control information from the network to the UE, for one, or several MTCHs. After establishing an RRC connection, this channel is only used by UEs that receive MBMS.[10]

- Dedicated Control Channel (DCCH): A point-to-point bidirectional channel that transmits dedicated control information between a UE and the network. Used by UEs having an RRC connection.

MAC Traffic Channels

Traffic channels are used for transferring user-plane information only. The traffic channels offered by MAC include:

- Dedicated Traffic Channel (DTCH): A DTCH is a point-to-point channel, dedicated to one UE, for the transfer of user information. A DTCH can exist in both UL and DL.

- Multicast Traffic Channel (MTCH): A point-to-multipoint DL channel for the transmission of traffic data from the network to the UE.

Mapping between logical channels and transport channels

Another important function in the E-UTRAN MAC sublayer is to perform the mapping between the logical channels and the transport channels.

The mapping in UL concerns the connections between logical channels and transport channels, to be explained as follows:

- CCCH can be mapped to RACH;[11]

- CCCH can be mapped to UL SCH;[12]

- DCCH can be mapped to RACH;

- DCCH can be mapped to UL SCH;

- DTCH can be mapped to UL SCH.

---

[8]This needs further study, depending on whether the access mechanism is contained in L1. If RACH is visible as a transport channel, CCCH will be used by the UEs when accessing a new cell or after cell reselection.

[9]This needs a study on whether it is distinct from CCCH.

[10]Note that the old version is MCCH + MSCH.

[11]This needs further study if the access procedure is not contained within L1.

[12]Further study is required to see if just a transient (random) ID is assigned for the resource request, if the actual RRC Connection Request message has still to contain a UE identifier and therefore such a message is considered to be a CCCH message, and even if it is transported on the UL_SCH. Also, the UE is not yet in a connected mode.

On the other hand, the mapping in DL concerns the connections between logical channels and transport channels, as explained below.

- BCCH can be mapped to BCH;

- PCCH can be mapped to PCH;[13]

- PCCH can be mapped to DL SCH;[14]

- CCCH can be mapped to DL SCH;

- DCCH can be mapped to DL SCH;

- DTCH can be mapped to DL SCH;

- MTCH can be mapped to DL SCH;[15]

- MTCH can be mapped to MCH;[16]

- MCCH can be mapped to DL SCH;[17]

- MCCH can be mapped to MCH.[18]

RLC sublayer and PDCP sublayer

The exact functionalities of the other two sublayers in the E-UTRAN Layer 2, RLC sublayer and PDCP sublayer, had not been determined at the time of writing this book.

Also, the proposals for the functions of the Layer 3 and many other detailed elements in the E-UTRAN protocol architecture will be collected and discussed in the subsequent 3GPP TSG RAN meetings scheduled in 2006, as shown in Figure 10.6.

10.7   E-UTRAN Physical Layer Aspects

As one of the most important part of the overall system architecture, the E-UTRAN physical layer aspects is discussed in this section [822]. The details of E-UTRAN physical layer aspects have been discussed in various 3GPP TSG RAN WG1 meetings and the discussions are continuing in the follow-up TSG RAN WG1 meetings, which are scheduled in 2006, as shown in Figure 10.6. Therefore, the information given in this section is only reflected from the proposals and discussions made before the time of writing this book.

Altogether six E-UTRAN Physical Layer proposals (all of which have claimed to satisfy the general technical features described in Section 10.3) have been discussed in 3GPP TR 25.814 [822], which include:

- FDD UL based on SC-FDMA, FDD DL based on OFDMA

- FDD UL based on OFDMA, FDD DL based on OFDMA

---

[13]Further study is needed to see if a separate PCH exists.

[14]Further study is needed to see if a separate PCH does not exist.

[15]Further study is needed to see if a separate MCH does not exist.

[16]Further study is needed to see if a separate MCH exists.

[17]Further study is needed to see if a separate MCCH exist.

[18]Further study is needed to see if a separate MCCH and MCH exist.

- FDD UL/DL based on MC-WCDMA

- TDD UL/DL based on MC-TD-SCDMA

- TDD UL/DL based on OFDMA

- TDD UL based on SC-FDMA, TDD DL based on OFDMA

which were proposed by various different parties (including vendors and service providers, etc.).

Because of the limited space, we introduce only one such proposed Physical Layer design scheme as an example, namely, the second scheme "FDD UL based on OFDMA, FDD DL based on OFDMA." However, it should be noted that E-UTRAN, similar to 3GPP UTRAN, will be designed based on either FDD or TDD operation modes, depending on the operational environment.

10.7.1   Downlink Aspects of FDD OFDMA

The DL design based on FDD OFDMA technology is one of the proposed DL physical layer architectures in 3GPP TSG RAN LTE WGs meetings. In this scheme, the DL transmission scheme is based on conventional OFDM using a cyclic prefix (CP), with a subcarrier spacing $\Delta f = 15$ kHz and a CP duration $T_{CP} \approx 4.7/16.7$ s (short / long CP).

Assuming that a 10 ms radio frame is divided into 20 equally sized subframes, this parameter set implies a subframe duration $T_{subframe} = 0.5\ ms$. The basic transmission parameters are then specified in more detail in Table 10.2. It may be noted that the information specified below is for the purpose of evaluation only.

It is noted that in the FDD OFDMA DL scheme, subcarrier spacing is constant regardless of the transmission bandwidth. To allow for operation in different spectrum allocation schemes, the transmission bandwidth can be varied by using different numbers of OFDM subcarriers. The need for supporting an additional longer cyclic-prefix duration, as shown in Table 10.2, may be necessary. The longer CP should then be more suitable for the applications in multicell broadcast and very-large-cell scenarios.

OFDM/OQAM modulation scheme

The FDD OFDMA[19] scheme should support two modulation schemes, one called the *basic modulation scheme* and the other called the *enhanced modulation scheme*. The DL basic modulation schemes include QPSK, 16QAM and 64QAM. It is also possible to use hierarchical modulation schemes for the purpose of broadcasting. The enhanced modulation scheme is referred to OFDM modulation with pulse shaping, namely, the OFDM/OQAM scheme.

Unlike conventional OFDM modulation, the OFDM/OQAM modulation does not require a guard interval (also called *CP*). For this purpose, the prototype function modulating each subcarrier must be accurately localized in the time domain, to limit the intersymbol interference for transmissions over multipath channels. This prototype function can also be accurately localized in the frequency domain, to limit the intercarrier interferences (due to Doppler effects, phase noise, etc.). This function must also guarantee orthogonality between subcarriers both in the time and frequency domains.

It is mathematically clear that when using complex valued symbols, the prototype functions guaranteeing perfect orthogonality at a critical sampling rate cannot be well localized both in time and frequency. For instance, the unity function used in conventional OFDM has weak frequency localization properties and we have to use a CP between the symbols to limit intersymbol interference.

---

[19]The principles of orthogonal frequency division multiple access (OFDMA) has been discussed in Section 7.5.4.

Table 10.2   Parameters for downlink transmission FDD OFDMA scheme

Transmission BW		1.25 MHz	2.5 MHz	5 MHz	10 MHz	15 MHz	20 MHz
Subframe duration		0.5 ms	0.5 ms	0.5 ms	0.5 ms	0.5 ms	0.5 ms
Subcarrier spacing		15 kHz	15 kHz	15 kHz	15 kHz	15 kHz	15 kHz
Sampling frequency		1.92 MHz (1/2 × 3.84 MHz)	3.84 MHz	7.68 MHz (2 × 3.84 MHz)	15.36 MHz (4 × 3.84 MHz)	23.04 MHz (6 × 3.84 MHz)	30.72 MHz (8 × 3.84 MHz)
FFT size		128	256	512	1024	1536	2048
Number of occupied subcarriers[a]		76	151	301	601	901	1201
Number of OFDM symbols per subframe (Short / Long CP)		7/6	7/6	7/6	7/6	7/6	7/6
CP length[b]	Short	(4.69/9) ×6, (5.21/10) ×1[c]	(4.69/18) ×5, (4.95/19) ×2	(4.69/36) ×3, (4.82/37) ×4	(4.75/73) ×6, (4.82/74) ×1	(4.73/109) ×2, (4.77/110) ×5	(4.75/146) ×5, (4.79/147) ×2
	Long	(16.67/32)	(16.67/64)	(16.67/128)	(16.67/256)	(16.67/384)	(16.67/512)

[a] This includes DC subcarrier which contains no data. This is the assumption for the baseline proposal. It may be possible for some more carriers to occupy a wider bandwidth.
[b] The unit of "CP length" is (µs/samples).
[c] $(x_1/y_1) \times n_1, (x_2/y_2) \times n_2$ means (x_1/y_1) for n_1 OFDM symbols and (x_2/y_2) for n_2 OFDM symbols.

To allow the use of accurately localized functions in the time-frequency domain, OFDM/OQAM scheme introduces a time offset between the real part and the imaginary part of the symbols. Orthogonality is then guaranteed only over real values. The corresponding multi-carrier modulation is an OFDM/OQAM. The OFDM/OQAM transmitted signal is expressed by

$$s(t) = \sum_n \sum_{m=0}^{M-1} a_{m,n} \underbrace{i^{m+n} e^{2i\pi m \nu_0 t} g(t - n\tau_0)}_{g_{m,n}(t)} \tag{10.1}$$

where $a_{m,n}$ denotes the real valued information (which can be the real part or the imaginary part of the offset complex QAM symbol) sent on the m-th subcarrier at the n-th symbol, M is the number of subcarriers, ν_0 is the intercarrier spacing, which is the same as the classical OFDM system. τ_0 is the OFDM/OQAM symbol duration, it is equal to $T_u/2$ (T_u is the OFDM symbol duration), and g is the prototype function.

It is important to note that the OFDM/OQAM symbol rate is twice the classical OFDM symbol rate without CP ($\tau_0 = N/2$); meanwhile, since the modulation used is a real one, the information amount sent by an OFDM/OQAM symbol is only half the information amount sent by an OFDM symbol. Figure 10.9 depicts the signal generation process of an OFDM/OQAM signal. The modulator generates N real valued symbols at each τ_0 where $\tau_0 = T_u/2$. The real valued symbols are then dephased, and are multiplied by i^{m+n} before the inverse fast Fourier transform (IFFT) as shown in Figure 10.9.

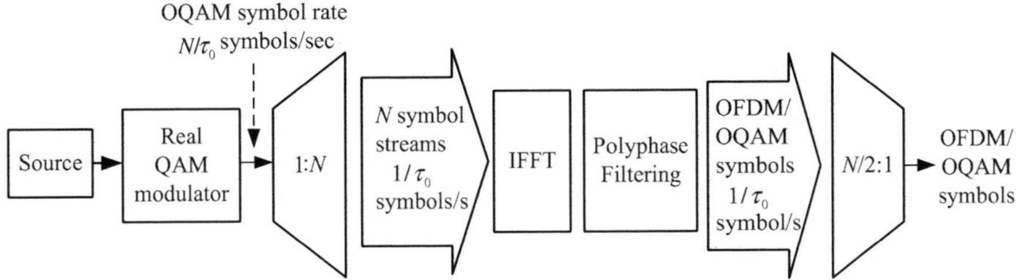

Figure 10.9 The OFDM/OQAM signal generation process for the FDD OFDMA downlink scheme in E-UTRAN architecture.

The main difference between OFDM/OQAM and conventional OFDM signal generation lies in the filtering by the prototype function g after the IFFT, instead of the CP addition.

Thanks to the IFFT, the prototype function g can be implemented in its polyphase form, which greatly reduces the complexity of the filtering. Moreover, the density 2 induces some more simplifications in the polyphase implementation. Figure 10.10 illustrates a possible polyphase implementation of both an OFDM/OQAM modulator and demodulator (G_i are the polyphase components of the prototype filter).

One possible candidate for the OFDM/OQAM filter (g) is the Isotropic Orthogonal Transform Algorithm (IOTA) prototype obtained by orthogonalizing the Gaussian function in both time and frequency domains according to the Gram-Schmidt algorithm. Another property of the IOTA is its spectrum. Thanks to its good frequency localization, the resulting spectrum is steeper than that generated from conventional OFDM.

A OFDM/OQAM transmitter (whose parameters are shown in Table 10.3) is very similar to the conventional OFDM transmitter, whose parameters have been listed in Table 10.2, with a subcarrier spacing $\Delta f = 15$ kHz. Assuming that a 10 ms radio frame is divided into 20 equally sized subframes, this parameter set implies a subframe duration $T_{subframe} = 0.5$ ms. As for conventional OFDM it may be noted that the numerology specified below are for the purposes of evaluation only. All remarks regarding the support of concatenated Transmission Time Interval (TTI) remain relevant.

Multiplexing and reference-signal structure

Both TDM and FDM are used in E-UTRAN FDD OFDMA DL design to map channel-coded, interleaved, and data-modulated information onto OFDM time/frequency symbols. The OFDM symbols can be organized into a number of resource blocks consisting of a number (M) of consecutive subcarriers for a number (N) of consecutive OFDM symbols. It should be possible to match the granularity of the resource allocation to the expected minimum payload. It also needs to take channel adaptation in the frequency domain into account.

The frequency and time allocations to map information for a certain UE to resource blocks are determined by the Node B scheduler and may depend on the frequency-selective CQI reported by the UE to the Node B. The channel coding rate and the modulation scheme (possibly different for different resource blocks) are also determined by the Node B scheduler and may also depend on the reported CQI based on a time/frequency-domain link adaptation algorithm.

In addition to block-wise transmission, transmission on nonconsecutive (scattered) subcarriers is also to be supported as a means of maximizing frequency diversity. Details of the multiplexing of lower-layer control signaling is still to be decided but may be based on time, frequency, and/or code multiplexing.

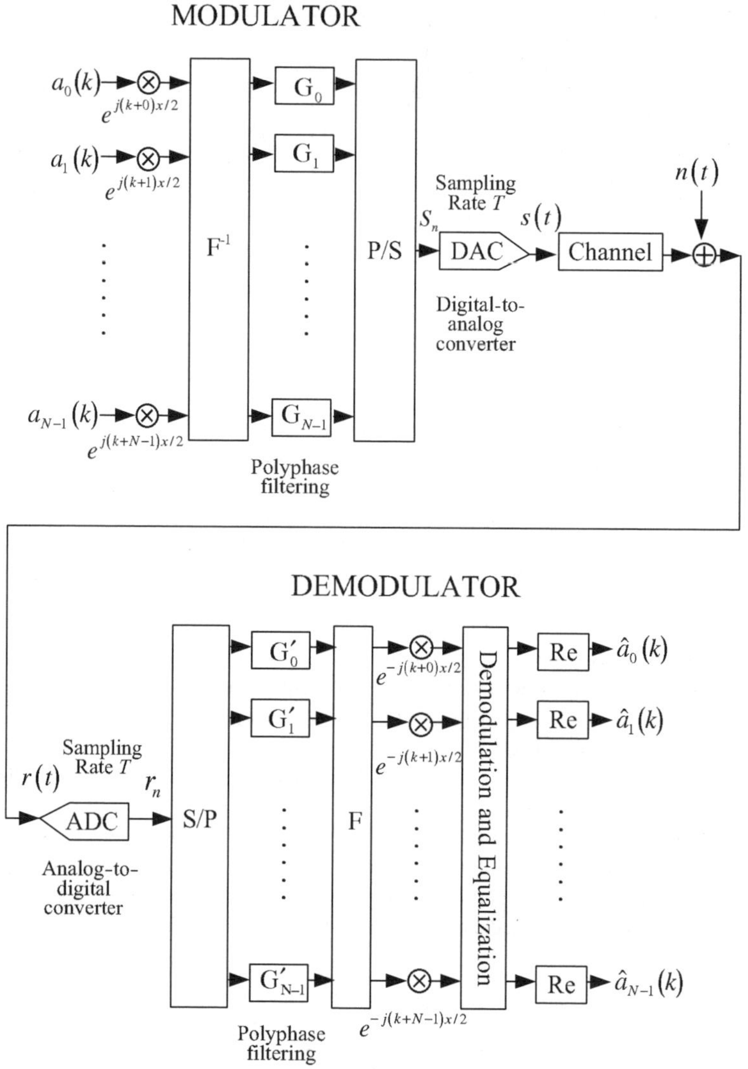

Figure 10.10 The OFDM/OQAM polyphase implementation for the FDD OFDMA downlink scheme in E-UTRAN architecture.

The functionalities of DL reference signal(s) can be summarized as follows: (1) DL-channel-quality measurements; (2) DL channel estimation for coherent demodulation/detection at the UE; and (3) Cell search and initial acquisition.

Reference symbols (also known as *"First reference symbols"*) are located in the first or second OFDM symbol of a subframe. Additional reference symbols (also known as *"Second reference symbols"*) may also be located in other OFDM symbols of a subframe. The position (in the frequency domain) of the reference symbols (the first as well as the second reference symbols) may vary from subframe to subframe. The first reference symbols are always transmitted from one or multiple Tx antennae. Currently, the issue on whether the "Second reference symbols" should be used is still open.

Table 10.3   OFDM/OQAM parameters for the downlink transmission scheme for E-UTRAN

Transmission BW	1.25 MHz	2.5 MHz	5 MHz	10 MHz	15 MHz	20 MHz
Subframe duration	0.5 ms	0.5 ms	0.5 ms	0.5 ms	0.5 ms	0.5 ms
Subcarrier spacing	15 kHz	15 kHz	15 kHz	15 kHz	15 kHz	15 kHz
Sampling frequency	1.92 MHz ($1/2 \times 3.84$ MHz)	3.84 MHz	7.68 MHz (2×3.84 MHz)	15.36 MHz (4×3.84 MHz)	23.04 MHz (6×3.84 MHz)	30.72 MHz (8×3.84 MHz)
FFT size	128	256	512	1024	1536	2048
Number of occupied subcarriers[a]	76	151	301	601	901	1201
Number of OQAM symbols per subframe	15[b]	15	15	15	15	15
CP length	0	0	0	0	0	0

a This includes the DC subcarrier which contains no data. This is the assumption for the baseline proposal. It may be possible for some more carriers to occupy a wider bandwidth.

b In OFDM/OQAM the symbol rate is twice higher than that of conventional OFDM (if no CP was included) and the amount of information transmitted per OFDM/OQAM symbol is only half the amount transmitted by one conventional OFDM symbol.

Channel coding scheme

On the channel coding scheme used in the OFDMA scheme, the current assumption for the study-item evaluations should be that channel coding for "normal" data is based on UTRA release 6 Turbo coding, possibly extended to lower rates by extension with additional code polynomials, extended longer code blocks, and modified by the removal of the tail. However, the use of alternative FEC encoding schemes could also be considered, especially if significant benefits in terms of complexity and/or performance can be shown. To achieve high processing gain, repetition coding can be used as a complement to FEC. Channel coding for lower-layer control signaling is the issue to be decided.

Downlink MIMO

The baseline antenna configuration for MIMO in E-UTRAN FDD OFDMA DL design is the use of two transmit antennae at the cell site and two receive antennae at the UE. The possibility for higher-order DL MIMO (more than two Tx / Rx antennae) should also be considered. Aspects to consider for the 3GPP LTE MIMO designs are given as follows: (1) Microcellular/Hot-spot and macrocellular environments should be considered in performance evaluation; (2) Not increasing the number of operation modes unnecessarily should be ensured. The impact on receiver architecture should also be considered; and (3) Realistic assumptions have to be taken into account when comparing different MIMO concepts, such as feedback errors and delays, which need to consider multiantenna reference signals overhead and its effect on performance, complexity, and signaling requirements, and so on. The resulting reference signal and signaling overheads in both UL and DL have to be justified by the shown improvements.

10.7.2   Uplink Aspects of FDD OFDMA

In this design scheme, proposed as a possible E-UTRAN UL architecture, the UL transmission scheme is based on conventional OFDM using a CP as described in Section 10.7.1. The basic transmission parameters such as subcarrier spacing, subframe duration and a CP duration are defined in Table 10.2 and are equally applicable to the UL. The need for longer CP durations is possible. It may be noted that the specified data shown in Table 10.2 is for the purpose of performance evaluation only. It is noted that the subcarrier spacing is constant, regardless of the transmission bandwidth. To allow for operation in different spectrum allocation schemes, the transmission bandwidth may vary in terms of different numbers of OFDM subcarriers.

Multiplexing and pilot structure

Two types of pilot symbols should be considered, including: (1) in band pilots, which are used for coherent data demodulation, for example, channel estimation. These pilots are transmitted in the part of the bandwidth used for data transmission; (2) out of band pilots, which are used for advanced frequency dependent scheduling and link adaptation. These pilots span a larger bandwidth than the one used for data transmission. Note that in-band pilots may also be used for frequency dependent scheduling and link adaptation.

It was suggested that orthogonal in-band pilot (IBP) symbol patterns are needed in the following cases: (1) If a UE transmits on two antennae (Antenna A and Antenna B) as in the case of MIMO or transmit diversity; and (2) If multiple UEs share the same time and frequency resource, and each of the UEs transmit on a single antenna, it is beneficial that UEs use orthogonal pilot patterns (this is described as virtual MIMO, a specific case of spatial division multiple access (SDMA)). The orthogonality of IBP symbol patterns can be achieved in the time and/or frequency domain.

Figure 10.11 shows an example of the IBP locations and overheads in the case that channel allocation to a UE in the time domain is done in multiples of seven symbols (a full subframe). The exact pilot locations and overhead are to be decided.

Figure 10.12 shows examples of the IBP location and overheads in the case that channel allocation in the time domain to a UE is done in multiples of six symbols, implying that the first symbol in a subframe may be used for other purposes (e.g. common control signaling). The exact pilot locations and overheads are not determined yet. In fact, Figure 10.12 exemplifies different cases of pilot pattern orthogonality.

Uplink MIMO

The baseline antenna configuration for UL single user MIMO is the use of two transmit antennae at the UE and two receiver antennae at the base station. The possibility for single user higher-order UL MIMO (more than two Tx / Rx antennae) should be considered. The possibility for SDMA should

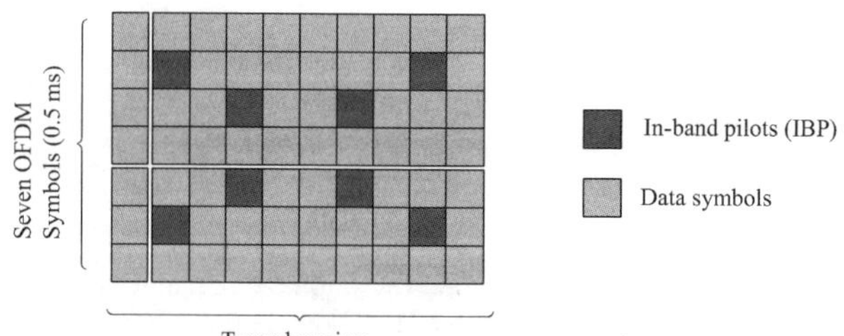

Figure 10.11 An example of the IBP locations and overheads in the case that channel allocation to a UE in the time domain is done in multiples of 7 symbols (a full subframe).

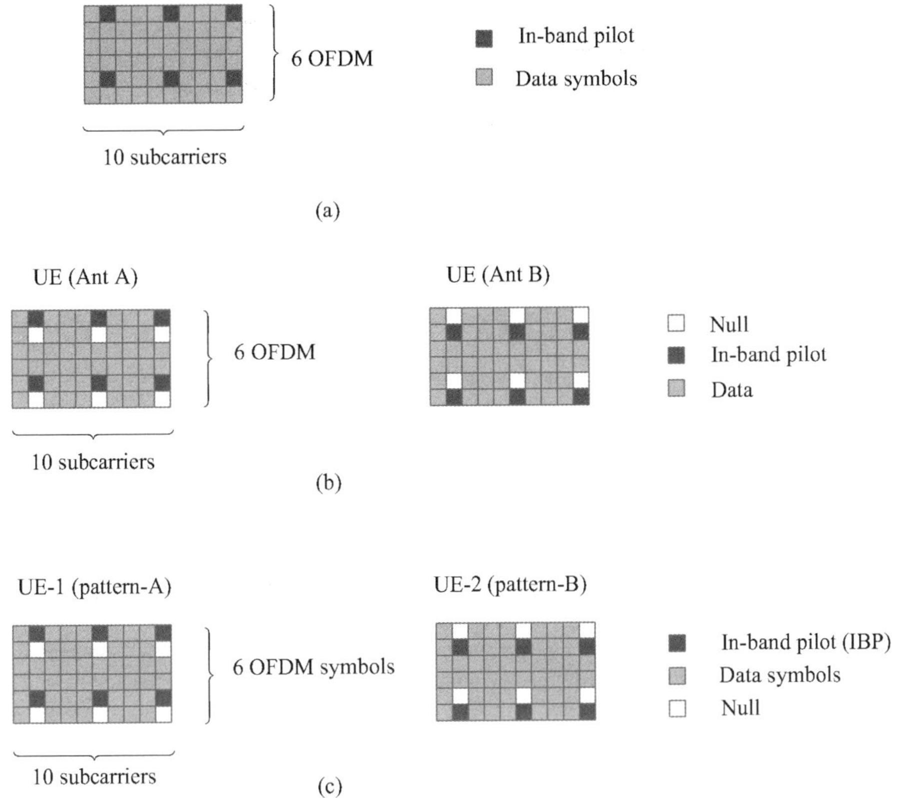

Figure 10.12 (a) Exemplifies the case of a single UE transmitting on a single antenna for which no orthogonal pilot is used. (b) Exemplifies the case of a UE transmitting on multiple antennae for which orthogonal pilot patterns are transmitted from multiple antennae. (c) Exemplifies the case of multiple UEs, each of which transmits on a single antenna, and shares the same time and frequency resource. Each UE transmits one orthogonal pilot pattern (virtual MIMO case).

also be considered. A specific example of SDMA corresponds to a (2×2) virtual MIMO, where two UEs, each of which transmits on a single antenna, and shares the same time and frequency resource allocation. These UEs apply mutually orthogonal pilot patterns in order to simplify cell site processing. Note that from the UE perspective, the difference between (2×2) virtual MIMO and single antenna transmission is only the use of a pilot pattern allowing for "pairing" with another UE.

PAPR reduction

The OFDMA-based UL transmission will lead to higher peak-to-average-power-ratio (PAPR) than the single carrier transmission schemes, the level of increase being dependent on the number of used subcarriers and/or presence of out of band pilots for the support of frequency based scheduling. However, several digital processing based PAPR reduction techniques can be employed to mitigate the higher PAPR for the OFDMA UL.

One approach to reduce the PAPR is the Tone Reservation (TR) method [823], in which both the transmitter and the receiver agree to reserve a subset of tones $\Re$ for generating PAPR reduction signals.

Assuming that a total of N available tones and K tones are reserved. Let $\mathbf{X}$ be frequency-domain data signal and $\mathbf{C} = [\mathbf{C}_0, \mathbf{C}_1, \ldots, \mathbf{C}_{K-1}]$ be a code on subset $\mathfrak{R}$. The goal of the TR method is to find the optimum code value $\mathbf{C}$ so that

$$\min_{\mathbf{C}}||\mathbf{x} + \mathbf{c}||_\infty = \min_{\mathbf{C}}||\mathbf{x} + \hat{\mathbf{Q}}\mathbf{C}||_\infty < ||\mathbf{x}||_\infty \qquad (10.2)$$

where $\mathbf{x}$ is the time domain signal of $\mathbf{X}$, $\hat{\mathbf{Q}}$ is a $N \times K$ submatrix of $\mathbf{Q}$, $\mathbf{Q}$ is the $N \times N$ inverse DFT matrix, and $||\mathbf{y}||_\infty$ is the ∞ norm of $\mathbf{y}$.

In the TR method, a simple gradient algorithm with fast convergence is proposed. The overall TR iterative algorithm can be defined as

$$\mathbf{x}^{i+1} = \mathbf{x}^i - \mu \cdot \sum_{\left|x_n^i\right| > A} \alpha_n^i \mathbf{p}_n \qquad (10.3)$$

where i is the iteration index variable, μ is the updating step size, and n is the index for which sample x_n is greater than the clipping threshold, which is defined as

$$\alpha_n^i = x_n^i - A \cdot \exp(j \cdot angle(x_n^i)) \qquad (10.4)$$

and $\mathbf{p}_n$ is called the *peak reduction kernel vector*. The kernel is a time domain signal that is as close as possible to the ideal impulse at the location where the sample amplitude is greater than the predefined threshold. This way the peak could be canceled as much as possible without generating secondary peaks. $\mathbf{p}_n$ is derived from original kernel $\mathbf{p}_0$ through the right circle shifting (by n-1 samples). The original kernel $\mathbf{p}_0$ can be calculated using 2-norm criteria and is given by the following formula:

$$\mathbf{p}_0 = \frac{\sqrt{N}}{K}\hat{\mathbf{Q}}\mathbf{1}_K \qquad (10.5)$$

where $\mathbf{1}_K$ is a vector of length K with all one elements.

In an example of the improved tone reservation with reduced complexity all tones except guard band [y] are used to calculate an original kernel. Then, α combined with μ is quantified to form derived reduction kernels. The phase is divided equally into s parts. The amplitude is divided into t parts represented by some special values according to different FFT sizes and step lengths. For example, if FFT size is 1024, the phase is divided equally into six parts represented by $\pm\pi/6$, $\pm\pi/2$, $\pm5\pi/6$ and the amplitude can be chosen among 0.01, 0.04, 0.08, 0.12, and 0.16. Thus, only 30 peak reduction kernels need to be stored.

In order to simplify the algorithm, we can only choose a fixed number of peaks to be canceled in one iteration instead of all the peaks that satisfies $\left|x_{n_i}\right| > A$. The steps of the improved TR method with reduced complexity is described as follows:

(A) Offline computation procedure:

- Calculate the original kernel vector $\mathbf{p}_0$ based on 2-norm criteria, which is the IFFT of $\mathbf{1}_K$ (all tones except the guard band);

- Quantify the original kernel to get derived kernels and store them in advance.

(B) Online computation procedure: [20]

- Select the target PAPR value and corresponding threshold A;

- Initially, set $\mathbf{x}^0 = \mathbf{x}$;

---

[20]This algorithm is based on each input OFDM symbol.

- Find a fixed number of samples (in order) with locations n_i in which $|x_{n_i}| > A$;

- If all the samples are below the target threshold, transmit $\mathbf{x}^i$. Otherwise, search among the derived kernels (stored in advance) to find matched ones according to Equation 10.4 and right circle shift them by n_i samples;

- Update $\mathbf{x}^i$ according to Equation 10.3;

- Repeat steps 3 to 5 until i reaches the maximum iteration limit. Transmit final $\mathbf{x}^i$.

10.8 Summary

Obviously, we can only see a very primitive skeleton of the E-UTRAN technology from the introduction covered in this chapter. At the time when this part of the book was being written, the work on 3GPP E-UTRAN architecture design was still going on in various TSG RAN and SA WGs meetings. Many new Working Group meetings have been scheduled in 2006 and later, as shown in 3GPP TSG RAN meeting schedule given in Figure 10.6. We have to wait for some more time before a final version of 3GPP E-UTRAN technical standards can be seen. However, one thing is for sure that the 3GPP Evolved UTRAN will definitely play an important role in the development of 4G technology in the world, as its predecessor, 3GPP UTRAN technology, has already been doing.

A

Orthogonal Complementary Codes (PG = 8 ∼ 512)

This Appendix presents the orthogonal complementary codes with their PG values from 8 to 512. It is noted that the commons "," are used to separate different element codes within a specific flock and blankets "()" are used to indicate flocks in an orthogonal complementary code set.

1. PG = 8

   Length of element codes = 4, flock size = 2

   (+ + + −, + − + +)

   (+ + − +, + − − −)

2. PG = 16

   Length of element codes = 8, flock size = 2

   (+ + + − + + − +, + − + + + − − −)

   (+ + + − − − + −, + − + + − + + +)

3. PG = 32

   Length of element codes = 16, flock size = 2

   (+ + + − + + − + + + + − − − + −, + − + + + − − − − + − + + − + + +)

   (+ + + − + + − + − − − + + + − +, + − + + + − − − − + − − + − − − −)

4. PG = 64

   (1) Length of element codes = 32, flock size = 2

   (+ + + − + + − + + + + − − − + − + + + − + + − + − − − + + + − +, + − + + + − − −
   + − + + − + + + + − + + + − − − − + − − + − − −)

   (+ + + − + + − + + + + − − − + − − − − − + − − + − + + + − − − + −, + − + + + − − −
   + − + + − + + + − + − − − + + + + − + + − + + +)

Next Generation Wireless Systems and Networks   Hsiao-Hwa Chen and Mohsen Guizani
© 2006 John Wiley & Sons, Ltd

(2) Length of element codes = 16, flock size = 4

(+ + + + + − + − + + − − + − − +, + − + − + + + + + − − + + + − −, + + − − + − −
+ + + + + + − + −, + − − + + + − − + − + − + + + +)

(+ + + + − + − + + + − − − + + −, + − + − − − − − + − − + − − + +, + + − − − + +
− + + + + − + − +, + − − + − − + + + − + − − − − −)

(+ + + + + − + − − − + + − + + −, + − + − + + + + − + + − − − − + +, + + − − − + − −
+ − − − − − + − +, + − − + + + − − − + − + − − − −)

(+ + + + − + − + − − + + + − − +, + − + − − − − − − + + − + + − −, + + − − − + + −
− − − − − + − + −, + − − + − − + + − + − + + + + +)

5. PG = 128

Length of element codes = 64, flock size = 2

(+ + + − + + + − + + + + + − − − + − + + + − + + − + − − − + + + − + + + + − + + − +
+ + + − − − + − − − − − + − − + − + + + − − − + −, + − + + + − − − + − + + − + + +
+ − + + + − − − − + − − + − − − + − + + + − − − + − + + − + + + − + − − − + + + +
− + + − + + +)

(+ + + − + + + − + + + + + − − − + − + + + − + + − + − − − + + + − + − − − + − − + −
− − − + + + − + + + + − + + − + − − − + + + − +, + − + + + − − − + − + + − + + +
+ − + + + − − − − + − − + − − − − + − − − + + + − + − − + − − − + − + + + − − − −
+ − − + − − −)

6. PG = 256

(1) Length of element codes = 128, flock size = 2

(+ + + − + + + − + + + + + − − − + − + + + − + + − + − − − + + + − + + + + − + + − +
+ + + − − − + − − − − − + − − + − + + + − − − + − + + + − + + − + + + + − − − + − +
+ + − + + − + − − − + + + − + − − − + − − + − − − − + + + − + + + + + − + + − + − −
− + + + − +, + − + + + + − − − + − + + − + + + + − + + + − − − − + − − + − − − + −
+ + + − − − + − + + − + + + − + − − − + + + + − + + − + + + + − + + + − − − + − +
+ − + + + + + − + + + − − − − + − − + − − − − − + − − − + + + − + − − + − − − + − + +
+ − − − − + − − + − − −)

(+ + + − + + + − + + + + + − − − + − + + + − + + − + − − − + + + − + + + + − + + − +
+ + + − − − + − − − − − + − − + − + + + − − − + − − − + − − − + − − + − − − − + + + − + −
− − + − − + − + + + − − − + − + + + − + + − + + + + − − − − + − − + − − + − + +
+ + − − − + −, + − + + + − − − + − + + − + + + + − + + + − − − + − − + − − − +
− + + + − − − + − + + − + + + − + − − − + + + + − + + − + + + − + − − − + + + − +
− − + − − − − − + − − − + + + + − + + − + + + + + + + − − − + − + + − + + + − + −
− − + + + + + − + + − + + +)

(2) Length of element codes = 64, flock size = 4

(+ + + + + − + − + + − − + − − + + + + + − + − + + + − − − + + − + + + + + + − + −
− − + + − + + − + + + + − + − + − − + + + − − +, + − + − + + + + + + − − + + + − −
+ − + − − − − − + − − + − − + + + − + + + + − + + − − − + + + − + − − − − − − −
+ + − + + − −, + + − − + − − + + + + + + + − + − + + − − + + − + + + + + − + − + +
+ − − + − − + − − − − − − + − + + + − − − + + − − − − − − + − + −, + − − + + + − − +
− + − + + + + + + − − + − − + + + − + − − − − − + − − + + + − − − + − + − − − − − +
− − + − − + + − + − + + + + +)

(+++++-+-++--+--+----+-+---+++--++++++-+-
--++-++------+-+-++---++-,+-+-++++++--+++--
-+-+++++++-++-++--+-+-++++-++---++-+-++++++
--+--++,++--+--+++++++-+---+++--+----+-+-+
+--+--+-----+-+--+++--++++++-+-+,+--+++--+
-+-+++++-++-++---+-+++++++--+++---+-+-----
++-++--+-+-----)

(+++++-+-++--+--+----+-+-+++--++-+++--++-+
++--+--+----+-+-++---++-,+-+-++++++--+++--
+-+------+-+--++-+-+----+--+++---+-++++++
--+--++,++--+--+++++++-+-++---++-++++-+-+-
-++-++-+++++-+--+++--++++++-+-+,+--+++--+
-+-+++++--+--+++-+-----++---+++-+-+++++-+
+-++--+-+-----)

(+++++-+-++--+--+----+-+---+++--+------+-+
++--+--+++++-+-+-+--+++--+,+-+-+++++--+++--
-+-+++++++-++-++---+-+----+--+++--+-+------
++-++--,++--+--+++++++-+---+++--+----+-+--
-++-++-+++++-+-++---++------+-+-,+--+++--+
-+-+++++-++-++---+-+++++-++---+++-+-+++++
--+--++-+-+++++)

7. PG = 512

(1) Length of element codes = 256, flock size = 2

(+++-++-++++---+-+++-++-+---+++-++++-++-+
+++---+----+--+-+++---+-+++-++-++++---+-
+++-++-+---+++-+---+--+----+++-++++-++-+
---+++-++++-++-++++---+-+++-+++-+---++++-+
+++-++-++++---+----+-+-+++--+----+--+--
--+++-+---+--+-+++--+-+++-++-+++-+-+-+--
--+-+-+++---+-,+-+++---+-++-++++-+++----
+--+---+-+++---+-++-+++-+---+++++-++-++++
-+++---+-++-++++-+++----+-+-+----+--+++
-+--+---+-+++---+--+---+-+++---+-++-+++
+-+++-----+--+---+-+++--+-++-+++-+---+++
+-++-+++-+---+++-+--+----+---++++-++-+++
+-++++---+-++-+++-+---++++-++-+++++)

(+++-++-++++---+-+++-++-+---+++-++++-++-+
+++---+----+--+-+++---+-+++-++-++++---+-+
++-++-+---+++-+---+--+----+++-+++++-++-+--
-+++-+---+-----+++-+---+--+-+++-+++-+---
+--+-----+++-++++-++-+---+++-++++-+-+-++++-
--+-+++-++-+---+++-+---+--+----+++-+++++-+
+-+---+++-+,+-+++---+-++-++++-+++----+--+
---+-++++---+-++-+++++++-+---+++++-++-++++
---+-++-+++++-+++---++--+---+----+++-+--+
---+-+++-++++-+++---+--+----+--+++-+--+
+++-++-+++-+---+++-+--+----+---+--+--

−+−+++−−−+−++−++++−+++−−−−+−−+−−−−−+−−−+++
−+−−+−−−+−+++−−−−+−−+−−−−)

(2) Length of element codes = 64, flock size = 8

(+++++++++−+−+−+−++−−++−−+−−++−−+++++−−−−
+−+−−+−+++−−−−+++−−+−++−,+−+−+−+−+++++++++
+−−++−−+++−−++−−+−+−−+−+++++−−−−+−−+−++−+
+−−−−++,++−−++−−+−−++−−+++++++++++−+−+−+−+
+−−−−+++−−+−++−++++−−−−+−+−−+−+,+−−++−−++
+−−++−−+−+−+−+++++++++−−+−++−++−−−−+++−
+−−+−++++−−−−,++++−−−−+−+−−+−+++−−−−+++−
−+−++−+++++++++++−+−+−+−++−−++−−+−−++−−+,+−
+−−+−+++++−−−−+−−+−++−++−−−+++−+−+−+−++
+++++++−−++−−+++−−++−−,++−−−−+++−−+−++−++
++−−−−+−+−−+−+++−−++−+−−++−−+++++++++++−+
−+−+−,+−−+−++−++−++−−−−+++−+−−+−+++++−−−−+−−
++−−+++−−++−−+−+−+−+−+++++++)

(+++++++++−+−+−+−+++−−++−−−++−−++−++++−−−−
−+−++−+−+++−−−−++−++−+−−+,+−+−+−+−−−−−−−−−−
+−−++−−+−−++−−+++−+−−+−+−−−++++++−+−++−
−−++++−−,++−−++−−−++−−++−+++++++++−+−+−+−+
++−−−−+++−++−+−−+++++−−−−−+−++−+−,+−−++−−
+−−++−−+++−+−+−+−−−−−−−−−−+−−+−++−−++++−
−+−+−−+−+−−−−++++,++++−−−−−+−++−+−++−−−−+
+−++−+−−+++++++++++−+−+−+++−−++−−−++−−++
−,+−+−−+−+−−−−+++++−−+−++−−−++++−−+−+−+−+
−−−−−−−−−+−−++−−+−−++−−++,++−−−−++−++−+−−
+++++−−−−−+−++−+−++−−++−+++++++++−+−+−+−+
+−+−+−+,+−−+−++−−−+++++−+−+−+−+−−−−+++
++−−++−−+−−++−−+++−+−+−+−−−−−−−−)

(+++++++++−+−+−+−−−++−−++−+++−++++−−−−
+−+−−+−+−++++−−−++−+−−+,+−+−+−+−++++++++
−++−−++−−−++−−+++−+−−+−+−−−−++++++−++−+−−+−
−++++−−,++−−++−−+−−++−−+−−−−−−−−−+−+−+−++
+−−−−+++−−+−++−−−−−++++−+−++−+−,+−−++−−+
++−−++−−−+−+−+−+−−−−−−−−−−+−−+−++−++−−−−++
−+−++−+−−−−−++++,++++−−−−+−+−−+−+−−++++−−
−++−+−−+++++++++++−+−+−+−−++−−++−++−−++−−,
+−+−−+−+++++−−−−−++−+−−+−−++++−−+−+−+−+−+
+++++++−++−−++−−−++−−++,++−−−−+++−−+−++−−
−−++++−+−++−+−++−−++−−+−−−−+−−−−−−−−−
+−+−+−+,+−−+−++−++−−−−++−+−++−+−−−−+++++
−−++−−+++−−++−−−+−+−+−+−−−−−−−−)

(+++++++++−+−+−+−+−−++−−+++−−+++−−+++++−−−−
−+−++−+−−−++++−−+−−+−++−,+−+−+−+−−−−−−−−−−
−++−−++−+++−−++−−+−+−−+−+−−−−++++−++−+−−+
++−−−−++,++−−++−−−++−−++−−−−−−−−−−+−+−+−+−
++−−−−++−++−+−−+−−−−++++−+−++−+−,+−−++−−+
++−−++−−−+−+−+−+−−−−−−−−−−+−−+−++−++−−−−++
−+−++−+−−−−−++++,++++−−−−−+−++−+−−++++−−
−−++−−++−+−+−+−++++++++++++−−+−++−−−++++−−
−+−++−+−++++−−−,++++−−−−−+−++−+−−−++++−−

```
+ - - + - + + - + + + + + + + + - + - + - + - + - - + + - - + + + - - + + - - +
+ - + - - + - + - - - - + + + + - + + - + - - + + + - - - - + + + - + - + - + -
- - - - - - - - + + - - + + - + + - - + + - -, + + - - - - + + - + + - + - - +
- - - - + + + + + - + - - + - + + + - - + + - - - + + - - + + - - - - - - - -
+ - + - + - + -, + - - + - + + - - - + + + + - - - + - + + - + - + + + + + - - -
+ - - + + - - + - - + + - - + + - + - + - + - + + + + + + + + +)

(+ + + + + + + + + - + - + - + - + + - - + + - - + - - + + - - + - - - - + + + +
- + - + + - + - - - + + + + - - - + + - + - - +, + - + - + - + - + + + + + + + +
+ - - + + - + + + - + + - - - + - + + - + - - - - + + + + - + + - + - - +
- - + + + + - -, + + - - + + - - + - - + + - + + + + + + + + + + + - + - + - + -
- - + + + + - - - + + - + - - + - - - - + + + - + - + + - + -, + - - + + - - +
+ + - - + + - + - + - + - + + + + + + + + - + + - + - - + - - + + + + - -
- + - + + - + - - - - - + + + +, + + + + - - - - + - + - - + - - + - + + + - - - + +
+ - - + - + + - - - - - - - - - - + - + - + - + - + + - - + + - + + - + + - - + + -
+ - + - - + - + + + + + - - - - + - + - + + - + + + + - - - - + + - + - + - + - + +
- - - - - - - - + + - - + + - - + + - - + +, + + - - - - + + + - - + - + + -
+ + + + - - - - + - + - - + - + - + + - + + - + + - + + - + + - - - - - - - -
- + - + - + - +, + - - + - + + - + + - - - - + + + - + - - + - + + + + + - - - -
- + + - - + + - - - + + - - + + - + - + - + - + - - - - - - - - -)

(+ + + + + + + + + - + - + - + - + + + - - + + - - - + + - - + + - - - - - + + + +
+ - + - - + - + - - + + + + - - + - - + - + + -, + - + - + - + - - - - - - - - -
+ - - + + - - + - + + - - + + - + - + + + + - - - - - + + - + - - +
+ + - - - - + + + + - - + + - - - + + - + + - + + + + + + + + + - + - + - + - +
- - + + + + - - + - - + - + + - - - - - + + + + + - + - - + - +, + - - + + - - +
- - + + - - + + + - + - + - + - - - - - - - - - - - + + - + - - + + + - - - - + +
- + - + + - + - + + + + - - - -, + + + + - - - - + - + + - + - + + - + + - - - - + +
- + + - + - - + - - - - - - - - + - + - + - + - - + + - + + + - - + + - - +
+ - - + - + - - - - - - - + + + + + - + - + + - - + + + + - - + - + - + - +
+ + + + + + + + + - + + - - + + - + + - + + - - + + - -, + + - - - + + - + + - + - - +
+ + + + - - - - + - + + - + - - + + - + + + - - + + - - + - - - - - - - -
+ - + - + - + - + - - + + - - + + - - + + + - - + - + - - + - + - - - + + + +
- + + - - + + - + + - - + + - - - + - + - + - + - + + + + + + + + +)

(+ + + + + + + + + - + - + - + - - - + + - - + + - + + - + + - - + + - - - - - + + + +
- + - + + - + - + + - - - + + + - - + - + + -, + - + - + - + - + + + + + + + +
- + + - - + + - - + + - - + + - + - + + - - - - + + + + + - - + - + + -
+ + - - - - + +, + + - - + + - - + - - + + - + - - - - - - - - - - + - + - + - +
- - + + + + - - + + - + - - + + + + + - - - + - + - - + - +, + - - + + - - +
+ + - - + + - - - + - + - + - - - - - - - - - + + - + - - + - - + + + + - -
+ - + - - + + + + + + - - -, + + + + - - - - + - + - - + - + - - + + + + - -
- + + - + - - + - - - - - - - - - + - + - + + + + - - + + - + - - + + - - +
+ - + - - + - + + + + + + - - - - - + + - + - - + - - - + + - - + + - - +
- - - - - - - + - - + + - - + + + - + + - - + + -, + + - - - + + + - - + - + + -
- - - - + + + + - + - + + - + - - - - + + - - + + - + + - - + + + + + + + + +
+ - + - + - +, + - - + - + + - + + - - - + + - + - + + - + - - - - - + + + +
- + + - - + + - - - + + - - + + + - + - + - + - + + + + + + + + +)

(+ + + + + + + + + - + - + - + - + - - + + - - + + + - - + + - - + - - - - + + + +
+ - + - - + - + + + - - - - + + - + + - + - - +, + - + - + - + - - - - - - - - -
-　 + + - - + + - + + - - + + - - + - + + - + - + + + + - - - - + - - + - + + -
- - + + + + - -, + + - - + + - - + + - - + + - - + + - - - - - - - - - - + - + - + - +
```

--++++--+--+-++-++++-----+-++-+-,+--++--+
--++--++-+-+-+-+++++++++++-++-+--+++----++
+-+--+-+----++++,++++-----+-++-+---++++--
+--+-++----------+-+-+-+-++--++---++--++-
+-+--+-+----++++-++-+--+++----++-+-+-+-+
+++++++++--++--+--++--++,++----++-++-+--+
----++++++-+--+-+--++--+++--++--+++++++++
-+-+-+-+,+--+-++---++++---+-++-+-++++-----
-++--++-++--++--+-+-+-+-+---------)

B

MAI in Asynchronous Flat Fading UWB Channel

Starting with Equation (7.17), we demonstrate how to derive multiple access interference (MAI) expression for a direct sequence-code division multiple access (DS-CDMA) ultra-wideband (UWB) system working in an asynchronous flat fading channel. Equation (7.17) is rewritten below as

$$
I_k = \alpha \left[b_{j-1}^{(k)} \int_0^{\tau_k} \sum_{n=0}^{N-1} a_n^{(k)} g(t - \tau_k + T_b - nT_c) \sum_{n=0}^{N-1} a_n^{(1)} g(t - nT_c) \, dt \right.
$$

$$
\left. + b_j^{(k)} \int_{\tau_k}^{T_b} \sum_{n=0}^{N-1} a_n^{(k)} g(t - \tau_k - nT_c) \sum_{n=0}^{N-1} a_n^{(1)} g(t - nT_c) \, dt \right] \tag{B.1}
$$

If $b_{j-1}^{(k)} = b_j^{(k)}$, I_k will be dependent on the sum of two partial aperiodic cross-correlation functions given in the above equation. On the other hand, if $b_{j-1}^{(k)} \neq b_j^{(k)}$, the value of I_k will depend on the subtraction between the two partial aperiodic cross-correlation functions given in Equation (B.1). The relative delay between the first and kth users' transmissions can be ordinarily expressed as $\tau_k = i_k T_c + \gamma_k T_c$ for $0 \leq i_k \leq N - 1$. Assume the signature codes for the first and kth users are $c^{(1)} = (a_0^{(1)}, a_1^{(1)}, a_1^{(1)}, \ldots, a_{N-1}^{(1)})$ and $c^{(k)} = (a_0^{(k)}, a_1^{(k)}, a_1^{(k)}, \ldots, a_{N-1}^{(k)})$. The integral in Equation (7.17) can be divided into two sections, one being from 0 to τ_k and the other being from τ_k to T_b, as shown in Figure B.1.

To facilitate the calculation of chipwise pulse autocorrelation function, the first integral section can also be divided into the following subsections as:

$$
\{0, \tau_k\} = \{0, r_k T_c\} + \{\gamma_k T_c, T_c\} + \{T_c, (1 + \gamma_k) T_c\} + \{(1 + \gamma_k) T_c, 2T_c\}
$$

$$
+ \cdots + \{i_k T_c, (i_k + \gamma_k) T_c\} \tag{B.2}
$$

Now, we can proceed with the integral subsection by subsection in the sequel.

Next Generation Wireless Systems and Networks   Hsiao-Hwa Chen and Mohsen Guizani

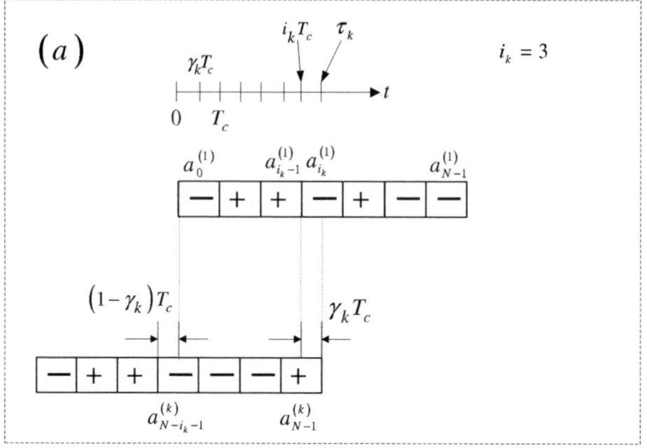

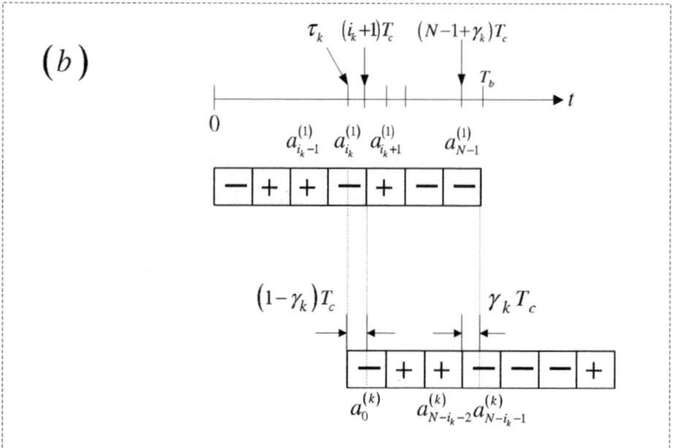

Figure B.1 This figure illustrates how to divide the integral given in Equation (7.17) into two sections, that is, (a) the integral section from 0 to τ_k; (b) the integral section from τ_k to T_b, where $i_k = 3$ and $N = 7$ are assumed.

For the subsection $\{0, \gamma_k T_c\}$, we have

$$\int_0^{\gamma_k T_c} \sum_{n=0}^{N-1} a_n^{(k)} g(t - \tau_k + T_b - nT_c) \sum_{n=0}^{N-1} a_n^{(1)} g(t - nT_c)\, dt$$

$$= \int_0^{\gamma_k T_c} a_{N-i_k-1}^{(k)} g(t - \tau_k + T_b - (N - i_k - 1)T_c) a_0^{(1)} g(t)\, dt$$

$$= \int_0^{\gamma_k T_c} a_{N-i_k-1}^{(k)} a_0^{(1)} g(t - \gamma_k T_c + T_c) g(t)\, dt \tag{B.3}$$

For the subsection $\{\gamma_k T_c, T_c\}$, we have

$$\int_{\gamma_k T_c}^{T_c} \sum_{n=0}^{N-1} a_n^{(k)} g(t - \tau_k + T_b - nT_c) \sum_{n=0}^{N-1} a_n^{(1)} g(t - nT_c)\, dt$$

$$= \int_{\gamma_k T_c}^{T_c} a_{N-i_k}^{(k)} g(t - \tau_k + T_b - (N - i_k)T_c) a_0^{(1)} g(t)\, dt$$

$$= \int_{\gamma_k T_c}^{T_c} a_{N-i_k}^{(k)} a_0^{(1)} g(t - \gamma_k T_c) g(t)\, dt \tag{B.4}$$

For the subsection $\{T_c, (1 + \gamma_k) T_c\}$, we have

$$\int_{T_c}^{(1+\gamma_k)T_c} \sum_{n=0}^{N-1} a_n^{(k)} g(t - \tau_k + T_b - nT_c) \sum_{n=0}^{N-1} a_n^{(1)} g(t - nT_c)\, dt$$

$$= \int_{T_c}^{(1+\gamma_k)T_c} a_{N-i_k}^{(k)} g(t - \tau_k + T_b - (N - i_k) T_c) a_1^{(1)} g(t - T_c)\, dt$$

$$= \int_{0}^{\gamma_k T_c} a_{N-i_k}^{(k)} a_1^{(1)} g(t - \gamma_k T_c + T_c) g(t)\, dt \tag{B.5}$$

For the subsection $\{(1 + \gamma_k) T_c, 2T_c\}$, we have

$$\int_{(1+\gamma_k)T_c}^{2T_c} \sum_{n=0}^{N-1} a_n^{(k)} g(t - \tau_k + T_b - nT_c) \sum_{n=0}^{N-1} a_n^{(1)} g(t - nT_c)\, dt$$

$$= \int_{(1+\gamma_k)T_c}^{2T_c} a_{N-i_k+1}^{(k)} g(t - \tau_k + T_b - (N - i_k + 1) T_c) a_1^{(1)} g(t - T_c)\, dt$$

$$= \int_{\gamma_k T_c}^{T_c} a_{N-i_k+1}^{(k)} a_1^{(1)} g(t - \gamma_k T_c) g(t)\, dt \tag{B.6}$$

and so on until we reach the last subsection $\{i_k T_c, (i_k + \gamma_k) T_c\}$, whose integral yields

$$\int_{i_k T_c}^{(i_k+\gamma_k)T_c} \sum_{n=0}^{N-1} a_n^{(k)} g(t - \tau_k + T_b - nT_c) \sum_{n=0}^{N-1} a_n^{(1)} g(t - nT_c)\, dt$$

$$= \int_{i_k T_c}^{(i_k+\gamma_k)T_c} a_{N-1}^{(k)} g\left(t - \tau_k + T_b - (N-1)T_c\right) a_{i_k}^{(1)} g\left(t - i_k T_c\right) dt$$

$$= \int_0^{\gamma_k T_c} a_{N-1}^{(k)} a_{i_k}^{(1)} g\left(t - \gamma_k T_c + T_c\right) g\left(t\right) dt \tag{B.7}$$

Therefore, we obtain the first integral over $\{0, \tau_k\}$ in Equation (7.17) or Equation (B.1) as

$$\int_0^{\tau_k} \sum_{n=0}^{N-1} a_n^{(k)} g(t - \tau_k + T_b - nT_c) \sum_{n=0}^{N-1} a_n^{(1)} g(t - nT_c)\, dt$$

$$= \left[a_0^{(1)} a_{N-i_k-1}^{(k)} + a_1^{(1)} a_{N-i_k}^{(k)} + a_2^{(1)} a_{N-i_k+1}^{(k)} + \cdots + a_{i_k}^{(1)} a_{N-1}^{(k)} \right]$$

$$\times \int_0^{\gamma_k T_c} g(t)g(t + (1 - \gamma_k)T_c)\, dt$$

$$+ \left[a_0^{(1)} a_{N-i_k}^{(k)} + a_1^{(1)} a_{N-i_k+1}^{(k)} + a_2^{(1)} a_{N-i_k+2}^{(k)} + \cdots + a_{i_k-1}^{(1)} a_{N-1}^{(k)} \right]$$

$$\times \int_{\gamma_k T_c}^{T_c} g(t)g(t - \gamma_k T_c)\, dt$$

$$= C_{k,1}(i_k - N + 1) \int_0^{\gamma_k T_c} g(t)\, g(t + (1 - \gamma_k)T_c)\, dt$$

$$+ C_{k,1}(i_k - N) \int_{\gamma_k T_c}^{T_c} g(t)g(t - \gamma_k T_c)\, dt \tag{B.8}$$

Similarly, the second integral section over $\{\tau_k, T_b\}$ can also be divided into the following subsections:

$$\{\tau_k, T_b\} = \{(i_k + \gamma_k)T_c, (i_k + 1)T_c\}$$

$$+ \{(i_k + 1)T_c, (i_k + 1 + \gamma_k)T_c\} + \{(i_k + 1 + \gamma_k)T_c, (i_k + 2)T_c\}$$

$$+ \cdots + \{(N - 1)T_c, (N - 1 + \gamma_k)T_c\} + \{(N - 1 + \gamma_k)T_c, NT_c\} \tag{B.9}$$

For the subsection $\{(i_k + \gamma_k)T_c, (i_k + 1)T_c\}$, we have

$$\int_{(i_k+\gamma_k)T_c}^{(i_k+1)T_c} \sum_{n=0}^{N-1} a_n^{(k)} g(t - \tau_k - nT_c) \sum_{n=0}^{N-1} a_n^{(1)} g(t - nT_c)\, dt$$

$$= \int_{(i_k+\gamma_k)T_c}^{(i_k+1)T_c} a_0^{(k)} g\,(t - \tau_k)\, a_{i_k}^{(1)} g\,(t - i_k T_c)\, dt$$

$$= \int_{\gamma_k T_c}^{T_c} a_0^{(k)} a_{i_k}^{(1)} g\,(t - \gamma_k T_c)\, g\,(t)\, dt \tag{B.10}$$

For the subsection $\{(i_k + 1)\, T_c,\, (i_k + 1 + \gamma_k)\, T_c\}$, we have

$$\int_{(i_k+1)T_c}^{(i_k+1+\gamma_k)T_c} \sum_{n=0}^{N-1} a_n^{(k)} g\,(t - \tau_k - nT_c) \sum_{n=0}^{N-1} a_n^{(1)} g\,(t - nT_c)\, dt$$

$$= \int_{(i_k+1)T_c}^{(i_k+1+\gamma_k)T_c} a_0^{(k)} g\,(t - \tau_k)\, a_{i_k+1}^{(1)} g\,(t - (i_k + 1)\, T_c)\, dt$$

$$= \int_{0}^{\gamma_k T_c} a_0^{(k)} a_{i_k+1}^{(1)} g\,(t - \gamma_k T_c + T_c)\, g\,(t)\, dt \tag{B.11}$$

For the subsection $\{(i_k + \gamma_k + 1)\, T_c,\, (i_k + 2)\, T_c\}$, we have

$$\int_{(i_k+\gamma_k+1)T_c}^{(i_k+2)T_c} \sum_{n=0}^{N-1} a_n^{(k)} g\,(t - \tau_k - nT_c) \sum_{n=0}^{N-1} a_n^{(1)} g\,(t - nT_c)\, dt$$

$$= \int_{(i_k+\gamma_k+1)T_c}^{(i_k+2)T_c} a_1^{(k)} g\,(t - \tau_k - T_c)\, a_{i_k+1}^{(1)} g\,(t - (i_k + 1)\, T_c)\, dt$$

$$= \int_{\gamma_k T_c}^{T_c} a_1^{(k)} a_{i_k+1}^{(1)} g\,(t - \gamma_k T_c)\, g\,(t)\, dt \tag{B.12}$$

The same process continues until we reach the last subsection $\{(N - 1 + \gamma_k)\, T_c,\, NT_c\}$ to yield

$$\int_{(N-1+\gamma_k)T_c}^{NT_c} \sum_{n=0}^{N-1} a_n^{(k)} g\,(t - \tau_k - nT_c) \sum_{n=0}^{N-1} a_n^{(1)} g\,(t - nT_c)\, dt$$

$$= \int_{(N-1+\gamma_k)T_c}^{NT_c} a_{N-1-i_k}^{(k)} g\,(t - \tau_k - (N - 1 - i_k)\, T_c)\, a_{N-1}^{(1)} g\,(t - (N - 1)\, T_c)\, dt$$

$$= \int_{\gamma_k T_c}^{T_c} a_{N-1-i_k}^{(k)} a_{N-1}^{(1)} g\left(t - \gamma_k T_c\right) g\left(t\right) dt \tag{B.13}$$

Thus, the integral over the section $\{\tau_k, T_b\}$ becomes

$$\int_{\tau_k}^{T_b} \sum_{n=0}^{N-1} a_n^{(k)} g\left(t - \tau_k - nT_c\right) \sum_{n=0}^{N-1} a_n^{(1)} g\left(t - nT_c\right) dt$$

$$= \left[a_{i_k}^{(1)} a_0^{(k)} + a_{i_k+1}^{(1)} a_1^{(k)} + \cdots + a_{N-1}^{(1)} a_{N-i_k-1}^{(k)} \right] \int_{\gamma_k T_c}^{T_c} g\left(t\right) g\left(t - \gamma_k T_c\right) dt$$

$$+ \left[a_{i_k+1}^{(1)} a_0^{(k)} + a_{i_k+2}^{(1)} a_1^{(k)} + \cdots + a_{N-1}^{(1)} a_{N-i_k-2}^{(k)} \right] \int_0^{\gamma_k T_c} g\left(t\right) g\left(t + (1 - \gamma_k) T_c\right) dt$$

$$= C_{k,1}\left(i_k\right) \int_{\gamma_k T_c}^{T_c} g\left(t\right) g\left(t - \gamma_k T_c\right) dt + C_{k,1}\left(i_k + 1\right) \int_0^{\gamma_k T_c} g\left(t\right) g\left(t + (1 - \gamma_k) T_c\right) dt \tag{B.14}$$

Finally, we obtain the kth MAI term with respect to the first user as

$$I_k = \int_{jT_b}^{(j+1)T_b} \sum_{j=-\infty}^{\infty} \sum_{n=0}^{N-1} \alpha b_j^{(k)} a_n^{(k)} g\left(t - \tau_k - jT_b - nT_c\right)$$

$$\times \sum_{n=0}^{N-1} a_n^{(1)} g(t - jT_b - nT_c)\, dt$$

$$= \alpha \Bigg\{ b_{j-1}^{(k)} \Bigg[C_{k,1}\left(i_k - N + 1\right) \int_0^{\gamma_k T_c} g\left(t\right) g\left(t + (1 - \gamma_k) T_c\right) dt$$

$$+ C_{k,1}\left(i_k - N\right) \int_{\gamma_k T_c}^{T_c} g\left(t\right) g\left(t - \gamma_k T_c\right) dt \Bigg]$$

$$+ b_j^{(k)} \Bigg[C_{k,1}(i_k) \int_{\gamma_k T_c}^{T_c} g(t) g(t - \gamma_k T_c)\, dt$$

$$+ C_{k,1}(i_k + 1) \int_0^{\gamma_k T_c} g(t) g(t + (1 - \gamma_k) T_c)\, dt \Bigg] \Bigg\} \tag{B.15}$$

which just gives Equation (7.18).

C

MI in Asynchronous Modified S-V UWB Channel

Starting with

$$I_{L,1} = \sum_{l=2}^{L_1} I_{1,l,1} + \sum_{q=2}^{Q_1} \sum_{l=1}^{L_1} I_{q,l,1} \tag{C.1}$$

we can derive the multipath interference (MI) component caused by the lth ray in the first cluster of the first user as

$$I_{1,l,1} = \int_{jT_b}^{(j+1)T_b} \beta_1 w_{1,l} \alpha_{1,l} \sum_{j=-\infty}^{\infty} \sum_{n=0}^{N-1} b_j^{(1)} a_n^{(1)} g(t - \tau_{1,l} - jT_b - nT_c)$$

$$\times \sum_{n=0}^{N-1} a_n^{(1)} g(t - jT_b - nT_c) \, dt \tag{C.2}$$

It is assumed that the relative delay between the first and lth ray, $\tau_{1,l}$, is uniformly distributed over $(0, T_b)$. b_{j-1}^1 and b_j^1 are two consecutive bits from the first user. From Equation (7.83) we can obtain

$$I_{1,l,1} = \int_{jT_b}^{(j+1)T_b} \beta_1 w_{1,l} \alpha_{1,l} \sum_{j=-\infty}^{\infty} \sum_{n=0}^{N-1} b_j^{(1)} a_n^{(1)} g(t - \tau_{1,l} - jT_b - nT_c)$$

$$\times \sum_{n=0}^{N-1} a_n^{(1)} g(t - jT_b - nT_c) \, dt$$

$$= \beta_1 w_{1,l} \alpha_{1,l} \Bigg[b_{j-1}^{(1)} \int_0^{\tau_k} \sum_{n=0}^{N-1} a_n^{(1)} g(t - \tau_{1,l} + T_b - nT_c) \sum_{n=0}^{N-1} a_n^{(1)} g(t - nT_c) \, dt$$

$$+ b_j^{(1)} \int_{\tau_k}^{T_b} \sum_{n=0}^{N-1} a_n^{(1)} g(t - \tau_{1,l} - nT_c) \sum_{n=0}^{N-1} a_n^{(1)} g(t - nT_c) \, dt \Bigg] \tag{C.3}$$

Next Generation Wireless Systems and Networks   Hsiao-Hwa Chen and Mohsen Guizani
© 2006 John Wiley & Sons, Ltd

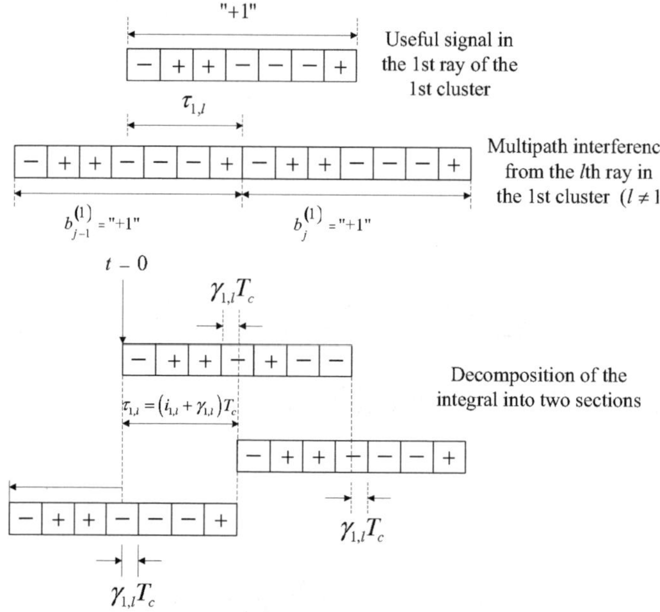

Figure C.1 This figure illustrates how the multipath interference between the first and lth rays in the same cluster (the first cluster in this case) is related to two partial aperiodic auto-correlation functions, where the relative delay between the first and lth rays can be any real value denoted by $\tau_{1,l} = (i_{1,l} + \gamma_{1,l})T_c$.

which shows that the MI component is related to the two partial auto-correlation functions, as shown in Figure C.1. Clearly, $I_{1,l,1}$ depends on the values of two consecutive bits appearing in the lth ray of the first cluster for the first user. If $b_{j-1}^{(1)} = b_j^{(1)}$, the calculation of $I_{1,l,1}$ can be simplified into the calculation of a periodic auto-correlation function; otherwise, $I_{1,l,1}$ depends on the difference between two partial aperiodic auto-correlation functions. The relative delay between the first and lth rays in the first cluster can be written as $\tau_{1,l} = i_{1,l}T_c + \gamma_{1,l}T_c$ for $0 \leq i_{1,l} \leq N - 1$.

The discrete partial aperiodic auto-correlation function for the spread code $c^{(1)} = \{a_n^{(1)}\}_{n=0}^{N-1}$ is defined by

$$
C_1(i) = \begin{cases}
\displaystyle\sum_{j=0}^{N-1-i} a_j^{(1)} a_{j+i}^{(1)}, & 0 \leq i \leq N - 1 \\[4mm]
\displaystyle\sum_{j=0}^{N-1+i} a_{j-i}^{(1)} a_j^{(1)}, & -(N - 1) \leq i \leq 0 \\[4mm]
0, & \text{otherwise}
\end{cases}
\tag{C.4}
$$

The two partial aperiodic auto-correlation functions in Equation (C.3) can be written as

$$
\int_0^{\tau_{1,l}} \sum_{n=0}^{N-1} a_n^{(1)} g\left(t - \tau_{1,l} + T_b - nT_c\right) \sum_{n=0}^{N-1} a_n^{(1)} g\left(t - nT_c\right) dt
$$

$$
= C_1\left(i_{1,l} - N + 1\right) R_p\left(-\left(1 - \gamma_{1,l}\right) T_c\right) + C_1\left(i_{1,l} - N\right) R_p\left(\gamma_{1,l} T_c\right)
\tag{C.5}
$$

$$\int_{\tau_{1,l}}^{T_b} \sum_{n=0}^{N-1} a_n^{(1)} g\left(t - \tau_{1,l} - nT_c\right) \sum_{n=0}^{N-1} a_n^{(1)} g\left(t - nT_c\right) dt$$

$$= C_1\left(i_{1,l}\right) R_p\left(\gamma_{1,l} T_c\right) + C_1\left(i_{1,l} + 1\right) R_p\left(-\left(1 - \gamma_{1,l}\right) T_c\right) \tag{C.6}$$

Thus, the term $I_{1,l,1}$ can be rewritten as

$$I_{1,l,1} = \beta_1 w_{1,l} \alpha_{1,l} \left\{ b_{j-1}^{(1)} \left[C_1\left(i_{1,l} - N + 1\right) R_p\left(-\left(1 - \gamma_{1,l}\right) T_c\right) \right. \right.$$

$$+ C_1(i_{1,l} - N) R_p(\gamma_{1,l} T_c) \big]$$

$$+ b_j^{(1)} \left[C_1\left(i_{1,l}\right) R_p\left(\gamma_{1,l} T_c\right) + C_1\left(i_{1,l} + 1\right) R_p\left(-\left(1 - \gamma_{1,l}\right) T_c\right) \right] \right\} \tag{C.7}$$

The MI component caused by the lth ray in the qth cluster with respect to the first ray in the first cluster can be expressed by

$$I_{q,l,1} = \int_{jT_b}^{(j+1)T_b} \beta_1 w_{q,l} \alpha_{q,l} \sum_{j=-\infty}^{\infty} \sum_{n=0}^{N-1} b_j^{(1)} a_n^{(1)} g\left(t - T_q - \tau_{q,l} - jT_b - nT_c\right)$$

$$\times \sum_{n=0}^{N-1} a_n^{(1)} g\left(t - jT_b - nT_c\right) dt \tag{C.8}$$

For noninteger chip relative delay, we always have

$$T_q + \tau_{q,l} = i_{q,l} T_c + \gamma_{q,l} T_c, \qquad 0 \le i_{1,l} \le N - 1 \tag{C.9}$$

where $i_{q,l}$ is the integer chip relative delay between the lth and first rays in the qth cluster, $\gamma_{q,l}$ is the fractional chip delay between the lth and first rays in the qth cluster. Therefore, Equation (C.8) can be written as

$$I_{q,l,1} = \beta_1 w_{q,l} \alpha_{q,l} \left\{ b_{j-1}^{(1)} \left[C_1\left(i_{q,l} - N + 1\right) R_p\left(-\left(1 - \gamma_{q,l}\right) T_c\right) \right. \right.$$

$$+ C_1(i_{q,l} - N) R_p(\gamma_{q,l} T_c) \big]$$

$$+ b_j^{(1)} \left[C_1\left(i_{q,l}\right) R_p\left(\gamma_{q,l} T_c\right) + C_1\left(i_{q,l} + 1\right) R_p\left(-\left(1 - \gamma_{q,l}\right) T_c\right) \right] \right\} \tag{C.10}$$

Inserting Equation (C.7) and Equation (C.10) into Equation (C.1), we obtain

$$I_{L,1} = \sum_{l=2}^{L_1} \beta_1 w_{1,l} \alpha_{1,l} \left\{ b_{j-1}^{(1)} \left[C_1\left(i_{1,l} - N + 1\right) R_p\left(-\left(1 - \gamma_{1,l}\right) T_c\right) \right. \right.$$

$$+ C_1(i_{1,l} - N) R_p(\gamma_{1,l} T_c) \big]$$

$$+ b_j^{(1)} \left[C_1\left(i_{1,l}\right) R_p\left(\gamma_{1,l} T_c\right) + C_1\left(i_{1,l} + 1\right) R_p\left(-\left(1 - \gamma_{1,l}\right) T_c\right) \right] \right\}$$

$$+ \sum_{q=2}^{Q_1} \sum_{l=1}^{L_1} \beta_1 w_{q,l} \alpha_{q,l} \left\{ b_{j-1}^{(1)} \left[C_1\left(i_{q,l} - N + 1\right) R_p\left(-\left(1 - \gamma_{q,l}\right) T_c\right) \right. \right.$$

$$+ C_1(i_{q,l} - N) R_p(\gamma_{q,l} T_c) \big]$$

$$+ b_j^{(1)} \left[C_1\left(i_{q,l}\right) R_p\left(\gamma_{q,l} T_c\right) + C_1\left(i_{q,l} + 1\right) R_p\left(-\left(1 - \gamma_{q,l}\right) T_c\right) \right] \right\} \tag{C.11}$$

Similarly, we can finally obtain

$$
\begin{aligned}
I_{L,p} = {} & \sum_{l=1,l\neq p}^{L_1} \beta_p w_{1,l}\alpha_{1,l}\left\{b_{j-1}^{(1)}\left[C_1\left(i_{1,l,p}-N+1\right)R_p\left(-\left(1-\gamma_{1,l,p}\right)T_c\right)\right.\right. \\
& \left.+ C_1(i_{1,l,p}-N)R_p(\gamma_{1,l,p}T_c)\right] \\
& + b_j^{(1)}\left[C_1\left(i_{1,l,p}\right)R_p\left(\gamma_{1,l,p}T_c\right)+C_1\left(i_{1,l,p}+1\right)R_p\left(-\left(1-\gamma_{1,l,p}\right)T_c\right)\right]\Bigg\} \\
& + \sum_{q=2}^{Q_1}\sum_{l=1}^{L_1}\beta_p w_{q,l}\alpha_{q,l}\left\{b_{j-1}^{(1)}\left[C_1\left(i_{q,l,p}-N+1\right)R_p\left(-\left(1-\gamma_{q,l,p}\right)T_c\right)\right.\right. \\
& \left.+ C_1(i_{q,l,p}-N)R_p(\gamma_{q,l,p}T_c)\right] \\
& + b_j^{(1)}\left[C_1\left(i_{q,l,p}\right)R_p\left(\gamma_{q,l,p}T_c\right)+C_1\left(i_{q,l,p}+1\right)R_p\left(-\left(1-\gamma_{q,l,p}\right)T_c\right)\right]\Bigg\}
\end{aligned} \qquad \text{(C.12)}
$$

D

Proof of Equation (8.44)

In order to simplify the proof of Equation (8.44), let us first look at no more than the first term of the received signal in Equation (8.43), $\mathbf{r}_{11}$, which can be written as

$$\mathbf{r}_{11} = (b_{1,o}\mathbf{c}_{o,11} + b_{1,e}\mathbf{c}_{e,11})h_{1,1} + (b_{1,e}\mathbf{c}_{o,11} - b_{1,o}\mathbf{c}_{e,11})h_{1,2} + \mathbf{n}_{1,1} \tag{D.1}$$

where we have assumed

$$\mathbf{c}_{o,11} = [\mathbf{c}_{1,1}, 0, 0, \ldots, 0] \tag{D.2}$$

$$\mathbf{c}_{e,11} = [0, 0, \ldots, 0, \mathbf{c}_{1,1}] \tag{D.3}$$

Using the above relation we can obtain the output signal from the correlator, if only the signal transmitted from the first user is present, as

$$
\begin{aligned}
[d_{1,11}, d_{2,11}] &= [(\mathbf{r}_{11} \otimes [\mathbf{c}_{o,11} + \mathbf{c}_{e,11}]) \oplus (\mathbf{r}_{11} \otimes [\mathbf{c}_{o,11} + \mathbf{c}_{e,11}])] \\
&= [\mathbf{r}_{11}\mathbf{c}_{o,11}^{H}, \mathbf{r}_{11}\mathbf{c}_{e,11}^{H}]
\end{aligned}
\tag{D.4}
$$

where $\mathbf{x}^{H}$ is the Hermitian form of $\mathbf{x}$. Therefore, based on the definitions given from Equation (8.37) to Equation (8.39) we can readily obtain

$$
\begin{aligned}
[d_{1,11}, d_{2,11}] &= [(\mathbf{r}_{11} \otimes [\mathbf{c}_{o,11} + \mathbf{c}_{e,11}]) \oplus (\mathbf{r}_{11} \otimes [\mathbf{c}_{o,11} + \mathbf{c}_{e,11}])] \\
&= [\mathbf{r}_{11}\mathbf{c}_{o,11}^{H}, \mathbf{r}_{11}\mathbf{c}_{e,11}^{H}] \\
&= \left[\begin{array}{c} \left((b_{1,o}\mathbf{c}_{o,11} + b_{1,e}\mathbf{c}_{e,11})h_{1,1} + (b_{1,e}\mathbf{c}_{o,11} - b_{1,o}\mathbf{c}_{e,11})h_{1,2} + \mathbf{n}_{1,1}\right)\mathbf{c}_{o,11}^{H} \\ \left((b_{1,o}\mathbf{c}_{o,11} + b_{1,e}\mathbf{c}_{e,11})h_{1,1} + (b_{1,e}\mathbf{c}_{o,11} - b_{1,o}\mathbf{c}_{e,11})h_{1,2} + \mathbf{n}_{1,1}\right)\mathbf{c}_{e,11}^{H} \end{array} \right]^{T} \\
&= \left[(h_{1,1}b_{1,o} + h_{1,2}b_{1,e} + \mathbf{c}_{o,11}^{H}\mathbf{n}_{1,1}), (-h_{1,2}b_{1,o} + h_{1,1}b_{1,e} + \mathbf{c}_{e,11}^{H}\mathbf{n}_{1,1}) \right]
\end{aligned}
\tag{D.5}
$$

where we have taken into account the following relation

$$\mathbf{c}_{o,11}\mathbf{c}_{e,11}^{H} = \mathbf{c}_{e,11}\mathbf{c}_{o,11}^{H} = \mathbf{0} \tag{D.6}$$

Next Generation Wireless Systems and Networks   Hsiao-Hwa Chen and Mohsen Guizani

Therefore, starting with the summation of all transmitting signals from different users $\mathbf{r}_1$, given in Equation (8.43), we can readily obtain

$$
\begin{aligned}
\left[d_{1,1}, d_{1,2}\right] &= \left[\left(\mathbf{r}_1 \otimes \left[\mathbf{c}_{o,1} + \mathbf{c}_{e,1}\right]\right) \oplus \left(\mathbf{r}_1 \otimes \left[\mathbf{c}_{o,1} + \mathbf{c}_{e,1}\right]\right)\right] \\
&= \left[(h_{1,1}b_{1,o} + h_{1,2}b_{1,e} + \mathbf{c}_{o,1}^H \mathbf{n}_1), (-h_{1,2}b_{1,o} + h_{1,1}b_{1,e} + \mathbf{c}_{e,1}^H \mathbf{n}_1)\right] \\
&\quad + \sum_{k=2}^{K} \left[\left(\mathbf{t}_k \otimes \left[\mathbf{c}_{o,1} + \mathbf{c}_{e,1}\right]\right) \oplus \left(\mathbf{t}_k \otimes \left[\mathbf{c}_{o,1} + \mathbf{c}_{e,1}\right]\right)\right]
\end{aligned}
\tag{D.7}
$$

E

Properties of Orthogonal Complementary Codes

Tables 7.2 and 7.3 give two typical examples of the orthogonal complementary (OC) code sets, which have the same processing gain (PG $= MN$) but differ in their individual parameters K, M and N. The first OC code set has $K = 8$, $M = 8$ and $N = 4$; while the second one has $K = 4$, $M = 4$ and $N = 8$. Using the information given in the two tables, we can easily observe the following orthogonality properties of an OC code set.

Take any two flocks of element codes from the tables, say

$$\begin{cases} \mathbf{c}_x = [\mathbf{c}_{x,1}(n), \mathbf{c}_{x,2}(n), \ldots, \mathbf{c}_{x,M}(n)] \\ \mathbf{c}_y = [\mathbf{c}_{y,1}(n), \mathbf{c}_{y,2}(n), \ldots, \mathbf{c}_{y,M}(n)] \end{cases} \tag{E.1}$$

where $\mathbf{c}_{x,m}(n)$ and $\mathbf{c}_{y,m}(n)$ are column vectors, and x and y are the flock indexes k defined in Tables 7.2 and 7.3, and chip index is n, $n \in (1, N)$. The in-phase auto-correlation and cross-correlation functions of an OC code set are always perfect (as shown by the code sets given in Tables 7.2 and 7.3), such that

$$\sum_{m=1}^{M} \mathbf{c}_{x,m}^T(n) \mathbf{c}_{y,m}(n - i) = 0, \qquad x \neq y \tag{E.2}$$

and

$$\sum_{m=1}^{M} \mathbf{c}_{x,m}^T(n) \mathbf{c}_{x,m}(n - i) = \sum_{m=1}^{M} \mathbf{c}_{y,m}^T(n) \mathbf{c}_{y,m}(n - i)$$

$$= \begin{cases} NM & \text{for } i = 0 \\ 0 & \text{for } i \neq 0 \end{cases} \tag{E.3}$$

where $\mathbf{c}_{x,m}^T(n) \mathbf{c}_{y,m}(n)$ stands for the inner product between two element codes $\mathbf{c}_{x,m}(n)$ and $\mathbf{c}_{y,m}(n)$. Now let us form two extended element codes as

$$\begin{cases} \mathbf{c}_{o,x,m} = [\mathbf{c}_{x,m}, 0, 0, \ldots, 0] \\ \mathbf{c}_{e,x,m} = [0, 0, \ldots, 0, \mathbf{c}_{x,m}] \end{cases} \tag{E.4}$$

Next Generation Wireless Systems and Networks   Hsiao-Hwa Chen and Mohsen Guizani
© 2006 John Wiley & Sons, Ltd

and

$$
\begin{cases}
\mathbf{c}_{o,y,m} = [\mathbf{c}_{y,m}, 0, 0, \ldots, 0] \\
\mathbf{c}_{e,y,m} = [0, 0, \ldots, 0, \mathbf{c}_{y,m}]
\end{cases}
\tag{E.5}
$$

Thus, we can have

$$
\begin{cases}
\tilde{\mathbf{c}}_{x,m} = \mathbf{c}_{o,x,m} + \mathbf{c}_{e,x,m} \\
\tilde{\mathbf{c}}_{y,m} = \mathbf{c}_{o,y,m} + \mathbf{c}_{e,y,m}
\end{cases}
\tag{E.6}
$$

The block-wise cross-correlation between $\tilde{\mathbf{c}}_{x,m}$ and $\tilde{\mathbf{c}}_{y,m}$ becomes

$$
\begin{aligned}
f_{bwc}(\tilde{\mathbf{c}}_{x,m}, \tilde{\mathbf{c}}_{y,m}) &= \left[\tilde{\mathbf{c}}_{x,m} \otimes \tilde{\mathbf{c}}_{y,m}\right] \oplus \left[\tilde{\mathbf{c}}_{x,m} \otimes \tilde{\mathbf{c}}_{y,m}\right] \\
&= \left[(\mathbf{c}_{o,x,m} + \mathbf{c}_{e,x,m}) \otimes (\mathbf{c}_{o,y,m} + \mathbf{c}_{e,y,m})\right] \oplus \left[(\mathbf{c}_{o,x,m} + \mathbf{c}_{e,x,m}) \otimes (\mathbf{c}_{o,y,m} + \mathbf{c}_{e,y,m})\right]
\end{aligned}
\tag{E.7}
$$

which will give block-wise auto-correlation function if $\tilde{\mathbf{c}}_{x,m} = \tilde{\mathbf{c}}_{y,m}$ and block-wise cross-correlation function otherwise. From the orthogonality property of an OC code set, we have

$$
\begin{aligned}
\sum_{m=1}^{M} f_{bwc}(\tilde{\mathbf{c}}_{x,m}, \tilde{\mathbf{c}}_{y,m}) &= \sum_{m=1}^{M} \left[(\mathbf{c}_{o,x,m} + \mathbf{c}_{e,x,m}) \otimes (\mathbf{c}_{o,y,m} + \mathbf{c}_{e,y,m})\right] \\
&\quad \oplus \left[(\mathbf{c}_{o,x,m} + \mathbf{c}_{e,x,m}) \otimes (\mathbf{c}_{o,y,m} + \mathbf{c}_{e,y,m})\right] \\
&= \begin{cases} [NM, NM] & x = y \\ [0, 0] & x \neq y \end{cases}
\end{aligned}
\tag{E.8}
$$

It can be readily shown that weighted block-wise correlation (BWC) between two different extended element codes $\tilde{\mathbf{c}}_{x,m}$ and $\tilde{\mathbf{c}}_{y,m}$ should also be equal to a null vector or

$$
\begin{aligned}
\sum_{m=1}^{M} &f_{bwc}(b_{x,o}\mathbf{c}_{o,x,m} + b_{x,e}\mathbf{c}_{e,x,m}, b_{y,o}\mathbf{c}_{o,y,m} + b_{y,e}\mathbf{c}_{e,y,m}) \\
&= \sum_{m=1}^{M} \left\{ \left[(b_{x,o}\mathbf{c}_{o,x,m} + b_{x,e}\mathbf{c}_{e,x,m}) \otimes (b_{y,o}\mathbf{c}_{o,y,m} + b_{y,e}\mathbf{c}_{e,y,m})\right] \right. \\
&\quad \left. \oplus \left[(b_{x,o}\mathbf{c}_{o,x,m} + b_{x,e}\mathbf{c}_{e,x,m}) \otimes (b_{y,o}\mathbf{c}_{o,y,m} + b_{y,e}\mathbf{c}_{e,y,m})\right] \right\} \\
&= [0, 0], \quad x \neq y
\end{aligned}
\tag{E.9}
$$

where $b_{x,o}$, $b_{x,e}$, $b_{y,o}$ and $b_{y,e}$ are constants with the same absolute value. In the analysis given here, we are particularly interested in the cases that take only binary values of either $+1$ or -1.

F

Proof of Equation (8.66)

In this Appendix we will show the validity of Equation (8.66), which is rewritten for easy reference as follows:

$$\left\{\sum_{m=1}^{M}\mathbf{I}_{1,m}\otimes(\mathbf{c}_{o,1,m}+\mathbf{c}_{e,1,m})\right\}\oplus\left\{\sum_{m=1}^{M}\mathbf{I}_{1,m}\otimes(\mathbf{c}_{o,1,m}+\mathbf{c}_{e,1,m})\right\}=[0,0] \qquad (\text{F.1})$$

where as also given in (8.31) and (8.63) we have

$$\begin{cases}\mathbf{c}_{o,1,m}=[\mathbf{c}_{1,m},0,0,\ldots,0]\\ \mathbf{c}_{e,1,m}=[0,0,\ldots,0,\mathbf{c}_{1,m}]\end{cases} \qquad (\text{F.2})$$

and

$$\mathbf{I}_{1,m}=\sum_{k=2}^{K}\left[(b_{k,o}\mathbf{c}_{o,k,m}+b_{k,e}\mathbf{c}_{e,k,m})h_{k,(2m-1)}+(b_{k,e}\mathbf{c}_{o,k,m}-b_{k,o}\mathbf{c}_{e,k,m})h_{k,2m}\right] \qquad (\text{F.3})$$

Thus, we get

$$\sum_{m=1}^{M}\mathbf{I}_{1,m}\otimes(\mathbf{c}_{o,1,m}+\mathbf{c}_{e,1,m})$$

$$=\sum_{m=1}^{M}\sum_{k=2}^{K}\left[(b_{k,o}\mathbf{c}_{o,k,m}+b_{k,e}\mathbf{c}_{e,k,m})h_{k,(2m-1)}+(b_{k,e}\mathbf{c}_{o,k,m}-b_{k,o}\mathbf{c}_{e,k,m})h_{k,2m}\right]$$

$$\otimes(\mathbf{c}_{o,1,m}+\mathbf{c}_{e,1,m})$$

$$=\sum_{k=2}^{K}\sum_{m=1}^{M}h_{k,(2m-1)}(b_{k,o}\mathbf{c}_{o,k,m}+b_{k,e}\mathbf{c}_{e,k,m})\otimes(\mathbf{c}_{o,1,m}+\mathbf{c}_{e,1,m})$$

$$+\sum_{k=2}^{K}\sum_{m=1}^{M}h_{k,2m}(b_{k,e}\mathbf{c}_{o,k,m}-b_{k,o}\mathbf{c}_{e,k,m})\otimes(\mathbf{c}_{o,1,m}+\mathbf{c}_{e,1,m})$$

$$=\mathbf{A}+\mathbf{B} \qquad (\text{F.4})$$

Next Generation Wireless Systems and Networks   Hsiao-Hwa Chen and Mohsen Guizani
© 2006 John Wiley & Sons, Ltd

where

$$
\begin{cases}
\mathbf{A} = \displaystyle\sum_{k=2}^{K}\sum_{m=1}^{M} h_{k,(2m-1)}(b_{k,o}\mathbf{c}_{o,k,m} + b_{k,e}\mathbf{c}_{e,k,m}) \otimes (\mathbf{c}_{o,1,m} + \mathbf{c}_{e,1,m}) \\[4mm]
\mathbf{B} = \displaystyle\sum_{k=2}^{K}\sum_{m=1}^{M} h_{k,2m}(b_{k,e}\mathbf{c}_{o,k,m} - b_{k,o}\mathbf{c}_{e,k,m}) \otimes (\mathbf{c}_{o,1,m} + \mathbf{c}_{e,1,m})
\end{cases}
\tag{F.5}
$$

and $(b_{k,o}\mathbf{c}_{o,k,m} + b_{k,e}\mathbf{c}_{e,k,m}) \otimes (\mathbf{c}_{o,1,m} + \mathbf{c}_{e,1,m})$ represents the elementa wise product (EWP) operation between two extended element codes belonging to two different flocks, and the same applies to $(b_{k,e}\mathbf{c}_{o,k,m} - b_{k,o}\mathbf{c}_{e,k,m}) \otimes (\mathbf{c}_{o,1,m} + \mathbf{c}_{e,1,m})$. $b_{k,o}$ and $b_{k,e}$ take either $+1$ or -1, which will not affect the orthogonality property of the OC codes sets involved in the above equations. Therefore, applying the half a length addition (HLA) operation to $\sum_{m=1}^{M} \mathbf{I}_{1,m} \otimes (\mathbf{c}_{o,1,m} + \mathbf{c}_{e,1,m})$ is just equal to applying the HLA operation to $\mathbf{A}$ and $\mathbf{B}$ individually, followed by a summation, or

$$
\left\{ \sum_{m=1}^{M} \mathbf{I}_{1,m} \otimes (\mathbf{c}_{o,1,m} + \mathbf{c}_{e,1,m}) \right\} \oplus \left\{ \sum_{m=1}^{M} \mathbf{I}_{1,m} \otimes (\mathbf{c}_{o,1,m} + \mathbf{c}_{e,1,m}) \right\}
$$

$$
= (\mathbf{A} + \mathbf{B}) \oplus (\mathbf{A} + \mathbf{B})
$$

$$
= (\mathbf{A} \oplus \mathbf{A}) + (\mathbf{B} \oplus \mathbf{B})
$$

$$
= [0, 0] + [0, 0]
$$

$$
= [0, 0]
\tag{F.6}
$$

where $\mathbf{A} \oplus \mathbf{A}$ and $\mathbf{B} \oplus \mathbf{B}$ just calculate the weighted in-phase cross-correlation functions between two different flocks of extended element codes, both of which should be equal to a null vector or $[0, 0]$, as shown in Appendix E.

Bibliography

[1] R. Peterson, R. E. Zimmer, and D. E. Borth, *Introduction to spread spectrum communications*, Prentice Hall International, 1995 Edition.

[2] K. S. H. Zigangirov, *Theory of Code Division Multiple Access Communications*, Wiley, 2004 Edition.

[3] J. G. Proakis, *Digital Communications*, 3rd Edition, McGraw-Hill, New York, 1995.

[4] M. J. C. Martin, *Managing Innovation and Entrepreneurship in Technology-Based Firms*, Wiley, 1994 Edition.

[5] H. P. Hsu, *Analog and Digital Communications*, McGraw-Hill, 1993.

[6] A. V. Oppenheim, A. S. Willsky, and I. T. Young, *Signals and Systems*, Prentice Hall, 1983.

[7] A. V. Oppenheim and R. W. Schafer, *Discrete-Time Signal Processing*, Prentice Hall, 1989.

[8] J. Mills, editor, *Radio Communication Theory and Methods*, McGraw-Hill, 1917.

[9] E. Kreyszic, *Advanced Engineering Mathematics*, John Wiley & Sons, 1988.

[10] C. Flammer, *Spheroidal Wave Functions*, Stanford University Press, 1957.

[11] M. I. Skolnik, *Introduction to Radar Systems*, McGraw-Hill Book Co, 1962.

[12] D. G. Fink and D. Christiansen, *Electronics Engineers Handbook*, McGraw-Hill Book Co, 1975.

[13] K. Pahlavan and A. Levesque, *Wireless Information Networks*, John Wiley & Sons, 1995.

[14] S. L. Marple, *Digital Spectral Analysis*, Prentice Hall, 1987.

[15] J. G. Proakis, *Digital Communications*, 4th Edition, Addison-Wesley, 2000.

[16] G. Arfken, *Mathematical Methods for Physicists, Chapter 12, Legendre Functions*, 3rd Edition, Academic Press, 1985.

[17] D. K. Cheng, *Field and Wave Electromagnetics*, Addison-Wesley, 1989.

[18] C. Balanis, *Antenna Theory*, John Wiley & Sons, 1997.

[19] S. Ramo and J. Whinnery, *Fields and Waves in Modern Radio*, John Wiley & Sons, 1962.

[20] S. Haykin, *Adaptive Filtering Theory*, 3rd Edition, 1996.

[21] D. Bertsekas and R. Gallager, *Data Networks*, 2nd Edition, Prentice Hall, Englewood Cliffs, NJ, 1992.

[22] A. S. Tanenbaum, *Computer Networks*, Prentice Hall, 1981.

[23] G. J. Bierman, *Factorization Method for Discrete Sequential Estimation*, Academic Press, New York, 1977.

[24] J. A. C. Bingham, *The Theory and Practice of Modem Design*, John Wiley & Sons, New York, 1988.

[25] L. Boithias, *Radio Wave Propagation*, McGraw-Hill Inc., New York, 1987.

[26] R. E. Ziemer and R. L. Peterson, *Digital Communications*, Prentice Hall, Englewood Cliffs, NJ, 1990.

[27] R. E. Ziemer and R. L. Peterson, *Introduction to Digital Communications*, Macmillan Publishing Company, 1992.

[28] M. D. Yacoub, *Foundations of Mobile Radio Engineering*, CRC Press, 1993.

[29] G. Calhuon, *Digital Cellular Radio*, Artech House Inc., 1988.

[30] K. W. Cattermole, *Principles of Pulse-Code Modulation*, Elsevier, New York, 1969.

[31] R. Steedman, The common air interface MPT 1375. In W. H. W. Tuttlebee, editor, *Cordless Telecommunications in Europe*, Springer-Verlag, 1990.

[32] R. Steele, *Mobile Radio Communications*, IEEE Press, 1992.

[33] R. Steele, editor, *Mobile Radio Communications*, IEEE Press, 1994.

[34] W. L. Stutzman and G. A. Thiele, *Antenna Theory and Design*, John Wiley & Sons, New York, 1981.

[35] W. L. Stutzman, *Polarization in Electromagnetic Systems*, Artech House, Boston, 1993.

[36] S. Baase and A. Van Gelder, *Computer Algorithms, Introduction to Design and Analysis*, 3rd Edition, Addison Wesley, Longman, 2000.

[37] K. Pahlavan and A. H. Levesque, *Wireless Information Networks*, John Wiley & Sons, 1995.

[38] L. Hanzo, W. Webb, and T. Keller, *Single and Multi-Carrier Quadrature Amplitude Modulation*, 1st Edition, John Wiley & Sons, 2000.

[39] A. Jamalipour, *The Wireless Mobile Internet–Architectures, Protocols and Services*, Wiley, 2003.

[40] G. L. Stuber, *Principle of Mobile Communications*, Kluwer, 1996.

[41] J. Craig, *Marconi's First Transatlantic Wireless Experiment*, http://www.ucs.mun.ca/jcraig/marconi.html, accessed 2001.

[42] J. P. Rybak, *Alexander Popov: Russia's Radio Pioneer*, http://www.ptti.ru/eng/forum/article2.html, accessed 1992.

[43] M. Hauben, *History of ARPANET Behind the Net – The Untold History of the ARPANET*, http://www.dei.isep.ipp.pt/docs/arpa.html, accessed 1993.

[44] H. Meyr, M. Moeneclaey, and S. Fechtel, *Digital Communication Receivers: Synchronization, Channel Estimation and Signal Processing*, John Wiley & Sons, New York, 1998 Edition.

[45] S. M. Kay, *Fundamentals of Statistical Signal Processing – Estimation Theory*, Prentice Hall, Englewood Cliffs, NJ, 1993 Edition.

[46] J. Baltersee, G. Fock, P. Schulz-Rittich, and H. Meyr, *Performance Analysis of Phasor Estimation Algorithms for a FDD-UMTS RAKE Receiver*, ISSSTA2000, Parsippany, NJ, September 2000.

[47] G. Fock, P. Schulz-Rittich, J. Baltersee, and H. Meyr, *Multipath Resistant Coherent Timing-Error-Detector for DS-CDMA Applications*, ISSSTA2000, Parsippany, NJ, September 2000.

[48] P. Schulz-Rittich, G. Fock, J. Baltersee, and H. Meyr, *Low Complexity Adaptive Code Tracking with Improved Multipath Resolution for DS-CDMA Communications ovdf Fading Channels*, ISSSTA, 2000.

[49] B. Middleton, *An Introduction to Statistical Communication Theory*, John Wiley & Sons, 1996 Edition.

[50] P. Y. C. Hwang, *Introduction to Random Signals and Applied Kalman Filtering with Matlab Exercises and Solutions*, 3rd Edition, Robert Grover Brown, Rockwell International Corporation, ISBN: 0-471-128, 1997 Edition.

[51] M. G. Cotton, R. J. Achatz, Y. Lo, and C. L. Holloway, *Indoor Polarization and Directivity Measurements at 5.8 GHz*, NTIA-Report 00–372, November 1999.

[52] P. B. Papazian, K. Allen, and M. Cotton, A comparison of 1920 MHz diversity gain using horizontally and vertically spaced antenna elements, in *Proceedings of 1999 IEEE Radio and Wireless Conference*, Denver, CO, August 1999, paper No. 161.

[53] P. Wilson, P. Papazian, M. Cotton, and Y. Lo, Advanced antenna test bed characterization for wideband wireless communication systems, NTIA Report, *Vehicular Technology Conference. (VTC 1999) – Fall. IEEE VTS 50th*, vol. 2, pp. 1048–1052, 19–22 September 1999.

[54] J. A. Wepman, J. R. Hoffman, and L. H. Loew, Analysis of impulse response measurements for PCS channel modeling applications, *IEEE Trans. Veh. Techno.*, vol. 44, no. 3, August 1995.

[55] J. A. Wepman, J. R. Hoffman, and L. H. Loew, *Impluse Response Measurements in the 1850–1990 MHz Band in Large Outdoor Cells*, NTIA Report 94–309, (NTIS Order No. PB 94204906), June 1994.

[56] J. A. Wepman, J. R. Hoffman, and L. H. Loew, Characterization of macrocellular PCS propagation channels in the 1850-1990 MHz bands, in *Proceedings of the 3rd International Conference on Universal Personal Communications*, San Diego, CA, September 1994.

[57] J. A. Wepman, J. R. Hoffman, L. H. Loew, and V. S. Lawrence, *Comparison of Wideband Propagation in the 902–928 and 1850–1990 MHz Bands in Various Macrocellular Environments*, NTIA Report 93–299, September 1993, Click here for Abstract.

[58] J. A. Wepman, J. R. Hoffman, L. H. Loew, W. J. Tanis, and M. Hughes, Impulse response measurements in the 902–928 and 1850–1990 MHz bands in macrocellular environments, in *Proceedings of the 2nd International Conference on Universal Personal Communications*, October 1993.

[59] U. Charash, *A Study of Multipath Reception with Unknown Delays*. Ph.D. dissertation, University of California, Berkeley, CA, January 1974.

[60] J. Luo, J. R. Zeidler, and S. McLaughlin, Performance Analysis of Compact Antenna Arrays with MRCin Correlated Nakagami Fading Channels, *IEEE Trans. Veh. Technol.*, vol. 50, pp. 267–277, January 2001.

[61] M. K. Simon and M.-S. Alouini, A simple integral representation of the bivariate Rayleigh distribution, *IEEE Commun. Lett.*, vol. 2, pp. 128–130, May 1998.

[62] G. K. Karagiannidis, D. A. Zogas, and S. A. Kotsopoulos, On the multivariate Nakagami-m distribution with exponential correlation, *IEEE Trans. Commun.*, vol. 51, no. 8, pp. 1240–1244, August 2003.

[63] H. Suzuki, A statistical model for urban multipath propagation, *IEEE Trans. Commun.*, vol. COM-25, pp. 637–680, July 1977.

[64] G. L. Turin, F. D. Clapp, T. L. Johnston, S. B. Fine, and D. Lavry, A statistical model of urban multipath propagation, *IEEE Trans. Veh. Technol.*, vol. VT-21, pp. 1–9, February 1972.

[65] H. Hashemi, Simulation of the urban radio propagation channel, *IEEE Trans. Veh. Technol.*, vol. VT-28, pp. 213–225, August 1979.

[66] T. S. Rappaport, S. Y. Seidel, and K. Takamizawa, Statistical channel impulse response models for factory and open plan building radio communication system design, *IEEE Trans. Commun.*, vol. COM-39, pp. 794–807, May 1991.

[67] P. Yegani and C. McGlilem, A statistical model for the factory radio channel, *IEEE Trans. Commun.*, vol. COM-39, pp. 1445–1454, October 1991.

[68] H. Hashemi, Impulse response modeling of indoor radio propagation channels, *IEEE J. Sel. Areas Commun.*, vol. SAC-11, pp. 967–978, September 1993.

[69] S. A. Abbas and A. U. Sheikh, A geometric theory of Nakagami fading multipath mobile radio channel with physical interpretations, in *Proc. IEEE Veh. Technol. Conf. (VTC'96)*, Atlanta, GA, April 1996, pp. 637–641.

[70] M. Abramowitz and I. A. Stegun, *Handbook of Mathematical Functions with Formulas, Graphs, and Mathematical Tables*, Dover Publications, New York, 1970.

[71] T. Aulin, Characteristics of a digital mobile radio channel, *IEEE Trans. Veh. Technol.*, vol. VT-30, pp. 45–53, May 1981.

[72] R. M. Baits and W. L. Stutzman, Modeling and simulation of mobile satellite propagation, *IEEE Trans. Ant. Propagat.*, vol. AP-40, pp. 375–382, April 1992.

[73] P. K. Banerjee, R. S. Dabas, and B. M. Reddy, C-band and L-band transionospheric scintillation experiment: Some results for applications to satellite radio systems, *Radio Sci.*, vol. 27, pp. 955–969, June 1992.

[74] S. Basu, E. M. MacKenzie, S. Basu, E. Costa, P. F. Fougere, H. C. Carlson, and H. E. Whitney, 250 MHz/GHz scintillation parameters in the equatorial, polar, and aural environments, *IEEE J. Sel. Areas Commun.*, vol. SAC-5, pp. 102–115, February 1987.

[75] K. Bischoff and B. Chytil, A note on scintillation indices, *Planet. Space Sci.*, vol. 17, pp. 1059–1066, 1969.

[76] W. R. Braun and U. Dersch, A physical mobile radio channel model, *IEEE Trans. Veh. Technol.*, vol. VT-40, pp. 472–482, May 1991.

[77] R. J. C. Bultitude, S. A. Mahmoud, and W. A. Sullivan, A comparison of indoor radio propagation characteristics at 910 MHz and 1.75 GHz, *IEEE J. Sel. Areas Communi.*, vol. SAC-7, pp. 20–30, January 1989.

[78] U. Charash, *"A Study of Multipath Reception with Unknown Delays."* Ph.D. dissertation, University of California, Berkeley, CA, January 1974.

[79] U. Charash, Reception through Nakagami fading multipath channels with random delays, *IEEE Trans. Commun.*, vol. COM-27, pp. 657–670, April 1979.

[80] B. Chytil, The distribution of amplitude scintillation and the conversion of scintillation indices, *J. Atmos. Terr. Phys.*, vol. 29, pp. 1175–1177, September 1967.

[81] G. Corazza and F. Vatalaro, A statistical model for land mobile satellite channels and its application to nongeostationary orbit systems, *IEEE Trans. Veh. Technol.*, vol. VT-43, pp. 738–742, August 1994.

[82] COST 207 TD(86)51-REV 3 (WG1), *Proposal on Channel Transfer Functions to be Used in GSM Test Late 1986*, Tech. Rep., Office Official Publications European Communities, September 1986.

[83] T. Eng and L. B. Milstein, Coherent DS-CDMA performance in Nakagami multi-path fading, *IEEE Trans. Commun.*, vol. COM-43, pp. 1134–1143, February-March-April 1995.

[84] E. J. Fremouw and H. F. Bates, Worldwide behavior of average VHF-UHF scintillation, *Radio Sci.*, vol. 6, pp. 863–869, October 1971.

[85] E. J. Fremouw, R. C. Livingston, and D. A. Miller, On the statistics of scintillating signals, *J. Atmos. Terr. Phys.*, vol. 42, pp. 717–731, August 1980.

[86] B. Glance and L. J. Greenstein, Frequency-selective fading effects in digital mobile radio with diversity combining, *IEEE Trans. Commun.*, vol. COM-31, pp. 1085–1094, September 1983.

[87] F. Hansen and F. I. Meno, Mobile fading-Rayleigh and lognormal superimposed, *IEEE Trans. Veh. Technol.*, vol. VT-26, pp. 332–335, November 1977.

[88] H. Hashemi, Impulse response modeling of indoor radio propagation channels, *IEEE J. Sel. Areas Commun.*, vol. SAC-11, pp. 967–978, September 1993.

[89] H. Hashemi, Simulation of the urban radio propagation channel, *IEEE Trans. Veh. Techno.*, vol. VT-28, pp. 213–225, August 1979.

[90] M. J. Ho and G. L. Stüber, Co-channel interference of microcellular systems on shadowed Nakagami fading channels, in *Proc. IEEE Veh. Technol. Conf. (VTC'93)*, Secaucus, NJ, pp. 568–571, May 1993.

[91] R. S. Hoyt, Probability functions for the modulus and angle of the normal complex variate, *Bell Syst. Tech. J.*, vol. 26, pp. 318–359, April 1947.

[92] S.-H. Hwang, K.-J. Kim, J.-Y. Ahn, and K.-C. Wang, A channel model for nongeostationary orbiting satellite system, in *Proc. IEEE Veh. Technol. Conf. (VTC'97)*, Phoenix, AZ, pp. 41–45, May 1997.

[93] S. Ichitsubo, T. Furuno, and R. Kawasaki, A statistical model for microcellular multipath propagation environment, in *Proc. IEEE Veh. Technol. Conf. (VTC'97)*, Phoenix, AZ, pp. 61–66, May 1997.

[94] H. B. James and P. I. Wells, Some tropospheric scatter propagation measurements near the radio-horizon, in *Proc. IRE*, pp. 1336–1340, October 1955.

[95] C. Loo, A statistical model for a land-mobile satellite link, *IEEE Trans. Veh. Technol.*, vol. VT-34, pp. 122–127, August 1985.

[96] E. Lutz, D. Cygan, M. Dippold, F. Dolainsky, and W. Papke, The land mobile satellite communication channel: Recording, statistics, and channel model, *IEEE Trans. Veh. Technol.*, vol. VT-40, pp. 375–386, May 1991.

[97] L. J. Mason, Error probability evaluation of systems employing differential detection in a Rician fading environment and Gaussian noise, *IEEE Trans. Commun.*, vol. COM-35, pp. 39–46, May 1987.

[98] D. Molkdar, Review on radio propagation into and within buildings, *IEE Proc. H*, vol. 138, pp. 61–73, February 1991.

[99] G. H. Munro, Scintillation of radio signals from satellites, *J. Geophys. Res.*, vol. 68, p. 1851, April 1963.

[100] M. Nakagami, The m-distribution: A general formula of intensity distribution of rapid fading, *Statistical Methods in Radio Wave Propagation*, Pergamon Press, Oxford, pp. 3–36, 1960.

[101] J. G. Proakis, *Digital Communications*, 3rd Edition, McGraw-Hill, New York, 1995.

[102] T. S. Rappaport, *Wireless Communications: Principles and Practice*, Prentice Hall, Upper Saddle River, NJ, 1996.

[103] T. S. Rappaport, S. Y. Seidel, and K. Takamizawa, Statistical channel impulse response models for factory and open plan building radio communication system design, *IEEE Trans. Commun.*, vol. COM-39, pp. 794–807, May 1991.

[104] T. S. Rappaport and C. D. McGillem, UHF fading in factories, *IEEE J. Sel. Areas Commun.*, vol. SAC-7, pp. 40–48, January 1989.

[105] M. Rice and B. Humphreys, Statistical models for the ACTS K-band land mobile satellite channel, in *Proc. IEEE Veh. Technol. Conf. (VTC'97)*, Phoenix, AZ, pp. 46–50, May 1997.

[106] S. O. Rice, Statistical properties of a sine wave plus random noise, *Bell Syst. Tech. J.*, vol. 27, pp. 109–157, January 1948.

[107] P. D. Shaft, On the relationship between scintillation index and Rician fading, *IEEE Trans. Commun.*, vol. COM-22, pp. 731–732, May 1974.

[108] A. U. Sheikh, M. Handforth, and M. Abdi, Indoor mobile radio channel at 946 MHz: Measurements and modeling, in *Proc. IEEE Veh. Technol. Conf. (VTC'93)*, Secaucus, NJ, pp. 73–76, May 1993.

[109] P. F. M. Smulders and A. G. Wagemans, Millimetre-wave biconical horn antennas for near uniform coverage in indoor picocells, *Electron. Lett.*, vol. 28, pp. 679–681, March 1992.

[110] T. L. Staley, R. C. North, W. H. Ku, and J. R. Zeidler, Performance of coherent MPSK on frequency selective slowly fading channels, in *Proc. IEEE Veh. Technol. Conf. (VTC'96)*, Atlanta, GA, pp. 784–788, April 1996.

[111] K. A. Stewart, G. P. Labedz, and K. Sohrabi, Wideband channel measurements at 900 MHz, in *Proc. IEEE Veh. Technol. Conf. (VTC'95)*, Chicago, pp. 236–240, July 1995.

[112] G. L. Stuber, *Principles of Mobile Communications*, Kluwer Academic Publishers, Norwell, MA, 1996.

[113] G. R. Sugar, Some fading characteristics of regular VHF ionospheric propagation, in *Proc. IRE*, pp. 1432–1436, October 1955.

[114] H. Suzuki, A statistical model for urban multipath propagation, *IEEE Trans. Commun.*, vol. COM-25, pp. 673–680, July 1977.

[115] G. L. Turin, F. D. Clapp, T. L. Johnston, S. B. Fine, and D. Lavry, A statistical model of urban multipath propagation, *IEEE Trans. Veh. Technol.*, vol. VT-21, pp. 1–9, February 1972.

[116] H. E. Whitney, J. Aarons, R. S. Allen, and D. R. Seeman, Estimation of the cumulative probability distribution function of ionospheric scintillations, *Radio Sci.*, vol. 7, pp. 1095–1104, December 1972.

[117] M. Wittmann, J. Marti, and T. Kiirner, Impact of the power delay profile shape on the bit error rate in mobile radio systems, *IEEE Trans. Veh. Technol.*, vol. VT-46, pp. 329–339, May 1997.

[118] P. Yegani and C. McGlilem, A statistical model for the factory radio channel, *IEEE Trans. Commun.*, vol. COM-39, pp. 1445–1454, October 1991.

[119] C. B. Emmanuel and P. A. Mandics, *A Feasibility Study for the Remote Measurement of Underwater Currents Using Acoustic Doppler Techniques*, NOAA Tech. Rep. ERL 278-WPL25, August 1973.

[120] R. Pinkel and F. N. Spiess, Space-time Measurement of Oceanic Motions from a Range-Gated Doppler Sonar, *J. Acoustical Soc. Am.*, vol. 59(Suppl. 1), 1976.

[121] W. D. Scherer, K. A. Sage, and D. E. Pryor, An intercomparison of an acoustic remote current sensor and Aanderaa current meters in an estuary, in Presented at *98th Meeting of the Acoustical Society of America*, Salt Lake City, Utah, November 1979.

[122] K. S. Miller and M. M. Rochwarger, A covariance approach to spectral moment estimation, *IEEE Trans. Inform. Theory*, vol. IT-18, no. 5, pp. 588–596, 1972.

[123] R. J. Doviac and D. S. Zrnic, *Doppler Radar and Weather Observations*, Academic Press, Orlando, FL, pp.103–107, 1984.

[124] K. B. Theriault, Incoherent Multibeam Doppler current profiler performance. Part I: Estimate variance, *J. Ocean Eng.*, vol. 0E-11, no. 1, pp. 7–15, 1986.

[125] A. W. Rihaczek, *Principles of High Resolution Radar*, McGraw-Hill, New York, p. 328, 1969.

[126] J. A. Edwards, Remote measurement of water currents using correlation sonar, in Presented at *98th Meeting of the Acoustical Society of America*, Salt Lake City, Utah, November 1979.

[127] J. A. Smith, Doppler sonar and surface waves: Range and resolution, *J. Atm and Oceanic Tech.*, vol. 6, no. 4, pp. 680–696, 1989.

[128] R. Pinkel and J. A. Smith, Repeat sequence codes for improved performance of doppler sounders, *J. Atm. Oceanic Tech.*, vol. 9, no. 2, pp. 149–163, 1991.

[129] W. D. Rummler, *Introduction of a New Estimator for Velocity Spectral Parameters*. Technical Memo MM-68-4141-5, Bell Telephone Labs, no. 24, 1968.

[130] R. Pinkel, On the use of Doppler sonar for internal wave measurements, *Deep Sea Res.*, vol. 28A, pp. 269–289, 1981.

[131] D. S. Hanson, Oceanic incoherent Doppler sonar spectral analysis by conventional and finite parameter modeling methods, *IEEE J. Ocean Eng.*, vol. 0E-11, no. 1, pp. 26–40, 1986.

[132] X. H. Chen, T. Lang, and J. Oksman, Searching for quasi-optimal subfamilies of m-sequences for CDMA systems, in *Seventh IEEE International Symposium on Personal, Indoor and Mobile Radio Communications (PIMRC'96)*, vol. 1, pp. 113–117, 15–18 October 1996.

[133] A. Z. Tirkel, Cross correlation of m-sequences-some unusual coincidences, in *Spread Spectrum Techniques and Applications Proceedings, 1996 IEEE 4th International Symposium*, vol.3, Mainz, Germany, pp. 969–973, 22–25 September 1996.

[134] T. Ito, S. Sampei, and N. Morinaga, M-sequence based M-ary/SS scheme for high bit rate transmission in DS/CDMA systems, *Electron. Lett.*, vol. 36, no. 6, pp. 574–576, 16 March 2000.

[135] A. Z. Tirkel, Cross correlation of m-sequences-some unusual coincidences, in *Spread Spectrum Techniques and Applications Proceedings, 1996 IEEE 4th International Symposium*, vol. 3, Mainz, Germany, pp. 969–973, 22–25 September 1996.

[136] K. Imamura and G.-Z. Xiao, On periodic sequences of the maximum linear complexity and M-sequences, Singapore ICCS/ISITA '92, *Commun. Move*, vol. 3, pp. 1219–1221, 16–20 November 1992.

[137] S. Uehara and K. Imamura, Some properties of the partial correlation of M-sequences, Singapore ICCS/ISITA '92, *Commun. Move*, vol. 3, pp. 1222–1223, 16–20 November 1992.

[138] A. M. D. Turkmani and U. S. Goni, Performance evaluation of maximal-length, Gold and Kasami codes as spreading sequences in CDMA systems, Universal Personal Communications, 1993. *Personal Communications: Gateway to the 21st Century Conference Record, 2nd International Conference on*, vol. 2, pp. 970–974, 12–15 October 1993.

[139] C. Balza, A. Fromageot, and M. Maniere, Four-level pseudo-random sequences, *Electron. Lett.*, vol. 3, pp. 313 315, 1967.

[140] P. A. N. Briggs and K. R. Godfrey, Autocorrelation function of a 4-level m-sequence, *Electon. Lett.*, vol. 4, pp. 232–233, 1963.

[141] J. Granlund, A. R. Thompson, and B. G. Clark, An application of walsh functions in radio astronomy instrumentation, *IEEE Trans. Electromagn. Comput.*, vol. EMC-20, pp. 451–453, 1978.

[142] H. F. Harmuth, *Transmission of Information by Orthogonal Functions*, Springer-Verlag, Berlin, 1970.

[143] M. A. Ryle, A new radio interferometer and it application to the observation of weak radio stars, *Proc. R. Soc. A*, vol. 211, pp. 351–375, 1952.

[144] W. J. Welch, *et al.*, The Berkeley-Illinois-Maryland-Association Millimeter Array, *Publ. Astron. Soc. Pacific*, vol. 108, pp. 93–103, 1996.

[145] M. C. H. Wright, B. G. Clark, C. H. Moore, and J. Coe, Hydrogen-line aperture synthesis at the National Radio Astronomy Observatory: Techniques and data reduction, *Radio Sci.*, vol. 8, p. 763, 1973.

[146] N. Zierler, Linear recurring sequences, *J Soc. Indust. Appl. Math.*, vol. 7, pp. 31–48, 1959.

[147] S. W. Golomb, *Shift Register Sequences*, Aegean Press, California, 1982.

[148] B. Gordon, W. H. Mills, and L. R. Welch, Some new difference sets, *Canad. J. Math.*, vol. 14, pp. 614–625, 1962.

[149] A. H. Chan and R. A. Games, On the linear span of Binary sequences obtained from finite geometry, *IEEE Trans. Inform. Theory*, vol. IT-36, pp. 548–552, 1976.

[150] E. L. Key, An analysis of the structure and camplexity of nonlinear binary sequence generators, *IEEE Trans. Inform. Theory*, vol. IT-33, pp. 124–129, 1987.

[151] A. H. Chan and R. A. Games, On the quadratic spans of DeBruijn sequences, *IEEE Trans. Inform. Theory*, vol. IT-36, pp. 822–829, 1990.

[152] L. H. Khachatrian, *The Lower Bound of the Quadratic Spans of DeBruijn Sequences, Designs, Codes and Crytography*, vol. 3, 1993.

[153] X. H. Chen, and J. Oksman, A new algorithm to optimize Barker code sidelobe suppression filters, *IEEE Trans. Aero. Electron. Syst.*, vol. 26, pp. 673–677, July 1990.

[154] I. Bar-David and R. Krishnamoorthy, Barker code position modulation for high-rate communication in the ISM bands, *Bell Labs Tech. J.*, vol. 1, no. 2, pp. 21–40, August 2002.

[155] R. H. Barker, Group synchronization of binary digital systems. In W. Jackson, editor, *Communication Theory*, Butterworths, London, 1953.

[156] G. F. M. Beenker, T. A. C. M. Claasen, and P. W. C. Heime, Binary sequences with a maximally flat amplitude spectrum, *Phillips J. Res.*, vol. 40, pp. 289–304, 1985.

[157] J. Bernasconi, Low autocorrelation binary sequences: Statistical mechanics and configuration space analysis, *J. Phys.*, vol. 48, pp. 559–567, 1987.

[158] J. Bernasconi, Optimization problems and statistical mechanics, in *Proceedings of Workshop on Chaos and Complexity*, World Scientific, Torino, 1987.

[159] L. D. Baumert, *Cyclic Difference Sets*, Springer-Verlag, Berlin, 1971.

[160] C. E. Cook and M. Bernfeld, *Radar Signals*, Academic Press, New York, 1967.

[161] C. De-Groot, D. Wurtz, and K. H. Hoffman, Low autocorrelation binary sequences: Exact enumeration and optimization by evolutionary stratigies, *Optimization*, vol. 23, pp. 369–384, 1992.

[162] K. Deergha Rao and G. Sridhar, Improving performance in pulse radar detection using neural networks, *IEEE Trans. Aero. Electron. Syst.*, vol. 30, pp. 1193–1198, 1995.

[163] M. J. E. Golay, A class of finite binary sequences with alternate autocorrelation values equal to zero, *IEEE Trans. Inform. Theory*, vol. IT-18, pp. 449–450, 1972.

[164] M. J. E. Golay, Sieves for low autocorrelation binary sequences, *IEEE Trans. Inform. Theory*, vol. IT-23, pp. 43–51, 1977.

[165] M. J. E. Golay, The merit factor of long low autocorrelation binary sequences, *IEEE Trans. Inform. Theory*, vol. IT-28, pp. 543–549, 1982.

[166] M. J. E. Golay, The merit factor of Legendrea sequences, *IEEE Trans. Inform. Theory* vol. IT-29, pp. 934–936, 1983.

[167] M. J. E. Golay and D. Harris, A new search for skew-symmetric binary sequences with optimal merit factors, *IEEE Trans. Inform. Theory*, vol. 36, pp. 1163–1166, 1990.

[168] T. Hoholdt and J. Justesen, Determination of the merit factor of Legendrea sequences, *IEEE Trans. Inform. Theory* vol. IT-34, pp. 161–164, 1988.

[169] T. Hoholdt, H. E. Jensen, and J. Justesen, Aperiodic correlations and the merit factor of a class of binary sequences, *IEEE Trans. Inform. Theory* vol. IT-31, pp. 549–552, 1985.

[170] J. M. Jensen, H. E. Jensen, and T. Hoholdt, The merit factor of binary sequences related to difference sets, *IEEE Trans. Inform. Theory*, vol. IT-37, pp. 617–626, 1991.

[171] A. M. Kerdock, R. Meyar, and D. Bass, Longest binary pulse compression codes with given peak side lobe levels, *Proc. IEEE*, vol. 74, p. 366, 1986.

[172] H. K. Kwan and C. K. Lee, Pulse radar detection using a multilayer neural network, in *Proc. Int. Joint Conf. Neural Networks*, Washington, DC, vol. 2, pp. 75–85, 1989.

[173] H. K. Kwan and C. K. Lee, A neural network approach to pulse radar detection, *IEEE Trans. Aero. Electron. Syst.*, vol. 29, pp. 9–21, 1993.

[174] P. S. Moharir, Generation of the approximation to binary white noise, *J. IETE*, vol. 21, pp. 5–7, 1975.

[175] P. S. Moharir, Non-linear non gaussian inversion. In N. K. Indira and P. K. Gupta, editors, *Inverse Methods: General Principles and Applications to Earth Sciences*, Narosa, New Delhi, 1998.

[176] P. S. Moharir and K. Subba Rao, Nonbinary sequences with superior merit factors, *IETE J. Res.*, vol. 1, pp. 49–53, 1997.

[177] P. S. Moharir, V. M. Maru, and R. Singh, S-K-H algorithm for signal design, *Electron. Lett.*, vol. 32, pp. 1648–1649, 1996.

[178] P. S. Moharir, V. M. Maru, and R. Singh, Bi-parental product algorithm for coded wave form design in radar, *Sadhana*, vol. 22, pp. 589–599, 1997a.

[179] P. S. Moharir, V. M. Maru, and R. Singh, Untrapping techniques for radar signal design, *Electron. Lett.*, vol. 33, pp. 631–633, 1997b.

[180] P. S. Moharir, V. M. Maru, and R. Singh, Simonization for signal design, *Sadhana*, vol. 23, pp. 351–358, 1998.

[181] D. J. Newmann and J. S. Byrnes, The L norm of polynomial with coefficients. 1, *Am. Math. Mon.*, vol. 97, pp. 42–45, 1990.

[182] R. Singh, P. S. Moharir, and V. M. Maru, Eugenic algorithm-based search for ternary pulse compression sequences, *J. Inst. Electron. Telecommun. Eng.*, vol. 42, pp. 11–19, 1996.

[183] R. Turyn, *Optimum code study*. Sylvania Electric Systems Report F 437–1, 1963.

[184] R. Turyn, Sequences with small correlation. In H. B. Mann, editor, *Error Correcting Codes*, Wiley, New York, pp. 195–228, 1968.

[185] K. H. A. Karkkainen, Linear complexity of Kronecker sequences, *IEICE Trans. Fundam.*, vol. E84-A, no. 5, pp. 1348–1351, May 2001.

[186] W. E. Stark and D. V. Sarwate, Kronecker sequences for spread spectrum communication, *IEE Proc. Part F*, vol. 128, no. 2, pp. 104–109, April 1981.

[187] M. Beale and T. C. Tozer, A class of composite sequences for spread-spectrum communications, *IEE J. Comput. Dig. Technol.*, vol. 2, no. 2, pp. 87–92, April 1979.

[188] S. A. Faulkner and J. S. Wight, Structure of composite codes for rapid acquisition of DS-SS signals, *Proceedings Spread Spectrum – Potential Commercial Applications Myth or Reality? A Workshop Held in Montebello*, pp. 7.3.1–7.3.3, Quebec, Canada, May 1991.

[189] S. Uehara and K. Imamura, Characteristic polynomials of binary complementary sequences, *IEICE Trans. Fundam.*, vol. E80-A, no. 1, pp. 193–196, January 1997.

[190] X. H. Chen and J. Oksman, BER performance analysis of 4-CCL and 5-CCL codes in slotted indoor CDMA systems, *IEE Proc. – I*, vol. 139, pp. 79–84, February 1992.

[191] R. A. Scholtz and L. R. Welch, GMW Sequences, *IEEE Trans. Inform. Theory*, vol. 30, no. 3, pp. 548–553, May 1984.

[192] H. H. Chen, T. Lang, and J. Oksman, Constructing quasi-optimal subfamilies of GMW sequences suitable for CDMA applications, *IEE Proc-Commun.*, vol. 144, no. 2, pp. 99–106, April 1997.

[193] H. H. Chen, T. Lang, and J. Oksman, Constructing quasi-optimal GMW & M-sequence subfamilies with minimized bit error rate, *IEICE Trans. Commun.*, vol. E79-B, no. 7, pp. 963–973, July 1996.

[194] J.-S. No and P. V. Kumar, A new family of binary pseudorandom sequences having optimal periodic correlation properties and large linear Span, *IEEE Trans. Inform. Theory*, vol. 35, no. 2, pp. 371–379, March 1989.

[195] R. Gold, Maximal recursive sequences with 3-valued recursive cross-correlation functions, *IEEE Trans. Inform. Theory*, vol. IT-14, pp. 154–156, January 1968.

[196] J. Lahtonen, On the odd and the aperiodic correlation properties of the Kasami sequences, *IEEE Trans. Inform. Theory*, vol. 41, no. 5, pp. 1506–1508, September 1995.

[197] O. N. Lebedev and I. L. Poliakov, Properties of composite Kasami sequence sets for wideband signals, 10th International Microwave Conference, 2000. *Microwave and Telecommunication Technology*, pp. 234–235, 2000.

[198] R. T. Barghouthi and G. L. Stuber, Rapid sequence acquisition for DS/CDMA systems employing Kasami sequences, *IEEE Trans. Commun.*, vol. 42, no. 2, pp. 1957–1968, Feb/Mar/Apr 1994.

[199] J. J. Komo and S.-C. Liu, Modified Kasami sequences for CDMA System Theory, *Twenty-Second Southeastern Symposium*, USA, pp. 219–222, 11–13 March 1990.

[200] D. Lee, H. Lee, and K. B. Milstein, Direct sequence spread spectrum Walsh-QPSK modulation, *IEEE Trans. Commun.*, vol. 46, no. 9, pp. 1227–1232, September 1998.

[201] D. Lee, H. Lee, and K. B. Milstein, Direct sequence spread spectrum Walsh-QPSK modulation, *IEEE Trans. Commun.*, vol. 46, no. 9, pp. 1227–1232, September 1998.

[202] J. Cho, Y. Kim, and K. Cheun, A novel FHSS multiple-access network using M-ary orthogonal Walsh modulation, VTC 2000, *IEEE VTS-Fall VTC 2000, 52nd*, vol. 3, pp. 1134–1141, 2000.

[203] S. Tsai, F. Khaleghi, S.-J. Oh, and V. Vanghi, Allocation of Walsh codes and quasi-orthogonal functions in cdma2000 forward link, VTC 2001 Fall. *IEEE VTS 54th*, vol.2, pp. 747–751, 2001.

[204] P. V. Kumar and R. A. Scholtz, Bounds on the linear span of Bent sequences, *IEEE Trans. Inform. Theory*, vol. 29, pp. 854–862, 1983.

[205] H. H. Chen, Multi-Band wavelet packet spreading codes with intra-code subband diversity for communications in multipath fading channels, *IEICE Trans. Commun.*, vol. E84-B, no. 7, pp. 1876–1884, July 2001.

[206] M. J. E. Golay, Complementary series, *IRE Trans. Inform. Theory*, vol. IT-7, pp. 82–87, 1961.

[207] R. Turyn, Ambiguity function of complementary sequences, *IEEE Trans. Inform. Theory*, vol. IT-9, pp. 46–47, January 1963.

[208] N. Suehiro, Complete complementary code composed of N-multiple-shift orthogonal sequences, *Trans. IECE of Japan* (in Japanese), vol. J65-A, pp. 1247–1253, December 1982.

[209] N. Suehiro and M. Hatori, N-Shift Cross-orthogonal sequences, *IEEE Trans. Inform. Theory*, vol. IT-34, no. 1, pp. 143–146, January 1988.

[210] H. H. Chen, J. F. Yeh, and N. Seuhiro, A multi-carrier CDMA architecture based on orthogonal complementary codes for new generations of wideband wireless communications, *IEEE Commun. Mag.*, vol. 39, no. 10, pp. 126–135, October 2001.

[211] H. H. Chen and Y.-C Yeh, Capacity of space-time block-coded CDMA systems: Comparison of unitary and complementary codes, *IEE Proc- Commun.*, vol. 152, no. 2, pp. 203–214, 8 April 2005.

[212] H. H. Chen, Y.-C. Yeh, C.-Y. Chao, and J.-F. Yeh, A Pilot-added signal detection algorithm and its application in OCC-CDMA systems under multipath interference, *IEE Electron. Lett.*, vol. 40, no. 8, pp. 488–489, 15th April 2004.

[213] H. H. Chen, On next generation CDMA technology for future wireless networking (Invited Paper), in *Wireless Ad Hoc and Sensor Networks Workshop, IEEE Globecom 2004*, Dallas, TX, 29 November–3 December, 2004.

[214] H. H. Chen and H.-W. Chiu, Generation of super-set of perfect complementary codes for next generation CDMA systems, in *IEEE Military Communication Conference (IEEE MILCOM) 2004*, Monterey, CA, October 31–November 3, 2004.

[215] H. H. Chen and H.-W. Chiu, Design of perfect complementary codes To implement an interference-free CDMA system, *IEEE Globecom 2004*, Dallas, TX, 29 November–3 December, 2004.

[216] H. H. Chen and H.-W. Chiu, Generation of perfect orthogonal complementary codes for their applications in interference-free CDMA systems, accepted for publication in the record of PIMRC 04, *15th IEEE Int. Symp. on Personal, Indoor and Mobile Radio Commun.*, 05.09.2004–08.09.2004, Barcelona, Spain, 2004.

[217] H. H. Chen, Y.-C. Yeh, C.-Y. Chao, and K.-S. Chen, Interference-free CDMA air-link technology promising noise-limited performance, *Proc. IEEE VTC 2003-Fall*, Orlando, USA, October 4–9, 2003.

[218] H. H. Chen, J.-X. Lin, S.-W. Chu, C.-F. Wu, and G.-S. Chen, Isotropic air-interface technologies for fourth generation wireless communications, *Wireless Commun. Mobile Comput. (WCMC) J.*, Wiley InterScience, John Wiley & Sons, vol. 3, no. 6, pp. 687–704, September 2003.

[219] H. H. Chen and J.-F. Yeh, A complementary codes based CDMA architecture for wideband mobile Internet with high spectral efficiency and exact rate-matching, *Int. J. Commun. Syst.*, John Wiley & Sons, vol. 16, pp. 497–512, 2003.

[220] H. H. Chen and Y.-C. Yeh, Capacity of space-time block-coded CDMA systems: Comparison of unitary and complementary codes, *IEE Proc- Commun.*, vol. 152, no. 2, pp. 203–214, 8 April 2005.

[221] L. R. Welch, Lower bounds on the maximum cross-correlation of signals, *IEEE Trans. Inform. Theory*, vol. IT-20, pp. 397–399, 1974.

[222] M. B. Pursley, Performance evaluation for phase-coded spread-spectrum multiple-access communications – Part I: System analysis, *IEEE Trans. Commun.*, vol. COM-25, no. 8, pp. 795–799, August 1977.

[223] M. B. Pursley and D. V. Sarwate, Performance evaluation for phase-coded spread-spectrum multiple-access communications – Part II: Code sequence analysis, *IEEE Trans. Commun.*, vol. COM-25, no. 8, pp. 800–803, August 1977.

[224] D. V. Sarwate and M. B. Pursley, Cross-correlation properties of pseudorandom and related sequences, *Proc. IEEE*, vol. 68, no. 5, pp. 593–620, May 1980.

[225] M. B. Pursley, D. V. Sarwate and W. E. Stark, Error probability for direct-sequence spread spectrum multiple-access communications – Part I: Upper and lower bounds, *IEEE Trans. Commun.*, vol. COM-30, no. 5, pp. 975–984, May 1982.

[226] E. A. Geraniotis and M. B. Pursley, Error probability for direct-sequence spread spectrum multiple-access communications – Part II: Approximations, *IEEE Trans. Commun.*, vol. COM-30, no. 5, pp. 985–995, May 1982.

[227] K. Yao, Error probability of asynchronous spread spectrum multiple access communication systems, *IEEE Trans. Commun.*, vol. 25, no. 8, pp. 803–809, August 1977.

[228] J. M. Holtzman, On calculating DS/SSMA error probabilities, *Proc. IEEE 2nd Int. Symp. Spread Spect. Techn. Appl. (ISSSTA' 92)*, Yokohama, Japan, pp. 23–26, December 1992.

[229] Ross, A. H. M., and K. L. Gilhousen, CDMA Technology and the IS-95 North American Standard, In J. D. Gipson, editor, *The Mobile Communications Handpaper*, CRC Press, pp. 430–448, 1996.

[230] N. Guo and L. B. Milstein, On rate-variable multidimensional DS/SSMA with dynamic sequence sharing, IEEE J. Select. Areas Commun., vol. 17, May 1999.

[231] C.-L. I and R. D. Gitlin, Multi-code CDMA wireless personal communications networks, in *Proc. IEEE Int. Conf. Commun. (ICC'95)*, Seattle, WA, vol. 2, pp. 1060–1064, 1999.

[232] S. Sasaki, H. Kikuchi, H. Watanabe, and J. Zhu, Performance evaluation of parallel combinatory SSMA systems in Rayleigh fading channel, in *Proc. IEEE 3rd Int. Symp. Spread Spectrum Techn. Appl. (ISSSTA'94)*, Oulu, Finland, vol. 1, pp. 198–202, 1994.

[233] S. Baey, M. Dumas, and M.-C. Dumas, QoS tuning and resource sharing for UMTS WCDMA multiservice mobile, *IEEE Trans. Mobile Comput.*, vol. 1, no. 3, pp. 221–235, Jul-Sep 2002.

[234] S. Insoo and B. S. Chan, Performance studies of rate matching for WCDMA mobile receiver, *Veh. Technol. Conf., 2000. IEEE VTS-Fall VTC 2000. 52nd*, vol. 6, pp. 2661–2665, 2000.

[235] A. C. Kam, T. Minn, and K.-Y. Siu, Supporting rate guarantee and fair access for bursty data traffic in WCDMA, *IEEE J. Sel. Areas Commun.*, vol. 19, no. 11, pp. 2121–2130, November 2001.

[236] M. Thit and K.-Y. Siu, Dynamic assignment of orthogonal variable-spreading-factor codes in WCDMA, *IEEE J. Sel. Areas Commun.*, vol. 18, no. 8, pp. 1429–1440, August 2000.

[237] R. Fantacci and S. Nannicini, Multiple access protocol for integration of variable bit rate multimedia traffic in UMTS/IMT-2000 based on wideband CDMA, *IEEE J. Sel. Areas Commun.*, vol. 18, no. 8, pp. 1441–1454, August 2000.

[238] L. Tao and X.-H. Chen, Comparison of correlation parameters of binary codes for DS/CDMA systems, *Singapore ICCS '94. Conf. Proc.*, Singapore, vol. 3, pp. 1059–1063, 14–18 November 1994.

[239] H. H. Chen, Spreading code dependent bit error rate and capacity analysis for finite asynchronous CDMA systems, *Int. J Commun. Syst.*, John Wiley & Sons, vol. 12, pp. 49–64, 1999.

[240] H. H. Chen, T. Lang, and J. Oksman, Multiple chip rate DS/CDMA system and its spreading code dependent performance analysis, *IEE Proc. Commun.*, vol. 145, no.5, pp. 371–377, October 1998.

[241] H. H. Chen, T. Lang, and J. Oksman, Performance analysis based on co-channel interference statistics of indoor CDMA systems with RAKE receiver & power control under multipath fading, *IEE Proc. Commun.*, vol. 144, no. 3, pp. 173–179, June 1997.

[242] H. H. Chen, T. Lang, and J. Oksman, Correlation statistics distribution convolution (CSDC) algorithm for studying CDMA indoor wireless systems with RAKE receiver, power control and multipath fading, *IEICE Trans. Commun.*, vol. E79-B, no. 10, October 1996.

[243] H. H. Chen, Adaptive traffic load shedding and its capacity gain in CDMA cellular, *IEE Proc- Commun.*, vol. 142, no. 3, pp. 186–192, June 1995.

[244] *1999 Federal Radionavigation Plan*, Washington, DC, U.S. Department of Transportation and Department of Defense. Available on line from United States Coast Guard Navigation Center, February 2000.

[245] Annex A, *Global Positioning System Standard Positioning Service Specification*, 2nd Edition, Available on line from United States Coast Guard Navigation Center, June 2, 1995.

[246] NATO and NAVSTAR-GPS Joint Program Office. '*NAVSTAR GPS User Equipment Introduction*. Available on line from United States Coast Guard Navigation Center, 1996.

[247] *GPS Joint Program Office. ICD-GPS-200: GPS Interface Control Document. ARINC Research.* Available on line from United States Coast Guard Navigation Center, 1997.

[248] B. Hoffmann-Wellenhof, H. Lichtenegger, and J. Collins, *GPS: Theory and Practice*. 3rd Edition, New York, Springer-Verlag, 1994.

[249] Institute of Navigation, 1980, 1884, 1986, 1993, *Global Positioning System Monographs*. Washington, DC, The Institute of Navigation.

[250] E. D. Kaplan, editor, *Understanding GPS: Principles and Applications*. Artech House, Boston, 1996.

[251] A. Leick, *GPS Satellite Surveying*, 2nd Edition. New York: John Wiley & Sons, 1995.

[252] National Imagery and Mapping Agency, *Department of Defense World Geodetic System 1984: Its Definition and Relationship with Local Geodetic Systems*. NIMA TR8350.2, 3rd Edition, 4 July 1997. Bethesda, MD: National Imagery and Mapping Agency, Available on line from National Imagery and Mapping Agency, at http://www.colorado.edu/geography/gcraft/notes/gps/gps$_f$.html, accessed 1997.

[253] B. W. Parkinson and J. J. Spilker, editors, *Global Positioning System: Theory and Practice*. vol. I and II, Washington, DC, American Institute of Aeronautics and Astronautics, 1996.

[254] D. Wells, editor, *Guide to GPS positioning*. Fredericton, NB, Canada, Canadian GPS Associates, 1989.

[255] C. R. Cahn, *Spectrum Reduction of Biphase Modulated (2-PSK) Carrier*, Magnavox Research Laboratories, MX-TM-3103-71.

[256] C. E. Gilchreist, *Pseudonoise System Lock-In*, JPL Research Summary No. 36–9.

[257] P. W. Nilsen, *PN Receiver Carrier and Code Tracking Performance*, Magnavox Research Laboratories, MX-TM-8-674-3043-68.

[258] G. F. Sage, Serial synchronization of pseudonoise systems, *IEEE Trans. Commun. Technol.*, vol. 12, pp. 123–127, December 1964.

[259] R. B. Ward, "Acquisition of pseudonoise signals by sequential estimation," *IEEE Trans. Commun. Technol.*, vol. 13, pp. 475–483, December 1965.

[260] L. E. Zegers, Common bandwidth transmission of information signals and pseudonoise synchronization waveforms, *IEEE Trans. Commun. Technol.*, vol. 16, pp. 796–807, December 1968.

[261] M. Lewis, PLLs Upconvert Chirp Radar Signals, *Microwaves*, June 1981.

[262] D. L. Schilling, L. B. Milstein, R. L. Pickholtz, and R. W. Brown, Optimization of the processing gain of an M-ary direct sequence spread spectrum communication system, *IEEE Trans. Commun.*, vol. 28, p. 1944, August 1980.

[263] M. B. Pursley, Performance evaluation for phase-coded spread spectrum multiple access communications–Part I: System analysis, *IEEE Trans. Commun.*, vol. COM-25, pp. 795–799, August 1977.

[264] F. D. Garber, Analysis of generalized quadriphase spread-spectrum communication, *Natl. Telec. Conf. Proc.*, November 1980.

[265] L. B. Milstein, R. L. Pickholtz, D. L. Schilling, and R. Brown, Optimization of the processing gain of an M-ary direct sequence spread spectrum communication system, *Int. Conf. Commun. Proc.*, June 1980.

[266] B. K. Levitt, On direct sequence spread spectrum systems, *Natl. Telec. Conf. Proc.*, November 1980.

[267] J. Low and S. M. Waldstein, "A direct sequence spread-spectrum modem for wideband HF channels," *IEEE Milcom. Proc. Conf.*, October 1981.

[268] D. L. Schilling, L. B. Milstein, R. L. Pickholtz, and R. W. Brown, Optimization of the processing gain of an M-ary direct sequence spread spectrum communication system, *IEEE Trans. Commun.*, vol. 28, p. 1944, August 1980.

[269] N. Abramson, Bandwidth and spectra of phase- and frequency-modulated waves, *IEEE Trans. Commun. Syst.*, vol. 11, pp. 407–414, December 1963.

[270] C. R. Cahn, *Noncoherent frequency hop sync mode performance*, Mag-navox Research Laboratories, STN-12, March 1964.

[271] O. H. George, Performance of noncoherent M-ary FSK systems with diversity under the influence of rician fading, *IEEE Int. Conf. Commun.*, June 1968.

[272] G. K. Huth, *Detailed frequency hopper analysis*, Magnavox Research Laboratories, STN-29, August 1966.

[273] A. Kaplan, Detection and analysis of frequency hopping radar signals, *Sylvania Elect. Syst.*, W.D.L. Mountain View, California.

[274] H. H. Schreiber, Self-noise of frequency hopping signals, *IEEE Trans. Commun. Techol.*, vol. 17, pp. 588–590, October 1969.

[275] F. G. Splitt, Combined frequency and time-shift keyed transmission systems, *IEEE Trans. Commun. Syst.*, vol. 11, pp. 414–421, December 1963.

[276] C. M. Thomas, *A Matched Filter Concept for Frequency Hopping Synchronization*, TRW 10C 7353.6-05.

[277] R. Malm and K. Schreder, Fast frequency hopping techniques, *Proc. Symp. Spr. Spec. Commun.*, March 1973.

[278] E. J. Nossen, Fast frequency hopping synthesizer, *Proc. Symp. Spr. Spec. Comm.*, March 1973.

[279] P. S. Henry, Spectrum efficiency of a frequency-hopped-DPSK spread spectrum mobile radio system, *IEEE Trans. Veh. Tech.*, vol. VT-28, November 1979.

[280] J. D. Edell, Wideband, noncoherent, frequency-hopped waveforms and their hybrids in low probability-of-intercept communications, *Naval Research Lab.*, Washington, DC, NRL Rep. 8025, November 8, 1976.

[281] D. J. Goodman, P. S. Henry, and V. K. Prabhu, Frequency-hopped multilevel FSK for mobile radio, *Bell Syst. Tech. J.*, vol. 59, pp. 1257–1275, September 1980.

[282] R. F. Pawula and R. F. Mathis, A spread spectrum system with frequency hopping and sequentially balanced modulation–parts one and two, *IEEE Trans. Commun.*, Part I, vol. 28, pp. 682–688, Part II, vol. 28, pp. 1785–1793, May 1980.

[283] M. K. Simon and A. Polydoros, Coherent detection of frequency-hopped quadrature modulations in the presence of jamming-parts I and II, *IEEE Trans. Commun.*, vol. 29, pp. 1644–1668, November 1981.

[284] M. K. Simon, G. K. Huth, and A. Polydoros, Differentially coherent detection of QASK for frequency-hopping systems, Parts I and II, *IEEE Trans. Commun.*, vol. COM-30, no. 1, pp. 158–172, January 1982.

[285] O.-C. Yue, Performance of frequency hopping multiple-access multilevel FSK systems with hard-limited and linear combining, *IEEE Trans. Commun.*, vol. 29, pp. 1687–1694, November 1981.

[286] O. C. Yue, Hard-limited versus linear combining for frequency hopping multiple-access systems in a Rayleigh fading environment, *IEEE Trans. Veh. Technol.*, vol. VT-30, pp. 10–14, February 1981.

[287] L. B. Milstein, R. L. Pickholtz, and D. L. Schilling, Optimization of the processing gain of an FSK-FH system, *IEEE Trans. Commun.*, vol. COM-28, pp. 1062–1079, July 1980.

[288] J. K. Omura, B. Levitt, and Stokey, R. FH/MFSK performance in a partial band jamming environment, *IEEE Trans. Commun.*, To be published.

[289] D. Avidor, Anti-jam analysis of frequency hopping M-ary FSK communication systems in HF Rayleigh fading channels, *Doctoral dissertation, School Engineering Applications Science*, University of California, Los Angeles, 1981.

[290] D. V. Sarwate and M. B. Pursley, Hopping patterns for frequency-hopped multiple-access communication, in *Proc. 1978 IEEE Int. Conf. Commun.*, vol. 1, pp. 7.4.1–7.4.3, 1978.

[291] P. S. Henry, Spectrum efficiency of a frequency-hopped-DPSK spread spectrum mobile radio system, *IEEE Trans. Veh. Technol.*, vol. VT-28, pp. 327–329, November 1979.

[292] R. W. Nettleton and G. R. Cooper, Performance of a frequency-hopped differentially modulated spread-spectrum receiver in a Rayleigh fading channel, *IEEE Trans. Veh. Technol.*, vol. VT-30, pp. 14–29, February 1981.

[293] E. A. Geraniotis and M. B. Pursley, Error probability bounds for slow frequency Hopped spread-spectrum multiple access communications over fading channels, in *Proceedings of 1981 IEEE International Conference on Communications*, 1981.

[294] A. J. Budreau, A. J. Slobodnick Jr, and P. H. Carr, Fast frequency hopping achieved with SAW synthesizers, *Microwave J.*, February 1982.

[295] E. Ribchester, Frequency hopping techniques vary with frequency, *Microwaves RF*, March 1983.

[296] S. M. Sussman and P. Kotiveeriah, Partial processing satellite relays for frequency hop antijam communications, *IEEE Trans. Commun.*, vol. 30, pp. 1929–1937, August 1982.

[297] A. K. Elhakeem, Overall SNR optimization of a FH/MFSK pulse code and adaptive data modulation systems in mixed jamming, *IEEE I.C.C. Proceedings*, 1982.

[298] S. M. Elnoubi, Error rate performance of frequency hopped MSK spread spectrum mobile radio system with differential detection, *I.C.C. Proceedings*, 1982.

[299] J. E. Blanchard, Performance of M-Ary FSK/FH against optimum multitone jamming, *I.C.C. Proceedings*, 1982.

[300] C. Niyonizeye, M. Lecours, and H. T. Huynh, Address assignment in a multiple access FH-FSK system, *I.C.C. Proceedings*, 1982.

[301] R. Muammar and S. C. Gupta, Performance of a frequency-hopped multilevel FSK spread spectrum receiver in a Rayleigh fading and log-normal shadowing channel, *I.C.C. Proc.*, 1982.

[302] M. Mizuno, Randomization effect of errors by means of frequency hopping techniques in a fading channel, *IEEE Trans. Commun.*, vol. 30, pp. 1052–1056, May 1982.

[303] M. B. Pursley and D. V. Sarwate, New results on frequency hop, spread-spectrum, multiple access communications, *Natl. Telec. Conf. Proc.*, November 1980.

[304] I. M. Jacobs, Dama-frequency hopping and pre-correction for a processing satellite, *Int. Conf. Comm. Proc.*, June 1980.

[305] W. C. Lindsey, L. Beiderman, and R. P. Sherwin, Coding and modulation tradeoffs for frequency-hopped channels, *Int. Conf. Comm. Proc*, June 1980.

[306] R. C. Dixon, Frequency hopping synthesizers employing conventional commercially-available integrated circuits, *ITC Conf. Proc.*, October 1981.

[307] B. G. Haskell, Computer simulation results on frequency hopped MFSK mobile radio-noiseless case, *Natl. Telec. Conf. Proc*, Texas, November 1980.

[308] http://www.itu.int/osg/imt-project, accessed 1865.

[309] http://www.3gpp.org, accessed 1998.

[310] http://www.3gpp2.org, accessed 2000.

[311] http://www.arib.or.jp/IMT-2000/ARIB/Document, accessed 1999.

[312] http://www.cdg.org, accessed 1993.

[313] http://www.umts-forum.org, accessed 1998.

[314] http://www.umtsuniversity.com, accessed 2004.

[315] http://www.cdmauniversity.com, accessed 2004.

[316] W. R. Young, Advanced mobile phone service : Introduction, background, and objectives, *Bell Syst. Tech. J.*, vol. 58, pp. 1–14, January 1979.

[317] TIA/EIA IS-19-B, *Recommended Minimum Standards for 800-MHZ Cellular Subscriber Units*, May 1988.

[318] TIA/EIA IS-20-A, *Recommended Minimum Standards for 800-MHZ Cellular Land Stations*, May 1988.

[319] EIA/TIA IS-41-B, *Cellular Radio-Telecommunications Intersystem Operations*, December 1992.

[320] TIA/EIA IS-41-C, *Cellular Radiotelecommunications Intersystem Operations*, January 1996.

[321] EIA/TIA IS-52, *Uniform Dialing Procedures and Call Processing Treatment for Use in Cellular Radio Telecommunications*, November 1989.

[322] EIA/TIA IS-53, *Cellular Features Description*, August 1991.

[323] EIA/TIA IS-54-B, *Cellular System Dual-Mode Mobile Station – Base Station Compatibility Standard*, April 1992.

[324] TIA/EIA IS-91, *Mobile Station-Base Station Compatibility Standard for 800 MHz Analog Cellular*, October 1994.

[325] TIA/E1A IS-95, *Mobile Station – Base Station Compatibility Standard for Dual-Mode Wideband Spread Spectrum Cellular System*, July 1993.

[326] TIA/EIA IS-95-B, *Mobile Station-Base Station Compatibility Standard for Dual-Mode Wideband Spread Spectrum Cellular Systems*, Baseline Version, July 31, 1997.

[327] Telecommunications Industry Association, TIA/EIA/IS-96-A, *Speech Service Option Standard for Wideband Spread Spectrum Digital Cellular System*, 1995.

[328] TIA/EIA IS-99, *Data Services Option Standard for Wideband Spread Spectrum Digital Cellular System*, 1995.

[329] TIA/EIA IS-124, *Cellular Radio Telecommunications Intersystem Non-Signaling Data Message Handlers (DMH)*, 1994.

[330] TIA/EIA IS-125, *Recommended Minimum Performance Standard for Digital Cellular Wideband Spread Spectrum Speech Service Option*, 1995.

[331] TIA/EIA IS-126, *Service Option 2: Mobile Station Loopback Service Option Standard*, Page G-2 TIA/EIA/IS-95-A, December 1994.

[332] TIA/EIA IS-634, *MSC-BS Interface for Public 800 MHz*, Revision 0, 1995.

[333] TIA/EIA IS-637, *Short Message Services for Wideband Spread Spectrum Cellular Systems*, 1997.

[334] TIA/EIA IS-657, *Packet Data Services Option for Wideband Spread Spectrum Cellular System*, 1996.

[335] TIA/EIA IS-687, *Data Services Inter-Working Function Interface Standard for Wideband Spread Spectrum Digital Cellular System*, 1995.

[336] TIA/EIA IS-707-A, *Data Services Options for Spread Spectrum Digital Cellular 24 Systems*, 1999.

[337] TIA 232E, *Interface Between DTE and DCE Employing Serial Binary Data Interchange*, 1991.

[338] TIA/EIA SP-2977, *Cellular Features Description*, Prepublication Version, March 14, 1995.

[339] TIA/EIA SP-3693, *Mobile Station-Base Station Compatibility Standard for Dual-Mode Wideband Spread Spectrum Cellular Systems*, November 18, 1997.

[340] TIA TR-45, *Reference Model*, 1990.

[341] TIA TR-46, *Reference Model*, 1991.

[342] JTC(AIR)/94.08.01-022R2, PN-3384, *Personal Station – Base Station Compatibility Requirements for 1.8 to 2.0 GHz Code Division Multiple Access (CDMA) Personal Communications Systems*, 1 August 1994.

[343] Qualcomm Inc., *An Overview of the Application of Code Division Multiple Access (CDMA) to Digital Cellular Systems and Personal Cellular Networks*, Qualcomm Inc., Doc. No. EX60–10010, 21 May 1992.

[344] R. Padovani, Reverse link performance of IS-95 based cellular systems, *IEEE Pers. Commun.*, Third Quarter, pp. 28–34, 1994.

[345] TIA TR 45.5, *The cdma2000 ITU-RTT Candidate Submission*, TR 45-ISD/98.06.02.03, May 15, 1998.

[346] 1xEV-DO Inter-Operability Specification (IOS), *For CDMA 2000 Access Network Interfaces*, Release 0, 3GPP2 A.S0007, Ballot Version, June 14, 2001.

[347] 3GPP2 WG5 Evaluation Ad Hoc, *1xEV-DV Evaluation methodology – Addendum (V6)*, July 25, 2001.

[348] 3GPP2, C.S0001-0, *Introduction to Cdma2000 Standards for Spread Spectrum Systems*, Version 1.0, Version Date, July 1999.

[349] 3GPP2 C.S0002-0, *Physical Layer Standard for cdma2000 Spread Spectrum Systems*, Version 1.0, Version Date, July 1999.

[350] 3GPP2 C.S0003-0, *Medium Access Control (MAC) Standard for cdma2000 Spread Spectrum Systems*, Version 1.0, Version Date, October 1999.

[351] 3GPP2, C.S0004-0, *Signaling Link Access Control (LAC) Standard for cdma2000 Spread Spectrum Systems*, Version 1.0, Version Date, July 1999.

[352] 3GPP2 C.S0005-0, *Upper Layer (Layer 3) Signaling Standard for cdma2000 Spread Spectrum Systems*, Version 1.0, Version Date, July 1999.

[353] 3GPP2 C.S0006-0, *Analog Signaling Standard for cdma2000 Spread Spectrum Systems*, Version 1.0, Version Date, July 1999

[354] 3GPP2 C.S0001-D, *Introduction to cdma2000 Spread Spectrum Systems Revision D*, Version 1.0, Date, February 2004.

[355] 3GPP2 C.S0002-D, *Physical Layer Standard for cdma2000 Spread Spectrum Systems Revision D*, Version 1.0, Date, February 13, 2004.

[356] Medium Access Control (MAC), *Standard for Cdma2000 Spread Spectrum Systems Release D*, 3GPP2 C.S0003-D, Version 1.0, Date, February 13, 2004.

[357] Signaling Link Access Control (LAC), *Standard for Cdma2000 Spread Spectrum Systems Release D*, 3GPP2 C.S0004-D Version 1.0 Date, February 13, 2004.

[358] Upper Layer (Layer 3), *Signaling Standard for Cdma2000 Spread Spectrum Systems Release D*, 3GPP2 C.S0005-D, Version 1.0, Date, February 2004.

[359] 3GPP2 C.S0006-D, *Analog Signaling Standard for Cdma2000 Spread Spectrum Systems Release D*, Version 1.0, Date, February 2004.

[360] *cdma2000 High Rate Packet Data Air Interface Specification*, 3GPP2 C.S20024 v2.0. October 2000.

[361] *cdma2000 High Rate Packet Data Air Interface Specification*, 3GPP2 C.S0024-A, Version 1.0, Date, March 2004.

[362] Rec.ITU-R M.1225, *Guidelines for Evaluation of Radio Transmission Technologies for IMT-2000*, 2000.

[363] P. Bender, M. Black, M. Grop R. Padovani, N. Sindhushyana, and S. Viterbi, CDMA/HDR: A bandwidth efficient high-speed data service for nomadic users, *IEEE Commun. Mag.*, vol.38, pp.70–77, July 2000.

[364] E. Esteves, The high data rate evolution of the cdma2000 Cellular System, In G. Stuber and B. Jabbari, editors, *Multiaccess, Mobility and Teletraffic in Wireless Communications*, vol. 5, Kluwer Academic Publishers, Norwell MA, 2000.

[365] A. Jalali, R. Padovani, and R. Pankaj, Data throughput of CDMA/HDR a high efficiency high data rate personal communication wireless system, *Proceedings of IEEE 51st Vehicular Technology Conference*, Tokyo, Japan, May 2000.

[366] P. J. Black and M. I. Gurelli, Capacity simulation of cdma2000 1xEV wireless internet access system, *The 3rd IEEE International Conference on Mobile and Wireless Communications Networks*, Recife, Brazil, August 2001.

[367] E. Esteves, P. J. Black, and M. I. Gurelli, Link adaptation techniques for high-speed packet data in third generation cellular systems, *European Wireless Conference*, 2002.

[368] Y.-K. Kim and B. K. Yi, 3G wireless and cdma2000 1x evolution in Korea, *IEEE Commun. Mag.*, vol. 43, no.4, pp. 36–40, April 2005.

[369] R. T. Derryberry and Z. Pi, Reverse high-speed packet data pyhsical layer enhancements in cdma2000 1xEV-DV, *IEEE Commun. Mag.*, vol. 43, no.4, pp. 41–47, April 2005.

[370] D. Comstock, R. Vannithamby, S. Balasubbamanin, L. A. Hsu and M. W. Cheng, Reverse high-speed packet data support in cdma2000 1xEV-DV:upper layer protocols, *IEEE Commun. Mag.*, vol. 43, no.4, pp. 48–56, April 2005.

[371] Y. Kim, J. Jung, B. Bae, D. Kim, P. R. Rajkotia, and Y. K. Kim, Upper layer enhancements for fast call setup in cdma2000 Revision D, *IEEE Commun. Mag.*, vol. 43, no.4, pp. 57–66, April 2005.

[372] H. Kwon, Y. Kim, J.-K. Han, D. Kim, H. W. Lee, and Y. K. Kim, Performance evaluation of high-speed packet enhancement of cdma2000 1xEV-DV, *IEEE Commun. Mag*, vol. 43, no.4, pp. 67–76, April 2005.

[373] S. Kwon, K. Kim, Y. Yun, S. G. Kim, and B. K. Yi, Power controlled H-ARQ in cdma2000 1xEV-DV, *IEEE Commun. Mag.*, vol. 43, no.4, pp. 77–81, April 2005.

[374] Y.-H. Choi, L. Park, B. Kim, and M. A. Shayman, A framework for elastic QoS provisioning in the cdma2000 1xEV-DV packet core network, *IEEE Commun. Mag.*, vol. 43, no.4, pp. 82–88, April 2005.

[375] J. A. Audestad, *Network aspects of the GSM system*, In EUROCON 88, June 1988.

[376] D. M. Balston, The Pan-European system: GSM. In D. M. Balston and R.C.V. Macario, editors, *Cellular Radio Systems*. Artech House, Boston, 1993.

[377] D. M. Balston, The pan-European cellular technology. In R. C. V. Macario, editor, *Personal and Mobile Radio Systems*, Peter Peregrinus, London, 1991.

[378] J. Varin, M. Bezler, R. Hofmans and K. Van den Bosse, GSM base station system, *Electrical Communication*, 2nd Quarter, 1993.

[379] D. Cheeseman. The pan-European cellular mobile radio system. In R. C. V. Macario, editor, *Personal and Mobile Radio Systems*, Peter Peregrinus, London, 1991.

[380] C. Dechaux and R. Scheller, What are GSM and DCS, *Electrical Communication*, 2nd Quarter, 1993.

[381] M. Feldmann and J. P. Rissen, GSM network systems and overall system integration, *Electrical Communication*, 2nd Quarter, 1993.

[382] J. M. Griffiths, *ISDN Explained: Worldwide Network and Applications Technology*, 2nd Edition, John Wiley & Sons, Chichester, 1992.

[383] I. Harris, Data in the GSM cellular network. In D. M. Balston and R. C. V. Macario, editors, *Cellular Radio Systems*. Artech House, Boston, 1993.

[384] I. Harris, Facsimile over cellular radio. In D. M. Balston and R. C. V. Macario, editors, *Cellular Radio Systems*. Artech House, Boston, 1993.

[385] T. Haug, Overview of the GSM project. In *EUROCON 88*, June 1988.

[386] J.-F. Huber, Advanced equipment for an advanced network, *Telcom Report International*, vol. 15, no. 3–4, 1992.

[387] H. Lobensommer and H. Mahner, GSM – A European mobile radio standard for the world market, *Telcom Report International*, vol. 15, no. 3–4, 1992.

[388] B. J. T. Mallinder, Specification methodology applied to the GSM system. In *EUROCON 88*, June 1988.

[389] S. Mohan and R. Jain, Two user location strategies for personal communication services, *IEEE Pers. Commun.*, vol. 1 no. 1, pp. 42–50, 1994.

[390] M. Mouly and M.-B. Pautet, *The GSM System for Mobile Communications*. Published by the authors, ISBN: 2-9507190-0-7, 1992.

[391] J. E. Natvig, S. Hansen, and J. de Brito, Speech processing in the pan-European digital mobile radio system (GSM) – system overview. In *IEEE GLOBECOM 1989*, November 1989.

[392] T. Nilsson, *Toward a new era in mobile communications*. http://193.78.100.33/ (Ericsson WWW server).

[393] M. Rahnema, Overview of the GSM system and protocol architecture, *IEEE Commun. Mag.*, vol. 31, pp. 92–100, April 1993.

[394] E. H. Schmid and M. Kahler, GSM operation and maintenance, *Electrical Communication*, 2nd Quarter, 1993.

[395] M. Silventoinen, *Personal email, quoted from European Mobile Communications Business and Technology Report*, March 1995, and December 1995.

[396] C. B. Southcott, D. Freeman, G. Cosier, D. Sereno, A. van der Krogt, A. Gilloire and H. J. Braun, Voice control of the pan-European digital mobile radio system. In *IEEE GLOBECOM 1989*, November 1989.

[397] K. Hellwig, P. Vary, D. Massaloux, J. P. Petit, C. Galand and M. Rosso, Speech codec for the European mobile radio system. In *IEEE GLOBECOM 1989*, November 1989.

[398] C. Watson, Radio equipment for GSM. In D. M. Balston and R. C. V. Macario, editors, *Cellular Radio Systems*, Artech House, Boston, 1993.

[399] R. G. Winch, *Telecommunication Transmission Systems*. McGraw-Hill, New York, 1993.

[400] J. A. Audestad. Network aspects of the GSM system. In *EUROCON 88*, June 1988.

[401] D. M. Balston, The pan-European system: GSM. In D. M. Balston and R. C. V. Macario, editors, *Cellular Radio Systems*, Artech House, Boston, 1993.

[402] D. M. Balston. The pan-European cellular technology. In R. C. V. Macario, editor, *Personal and Mobile Radio Systems*, Peter Peregrinus, London, 1991.

[403] J. Varin, M. Bezler, R. Hofmans and K. Van den Bosse, GSM base station system, *Electrical Communication*, 2nd Quarter, 1993.

[404] D. Cheeseman, The pan-European cellular mobile radio system. In R. C. V. Macario, editor, *Personal and Mobile Radio Systems*, Peter Peregrinus, London, 1991.

[405] C. Dechaux and R. Scheller, What are GSM and DCS, *Electrical Communication*, 2nd Quarter, 1993.

[406] M. Feldmann and J. P. Rissen, GSM network systems and overall system integration, *Electrical Communication*, 2nd Quarter, 1993.

[407] J. M. Griffiths, *ISDN Explained: Worldwide Network and Applications Technology*, 2nd Edition, John Wiley & Sons, Chichester, 1992.

[408] I. Harris, Data in the GSM cellular network. In D. M. Balston and R. C. V. Macario, editors, *Cellular Radio Systems*, Artech House, Boston, 1993.

[409] I. Harris, Facsimile over cellular radio. In D. M. Balston and R. C. V. Macario, editors, *Cellular Radio Systems*, Artech House, Boston, 1993.

[410] T. Haug, Overview of the GSM project. In *EUROCON 88*, June 1988.

[411] J.-F. Huber, Advanced equipment for an advanced network, *Telecom Report International*, vol. 15, no. 3–4, 1992.

[412] H. Lobensommer and H. Mahner, GSM – a European mobile radio standard for the world market, *Telcom Report International*, vol. 15, no. 3–4, 1992.

[413] B. J. T. Mallinder, Specification methodology applied to the GSM system. In *EUROCON 88*, June 1988.

[414] S. Mohan and R. Jain, Two user location strategies for personal communication services, *IEEE Pers. Commun.*, vol. 1 no. 1, pp. 42–50, 1994.

[415] M. Mouly and M.-B. Pautet, *The GSM System for Mobile Communications*, Published by the authors, 1992.

[416] J. E. Natvig, S. Hansen, and J. de Brito, Speech processing in the pan-European digital mobile radio system (GSM) – system overview. In *IEEE GLOBECOM 1989*, November 1989.

[417] T. Nilsson, *Toward a new era in mobile communications*, http://193.78.100.33/ (Ericsson WWW server).

[418] M. Rahnema, Overview of the GSM system and protocol architecture, *IEEE Communications Magazine*, vol. 31, pp. 92–100, April 1993.

[419] E. H. Schmid and M. Kahler, GSM operation and maintenance, *Electrical Communication*, 2nd Quarter, 1993.

[420] M. Silventoinen, *Personal email, quoted from European Mobile Communications Business and Technology Report*, March 1995, and December 1995.

[421] C. B. Southcott *et al.*, Voice control of the pan-European digital mobile radio system. In *IEEE GLOBECOM 1989*, November 1989.

[422] P. Vary *et al.*, Speech codec for the European mobile radio system. In *IEEE GLOBECOM 1989*, November 1989.

[423] C. Watson, Radio equipment for GSM. In D. M. Balston and R. C. V. Macario, editors, *Cellular Radio Systems*, Artech House, Boston, 1993.

[424] R. G. Winch, *Telecommunication Transmission Systems*, McGraw-Hill, New York, 1993.

[425] ETSI, *The ETSI UMTS Terrestrial Radio Access (UTRA) ITU-R RTT Candidate Submission*, January 29, 1998.

[426] H. Holma and A. Toskala, Editors, *WCDMA for UMTS: Radio Access for Third Generation Mobile Communications*, New York, Wiley, 2000.

[427] H. Kaaranen, S. Naghian, L. Laitinen, A. Ahtiainen and V. Niemi, *UMTS Networks. Architecture, Mobility and Services*, John Wiley & Sons, 2001.

[428] J. Laiho, A. Wacker and T. Novosad, *Radio Network Planning and Optimisation for UMTS*, John Wiley & Sons, 2002.

[429] J. P. Castro, *The UMTS Network and Radio Access Technology*, John Wiley & Sons, 2001.

[430] J. Korhonen, *Introduction to 3G Mobile Communications*, 2nd Edition, Artech House, 2001.

[431] ARIB, *Japan's Proposal for Candidate Radio Transmission Technology on IMT-2000: WCDMA*, June 26, 1998.

[432] CATT, *TD-SCDMA Radio Transmission Technology For IMT-2000 Candidate submission*, Draft V.0.4, September 1998.

[433] H.-H. Chen, C. X. Fan, and W. W. Lu, China's perspectives on 3G mobile communications and beyond: TD-SCDMA technology, *IEEE Wireless Communications*, pp. 48–59, April, 2002.

[434] CWTS-WG1, *Pyhycial Layer – General Description*, TS C101, V3.1.1, September 2000.

[435] CWTS-WG1, *Physical channels and mapping of transport channels onto physical channels*, TS C102, V3.3.0, September 2000.

[436] CWTS-WG1, *Multiplexing and channel coding*, TS C103, V3.1.0, September 2000.

[437] CWTS-WG1, *Spreading and modulation*, TS C104, V3.3.0, September 2000.

[438] CWTS-WG1, *Pyhycial layer procedures*, TS C105, V3.2.0, September 2000.

[439] CWTS-WG1, *Pyhycial layer – Measurements (TD-SCDMA)*, TS C106, V3.0.0, May 2000.

[440] CWTS WG1 LAS-CDMA, 2001 *Physical Channels and Mapping of Transport Channels onto Physical Channels*, LAS TS 25.221, V1.0.0, July 17–17, 2001.

[441] CWTS-SWG2, LAS-CDMA, *Pyhycial Layer Apects of TD-LAS High Speed Packet Technology*, LAS-TR 25.951, V1.0.0, July 2001.

[442] CWTS-SWG2, LAS-CDMA, *Physical Channels and Mapping of Transport Channels onto Physical Channels*, LAS-TS 25.221, V1.0.0, July 2001.

[443] CWTS-SWG2, LAS-CDMA, *Multiplexing and Channel Coding*, LAS-TS 25.222, V1.0.0, July 2001.

[444] CWTS-SWG2, LAS-CDMA, *Spreading and Modulation*, LAS-TS 25.223, V1.0.0, July 2001.

[445] CWTS-SWG2, LAS-CDMA, *Pyhycial Layer Procedures*, LAS-TS 25.224, V1.0.0, July 2001.

[446] CWTS-SWG2, LAS-CDMA, *Pyhycial Layer – Measurements*, LAS-TS 25.225, V1.0.0, July 2001.

[447] CWTS-SWG2, LAS-CDMA, *TD-LAS High Level System Design Document*, LAS-TR 25.960, V1.0.0, July 2001.

[448] 3GPP, *Technical Specification Group Radio Access Network; Physical layer – General description*, (3G TS 25.201 version 3.0.0), 1999.

[449] 802.11 Standard, *Draft supplement to standard for telecommunications and information exchange between systems – LAN/MAN specific requirements-part 11: Wireless MAC and PHY specifications: High speed physical layer in the 5 GHz band*, P802.11a/D6.0, May 1999.

[450] IEEE Std 802.11b, *Part 11: Wireless LAN Medium Access Control (MAC) and Physical Layer (PHY) specifications: Higher-Speed Physical Layer Extension in the 2.4 GHz Band*, ISBN 0-7381-1811-7 SH94788, approved 16 September 1999.

[451] IEEE Std 802.16.2, *IEEE Recommended Practice for Local and Metropolitan Area Networks, Coexistence of Fixed Broadband Wireless Access Systems*, IEEE-SA Standards Board. ISBN 0-7381-3985-8 SH95215, approved 9 February 2004.

[452] IEEE 802.11, 1999 Edition. http://standards.ieee.org/getieee802/802.11.html, accessed 1999.

[453] B. Walter, *Wireless LANs End to End: Installing a Wireless Network*, ISBN: 0764548883, 2002.

[454] P. Kaveh and K. Prashant, *Principles of Wireless Networks: A Unified Approach*, Prentice Hall, 2002.

[455] F. Behrouz, *Data Communications and Networking*, McGraw-Hill, 2001.

[456] C. Eduardas, *White Paper on IEEE 802.11: Part 2*, http://gauge.upb.de/pgmanet/seminar/workout/ieee80211part2.pdf, accessed January, 2004.

[457] E. John and A. William, *Real 802.11 Security: Wi-Fi Protected Access and 802.11i*, Addison-Wesley, 2004.

[458] M. Stewart, *Wi-Fi Security*, McGraw-Hill, 2003.

[459] M. Stuart, S. Joel, and K. George, *Hacking Exposed: Network Security Secrets and Solutions*, McGraw-Hill, 2003.

[460] K. Jason, *An IEEE 802.11 Wireless LAN Security White Paper*, http://www.llnl.gov/asci/discom/ucrl-id-147478.html,2001, accessed 2002, 2001.

[461] Silicon Wave, *Bluetooth and 802.11 Compared*, http://www.siliconwave.com/pdf/73_0005_R00A_Bluetooth_802_11.pdf, accessed 2001.

[462] M. Heidi, *Bluetooth Technology and Implications*, http://www.sysopt.com/articles/bluetooth/index1.html, accessed 1999.

[463] K. Janne, *HIPERLAN/2*, http://www.tml.hut.fi/studies/Tik-100.300/1999/Essays/hiperlan2.html, accessed 1999.

[464] Ministry of Posts and Telecommunications (Japan), *Status of Efforts to Promote Multimedia Mobile Access Communication (MMAC) Systems*, http://www.soumu.go.jp/joho_tsusin/pressrelease/english/telecomm/news8-16-2.html, accessed 1996.

[465] Black Box Networking Services, *802.11: Wireless Networking: A White Paper*, http://www.blackbox.com/homenetworking/wireless_white_paper.pdf, accessed 2002.

[466] G. Paul, *802.11: A Standard for the Present and Future*, http://www.mtghouse.com/MDC_8021X_White_Paper.pdf, accessed 2003.

[467] IEEE 802.11b, 1999 Edition, http://standards.ieee.org/getieee802/802.11.html, accessed 1999.

[468] Dell Technology White Paper, *Wireless Security in 802.11 (Wi-Fi) Networks*, http://www.dell.com/downloads/global/vcctors/wirclcss_sccurity.pdf, acccsscd 2003.

[469] Intel, *Intel Building Blocks for Wireless LAN Security*, http://www.intel.com/network/connectivity/resources/doc_library/white_papers/WLAN_Security_WP.pdf, accessed 2003.

[470] Atheros, *Building a Secure Wireless Network: How Atheros Defines Wireless Network Security Today and in the Future*, http://www.atheros.com/pt/atheros_security_whitepaper.pdf, accessed April, 2004.

[471] G. Matthew, *802.11 Wireless Networks: The Definitive Guide*, O'Reilly, 2002.

[472] B. Benn editor, *Wireless Local Area Networks: The New Wireless Revolution*, Wiley, 2002.

[473] T. K. Tan, and B. Benny, *World Wide Wi-Fi: Technological Trends and Business Strategies*, Wiley, 2003.

[474] N. Borisov, I. Goldberg, and D. Wagner, *Intercepting Mobile Communications: The Insecurity of 802.11*, http://delivery.acm.org/10.1145/390000/381695/p180-borisov.pdf?key1 =381695&key2=0707229211& coll=GUIDE&dl=GUIDE&CFID=55388121&CFTOKEN=8228593, accessed 2001.

[475] I. Mantin and A. Shamir, *A Practical Attack on Broadcast RC4*, http://www.wisdom.weizmann.ac.il/itsik/RC4/Papers/bc_rc4.ps, accessed 2001.

[476] S. Fluhrer, I. Mantin, and A. Shamir, *Weaknesses in the Key Scheduling Algorithm of RC4*, www.wisdom.weizmann.ac.il/itsik/RC4/Papers/Rc4_ksa.ps, accessed 2001.

[477] W. Arbaugh, N. Shankar, and Y. C. J. Wan, *Your 802.11 Wireless Network has No Clothes*, http://www.cs.umd.edu/waa/wireless.pdf, accessed 2001.

[478] IEEE, *Wireless LAN Medium Access Control (MAC) and Physical Layer (PHY) specifications: Amendment 4: Further Higher Data Rate Extension in the 2.4 GHz Band*, 2003.

[479] Cisco White Paper, *Capacity, Coverage and Deployment Considerations for IEEE 802.11g*, http://www.cisco.com/en/US/products/hw/wireless/ps4570/products_white_paper09186a0, accessed 2003.

[480] Broadcom White Paper, *IEEE 802.11g: The New Mainsteam Wireless LAN Standard*, 2003.

[481] U. S. Robotics, *802.11g Wireless Turbo White Paper*, http://www.usr.com/download/whitepapers/80211g-wp.pdf, accessed 2003.

[482] IEEE, *Wireless Medium Access Control (MAC) and Physical Layer (PHY) Specifications for Wireless Personal Area Networks (WPANs)*, 2002.

[483] IEEE, *Wireless Medium Access Control (MAC) and Physical Layer (PHY) Specifications for High Rate Wireless Personal Area Networks (WPANs)*, 2002.

[484] IEEE, *Wireless Medium Access Control (MAC) and Physical Layer (PHY) Specifications for Low-Rate Wireless Personal Area Networks (LR-WPANs)*, 2003.

[485] IEEE, *Air Interface for Fixed Broadband Wireless Access Systems*, http://standards.ieee.org/getieee802/download/802.16-2004.pdf, accessed 2002.

[486] S. Andre, *IEEE 802.11*, http://www.tehnicom.net/clanci/pdf/802.11Overview.pdf, accessed 2003.

[487] K. Murakami *et al.*, *Mobility Management Alternatives for Migration to Mobile Internet Session-Based Services*, http://www.bell-labs.com/user/oli/papers/jsac04.pdf, accessed 2004.

[488] National Chiao Tung University, *Advanced Technologies and Application for Next Generation Information Networks (II)*, http://www.csie.nctu.edu.tw/b3g/Documents/TH0_20050301.pdf, accessed 2005.

[489] Bell Labs Research China, *Research Projects of BLRC*, http://blrc.edu.cn/research/project_description.htm, accessed 2003.

[490] Wirelab, *Research*, http://wire.cs.nthu.edu.tw/research.php, accessed 2001.

[491] Network Reliability and Interoperability Council, *Network Interoperability*, http://www.nric.org/fg/charter_vi/fg3/NRIC6FG3FinalReport.pdf, accessed 2003.

[492] A. Kupetz and K. T. Brown, *4G-A Look Into the Future of Wireless Communications*, http://www.crummer. rollins.edu/journal/articles/2004_1_4G.pdf, accessed 2004.

[493] C. Perkins, *Mobile Networking Through Mobile IP. IEEE Internet Computing*, http://www.computer.org/ internet/v2n1/perkins.htm, accessed January, 1998.

[494] Sun Microsystems, *How Mobile IP Works*, http://docs.sun.com/app/docs/doc/806-7600/6jgfbep13?a=view, accessed 2002.

[495] Cisco Systems, *Introduction to Mobile IP*, http://www.cisco.com/en/US/tech/tk827/tk369/technologies_ white_paper09186a00800c9906.shtml#1030824, accessed 2001.

[496] The TCP/IP Guide, *Internet Protocol Mobility Support (Mobile IP)*, http://www.tcpipguide.com/free/t_ InternetProtocolMobilitySupportMobileIP.htm.

[497] L. Mittag, *Mobile IP*, http://www.embedded.com/story/OEG20010628S0054, accessed 2001.

[498] Sun Microsystems, *System Administration Guide: IP Services*, http://docsun.cites.uiuc.edu/sun_docs/C/ solaris_9/SUNWaadm/SYSADV3/toc-chapter-23.html, accessed 2004.

[499] Sun Microsystems, *Mobile IP (Overview)*, http://docsun.cites.uiuc.edu/sun_docs/C/solaris_9/SUNWaadm/ SYSADV3/p91.html#MIPOVERVIEW-2, accessed 2004.

[500] J. Li and H. H. Chen, Mobility support for IP-based networks, *IEEE Commun. Mag.*, vol. 43, no.10, pp. 127–132, October 2005.

[501] G. Kessler, *Mobile IP: Harbinger of Untethered Computing*, http://www.garykessler.net/library/mobileip. html, accessed 1998.

[502] *Mobile IPv6 Issue List*, http://users.piuha.net/jarkko/publications/mipv6/MIPv6-Issues.html.

[503] K. Zhigang *et al*, *Mobile IPv6 and Some Issues for QoS*, http://www.isoc.org/isoc/conferences/inet/01/ CD_proceedings/T28/T28.htm, accessed 2001.

[504] A. Yegin and C. Williams, *IPv6: Necessary for Mobile and Wireless Internet*, http://www.isoc.org/briefings/ 014, accessed 2003.

[505] W. Fritsche and F. Heissenhuber, *Mobility Support for the Next Generation Internet*, http://www.6bone.sk/ zaujim/MobileIPv6_Whitepaper.pdf, accessed 2000.

[506] Nokia White Paper, *Introducing Mobile IPv6 in 2G and 3G Mobile Networks*, http://www.nokia.com/ BaseProject/Sites/NOKIA_MAIN_18022/CDA/Categories/Business/NetworkSecurity/Firewalls/IPv6/ _Content/_Static_Files/mobileipv6in3gnetworks.pdf, accessed 2001.

[507] The Harm of the Wireless Application Protocol (WAP), http://www.freeprotocols.org/harmOfWap/main. html, accessed 2000.

[508] Javvin Network Management and Security, *WAP: Wireless Application Protocol and WAP Architecture*, http://www.javvin.com/protocolWAP.html, accessed 2001.

[509] J. Tyson, *How WAP Works*, http://electronics.howstuffworks.com/wireless-internet3.htm.

[510] Wireless Developer Network, *Introduction to the Wireless Application Protocol*, http://wireless.ittoolbox. com/documents/document.asp?i=3291, accessed 2005.

[511] G. Q. Maguire, *WAP, Heterogeneous PCS, 3G*, http://www.imit.kth.se/courses/2G1330/Lectures-2002/P4- Lecture5-2002.pdf, accessed 2002.

[512] Wireless Developer Network, *Introduction to the Wireless Application Protocol*, http://www.wirelessdevnet. com/channels/wap/training/wapoverview.html, accessed 2005.

[513] Open Mobile Alliance, http://www.openmobilealliance.org/tech/affiliates/wap/wapindex.html, accessed 2006.

[514] Open Mobile Alliance, *WAP Forum*, http://www.wapforum.org, accessed 2003.

[515] R. Lanka, *MIPMANET-Mobile IP for Mobile Ad Hoc Networks*, http://www.cs.umn.edu/research/mobile/ seminar/FALL02/WNfiles/MIPMANET.ppt#1, accessed 2002.

[516] M. Mohsin and R. Prakash, *IP Address Assignment in a Mobile Ad Hoc Network*, http://www.utdallas.edu/ mmohsin/publications/IPAssignment.pdf IPAssignment.pdf, accessed 2002.

[517] D. Zeinalipour-Yazti, *A Glance at Quality of Services in Mobile Ad-Hoc Networks*, http://www.cs.ucr.edu/ csyiazti/courses/cs260/manetqos.pdf, accessed 2001.

[518] P. Papadimitratos and Z. Haas, *Secure Routing for Mobile Ad Hoc Networks* http://people.ece.cornell.edu/ haas/wnl/Publications/cnds02.pdf, accessed 2002.

[519] Computer Security Resource Center, *Mobile Ad Hoc Network Security*, http://csrc.nist.gov/manet, accessed 2005.

[520] J. Moore, *MANET and the Art of Communication*, http://www.fcw.com/supplements/homeland/2003/sup3/ hom-manet-08-25-03.asp, accessed 2003.

[521] S. Ding, A. Dadej, and S. Gordon, *Internet Integrated MANETs using Mobile IP*, http://www.itr.unisa.edu. au/sgordon/doc/ding2004-internet.pdf, accessed 2004.

[522] M. Conti, E. Gregori, and G. Maselli, *Cooperation Issues in Mobile Ad Hoc Networks*, http://www.di.unipi. it/maselli/WWAN_Maselli_G.pdf, accessed 2004.

[523] Overview of Ad-Hoc Networking, http://www.it.iitb.ac.in/it644/lectures/notes/manet-notes/adhoc/index. html.

[524] M. S. Corson, *An Overview of Mobile Ad Hoc Networking*, http://inet2002.org/CD-ROM/lu65rw2n/papers/ t13-a.pdf, accessed 2002.

[525] P. Nicopolitidis, M. S. Obaidat, G. I. Papadimitriou and A. S. Pomportsis, *Wireless Networks*, Chichester, John Wiley & Sons, 2003.

[526] AODV, http://moment.cs.ucsb.edu/AODV/aodv.html#Description, accessed 2003.

[527] J. Schaumann, *Analysis of the Zone Routing Protocol*, http://www.netmeister.org/misc/zrp/zrp.html, accessed 2002.

[528] P. Samar, *Independent Zone Routing: An Adaptive Hybrid Routing Framework for Mobile Ad Hoc Networks*, http://wisl.ece.cornell.edu/ECE794/Mar26/IZR_ECE794.ppt#5, accessed 2004.

[529] N. Beijar, *Zone Routing Protocol (ZRP)*, http://www.netlab.hut.fi/opetus/s38030/k02/Papers/08-Nicklas. pdf, accessed 2002.

[530] K. Leppanen, *Alustus: 4G*, http://websrv2.tekes.fi/opencms/opencms/OhjelmaPortaali/Kaynnissa/NETS/fi/ Dokumenttiarkisto/Viestinta_ja_aktivointi/Seminaarit/Aiheryhmat/Aiheryhmx1d25112003_4Galustus-LeppxnenNRC.pdf.

[531] Wireless Networks Architecture and Standards, http://www.enel.ucalgary.ca/People/fapojuwo/619.96/ topic2_04_content.pdf.

[532] M. Katz and F. Fitzek, *On the Definition of the Fourth Generation Wireless Communication Networks: The Challenges Ahead*, http://kom.aau.dk/ff/documents/IWCTKatzFitzek2005.pdf, accessed 2005.

[533] Wireless World Research Forum, *Cognitive Radio, Spectrum and Radio Resource Management*, http://wg6.ww-rf.org/images/pdfs/WG6_WP4_CogRaSpeRRM-20041208.pdf, accessed 2004.

[534] G. Oien, *Flexible and Heterogeneous: Radio Access Beyond 3G*, http://www.telenor.com/telektronikk/Oien_ Beyond3G.pdf, accessed 2004.

[535] MobileInfo.com, 4G-Beyond 2.5G and 3G Wireless Networks, http://www.mobileinfo.com/3G/ 4GVision&Technologies.htm, accessed 2002.

[536] R. Hurwitz and B. Peebler, *Overview: The Future of Wireless Handsets*, http://www.deviceforge.com/ articles/AT7085477626.html, accessed 2003.

[537] G. Legg, *Beyond 3G: The Changing Face of Cellular* http://www.techonline.com/community/tech_group/ 37977, accessed 2005.

[538] V. Gurbani and X. Sun, *A Systematic Approach for Closer Integration of Cellular and Internet Services*, http://www.cs.iit.edu/scs/psfiles/IEEE-Network-2005.pdf, accessed 2005.

[539] M. Abualreesh, *4G*, http://www.comlab.hut.fi/opetus/333/2004_2005_slides/4G.pdf, accessed 2005.

[540] S. Denno, *Recursive Vector Algorithm for Multibeam Interference Cancellers*, http://www.docomoeurolabs. de/pdf/publications/WSL-viterbi_vtc_spring_03.pdf, accessed 2003.

[541] Wireless World Research Forum, *Cognitive Radio, Spectrum and Radio Research Management*, http://wg6.ww-rf.org/images/pdfs/WG6_WP4_CogRaSpeRRM-20041208.pdf, accessed 2004.

[542] P. Demestichas, *Design of Wireless Networks in a B3G Reconfigurable Radio Context*, http://www.ait.gr/ Publevents/01042005_Reconfigurability_Demestichas.pdf, accessed 2005.

[543] Wireless World Research Forum, *Cooperative Networks for the Future Wireless World*, http://dmc.ajou.ac. kr/paper/wwrf_ieee_com_2004_sep.pdf, accessed 2004.

[544] Ad Hoc Networking Protocols, http://ntrg.cs.tcd.ie/adhoc.php, accessed 2001.

[545] F. Fitzek and M. Reisslein, *Ad-hoc Technology in Future IP based Mobile Communication Systems*, http://www.acticom.de/fileadmin/data/publications/WWRF5_Contribution.pdf, accessed 2002.

[546] J. Hoebeke, I. Moerman, B. Dhoedt, and P. Demeester, *An Overview of Mobile Ad Hoc Networks: Applications and Challenges*, http://www.ist-magnet.org/private/files/Dissemination/WP2/1%20An%20Overview%20of%20Mobile%20Ad%20Hoc%20Networks%20%20Applications%20and%20Challenges.pdf, accessed 2004.

[547] E2R White Paper, *Dynamic Network Planning and Management*, http://e2r.motlabs.com/whitepapers, accessed 2005.

[548] Wireless World Research Forum, *Cognitive Radio, Spectrum and Radio Resource Management in Reconfigurable Networks*, http://wg6.ww-rf.org/images/pdfs/WG6_WP4_CogRaSpeRRM-20041208.pdf, accessed 2004.

[549] M. Odroma, *Wireless, Mobile and Always Best Connected*, http://www.ul.ie/odroma/odroma_p_p.pdf, accessed 2003.

[550] M. Abualreesh, *4G*, http://www.comlab.hut.fi/opetus/333/2004_2005_slides/4G_text.pdf, accessed 2005.

[551] R. Grunheid and H. Rohling, Adaptive modulation and multiple access for the OFDM transmission technique, *Wireless Pers. Commun.*, vol. 13, pp. 5–13, 2000.

[552] A. Jamalipour, T. Wada, and T. Yamazato, A Tutorial on Multiple Access Technologies for Beyond 3G Mobile Networks, *IEEE Commun. Mag.*, vol. 43, pp. 110–117, February 2005.

[553] L. Bos and S. Leroy, Toward an all-IP UMTS System Architecture, *IEEE Network*, vol. 15, no. 1, pp. 36;V45, 2001.

[554] H. Muramatsu M. Harada, T. Yamazato, H. Okada, and M. Katayama, Effect of Nonlinear Amplifiers of Transmitters in Multicarrier CDMA Systems, *IEICE Trans. Fundam.*, vol. J85-A, no. 3, pp. 340;V48, Mar. 2002.

[555] Z. Dawy and A. Seeger, Coverage and capacity enhancement of multiservice WCDMA cellular systems via serial interference cancellation, *Proceedings of the ICC 2004*, Paris, France, June 2004.

[556] D. Yu, H. Li, and H. Hagenauer, Multihop Network Capacity Estimation, *Proceedings of the ICC 2004*, Paris, France, June 2004.

[557] R. Esmailzadeh and M. Nakagawa, TDD-CDMA for the 4th generation of wireless communications, *IEEE Commun. Mag.*, vol. 41, no. 8, pp. 8;V15, August 2003.

[558] F. Piolini and A. Rlolando, smart channel-assignment algorithm for SDMA systems, *IEEE Trans. Microwave Theory Techn.*, vol. 47, no. 6, pp. 693;V99, June 1999.

[559] S. Suwa, H. Atarashi, and M. Sawahashi, Performance comparison between MC/DS-CDMA and MC-CDMA for reverse link broadband packet wireless access, *Proceedings of VTC- 2002 Fall*, Vancouver, Canada, pp. 2076;V80, September 2002.

[560] L.-L. Yang and L. Hanzo, Multicarrier DS-CDMA: A Multiple Access Scheme for Ubiquitous Broadband Wireless Communications, *IEEE Commun. Mag.*, vol. 41, no. 10, pp. 116;V24, October 2003.

[561] P. Xia, S. Zhou, and G. B. Giannakis, bandwidth- and power-efficient multicarrier multiple access, *IEEE Trans. Commun.*, vol. 51, no. 11, pp. 1828;V37, November 2003.

[562] H.-H. Chen and M. Guizani, Guest Editorial, Multiple access technologies for B3G wireless communications, *IEEE Commun. Mag.*, vol. 43, pp. 65–67, February 2005.

[563] H. Wei, L.-L. Yang, and L. Hanzo, Interference-free broadband single- and multicarrier DS-CDMA, *IEEE Commun. Mag.*, vol. 43, pp. 68–73, February 2005.

[564] C. William and Y. Lee, CS-OFDMA: A new wireless CDD physical layer scheme, *IEEE Commun. Mag.*, vol. 43, pp. 74–79, February 2005.

[565] R. C. Qiu, H. Liu, and X. (Sherman) Shen, Ultra-wideband for multiple access communications, *IEEE Commun. Mag.*, vol. 43, pp. 80–87, February 2005.

[566] F. Khan, A time-orthogonal CDMA high-speed uplink data transmission scheme for 3G and beyond, *IEEE Commun. Mag.*, vol. 43, pp. 88–94, February 2005.

[567] R. Fantacci, F. Chiti, D. Marabissi, G. Mennuti, S. Morosi, and D. Tarchi, Perspectives for present and future CDMA-based communications systems, *IEEE Commun. Mag.*, vol. 43, pp. 95–100, February 2005.

[568] S. Nanda, R. Walton, J. Ketchum, M. Wallace, and S. Howard, A high-performance MIMO OFDM wireless LAN, *IEEE Commun. Mag.*, vol. 43, pp. 101–109, February 2005.

[569] A. Jamalipour, T. Wada, and T. Yamazato, A tutorial on multiple access technologies for beyond 3G mobile networks, *IEEE Commun. Mag.*, vol. 43, pp. 110–117, February 2005.

[570] M. Juntti, M. Vehkapera, J. Leinonen, Z. Li, and D. Tujkovic, S. Tsumura, and S. Hara, MIMO MC-CDMA communications for future cellular systems, *IEEE Commun. Mag.*, vol. 43, pp. 118–124, February 2005.

[571] B. M. Popovic, Spreading sequences for multicarrier CDMA systems, *IEEE Trans. Commun.*, vol. 47, no. 6, pp. 918–926, June 1999.

[572] A. M. Tulino, Linbo Li, and S. Verdu, Spectral efficiency of multicarrier CDMA, *IEEE Trans. Inform. Theory*, vol. 51, no. 2, pp. 479–505, February 2005.

[573] S. Hara and R. Prasad, Overview of multicarrier CDMA, *IEEE Commun. Mag.*, vol. 35, pp. 126–133, Decmber 1997.

[574] N. Yee, J. P. M. G. Linnartz, and G. Fettweis, Multi-carrier CDMA in indoor wireless radio networks, *IEEE Personal Indoor and Mobile Radio Communications (PIMRC) International Conference*, Yokohama, Japan, pp. 109–113, September 1993.

[575] L. Yun, M. Couture, J. R. Camagna, and J. P. M. G. Linnartz, BER for QPSK DS-CDMA indoor downlink to Rician dispersive channels, *Asilomar Conference*, Monterey, CA, 1993, pp. 1417–1421, November 1–3.

[576] N. Yee and J. P. M. G. Linnartz, BER for multi-carrier CDMA in indoor Rician-fading channel, *Asilomar Conference*, Monterey, CA, pp. 426–430, November 1–3, 1993.

[577] N. Yee and J. P. M. G. Linnartz, Controlled equalization for MC-CDMA in Rician fading channels, *44th IEEE Vehicular Technology Conference*, Stockholm, pp. 1665–1669, June 1994.

[578] N. Yee and J. P. M. G. Linnartz, Wiener filtering for multi-carrier CDMA, *IEEE / ICCC Conference on Personal Indoor Mobile Radio Communications (PIMRC) and Wireless Computer Networks (WCN)*, The Hague, vol. 4, pp. 1344–1347, September 19–23, 1994.

[579] N. Yee, J. P. M. G. Linnartz, and G. Fettweis, Multi-Carrier-CDMA in indoor wireless networks, *IEICE Trans. Commun. Japan*, vol. E77-B, no. 7, pp. 900–904, July 1994.

[580] J. P. M. G. Linnartz, Performance analysis of synchronous MC-CDMA in mobile Rayleigh channels with both delay and doppler spreads, *IEEE VT*, vol. 50, no. 6, pp. 1375–1387, November 2001.

[581] A. Gorokhov, J. P. M. G. Linnartz, Robust OFDM receivers for dispersive time varying channels: Equalization and channel acquisition, *IEEE Trans. Commun.*, vol. 52, no. 4, pp. 572–583, April 2004.

[582] S. Tomasin, A. Gorokhov H. Yang, and J. P. M. G. Linnartz, Iterative interference cancellation and channel estimation for mobile OFDM, accepted for *IEEE Trans. Wireless Commun.*, vol. 4, pp. 238–245, TW-3-038, 2004.

[583] S. Tomasin, A. Gorokhov, H. Yang, and J.-P. Linnartz, *Reduced Complexity Doppler Compensation for Mobile DVB-T*, PIMRC, Lisbon, 2002.

[584] J.-P. Linnartz, A. Gorokhov, S. Tomasin, and H. Yang, "Achieving mobility for DVB-T by signal processing for doppler compensation", in session: *Cutting Edge, the Latest From the Labs*, IBC, Amsterdam, pp. 412–420, September 14th, 2002.

[585] H. H. Chen and X. D. Cai, Optimization of transmitter and receiver filters for the OQAM-OFDM systems by using nonlinear programming algorithms, *IEICE Trans. Commun.*, vol. E80-B, no. 11, November 1997.

[586] H. H. Chen, Performance analysis of an improved multi-carrier CDMA system under frequency-selective Rayleigh fading channels, *Int. J. Commun. Syst.*, vol. 16, no. 7, John Wiley & Sons, pp. 267–646, September 2003.

[587] E. A. Sourour and M. Nakagawa, Performance of orthogonal multicarrier CDMA in a multipath fading channel, *IEEE Trans. Commun.*, vol. 44, no. 3, pp. 356–367, March 1996.

[588] S. Kondo and L. B. Milstein, Performance of multicarrier DS CDMA systems, *IEEE Trans. Commun.*, vol. 44, no. 2, pp. 238–246, February 1996.

[589] A. C. McCormick and E. A. Al-Susa, Multicarrier CDMA for future generation mobile communication, *Electron. Commun. Eng. J.*, vol. 14, no. 2, pp. 52–60, April 2002.

[590] L. Loyola and T. Miki a new transmission and multiple access scheme based on multicarrier cdma for future highly mobile network, *IEEE Proc. Pers. Indoor Mobile Radio Commun. (PIMRC) 2003*, vol. 2, pp. 1944–1948, September 7–10, 2003.

[591] Q. Shi and M. Latva-aho, Simple spreading code allocation scheme for downlink MC-CDMA, *Electron. Lett.*, vol. 38, no. 15, pp. 807–809, 18 July 2002.

[592] L.-L. Yang and L. Hanzo, Multicarrier DS-CDMA: A multiple access scheme for ubiquitous broadband wireless communications, *IEEE Commun. Mag.*, vol. 41, no. 10, pp. 116–124, October 2003.

[593] F. Frederiksen and B. Prasad, An overview of OFDM and related techniques towards development of future wireless multimedia communications, *Radio and Wireless Conference, 2002*. IEEE RAWCON 2002, USA, pp. 19–22, 11–14, August 2002.

[594] Intel Corporation, http://www.intel.com/technology/ultrawideband/.

[595] Motorola Corporation, http://www.motorola.com, accessed since 1994.

[596] Communication Research Laboratory, http://www2.crl.go.jp/.

[597] General Atomics, http://www.fusion.gat.com/photonics/uwb/.

[598] Wisair, http://www.wisair.com.

[599] Time Domain, http://www.timedomain.com.

[600] XtremeSpectrum, http://www.xtremespectrum.com.

[601] FCC regulations, 47CFR Section 15.5 (d). http://ftp.fcc.gov, accessed 1998.

[602] Farr Research Inc, http://www.farr-research.com, accessed 2004.

[603] J. McCorkle, *Why such uproar over ultrawideband? Communication Systems Design Website*, http://www.commsdesign.com/csdmag/sections/feature_article/OEG20020301S0021, accessed March 2002.

[604] J. D. Taylor, editor, *Ultra-Wideband Radar Technology*, CRC Press, 2001.

[605] J. D. Taylor, editor, *Introduction to Ultra-Wideband Radar Systems*, CRC Press, 1995.

[606] X. Li, *Super-Resolution to a Estimation with Diversity Techniques for Indoor Geolocation Applications*. PhD thesis, 2003.

[607] K. Y. Yazdandoost and R. Kohno, Ultra Wideband Antenna. CRL report. *IEEE, Commun. Mag.* vol. 42, no. 6, pp. S29–S32, June 2004.

[608] G. F. Ross, *Transmission and Reception System for Generating and Receiving Base-Band Duration Pulse Signals Without Distortion for Short Base-Band Pulse Communicaton System*, U.S. Patent 3,728,632, April 1973.

[609] T. W. Barrett, History of ultra wideband (UWB) radar and communications: Pioneers and innovators. In *Proceedings of Progress in Electromagnetics Symposium 2000 (PIERS2000)*, July 2000.

[610] C. L. Bennett and G. F. Ross, Time-domain electromagnetics and its applications, *Proc. IEEE*, vol. 66, pp. 299–318, March 1978.

[611] J. Williams, *The IEEE 802.11b Security Problem, part 1. IT Professional*, pp. 91–96, November 2001.

[612] F. Ramirez-Mireles and R. A. Scholtz, Wireless multiple-access using SS time-hopping and block waveform pulse position modulation, part 2: Multiple-access performance. In *Proceedings ISITA Symposium*, October 1998.

[613] M. Z. Win and R. A. Scholtz, Ultra-wide bandwidth time-hopping spread spectrum impulse radio for wireless multiple-access communication, *IEEE Trans. Commun.*, vol. 48, no. 4, pp. 679–691, April 2000.

[614] D. G. Leeper, *Wireless Data Blaster*. Scientific American, May 2002.

[615] H. Kikuchi, *UWB Arrives in Japan*, Nikkei Electronics, pp. 95–122, February 2003.

[616] R. Mark, *XtremeSpectrum Rolls out First UWB Chipset*, InternetNews Website, June 2002.

[617] R. A. Scholtz, Multiple access with time-hopping impulse modulation. In *IEEE MILCOM93*, vol. 2, October 1993.

[618] J. T. Conroy, J. L. LoCicero, and D. R. Ucci, Communication techniques using monopulse waveforms. In *IEEE MILCOM99*, vol. 2, November 1999.

[619] M. Ghavami, L. B. Michael, S. Haruyama, and R. Kohno. A novel UWB pulse shape modulation system, *Kluwer Wireless Pers. Commun. J.*, vol. 23, pp. 105–120, 2002.

[620] J. B. Martens, The hermite transform-theory, *IEEE Trans. Acoust. Speech Signal Process.*, vol. 38 pp. 1595–1606, 1990.

[621] M. R. Walton and H. E. Hanrahan. Hermite wavelets for multicarrier data transmission. In *South African Symposium on Communications and Signal Processing ComSIG 93*, South African, August 1993.

[622] J. M. Cramer, R. A. Scholtz, and M. Z. Win, On the analysis of UWB communication channels. In *IEEE MILCOM99*, November 1999.

[623] M. Ghavami, L. B. Michael, and R. Kohno. Hermite function-based orthogonal pulses for ultra wideband communication. In *WPMC'01*, September 2001.

[624] M. Z. Win and R. A. Scholtz, Impulse radio: How it works, *IEEE Commun. Lett.*, vol. 2, pp. 36–38, 1998.

[625] D. Slepian, Prolate spheroidal wave functions, fourier analysis and uncertainty V: The discrete case, *Bell Syst. Techn. J.*, vol. 57, pp. 1371–1429, 1978.

[626] R. S. Dilmaghani, M. Ghavami, B. Allen, and H. Aghvami. Novel pulse shaping using prolate spheroidal wave functions for UWB. In *IEEE PIMRC 2003*, Beijing, China, 2003.

[627] N. W. Bailey, On the product of two Legendre polynomials, *Proc. Cambridge Philos. Soc.*, vol. 29, pp. 173–177, 1933.

[628] J. M. Wilson, Ultra wideband technology update at spring 2003, *Intel Developer UPDATE Magazine*, pp. 1–9, 2003.

[629] New Ultra-Wideband Technology, White Paper, *Discrete Time Communications*, pp. 1–8, 2002.

[630] H. F. Harmuth, Radio signals with large relative bandwidth for over-the-horizon radar and spread spectrum communications, *IEEE Trans. Electromag. Compat.*, vol. 20, pp. 501–512, 1978.

[631] J. R. Davis, D. J. Baker, J. P. Shelton, and W. S. Ament, Some physical constraints on the use of carrier free waveforms in the radio-wave transmission systems, *Proc. IEEE*, vol. 67, pp. 884–890, June 1979.

[632] P. P. Newaskar, R. Blazquez, and A. P. Chandrakasan, A/D precision requirements for an ultra-wideband radio receiver. In *SIPS 02*, October 2002.

[633] W. Ellersick, C. K. Ken Yang, W. Horowitz, and W. Dally. Gad: A 12gs/s cmos 4-bit A/D converter for an equalized multi-level link. In *Symposium on VLSI Circuits*, Digest of Technical Papers, 1999.

[634] T. E. McEwan, *Ultra-Wideband Radar Motion Sensor*, US Patent 5,361,070, 1994.

[635] J. R. Foerster, The effects of multipath interference on the performance of UWB systems in an indoor wireless channel. In *Spring Vehicular Technology Conference*, Rhodes Island, Greece, May 2001.

[636] H. Hashemi, Impulse response modeling of indoor radio propagation channels, *IEEE J. Sel. Areas Commun.*, vol. 11, pp. 967–978, 1993.

[637] M. Z. Win and R. A. Scholtz, On the robustness of ultra-wide bandwidth signals in dense multipath environments, *IEEE Commun. Lett.*, vol. 2, pp. 10–12, 1998.

[638] A. A. Saleh and R. A. Valenzuela, A statistical model for indoor multipath propagation, *IEEE J. Sel. Areas Commun.*, vol. 5, pp. 128–137, 1987.

[639] H. Suzuki, A statistical model for urban radio propagation, *IEEE Trans. Commun.*, vol. 25, pp. 673–680, 1977.

[640] R. Ganesh and K. Pahlavan, Statistical modeling and computer simulation of indoor radio channel, *IEE Proc.*, vol. 138, part 1, no. 3, pp. 153–161, 1991.

[641] S. S. Ghassemzadeh, R. Jana, C. Rice, W. Turin, and V. Tarokh. A statistical. path loss model for in-home UWB channels. In *IEEE UWBST*, May 2002.

[642] J. Foerster and Q. Li, UWB channel modeling contribution from Intel, *Technical report, IEEE document*, 2002.

[643] K. Siwiak and A. Petroff, A path link model for ultra wide band pulse transmission. In *IEEE Vehicular Technology Conference 2001*, Rhodes, pp. 1173–1175, May 2001.

[644] D. Cassioli, M. Z. Win, and A. R. Molisch, The ultra-wide bandwidth indoor channel: From statistical model to simulations, *IEEE J. Sel. Areas Commun.*, vol. 20, pp. 1247–1257, 2002.

[645] A. Armogida, B. Allen, M. Ghavami, M. Porretta, and H. Aghvami, Path loss modeling in short-range UWB transmissions. In *International Workshop on UWB Systems*, IWUWBS2003, Oulu, Finland, June 2003.

[646] W. C. Stone, *Nist Construction Automation Report No. 3: Electromagnetic Signal Attenuation in Construction Materials. Technical report*, BFRL Publications, 1997.

[647] T. S. Rappaport, *Wireless Communications: Principles and Practice*, Prentice Hall, 1996.

[648] W. Turin, R. Jana, S. S. Ghassemzadeh, C. W. Rice, and V. Tarokh, Autoregressive modeling of an indoor UWB channel. In *IEEE UWBST*, May 2002.

[649] S. Howard and K. Pahlavan. Autoregressive modeling of wide-band indoor radio propagation, *IEEE Trans. Commun.*, vol. 40, pp. 1540–1552, September 1992.

[650] L. Zhao and A. M. Haimovich, The capacity of a UWB multiple access communication system. In *IEEE International Conference on Communications, ICC '02*, pp. 1964–1968, May 2002.

[651] K. Eshima, K. Mizutani, R. Kohno, Y. Hase, S. Oomori, and F. Takahashi, Comparison of ultra-wideband (UWB) impulse radio with DS-CDMA and FH-CDMA. In *Proceedings of 24th Symposium on Information Theory and Applications (SITA)*, Kobe, Japan, pp. 803–806, 2001, In Japanese.

[652] T. Ikegami and K. Ohno. Interference mitigation study for UWB impulse radio. In *IEEE PIMRC 2003*, vol. 1, pp. 583–587, September 2003.

[653] M. Hamalainen, J. Saloranta, J. P. Makela, I. Opperman, and T. Pantana, Ultra wideband signal impact on IEEE 802.11b and bluetooth performance. In *IEEE PIMRC 2003*, vol. 1, pp. 280–284, September 2003.

[654] M. Luo, M. Koenig, D. Akos, S. Pullen, and P. Enge, Potential interference to GPS from UWB transmitters phase II test results accuracy, loss-of-lock, and acquisition testing for GPS receivers in the presence of UWB signals, *Technical Report 3.0*, Stanford University, March 2001.

[655] J. P. Van't Hof and D. D. Stancil, Ultra-wideband high data rate short range wireless links. In *IEEE Vehicular Technology Conference 2002*, pp. 85–89, 2002.

[656] I. I. Immoreev and A. N. Sinyavin, Features of ultra-wideband signals' radiation. In *UWBST 2002 IEEE Conference on Ultra Wideband Systems and Technologies*, May 2002.

[657] F. Sabath, Near field dispersion of impulse radiation. In *URSI General Assembly 2002*, August 2002.

[658] B. Widrow, P. E. Mantey, L. J. Griffiths, and B. B. Goode, Adaptive antenna systems, *Proc. IEEE*, vol. 55, pp. 2143–2159, December 1967.

[659] M. G. M. Hussain, An overview of the principle of ultra-wideband impulse radar. In *CIE 1996 International Conference of Radar*, November 1996.

[660] M. G. M. Hussain, Antenna patterns of nonsinusoidal waves with the time variation of a gaussian pulse – part I, *IEEE Trans. Electromag. Compat.*, vol. 30, pp. 504–512, 1988.

[661] CDMA Development Group, *CDG: Test plan document for location determination technologies evaluation*, 2000.

[662] N. Lenihan and S. McGrath, *REALM: Analysis of alternatives for location positioning*.

[663] K. Pahlavan, X. Li, and J. Makela. Indoor geolocation science and technology, *IEEE Commun. Soc. Mag.*, vol. 40, pp. 112–118, February 2002.

[664] M. O. Sunay and I. Tekin. Mobile location tracking for IS-95 using the forward link time difference of arrival techniques and its application to zone-based billing. In *IEEE GLOBECOM Conference*, Brazil, 1999.

[665] R. J. Fontana, Experimental results from an ultra wideband precision geolocation system, *Ultra- Wideband, Short-Pulse Electromagnetics IV*, Kluwer Academic/Plenum Publishers, May 2000.

[666] R. J. Fontana and S. Gunderson, Ultra-wideband precision asset location system. In *UWBST 2002 IEEE Conference on Ultra Wideband Systems and Technologies*, May 2002.

[667] I. Maravic, M. Vetterli, and K. Ramchandran, Channel estimation and synchronization with sub-Nyquist sampling and application to ultra-wideband systems. In *ISCAS*, 2004.

[668] R. Fleming and C. Kushner, Low-power miniature distributed position location and communication devices using ultra wideband, nonsinusoidal communication technology, *Technical Report*, AetherwireLocation Inc., July 1995.

[669] D. Porcino and W. Hirt. Ultra-wideband radio technology: Potential and challenges ahead, *IEEE Commun. Mag.*, vol. 41, pp. 66–74, July 2003.

[670] M. Nakagawa, H. Zhang, and H. Sato, Ubiquitous homelinks based on IEEE 1394 and ultra wideband solutions, *IEEE Commun. Mag.*, vol. 41, no. 4, pp. 74–82, April 2003.

[671] J. C. Harrtsen, The bluetooth radio system, *IEEE Pers. Commun.*, vol. 7, no. 1, pp. 28–36, February 2000.

[672] K. J. Negus, A. P. Stephens, and J. Landsford. Homerf: Wireless networking for the connected home, *IEEE Pers. Commun.*, vol. 7, no. 1, pp. 20–27, February 2000.

[673] R. J. Fontana, E. Richley, and J. Barney, Commericalization of an ultra wideband precision asset location system. In *UWBST 2003 IEEE Conference on Ultra Wideband Systems and Technologies*, November 2003.

[674] L. Fullerton, UWB waveforms and coding for communications and radar, *Telesystems Conference, 1991. Proceedings. vol. 1, NTC '91, National*, pp. 139–141, 26–27 March 1991.

[675] M. Z. Win and R. A. Scholtz, Ultra-wide bandwidth time-hopping spread-spectrum impulse radio for wireless multiple-access communications, *IEEE Trans. Commun.*, vol. 48, no. 4, pp. 679–689, April 2000.

[676] K. Hase, Y. Oomori, S. Takahashi, R. Kohno, Performance analysis of interference between UWB and SS signals, Eshima, Spread Spectrum Techniques and Applications, *2002 IEEE Seventh International Symposium on*, Prague, Czech Republic, vol. 1, pp. 59–63, 2002.

[677] V. S. Somayazulu, Multiple access performance in UWB systems using time hopping vs. direct sequence spreading, *Wireless Communications and Networking Conference, 2002, WCNC2002. 2002, IEEE*, vol. 2, pp. 522–525, 17–21 March 2002.

[678] J. R. Foerster, The performance of a direct-sequence spread ultrawideband system in the presence of multipath, narrowband interference, and multiuser interference, Ultra Wideband Systems and Technologies, 2002, *Digest of Papers. 2002 IEEE Conference*, Intel Labs., Hillsboro, OR, pp. 87–91, on 21–23 May 2002.

[679] M. Welborn, T. Miller, J. Lynch, and J. McCorkle, Multi-user perspectives in UWB communications networks, Ultra Wideband Systems and Technologies, 2002, *Digest of Papers. 2002 IEEE Conference*, Vienna, VA, pp. 271–275, on 21–23 May 2002.

[680] Q. Li and L. A. Rusch, Multiuser receivers for DS-CDMA UWB, Ultra Wideband Systems and Technologies, 2002, *Digest of Papers. 2002 IEEE Conference*, Baltimore, MD, pp. 163–167, on 21–23 May 2002.

[681] Q. Li and L. A. Rusch, Multiuser detection for DS-CDMA UWB in the home environment, Selected Areas in Communications, *IEEE J.*, vol. 20, no. 9, pp. 1701–1711, December. 2002.

[682] B. M. Sadler and Swami A. On the performance of UWB and DS-spread spectrum communication systems, Ultra wideband systems and technologies, 2002, *Digest of Papers. 2002 IEEE Conference*, Adelphi, MD, pp. 289–292, on 21–23 May 2002.

[683] C. M. Canadeo, M. A. Temple, R. O. Baldwin, and R. A. Raines, Code selection for enhancing UWB multiple access communication performance using TH-PPM and DS-BPSK modulations, wireless communications and networking, 2003. WCNC 2003, *2003 IEEE*, vol. 1, pp. 678–682, 16–20 March 2003.

[684] N. Boubaker and K. B. Letaief, Ultra wideband DSSS for multiple access communications using antipodal signaling, communications, 2003. ICC '03, *IEEE Int. Conf.*, vol. 3, pp. 2197–2201, 11–15 May 2003.

[685] C. R. Nassar, F. Zhu, and Z. Wu, Direct sequence spreading UWB systems: frequency domain processing for enhanced performance and throughput, Communications, 2003. ICC '03, *IEEE Int. Conf.*, vol. 3, pp. 2180–2186, 11–15 May 2003.

[686] V. Venkatesan, H. Liu, C. Nilsen, R. Kyker, M. E. Magana, Performance of an optimally spaced PPM ultra-wideband system with direct sequence spreading for multiple access, vehicular technology conference, 2003, *VTC 2003-Fall. 2003 IEEE 58th*, vol. 1, pp. 602–606, 6–9 October 2003.

[687] R. A. Jones, D. H. Smith, and S. Perkins, Assignment of spreading codes in DS-CDMA UWB systems, ultra wideband systems and technologies, 2003, *IEEE Conference*, Reston, Virginia, pp. 359–363, on November 16–19, 2003.

[688] P. Runkle, J. McCorkle, T. Miller, and M. Welborn, DS-CDMA: the modulation technology of choice for UWB communications, ultra wideband systems and technologies, 2003, *IEEE Conference*, Reston, VA, pp. 364–368, on November 16–19, 2003.

[689] R. D. Wilson and R. A. Scholtz, Comparison of CDMA and modulation schemes for UWB radio in a multipath environment, *Global Telecommunications Conference, 2003. GLOBECOM '03. IEEE*, vol. 2, pp. 754–758, 1–5 December 2003.

[690] A. Saleh and R. Valenzuela, A statistical model for indoor multipath propagation, *IEEE J. Sel. Areas Commun.*, vol. 5, no. 2, pp. 128–137, February 1987.

[691] J. Foerster and Q. Li, Intel research and development, *UWB Channel Modeling Contribution from Intel*, submission to IEEE 802.15.3a Working Group, June, 2002.

[692] V. Tarokh, A. Naguib, N. Seshadri, and A. R. Calderbank, Space-time codes for high data rate wireless communication: Performance criterion and code construction, *IEEE Trans Inform. Theory*, vol. 44, no. 2, pp. 744–765, 1998.

[693] S. M. Alamouti, A simple transmit diversity technique for wireless communications, *IEEE J. Sel. Areas Commun.*, vol. 16, no. 8, pp. 1451–1458, October 1998.

[694] V. Tarokh, H. Jafarkhani, and A. R. Calderbank, Space-time block codes from orthogonal designs, *IEEE Trans. Inform. Theory*, vol. 45, no. 5, pp.1456–1467, July 1999.

[695] B. M. Hochwald, T. L. Marzetta, and C. B. Papadias, A transmitter diversity scheme for wideband CDMA systems based on space-time spreading, *IEEE J. Sel. Areas Commun.*, vol. 19, no. 1, pp. 48–60, January 2001.

[696] C. B. Papadias and H. Huang, Linear space-time multiuser detection for multipath CDMA channels, *IEEE J. Sel. Areas Commun.*, vol. 19, no. 2, pp. 264–265, February 2001.

[697] M. O. Damen, A. Safavi, and K. Abed-Meraim, On CDMA with space-time codes over multipath fading channels, *IEEE Trans. Wireless Commun.*, vol. 2, no. 1, pp. 11–19, January 2003.

[698] F. Petre, G. Leus, L. Deneire, M. Engels, M. Moonen, and H. De Man, Space-time block coding for single-carrier block transmission DS-CDMA downlink, *IEEE J. Sel. Areas Commun.*, vol. 21, no. 3, pp. 350–361, April 2003.

[699] G. L. Stuber, *Principles of Mobile Communication*, 2nd Edition, Kluwer Academic Publishers, pp. 280–284, 2001.

[700] N. Al-Dhahir, C. Fragouli, A. Stamoulis, W. Younis, and R. Calderbank, Space-time processing for broadband wireless access, *IEEE Commun. Mag.*, vol. 40, no. 9, pp. 136–142, 2002.

[701] I. E. Telatar, *Capacity of Multi-Antenna Gaussian Channels*, No. BL0112170-950615-07TM, AT&T Bell Laboratories Technical Report, 1995.

[702] G. J. Foschini and M. J. Gans, On limits of wireless communications in a fading environment when using multiple antennas, *Wireless Pers. Commun.*, vol. 6, pp. 311–335, 1998.

[703] V. Erceg, P. Soma, D. S. Baum, and A. J. Paulraj, Capacity obtained from multi-input-multi-output channel measurements in fixed wireless environments at 2.5GHz, ICC 2002, *IEEE Int. Conf. Commun.*, vol. 1, pp. 396;V400, 2002.

[704] P. K. Enge and D. V. Sarwate, Spread-spectrum multiple-access performance of orthogonal codes: Linear receivers, *IEEE Trans. Commun.*, vol. COM-35, pp. 1309–1319, December 1987.

[705] C. L. I and R. D. Gitlin, Multi-code CDMA wireless personal communication networks, in *Proceedings of IEEE International Conference on Communications, (ICC 1995)*, Seattle, WA, vol. 2, pp. 1060–1064, 1995.

[706] T. F. Wang and T. M. Lok, Transmitter adaptation in multicode DS-CDMA systems, *IEEE J. Sel. Areas Commun.*, vol. 19, no. 1, January 2002.

[707] H. H. Chen, Y.-T. Wu, and C.-Y. Chao, Unified approach for BER analysis of a generic multi-code CDMA with optimized decision thresholds, *IEE Electron. Lett.*, vol. 39, no. 22, October 30, 2003.

[708] V. D. Pham and T. B. Vu, Adaptive space-time MMSE receivers in DS/CDMA systems, *Proceedings of ICSP*, pp. 470–473, 1998.

[709] S. L. Miller, An adaptive DS-CDMA receiver for multiuser interference rejection, *IEEE Trans. Commun.*, vol. 43, no. 2/3/4, pp. 1746–1755, February/March/April 1995.

[710] M.-C. Chin and C.-C. Chao, Analysis of LMS-adaptive MLSE equalization on multipath fading channels, *IEEE Trans. Commun.*, vol. 44, no. 12, pp. 1684–1692, December 1996.

[711] J. Razavilar, F. Rashid-Farrokhi, and K. J. Ray Liu, Software radio architecture with smart antennas: A tutorial on algorithms and complexity, *IEEE J. Sel. Areas Commun.*, vol. 46, no. 4, pp. 1313–1324, October 1998.

[712] R. Lupas and S. Verdu, Linear multi-user detectors for synchronous code-division multiple-access channels, *IEEE Trans. Inform. Theory*, vol. 35, pp. 123–136, January 1989.

[713] R. Lupas and S. Verdu, Near-far resistance of multi-user detectors in asynchronous channels, *IEEE Trans. Commun.*, vol. 38, pp. 496–508, April 1990.

[714] H.-H. chen and Z.-Q. Liu, Zero-insertion adaptive minimum mean square error (MMSE) receiver for asynchronous CDMA multiuser detection, *IEEE Trans. Veh. Technol.*, vol. 50, no. 2, pp. 557–569, March 2001.

[715] H.-H. Chen, Y.-N. Chang, and Y.-B. Wu, Single code cyclic shift detection – A pilot aided CDMA multiuser detector without using explicit knowledge of signature codes, *IEICE Trans. Commun.*, vol. E86-B, no. 4, pp. 1286–1296, April 2003.

[716] F. Rashid-Farrokhi, L. Tassiulas, and K. J. Ray Liu, Joint optimal power control and beamforming for wireless networks with antenna arrays, *IEEE Trans. Commun.*, vol. 17, no. 4, pp. 662–676, April 1999.

[717] H. V. Poor and S. Verdu, Probability of error in MMSE multiuser detection, *IEEE Trans. Inform. Theory*, vol. 43, no. 3, pp. 858–871, May 1997.

[718] H. H. Chen, Quasi-decorrelating detector – A non-matrix inversion based decorrelating detector with near-far resistance and complexity trade-off, accepted for publication in *European Transactions on Telecommunications (ETT)*.

[719] H. H. Chen and C.-F. Wu, A novel approach to enable drecorrelating multiuser detection without matrix inversion operations, accepted for publication, in *Int. J. Commun. Syst.*, John Wiley & Sons, 2004.

[720] H. H. Chen and H. K. Sim, A new CDMA multiuser detector – orthogonal decision-feedback detector for asynchronous CDMA systems, *IEEE Trans. Commun.*, vol. 49, no. 9, pp. 1649–1658, September 2001.

[721] H. H. Chen, Asynchronous orthogonal decision-feedback multi-user detector (AODFD) and its alternative decoding strategies, *Int. J. Commun. Syst.*, John Wiley & Sons, vol. 14, no. 6, pp. 561–574, June 2001.

[722] H. H. Chen and Z.-Q. Liu, Zero-insertion adaptive minimum mean square error (MMSE) receiver for asynchronous CDMA multiuser detection, *IEEE Trans. Veh. Technol.*, vol. 50, no. 2, pp. 557–569, March 2001.

[723] H. H. Chen and Z. Q. Liu, Asynchronous block-based minimum mean square error (B-MMSE) CDMA multiuser detection, *Int. J. Commun. Syst.*, John Wiley & Sons, vol. II, pp. 395–401, 1998.

[724] H. H. Chen and Z. Q. Liu, A CDMA multiuser detector with block channel coding and its performance analysis under multiple access interference, *IEICE Trans. Commun.*, vol. E81-B, no. 5, pp. 1095–1101, May 1998.

[725] H. H. Chen and H. K. Sim, Novel synchronous CDMA multiuser detection scheme: Orthogonal decision-feedback detection and its performance study, *IEE Proc- Commun.*, vol. 144, no. 3, pp. 275–280, August 1997.

[726] H. H. Chen and H. K. Sim, Quasi-decorrelating detector (QDD) and its spreading code dependent performance analysis, *IEICE Trans. Commun.*, vol. E80-B, no. 9, pp. 1337–1344, September 1997.

[727] H. H. Chen, H. K. Sim, and P. K. Kooi, An effective CDMA multi-user detection scheme – orthogonal decision-feedback and its performance analysis, *IEICE Trans. Commun.*, vol. E80-B, no.1, pp. 145–155, January 1997.

[728] A. Duel-Hallen, Decorrelating decision-feedback multiuser detector for synchronous code-division multiple-access channel, January 1997, *IEEE Trans. Commun.*, vol. 41, pp. 285–290, February 1993.

[729] L. Wei and C. Schlegel, Synchronous DS-SSMA system with improved decorrelating decision-feedback multiuser detection, *IEEE Trans. Veh. Technol.*, vol. 43, pp. 767–772, August 1994.

[730] A. Duel-Hallen, A family of multiuser decision-feedback detectors for asynchronous code-division multiple-access channels, *IEEE Trans. Commun.*, vol. 43, pp. 421–434, February, March, April 1995.

[731] A. Duel-Hallen, Equalizers for multiple input-multiple output channels and PAM systems with cyclostationary input sequences, *IEEE JSAC*, vol. 10, no. 3, pp. 630–639, April 1992.

[732] G. D. Forney, Maximum likelihood sequence estimation of digital sequences in the presence of intersymbol interference, *IEEE Trans. Inform. Theory*, vol. IT-18, pp. 363–378, May 1972.

[733] L. Wei and L. K. Rasmussen, A near ideal whitening filter for an asynchronous time-variant CDMA system, *IEEE Trans. Commun.*, vol. 44, no. 10, pp. 1355–1361, October 1996.

[734] C. Schlegel, S. Roy, P. D. Alexander, and Z. J. Xiang, Multiuser projection receivers, *IEEE JSAC*, vol. 14, no. 8, October 1996.

[735] K. B. Lee, Orthogonalization based adaptive interference suppression for DS-CDMA systems, *IEEE Trans Commun.*, vol. 44, no. 9, pp. 1082–1085, September 1996.

[736] Z. Xie, R. T. Short, and C. K. Rushforth, A family of suboptimal detectors for coherent multiuser communications, *IEEE J. Sel. Areas Commun.*, vol. 8, no. 4, pp. 683–690, May 1990.

[737] H.-H. Chen, T. Lang, and J. Oksman, Correlation statistics distribution convolution (CSDC) algorithm for studying CDMA indoor wireless systems with RAKE receiver, power control and multipath fading, *IEICE Trans. Commun.*, vol. E79-B, no. 10, October 1996.

[738] N. Abramson, Multiple access in wireless digital networks, *Proc. IEEE*, vol. 82, no. 9, pp. 1360–1370, September 1994.

[739] N. Abramson, The throughput of packet Broadcasting channels, in N. Abramson, editor, *Multiple Access Communications-Foundations for Emerging Technologies*, pp. 233–244, 1992.

[740] J. Ackermann, *Getting Started with TCP/IP on Packet Radio*, available through anonymous ftp at ftp.ucsd.edu as file intronos.zip, accessed 1992.

[741] AX.25 Amateur Packet-Radio Link-Layer Protocol, Version 2.0, *American Radio Relay League*, October 1984.

[742] R. Gerhards and P. Dupont, The RD-LAP Air interface protocol, *INTERCOMM 93*, 1993.

[743] L. Kleinrock and F. A. Tobagi, Packet switching in radio channels: Part I carrier sense multiple access modes and their throughput-delay characteristics, *IEEE Trans. Commun.*, vol. COM 23, no. 12, pp. 1400–1416, December 1975.

[744] K. Pahlavan and A. H. Levesque, Wireless Data Communications, *Proc. IEEE*, vol. 82, pp. 1398–1430, September 1994.

[745] RAM Mobile Data-System Overview, Release 5.3, *RAM Mobile Data*, Woodbridge, NJ, November 1994.

[746] B. Leiner and D. Neilson, Special issue on Packet Radio Network, *Proc. IEEE* vol. 75, no. 1, pp. 6–20, January 1987.

[747] F. A. Tobagi and L. Kleinrock, Packet switching in radio channels: Part II the hidden terminal problem in carrier sense multiple-access and the busy-tone solution, *IEEE Trans. Commun.*, vol. COM 23, no. 12 December 1975.

[748] I. Wade, NOSintro, TCP I IP over packet radio, *An Introduction to the KA9Q Network Operating System*, Dowermain LTD, Luton, Bedfordshire: United Kingdom, 1992.

[749] E. S. Sousa and J. A. Silvester, Spreading code protocols for distributed spread-spectrum packet radio networks, *IEEE Trans. Commun.*, vol. 36, no. 3, pp. 272–281, March 1988.

[750] X. H. Chen, W. X. Lu, and J. Oksman, Use of code sensing technique in the receiver-based spreading code protocol and its performance analysis, *IEE Proc.-I*, vol. 139, no. 1, pp. 85–90, February 1992.

[751] X. H. Chen and N. C. Lim, Triple-receiver-based code protocol for unslotted DS/SSMA packet-radio networks and its performance analysis, *IEE Proc- Commun.*, vol. 142, no. 3, pp. 193–200, June 1995.

[752] S. Jiang, S. Man-Tung, and T. Hsiao, Performance evaluation of a receiver-based handshake protocol for CDMA networks, *IEEE Trans. Commun.*, vol. 43, no. 6, pp. 2127–2138, June 1995.

[753] J. H. Huang and L. Kleinrock, Throughput analysis and protocol design for CSMA and BTMA protocols under noisy environments, *IEE Proc.-I*, vol. 139, no. 3, pp. 289–296, June 1992.

[754] D.-M. Lim and H.-S. Lee, Throughput-delay and stability analysis of an asynchronous spread spectrum packet radio network, *IEEE Trans. Veh. Technol.*, vol. 41, no. 4, pp. 469–478, November 1992.

[755] D. Raychaudhuri, Performance analysis of random access packet-switched code division multiple access systems, *IEEE Trans. Commun.*, vol. COM-23, no. 6, pp. 895–901, June 1981.

[756] D. H. Davis and S. A. Gronemeyer, Performance of slotted ALOHA random access with delay capture and randomised time of arrival, *IEEE Trans. Commun.*, vol. COM-28, no. 5, pp. 703–710, May 1980.

[757] X. H. Chen and J. Oskman, Busy code broadcasting and sensing protocol for collision-free CDMA packet radio networks and its performance analysis, *IEE Proc.-I*, vol. 139, no. 6, pp. 613–619, December 1992.

[758] M. C. Yuang and P. L. Tien, Multiple access control with intelligent bandwidth allocation for wireless ATM networks, *IEEE J. Sel. Areas Commun.*, vol. 18 no. 9 , pp. 1658–1669, September 2000.

[759] R. Srinivasan and A. K. Somani, On achieving fairness and efficiency in high-speed shared medium access, *IEEE/ACM Trans. Network.*, vol. 11 no. 1 , pp. 111–124, February 2003.

[760] H. H. Chen and W.-T. Tea, Hierarchy Schedule Sensing protocol for CDMA wireless networks – performance study under multipath, multi-user interference and collision-capture effect, to appear in *IEEE Trans. Mobile Comput.*, vol. 4, no. 2, pp. 178–188, 2005.

[761] H. H. Chen and W.-T. Tea, Hierarchy schedule sensing protocol–its performance analysis considering packet collision, multipath and channel coding, accepted for publication in *Eur. Trans. Telecommun. (ETT)*, 2004.

[762] H. H. Chen and W. T. Tea, Performance of hierarchy schedule sensing protocol for ad-hoc CDMA networks under multiple packet collision and capture effect, accepted for publication in *IEEE/ACM Trans. Network.*, vol. 12, no. 6, pp. 1036–1048, 2004.

[763] H. H. Chen and W. T. Tea, A comparison study of HSS protocol versus Triple-R protocol for CDMA wireless data networking with packet collision and capture effect, *Eur. Trans. Telecommun. (ETT)*, vol. 14, no. 3, pp. 279–291, 2003.

[764] H. H. Chen, Simultaneous multiple packet capture based on SIR levels and arrival delay offsets, In *CDMA Packet Networks, IEEE Trans. Veh. Technol.*, vol. 51, no. 6, pp. 1560–1568, November 2002.

[765] H. H. Chen, On joint power-delay double packet capture in an SSMA network with Rayleigh fading, shadowing and propagation path loss, *IEEE Trans. Veh. Technol.*, vol. 50, no. 6, pp. 1388–1402, November 2001.

[766] H. H. Chen, Group-based common-receiver (GBCR) code protocol for DS/CDMA wireless networks and its performance study, *Int. J. Commun. Syst.*, John Wiley & Sons, vol. 14, no. 7, pp. 725–734, June 2001.

[767] H. H. Chen, Hierarchy schedule sensing protocol for CDMA wireless networks and its performance under multiple collision and capture effect, *IEICE Trans. Commun.*, vol. E83-B, no. 3, pp. 703–712, March 2000.

[768] H. H. Chen and W. T. Tea, Novel group-based spreading code protocol: Hierarchy schedule sensing protocol for CDMA wireless networks, *IEE Proc.- Commun.*, vol. 146, no. 1, pp. 15–22, February 1999.

[769] H. H. Chen, N. C. Lim, and J Oksman, Spreading code protocol enabling programmable complexity/performance for CDMA local wireless networks, *IEE Proc.- Commun.*, vol. 144, no. 6, pp. 395–401, December 1997.

[770] H. H. Chen and J. Oksman, Destructive collision free protocol for distributed DS/SSMA wireless networks using code sensing and chip rate division techniques, *IEE Proc.-Commun.*, vol. 143, no. 1, pp. 101–120, February 1996.

[771] H. H. Chen and J. Oksman, Performance bound analysis of the new DS/SSMA protocol for wireless data network, *IEE Proc.-Commun.*, vol. 142, no. 4, pp. 255–262, August 1995.

[772] H. H. Chen and N. C. Lim, Triple-receiver-based code protocol for unslotted DS/SSMA packet radio networks and its performance analysis, *IEE Proc.-Commun.*, vol. 142, no. 3, pp. 193–200, June 1995.

[773] H. H. Chen and J. Oksman, Using code sensing and chip rate division techniques to improve stability & throughput-delay performance for distributed DS/SSMA wireless networks, *Wireless Pers. Commun., Int. J.*, Kluwer Academic Publisher, vol. 1, no. 3, pp. 191–209, 1994/1995.

[774] H. H. Chen and J. Oksman, Busy code broadcasting and sensing protocol for collision-free CDMA packet radio networks and its performance analysis, *IEE Proc.-I*, vol. 139, no. 6. pp. 613–619, December 1992.

[775] H. H. Chen, WX Liu, and J. Oksman, Use of code sensing technique in the receiver-based spreading code protocol and its performance analysis, *IEE Proc.-I*, vol. 139, no. 1, pp. 85–90, February 1992.

[776] H. H. Chen and J. Oksman, New collision-channel model for packet-switched CDMA networks, *IEE Electron. Lett.*, vol. 27, no. 20, pp. 1792–1793, September 1991.

[777] H. H. Chen, Y.-C. Yeh, C.-Y. Chao, and J.-F. Yeh, A pilot-added signal detection algorithm and its application in OCC-CDMA systems under multipath interference, *IEE Electron. Lett.*, vol. 40, no. 8, pp. 488–489, 15th April 2004.

[778] H. H. Chen, Y.-N. Chang, and Y.-B. Wu, Single code cyclic shift detection – A pilot aided CDMA multiuser detector without using explicit knowledge of signature codes, *IEICE Trans. Commun.*, vol. E86-B, no. 4, pp. 1286–1296, April 2003.

[779] H. H. Chen and J.-S. Lee, On adaptive joint beamforming and B-MMSE detection under multipath interference, accepted for publication in *IEE Proc. Commun.*, vol. 151, pp. 605–612, 2004.

[780] H. H. Chen and J.-S. Lee, Adaptive joint beamforming and B-MMSE detection for CDMA signal reception under multipath interference, accepted for publication in *Int. J. Commun. Syst.*, John Wiley & Sons, 2004.

[781] H. H. Chen, Y.-C. Yeh, C.-H. Tsai, and W.-H. Chang, Uplink synchronization control technique and its environment-dependent performance analysis, *IEE Electron. Lett.*, vol. 39, no. 24, pp. 1755–1757, 27th November 2003.

[782] H. H. Chen, Y.-C. Yeh, C.-Y. Chao, C.-H. Tsai, and W.-H. Chang, Isotropic air-interface in TD-SCDMA: Uplink synchronization control and its environment-dependent performance analysis, *Proc. IEEE VTC 2003-Fall*, Orlando, USA, October 4–9, 2003.

[783] L. Tong, Q. Zhao, and G. Mergen, Multipacket reception in random access wireless networks: From signal processing to optimal medium access control, *IEEE Communications Magazine*, vol. 39, pp. 108–112, November 2001.

[784] S. Shakkottai, T. S. Rappaport, and P. C. Karlsson, Cross-layer design for wireless networks, *IEEE Commun. Mag.*, vol. 41, no. 10, pp. 74–80, October 2003.

[785] H. H. Chen, Convoluted raised-cosine and triangular-cosine (RCTC) modulation and its performance against timing jitters over noisy channels, *IEE Electron. Lett.*, vol. 29, no. 14, pp. 1239–1240, 8th July 1993.

[786] H. H. Chen and J. Oksman, A quasi-constant envelope quadrature overlapped modulation and its performance over nonlinear bandlimited satellite channels, *Int. J. Sat. Commun.*, John Wiley & Sons, vol. 14, pp. 351–359, 1996.

[787] H. H. Chen and S. Y. Wong, Spectral efficiency analysis of new quadrature overlapped modulations over band-limited non-linear channels, *Int. J. Commun. Syst.*, John Wiley & Sons, vol. 10, no. 1–11, pp. 1–11, 1997.

[788] H. H. Chen and S. Y. Wong, Four novel quadrature overlapping modulations and their spectral efficiency analysis over band-limited non-linear channels, *Int. J. Sat. Commun.*, John Wiley & Sons, vol. 15, pp. 117–127, 1997.

[789] J. Mitola III, Software radios-survey, critical evaluation and future directions. In *IEEE National Telesystems Conference*, New York, pp. 13/15;V13/23, 1992.

[790] J. Mitola III, Cognitive radio for flexible mobile multimedia communications. In *Sixth International Workshop on Mobile Multimedia Communications (MoMuC'99)*, San Diego, CA, 1999.

[791] J. Mitola III, Software radio architecture: A mathematical perspective, *IEEE J. Sel. Areas Commun.*, vol. 17, no. 4, pp. 514–538, April 1999.

[792] http://www.ntia.doc.gov/osmhome/allochrt.html, accessed 1978.

[793] J. Mitola, III, G. Q. Maguire Jr, Cognitive radio: Making software radios more personal, *IEEE Pers. Commun.*, vol. 6, no. 4, pp. 13–18, August 1999.

[794] J. Mitola III, *Cognitive Radio an Integrated Agent Architecture for Software Defined Radio Dissertation*, Royal Institute of Technology, http://web.it.kth.se/jmitola/Mitola_Dissertation8_Integrated.pdf, accessed May 8, 2000.

[795] Federal Communications Commission, *Unlicensed Operation in the TV Broadcast Bands*, ET Docket No. 04–186, 2004.

[796] J. Mitola III, Cognitive radio for flexible mobile multimedia communications, Mobile Multimedia Communications, *IEEE International Workshop*, 1999.

[797] Federal Communications Commission, *Spectrum Policy Task Force Report*, ET Docket No. 02–135, November 2002.

[798] W. Lehr, The economic case for dedicated unlicensed spectrum below 3GHz, *Spectrum Policy Program White Paper*, New America Foundation, Spectrum Series Issue Brief No.16, July 2004.

[799] J. Notor, Radio architectures for unlicensed reuse of broadcast TV channels, *Communications Design Conference*, September, 2003.

[800] D. Cabric, S. M. Mishra, and R. W. Brodersen, Signals, implementation issues in spectrum sensing for cognitive radios, *Conference Record of the Thirty-Eighth Asilomar Conference on Systems and Computers*, vol. 1, no. 7–10, pp. 772–776, November 2004.

[801] T. A. Weiss and F. K. Jondral, Spectrum pooling: An innovative strategy for the enhancement of spectrum efficiency, *IEEE Commun. Mag.*, vol. 42, no. 3, pp. S8–14, March 2004.

[802] IEEE 802.15 WPAN? Task Group 2 (TG2), http://ieee802.org/15/pub/TG2.html, accessed 2004.

[803] IEEE 802.19 Coexistence Technical Advisory Group (TAG), http://grouper.ieee.org/groups/802/19, accessed 2005.

[804] N. Hoven, R. Tandra, and A. Sahai, *Some Fundamental Limits on Cognitive Radio*, http://www.eecs.berkeley.edu/wireless/posters/WFW05_cognitive.pdf, accessed 2004.

[805] J. Hillenbrand, T. A. Weiss, and F. K. Jondral, Calculation of detection and false alarm probabilities in spectrum pooling systems, *IEEE Commun. Lett.*, vol. 9, no. 4, pp. 349–351, April 2005.

[806] IEEE 802.11h, Part 11: Wireless LAN Medium Access Control (MAC) and Physical Layer (PHY) specifications, *Amendment 5: Spectrum and Transmit Power Management Extensions in the 5 GHz band in Europe*, 14 October 2003.

[807] ITU Recommendation ITU-R M.1461, *Procedures for Determining the Potential for Interference Between Radars Operating in the Radiodetermination Service and Systems in Other Services*, at Annex 1, 2000.

[808] National Telecommunications and Information Administration, NTIA Report 82–100, *A Guide to the Use of the ITS Irregular Terrain Model in the Area Prediction Mode* April 1982.

[809] IEEE 802.22, *Standard for Wireless Regional Area Networks (WRAN)-Specific Requirements-Part 22: Cognitive Wireless RAN Medium Access Control (MAC) and Physical Layer (PHY) Specifications: Policies and Procedures for Operation in the TV Bands*, 2004.

[810] H. Holma and A. Toskala, *WCDMA for UMTS: Radio Access For Third Generation Mobile Communications*, Wiley, 2002.

[811] G. Patel and S. Dennett, The 3GPP and 3GPP2 Movements Toward an All-IP Mobile Network, *IEEE Pers. Commun.*, vol. 7, no. 4, pp. 62;V64, August 2002.

[812] Wireless World Research Forum, http://www.wireless-world-research.org, accessed 2001.

[813] ITU-R PDNR WP8F, *Vision, Framework and Overall Objectives of the Future Development of IMT-2000 and Systems beyond IMT-2000*, 2002.

[814] R. Nee and R. Prasad, *OFDM for Wireless Multimedia Communications*, Artech House, 2000.

[815] 3GPP TD RP-040461, *Proposed Study Item on Evolved UTRA and UTRAN*, 2004.

[816] 3GPP TR 21.905, *Vocabulary for 3GPP Specifications*, 2004.

[817] 3GPP TR 25.913, *Requirements for Evolved UTRA (E-UTRA) and Evolved UTRAN (E-UTRAN)*.

[818] 3GPP TR 25.814, *Physical layer aspect for evolved UTRA*, 2005.

[819] ftp://ftp.3gpp.org/tsg_ran/WG2_RL2/TSGR2_AHs/2005_06_LTE, accessed 2005.

[820] 3GPP TR 23.882, *3GPP System Architecture Evolution (SAE): Report on Technical Options and Conclusions*, 2005.

[821] 3GPP TR 25.813, *Evolved Universal Terrestrial Radio Access (UTRA) and Universal Terrestrial Radio Access Network (UTRAN); Radio Interface Protocol Aspects (Release 7)*, October 2005.

[822] 3GPP TR 25.814, *Physical Layer Aspects for Evolved UTRA (Release 7)*, October 2005.

[823] Motorola, 3GPP R1-040642, *Comparison of PAR and Cubic Metric for Power De-rating*, Motorola, 2005.

Index